Palgrave Studies in Environmental Sustainability

Series Editor
Robert Brinkmann, Hofstra University, Hempstead, USA

The series will take advantage of the growing interest in a number of environmental issues in sustainability, particularly in those that focus on interdisciplinary approaches to solving real-world problems. Unfortunately, many of the themes of sustainability address short- and long-term survivability of our planet and there is a pressing need for access to information to those who seek to solve sustainability problems. Thus, the scope of the book series seeks to be comprehensive within an environmental sustainability framework. The topics covered in the series range from climate change to public land management. While no series can ever be truly comprehensive, the series will provide one of the most definitive surveys of knowledge in the area of environmental sustainability.

Sandipan Ghosh
Editor

Reimagining Indian Rivers for Sustainability

palgrave
macmillan

Editor
Sandipan Ghosh
Department of Geography
Chandrapur College
Bardhaman, West Bengal, India

ISSN 2946-5311 ISSN 2946-532X (electronic)
Palgrave Studies in Environmental Sustainability
ISBN 978-3-032-04753-3 ISBN 978-3-032-04754-0 (eBook)
https://doi.org/10.1007/978-3-032-04754-0

Cover credit: Vinod Khapekar

This Palgrave Macmillan imprint is published by the registered company Springer Nature Switzerland AG
The registered company address is: Gewerbestrasse 11, 6330 Cham, Switzerland

Damodar River
(Bardhaman, West Bengal)
@ Rahul Singh

To the timeless rivers of India—
lifelines of our civilization,
keepers of culture, and
silent witnesses to our journey through time.
To the countless scholars, riverine communities, policymakers, and practitioners whose work, wisdom, and resilience continue to inspire a rethinking of how we engage with rivers—not as resources to be exploited, but as living systems to be understood, respected, and sustained. And to future generations, in the hope that the rivers they inherit will flow cleaner, freer, and with renewed vitality.

Foreword

While human interventions—whether through physical infrastructure, technological innovations, or extractive activities—have historically been motivated by immediate social and economic objectives, such actions frequently unsettle the delicate equilibrium that characterised river systems in their pristine, pre-industrial state. In the contemporary Anthropocene era, the functioning of riverine systems is profoundly altered by flow regulations, excessive sedimentation, nutrients, and pollutants introduced or removed through industrial, agricultural, and urban activities. Achieving sustainability in our interactions with river systems, therefore, demands a fundamental realignment between human needs and the ecological and geomorphological imperatives of rivers themselves.

This book offers a comprehensive synthesis of the current status of Indian rivers, with particular emphasis on environmental degradation, riverine processes, and the spectrum of risks manifest in the Anthropocene. The book is structured in two distinct yet interconnected sections, each designed to elucidate critical dimensions of riverine science and management. Part I, titled *Fluvial Aspects of Environmental Fragility and Sustainability*, lays the groundwork by engaging with the vulnerabilities and sustainability imperatives of Indian rivers. The section commences with a lead paper that appraises the anthropogenic pressures exerted upon river ecosystems, setting the stage for a suite of subsequent contributions. These papers investigate a diverse array of themes, including water quality assessment, watershed management strategies, engineering interventions, patterns of bioaccumulation, and the valuation of ecosystem services, each substantiated by case studies drawn from various regions of the country. Importantly, the analyses are anchored in multiple datasets and methodological approaches, encompassing water chemistry, ecological databases, and socio-economic metrics, thereby reflecting the inherently interdisciplinary nature of contemporary river studies.

Part II, entitled *Fluvial Dynamics and Risk in Anthropocene*, shifts focus to the consequences of human interventions on river processes and fluvial hazards. Contributions in this section investigate topics such as flow alterations induced by dams, flood susceptibility, spatiotemporal river dynamics, bank erosion, and other emergent riverine hazards. These studies are underpinned by a robust application of advanced tools and techniques, including remote sensing and GIS, hydrological and hydro-geomorphic modelling, as well as multivariate statistical analyses. Through this methodological rigor, the section provides a compelling array of case studies that collectively enrich our understanding of the present and future challenges faced by Indian rivers. By assembling these diverse and in-depth scholarly perspectives, this book aspires not only to document the extent and complexity of riverine change in India, but also to inform the development of more resilient, scientifically grounded, and adaptive river management strategies.

This book seeks to address the diversity and complexities of the Indian rivers, emphasising the urgent need for a multi-dimensional and multi-disciplinary approach as the bedrock for any meaningful advancement in river science and management. The integration of geomorphology, hydrology, ecology, and social sciences into a unified scientific paradigm is not only timely but also indispensable for designing sustainable river management strategies that are sensitive to local and regional specificities. Furthermore, this book acknowledges the increasingly vital role of inclusive, technology-enabled platforms that connect policymakers, researchers, civil society, and affected communities. By fostering participatory forums, it is important to ensure that a diverse spectrum of voices is actively engaged in dissecting, evaluating, and implementing alternative river management pathways. The effective utilisation of advanced computational modelling, geospatial analysis, and scenario planning tools emerges, therefore, as a cornerstone for achieving transparency, scientific rigor, and actionable policy outcomes. We hope that the insights and frameworks presented herein contribute to a stronger foundation for river health and resilience across India, equipping practitioners, scholars, and stakeholders alike to navigate the challenges of environmental and socio-economic change with a commitment to evidence-based, equitable, and forward-thinking solutions.

Human civilisations have, since antiquity, flourished along riverbanks, underscoring the foundational role rivers have played in sustaining societal development. Rivers have not merely shaped landscapes but have been intricately woven into the fabric of human existence. However, mounting anthropogenic pressures—principally stemming from burgeoning populations and escalating water demand—have precipitated a global decline in river health and availability of water resources. India, despite being endowed with several major river systems, is not immune to these challenges; many of its rivers now exhibit alarming signs of ecological distress. The ongoing degradation of riverine health, as manifested in diminishing water quality, loss of aquatic and riparian biodiversity, and disruptions to fundamental ecosystem processes, constitutes a matter of urgent concern with profound implications for both human welfare and the broader environment. These disturbing trends starkly illuminate the necessity for holistic, science-driven, and adaptive river management paradigms—ones capable of addressing complex ecological pressures and balancing the often-conflicting demands placed on freshwater systems.

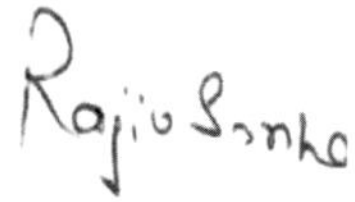

IIT Kanpur, Uttara Pradesh, India
July 2025

Dr. Rajiv Sinha

Dr. Rajiv Sinha is a professor at the Department of Earth Sciences, Indian Institute of Technology Kanpur, India. He stands as one of India's most eminent scholars in the field of large river research. Over his distinguished career, his pioneering contributions have transformed the scientific understanding of hydrological processes and the geomorphic diversity of major river systems, with particular emphasis on the interplay of hinterland controls, channel morphology, and river dynamics. Professor Sinha was the Founding Head of the Department of Earth Sciences at the Indian Institute of Technology Kanpur, (established in 2014), following his tenure as faculty in the Department of Civil Engineering at the same institution. With over 200 peer-reviewed publications and close to 12,000 citations (*h*-index: 62), Professor Sinha's scholarly output has received wide national and international recognition. He is an elected Fellow of the Indian Academy of Sciences, Bangalore, and the National Academy of Sciences, Allahabad.

Preface

India's rivers have long been the lifeblood of its civilization, shaping not only the nation's landscapes but also its socio-economic, cultural, and ecological foundations. From the fertile Indo-Gangetic plains to the intricate river networks of the peninsula, these waterways nourish agriculture, fuel industries, support biodiversity, and provide essential drinking water to millions. Yet, as we move deeper into the Anthropocene—a geological epoch defined by significant human impact—the rivers of India are under growing distress.

This edited volume, *Reimagining Indian Rivers for Sustainability*, emerges from the urgent need to understand and address the complex challenges confronting India's fluvial systems. Climate change, with its manifold expressions—rising temperatures, shifting precipitation regimes, retreating glaciers, and extreme hydrological events—is disrupting river hydrology, geomorphology, and ecological integrity. Compounded by anthropogenic pressures such as urbanization, pollution, damming, sand mining, and land use changes, these forces are reshaping riverine environments in unprecedented ways.

The objective of this book is twofold: first, to assess the current and emerging impacts of climate and human-induced changes on the rivers of India; and second, to identify pathways toward mitigation, adaptation, and long-term sustainability. In doing so, the volume brings together interdisciplinary perspectives and evidence-based research from diverse academic fields—geography, geomorphology, hydrology, environmental science, ecology, water economics, and sociology.

This book is divided into two principal sections (Part I and II), each of which focuses on a specific aspect of the Indian rivers. As an introductory note, the first chapter looks at how the regional fluvial ecosystems, geomorphology, sustainable development of fluvial environment and human populations are influenced and affected by climate change, geohazard, pollution, changes in land use and land cover, and infrastructural development. The foundation is laid by **Part I**, '*Fluvial Aspects of Environmental Fragility and Sustainability*' (collection of eleven book chapters), which discusses the sustainability imperatives and vulnerabilities of Indian rivers. With the support of case studies from around the nation, these articles explore a wide range of topics, such as environmental vulnerability, water quality evaluation, watershed management tactics, engineering solutions, bioaccumulation patterns, and the value of ecosystem services. The second part (collection of fourteen book chapters), *Fluvial Dynamics and Risk in Anthropocene*, is on how human activities affect river processes and fluvial hazard. This section's contributions explore issues including flood vulnerability, bank erosion, and other emerging riverine hazards, as well as dam-induced changes.

Through a collection of scholarly contributions, this book offers not only an analysis of river degradation but also a forward-looking vision for resilience—restoring wetlands, reviving riparian ecosystems, rewilding river channels, and adopting integrated river basin management strategies. It calls for a systemic rethinking of how rivers are valued, governed, and sustained in the face of climatic and societal transitions.

This work aspires to inform researchers, policymakers, practitioners, and citizens alike. By fostering dialogue and collaboration across disciplines and sectors, it aims to contribute meaningfully to the evolving discourse on river sustainability in India and inspire collective action toward preserving these irreplaceable natural assets for future generations.

Dr Sandipan Ghosh

Bardhaman, West Bengal, India
July 2025

Dr. Sandipan Ghosh

Acknowledgements

I am deeply grateful to the individuals of Springer Nature that have supported to the creation of this edited book. Their support, guidance, and expertise have been invaluable in helping us to bring this project to fruition. I am indebted to Prof. Racheal Ballard (Publishing Editor in Palgrave Studies in Environmental Sustainability, Springer Nature) for this book project. I owe a debt of gratitude to Prof. Rajiv Sinha (Department of Earth Sciences, Indian Institute of Technology Kanpur, India) and Prof. Sanat Kumar Guchhait (Department of Geography, The University of Burdwan, West Bengal, India) for their unwavering academic support and advice throughout the development of this book project.

I am deeply grateful to the esteemed authors who have contributed their expertise and knowledge to this book through their chapters. Their contributions have enriched our understanding of the Indian river dynamics and its complex challenges. We appreciate the time, effort, and expertise that each of these authors has brought to their chapters, and I am grateful for their contributions to this book.

I would like to express our sincere gratitude to the riverine communities of India, who have shared their stories, wisdom, and experiences with me. Your resilience, adaptability, and generosity of spirit are an inspiration to all.

Finally, I would like to acknowledge the beauty, majesty, and power of the Indian rivers which have inspired and awed us throughout this project. May this book contribute to a deeper understanding and appreciation of this incredible region and its people.

Contents

List of Contributors

Shukla Acharjee Centre for Studies in Geography, Dibrugarh University, Dibrugarh, Assam, India

Pratibha S. Agrawal Department of Applied Chemistry, Laxminarayan Innovation Technological University, Nagpur, India

Tarig Ali Department of Civil Engineering, American University of Sharjah, Sharjah, UAE

Gayatri Barabde Department of Environmental Science, The Institute of Science, Dr. Homi Bhabha State University, Mumbai, India

Shreya Basak Department of Geography, University of Calcutta, Kolkata, India

Subhankar Bera Burdwan Model School, Purba Bardhaman, West Bengal, India

Arup Bishram Centre for the Study of Regional Development, Jawaharlal Nehru University, New Delhi, India

Akshay Botle Department of Environmental Science, The Institute of Science, Dr. Homi Bhabha State University, Mumbai, India

Advait G. Burande Department of Applied Chemistry, Laxminarayan Innovation Technological University, Nagpur, India

Rabin Chakrabortty Department of Civil Engineering, American University of Sharjah, Sharjah, UAE

Pankaj Chakraborty 21St Century GIS Academy, Kolkata, India

Saptadipa Chakraborty Department of Geography, Vedanta College, Kolkata, India

Komal R. Chandankhede Department of Applied Chemistry, Laxminarayan Innovation Technological University, Nagpur, India

Akshay Chavan Department of Botany, The Institute of Science, Dr. Homi Bhabha State University, Mumbai, India

Nandini Choudhury Department of Earth Sciences, Indian Institute of Engineering Science and Technology, Shibpur, Howrah, West Bengal, India

Anup Das North Eastern Space Applications Centre, Umiam, Meghalaya, India

Biyas Dey Landscape Ecology Laboratory, Department of Geography, Jadavpur University, Kolkata, India

Shreyasee Das Mistry Post Graduate Department of Geography, Krishnagar Government College, Nadia, West Bengal, India

Debajit Datta Landscape Ecology Laboratory, Department of Geography, Jadavpur University, Kolkata, India

Asish Dhara Department of Geography, The University of Burdwan, Burdwan, West Bengal, India

Kishore Dhara Office of the Dy. Director of Fisheries, Gour Banga Zone, Malda, India

Olena Dubovyk Department of Earth System Sciences, Geography Institute, University of Hamburg, Hamburg, Germany

Lopamudra Ganguly Department of Geography, Vedanta College, Kolkata, India

Arnab Ghosh 21St Century GIS Academy, Kolkata, India

Debika Ghosh Post Graduate Department of Geography, Krishnagar Government College, Nadia, West Bengal, India

Sandipan Ghosh Department of Geography, Chandrapur College, Bardhaman, India

Santu Ghosh Publication Unit, Registrar Section, The University of Burdwan, Bardhaman, India

Surajit Ghosh Ecotoxicology Research Laboratory, Department of Zoology, The University of Burdwan, Bardhaman, West Bengal, India

Manisha Giri North Eastern Space Applications Centre, Umiam, Meghalaya, India

Ananta Gope Department of Geography, Vivekananda Mahavidyalaya, Sripally, Burdwan, India

Mohammad Irfan Department of Earth Sciences, University of Kashmir, Srinagar, India

Umesh Kakde Department of Botany, The Institute of Science, Dr. Homi Bhabha State University, Mumbai, India

Sandeep Kumar Department of Zoology, Doctor Harisingh Gour Vishwavidyalaya (A Central University), Sagar, India;
Aquatic Biodiversity Unit, Department of Zoology, Hemvati Nandan Bahuguna Garhwal University (A Central University), Srinagar, India

Prabir Kumar Kundu Department of Geography, Siliguri Mahila Mahavidyalaya, Siliguri, Darjeeling, West Bengal, India

Avijit Mahala Department of Geography, Swamy Shradhanand College, University of Delhi, Delhi, India

Ahamadul Hoque Mandal Ecotoxicology Research Laboratory, Department of Zoology, The University of Burdwan, Bardhaman, West Bengal, India

Atin Kumar Mitra Department of Earth Sciences, Indian Institute of Engineering Science and Technology, Shibpur, Howrah, West Bengal, India

Ayan Mondal Department of Zoology, Government General Degree College, Mohanpur, Paschim Medinipur, West Bengal, India

Ranjita Mondal Department of Geography, Vedanta College, Kolkata, India

Subhadip Mondal Department of Geography, Vedanta College, Kolkata, India

Neha Mourya Department of Botany, The Institute of Science, Dr. Homi Bhabha State University, Mumbai, India

Shambhu Nath Sing Mura Department of Geography, Vivekananda Mahavidyalaya, Sripally, Burdwan, India

Hasiba Nasrin Department of Geography, Vedanta College, Kolkata, India

Biswajit Nath Aerospace Information Research Institute (AIR), National Engineering Laboratory for Satellite Remote Sensing Applications, Chinese Academy of Sciences (CAS), Beijing, P. R. China; Lab of Geoinformatics and Earth Observation Research (LGEOR), Department of Geography and Environmental Studies, Faculty of Biological Sciences, University of Chittagong, Chittagong, Bangladesh

Prakash Nautiyal Aquatic Biodiversity Unit, Department of Zoology, Hemvati Nandan Bahuguna Garhwal University (A Central University), Srinagar, India

Alok Pandey Centre for Remote Sensing and GIS, School of Earth and Environmental Science, Central University of Himachal Pradesh, Shahpur, India

Kusum Pandey G. B. Pant National Institute of Himalayan Environment, Garhwal Regional Centre, Srinagar, Uttarakhand, India

Padmini Pani Centre for the Study of Regional Development, Jawaharlal Nehru University, New Delhi, India

Uttam Pawar Department of Geography, Shri Bhausaheb Vartak Arts, Commerce and Science College, Mumbai, India

Jitendra Rajput ICAR-Indian Agricultural Research Institute, New Delhi, India

Arka Prava Ray Mirjapur High School, Bankura, West Bengal, India

Paramita Roy Department of Geography, The University of Burdwan, Burdwan, West Bengal, India

Arpita Saha Department of Geography, Chanchal College, Malda, West Bengal, India

Arup Kr Saha Chandrapur College, Purba Bardhhaman, West Bengal, India

Nimai Chandra Saha School of Life Science, Swami Vivekananda University, North 24 Parganas, West Bengal, India;
Department of Zoology, Bidhannagar College, Kolkata, India

Shubhajit Saha Ecotoxicology Research Laboratory, Department of Zoology, The University of Burdwan, Bardhaman, West Bengal, India

Sayli Salgaonkar Department of Environmental Science, The Institute of Science, Dr. Homi Bhabha State University, Mumbai, India

Joy Sanyal Department of Geography, Presidency University, Kolkata, India

Joyeta Sarkar Department of Geography, Presidency University, Kolkata, India

Arindam Sarkar Department of Geography, Purash Kanpur Haridas Nandi College, University of Calcutta, Kanpur Puras, India

Ankita Singh Department of Environmental Science and Limnology, Barkatullah University, Bhopal, Madhya Pradesh, India

Nishant Singh Department of Civil Engineering, NRI Institute of Information Science and Technology (NIIST), Bhopal, Madhya Pradesh, India

Pratiksha Singh Department of Zoology and Applied Aquaculture, Barkatullah University, Bhopal, Madhya Pradesh, India

Rahul Tiwari Department of Chemistry, Institute of Basic Science, Dr. B. R. Ambedkar University, Agra, India

Junjun Wu Aerospace Information Research Institute (AIR), National Engineering Laboratory for Satellite Remote Sensing Applications, Chinese Academy of Sciences (CAS), Beijing, P. R. China

Shanlong Wu Aerospace Information Research Institute (AIR), National Engineering Laboratory for Satellite Remote Sensing Applications, Chinese Academy of Sciences (CAS), Beijing, P. R. China

Aixia Yang Aerospace Information Research Institute (AIR), National Engineering Laboratory for Satellite Remote Sensing Applications, Chinese Academy of Sciences (CAS), Beijing, P. R. China

Bo Zhong Aerospace Information Research Institute (AIR), National Engineering Laboratory for Satellite Remote Sensing Applications, Chinese Academy of Sciences (CAS), Beijing, P. R. China

List of Figures

Fig. 7.8 This image showcases a semi-arid riverine landscape at Chakdaha-Madhuban, Bankura (Latitude: 23° 5′39.39″ N, Longitude: 87°20′32.27″E), where the river Berai meets the Dwarkeshwar. It illustrates a seasonal stream highlighting essential river aggradation and ecological stress characteristics. A narrow, shallow stream winds through the scene, clearly showing reduced water flow typical of non-perennial rivers during post-monsoon or dry seasons. The stream's sinuous shape and exposed mid-channel bars reveal ongoing sediment deposition (aggradation). A dark area along the stream may indicate bank erosion, but the prevalence of sandy and silty bars points to sediment overload from upstream sources. The image features standard sandbars and point bars in aggradation zones where the sediment load surpasses the river's carrying capacity. These deposits are sparsely vegetated, indicating that aggradation has occurred and that the sites are nearing stability. The vast floodplain is devoid of vegetation or has scant grass, indicating periodic flooding, erosion, or human activity such as grazing or farming. On the left bank, grasses and reeds dominate the riparian area, which helps to stabilize the bank and indicates damaged riparian vegetation in transition. The far tree cover suggests a move towards more stable, less disturbed upland zones. The striking contrast between the lush uplands and sandy lowlands highlights the continuous fluvial changes. Aggradation Prevails: This landscape exemplifies aggradation dominance, where sediment deposition outpaces removal. Significant implications of the features depicted in this image include reduced water-holding capacity of the river, increased flood risk during monsoons due to choked channels, and a decline in aquatic biodiversity and ecosystem services 336

Fig. 7.9 This image displays the Berai riverine environment, highlighting elements that demonstrate seasonal hydrology and riparian vegetation dynamics, most likely within a degraded or aggradation fluvial situation. It has a shallow, softly flowing channel with brownish water, indicating a moderate to low discharge, probably after recent rains or during the post-monsoon season. The presence of floating aquatic plants, likely Eichhornia or other invasive species, suggests nutrient-rich stagnant zones, indicative of channel aggradation and reduced flow velocity. The expansive, flat area between the riverbanks and the islands, as well as the lateral bars, is covered with grass and aquatic weeds, suggesting sediment deposition. The vegetation-covered depositional zones confirm that aggradation has stabilized, indicating this river segment has been accumulating and retaining sediment over various seasons. On the right bank, a diverse community of tall grasses and shrubs typically shows early successional species establishing on degraded or aggraded banks. In contrast, the left bank has high tree cover, which could indicate a semi-stable natural levee or a largely undisturbed area. The right-side floodplain appears to be more open, possibly used for grazing or subjected to repeated human pressure. The dense vegetation in the channel, particularly floating weeds, suggests a diminished flushing capacity, often a result of reduced water discharge and sediment overload, indicating ongoing transformations in the riverine ecosystem driven by both natural deposition and human-altered catchment processes 337

Fig. 7.10 This image illustrates a riverbank slope that showcases distinct sedimentary and erosion features. The layered profile consists of a dark brown upper horizon, rich in organic matter, which is typical of topsoil. Beneath it lies a yellowish to reddish sandy-silt layer, which is looser and poorly compacted, typical of fluvial deposits. These layers likely indicate episodic sediment deposition, each stratum formed during high-flow flood events, shaping the river's aggradation history. The steep, eroded face displays crumbling blocks and loose clods, indicating active bank failure due to lateral erosion during high discharge, slumping, and sheet erosion, particularly following rainfall. The presence of angular clumps also suggests physical weathering, possibly caused by dry–wet or freeze–thaw cycles. Sparse grasses and pioneer vegetation struggle to establish themselves, especially at the slope's top and base. The absence of deep-rooted plants suggests an unstable substrate and ongoing soil erosion or human-induced clearance (e.g., grazing, land-use change). This exposure exemplifies classic riverbank degradation, characterized by the loss of cohesive vegetation cover, oversteepening of the bank (likely spurred by channel migration or monsoonal flash floods), and poor soil binding capacity in the alluvial material, resulting in accelerated slumping. Such degraded banks significantly contribute to increased sediment load downstream, the formation of mid-channel bars through redeposition (i.e., aggradation elsewhere), and potential badland expansion if left unchecked, particularly in lateritic and semi-arid regions 338

Fig. 22.7 Changes in the main course of the Kopai River from 2005 to 2024. Over the last twenty years, the Kopai River has drastically changed its morphodynamics and hydrology, including lateral shifts, degradation, sediment imbalance, and anthropogenic stress. These factors have combined to create a more seasonal, fragmented, and stressed river system. Furthermore, it is emphasized that sustainable development and the restoration of the river are urgent requirements to preserve its socio-economic and ecological balance. The Kopai River's main course changes from 2005 to 2024 have been analyzed using multi-temporal satellite imagery, field observations, and secondary data. Evidence shows lateral shifting of the river course at Labhpur and Chondrapur, highlighting significant meander migration due to high sediment loads and low bank resistance. The channel is widening in the upper reaches and narrowing downstream, especially in areas affected by embankments and human activities. Intermittent dry periods occur in the basin during the pre-monsoon months, particularly around Nanoor and *Surul*. Aggradation is evident in certain sections of the Kopai River, caused by reduced stream velocity and sediment trapping due to the construction of a dam and barrage for irrigation and water retention. Bed degradation and river incision in the middle and lower reaches of the Kopai are driven by sand and clay mining operations. Additionally, post-monsoon periods see the formation of mid-channel bars and shoals. The floodplain area faces encroachment, significantly reducing riparian vegetation, exacerbating bank erosion and sediment influx into the river. The peak flow magnitude has decreased, and monsoonal surges have become shorter, affecting the river's ability to reshape its channel. Irregular rainfall patterns and human intervention have contributed to the loss of connectivity with oxbow lakes and channel abandonment. The seasonal channel of the Kopai is currently facing a shortage of freshwater 805

List of Tables

CHAPTER 1

Sustainability of Rivers in India: Addressing Leading Concepts, Issues, Challenges and Solutions in the Anthropocene

Sandipan Ghosh

1.1 Introduction

The rivers of India have long been revered as sacred lifelines, nourishing civilizations, ecosystems, and economies across the subcontinent. However, in the current epoch—widely referred to as the Anthropocene, characterized by the dominant influence of human activity on the Earth's systems—these rivers face unprecedented threats. Pollution, over-extraction, habitat fragmentation, and climate change have pushed many river systems into ecological crisis. Traditional approaches to river management, often rooted in exploitation and fragmented governance, are proving inadequate in addressing these complex, interlinked challenges. To ensure the long-term health of Indian rivers, a fundamental rethinking is imperative. Reimagining rivers for sustainability in the

S. Ghosh (✉)
Department of Geography, Chandrapur College, Bardhaman, India
e-mail: sandipanghosh19@gmail.com

S. Ghosh (ed.), *Reimagining Indian Rivers for Sustainability*, Palgrave Studies in Environmental Sustainability,
https://doi.org/10.1007/978-3-032-04754-0_1

Anthropocene calls for a shift towards integrated, ecosystem-based, and community-driven approaches that respect the natural flow, function, and cultural significance of rivers. It demands the blending of modern science with traditional wisdom, and the creation of resilient policies that can withstand and adapt to ongoing environmental change. This vision not only safeguards water security but also affirms our ethical and ecological responsibility to future generations. Reimagining Indian rivers in the Anthropocene requires a paradigm shift—from exploitation to restoration, from isolation to integration. A sustainable future for rivers lies in blending modern science with traditional wisdom, strong policies with local action, and ecological awareness with collective responsibility. In the following discussion, leading issues, ideas, and concepts are analysed with examples from India.

1.2 The Rivers of Indian Civilization

In ancient India, rivers have long played a significant role in daily life. Rivers have been the lifeblood of several civilizations since the dawn of time. They served as supplies of water in the past, but they also aided in trade, farming, and cultural development. The historical foundation of India's hydrologic knowledge spans several millennia, including the Vedic Period (c. 1500–500 BCE) and the Harappan civilization (c. 3000–1500 BCE). In ancient India, the necessity to manage water drove the development of hydrologic science, as it did in other ancient civilizations throughout the world. The creation of intricate hydraulic structures, centralized and decentralized wastewater disposal systems, and wastewater treatment techniques are all examples of how advanced water sciences were in ancient India, and the Harappan civilization is a prime example of this. The building of reservoirs, canals, and dams with spillways is what defines the Mauryan Empire (c. 322–185 BCE), which is recognized as the first "hydraulic civilization". Significant Holocene climate changes differed from place to place, but floodplain environment dynamics were as variable, with changes in river channels having a major impact on ancient human settlements (Macklin & Lewin, 2015). These had an effect on the following five flooding and settlement scenarios: (i) alluvial fans and aprons; (ii) rivers that move laterally; (iii) rivers that have built levees and flood basins; (iv) river systems that are characterized by floodouts and avulsions; and (v) vast wetlands supplied by rivers (Macklin & Lewin, 2015).

Rakhal Das Banerjee and Daya Ram Sahni, two archaeological Survey of India personnel, explored the Mohenjodaro and Harappa archaeological sites in 1921–2022. On the banks of the River Indus and its tributaries, further villages and cities of this civilization were discovered prior to 1947 through archaeological expeditions and excavations. As a result, the "Indus Valley Civilization" (IVC) was adopted as a name. But with Pakistani archaeologist M.R. Mughal's studies and discoveries of 363 archaeological sites in Bahawalpur, on the dried-up bed of the River Saraswati, a major shift occurred. The dried-up bed of Saraswati-Drishadvati in the states of Rajasthan, Haryana, and Uttar Pradesh also yielded some 250 archaeological monuments. More than sixty-three per cent of the archaeological sites are situated along the banks of the Saraswati and Drishadvati rivers, according to recent study conducted by Michel Danino of the Indian Institute of Technology, Gandhinagar. Between the banks of the Indus River are around 21% of all archaeological sites. Consequently, the site distribution data show that the term "Indus-Saraswati Civilization" (ISC) is a suitable appellation. The Saraswati or Ghaggar-Hakra River, which formerly ran across northwest India and eastern Pakistan, and the Indus River, one of Asia's principal rivers, were the sites of the Indus-Sarasvati Civilization, which peaked around thirteen centuries ago. Most of the found sites are along trade routes connecting major metropolitan centres or along these important rivers and their tributaries. In the Indus-Sarasvati area, discoveries have been made from 7000 to 300 BCE.

There has long been discussion about the role that great rivers play in supporting civilizations across the world. A common method of supporting the so-called river-culture theory is the use of satellite imagery to map palaeochannels and superimpose known archaeological sites along them. Khan and Sinha (2019) found that the ISC, whose demise was ascribed to the deterioration of a sizable river (the Saraswati) that formerly traversed the area, was connected to a broad and extensive trace of a palaeochannel (the Ghaggar-Hakra channel) mapped from satellite images in northwest India. This vast palaeochannel was evidently abandoned soon after ~ 8 ka, and this was well before the Mature Harappan Civilization (~4.6–3.9 ka) emerged in this area. These results severely disputed the idea that the ISC and the huge river occurred at the same time. Rather, they pointed to river avulsion as a key element in the beginning and development of Harappan culture. Chronology of the ISC cultural periods include: (i) an Early Ravi Phase at ~ 5.7–4.8 ka BP, growth of agricultural communities, (ii) Transitional Kot Diji phase at ~ 4.8–4.6 ka BP, (iii) Mature phase at ~ 4.6–3.9 ka BP during which urbanization peaked, and (iv) Late declining (Painted Grey Ware) phase at 3.9–3.3 ka BP, which marked the collapse or deurbanization (Khan & Sinha, 2019). The trench sediment log architecture and archaeological sites of the Saraswati River palaeochannels in northern India (Haryana) were being characterized by Chaudhri et al. (2021), along with the potential palaeo-path of this ancient river, using remote sensing technology and ^{14}C dating, and optically stimulated luminescence chronology. The constant water flow in the Saraswati River input channels in the ISC before 14 ± 3 ka until 2 ± 0.3 ka AD 1402, the archaeological sites' close proximity to the Saraswati River channels, and the persistence of culture in Rakhigarhi and Bhirrana from the Mesolithic pre-formative stage to Harappan Urbanism in Mature Harappan Phase 2 period, all attest to the Saraswati River system's role in supporting ancient settlements in Haryana (Chaudhri et al., 2021).

Rivers were essential to the existence and prosperity of the Harappan city of Dholavira (an archaeological UNESCO world heritage site in Khadirbet, Gujarat, western India) because of their function in resource availability and water management (Fig. 1.1). Two seasonal streams, Manhar and Mansar, provided the most of the fresh water, even though the Arabian Sea was close by. Dholavira is the fifth-biggest Harappan site in Asia and one of the two largest in India. Dholavira offers a singular view of the layout of a whole city rather than simply a portion of it, in contrast to other Harappan sites in Gujarat like Lothal. After excavation, the site showed a carefully designed metropolis with separate areas for business, residential, and administrative use. The capacity of Dholavira can endure the severe desert climate is among its most amazing features. The Harappan civilization's inventiveness is demonstrated by the city's vast reservoirs, water conservation system, and well-planned buildings. One of the side chambers of the city's northern entryway, commonly referred to as the Dholavira Signboard, is where the most important findings at Dholavira were made. On a massive wooden board, the Harappans had organized and put fragments of the mineral gypsum to create 10 gigantic characters or symbols.

Fig. 1.1 The rivers have long played a significant role in evolution of ancient civilization, like the archaeological site of Dholavira (23°53'13.12"N, 70°12'48.19"E). Dholavira, on July 27, 2021, the ancient Gujarati city of Dholavira was included into the UNESCO World Heritage List. It is situated in the Kutch area. Situated in the ISC's five greatest Harappan sites, the city was inhabited from around 3500 BCE (pre-Harappan period) to 1800 BCE (late Harappan period)—(a) aerial plan view of the site, (b) field photograph of ancient urban plan, and (c) example of ancient seal (Bull Seal) of ISC discovered from this site

1.3 Salvation by Water in Riverine Society

In India, rivers are revered and worshipped, often personified as goddesses or deities. This reverence extends to their environmental importance, as these rivers are considered lifelines, sustaining the land, biodiversity, and the lives of millions. Since Vedic period, the rivers are connected to the philosophy of "divine ecology" which refers to the relationship between humanity, nature, and the divine, emphasizing the spiritual and ecological aspects of creation. For Indian society and the diaspora, the Ganga, the holiest of all rivers, is extremely important. Some attribute this significance to the societal and economic opportunities it offers, while others attribute it to religious and spiritual beliefs and mythology. There is no need for a formal introduction to the river's varied relationship with Indians and their culture, as this relationship is ingrained in every Indian's mind from an early age through folklore, references to scripture, and a variety of customs and traditions. These, in turn, encourage pilgrims to visit the river's banks throughout the year, from its source in the Himalayas to its mouth on the Bay of Bengal. Every year, tens of millions of pilgrims, motivated by religion, take a sacred dip in the Ganga, and the rites they follow are connected to some aspect of their life. Given that over 37% of Indians reside in its basin and the remainder are either directly or indirectly connected to it, the Ganga is undoubtedly an integral part of Indian culture. Ganga water, also called "Gangajal", is revered and is commonly available in most Hindu homes throughout the nation. As such, it is highly suitable for listing on the Geographical Indication registry.

Interestingly, Gangajal has an unlimited shelf life and is thought to contain hitherto unknown chemical and biological qualities, which contribute to its holiness. Since Gangajal is thought to have more medicinal qualities than any other river's water, ayurveda and naturopathic practitioners use it to treat a variety of illnesses due to its paranormal properties. This magical quality is ascribed to the abundance of minerals and botanicals found in the river's higher Himalayan catchments. In both water and sediment samples of the Ganga, more pathogenic bacteria than opportunistic pathogenic and non-pathogenic bacteria were found within the bacterial diversity detected, according to a study titled "Assessment of Water Quality and Sediment to Understand Special Properties of River Ganga". A comparable pattern was noted in bacteriophages. From Bijnore to Varanasi, the greatest number of bacteriophages was found in water and sediment samples. The flora and animals of Ganga Jal are diverse and contribute to the maintenance of a distinct environment. Efforts to save and restore the river have been sparked by the belief in its holiness, despite pollution issues.

In Hinduism and Indian culture, the Ganga River, often known as the "Mother Ganga", is extremely important. A vital part of Hindu spirituality and the socio-economic structure of India, the Ganga is revered as a goddess of forgiveness and purity. Bathing in its water, according to Hindu belief, cleanses sins and aids in achieving moksha, or freedom from the cycle of life and death. Pilgrimage sites like Varanasi, Haridwar, and Rishikesh along the river's bank are significant for rituals and spiritual practices (Fig. 1.2):

- Varanasi (25°18′10.56"N, 83° 0′19.51"E)—known for its ancient ghats (its history stretches back over 3,000 years), it is believed that dying here ensures liberation;
- Haridwar (29°57′12.83"N, 78°10′16.33"E)—the ancient site (700 BCE and 1200 BCE) where the Ganga enters the plains, famous for the Ganga Aarti Ceremony; and
- Rishikesh (30° 6′32.98"N, 78°18′23.66"E)—a hub for yoga and meditation along the river bank (Yoga Capital of the World), offering spiritual awakening.

Another key element of river worship is the "Mahakumbh Mela", which takes place in the Ganga valley and is a very important Hindu festival, particularly the one held at Prayagraj once every 144 years. With millions of visitors seeking spiritual freedom and cleansing by holy dips in the confluence of three rivers (the Ganga, Yamuna, and the legendary Saraswati), it's the greatest peaceful gathering in the world and a profound religious experience. The celebration serves as a forum for spiritual discussion, bringing together sages, academics, and followers to share thoughts and broaden their grasp of religion. After the Mahakumbh Mela in Prayagraj in 2025, the next one is expected to happen in 2169. From the above statements it is cleared that rivers are seen as being extremely sacred and life-giving in Indian culture, and they are widely used in daily life as well as in religious rituals and spiritual beliefs, as mentioned in table. The rivers' metaphysical, transcendental, and symbolic connotations submerge and frequently alter their corporeality. Within a holy discourse, the story of India's major and smaller rivers—including the

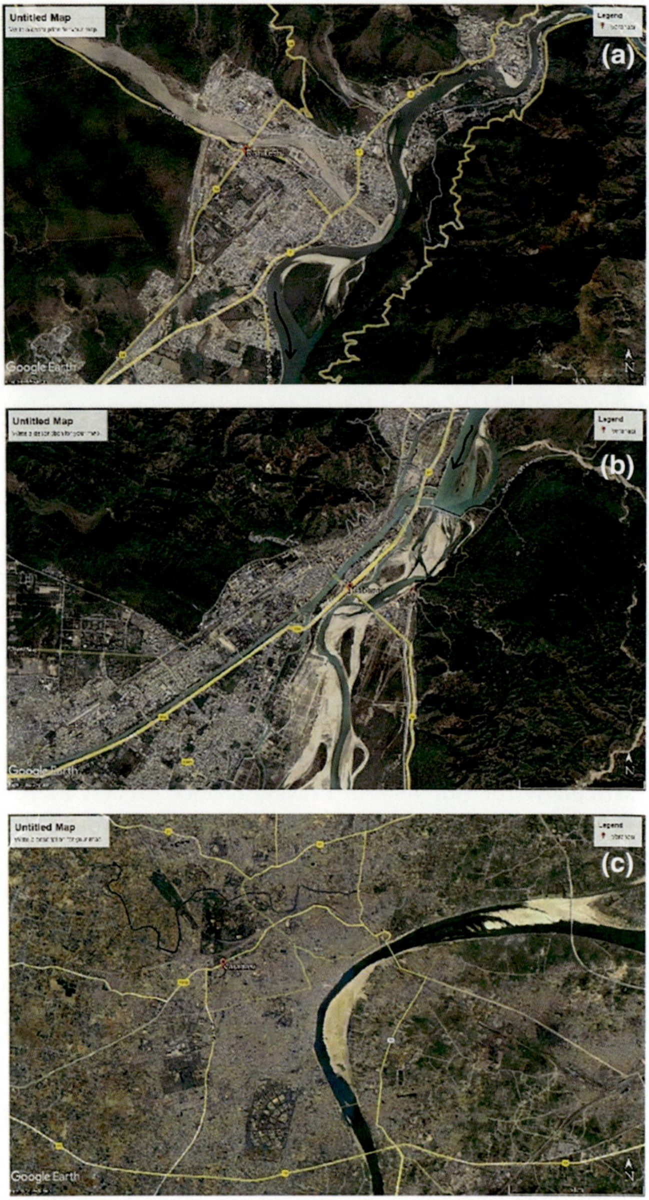

Fig. 1.2 The Ganga River has historically been a vital artery for several ancient cities, like (c) Varanasi (also known as Banaras or Kashi), (b) Haridwar, and (a) Rishikesh (a). The Ganga River‘s influence extends beyond these specific cities, shaping the broader cultural and economic landscape of the region

Yamuna, Sarasvat, Narmada, and Ganga (Ganges)—is told. The spiritual discourse of ancient faiths included the idea that the divine is present in nature (Fig. 1.3 and Table 1.1).

1.4 Environmental Philosophy and River

Within the framework of the Gaia hypothesis, rivers are crucial components of the Earth's self-regulating system. They act as a vital link between the atmosphere, lithosphere, biosphere, and human sphere, influencing the environment and supporting life. The Gaia hypothesis, also known as Gaia theory or Gaia principle, proposes that all organisms and their inorganic surroundings on Earth are closely integrated to form a single and self-regulating complex system, maintaining the conditions for life on the planet (Lovelock, 1972). River is an integral part of the Gaia hypothesis:–

- Rivers transport water, nutrients, and sediments, connecting different parts of the Earth system, including the atmosphere, land, and oceans.
- Rivers play a significant role in regulating global water cycles and influencing ocean salinity, contributing to the stability of the Earth's climate and environment.
- Rivers carry essential nutrients that support aquatic ecosystems and terrestrial ecosystems along riverbanks, influencing plant growth and biodiversity.
- Human activities, such as dam construction and pollution, can disrupt river ecosystems and affect the functioning of the Earth's self-regulating system.
- From a Gaia perspective, rivers are not just physical features, but dynamic systems that interact with the broader Earth system, contributing to its overall health and stability.

The fundamentals of environmental philosophy and ethics should be applied to the treatment of rivers. The health of rivers is essential to the sustainable development of the planet because they are fundamental natural resources that sustain ecosystems and human communities. Rivers sustain biodiversity in terms of biology and habitat. Biotopes have habitats, material pathways, and barriers. The most important habitats for an organism's ability to survive and reproduce are rivers. They provide

Fig. 1.3 The Ganga River, frequently referred to as the "Mother Ganga", is a very significant part of Indian culture and Hinduism. As a goddess of forgiveness and cleanliness, the Ganga is an important component of Hindu faith and India's socioeconomic system—(a) In Varanasi, Ganga Pranam is a significant practice in Hindu culture, highlighting the reverence for the river and its role in spiritual life; and (b) people chanting the mantra, bathing in the Ganges, or simply remembering Ganga with a pure heart are believed to bring divine grace and spiritual benefits

Table 1.1 The symbolic meaning of main Indian rivers in Hindu mythology

Scared rivers	*Symbolic identity*	*Spiritual significance*
Ganga	The Holiest River in Hinduism—maa ganga	Mythology asserts that the Ganges River fell from the heavens to Earth through the matted hair of Lord Shiva. Her water is regarded as sacred, and bathing in it is said to atone for sins. Additionally, Hindus carry out pinda daan (rituals for the departed) and tarpana (offering to ancestors) on her banks in the belief that these rites aid in achieving redemption for the souls of the dead
Yamuna	The River of Love and Devotion	The tale of Lord Krishna, one of the most adored gods in Hindu mythology, is intimately associated with the Yamuna River. Legend has it that Krishna performed heavenly deeds (leelas) while growing up near the Yamuna River. To carry out religious rites and ask for the blessings of the Yamuna, many devotees travel to its banks, particularly in locations like Vrindavan, Mathura, and Prayagraj
Saraswati	The Lost River of Knowledge	The vanishing of the Saraswati River is among its most fascinating features. Legend has it that Saraswati once swept gloriously over northern India's plains before gradually disappearing. The Saraswati was once a powerful river that served as the birthplace of the Vedic culture. Saraswati represents all types of knowledge in Hinduism, including spiritual, artistic, and intellectual understanding. The goddess Saraswati wisely feeds the intellect and spirit, just like the real river did for the first civilizations
Godavari	The Southern Ganga	Particularly in southern India, the river has great cultural and spiritual importance. The Godavari, which flows through Maharashtra, Telangana, Andhra Pradesh, and other states, is linked to a number of myths and tales that emphasize its heavenly origin and cleansing properties. One of India's most cherished epics, the Ramayana, also mentions the Godavari River. During their 14-year exile, Lord Rama, his brother Lakshmana, and his wife Sita are said to have resided close to the banks of the Godavari. Every twelve years, the banks celebrate a major celebration called Godavari Pushkaram

(continued)

Table 1.1 (continued)

Scared rivers	*Symbolic identity*	*Spiritual significance*
Narmada	The River of Redemption	The river is said to have great spiritual power and holiness because it was created by the highest deity. The pilgrimage known as Narmada Parikrama is among the most distinctive features of the Narmada. From the river's source in Amarkantak to its mouth at the Arabian Sea and back on the other side, devotees trek the whole distance. This trip, which is regarded as extremely fortunate, may take months or even years to finish
Kaveri	The Lifeline of South India	The river Kaveri, which supplies water to Tamil Nadu and Karnataka in the south, is worshipped as the goddess of fertility and plenty. In addition to being a physical reality, the Kaveri River is revered as a goddess and mother who cares for the land and its inhabitants. The river is revered with great fervour, especially during special occasions like the Kaveri Pushkaram. Kaveri is recognized as a more pragmatic life giver in addition to her spiritual importance. She plays a part in supplying water for irrigation and maintaining agriculture in the lush areas she travels through

Source Shankar (2002)

people with shelter, living space, and supplies of food and water. Additionally, the habitat function will be directly impacted by the breadth, bed composition, and hydrological rhythm of rivers. Large, connected rivers with rich environmental gradients tend to have more abundant life forms than small, scattered rivers with little change in environmental characteristics. The moral link between people and the environment is the main emphasis of environmental ethics, especially when it comes to matters like river protection. It highlights how crucial it is to uphold rivers' inherent worth, acknowledge their contribution to ecosystems and human welfare, and promote sustainable behaviours. This entails striking a balance between the need to preserve river ecosystems and economic and social growth, as well as incorporating river ethics into all phases of decision-making, from legislation to policy implementation. Key aspects of environmental ethics in river conservation and stream ecology include:

- Recognizing the intrinsic value of rivers: Rivers are not just resources for human use but possess intrinsic worth, deserving respect and protection for their own sake.
- Promoting sustainability: Environmental ethics encourages the use of rivers in a way that ensures their long-term health and productivity, considering the needs of both present and future generations.
- Advocating for equitable access and use: Ethical river conservation considers the needs of all stakeholders, including local communities, downstream users, and non-human species.
- Embracing a biocentric approach: Shifting from a purely anthropocentric view (humans as the primary beneficiaries) to a biocentric view that values the well-being of all living beings, including rivers and their ecosystems.
- Emphasizing river ecological aspect: The study of river ecological conservation focuses on the interactions that occur between living things and their surroundings in a river ecosystem. It is important because numerous stresses, such as pollution, climate change, and human activity, can affect the health and stability of river ecosystems.

The significance of river ethics, emphasizing the necessity of upholding rivers' inherent right to exist in the natural world, acknowledging them as living things, and establishing moral standards to preserve their well-being and guarantee peaceful cohabitation with people. The idea of river ethics highlighted that it entails recognizing rivers as living things and upholding their fundamental rights to health and life. In order to promote peaceful coexistence and keep rivers alive and active, it urges a reorientation of moral principles, values, obligations, and behaviour. This will enable the long-term growth of human civilization. Globally, floodplain river ecosystems have been subjected to several anthropogenic constraints. A paradigm shifts towards "living with water" is starting to replace the command-and-control approach to river flood management as flood hazards increase due to climate change. The freshwater biodiversity issue and hazard management as climate-driven extremes intensify can both benefit from giving rivers greater space to flow. The primary socioeconomic and ecological component of river conservation is that it would be able to protect those rivers that are unavoidably impacted by human activity in part by managing them to minimize harm and eventually restore them to the greatest extent feasible. The traits that give river habitats their pattern, process, connection, and biological variety also make them vulnerable to human use, misuse, and modification: in addition to sustaining natural flows, hydrological connection spreads stressor impacts through river ecosystems.

1.5 Stream Ecology and Ecohydrology

An essential component of the Earth's circulatory system is the stream. They transport fertilizers, sediments, and other elements into lakes, rivers, and eventually the ocean. Rivers, lakes, and estuaries all benefit from healthy streams. However, streams are also essential ecosystems in and of themselves, offering home for species, preventing flooding, and supplying a large portion of our drinking water. Because streams are essential ecosystems that sustain biodiversity, offer ecosystem services, and are sensitive to changes in the environment, stream ecology is essential (Fig. 1.4). Better management and preservation tactics are made possible by an understanding of how these ecosystems operate, which is gained via studying stream ecology which helps predict the impacts of environmental changes, such as climate change, pollution, and habitat loss, on stream ecosystems and their ability to provide services. The study of interactions between species and their physical and chemical surroundings within a stream or river system is known as stream ecology. The impact of human activity on stream ecosystems, the significance of energy flow via food webs, and the effect of flow on habitat and biotic composition are important concepts. Key principles of stream ecology can be categorized as follows:-

- Flow as a Crucial Determinant: The kinds of plants and animals that may live in a stream are greatly influenced by the quantity and velocity of water flow. The structure and function of stream ecosystems can be significantly impacted by changes in flow regimes, such as those brought about by dams or changed land use.
- Energy Flow and Trophic Dynamics: Streams are dynamic systems in which food webs and chains allow energy to flow. Algae are examples of primary producers that absorb solar energy and subsequently pass it to consumers and decomposers.
- Ecosystem Metabolism: Through processes including primary generation and respiration, energy and matter are continuously exchanged in stream ecosystems. Stream metabolism measurements reveal information about the health of ecosystems and how they react to changes in their surroundings.

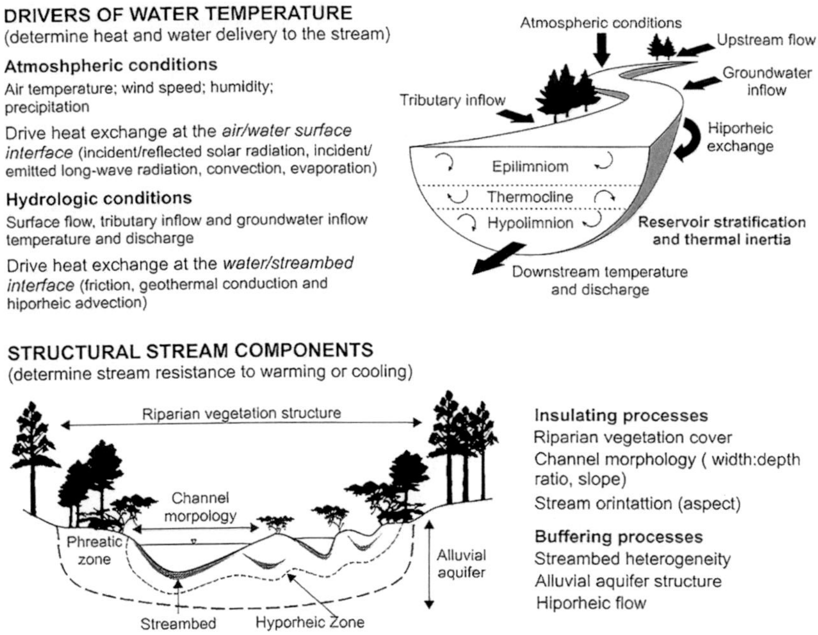

Fig. 1.4 Importance of stream ecology—rivers are ecologically vital, serving as habitats for diverse freshwater species, supporting food chains, and playing a crucial role in water hydro-thermal regulation and nutrient cycling

The new study, by Thompson et al. (2021), brings to light a significant problem regarding the importance of stream ecology. The ecological fragility and geographical variability of the threats that unchecked climate change presents to rivers throughout the world are highlighted in this study. In all river basins, there is a medium–high risk of change in high (low) flows in 21.4% (22.4%) of ensemble members for 1.0 C warming, rising to 61.5% (63.2%) for 3.0°C warming (Thompson et al., 2021). This indicates that, on a global scale, the likelihood of high risks of significant ecological change in both high and low flows increases with global warming. The research during COVID-19, by Singh and Pandey (2022), shown a remarkable result in the Ganga River to grasp sensitivity to minor changes in the environment. The COVID-19-induced lockout resulted in improvements in river ecology. The resulting limitations on human activity caused industrial operations to halt and the flow of cars, trains, and airplanes to decrease. Rivers showed noticeable changes as the water's colour became cleaner. There has been an increase in the number of migrating birds landing on riverbanks. While most people observed an increase in aquatic life in Krishna (75%), Kaveri (70%), Gandak (75%), Godavari (80%), and Gomati (78%), there has been a rise in aquatic species in Mahanadi, Narmada, Ganga, Yamuna, and Nagavali. The heavy metal index varied from 33.96 to 117.33 throughout the lockout, with 27.27% of the water being of good quality (Chakraborty et al., 2021). This indicates that the aquatic environment is not at high ecological danger because there is little mixing of hazardous metals in the river water of Damodar.

Many researchers are now concentrating on ecohydrology, which is the study of how biotic and hydrological processes affect water cycle components and how soil, soil moisture, and vegetation relate to one another at various scales, from the pore to the global level. Ecohydrology, which combines hydrology with ecology, will be useful in addressing a number of important issues pertaining to the Anthropocene's water-related sustainable development. Four main ideas highlight the characteristics of ecohydrology (Bridgewater et al., 2018): (a) combining the science of water and biodiversity at scales relevant to management; (b) being aware of how the environment is changing and how individuals may manage it; (c) recognizing the function of ecological services; and (d) using ecological characteristics as change markers. A specific area of ecohydrology called environmental flows or e-flows is essential to preserving the health of Indian rivers. The amount, time, length, frequency, and quality of water flow required to support terrestrial and aquatic ecosystems, as well as human livelihoods and well-being, are known as e-flows. Understanding biological and hydrological processes is used in ecohydrology to boost biodiversity, improve water quality, and advance sustainable development. A crucial component of this is e-flows, which guarantee that rivers have enough water to meet their biological requirements. Environmental flow requirements are now crucial because they establish the amount, time, and quality of water flow required to support aquatic ecosystems and the human lives that rely on them. The Indian government has announced minimum e-flows for the Ganga River, and the National Green Tribunal has ordered states to maintain minimum environmental flows in rivers. Numerous Indian River basins, including the Ganga, Yamuna, and Son rivers, have been the subject of e-flow studies. Maintaining a minimum flow in rivers is usually required by environmental flow requirements in India. This flow is usually represented as a percentage of the 90% reliable flow or the average lean season flow. The average lean season flow must include at least 15–25%, according to the National Green Tribunal (NGT). According to the 90% reliable year, the Ministry of Environment, Forest & Climate Change (MoEF&CC) establishes certain standards for each season, with 30% designated for the monsoon, 20% for the lean season, and 25% for the non-monsoon/non-lean seasons. According to the 1994 Memorandum of Understanding (MoU) between basin states and the National Green Tribunal (NGT), the Yamuna River in India must have a minimum flow of 10 cubic metres per second (10 cumec) downstream of the Tajewala and Okhla Headworks at all times of the year. The estimated environmental flow of the lower Damodar River (downstream of the Durgapur Barrage, West Bengal) is 1.14 to 0.79 m^3s^{-1}, which is derived using stochastic Flow Duration Curves of 60-day daily mean flows (Hoque et al., 2022).

1.6 Changing Climate and Fluvial System

A group of linked items and the procedures that connect them make up a system. The mechanisms that transport water and sediment between hillslopes, the channel network, and floodplains are all connected within fluvial systems. A network of rivers and streams known as a fluvial system carries dissolved materials, sediment, and water from highland regions to confined basins or the ocean. The drainage basin or catchment, which is a region of land that each river drains and that is enclosed by a drainage divide or watershed and that provides water and sediment to the channel, is the fundamental unit of study in fluvial systems. These systems are essential for maintaining ecosystems, supplying essential resources, and forming landscapes. By means of deposition, transport, and erosion, rivers and streams sculpt and mould the Earth's surface. In addition to transporting material downstream, fluvial channels are essential for linking highland regions to the shore. The main supply of fresh water for human usage comes from rivers and streams. Habitat for a wide range of plants and animals is provided by fluvial systems. The climatic variables listed in this section serve as the system's overall regulators. The fluvial system will undergo a complicated series of modifications and adjustments in response to any change in one of these factors. Temperature, precipitation, vegetation, sediment supply, and river discharge are all impacted by climate, which has a substantial impact on fluvial systems and alters their processes and landforms. These alterations may occur directly, as in the case of peak precipitation, or indirectly, as in the case of vegetation being impacted by permafrost. Increased erosion, deposition, and changed channel dimensions and patterns can all result from climate change, especially global warming (Fig. 1.5).

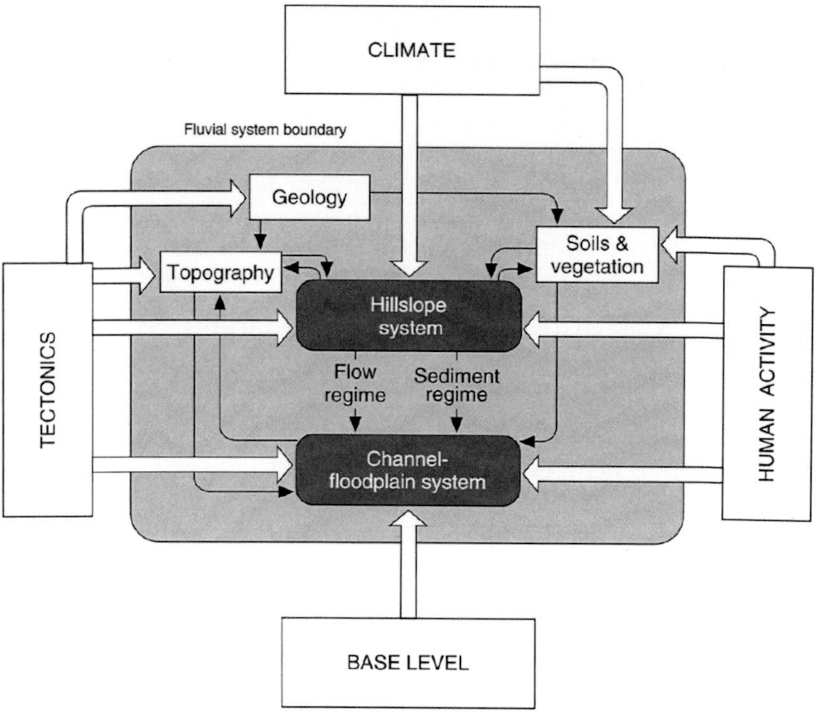

Fig. 1.5 Fluvial systems, which encompass rivers and their associated landforms, are environmentally important due to their role in shaping the landscape, supporting ecosystems, and providing essential resources. In the Anthropocene, fluvial systems are composed of natural components like rivers, streams, and floodplains, but also include human-made elements like dams, levees, and channels (after Charlton, 2007)

1.6.1 Fluvial Facies and Palaeoclimate

In order to answer the concerns on climate change analysis and prediction, it is important to rebuild the historical climates of different places and utilize the knowledge of these changes to offer the baseline data of climate variability (without human involvement) that may be used as future analogy. According to Singhvi and Kale's (2010) recent reconstructions of previous climates, significant variations in temperature and precipitation have taken place on almost all time periods, from decades to hundreds to millennia, and on virtually all geographical sizes, from tiny areas to much of the world. Any change in the climatic and environmental circumstances causes the fluvial systems to react quickly, making them the most sensitive features of the earth's surface. It is possible to reconstruct the long-term (millennial scale) and short-term (centennial scale) changes in the hydrology, morphology, and sedimentation pattern of the river basins using the palaeo-records of river planform modifications and sedimentary facies or deposits that take place during the Indian monsoon's flood events.

A study of slackwater flood deposits from the valley of Narmada River Basin yielded a continuous record of the biggest floods over the past two millennia (Kale et al., 1997). Two periods of severe flooding occur about 400–1000 AD and around 1900 AD. Again, Kale et al. (2003) found a 5000-year record of large floods ($>4500\ m^3/s$) on the Choral River of the central Narmada basin using radiocarbon dating of charcoal samples collected from a depth of 150 cm (5170 ± 135 years BP). Climate variations in the monsoonal wet-dry zone of the lower Damodar Basin (West Bengal, India) and the alluvial history from the Late Pleistocene to the Early Holocene have been reconstructed by Ghosh and Guchhait (2014) using stratigraphic exposures, fluvial archives, and borehole data. According to climatic proxies and facies structures, there were five to six climate shifts between around 14 and 6 ka. Most importantly, the Late Pleistocene—Early Holocene saw a decrease in fluvial activity of the Damodar River (meandering nature), which was primarily impacted by the post-glacial dry environment (yellowish to grey sediments and caliches) and sporadic severe floods (Ghosh & Guchhait, 2014).

1.6.2 Climatic Control on Fluvial Processes

Basins receiving more precipitation will reach equilibrium more quickly and have higher time-averaged erosion rates during landscape adjustment compared to basins receiving less precipitation, according to Marder and Gallen (2023), who noted that climate appears to have a measurable impact on bulk erodibility and landscape response time. The topography, not the long-term erosion rate, records the climatic signal at steady state. Systematic changes in the degree of nonlinearity of $E - K_{sn}$ (i.e., $E =$ long-term erosion rate and $K_{sn} =$ normalized channel steepness index) as a function of mean annual precipitation and aridity index, providing a distinct climatic signal in topographic form. Higher mean annual precipitation and decreased aridity are projected to exacerbate the nonlinearity between E and K_{sn} in bedrock channels when incision thresholds are significant. It appears that the impact of rainfall on erosion rates (22 and 3670 m My^{-1}) may be adequately represented by K_{sn} (Fig. 1.6), the increased channel steepness metric based on a proxy for discharge rather than just drainage area, given in the analytical observation that K grows monotonically with mean annual rainfall (ranges between 0.72 and 5.9 m $year^{-1}$) in the fluvial landscapes of the Bhutan Himalaya.

As a key factor in determining the availability and quality of habitat, geomorphology plays a significant role in the geomorphic response of fluvial systems to global warming in temperate regions, such as increased flood frequency and intensity. Potential geomorphic reactions include enhanced deposition and/or erosion in the floodplain, riparian zone, and river channel, with related habitat consequences. According to the current findings, the peak discharge for the future climate scenario will rise between 34.3% and 91.9% over the 100-year flood return period when compared to the current climatic scenario. Similarly, future forecasts based on GCMs Global Circulation Models indicate that the flooded area will expand by 12.6 to 26.4% in India. Pandey et al. (2025) found that the amount of water in the soil and river systems of southern India before to a rainstorm event can also have a big impact on how severe and widespread flooding is. A river's likelihood of overflowing its banks during periods of intense precipitation may be increased by higher antecedent moisture levels. The current climatic regimes of the Indian subcontinent are incredibly diverse, and this great diversity is also evident in the late Quaternary and Holocene time periods. Variable responses to monsoonal changes over time are documented (Table 1.2) in the

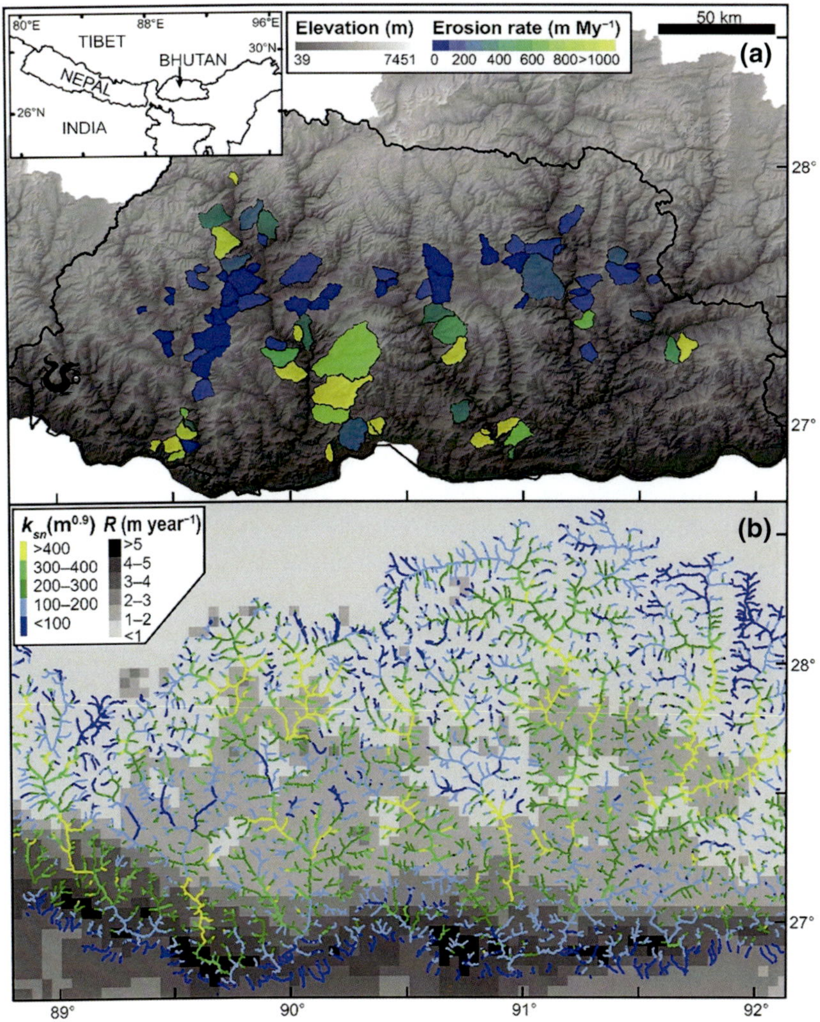

Fig. 1.6 Climatic control on fluvial erosion and channel steepness in the Bhutan Himalaya—climate significantly influences fluvial erosion in the Himalaya by controlling the amount and intensity of precipitation, glacial cover, and snowmelt, which all directly affect river discharge, fluvial kinetic energy and sediment transport

patterns of sedimentary sequences and distinctive facies seen in all of India's morpho-climatic zones' fluvial archives (Sinha et al., 2020). Rivers in various parts of India exhibit extensive floodplain aggradation during Marine Isotope Stage (MIS)-5 (105–74 ka), although they reacted quite differently during MIS-4 (74–59 ka), as evidenced by aeolian deposits in the western portion and deterioration in the Ganga Plains (Sinha et al., 2020). MIS-3 (59–28 ka) and MIS-2 (28–14 ka) saw the development of notable discontinuities in the Ganga Plains' interfluves, whereas the western Indian rivers showed erratic responses (Sinha et al., 2020). The Holocene monsoon changes are shown in the Ganga Plains through many valley filing episodes and extensive incision throughout western India. Floodplain, channel, and aeolian depositional settings were identified, by Singh and Sinha (2019), the Ghaggar-Hakra channel of the Indo-Gangetic plains (IGP)—(a) Floodplain deposition is seen in Period II (~62–30 ka); (b) while coarse-grained grey micaceous channel sand is seen in Period I (~86–64 ka); (c) Period III deposits of channel margin settings (~25–12 ka); and (d) present-day Ghaggar deposits and the great Himalayan river's discharge stopped during Period IV (~11–4.0 ka) (Fig. 1.7).

1.6.3 Changes in River Flow Patterns and Increased Flooding

Global heatwaves, dry spells, and severe glacier melting are all being exacerbated by climate change, which has also changed how rivers flow globally. The seasonality of river flow, especially in high northern latitudes, is being greatly impacted by climate change, according to recent research (Wang et al., 2024). India's rivers are experiencing both increased and decreased flows in different areas and at different times of the year as a result of climate change. Rainfall, temperature, and glacier melt variations are all connected to this unpredictability, which affects water supplies and raises the possibility of floods and droughts. About 40% of sites in northern North America displayed a notable decrease in the seasonality of river flow and in almost 25% of the stations in southeast Brazil, rising seasonality of river flow was noted (Wang et al., 2024). Seasonality of river flow may also result from human involvement, such as controlling flow through reservoirs. Human-caused greenhouse gas emissions may have an impact on the seasonality of monsoon precipitation and the dynamics of runoff that follow (Wang et al., 2024). According to some research, monsoon rainfall and runoff would rise, which might

Table 1.2 Examples of river dynamics due to past climate changes in Indian rivers (after Sinha et al., 2020)

Basin/Interfluve	*Processes active in time due to changing climate*				
	14 ka to present *MIS–1*	*28–14 ka* *MIS–2*	*59–28 ka* *MIS–3*	*74–59 ka* *MIS–4*	*105–74 ka* *MIS–5*
Yamuna valley	Aggradation—(4–7 ka), (3–2 ka) & (<2 ka); Incision—(7–11 ka), (3 ka) & (<2 ka)	Incision—(11 ka); Aggradation—(12–25 ka)	Incision—(24 ka); Aggradation—(24–37 ka)		
Alaknanda valley	Valley aggradation—(15–8 ka); Incision—(<11 ka)	Aggradation—(11–18 ka)	Aggradation—(25–49 ka); Floodplain aggradation and humid palaeosols—(40–60 ka)		
Ganga–Yamuna interfluve	Fluvial humid (6–11 ka); Incision (13–9 ka)	Aggradation Floodplain palaeosols—(20 ka)	Aggradation—(30–40 ka)	Drier palaeosols (75–60 ka)	Humid conditions; palaeosols showing extensive illuviation (90–80 ka)

(continued)

Table 1.2 (continued)

Basin/Interfluve	*Processes active in time due to changing climate*				
	14 ka to present *MIS–1*	*28–14 ka* *MIS–2*	*59–28 ka* *MIS–3*	*74–59 ka* *MIS–4*	*105–74 ka* *MIS–5*
Mahi Basin	Incision during Early Holocene	Fluvial dormancy and aeolian dunes	Mixed load meandering river; aggradation during 52–44 ka and 37–30 ka; warmer-wetter		mud deposits followed by gravel bed load braided stream deposits (34 ka); humid
Sabarmati Basin	regional aridity (6–3.5 ka); Aeolian (~12 ka)	aeolian (30–14 ka); aeolian dunes (22 ka) dry climate; fluvial aggradation & pedogenesis (18–12 ka) wet climate	mixed load and floodplain (30–80 ka)		fluvial aggradation (90–84 ka) & red soil deposition (prior to 63 ka)

Note Marine Isotope Stages—MIS

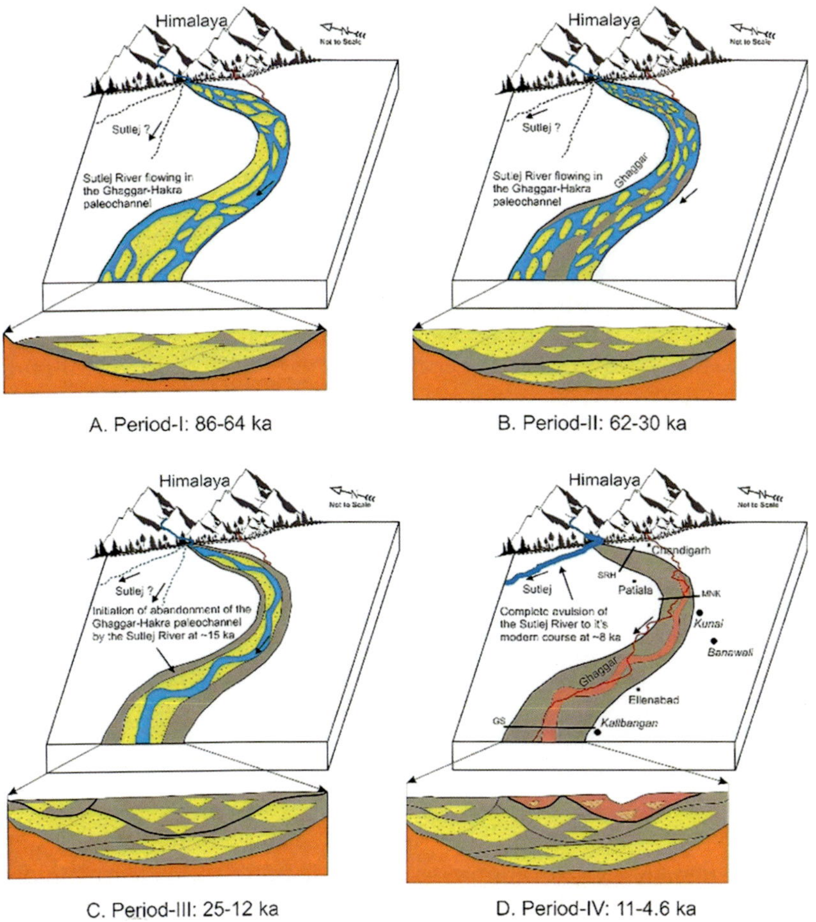

Fig. 1.7 Drainage evolution during different Late Quaternary climatic periods of fluvial aggradation in the Ghaggar–Hakra paleochannel (after Singh & Sinha, 2019)—(A) Widespread braided channel fluvial activity. (B) The river channel and the floodplain it is situated in, which is constantly changing throughout the Palaeovalley. (C) The start of the Sutlej River's avulsion and the shift from extensive riverine activity to carved-out channel activity. (D) Total abandonment of the paleo-Sutlej valley, which the present-day Ghaggar River once occupied

result in more frequent and severe floods, particularly in areas like Kerala. As an illustration of how climate change affects flood occurrences, the 2018 Kerala floods (Idamalayar catchment) were connected to extreme rainfall and altered flow patterns (Wang et al., 2024). Extreme precipitation is expected to rise significantly in both frequency and area as the climate warms. Additionally, hydrological model simulations indicate a considerable increase in widespread floods in Indian river basins as a result of a warmer climate. With a rise on the western, northwest, and eastern coastlines and a decrease in severe precipitation in the Gangetic and Brahmaputra plains, the daily annual maximum precipitation trend showed a similar regional pattern. Hvaoc floods were caused by a growing trend in the frequency of severe precipitation (>150 mm/day) in central India between 1950 and 2015, according to Nanditha and Mishra (2024). In the middle and end of the century, the peninsular river basins have a widespread flood chance of above 30% (Nanditha and Mishra, 2024). Anthropogenic activities and climate change are major concerns to the Ganga river basin hydrology. Numerous hydraulic constructions, both big and small, have altered the flow regime in the Upper Ganga Basin, especially in the Bhagirathi basin (western tributary) (Swarnkar et al., 2021). The eastern tributary, the Alaknanda basin, on the other hand, has seen an increase in the size of extreme rainfall events and frequency of flash floods between 1970 and 2019. In pre- and post-monsoon, the reservoir causes minor flow changes in the downstream reaches, and during the monsoon, the severe flood magnitudes increase (Swarnkar et al., 2021). Additionally, it is noted that the Alaknanda's downstream (Rudraprayag) section exhibits an incremental variation in extreme flows of up to 15%, rendering the entire downstream Alaknanda Basin (Uttarakhand) susceptible to severe flooding occurrences in the near future (Swarnkar et al., 2021). The whole Ganga river basin's flood hazard index (FHI), flood exposure index (FEI), and flood vulnerability index (FVI) are disclosed by Aggarwal et al. in 2025 (Fig. 1.8). Areas of Bihar, eastern Madhya Pradesh, eastern Uttar Pradesh, and the northern section of West Bengal are identified as high flood risk zones due to the notable rise in the flood risk trend in the eastern portion of the basin (Aggarwal et al., 2025).

Climate change is linked to increased cloudburst intensity and frequency due to rising global temperatures, leading to more atmospheric moisture and changes in rainfall patterns. A cloudburst can cause flooding because it produces a lot of precipitation in a short amount of

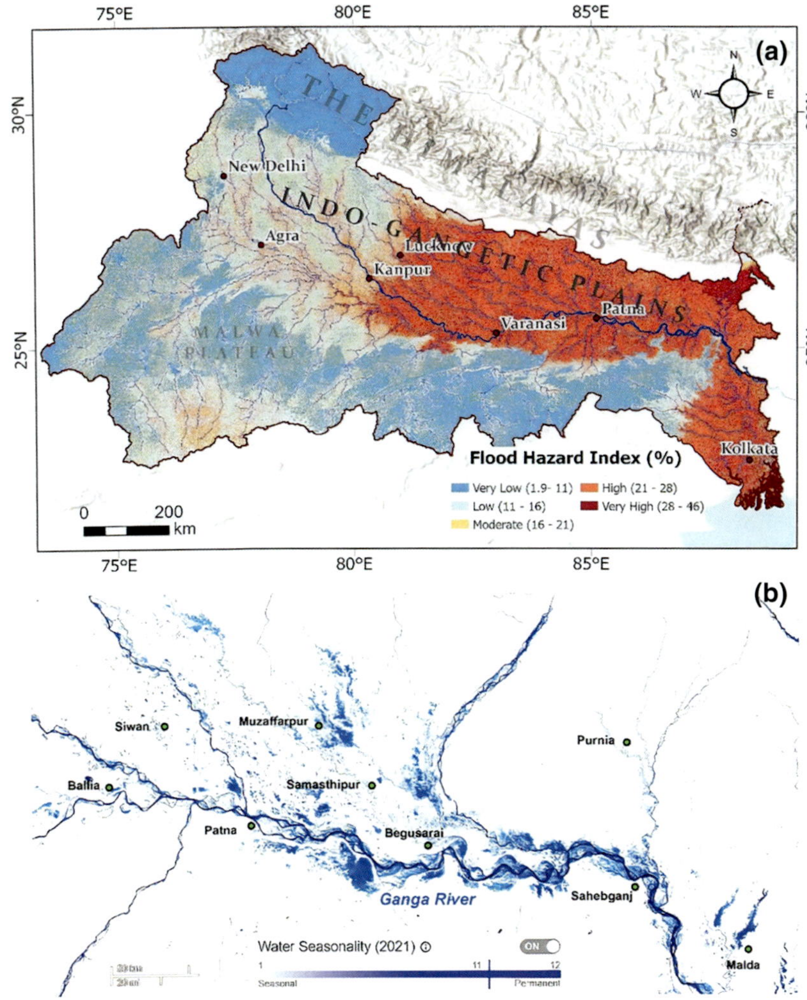

Fig. 1.8 The Ganga River Basin (GRB) is highly susceptible to floods, primarily due to heavy monsoon rainfall, siltation, and human interventions. The basin's flat terrain, high population density, and significant agricultural activities further exacerbate flood risks. Climate change is also impacting the basin, leading to more extreme rainfall events and increasing flood hazards. (a) Indo-Gangetic Plains of GRB is very much susceptible to high to very high flood hazard (after Aggarwal et al., 2025). (b) Flood inundation (https://global-surface-water.appspot.com/map) is a significant problem in the Indo-Gangetic Plains, a vast, densely populated and agriculture-dominated region encompassing parts of Uttar Pradesh and Bihar

time, occasionally with hail and thunder. Cloudbursts have the ability to suddenly release massive volumes of water; for example, 25 mm of precipitation is equivalent to 25,000 metric tonnes per square kilometre. Rising global temperatures, which increase atmospheric moisture and alter rainfall patterns, are connected to a rise in the severity and frequency of cloudbursts. Because the Himalayan region now experiences high temperatures, the atmosphere may not be as stable and may be more vulnerable to abrupt fluctuations in pressure and temperature, which may lead to the creation of cloudbursts. Over the past three years, 29 cloudburst occurrences have affected Himachal Pradesh. Cloudbursts are the primary cause of flash floods. Flash floods, which often affect small basins of less than 500 km^2, are characterized by fast flooding that follows an extreme storm event. Water levels rise from minutes to a few hours after heavy rainfall. A cloudburst on the night of July 20, 1970, on the southern mountain front in the Alaknanda valley (between Joshimath and Chamoli) was blamed for the significant flood that occurred in the Alaknanda valley in July 1970. In Shimla district, Himachal Pradesh's Chirgaon, a cloudburst on August 15, 1997, killed 115 people, leaving only a path of death in its wake. At Shilagarh in the Gursa region of Kullu district, Himachal Pradesh, flash floods brought on by a cloudburst killed over 40 people on July 16, 2003. On August 6, 2010, a series of cloudbursts in the Ladakh region's frontier town of Leh killed over 1,000 people (updated total) and wounded over 400 more. A cloudburst was observed in the Rudraprayag district of Uttarakhand on June 15, 2013, in the Kedarnath and Rambara area. More than 1,000 people have been killed thus far. One such incident was the Joshimath flash flood scenario on February 7, 2021, which resulted in extensive damage and the total washout of the several hydroelectric power facilities situated along the Dhauliganga River's path. On October 4, 2023, a glacial lake outburst flood occurred in Sikkim, a state in northeastern India, when the glacial South Lhonak lake broke its banks due to severe rainfall.

1.6.4 Declining Water Quality and Riverine Ecosystems

Important indicators for determining the ecosystem health of a river water body include river water quality metrics like dissolved oxygen (DO) and river water temperature (RWT). Rajesh and Rehana (2022) have conducted extensive study on the effects of climate change on dissolved oxygen (DO) saturation levels, while other recent studies have focused on the rising river water temperatures (RWTs) associated with climate change signals. India's riverine ecosystems and water quality are greatly impacted by climate change, which causes extreme weather events, higher temperatures, and changed rainfall patterns. These modifications cause variations in river flows, which impact water quality and disturb natural ecosystems. Aquatic life and human health are impacted by the ongoing degradation of water quality caused by the combination of contaminants from household and industrial waste and rising temperatures. Degradation of water quality and ecological distortion can result from the impact of climate change on DO in relation to RWT (Rajesh & Rehana, 2022). Because RWT and DO concentration are inversely correlated, any change in RWT reduces the quantity of oxygen that can dissolve and be used for biodegradation, which in turn impacts the river's capacity to self-purify (Rajesh & Rehana, 2022). The RWT drops DO by 2–12% and raises by up to 7 C in the summer, reaching nearly 35 C. This lowers the saturation capacity for DO between 2071 and 2100. The observed and predicted mean RWT levels of the catchments from 2071 to 2100 are represented by a decline in DO concentration (mgO_2/L) scale scores from 7.9 to 7.3 mgO_2/L (Rajesh & Rehana, 2022). It is predicted that in comparison with summer RWTs, the Narmada, Musi, and Ganga catchment sites have a larger proportion of DO drop during 2071–2100. In general, DO saturation level concentrations over Indian catchments will fall by around 2.3% for every 1 C RWT increase under climatic signals (Rajesh & Rehana, 2022).

Changes in river flow, water quality, and biodiversity can disrupt ecological processes, affecting the overall health and resilience of riverine ecosystems (Siddha & Sahu, 2022). As ectotherms, fish and other aquatic invertebrates may be directly affected by changes in RWT in their physiology. Organisms can't sustain essential survival functions over their critical thermal minimum and maximum temperatures. Thermal maxima, or temperature limits on oxygen consumption, have been found for a number of freshwater species. With an average temperature of 33.5 degrees Celsius, the 92 freshwater fish entries in the GlobTherm database have thermal maxima ranging from 20 to 43 degrees. Note that climate change, man-made changes to river systems (like reservoirs), and climatic variability (like La Nina and El Nino) may all contribute to temperatures over 30 degrees Celsius for longer periods of time (weeks). For example, over a few days in August 2023, the Amazon River's maximum daily RWT surpassed 30+ , resulting in widespread fish and animal mortality. Increased RWT could encourage disease transmission in wildlife. The establishment and spread of diseases with relatively high temperature thresholds, as well as bacterial and viral infections, may be more likely if RWT increases. Higher RWT, lower flows, and eutrophication are likely to result in a weakened immune system and increased susceptibility to illness in animals.

The issue of Anthropocene is to enhance "river resilience" which refers to a river's capacity to withstand and recover from disturbances, including natural events, and human-induced changes like dam construction or pollution. The intricacy of biological connections contributes to the uncertainty of the future, thus measures are required to make sure rivers remain resilient to whatever lies ahead. A river with no anthropogenic disturbance, complete groundwater and floodplain connection, an erodible corridor, longitudinal connectivity, and the capacity to react freely to shifts in climatic factors offers the most resilience. Channel modification, flow alteration, floodplain disconnection, and riparian land use change are examples of anthropogenic activities that affect climate resilience anytime they hinder or delay adaptation to changing circumstances and when they simplify habitats and diminish functional redundancy (Fig. 1.9). Freshwater resource dynamics, such as the movement of organic matter from terrestrial to freshwater systems, are anticipated to change as a result of warming. Ecological communities tend to reconstruct after disturbances, and the more easily they do so, the more robust they are. Individuals' capacity for dispersion and the damaged site's link to the local species pool are key factors in community reassembly. Consequently, previous management practices that have caused longitudinal river disconnection through dams, weirs, and other obstacles may have an effect on warming resilience. Because of the lack of riparian shade and the hydrological disconnection that has occurred longitudinally, laterally, and vertically as a result of historical management, rivers are particularly sensitive to heat. River restoration should be prioritized as part of global initiatives for climate adaptation and mitigation because it will increase resilience to a variety of anticipated co-occurring stresses, including warming, by restoring connection and diversification of conditions.

1.6.5 Salinity Intrusion in Delta

Over 500 million people may be accommodated in the world's major deltas, which are under growing stress due to a combination of human

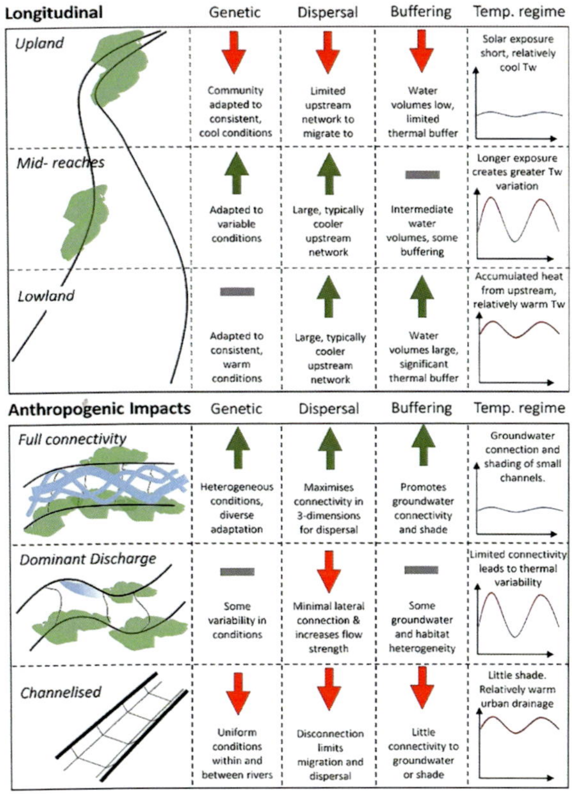

Fig. 1.9 River resilience and biological functions are shown in a schematic figure (after Johnson et al., 2024) based on (a) the catchment's longitudinal location and (b) human influences on channel connectivity. A green arrow indicates resilience to warming, whereas a red arrow indicates susceptibility. "Dispersal" refers to an animal's capacity to migrate or seek refuge; "Buffering" refers to the thermal cushioning against warming caused by landscape conditions; "Genetic" refers to the potential resilience bestowed by diversity in ecological communities; and "Temp regime" refers to the typical temperature profile for each condition

and environmental factors. One of the most significant effects of global warming is the rise in sea level (SLR), which is caused by changes in precipitation and evaporation, as well as anomalies in river discharges (Eslami et al., 2021). In growing economies in particular, upstream impoundments, sand mining, and groundwater extraction all show the simultaneous need on natural resources like freshwater and sand to sustain industrial and urban expansion, as well as agri-/aqua-cultural production (Eslami et al., 2021). Given the uncertainty in SLR forecasts, the sensitivity analysis reveals that a severe SLR of 60 cm by 2050 in response to Antarctic ice-sheet melting can significantly affect land use and adaptation plans by adding an extra 9–13% to the total area affected by saline water intrusion. Climate change, particularly rising sea levels and altered monsoon patterns, is making salinity intrusion in India's delta regions—such as the Ganges-Brahmaputra-Meghna (GBM) delta—an increasingly serious issue (Fig. 1.10). As a result, saltwater infiltrates freshwater sources, affecting ecosystems, agriculture, and drinking water. Rising sea levels brought on by climate change force saltwater farther inland into rivers and aquifers. Decreased freshwater discharge into rivers due to altered monsoon patterns might lessen the freshwater flow that naturally forces saltwater farther out to sea. Storm surges brought on by more frequent and powerful cyclones—which are also associated with climate change—can cause seawater to overwhelm coastal regions. According to the GBM delta's observed salinity value, the base year (2011) had a rather fresh upper eastern-central coast (0–1 ppt) (Akter et al., 2019). Given that its salinity magnitude was more than 15 ppt, the western shore had extremely high salinity. Depending on the discharge scenarios, the steady increase in salinity magnitude along the shore was caused by reduced upstream releases (Akter et al., 2019). Because the sea level rise circumstances were 0.25m and 0.50m, the salinity magnitude rose by less than 1 ppt. In the central-eastern coastlines, a 1m rise in sea level drove the saline front forward, and in the seaward portion, the salinity magnitude rose by 2–4 ppt (Akter et al., 2019).

1.7 River Metamorphosis due to Human Activity

River metamorphosis is the term, introduced by Schumm (1969), used to describe notable alterations in a river's shape and behaviour, frequently brought on by outside influences such as man-made actions like building dams or river impoundments or natural occurrences like floods and earthquakes. It includes modifications to bed morphology, general river system dynamics, and channel planform (such as meandering and braiding) (Fig. 1.11). Schumm's river metamorphosis is a theoretical framework

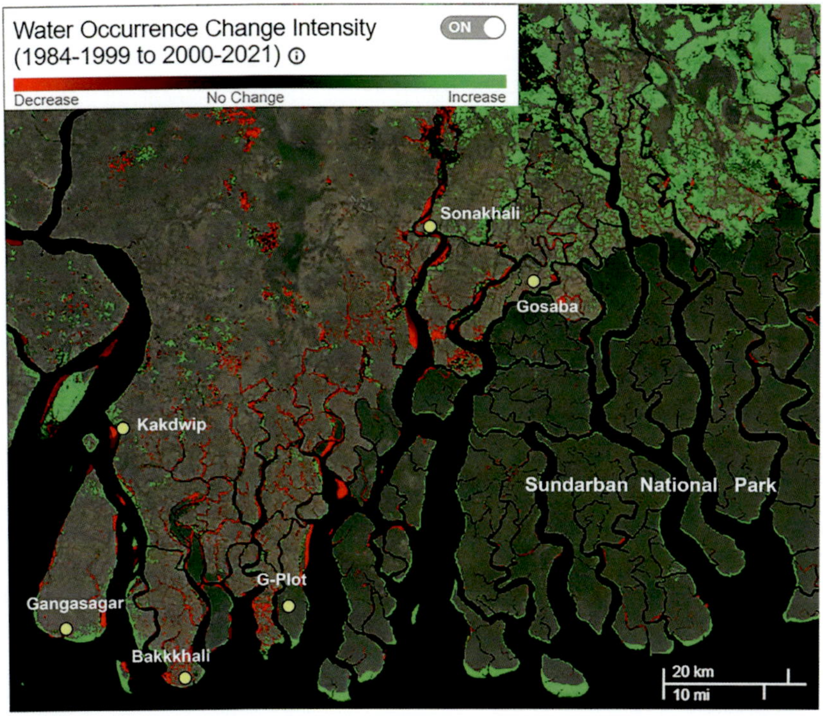

Fig. 1.10 The Water Occurrence Change Intensity map shows where high salt water occurrence (due to sea level rise and storm surges) increased (green colour) in space between 1984 and 2021 in the estuarine part of the Ganga Brahmaputra Meghna Delta (virtual map derived from https://global-surface-water.appspot.com/map)

that explains how rivers modify the size, form, and pattern of their channels in reaction to environmental shifts. In 1969, Stanley A. Schumm presented this model, which highlights the significance of sediment load and discharge in forming river channels. All things considered, river metamorphosis describes how rivers adjust to changed climate, hydrological circumstances, and other elements that affect sediment transport and deposition (Casado, 2013). River metamorphosis is mostly caused by human activity, which has significant effects on human society and the environment. Developing successful river management plans that strike a balance between the demands of people and the preservation of river ecosystems requires an understanding of these effects. The following human endeavours influence the transformation of rivers (Casado, 2013):

- Channel Modifications: A river's channel size, shape, and flow patterns can be drastically changed by operations including dam building, channel dredging, and straightening.
- Controlled Sediment Dynamics: While deforestation and agriculture can increase sediment runoff, dams trap sediment, lowering sediment loads downstream. The natural system and form of rivers may alter as a result.
- Water Flow: By altering natural flow regimes, dam building, water abstraction, and channel engineering can have an adverse effect on aquatic ecosystems and water quality.
- Land Use Changes: Intense agriculture, urbanization, and deforestation change the terrain by causing pollution, runoff, and sedimentation, which causes rivers to change.
- Pollution: Water quality and river ecosystems can be impacted by pollutants introduced into rivers by wastewater treatment, industrial runoff, and agricultural runoff.

The time when river processes and formations re-adjust to a shift in the sedimentologic and hydrologic regime [due to human controlled changes or natural event] is commonly regarded as the disequilibrium phase in a fluvial system (Miller & Miller, 2007). The driving and/or resistive forces acting in the channel were changed to the point where they surpassed the equilibrium limits, which Schumm (1973) referred to as thresholds. When a river system transitions from a balanced state to a transient state of disequilibrium, this is known as a threshold crossing event. A new state of balance that is adapted to a fresh set of external variables causes the system to progressively rectify this disequilibrium phase. A substantial shift in channel form and process connected to a threshold crossing event is represented by a change in channel form (e.g., width, depth, and slope), which fluctuates around the mean condition of a river in equilibrium (Miller & Miller, 2007). The recuperation or recovery time is the amount of time required for something to happen after triggering threshold (Casado, 2013). It is often believed that when a threshold is exceeded, the channel's shape would change quickly at

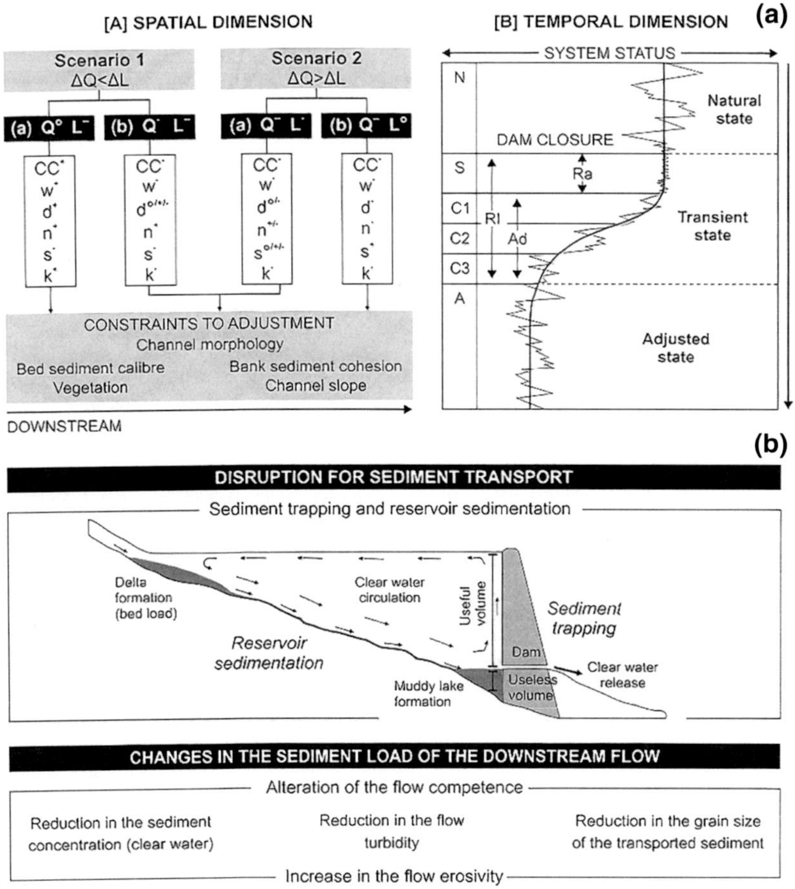

Fig. 1.11 River metamorphosis: (a) The widest foundation for forecasting the course of channel metamorphosis in response to variations (increase or reduction) in flow discharge and sediment load has been offered by Schumm's (1969) qualitative model (for details see Box 1.1); and (b) dams significantly alter sediment transport in rivers, primarily by trapping sediment within reservoirs, and downstream rivers and deltas experience a reduction in sediment supply, which can lead to erosion, channel instability, and habitat degradation

first and gradually until a new equilibrium state is reached. The ability to forecast how a river system will react to a certain disturbance and how this can impact future sediment loads is ideal for river scientists. Channel form (width, depth, hydraulic roughness, or width/depth ratio), gradient, planimetric characteristics (sinuosity, meander geometry), and pattern (straight, meandering, braided, anabranching) are just a few of the many variables that can be changed to adapt to a change [tend to river metamorphosis] (Miller & Miller, 2007). Sensitivity, or the propensity of landforms to adapt to an environmental perturbation by reaching a new equilibrium state, is a significant component in river metamorphosis. From a conceptual standpoint, sensitivity encompasses both the system's inclination to change and its capacity to withstand change by absorbing the forces of the disruption. Schumm (1968, 1969) introduced a conceptual model that is frequently used to explain how a channel adjusts to a change in water and sediment discharge in a series of publications outlining the idea of river metamorphosis. Schumm assembled a variety of empirical formulas designed to evaluate the connections between river shape and processes in order to provide a thorough, albeit qualitative, picture of the potential river adaptations to changed load and hydrology. When sediment and water discharge changes, due to dam construction, are taken into account, four distinct equations [Box 1.1] may be used to characterize the potential modifications of river (Schumm, 1969). Similar kind of model-based study was done by Graf, 2006, Brizga & Finlayson, 1990, Islam et al., 2025a and 2025b to assess human-induced changes in river basins.

Box 1.1: River Metamorphosis Model

- $Q_w + Q_t + = w + L + F + P - S \pm d \pm$
- $Q_w Q_t - = w - L - F + P + S \pm d \pm$
- $Q_w + Q_t - = d + P + S - F \pm w \pm L \pm$
- $Q_w + Q_t - = S + F \pm d - P - w \pm L \pm$

where Q_t is the percentage of the total load transported as bedload (sand-sized or larger), and Q_w can be either the mean annual discharge or the mean annual flood. The other variables are width (w), depth (d), slope (S), meander wavelength (L), width-depth ratio (F), and sinuosity (P). The exponents, expressed as either a plus or minus, indicate whether the dimensions of the variables are increasing or decreasing.

To illustrate the potential applications of these equations, let us consider a clear-cut forest area (deforestation). As a result of reduced infiltration rates and more direct runoff to the stream, Q_w is probably going to rise. Additionally, because plant roots were able to maintain the coarse material on the hillsides, it may now be eroded and enter the channel, increasing Q_t. In case dam-controlled river, regardless of the possible decrease in effective flow, clear water discharges from reservoir sedimentation have an unfulfilled capacity to carry sediment, which increases their erosive strength (valley incision). Within the stream by removing silt, sand mining immediately lowers the riverbed. As the river looks for a new equilibrium, the lowered sediment load may lead the river to expand. This may affect the general stability of the river morphologic system, accelerate bank erosion, and produce new incised channels.

Throughout the Anthropocene, human activity has drastically changed rivers, resulting in "novel ecosystems" that are structurally and functionally different from those seen in the natural world (Wohl, 2020). These changes are caused by climate change, river engineering, flow management, and changes in land cover (Figs. 1.12 and 1.13). Recognizing the cumulative effects of human activity and using multidisciplinary methods are necessary for comprehending and managing these modified river systems (Wohl, 2020). The natural equilibrium of sediment transport and flow dynamics is upset by sand mining, which drastically changes the shape of rivers. This may result in a number of changes, such as deterioration of the channel bed, increased erosion, and changes to the river's planform. Because of the changing flow and sediment transfer, the river may become more braided, meander, or change its course. Sand mining causes imbalances in the river's sediment budget by lowering the quantity of material available for natural transit. The aquatic and riparian environments, including fish breeding grounds, flora, and other species, can be adversely affected by altered river morphology. Sand mining, the construction of high embankments, and the release of pollutants are just a few of the ways that industrial operations may drastically change

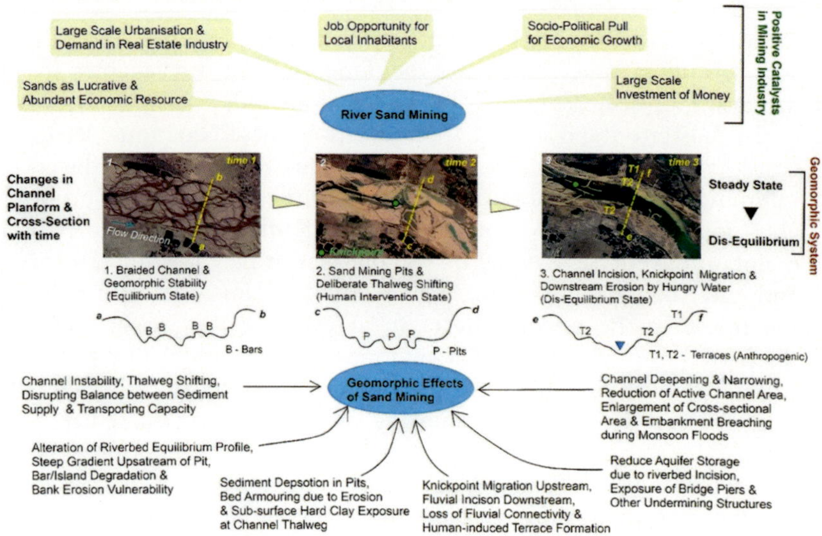

Fig. 1.12 Sand mining significantly alters river morphology (after Islam et al., 2025a, 2025b), leading to channel degradation, instability, and changes in water levels and flow patterns. The river cuts deeper into the valley floor, leading to a steeper channel. The removal of sediment destabilizes riverbanks, making them more susceptible to erosion

river patterns. These activities have the potential to alter river flow, water quality, and ecosystem health generally, which frequently has detrimental effects on both people and the environment.

1.8 Urban Rivers

India's urban rivers confront serious environmental problems, mostly because of water pollution from industrial effluents, municipal sewage, and agricultural runoff, among other causes. These contaminants damage aquatic life and have an adverse effect on human health by causing the quality of the water to deteriorate (Fig. 1.14). Finding the blue arteries of cities is further hampered by the deterioration and degeneration of channels brought on by urban encroachment. Urban rivers are known to be physically threatened by a number of factors, including

Fig. 1.13 Rivers in the Anthropocene: (a) concrete pier bridge (Krishak Setu built in 1978) and sand mining pits have changed river morphology in the Damodar River, Bardhaman, West Bengal (drone view); and (b) large scale paper industry unit (Shree Krishna Paper Mills & Industries Ltd. established in 2007) have degraded river water and its channel in the Khari River, Bardhaman, West Bengal

pollution, encroachment, solid waste dumping, damming, water diversions, groundwater overexploitation, catchment degradation, destruction of water bodies, wetlands, and forests, indiscriminate mining, and the effects of bridge, flyover, and metro construction. A worldview that views rivers as non-essential, disposable things and a total absence of institutional or legal protection are some of the main reasons of this predicament. Some examples of rivers that fall under this category are the Yamuna (Delhi, Agra, Mathura), the Ganga (Allahabad, Varanasi, Patna), the Gomati (Lucknow), the Mithi (Mumbai), the Mula-Mutha (Pune), the Sabarmati (Ahmedabad), the Dravyawati (Jaipur), Khan (Indore), the Kshipra (Ujjain), the Jhelum (Srinagar), the Mahi and Vishvamitri (Vadodara), the Tapi (Surat), and the Arkavathi & Virishabhavati (Bangalore), to mention a few. Floods are becoming more damaging, there is a shortage of water, people's quality of life is declining, and the economic, social, and cultural effects of this deteriorating urban river situation are already being felt in a number of ways. Urban India's fast growing per capita footprint and urbanization mean that these effects will only get worse. To evaluate the pollution in different river segments across the nation, the Central Pollution Control Board's (CPCB) report and the data it contains serve as a guide. In 2022, the CPCB identified 311 contaminated river sections on 279 rivers in 30 States and UTs in the nation based on the study of water quality data of 603 rivers on 1920 sites for the years 2019 and 2021 (Box 1.2). This was done using the Biochemical Oxygen Demand (BOD) (3 mg/L) indicator of organic pollution. A river is deemed severely polluted if its BOD (biochemical oxygen demand) level is more than 8 mg/L, and critically contaminated if its overall score (which is determined by the BOD and faecal coliform counts) is higher than 90.

Box 1.2: Pollution level of rivers in India

- Musi River—polluted stretch between Hyderabad and Rangareddy [Adhra Pradesh]—BOD 34 mg/L
- Kalong River—polluted stretch at Nagaon [Assam]—BOD 50 mg/L
- Yamuna River—polluted stretch between Wazirabad and Okhla [Delhi UT]—BOD 55 mg/L

- Sabarmati River—polluted stretch between Ahmedabad and Vautha [Gujarat]—BOD 48 mg/L
- Sukhna River—polluted stretch at Parwanoo [Himachal Pradesh]—BOD 36 mg/L
- Godavari River—polluted stretch at Nashik [Maharashtra]—BOD 36 mg/L
- Cauvery River—polluted stretch at Erode [Tamil Nadu]—BOD 38 mg/L
- Ganga River—polluted stretch between Kannauj and Kanpur [Uttar Pradesh]—BOD 21 mg/L
- Kosi River—polluted stretch at Kashipur [Uttarakhand]—BOD 13 mg/L
- Damodar River—polluted stretch Asansol [West Bengal]—BOD 6.8 mg/L

In Indian rivers, a BOD (Biochemical Oxygen Demand) level exceeding 6 mg/L is generally considered polluted and requires remedial action.

SOURCE—Central Pollution Control Board, Government of India; website [https://cpcb.nic.in/].

India's rivers are severely polluted and choked by development as a result of the country's fast growing metropolitan centres and burgeoning population. However, campaigners in Indian cities are going to court to compel governments and municipalities to act. In 2015, activists in Hyderabad, a centre for technology, petitioned the National Green Tribunal, a quasi-judicial body, to stop unauthorized building along the Musi River. In Chennai, South India, residents have petitioned the tribunal to prevent contamination of the Cooum River and to guarantee that a sizable canal is properly dredged in order to remove silt and enhance flow. To prevent several constructions, like as a road and subway depot, from encroaching on the floodplain and riverbed of the Yamuna, a significant tributary of the Ganges, campaigners in New Delhi have been engaged in a legal battle for years. Additionally, people and environmentalists who are angry about the government's inability to clean up the severely polluted Ganges have launched a court battle over the holy river, which flows through five Indian states. The National Mission for Clean Ganga (NMCG) has approved an annual master plan for the River Cities Alliance (RCA), including a dynamic and action-oriented roadmap of projects to be carried out throughout the year. This is a major step towards sustainable urban river restoration. The strategy includes a number of initiatives aimed at integrating river-sensitive urban design into India's expanding cities, including knowledge-sharing platforms, technical tool development, expert advice, and theme case studies.

Due to intentional neglect and haphazard development initiatives, India's urban rivers have been in increasingly dire circumstances. By diverting water for household use, they are losing freshwater flows on the one hand, and on the other, they have become a landfill for large volumes of generally untreated solid and liquid waste from both residential and industrial regions. Individuals, groups, and government agencies are nonetheless making some impressive attempts to restore urban rivers in various locations around the nation. Ahmedabad's Sabarmati Riverfront project is an example of river restoration that prioritizes environmental enhancement, social advancement, and urban renewal. A historic urban revitalization project in Ahmedabad, the Sabarmati Riverfront Development has created vibrant public spaces along 11 kilometres of once unutilized riverbanks. The project, which prioritizes sustainability and environmental restoration, installs sewage treatment facilities to enhance water quality and uses water from the Narmada Canal to ensure a steady river flow. The city's historical flood hazards are greatly decreased by building embankments and carefully managing river channels, while urban growth prospects are created by reclaiming floodplains. By creating social coherence through its parks, promenades, and plazas, the riverside has developed into a major meeting spot. Delhi's 22-kilometre section of the Yamuna between Wazirabad and Okhla barrage is the most poisonous of India's 311 contaminated river segments. It makes up 79% of the river's overall pollutant burden even though it only makes up only 2% of its length. In order to combat the pollution of the Yamuna River, a significant tributary of the Ganges, the Indian government initiated the multi-phase Yamuna Action Plan (YAP) in 1993. Untreated sewage, industrial waste, and other pollutants that enter the river are among the many causes of pollution that it seeks to lessen. The strategy aims to improve the quality of the water by combining various tactics, such as awareness campaigns, with the construction of infrastructure, such as sewage treatment facilities.

Fig. 1.14 The photographs showing significant environmental issues (like critical BOD level greater than 50 mg/L) plague India's metropolitan rivers (like Yamuna River, Delhi), mostly due to water contamination from municipal sewage, industrial effluents, and agricultural runoff, among other sources

1.9 SDGs and River Restoration

All United Nations member nations endorsed the 2030 Agenda for Sustainable Development in 2015, and it offers a common roadmap for world peace and prosperity both now and in the future. It is centred on the 17 Sustainable Development Goals (SDGs), which acknowledge that eradicating poverty and other forms of deprivation requires a combination of measures to enhance health and education, lower inequality, and promote economic growth—all while addressing climate change and protecting the environment. Particularly concerning is the status of the SDGs pertaining to water, climate change, and environment loss. SDG 6 demands for clean water and sanitation, for instance, yet just 15% of the planet's wetlands remain to offer services like groundwater and aquifer recharge, since 85% of them have been gone (Xue et al., 2020). Another concerning figure indicates that between 2015 and 2030, the number of medium-to-large-scale catastrophes (many of which are connected to water) will rise by 40% (Xue et al., 2020). This means that there will be just over five years to create resilient communities that are ready for more frequent calamities like droughts and floods. Furthermore, 40,000 species—including freshwater species that are essential to the economy and food supply—are in danger of going extinct.

Fortunately, rivers are a solution that is frequently disregarded but is an SDG hero. The lifeblood of civilizations worldwide has always been rivers and the marshes that surround them. Indeed, the benefits that rivers offer enable many of the pillars of a functional community. Nearly half of the 17 Sustainable Development Goals (SDGs) of the UN depend on healthy, linked rivers because of their immense value. By enhancing a river system's biological functions and natural processes, river restoration increases river resilience, enabling it to endure and bounce back from shocks like droughts and floods (Tullos et al, 2021; Xue et al., 2020). Restoring natural ecosystems, enhancing water flow, and re-establishing rivers' connections to their floodplains all help rivers become more resilient to climate change. River restoration and enhancing river resilience are crucial for achieving several Sustainable Development Goals (SDGs), particularly SDG 6 (Clean Water and Sanitation) and SDG 14 (Life Below Water). How healthy and connected rivers contribute towards 8 of the 17 SDGs is depicted in figure. Here, river restoration and river resilience connect to specific SDGs (Fig. 1.15):

Fig. 1.15 River restoration and resilience are crucial for achieving multiple Sustainable Development Goals (SDGs), particularly SDG 6 (Clean Water and Sanitation) and SDG 13 (Climate Action). By restoring natural river conditions, improving water quality, and increasing resilience to climate change, these efforts contribute to a more sustainable future

- SDG 6 Clean Water and Sanitation: By stopping pollutants from entering rivers and reestablishing the natural mechanisms that filter water, river restoration can aid in the reduction of pollution. River restoration may improve the resilience of water resources and increase their capacity to adapt to climate change by restoring natural river flows and ecosystems. The preservation and restoration of water-related ecosystems, such as rivers, wetlands, and other freshwater habitats, are directly aided by river restoration.

- SDG 14 Life Below Water: In order to maintain the health and productivity of freshwater ecosystems, river restoration can aid in their sustainable management and protection. River habitat restoration can contribute to biodiversity conservation and safeguard endangered species that depend on these environments. Since rivers are frequently a source of pollutants that enter the ocean, minimizing pollution from land-based activities through river restoration also helps avoid and reduce marine pollution.
- SDG 13 Climate Action: River restoration can enhance the resilience of river systems to climate change impacts, such as droughts, floods, and increased water temperatures.
- SDG 15 Life on Land: Restoring riparian areas and river corridors can help protect and restore terrestrial ecosystems that are linked to rivers.
- SDG 17 Partnerships for the Goals: Effective river restoration requires collaboration between various stakeholders, including governments, communities, and NGOs, to achieve sustainable outcomes.

The degree of change that a system can withstand in the face of disruption while preserving a distinctive, self-sustaining regime of operations, procedures, or sets of feedback loops is known as resilience, or absorbing capacity (Wohl et al., 2024). Although rivers have different degrees of resilience, industrialized human change has generally reduced river resilience. River form and connection are altered by human activity and climate change; therefore, restoration efforts are increasingly looking at ways to improve resilience (Wohl et al., 2024). River restoration and conservation are crucial for maintaining India's ecology and human population for both the present and future generations. Unexpected floods, droughts, and water-borne illnesses are the most frequent disasters brought on by distressed rivers, and their degradation in modern times has consequently had equally far-reaching and occasionally catastrophic effects on humans. India's river revitalization initiatives have mostly concentrated on big rivers overall, but they haven't made much progress, in part because the work is too big to be understood comprehensively and completed effectively in a short amount of time. In addition to being too big to be thoroughly seen and tracked, the anthropogenic causes influencing the river are frequently too varied and unequally dispersed throughout the river basin (cGanga, 2019). When the effort to reverse the degradation process is directed towards the smaller urban and semi-urban tributaries and drains, particularly those that are perennial or readily made so, the work becomes much easier (cGanga, 2019). The economic, environmental, aesthetic, and cultural advantages that come with this kind of rejuvenation are also instantaneous, and they affect sizable populations, which may have a domino effect on river revitalization projects throughout the basin. By taking a bottom-up approach, beginning with the rejuvenation of small perennial (or possibly perennial) rivers in and around urban areas before concentrating on the larger tributaries or the main river, river restoration and conservation can be a true success and ensure sustainable development in the nation due to its clear focus and objectives (Fig. 1.16). Given the previous discussions, the following succinctly describes the top priorities of the River Basin Organization of India for achieving the SDGs in the rivers of India (cGanga, 2019):

- Health of the river to guarantee its restoration and preservation as a perennial river or natural water channel: guaranteed the use of clean water (or uncontaminated water) to sustain the lowest acceptable flows throughout the year; keeping the flow's breadth and depth appropriate for aquatic life by using structural interventions (such as weirs and bounds) as necessary; keeping pollutants and waste from entering; and preserving the river's and floodplains' physical integrity.
- Conformity to the local environment and ecology, including: Unobstructed natural drainage pathways; connectivity with nearby water bodies; improving groundwater recharge by lowering watershed runoff rates.
- Maximizing the benefits of rivers, including increased public revenue from river-centric operations, fish and other aquatic farming, tourism and recreation, navigation and transportation, and surplus water storage for human use
- Reducing losses or damages brought on by rivers, including minimizing flooding and water logging as well as land reclamation.
- Increasing the downstream regions' benefits, including increased biodiversity and water quality.

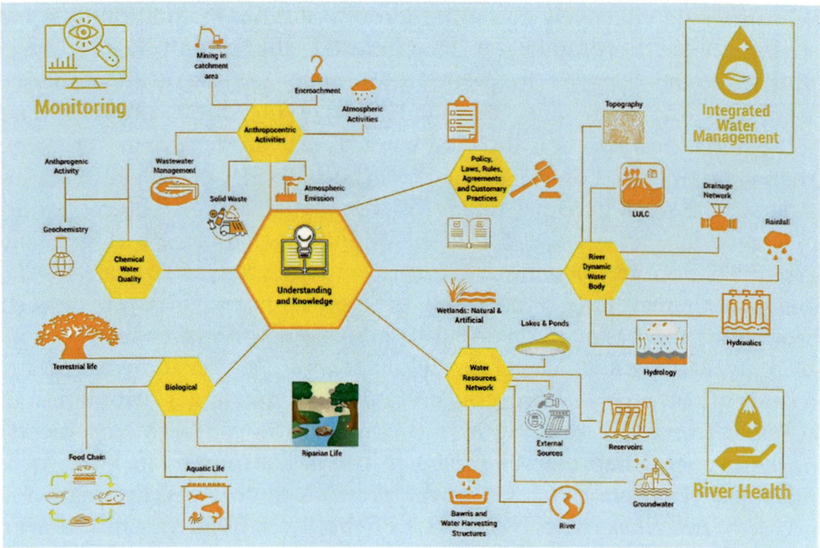

Fig. 1.16 Climate Resilient Integrated Water Management Plan: An Integrated Water Management Plan (IWMP), coupled with river restoration and resilience initiatives, provides a holistic approach to managing water resources and ensuring the health of rivers while enhancing their capacity to withstand environmental changes. IWMPs promote coordinated development and management of water, land, and related resources, balancing economic and social needs with the sustainability of ecosystems and legal rights to rivers

1.10 Importance of Education in River Conservation

Education is crucial for river conservation in India because it raises awareness about the importance of rivers, their ecosystems, and the threats they face. By educating individuals, communities, and stakeholders, it fosters a sense of responsibility and empowers people to take action in protecting rivers. Rivers nourish animals, supply clean water, and give recreational possibilities, making them essential ecosystems. Climate change, habitat damage, and pollution, however, are posing a growing threat to them. River protection is vital, but it also needs education in addition to reforming policies and conducting scientific studies. Communities benefit from education when they learn about the value of rivers, the dangers they face, and the personal contributions that each person can make. By empowering people to preserve and repair the rivers they depend on, river education promotes environmental responsibility. Numerous national and international organizations are leading the way in river education, enabling people and communities to take up the cause of river protection. Michigan Sea Grant which is part of NOAA Sea Grant, promotes sustainable water management in Michigan's rivers and Great Lakes. Volunteers are taught how to monitor water quality and safeguard nearby rivers through their Adopt-a-Stream programme. Connecting watersheds and cleaning up the Mississippi River are the main goals of Living Lands & Waters. Their Waterways Education Program teaches community people, educators, and students about conservation issues by taking them straight to the river. Missouri River Relief is committed to restoring the Missouri River and educating local populations about conservation. Students learn about the environmental issues facing the Missouri River and how to monitor water quality through their River Watch and Missouri River Field Days programmes.

NGOs like the Environmentalist Foundation of India, SARA (Sustainable Alternatives for Rural Accord), and Jal Bhagirathi Foundation are among the many Indian groups that play a crucial role in spreading awareness about and advocating for river protection. Through a number of programmes, such as the National Water Awards, the Ministry of Jal Shakti also plays an important role. Organizations Involved in River Conservation Education:

- Environmentalist Foundation of India: Focuses on holistic approaches including wildlife conservation and habitat restoration.
- SARA (Sustainable Alternatives for Rural Accord): Works on sustainable models through its "Swagrama" programme.
- Jal Bhagirathi Foundation: Addresses water scarcity and contamination in arid regions like the Marwar region of Rajasthan.
- Watershed Organisation Trust (WOTR): Works to improve potable and sanitation facilities across rural India, with initiatives like water budgeting.
- Centre for Aquatic Livelihood Jaljeevika: Supports fish farmers and promotes water conservation.
- Project Chirag: It is committed to empowering communities and individuals to conserve water and protect our precious water resources. Through education and awareness programmes, Project

Chirag aims to inspire people to take action and make a positive impact on the environment.
- Ministry of Jal Shakti: Organizes the National Water Awards to encourage a holistic approach to water resource management.
- National River Conservation Plan: Involves activities like municipal sewage treatment, River Front Development, and improved sanitation.

1.11 Legal Status and River Conservation Projects in India

India's indigenous and other nature-dependent cultures have questioned the core assumptions of Western thought, which hold that only people have rights and that everything else in the living world exists for human purpose. It is venerated in many religions and traditions of coexisting with the natural world. By fostering collaboration and honouring ecological boundaries, their daily lives have been steered through the rhythms and moods of nature. An increasing number of communities, organizations, and governments worldwide are advocating for the replacement of anthropocentric legal and governance systems with ecocentric ones as a result of the startling effects of climate change, the possibility of mass extinction of species, and the waning window of opportunity for significant action. In the past twenty years, the number of legislations based on ecological jurisprudence has dramatically increased. When a river or other natural entity is given the same rights and obligations as a "living person", what does it mean? What will happen to those who live alongside these ecosystems in terms of their rights? Are these discussions actually transforming? The current global crisis has evolved into a civilizational crisis. The essential aspects of life are deteriorating, leaving many communities in dire circumstances and with a diminished future. Globally, a new vocabulary is emerging that acknowledges the rest of nature as alive in formal legal systems and goes beyond the conventional bounds of classical environmental law. In India, a mix of legislative frameworks, governmental programmes, and court rulings are used to address river protection. Although rivers have some legal safeguards, especially when it comes to animals, the idea of giving them legal personality is still up for dispute and has been interpreted differently by courts. The National Water Mission concentrates on managing and conserving water resources, while National River Conservation Plans and other initiatives seek to reduce pollution and enhance water quality.

1.11.1 Legal Framework for River Conservation

India does not have a single consolidated law for river conservation; instead, a combination of environmental laws and judicial interventions shapes the legal status of rivers.

- Article 48A (Directive Principles of State Policy) mandates the state to protect and improve the environment.
- Article 51A (g) places a duty on citizens to protect natural resources, including rivers.
- The Water (Prevention and Control of Pollution) Act, 1974: The primary legislation for the prevention and control of water pollution and maintaining or restoring the wholesomeness of water.
- The Environment (Protection) Act, 1986: An umbrella legislation empowering the central government to take measures to protect and improve the environment.
- The National Green Tribunal Act, 2010: Provides for effective and expeditious disposal of cases relating to environmental protection, including river pollution.
- Indian courts have played a significant role in river conservation. Notably, in 2017, the Uttarakhand High Court declared the Ganga and Yamuna rivers as "living entities", granting them legal personhood. However, the Supreme Court later stayed the order, citing implementation concerns.

1.11.2 Major River Conservation Projects

Several government-led projects aim to address the multifaceted threats facing India's rivers. Some notable initiatives include:

- Namami Gange Programme (2014): This flagship initiative aims to rejuvenate the Ganga River by addressing pollution, promoting sustainable sanitation, and involving community participation. It includes components such as sewage treatment, river surface

cleaning, afforestation, and biodiversity conservation. It is one of the largest integrated river rejuvenation programmes in India.
- National River Conservation Plan (NRCP): Launched in 1995, this plan covers pollution abatement in 190 towns along 41 rivers spread over 20 states. The NRCP finances the creation of sewage treatment plants and sanitation infrastructure to reduce the pollution load.
- Yamuna Action Plan: Funded with assistance from the Japan International Cooperation Agency (JICA), this project focuses on improving water quality in the Yamuna through waste treatment and improved infrastructure in Delhi and nearby regions.
- Clean Ganga Fund: Introduced in 2014, this fund invites contributions from residents, NRIs, and corporations to support the Namami Gange objectives. It promotes community participation and CSR involvement in river conservation.

1.12 Concluding Remarks

Reimaging Indian rivers for sustainability in the Anthropocene necessitates a shift from a purely developmental view to a holistic approach that considers ecological, social, and economic dimensions. This requires understanding the intricate relationships between human activities, river ecosystems, and the broader environment. The Anthropocene is characterized by the profound and pervasive influence of human activities on the planet, including alterations to river systems. Along with the impact of climate change the crisis of river sustainability includes issues like deforestation, urbanization, industrial pollution, and unsustainable water management practices. The water life is supported by the rivers. The rivers are also essential to the vegetation and wildlife. Rivers are struggling to breathe. We must acknowledge the "Mother Earth" and provide her the legal rights in the Constitution. Rivers, forests, lakes, water bodies, air, and glaciers are all at jeopardy because of pollution, climate change, and global warming. India's river systems are under severe stress, but concerted legal, institutional, and societal efforts offer a pathway towards sustainable conservation. Strengthening the legal framework, empowering local stakeholders, and ensuring scientific and transparent project implementation will be key to securing the future of India's rivers, and with them, the health of the nation. It's anticipated that the current volume will address major issues and challenges hindering the sustainability of Indian rivers, focusing on areas like pollution, water management, and the impact of development. Specifically, it is expected to tackle issues related to sewage, industrial waste, and unsustainable water usage practices. This edited volume will likely examine the impact of rapid urbanization and industrialization on river ecosystems. It may also discuss the need for sustainable development practices that minimize environmental degradation.

References

Aggarwal, E., Ruiter, M. C., & Sangwan, R. J. et al. (2025). *Increasing flood risk in the Ganga Basin: A perspective from night-time lights*. EGUsphere [preprint]. https://doi.org/10.5194/egusphere-2024-3901

Akter, R., Asik, T. Z., & Sakib, M. et al. (2019). The dominant climate change event for salinity intrusion in the GBM Delta. *Climate, 7*(5), 69. https://doi.org/10.3390/cli7050069

Bridgewater, P., Guarino, E., & Thompson, R. M. (2018). Hydrology in the Anthropocene. *Encyclopedia of the Anthropocene, 1*, 87–92. https://doi.org/10.1016/B978-0-12-809665-9.09916-X

Brizga, S. O., & Finlayson, B. L. (1990). Channel avulsion and river metamorphosis: The case of the Thomson River, Victoria. *Australia. Earth Surface Processes and Landforms, 15*(5), 391–404.

Casado, A. (2013). *Human impacts and fluvial metamorphosis: the effects of flow regulation on the hydrology, morphology and water temperature of the Sauce Grande River, Argentina.. Geography*. Université Blaise Pascal—Clermont-Ferrand II, 2013. English. ffNNT: 2013CLF20013. https://theses.hal.science/tel-01077867v1/file/CASADO_2013CLF20013.pdf

cGanga (2019). *River restoration and conservation—A concise manual and guide*. Centre for Ganga River Basin Management and Studies, National Mission for Clean Ganga, Govt. Of India, New Delhi, p. 33.

Chakraborty, B., Bera, B., & Adhikary, P. P. et al. (2021). Positive effects of COVID-19 lockdown on river water quality: evidence from River Damodar, India. *Scientific Reports, 11,* 20140. https://doi.org/10.1038/s41598-021-99689-9

Charlton, R. (2007). *Fundamentals of fluvial geomorphology*. Routledge. https://doi.org/10.4324/9780203371084

Chaudhri, A. R., Chopra, S., & Kumar, P., et al. (2021). Saraswati River in northern India (Haryana) and its role in populating the Harappan civilization sites—A study based on remote sensing, sedimentology, and strata chronology. *Archaeological Prospection, 28*(4), 565–582.

Eslami, S., Hoekstra, P., & Minderhoud, P. S. J. et al. (2021). Projections of salt intrusion in a mega-delta under climatic and anthropogenic stressors. *Communications Earth & Environment, 2*, 142. https://doi.org/10.1038/s43247-021-00208-5

Ghosh, S., & Guchhait, S. K. (2014). Palaeoenvironmental significance of fluvial facies and archives of late quaternary deposits in the floodplain of Damodar River, India. *Arabian Journal of Geosciences, 7*(10), 4145–4161.

Graf, W. L. (2006). Downstream hydrologic and geomorphic effects of large dams on American Rivers. *Geomorphology, 79*, 336–360.

Hoque, M. M., Islam, A., & Ghosh, S. (2022). Environmental flow in the context of dams and development with special reference to the Damodar Valley Project, India: a review. Sustain. *Water Resources Management,* 8, 62. https://doi.org/10.1007/s40899-022-00646-9

Islam, A., Dey, S., & Mohinuddin, S. et al. (2025a). Planform metamorphosis of the rivers in the Ganga-Brahmaputra-Meghna Delta under variable damming and tidal effects. *Catena, 249*, 108645. https://doi.org/10.1016/j.catena.2024.108645

Islam, A., Das, B. C., & Ghosh, S. et al. (2025b). Role of riverbed sand mining on planform and cross-sectional morphology of Mayurakshi River, India. *Science of The Total Environment, 963* 178465. https://doi.org/10.1016/j.scitotenv.2025.178465

Johnson, M. F., Albertson, L. K., & Algar, A. C., et al. (2024). Rising water temperature in rivers: Ecological impacts and future resilience. *WIREs Water, 11*(4), e1724. https://doi.org/10.1002/wat2.1724

Kale, V. S., Baker, V. R., & Mishra, S. (2003). Flood hydrology and geomorphology of the monsoon-dominated Narmada basin, central India. *Journal of Hydrology, 275*(3–4), 249–268.

Kale, V. S., Ely, L. L., Baker, V. R., & Mishra, S. (1997). Palaeoflood evidence of changes in the magnitude and frequency of monsoon floods on the Narmada River, central India. *Bulletin of the Geological Society of America, 108*(9), 1134–1148.

Khan, I., & Sinha, R. (2019). Discovering 'buried' channels of the Palaeo-Yamuna River in NW India using geophysical evidence: Implications for major drainage reorganization and linkage to the Harappan Civilization. *Journal of Applied Geophysics, 167*, 128–139.

Lovelock, J. E. (1972). Atmospheric composition as a planetary property. *Environmental Science and Technology, 6*(10), 898–902.

Macklin, M. G., & Lewin, J. (2015). The rivers of civilization. *Quaternary Science Reviews, 114,* 228–244.

Miller, J. R., & Miller, S. M. D. (2007). *Contaminated Rivers—a geomorphological-geochemical approach to site assessment and reclamation* (pp. 235–270). Springer.

Nanditha, J. S., & Mishra, V. (2024). Projected increase in widespread riverine floods in India under a warming climate. *Journal of Hydrology, 630*, 130734. https://doi.org/10.1016/j.jhydrol.2024.130734
Rajesh, M., & Rehana, S. (2022). Impact of climate change on river water temperature and dissolved oxygen: Indian riverine thermal regimes. *Scientific Reports, 12*, 9222. https://doi.org/10.1038/s41598-022-12996-7
Reshma, C., & Arunkumar, R. (2023). Assessment of impact of climate change on the streamflow of Idamalayar River Basin, Kerala. *Journal of Water and Climate Change, 14*(7), 2133–2149.
Schumm, S. A. (1968). River adjustment to altered hydrologic regimen—Murrumbidgee River and paleochannels. *Australia. USGS Professional Paper, 598*. https://doi.org/10.3133/pp598
Schumm, S. A. (1969). River metamorphosis. *Journal of the Hydraulics Division, 95*(1). https://doi.org/10.1061/JYCEAJ.0001938
Schumm, S. A. (1973). Geomorphic thresholds and complex response of drainage systems. In M. Morisawa (Ed.), *Fluvial Geomorphology* (pp. 299–310). Publications of Geomorphology, State University of New York.
Shankar, A. (2002). The mighty and mystical rivers of India. Publications Division, Ministry of Information & Broadcasting, Govt. of India.
Siddha, S., & Sahu, P. (2022). Impact of climate change on the river ecosystem. In S. Madhav et al. (Eds.), *Ecological significance of riparian ecosystems: Challenges and management*, pp. 79–104. Elsevier. https://doi.org/10.1016/B978-0-323-85045-2.00014-5
Singh, A., & Sinha, R. (2019). Fluvial response to climate change inferred from sediment cores from the Ghaggar-Hakra palaeochannel in NW Indio-Gangetic Plains. *Palaeogeography Palaeoclimatology Palaeoecology, 532*, 109247. https://doi.org/10.1016/j.palaeo.2019.109247
Singh, M., & Pandey, J. (2022). COVID-19 lockdown-driven changes in the Ganga River ecosystem in response to human perturbations. *Environmental Monitoring and Assessment, 194*, 858. https://doi.org/10.1007/s10661-022-10507-9
Singhvi, A. K., & Kale, V. S. (2010). *Paleoclimate studies in India: Last ice age to the present* (Vol. 4, IGBP WCRP SCOPE Report Series). Indian National Science Academy.
Sinha, R., Singh, A., & Tandon, S. K. (2020). Fluvial archives of north and northwestern India as recorders of climatic signatures in the Late Quaternary: Review and assessment. *Current Science, 119*(2), 232–243.
Swarnkar, S., Mujumdar, P., & Sinha, R. (2021). Modified hydrologic regime of upper Ganga Basin induced by natural and anthropogenic stressor. *Scientific Reports, 11,* 19491. https://doi.org/10.1038/s41598-021-98827-7
Thompson, J. R., Gosling, S. N., Zaherpour, J., & Laize, C. L. R. (2021). Increasing risk of ecological change to major rivers of the world with

global warming. *Earth's Future, 9*(11), e2021EF002048. https://doi.org/10.1029/2021EF002048

Tullos, D., Baker, D. W., & Curran, J. C. et al. (2021). Enhancing resilience of river restoration design in systems undergoing change. *Journal of Hydraulic Engineering, 147*(3). https://doi.org/10.1061/(ASCE)HY.1943-7900.0001853

Wang, H., Liu, J., & Klaar, M., et al. (2024). Anthropogenic climate change has influenced global river flow seasonality. *Science, 383*(6686), 1009–1014.

Wohl, E. (2020). Rivers in the Anthropocene: The U.S. perspective. *Geomorphology, 366*. https://doi.org/10.1016/j.geomorph.2018.12.001

Wohl, E., Fryirs, K., Grabowski, R. C., Morrison, R. R., & Sear, D. (2020). Enhancing the natural absorbing capacity of rivers to restore their resilience. *Bioscience, 74*(11), 782—796, https://doi.org/10.1093/biosci/biae090

Wohl, E., Fryirs, K., Grabowski, R. C., Morrison, R. R., & Sear, D. (2024). Geomorphic context in process-based river restoration. *River Research and Applications, 40*(4), 434–449. https://doi.org/10.1002/rra.4236

Xue, C., Shao, C., & Chen, S. (2020). SDGs-based river health assessment for small-and medium-sized watersheds. *Sustainability, 12*(5), 1846. https://doi.org/10.3390/su12051846

PART I

Fluvial Aspects of Environmental Fragility and Sustainability

CHAPTER 2

Anthropogenic Pressures on Indian River Ecosystems: Strategies for Ecosystem Restoration and Sustainable Management

Akshay Botle, Sayli Salgaonkar, Rahul Tiwari, and Gayatri Barabde

2.1 Introduction

Rivers are prevalent and impact biodiversity. They support aquatic life and shape riparian and non-riparian ecosystems. Thus, rivers may impact aquatic and terrestrial biota origins, evolution, adaptability, and distribution (Naka et al., 2022). Rivers give fresh water and many environmental and social advantages (Tong, 2023). Healthy river ecosystems offer humanity essential materials and services and are vital dwellings for varied

A. Botle (✉) · S. Salgaonkar · G. Barabde
Department of Environmental Science, The Institute of Science, Dr. Homi Bhabha State University, Mumbai, India
e-mail: 23res.evs.akshay@iscm.ac.in

R. Tiwari
Department of Chemistry, Institute of Basic Science, Dr. B. R. Ambedkar University, Agra, India

S. Ghosh (ed.), *Reimagining Indian Rivers for Sustainability*, Palgrave Studies in Environmental Sustainability,
https://doi.org/10.1007/978-3-032-04754-0_2

species on Earth (Kaiser et al., 2020). Rivers have a vital role in many aspects of life, including energy generation, water supply, sediment movement, shipping, food production, sightseeing, and agricultural output (Pan et al., 2016). Freshwater from rivers sustains millions of species, including humans. They are crucial to managing the water cycle between the atmosphere, land, and seas. Rivers convey precipitation and surface runoff throughout the planet, replenishing groundwater supplies and supporting wetlands, lakes, and other aquatic ecosystems. Rivers also nourish terrestrial ecosystems, which sustain plant development and offer homes for many species. Rivers are important economically, socially, and ecologically. They provide freshwater to millions of people and irrigation for farmland. Hydropower, a clean, sustainable energy source, comes from rivers. They also facilitate trade by transporting products and people (Tong, 2023).

Rivers are among the most damaged ecosystems on the planet (Best, 2019). Humans play a significant role in the transformation of rivers (Gregory, 2006). A large geomorphological tradition has studied human effects on river systems, revealing that water quality is impaired by activities such as agriculture and urbanization and that currently, only 23% of Earth's most significant rivers flow uninterrupted to the ocean (García, 2021). However, human activities split river ecosystems, disrupt their structure, pollute the environment, and reduce biodiversity. Thus, protecting river ecosystems is a worldwide problem (Xi et al., 2023). The freshwater living planet index (LPI) and data from 6617 monitored populations of 1398 freshwater species show an average 83% reduction in global freshwater biodiversity between 1970 and 2018 (Valentim et al., 2024; WWF, 2022). There are growing worries regarding rivers ecological health and ability to offer ecosystem services worldwide (Meynell et al., 2021).

River Science is an interdisciplinary subject that studies physical, chemical, and biological interactions in riverscapes of various sizes. This discipline has advanced scientific understanding and applications in environmental fluxes, eco-geomorphology, river health, restoration, and dangers during the last several decades. The Indian subcontinent is drained by 18 primary river basins that provide vital water. Most Indian rivers are located in tropical regions, where the monsoon season brings an average annual rainfall of ~ 3893 billion cubic meters, affecting water availability (Jain et al., 2022). Himalayan and Peninsular rivers comprise

Indian rivers. These are further divided into six landscape, geological, and river types: “Large Himalayan rivers”, “Himalayan foothills-fed rivers”, “North-flowing cratonic rivers”, “West-flowing peninsular rivers”, “East-flowing peninsular rivers”, and “Coastal rivers” (Jain et al., 2020).

India has a vast river network, with twelve major, seven medium, and several small rivers and watercourses totalling around 16×10^4 km (Neelavannan & Sen, 2023). Indian rivers are sacred and ecological. Sacred rivers include the Ganga, Yamuna, and Sabarmati. All these rivers run through India’s most populous and polluted cities (Khan et al., 2021). As much as 70% of India’s water resources are contaminated (Dwivedi, 2017). Furthermore, 30% of waste is not collected, and the remaining 70% is deposited in landfills or neighboring residents spaces, which are swept away and mingled with local water bodies during monsoon season (Panda et al., 2018). Only 18% of collected waste is cleaned, and 38 billion liters of sewage are created daily. The installed capacity to treat wastewater is 12 billion liters; therefore, 26 billion liters are thrown into rivers (CAG, 2012). Dwivedi (2017) studied Indian rivers from the 1950s to the 1990s, finding worsening physico-chemical characteristics, abundant pathogenic microorganisms, and nutrient contamination. Moreover, open defecation in rural and urban India pollutes surface water due to a lack of toilets and sanitation (Panda et al., 2018). The major stressors of Indian rivers are highlighted in (Fig. 2.1), which increases the need to address river pollution. Thus, this chapter discusses major anthropogenic stressors affecting rivers, the ecological effects of human-induced disturbance, sustainable management strategies, and the need for a holistic and integrated approach to river sustainability.

2.2 Major Anthropogenic Stressors on Indian Rivers

2.2.1 *Land Use Changes and Urbanization*

2.2.1.1 Urban Expansion and River Encroachment

Due to the increase in natural population and economic difficulties, people have moved to cities (Björklund et al., 2012; Shukla & Gedam, 2018). Since 16% of the world’s population lives in 2.5% of the landmass, India’s urbanization issue is crucial. The population of India has doubled in 50 years. Many cities have grown rapidly unplanned (Nath et al., 2021). Urbanization involves chopping down trees and removing vegetation on

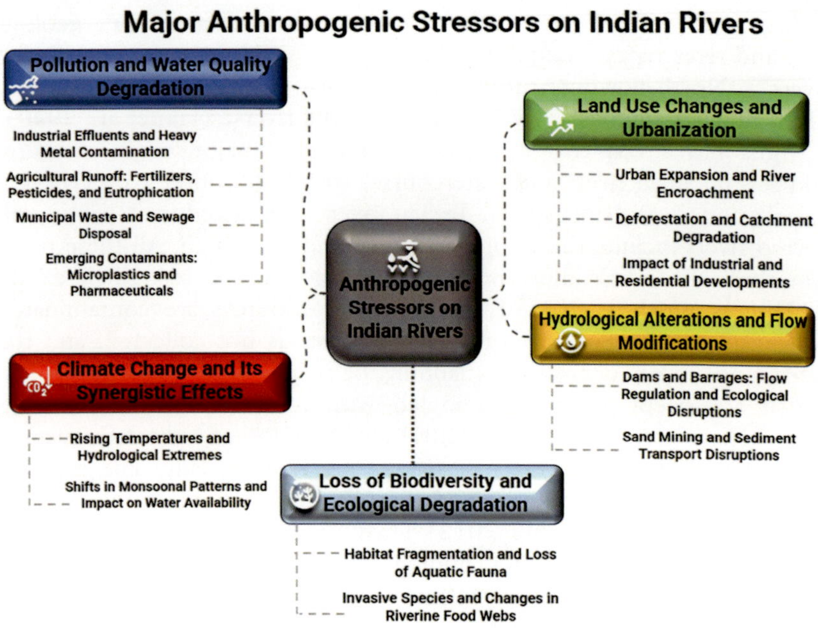

Fig. 2.1 Major anthropogenic stressors on Indian Rivers

forest land to create dwellings, factories, roads, and modern infrastructure (Dewan & Yamaguchi, 2009). Population movement from rural to urban regions has caused these changes (Sahana et al., 2018).

Urban pollution comes from landfill leachates, rainwater runoff, and garbage dumping (Shukla et al., 2018). Construction encroachment on natural drains and trash disposal impede storm water drains, causing congestion. Unplanned building and riverbed encroachment compromise river depths, possibly causing floods. River encroachment and urbanization-driven land use/cover (LULC) change promote urban catchment floods (Danegulu et al., 2024). Similarly, stormwater runoff is a significant non-point source of pollution in metropolitan areas, damaging rivers and inlets. Non-point sources of river contamination include septic tank effluent seepage and surface runoff (Arefin & Mallik, 2018). Sharma et al. (2014) used a WQI and statistical methodologies to study Ganga River water quality variations in Allahabad during the post-monsoon season. Findings revealed that the river's proximity to Allahabad's dense population caused water quality to deteriorate in many spots. Sinha and Eldho (2018) observed that LULC change boosted streamflow by 7.88% in the Netravati River Basin, India, from 1979 to 2012.

2.2.1.2 Deforestation and Catchment Degradation

Deforestation, including forest cover loss and land use change, affects water quality. First, widespread deforestation increases soil erosion and water turbidity. Forest cover loss exposes soil to rainfall and runoff, causing sedimentation in rivers and streams and lowering water clarity. In areas like the Amazon rainforest, deforestation for agriculture and logging has raised canal turbidity (Beuchle et al., 2021). Second, land use changes from deforestation may increase water nutrient contamination. When forests are removed for agriculture, fertilizers are used more, increasing nutrient runoff into rivers and lakes. Algal blooms and eutrophication from this inflow of nutrients, especially nitrogen and phosphorus, degrade water quality and threaten aquatic ecosystems. Thus, successful land management and conservation require knowing the relationship between deforestation and water quality (Ngwane, 2024).

Many researchers have examined how deforestation affects Indian rivers. Over the last several decades, human and natural factors have devastated the Man River Basin's enormous forest region in central India. Most of the region's forests are degraded by permanent agriculture, which is harder to correct since the state government offers local farmers land ownership permits (Thakur, 2022). Rajwant and Sharma (2005) examined the Baner Khad watershed (395 km^2) in Himachal Pradesh, part of the Beas River system. Erosion is common in the active Himalayas. However, tourism-driven development on steep slopes has accelerated deforestation. Behera et al. (2018) examined urbanization, industrialization, and agriculture-driven deforestation in the Brahmaputra (BRB) and Mahanadi (MRB) river basins. According to Landsat's pictures, agriculture covered 55% of MRB but just 16% of BRB. Both basins lost forests for agriculture, but BRB lost more. Matin and Behera (2019) examined forest cover changes (1975–2010), forest fragmentation, and rain-use efficiency (2000–2010) in the Ganga River Basin to estimate land degradation. They found active deterioration via green cover loss, forest fragmentation, and lower primary productivity. Various land use changes and urbanization studies have been reported in (Table. 2.1).

Table 2.1 Various studies indicating impact of urbanization on various rivers across India

Sr. No	*Study Area*	*Year*	*Findings*	*Reference*
1	Karamana River Basin, Kerala, India	2001–2020	Urbanization increased from 63.65 km^2 to 132.72 km^2; Vegetation cover decreased from 631 km^2 to 563.63 km^2; and Land surface temperature increased from 47.55% to 64.01%	(Prasood et al., 2021)
2	Upper Bhima River Basin, Maharashtra, India	1992–2009	Built-up area increased from 2.6% to 7.3%; Urbanization caused increased surface runoff (up to 13.6%) decreased baseflow & percolation. Evapotranspiration increased due to rise in water bodies (up to 59%)	(Samal & Gedam, 2021)
3	Kuttanad & Meenachil River Basin, Kerala, India	2005–2025 (including projections)	Built-up increased by 19.26%, Agri plantation decreased by 23.5%, Forest increased slightly (2005–2015); 53% of floodplain altered to built-up; Surface runoff increased from 856.83 mm (2005) to 937 mm (2025 projection); Vertical water rise increased from 0 mm (2005) to 850 mm (2025); and 83% built-up flooded	(T.S. et al., 2022)
4	Krishna River Basin, India	Predicted changes (2025–2085)	Cropland increases by 7.7%, Built-up increases by 2.5%, Fallow land decreases by 7%, and Wasteland decreases by 2.7%	(Chanapathi & Thatikonda, 2020)

(continued)

Table 2.1 (continued)

Sr. No	*Study Area*	*Year*	*Findings*	*Reference*
5	Brahmaputra Valley, Assam, India	1973–2021 (Past) and 2030–2050 (Future Projections)	Vegetation decreases by 19.48%, Water bodies decrease by 47.13%, Built-up increase by 384.99% (1973–2021); Built-up area increase to 18.09%, Cultivated land decrease to 32.31%, Vegetation decrease to 39.75% by (2050)	(Debnath et al., 2022)
6	Lower Son River Basin, India	1991–2020	Vegetation decreased by 6.74%; Agricultural land increased by 7%; Settlements increased by 2.6%; Barren land decreased from (2.95% to 0.25%)	(Singh et al., 2024)

2.2.1.3 Impact of Industrial and Residential Developments

Rapid urbanization and industrialization, mainly along riverbanks, have caused the discharge of untreated or partly treated sewage and eluents into the river, degrading river water quality worldwide (RamyaPriya & Elango, 2018). India generates 13,468 MLD of industrial wastewater. About 8080.8 MLD wastewater is processed and utilized for irrigation or released into nearby waterways. Partial treatment of wastewater pollutes surface and groundwater, degrades soil, and contaminates crops cultivated on contaminated water (Mukate et al., 2020). Sarker et al. (2021) estimate that 70% of Indian water is contaminated. Various industries for their massive water needs, are located along the river, and their effluents, which contain acids, alkalis, dyes, and other chemicals, are dumped into rivers. The Ganga basin covers 8,60,000 km^2 and 26.3% of India's landmass, home to 43% of the people (Dutta et al., 2020). The Ganga's bio-geomorphological functions have been altered by anthropogenic factors like habitat fragmentation, dams, wastewater discharge, and intensive agriculture using chemical fertilizers and pesticides (Bhardwaj et al., 2010; Sinha et al., 2017). Similarly, sugar and paper mills in Kanpur, tanneries, and distilleries in the Yamuna, Kosi, Ramganga, and Kali River catchments are important polluters (Dutta et al., 2020).

Common effluent treatment plants (CETP) have been built in industrial clusters to reduce the pollutants. These facilities provide water treatment infrastructure for various industrial complexes to protect adjacent communities and surroundings, but many industries do not use these facilities and release the wastewater untreated via trenches (Mukate et al., 2020). One example is the Punjab's Sutlej River which is polluted by home sewage, agricultural runoff, and industrial pollutants that are dumped into or on land nearby (Setia et al., 2020). Another study findings from Khurana et al. (2014) observed that industrial effluents, human waste, and sludge pollute Ludhiana's surface water. Prabha et al. (2013) examined how textile effluent affected water resources in Tamil Nadu's Tirupur industrial sector. Similar investigations from RamyaPriya and Elango (2018) investigated the Cauvery River's chemical load from natural and anthropogenic sources and recommended more burdensome discharge regulations to reduce anthropogenic pollution.

2.2.2 Pollution and Water Quality Degradation

2.2.2.1 Industrial Effluents and Heavy Metal Contamination

Rapid and indiscriminate industrial growth is the leading cause of global water pollution, especially in urban riverine environments where effluents from industries end up in rivers (Kumar et al., 2020a, 2020b; Pandey et al., 2019). Processing and manufacturing of raw materials produce industrial wastewater. Washing, cooking, chilling, heating, extraction, reaction by-products, separation, and quality control produce wastewater streams that adds to the pollution (Sathya et al., 2022). Industrial operations create vast amounts of liquid effluents and raw industrial wastewater with varying compositions (Ishaq et al., 2023). One of the major components of industrialization is the increasing heavy metal contamination worldwide (Saravanan et al., 2024). Heavy metals are those above 20 atomic masses (Barathi et al., 2023). Chromium, arsenic, cadmium, iron, copper, zinc, lead, and mercury are some widely used heavy metals (Singh & Steinnes, 2020). They are widely utilized in industry for many reasons; therefore, their effluents are often released into the environment (Upadhyay et al., 2023; Verma et al., 2023). Different water bodies have toxic heavy metals (Gupta et al., 2023).

Indian rivers are polluted with heavy metals (Botle et al., 2023). Since 1900, India had substantial growth in urban and industrial sectors, mining, smelting, and manufacturing processes which discharge heavy metals throughout the whole country (Barathi et al., 2023). Additionally, fertilizers and insecticides used in agriculture may potentially pollute with heavy metals (Botle et al., 2024b). These human activities pollute surface water bodies heavily, notably with heavy metals. Moreover, sediment transports 30–98% of heavy metals into rivers (Barathi et al., 2023; Wang et al., 2011). Thus, the continuous release of large amounts of untreated industrial effluents increases the concentration of heavy metals in the aquatic systems and poisons organisms and disrupting ecosystem functioning (Proshad et al., 2021). "The Central pollution Control Board (CPCB)" reports that Maharashtra and two other states create 80% of India's hazardous waste, primarily heavy metal pollution (Mawari et al., 2022). Ganga, the country's river, was polluted by five primary heavy metals: chromium, copper, nickel, lead, and iron (Lee, 2013; Mishra et al., 2022a, 2022b). Ganga receives 8250 MLD of municipal discharge. Many tanneries, synthetic rubber companies, distilleries, sugar mills, electro-processing businesses, pesticide plants, and textile manufacturers discharge effluents into the river (Kumar et al., 2020a, 2020b). Bisimwa et al. (2022) found that wastewater effluents degrade surface water, while industrial operations tend to be the primary source of heavy metals in surface water (Botle et al., 2025).

2.2.2.2 Agricultural Runoff: Fertilizers, Pesticides, and Eutrophication

Agriculture water pollution impacts developed and emerging nations (Zahoor & Mushtaq, 2023). Agriculture is one of the biggest drivers of environmental change worldwide. However, it is also one of the most vulnerable sectors to climate change, especially in India, which has only 2.4% of the global land, 18% of the population, and 4% of the water resources (Sharma et al., 2024). As one of the world's fastest-growing economies, India puts much strain on water, land, and air resources. The high-yield aquifers of the Indo-Gangetic and Brahmaputra alluvial plains provide India with one of its primary water resources. The nation uses 87% of freshwater for irrigation, followed by home and industrial consumption (Mukherjee et al., 2015). The causes and effects of agriculture-related water pollution differ by nation. Each country's population and land area need distinct livestock and agricultural management practices. Traditional agricultural methods, poor infrastructure, and illiterate farmers are the primary sources of agricultural pollution in India and other developing nations (Zahoor & Mushtaq, 2023).

The "Food And Agricultural Organisation Statistics" (FAOSTAT) reports that the US, China, and India use more than 50% of fertilizer, which improves their agriculture and finances (Rashmi et al., 2020). Farms dump massive volumes of agrochemicals, drug residues, organic materials, sediments, and salty drainage into waterways. Agriculture is the primary source of revenue and water pollution in rural regions. Since 1970, the global population has grown by 78% and fertilizer consumption by 120%. Nitrogen fertilizer consumption climbed by 630 million between 2014 and 2018, with 58% from Asia (primarily China 18% and India 17%) (Sharma et al., 2024). Numerous studies have shown that excessive fertilizer usage degrades environmental components, particularly the lithosphere and hydrosphere (Almasri & Kaluarachchi, 2004; Ning et al., 2017; Savci, 2012). The main contributors are nitrogen and phosphorus. Synthetic fertilizers unintentionally pollute waterways with nitrogen and phosphorus (Saritha Vara, 2023). Due to river pollution, phosphorus fractions in freshwater systems worldwide have been examined (Varma & Jha, 2024).

Similarly, pesticides are classified by their target pests/weeds, manner of application, and chemical composition (Srivastav, 2020). Pesticides protect crops, preserve food and materials, and prevent diseases, yet they may harm humans and the environment (Kaur & Sinha, 2019). Most of the 500 pesticides listed in the literature are used in agriculture (Arias-Estévez et al., 2008). These chemicals are used to control weeds, pests, and insects and are frequently utilized pollutants in biological systems at elevated levels (Alvarez et al., 2017). Organochlorine insecticides like Aldrin, Chlordane, Dichlorodiphenyltrichloroethane (DDT), Endosulfan, Endrin, Heptachlor, Hexachlorobenzene, Hexachlorocyclohexane (HCH), Metolachlor, Methoxychlor, Mirex, Toxaphene protect crops. Due to their inexpensive cost and ability to kill several pests, they are widely used in Asia (Srivastav, 2020). Yadav et al. (2015) reported over 3 million instances of hazardous pesticide poisoning annually. According to Chakraborty et al. (2016), Σ OCP concentrations in the Brahmaputra and Hooghly rivers in India ranged from 2 to 245 ng/L and 12 to 154 ng/L, respectively.

Agricultural non-point source pollution has long been linked to eutrophication (Xia et al., 2020). Since the beginning of the twentieth century, eutrophication has been widely acknowledged and ascribed to specific causes. Currently, dispersed nitrogen and phosphorus sources are the primary variables. Nutrient accumulation leads to cyanobacteria blooms, eutrophication, fish death, oxygen depletion, poor water quality, and red tides (García, 2021). Cyanobacteria produce poisonous secondary metabolites/cyanotoxins when given enough light and nutrition, causing aquatic life sickness and death, including humans, by ingestion, inhalation, or direct contact (Kakade et al., 2021). Excess phosphorous and nitrogen enrichment in water bodies causes nutritional imbalance, bloom development, and anoxic conditions (Carpenter, 2005; Häder et al., 2020; Liang et al., 2021). "Algal bloom" is the fast development and accumulation of phytoplankton (cyanobacteria, algae, diatoms, dinoflagellates) in waterways, causing hypoxia and the death of fish and other aquatic species. Similarly, a "Red Tide" occurs when red or brown algae accumulate in significant amounts. Harmful Algal Blooms (HAB) include "red tides" and other algal blooms. Many Indian researchers have found algal blooms in inland waterways (Prasad and Prasad, 2019). Moreover, dams slow river flow, enhance eutrophication and promote blooms (Domingues et al., 2012; Morais et al., 2009). The researchers observed that high nitrogen loading caused eutrophication, groundwater pollution, soil pH alterations, and nutrient fluctuation in surface water reservoirs, compromising soil health (Cui et al., 2014; Liu et al., 2011). The Ganges in Kanpur has seen regular fish mortality. In June 2014, thousands of dead fish were observed floating in the Ganga near Ganga Barrage and on city ghats (Prasad and Prasad, 2019).

2.2.2.3 Municipal Waste and Sewage Disposal

Worldwide sewage pollution models predict profound deterioration of surface water caused by excessive human waste entering streams. Wastewater pollution of surface waterways threatens human health worldwide, but recent investigation reveals it may also impact species and natural environments (Wear et al., 2021). "Class I and Class II" cities in India create 29,129 MLD of sewage (Varma et al., 2022). In India, Maharashtra produces (3.32 km^3) sewage water, whereas Lakshadweep generates just 0.47×10^{-2} km^3. The southern region of India produces the most sewage water (7.63 km^3), followed by the northern (7.33 km^3), western (5.2 km^3), eastern (3.83 km^3), central (1.77 km^3), and north-eastern (0.61 km^3). Class I cities and class II towns in the Ganga basin produce 4.16 km^3 of wastewater while, Union territories like Lakshadweep, Delhi, and Chandigarh generate more sewage water per person (178–202 L/d^{-1}) than the national average (72 L/d^{-1}). India produces > 2.26 km^3 of industrial wastewater (Ranade & Bhandari, 2014). Only 7.38 km^3 (28%)

of wastewater produced is treated by 1093 STPs, leaving 19.03 km^3 (72%) untreated (Minhas et al., 2022). Operational STPs process just 37% of 72,368 MLD of produced sewage and perform poorly, with 35–50% not fulfilling discharge limits. Recent studies indicate poor compliance rates with industrial effluent limits (<35% for ETPs and < 10% for CETPs) (Breitenmoser et al., 2022).

Poor sewage treatment in Delhi causes 19 drainage routes to empty into the Yamuna. These pipes carry wet sewage to rivers. About 22 industrial sectors in Haryana, 42 in Delhi, and 17 in Uttar Pradesh, plus many smaller ones, dump trash into the river. Factory pollution in Delhi accounts for 80% of river pollution (Mantey, 2017; Vashisht et al., 2020). The Bharalu River in Assam, India, is one of the most polluted Brahmaputra tributaries. Most municipal solid garbage and other city wastes harm the Brahmaputra River. The main Bharalu River pollutants include BOD, fluoride, pH, and dissolved oxygen (Tariq & Mushtaq, 2023). Another study carried out in Lucknow, 40% of the population lacks adequate sanitation, which leads to open defecation by slum dwellers along both banks of the Gomti River, which pollutes the river by releasing untreated feces, sewage, plastics, municipal waste, and industrial toxins. This feces contamination introduces harmful pathogens—including bacteria, viruses, fungi, and protozoa—into the water, posing serious health risks to individuals (Singh et al., 2020).

2.2.2.4 Emerging Contaminants: Microplastics and Pharmaceuticals

Emerging contaminants (ECs) are chemical and biological agents that endure, affect biotic/abiotic indices, impair human immune systems, and cause ailments (Sudarsan et al., 2024). Concerns over "emerging pollutants" and their impact on the environment have been raised by scientists, engineers, and the general public (Botle et al., 2024a). EC concerns have become a worldwide issue in recent decades, with India being the most susceptible because of its exponential production rate (Tripathi et al., 2023). India, China, the UK, Japan, the US, and Germany have reported more than 70 complex ECs mixtures, mostly in urban water bodies, including propranolol, codeine, and diclofenac (Sudarsan et al., 2024). Due to improved detection technology, recent investigations have detected emerging pollutants (EPs) in water. The main EP classes include fire retardants, nanomaterials, pesticides, pharmaceuticals and personal care products (PPCPs), plasticizers, and surfactants (Anh et al., 2023; Kumar et al., 2022). Due to treated wastewater effluents in rivers, medicines, steroid hormones, and personal care goods were found (De Sousa et al., 2018; Ngo et al., 2021). Nonetheless, river water contains emerging and persistent organic pollutants like organochlorine pesticides, perfluoroalkyl and polyfluoroalkyl substances, polychlorinated biphenyls, and polycyclic aromatic hydrocarbons, which were discharged from industrial and agricultural activities in river basins (Anh et al., 2023). Countrywide research in Indian rivers found that most aquatic systems are tainted with ECs contaminations (Sudarsan et al., 2024). ECs in rivers are caused by uncontrolled dumping of agricultural, residential, and industrial wastewater (Khalid et al., 2018). Williams et al. (2019) reported that untreated wastewater effluents cause ECs in Indian rivers. For example, rivers such as the Brahmaputra, Ganga, Kaveri, Sabarmati, Tamirabarani, Vellar, and Yamuna consist of phthalates, pharmaceuticals, triclosan, carbamazepine, and parabens (Tripathi et al., 2023).

Plastic wastes accumulate in the environment because they are chemically stable, corrosion-resistant, and difficult to biodegrade (Salgaonkar et al., 2024). India generates the third most plastic garbage. India generates 4 kilograms of plastic garbage per person (Randhawa, 2025). The Ganga River, the second most polluted river in the world, releases around 1 lakh tons of plastic into the Indian Ocean annually (Simon & Joshi, 2022). Microplastics (MPs) are small pieces of plastic trash made from a variety of sources and have a variety of chemical properties. Their environmental and eco-toxicological effects are becoming more prevalent and concerning, and they are abundant in beach and river sediments in India (Patidar et al., 2023). Napper et al. (2021) found that the Ganges might discharge 1–3 billion MPs into the Indian Ocean every day. Moreover, pharmaceuticals in aquatic environments may cause antibiotic resistance, endocrine disruption in fish, and cell proliferation suppression, impacting physiology and morphology (Konda et al., 2024). Lübbert et al. (2017) found fluconazole, voriconazole, and ciprofloxacin in Hyderabad urban wastewater samples. Williams et al. (2019) found caffeine, carbamazepine, diclofenac, naproxen, and venlafaxine in the Ahar River in Udaipur, India. Konda et al. (2024) examined 46 PPCPs in wastewater samples from 11 Hyderabad Musi River locations. Biswas and Vellanki (2021) detected medications, personal care items, and hormones discharged by people and animals in the Yamuna River. Similarly, numerous investigations have discovered active pharmaceutical ingredients, personal care products, pesticides, endocrine-disrupting chemicals, and artificial sweeteners in surface water bodies (Dwivedi et al., 2018). Sharma et al. (2019) found medicines, personal care items, and artificial sweeteners in surface and groundwater in the Ganges River Basin, India.

2.3 Hydrological Alterations and Flow Modifications

2.3.1 Dams and Barrages: Flow Regulation and Ecological Disruptions

Growing human requirements have prompted dam construction worldwide. Dams have increased economic progress and human well-being by providing water, electricity, and food (Ekka et al., 2022). Nearly 63% of free-flowing rivers worldwide have reservoir activities, affecting basin ecology (Grill et al., 2019). Over 16 million reservoirs globally store 65 Gt/year of river sediments, impacting river geomorphology, ecology, and livelihoods (Ghosh & Muñoz-Arriola, 2023). Considering only water's natural flow arteries, direct impoundments (transverse intervention) increased to a point where two new dams were built each day in the latter half of the twentieth century (Richter & Thomas, 2007). While dams and river water storage structures provide significant advantages to civilization, they can degrade geomorphic, hydrological, and ecological health (Ghosh & Muñoz-Arriola, 2023). Dam-building causes the river to become unstable and preserve hydrogeomorphic equilibrium until the human-induced threshold is reached (Ghosh & Guchhait, 2014). Dams, reservoirs, or levees may restrict river flow. Variations in flow extremes disrupt sediment movement, river connection, nutrient availability, and species growth and abundance, which may harm delicate biological ecosystems (Jain et al., 2022). River flow may be restricted, and channels may be downcast by artificial levees and other channelization devices.

River regulation began with ancient Indian civilizations. India operates around 5000 dams in various rivers, making it the third biggest dam-building country. New research predicts that India's energy consumption will expand by 4.2% per year for the next 30 years, faster than any other large country (Pradhan et al., 2023). Thus, the dam's upstream and downstream ecosystems are affected (Crossman & Pollino, 2018). Dams turn rivers into stagnant water (Gopal, 2016). This affects river flow amplitude, frequency, duration, timing, and rate of change. Changes in river regimes affect biological processes that affect ecosystem services (Brauman et al., 2007; Renöfält et al., 2010). According to Dhanakumar et al. (2015) and Venkatachalapathy and Karthikeyan (2015), reservoir development has changed sediment loads and aquatic species composition. Many dam-induced alterations studies have used "Indicators of Hydrological Alterations (IHA)" to assess dams' hydrological effects on flow regime characteristics (Ekka et al., 2022).

To harness water for development, India has built Bhakra Nangal, Chambal, Hirakud, Kakrapar, Kosi, Nagarjunasagar, and Tungabhadra dams. Development and economic prosperity have been associated with these dams, which provide hydropower, irrigation, and residential water (Klingensmith, 2003). "The National Registry for Large Dams (NRLD)" reported 4877 dams constructed in India in 2016 and 313 under development (Ekka et al., 2022). Currently, as per the reports published by India WRIS (2025), there are fifty large dams completed from 1926 to 2014. Maximum dams are constructed in the state of "Maharashtra and Kerala" (7), "Tamil Nadu and Uttarakhand" (5), "Himachal Pradesh" (4), "(Gujarat, Jammu and Kashmir, and Karnataka)" (3), "Arunachal Pradesh and Telangana" (2), and rest all with (1) (Fig. 2.2). Dam-building on Indian rivers will grow to provide renewable energy and flood and irrigation control in the near future, though it will alter sediment movement across river segments (Pradhan et al., 2023). Pal (2017) found that tributary channel sedimentation increases backwater reach flooding, while sand bars were gradually formed near the Ganga River's Farakka barrage, which affects the upstream reach (Mirza, 2004; Parua, 2010) and braids the river (Banerjee, 1999; Rudra, 2006). A Giri River barrage has also reduced the fish diversity and resulted to fragmented habitat (Rumana et al., 2015). Another study by Pal, (2016), on the Post-Massanjore reservoir building indicated narrowed active channel, coarsened bed material, lowered lateral stability, and hastened braiding on the Mayurakshi River.

2.3.2 Sand Mining and Sediment Transport Disruptions

A major sustainability problem of the twenty-first century is riverbed mining (Gruel & Latrubesse, 2021). River ecosystems are vital to human cultures, but widespread exploitation and mining of riverbed materials have damaged them (Rentier & Cammeraat, 2022). Overexploitation

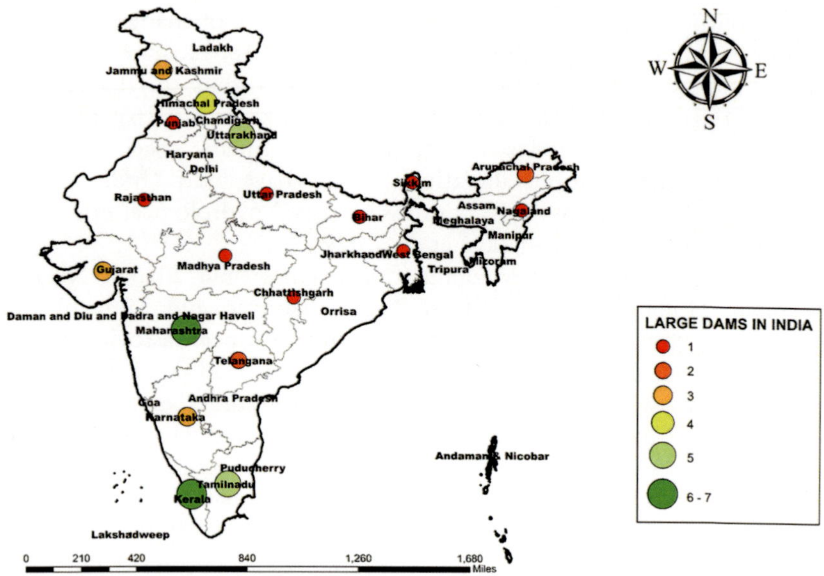

Fig. 2.2 Large dams in India

of materials distorts river ecosystems natural equilibrium (Banjoko & Eslamian, 2016), but excessive removal disrupts sediment transport, channel incisions, and sediment mass balance (Dar et al., 2023; Ghani et al., 2011). Global sand consumption rose 23-fold between 1900 and 2010 and is anticipated to reach 82 billion tons by 2060. About 50 billion tons of sand are utilized yearly, twice the quantity created by nature (Rentier & Cammeraat, 2022). Sand is solid particles 0.2 to 2.0 mm in diameter. Larger particles are classified as gravel (2–63 mm), cobble (64–256 mm), and boulder (>256 mm), while smaller particles are silt (2–0.002 mm) and clay (< 2 microns) (Gopal & Chauhan, 2025).

Unscientific river bed mining often destroys river aquatic habitats, requiring immediate remediation (Dinesh Kamboj et al., 2017; Kumar et al., 2019). Problems arise when river bed mining materials are extracted faster than natural replenishment, and their severity varies depending on the pace, kind, and execution of the extraction (Hackney et al., 2020; Malik et al., 2021). The activities may contaminate periphery communities waterways and ecosystems (Zou et al., 2019). River mining provides money and livelihood for many people, including the government, via royalties or taxes (Shymbin & Nongbri, 2022). As a result, boatmen, brick and block producers, drivers, miners, truck owners, and others depend on mining (Mngeni et al., 2017). Mazumder et al. (2014) found that river sand mining endangered the Ganges river dolphin. Prasad and Nair (2004) studied the Kerala Achankovil River basin where indiscriminate sand mining is the leading cause of drinking water concerns there.

2.4 Climate Change and Its Synergistic Effects on River Ecosystems

2.4.1 Rising Temperatures and Hydrological Extremes

Climate is defined by "change", but the speed is what matters. Hydrological extremes, including floods and droughts, have increased substantially in the recent decade. Climate change affects more than hydrological extremes (Panday & Kumar, 2022). Climate change and human activities have substantially altered hydrological regimes worldwide, altering the magnitude and timing of flood and drought events, affecting river discharge and other water-related hazards and water and food security. In recent decades, hydrological extremes and their devastation have escalated in India, affecting a significant population's livelihood (Mishra et al., 2024). Droughts, floods, and heatwaves may damage ecosystems and pose major social issues (Kumar et al., 2021). Anthropogenic greenhouse gas emissions have grown since the pre-industrial period, warming the Earth's atmosphere, according to the IPCC (Kumar et al., 2021). Low water quantity in low-precipitation places causes other problems, such as poor water quality. Consequently, pollution is growing in specific locations, followed by population increase and limited water supply per person (Botle et al., 2024c).

Two-thirds of Indians work in agriculture and related industries (Pathak, 2023). Hydrological extremes have ripple effects on socioeconomic development and may worsen social inequality in developing countries like India (V. Mishra et al., 2022a, 2022b). Anthropogenic climate change, ocean and atmospheric circulation dynamics, topography patterns, and aerosols affect hydrological extremes in India (Mishra et al., 2024). India's floods and droughts killed hundreds, displaced millions, damaged infrastructure and crops, and disrupted market and social operations (Ray et al., 2019; Sharma et al., 2020; Zhang et al., 2017).

Floods are the most common natural disaster. India's annual floods kill people and destroy infrastructure and ecosystems. Floods in 1998–2017 impacted almost two billion people. Over 4.84 million Indians were exposed to riverine floods, ranking highest among 164 nations (Pandey et al., 2022). India has the greatest flood societal vulnerability (Wasson, 2018). About 11% of India's land is prone to flooding (Pandey et al., 2022). Floods may be riverine, flash, urban, pluvial, or coastal due to excessive rainfall, storm surge, poor drainage, dam collapses, and reservoir malfunctions (Dhar & Nandargi, 2003; Mujumdar et al., 2020). Riverine floods in India peak in the mid- and late-summer monsoon season due to excessive precipitation and wet antecedent moisture (Nanditha et al., 2022; Nanditha and Mishra, 2022). Indian pluvial floods, caused by excessive precipitation, have also risen (Ali et al., 2019).

Meteorological, agricultural, hydrological, and socioeconomic droughts are characterized by precipitation deficit, soil moisture, streamflow or groundwater, and socioeconomic consequences. Flash droughts develop, proliferate, and dissipate quickly, causing irreparable harm to agriculture and vegetation. Instead of months or years, flash droughts arise and end in weeks (Mishra et al., 2024). Flash droughts are unpredictable because of their fast start and brief duration (Pendergrass et al., 2020). Mujumdar et al. (2020) found that central Indian droughts increased in size and severity from 1951 to 2015. After 1950, the central and northeast Indian areas had numerous droughts (Mishra et al., 2024). The climate has increased dryness in India's Gangetic plains (Aadhar & Mishra, 2019).

2.4.2 *Shifts in Monsoonal Patterns and Impact on Water Availability*

Most hydrological processes depend on rainfall, and its spatiotemporal distribution and changing characteristics can affect agriculture, ecosystems, and water resource management (Mathew et al., 2021). Global rainfall patterns altered dramatically in the twentieth century (New et al., 2001). Alterations in rainfall patterns cause frequent floods and drought, biodiversity loss, and river ecosystem disruption. Many studies have revealed global and regional rainfall changes due to global climate changes. These changes are especially critical for tropical nations like India, whose economy, agriculture, and food security rely on monsoon rains (Mathew et al., 2021).

Different parts of India have different climates. The south has distinct meteorological characteristics from the north, while varied places have varied rainfall quantities (Purnadurga et al., 2018). The Banas basin gets 600 mm of rainfall per year, ranging from 314 to 1167 mm from 1970 to 2013 (Rashmi et al., 2020). This rainfall pattern illustrates India's climatic patterns, which range from highly desert in Rajasthan to pleasantly moist in the northeast. Indian climates range from highly dry in Rajasthan to pleasantly rainy in the northeast, which explains this rainfall pattern (Singh & Kumar, 2015). The Western Ghats in South and North Eastern India get the most monsoon rainfall in India. Monsoon rainfall trends in the twentieth century have been thoroughly studied. Rainfall patterns in India varied by regional and temporal scale (Mathew et al., 2021). Most 1901–2015 research employing yearly rainfall averaged across the Indian landmass found trendless and unpredictable rainfall. However, rainfall in the Indian subcontinent declined from 1951 to 2004, and the decline accelerated from 1971 to 2002 (Mooley & Parthasarathy, 1983; Rao & Jagannathan, 1963; Srivastava et al., 1992; Thapliyal & Kulshrestha, 1991).

Summer monsoons deliver heavy rain to the Ganga, Indus, and Brahmaputra basins in northern India (Immerzeel et al., 2009). Over 700 million people reside in these river basins, with 60% in the Ganga (Nepal & Shrestha, 2015). The Ganga basin's direct and indirect freshwater needs depend on the "Indian summer monsoon (ISM)" (June–September), which provides 80% of rainfall (Auffhammer et al., 2012; Gadgil & Gadgil, 2006). "The Indo-Gangetic Plains (IGP)", which gets over 80% rainfall over ISM months, is crucial for agro-industrial development (Bhatla et al., 2015). Hydrological extremes throughout the nation, particularly the densely packed Ganga basin, might increase with considerable monsoon rainfall departures. Thus, long-term rainfall data are essential for understanding monsoon rainfall. Most studies have examined rainfall patterns and seasonality throughout the nation (Maurya et al., 2024). Central and North India have had considerable annual and seasonal rainfall declines (Ranade & Gurrapu, 2025). Two studies indicated a substantial decrease in seven subdivisions between 1901 and 2015: Praveen et al. (2020) and Sahoo and Yadav (2022) in the north Indian homogenous zone. River basin studies show that altering rainfall patterns has complicated and diverse consequences, reflecting climate change and regional variability (Ranade & Gurrapu, 2025). Over 135 years, Kumar and Jain (2011) found declining rainfall in 15 river basins. Deka et al.

(2013) found a substantial drop in monsoon and post-monsoon rainfall in northeast India's Brahmaputra and Barak basins between 1901 and 2010, while Taxak et al. (2014) found an 8.4% decline in Wainganga basin rainfall from 1901 to 2012. The Satluj, Ganga, and Kosi basins showed declining trends from 1984 to 2010 (Gajbhiye et al., 2016; Kumar & Gill, 2015; Srivastava et al., 2021).

2.5 Loss of Biodiversity in Riverine Ecosystem

2.5.1 *Habitat Fragmentation and Loss of Aquatic Fauna*

Several permanent and seasonal rivers cross India. Fisheries are the main contribution to India's food security and livelihood from most of its rivers rich biodiversity (A. M. et al., 2023). About 45,000 kilometers of riverine resources run across India. Freshwater habitats sustain about 900 endemism-rich fish species. Alterations in water flow, abstraction of water and sand, climate change impacts, the influx of pollutants, poaching and unsustainable fish exploitation, and the spread of invasive species are reducing freshwater biodiversity worldwide (Kantharajan et al., 2022). One of the world's richest fish biodiversity reserves, the Indian rivers feed millions of impoverished people, particularly in rural regions (Vass et al., 2011).

India has 2801 fish species and 999 freshwater species (Koushlesh et al., 2021). River fish species diversity relies on ecological characteristics such as river size, surface area, mean annual river flow, temperature, depth, water velocity, channel shape, substrate, and climate (Poff & Zimmerman, 2010). In recent years, water pollution, other anthropogenic stress, habitat degradation, overfishing, and other factors have threatened the aquatic ecosystem ichthyofaunal variety (Bhattacharjya et al., 2017; Pathak et al., 2020). River fragmentation is a significant cause of biodiversity changes in the Neotropics (Su et al., 2021), and a study by Sajina et al. (2022) on Mahanadi's fish assemblages showed that stressors have impaired the river's ecological health in some areas. Likewise, Das et al. (2013) assessed Ganga's ecological integrity, fish diversity, distribution, and community structure throughout its longitudinal gradient. *Cyprinidae* dominated the river's 143 fish species. The middle length of the river at Allahabad had a considerable catch drop from 1961 to 2010. Indian major carps (IMC) and the *Tenualosa Elisha* were declining, but foreign fish *Cyprinus carpio* and *Oreochromis niloticus* increased (43–48% of the total capture).

2.5.2 *Invasive Species and Changes in Riverine Food Webs*

Multiple factors, such as habitat alteration, overexploitation, and invasive alien species (IAS), contribute to biodiversity loss and its impact on life and livelihoods (Dueñas et al., 2018; Sreekanth et al., 2022). IAS is considered the second most significant threat to global biodiversity. Introductions and expansion of foreign species beyond their natural habitats also cause species extinctions (Dudgeon et al., 2006; Puth & Post, 2005). Invasive species develop, breed, and spread rapidly, potentially harming others. A non-native organism is an invading species. It may impact the environment or economy in a new setting. Invasive species reduce biodiversity, compete with native species for resources, and modify ecosystems. Oceangoing ships bilge water, aquaculture specimens or bait, and other methods may introduce invasive species (Mahapatra & Mohanty, 2023). Functional alien species may also cause biodiversity loss, habitat adjustments, parasite propagation, ecological food web changes, and trophic interactions in the receiving community (Thomsen et al., 2014). In imported habitats, non-native species are usually hardy and may disperse (Tierney et al., 2020).

Over the past century, foreign invading species have threatened India's mega diversity of floral and faunal resources. The aquatic world has been severely affected by biological invasions, with over 300 alien species recognized and many more predicted (Bijukumar, 2019; Singh & Lakra, 2011). Invasive foreign fish species threaten the Indian Inland Aquatic Ecosystem (Mahapatra & Mohanty, 2023). Due to invading alien species, the freshwater ecology that sustains 40% of all fish species is vulnerable. They affect indigenous aquatic biodiversity, especially fish (Sandilyan, 2016). Indian exotic fish include 31 aquaculture species, 600 decorative variations, and three larvivorous species. Some of the invasive species found in Indian Rivers are highlighted in (Table 2.2). Most intentional introductions aimed to improve fishery management, aquaculture, and fish pond production or to fill niches, increase production, and provide new sport fishing and ornamental fish objects in natural ecosystems (Singh & Lakra, 2011). Mahapatra and Mohanty (2023) identified various invasive fish and shellfish that threaten India's freshwater ecosystems. These invasive species prey on local wildlife, compete for resources, spread illnesses, and hinder reproduction. Indirectly, they impact food webs, habitat structure, and native species abundance or diversity. *Aristichthys nobilis* (Bighead carp), *Clarias gariepinus* (African catfish), *Mylopharyngodon piceus* (Black carp), *Oreochromis mossambicus* and *O. niloticus* (Mozambique and Nile tilapia), *Cyprinus carpio* (Common carp), and several *Pterygoplichthys* species are invasive. *Gambusia affinis* and *Holbrooki* (Western and Eastern mosquito fish), *Poecilia reticulata* (Guppy), and *Pomacea canaliculata* (Golden Apple snail) are also intruders. This research emphasizes the need to manage these species to reduce ecological impact. Lija and Radhamany (2023) assessed invasive alien plant species along the Neyyar River in Poovar, a popular tourist spot in Thiruvananthapuram, Kerala. The survey found 41 invasive species in 34 genera and 20 families, including five aquatic species. Fourteen were very toxic, threatening native biodiversity and ecological functioning. Aquatic invasives disrupted water movement and ecology. The authors stressed the need to prevent these species from spreading to protect local flora and promote sustainable ecotourism.

2.6 Ecological Consequences of Anthropogenic Impacts

2.6.1 Loss of Biodiversity: Endangered Aquatic Species

The worldwide freshwater species have fallen to 83% since 1970 (Grooten & Almond, 2018). Human activity significantly reduced freshwater ecosystem biodiversity (Grizzetti et al., 2016; Jenkins, 2003). Freshwater biodiversity is threatened by climate change and human river hydrological regime alteration (Pownkumar et al., 2022). The Western Ghats, North East Region, Himalayas, and Nicobar Islands—four of the 34 global biodiversity hotspots—provide a large portion of the world's biological resources, notably teleosts. Although biodiversity is vital to our existence, all species confront danger, and it is crucial to distinguish between various forms of biodiversity threats. Multiple methods and priorities have been presented to address the aquatic environment's biodiversity and ecological stability challenges. Artificial or natural dangers may have cascading effects. Threats include habitat changes, overexploitation, reduction of natural habitat area, dam construction, diversion or reclamation of river beds for urbanization, siltation, overfishing, pollution of water bodies, introduction/entry of non-native species, and global climatic variations. "The International Union for Conservation of

Table 2.2 Invasive species found in Indian rivers

Sr. No	*Name*	*Abundance in river streams*	*Native*
1	*Amphilophus trimaculatum* (Three spot cichlid)	+ + +	USA
2	*Aristichthys nobilis* (Bighead carp)	+ + +	China
3	*Clarias gariepinus* (North African catfish)	+ + +	Africa
4	*Ctenopharyngodon Idella* (Grass carp)	+ + +	Eastern Asia
5	*Cyprinus carpio* (Common carp)	+	Europe
6	*Gambusia affinis* (Topminnow)	+ +	USA
7	*Hypophthalmichthys molitrix* (Silver carp)	+ + +	China
8	*Molypharyngodon piceus* (Black carp)	+	China
9	*Oreochromis mossambicus* (Mozambique tilapia)	+ +	Nilo Sudanic Region
10	*Oreochromis niloticus* (Nile tilapis)	+ +	South Africa
11	*Poecilia reticulata* (Guppy)	+ +	South America
12	*Pterygoplichthys anisitsi* (Parana sailfin catfish)	+ +	South America
13	*Pterygoplichthys disjunctivus* (Vermi miculated sailfin catfish)	+ +	South America
14	*Pterygoplichthys multiradiatus* (Suckermouth armoured catfish)	+ +	South America
15	*Pterygoplichthys pardalis* (Amazon sailfn catfish)	+	South America
16	*Salmo trutta* (Brown trout)	+ +	Europe

Source (Nobinraja et al., 2023; Singh, 2025)

Nature" (IUCN), "World Wildlife Fund" (WWF), Conservation International, and "United Nations Environment Programme" (UNEP) have warned about rising species extinctions for years. IUCN, the World Conservation Union, regularly assesses globally threatened species. The

Red List 2011, version 2 lists 2028 endangered fish species from 9,554 assessed fishes (Jena & Gopalakrishnan, 2012).

River water manipulation has helped human progress, but it has damaged river ecosystems (Pownkumar et al., 2022). The Ganga Basin has several rare and vulnerable species, such as the Critically Endangered Gharial (*Gavialis gangeticus*), Mugger or Indian marsh crocodile (*Crocodylus palustris*), Estuarine crocodile (*Crocodylus porosus*), Ganges shark (*Glyphis gangeticus*), Gangetic stingray (*Himantura fluviatilis*), Golden mahseer (*Tor putitora*), and Hilsa (*Tenualosa ilisha*) (NMCG, 2025). Moreover, from research, it was found that Chinese *hypophthalmichthys nobilis* thrives in the Damodar, Ganges, Godavari, Irrawaddy, Kaladan, Mahi, Mahanadi, and Narmada river basins (Kulkarni, 2023). Moreover, dams block upstream fish spawning movement, displace populations, and prevent gene flow. The upstream migration of anadromous hilsa (*T. Elisha*) after the construction of the Farrakka barrage on River Ganga in India reduced the catch of the species above the barrage from 19.3 tons to 1.04 tons at Allahabad and from 32 tons to 0.6 tons at Buxar, demonstrating water obstruction (Gopal, 2004). Silver carp have displaced Indian main carp, notably *C. catla*, in Govindsagar Dam (Himachal Pradesh) (Sugunan, 1995). A. K. Singh and Lakra (2011) found evidence that *Clarias gariepinus*, *Cyprinus carpio*, and *Pangasianodon hypophthalmus* which harmed Indian river fish biodiversity.

2.6.2 Eutrophication and Dead Zones

Eutrophication pollutes water bodies and threatens aquatic organisms by causing algae blooms due to human dependence on nutrient fertilizers for food production (Akinnawo, 2021; Carpenter, 2008; Zhang, 2022). Increased nutrients lead to eutrophication, cyanobacteria blooms, fish death, oxygen depletion, poor water quality, and red tides (García, 2021). Eutrophication causes algal blooms by increasing the growth of phytoplankton like green algae, diatoms, and cyanobacteria, also known as blue-green algae, and poisoning water bodies, making them unfit for drinking, recreation, and aquatic life (Costa et al., 2018). Harmful algal blooms may upset aquatic environments and cause ecological damage. An overabundance of microalgae or phytoplankton may alter aquatic food web dynamics. Additionally, algal blooms may physically modify aquatic ecosystems (Vasseghian et al., 2024).

Cyanobacteria produce harmful secondary metabolites/cyanotoxins when given enough light and nutrition, which may kill aquatic life and people by direct touch, inhalation, or ingestion (Kakade et al., 2021). Cyanobacteria dominance in eutrophic water bodies causes toxins and a dead zone in aquatic ecosystems (Rathore et al., 2016). Dead zones caused by eutrophication have been shown to reduce aquatic life biodiversity. Depending on the colour of the algae, cyanobacteria cause "red or brown tides" (Ali et al., 2019). Red tides occur when fish, birds, and other fish-eating animals perish. Larger species that devour this marine life may die if they swallow enough poison (El-Sheekh et al., 2021). Cyanotoxins are produced by HABs, notably those caused by blue-green algae. The effects of microcystins, anatoxins, saxitoxins, and cylindrospermopsins vary. Cyanobacteria toxins pollute water and harm humans (Yang et al., 2022). Toxins harm fish growth, reproduction, and histopathology, lowering their survival chances (Drobac et al., 2016). Algal mats on water surfaces reduce light penetration and create hypoxia, killing vegetation and wildlife. These blooms discharge poisons that harm people. Microcystin-LR (MC-LR) and Cylindrospermopsin may induce tumors (Szlag et al., 2015). Abdominal discomfort, diarrhea, dry cough, headache, mouth blisters, nausea, pneumonia, sore throat, and vomiting are common acute health consequences of MC-LR. Anatoxin-a produces burning, confused speech, death, numbness, respiratory paralysis, salivation, sleepiness, and tingling, whereas cyclopsin induces bloody diarrhea, fever, headache, and vomiting (Kakade et al., 2021). Cyanotoxin-contaminated water may cause gastrointestinal issues, liver damage, neurological issues, and skin irritation (Nielsen & Jiang, 2020). The ecological effects of algal toxins may kill fish, cause aquatic illnesses, and harm other aquatic creatures (Zhang et al., 2023).

2.7 River Fragmentation

The dwelling space of biota called a "habitat", is spatially and temporally variable, indicating that integrated habitat information might be used for river health assessments and management (Kuo et al., 2021). Intermittent rivers biodiversity, biogeochemical cycle, and habitat variety are maintained by natural hydrology (Datry et al., 2016). Waterfalls, steep cascades, and canyons are physical obstacles in high-relief landscapes or when streams pass refractory geologic layers (e.g., Niagara Falls, the Fall Line in the Southeast) (Fuller et al., 2015). Beaver and big woody debris dams are comparable physical barriers in material composition and partial permeability. Both are often modest-scale (<10 m of stream channel) obstacles on small streams (Montgomery et al., 2003; Pollock et al., 2003)—anthropogenic fragments. Large numbers and significant biological consequences make dams the major way humans fracture river networks globally (Dynesius & Nilsson, 1994; Van Looy et al., 2014).

Ecologists have extensively investigated dams upstream and downstream edge impacts (Fuller et al., 2015). The dam's reservoir, which may be hundreds of kilometers long, produces a lake-like habitat barrier instead of a river. Riverine populations in reservoir tributaries, newly separated by the reservoir habitat barrier, may extirpate rapidly (Matthews & Marsh-Matthews, 2007). Weirs, bridge aprons, and sluices help in flood control, irrigation, navigation, water level measurement, and water supply. Human-engineered constructions, big and small, harm aquatic ecosystems, changing the environment and reducing biodiversity (Sun et al., 2023).

2.8 Socioeconomic Impacts: Decline of Fisheries and Livelihoods

Socioeconomic status includes demographics, income, living costs, boat transport, fishing gear, marketing infrastructure, and more, and it helps understand people, households, and communities social, cultural, and economic conditions (Parashar et al., 2016). Fishery supports livelihoods and feeds approximately one billion people. Inland fisheries provide 15% of worldwide employment and are important to the rural poor (Gogoi et al., 2015). In addition to earning foreign cash, it boosts the establishment of other subsidiary businesses and provides affordable, healthy food (Manasi et al., 2009). Large-scale flow modifications, water pollution, and resource exploitation have endangered river floodplain fish diversity and productivity worldwide for decades (Dudgeon, 2011; Welcomme, 1995). To avoid overfishing, riverine fisheries management faces scarcity, unpredictability, and dramatic changes due to hydro-climatic variability (Dey et al., 2020). When river pollution is severe and persistent, it hurts the economy, society, and livelihood of river users (Shameem et al., 2014). Massive water contamination and dramatic declines in the market price of fish taken from the contaminated river cripple fisherman communities' economies (Sarkar & Islam, 2022).

India's interior fisheries and indigenous fishing variety are valued as a growing nation. These species provided food and income to rural communities. In India, the fisheries industry accounts for 60% of fish output and employs 23 lakh households. Providing food security and employment to landless, small, and marginal farmers is more productive and sustainable. Andhra Pradesh, Bihar, Chhattisgarh, Gujarat, Kerala, Karnataka, Maharashtra, Orissa, Tamil Nadu, Uttar Pradesh, West Bengal, etc., produce fish. About 12 million Indians work directly in fishing, and 60 million rely on it (Paul & Chakraborty, 2016). Das et al. (2022) assessed fisher communities in 24 districts of Uttarakhand, Uttar Pradesh, Bihar, and West Bengal on the Ganga's upper, middle, and lower sections. The survey found that fishermen in the lower stretch earned the most per month (₹7,283). Fish supplies were falling owing to dam and barrage construction and overfishing, compounded by insufficient compensation during the fishing ban. Pandit et al. (2019) examined 500 fisher families on the Ganga's Bhagirathi–Hooghly length (Sagar to Farakka, 560 km) in West Bengal's socioeconomic circumstances and livelihood diversification. Fishing was the principal employment for 88.6% of respondents but only 70.3% of family income. The average monthly income was ₹9,391, with lower diversification levels. The report advocated government-supported fish selling, agro-processing, and skill development for alternative revenue streams. The Tungabhadra River Basin's large-scale and small-scale fishermen are trapped in poverty, according to Manasi (2013). Das (2015) also noticed fishermen's professional shifts, declining per-capita income, and persistent poverty owing to the Jalangi River deterioration.

2.9 Mitigation and Sustainable Management Strategies

2.9.1 Policy and Governance

2.9.1.1 Strengthening the National River Conservation Plan (NRCP)

Water is critical for all life, supporting healthy ecosystems, societies, and economies (Sinha & Sedai, 2025). Human civilization relies on rivers and water management (Azhoni et al., 2022). River conservation must be inclusive and backed by the finest knowledge for policy and management (Hussain et al., 2020). Large river systems like the Ganga, Yangtze, and Amazon span extensive geo-climatic and administrative territories, making it impossible to conserve them as a single system (Abell, 2002). Social and economic limitations by dependent human populations make it impossible to safeguard whole rivers and their catchments or all habitat elements in significant protected areas in human-dominated regions (Chessman, 2013; Nel et al., 2009). However, comprehensive river conservation planning is needed to guide policy and execute targeted actions to safeguard riverine ecosystems and biodiversity legally (Hermoso et al., 2011).

The national and state governments set river water management policies in India to handle water resource demands and difficulties. Nationally, the Ministry of Jal Shakti manages water resources, including river water. The ministry creates policies, rules, and frameworks for sustainable and equitable water use nationwide. The National Water Policy underpins water management. State governments must establish and enforce water management policies (Kumar et al., 2024). The officially financed "National River Conservation Plan (NRCP)" helps states improve river water quality by encompassing 38 rivers in 20 states (Benny & Kumar, 2024). India's CPCB found that 817 river sites surpassed the Primary Water Quality Criteria for outdoor swimming in 1920 locations evaluated between 2019 and 2021 (Sinha & Sedai, 2025). The NRCP aims to enhance water quality via pollution abatement. Pollution abatement techniques focus on sewage interception and treatment. The Ganga, Yamuna, Gomti, and Damodar rivers in North and Eastern India are not covered by the NRCP. Benny and Kumar (2024) state that other important river-specific systems cover them individually. NRCP of 1995 replaced the "Ganga Action Plan (GAP-I and II)" and encompassed the most important Indian rivers. In 1985, the Central Government launched the "Ganga Action Plan (GAP-I)" to reduce pollution. After many years, the "Ganga Action Plan (GAP-II)" was initiated in 1993 and expanded to Yamuna and Gomti, significant Ganga tributaries. NRCP extended this to encompass the other main rivers of the nation in 1995 and integrated the previous systems. NRCP conducted pollution abatement activities to improve river water quality. The NRCP intercepted, diverted, and treated sewage to decrease river pollution. Crematoria, low-cost sanitation, riverside development, and public awareness were non-core initiatives. The following discussion shows that NRCP is a significant intervention in improving Indian River water quality. The primary goal of this study is to evaluate NRCP's effect (Sinha & Sedai, 2025). "The Central Water Commission (CWC)" is vital to analyzing and monitoring main river water supplies. Over the years, CWC data has shown diminishing per-capita water supply in India, prompting worries about water shortage. "The National River Conservation Plan" and "Namami Gange program" clean and rejuvenate key rivers like the Ganges to solve these issues. These programs enhance water quality, encourage sustainable water usage, and protect river ecosystems (Kumar et al., 2024).

2.9.1.2 Legal Frameworks: Water (Prevention and Control of Pollution) Act, 1974

India enacted the Water Act of 1974 to address water pollution. The first effort is to address environmental problems thoroughly. The 1972 "UN Conference on the Human Environment" in Stockholm, when Indian delegations represented Late Smt. Indira Gandhi inspired this act. The law prevents and controls water contamination and maintains and restores water quality. The main goals are: (1) To prevent and regulate water contamination. (2) Keep or restore water purity. (3) Create Central and State Water Pollution Control Boards. (4) To provide such Boards authorities and duties related to them and related concerns. (5) Provide penalties for Act violations. (6) Establish Central and State Water Testing Laboratories to help the Board analyze pollution, set standards, and determine culpability (Verma, 2016).

2.10 Technological and Community-Based Solutions

2.10.1 Wastewater Treatment Innovations

We can limit growing pollution concerns and preserve water system sustainability by implementing adequate wastewater management methods, including contemporary technologies for treatment and robust regulatory frameworks (Dubey et al., 2023). Innovations in filtration technology have improved water treatment while conserving resources. Membrane filtration can recover energy and minerals from wastewater. Forward osmosis, a membrane-based process that separates water from solutes using osmotic pressure, saves energy (Maroufi & Hajilary, 2023). Over 450 angiospermic plants are metal hyperaccumulators, accumulating potentially phytotoxic metals up to 1000 times greater than typical plants. Plants remove hazardous components from wastewater through phytodegradation, phytoextraction, phytovolatilization, hemofiltration, and phytostabilization (Singh et al., 2019). Aquatic plants remove metals from wastewater without apparent damage (Lesage et al., 2007; Mishra & Tripathi, 2009).

Nanotechnology also advances water treatment. High surface area, adsorption capacity, and antibacterial characteristics make carbon nanotubes and graphene oxide ideal for water filtration (Das et al., 2023). These materials may remove heavy metals and organic pollutants from water more effectively and sustainably than current approaches (Sodhi & Singh, 2023). Decentralized water treatment techniques are being used to enhance local water management without huge, centralized treatment facilities (Das et al., 2023). These systems treat water near consumption, decreasing distribution losses and encouraging reuse. Decentralized technology like solar-powered desalination machines and small membrane bioreactors lets communities treat water independently. Resource efficiency, waste reduction, and water shortage resistance are promoted by this circular economy strategy (Guven et al., 2023).

2.10.2 Community-Based Solutions

Community-based conservation often links conservation to livelihood. As species and preserved areas gain in value, they boost local livelihoods and well-being (Kumar et al., 2011). However, communities respond immediately to life-support system threats. India's river restoration efforts fail due to their size and diverse (and frequently uninterested) stakeholders (Brugnach & Ingram, 2011). Civil society, communities, and the state and national governments have made localized freshwater conservation initiatives, but scaling up is needed. Only by harmonizing the fragmented planning and execution of conservation measures will protect fresh water ecosystems dependent species, ecosystem services, and thus people will get permanent benefits. Systematic conservation planning can help develop collaborative basin/sub-basin conservation plans based on the national vision to target conservation interventions at places with the greatest return on investment. The Seechewal model rejuvenation of the Kali Bein River in Punjab is a typical community-based Conservation Initiative in river management (Vazhayil & Wantzen, 2023). Demand- and supply-side irrigation water management improved Karula river flows in Uttar Pradesh's Ramganga river system (Arora et al., 2024; Nel et al., 2009). By 2023, nearly 1,00,000 farmers in 500 villages in four Uttar

Pradesh districts are restoring Karula, Gaagan (Ramganga basin), and Kali (Ganga basin) by improving flows. Over 2 billion liters of water have been restored to these rivers. Another restoration effort of the Arvari River which runs through Rajasthan's Alwar District from the Aravalli mountains was carried out, where local organizations and people recovered this 1985-dried river by 1996 using water gathering systems and groundwater recharge (DTE, 1999).

Similarly, the Naganathi River in Vellore and Tiruvannamalai districts dried up years ago due to groundwater depletion. Over 1,000 women from 21 villages in a Vellore revenue block constructed recharge wells and check dams to replenish groundwater and revitalize the river. Nine districts headed by over 20,000 women have joined the groundwater recharge project (Menon, 2021). Similar efforts were taken up by farmers who used the Kalyani River in Barabanki, Uttar Pradesh, for irrigation, which was silt dried. The river restoration effort began in 2013. Administrative authorities of the district sensitized locals, eliminated encroachments, reduced waste dumping, de-silted, and banned open defecation on the riverside to revitalize the river.

2.10.3 Role of School Education in River Conservation.

Role of schools, higher educational institutions in educating students for water conservation and pollution prevention practices play an important role as it fosters environmental stewardship. Education for sustainable development strives to help students acquire the mindsets, abilities, and understanding needed to make informed choices for betterment of others as well as themselves, now and in the future, and then act on these decisions (Ramadoss & Poyyamoli, 2011). In 1976, India added the 42 amendment to its constitution to address environmental issues. Following the introduction of a full-fledged Ministry of Environment and Forests, the environment has been a top concern in several regional and national policies, plans, and strategies. Growing worries about the environment in India prompted the government to develop a strategy for introducing Environmental Education in schools (Bhat et al., 2017).

The Ganga River Dolphin Conservation Education Program, implemented in Uttar Pradesh and Bihar explains the comprehensive approach where 20 clustered schools where each cluster involved 30 to 35 schools carried out an awareness drive with the help of local NGO's for the protection of Dolphins (NMCG, 2017). Similarly, the "My River My

Life" project for the Chalakudy River, in Kerala was carried out as a conservation strategy. Training was provided to 100 student and 2 teachers from various schools. Students carried out various field surveys and interviewed various members regarding the health conditions of the river. Thus the project brought an insight regarding the conservation of river and its ecology (TIES, 2017).

2.11 Conclusion

Rivers and fluvial ecosystems are highly affected by human activity, affecting biodiversity, water quality, and livelihoods. This chapter shows that land use changes, pollution, hydrological changes, and climate change put tremendous strain on river systems, creating permanent ecological deterioration. The loss of biodiversity, eutrophication, and fragmentation of river habitats need immediate intervention to safeguard these essential ecosystems. However, obstacles can be overcome. Effective legislative frameworks, wastewater treatment technology, and community-led efforts may alleviate these consequences and restore river ecosystems. Governance via national initiatives like the National River Conservation Plan and strong legislative regulations can lead to sustainable water management. Cultural initiatives and river monitoring networks let communities take responsibility for river health. River ecosystems need scientific investigation, legislative action, and community participation to survive. Addressing anthropogenic stresses and using adaptive management techniques may protect rivers for future generations, ensuring they continue to offer ecosystem services and sustain ecological and human well-being.

References

AM., S., Nair, S. M., Sudheesan, D., Samanta, S., Paul, S. K., Bhowmick, S., Kumar, V., & Das, B. K. (2023). Reconnoitre on ichthyofauna of Mahanadi River of India: shifting diversity down the river continuum and linking ecological traits with patterns in biodiversity. *Environmental Science* and *Pollution Research, 31*, 5684–5698. https://doi.org/10.1007/s11356-023-31411-w

Aadhar, S., & Mishra, V. (2019). A substantial rise in the area and population affected by dryness in South Asia under 1.5 °C, 2.0 °C and 2.5 °C warmer worlds. *Environmental Research Letters, 14*, 114021. https://doi.org/10.1088/1748-9326/ab4862

Ab. Ghani, A., Azamathulla, H. M., Chang, C. K., Zakaria, N. A., & Hasan, Z. A. (2011). Prediction of total bed material load for rivers in Malaysia: A case study of Langat, Muda and Kurau Rivers. *Environmental Fluid Mechanics, 11*, 307–318. https://doi.org/10.1007/s10652-010-9177-9

Abell, R. (2002). Conservation biology for the biodiversity crisis: A freshwater follow-up. *Conservation Biology, 16*, 1435–1437. https://doi.org/10.1046/j.1523-1739.2002.01532.x

Akinnawo, S. O. (2021). Chemical precipitation and reduction methods for the restoration of water from aquaculture operation. *Coast Journal of the School of Science Oaustech Okitipupa, 3*(1).

Ali, H., Modi, P., & Mishra, V. (2019). Increased flood risk in Indian subcontinent under the warming climate. *Weather and Climate Extremes, 25*, Article 100212. https://doi.org/10.1016/j.wace.2019.100212

Almasri, M. N., & Kaluarachchi, J. J. (2004). Assessment and management of long-term nitrate pollution of ground water in agriculture-dominated watersheds. *Journal of Hydrology, 295*, 225–245. https://doi.org/10.1016/j.jhydrol.2004.03.013

Alvarez, A., Saez, J. M., Costa, J. S. D., Colin, V. L., Fuentes, M. S., Cuozzo, S. A., Benimeli, C. S., Polti, M. A., & Amoroso, M. J. (2017). Actinobacteria: Current research and perspectives for bioremediation of pesticides and heavy metals. *Chemosphere, 166*, 41–62.

Anh, N. T., Can, L. D., Nhan, N. T., Schmalz, B., & Luu, T. L. (2023). Influences of key factors on river water quality in urban and rural areas: A review. *Case Studies in Chemical and Environmental Engineering*, *8*, Article 100424. https://doi.org/10.1016/j.cscee.2023.100424

Arefin, M. A., & Mallik, A. (2018). Sources and causes of water pollution in Bangladesh: A technical overview. *Bibechana, 15*, 97–112.

Arias-Estévez, M., López-Periago, E., Martínez-Carballo, E., Simal-Gándara, J., Mejuto, J.-C., & García-Río, L. (2008). The mobility and degradation of pesticides in soils and the pollution of groundwater resources. *Agriculture, Ecosystems & Environment, 123*, 247–260.

Arora, R., Balachander, T., Agrawal, I., Panda, R., Gupta, D., Kasturirangan, A., Vencatesan, J., Sivakumar, K., Kumar, R., Mathur, V. B., Chokkakula, S., Lal, T., Kaushal, N., Babu, S., & Kiesecker, J. (2024). Conserving Freshwater Ecosystems in India: A call to action. *Aquatic Conservation: Marine and Freshwater Ecosystems, 34*, Article e4165. https://doi.org/10.1002/aqc.4165

Auffhammer, M., Ramanathan, V., & Vincent, J. R. (2012). Climate change, the monsoon, and rice yield in India. *Climate Change, 111*, 411–424. https://doi.org/10.1007/s10584-011-0208-4

Azhoni, A., Holman, I., & Grabowski, R. C. (2022). Identifying evolving priorities in national river governance from Parliamentary Questions. *Water Policy, 24*, 1414–1433.

Banerjee, M. (1999). *A report on the impact of Farakka Barrage on the human fabric*. New Delhi South Asian Netw. Dams Rivers People.

Banjoko, B., & Eslamian, S. (2016). Environmental impact assessment: An application to urban water reuse. In *Urban water reuse handbook* (pp. 265–278). CRC Press.

Barathi, S., Lee, J., Venkatesan, R., & Vetcher, A. A. (2023). Current status of biotechnological approaches to enhance the phytoremediation of heavy metals in India—A Review. *Plants, 12*, 3816. https://doi.org/10.3390/plants12223816

Behera, M. D., Tripathi, P., Das, P., Srivastava, S. K., Roy, P. S., Joshi, C., Behera, P. R., Deka, J., Kumar, P., Khan, M. L., Tripathi, O. P., Dash, T., & Krishnamurthy, Y. V. N. (2018). Remote sensing based deforestation analysis in Mahanadi and Brahmaputra river basin in India since 1985. *Journal of Environmental Management, 206*, 1192–1203. https://doi.org/10.1016/j.jenvman.2017.10.015

Benny, C. S., & Kumar, S. (2024). Pilgrimage by Pamba: Sustainable planning for mitigating environmental impact of tourism along the Pamba River. *International Journal of Advances in Engineering and Management*, *6*, 274–281.

Best, J. (2019). Anthropogenic stresses on the world's big rivers. *Nature Geoscience, 12*, 7–21.

Beuchle, R., Frédéric, A., Bourgoin, C., Vancutsem, C., Eva, H., & Follador, M. (2021). *Deforestation and forest degradation in the Amazon—Status and trends up to year 2020*. https://doi.org/10.2760/61682

Bhardwaj, V., Singh, D. S., & Singh, A. K. (2010). *Water quality of the Chhoti Gandak River using principal component analysis*. Ganga Plain.

Bhat, S. A., Zahid, A. T., Sheikh, B. A., & Parrey, S. H. (2017). Environmental education in India: An approach to sustainable development. *FIIB Business Review, 6*(1), 14. https://doi.org/10.29368/FIIB.6.1.2017.14-21

Bhatla, R., Singh, M., Mall, R. K., Tripathi, A., & Raju, P. V. S. (2015). Variability of summer monsoon rainfall over Indo-Gangetic plains in relation to El-Nino/La-Nina. *Natural Hazards, 78*, 837–853. https://doi.org/10.1007/s11069-015-1746-2

Bhattacharjya, B. K., Bhaumik, U., & Sharma, A. P. (2017). Fish habitat and fisheries of Brahmaputra River in Assam, India. *Aquatic Ecosystem Health & Management*, *20*, 102–115.

Bijukumar, A. (2019). *Survey and mapping of exotic freshwater biodiversity in Kerala using Geographic Information System*. Final Project Report submitted to the Directorate of Environment and Climate Change, Government of Kerala, India.

Bisimwa, A. M., Amisi, F. M., Bamawa, C. M., Muhaya, B. B., & Kankonda, A. B. (2022). Water quality assessment and pollution source analysis in Bukavu urban rivers of the Lake Kivu basin (Eastern Democratic Republic of Congo). *Environmental and Sustainability Indicators*, *14*, Article 100183.

Biswas, P., & Vellanki, B. P. (2021). Occurrence of emerging contaminants in highly anthropogenically influenced river Yamuna in India. *Science of the Total Environment, 782*, Article 146741.

Björklund, G., Connor, R., Goujon, A., Hellmuth, M., Moriarty, P., Rast, W., Warner, K., Winpenny, J., Hoekstra, A., & Wiberg, D. (2012). Demographic, economic and social drivers. In *The United Nations World Water Development Report 3* (pp. 29–40). Routledge.

Botle, A., Salgaonkar, S., Barabde, G., & Herlekar, M. (2024a). Abatement of pharmaceutical compounds in wastewater using green nanomaterials: an eco-friendly alternative to conventional nanomaterials. In S. A. Bhat, V. Kumar, F. Li, & P. Verma (Eds.), *Detection and treatment of emerging contaminants in wastewater* (pp. 227–251). IWA Publishing. https://doi.org/10.2166/9781789063752_0227

Botle, A., Salgaonkar, S., Tiwari, R., Ambadekar, S., & Barabde, G. R. (2023). Brief status of contamination in surface water of rivers of India by heavy metals: A review with pollution indices and health risk assessment. *Environmental Geochemistry and Health, 45*, 2779–2801. https://doi.org/10.1007/s10653-022-01463-x

Botle, A., Salgaonkar, S., Tiwari, R., & Barabde, G. (2025). Heavy metal pollution in River Ulhas, Maharashtra, India: Unraveling contamination dynamics through pollution indices, multivariate analysis, and health risk assessment. *International Journal of Environmental Research*, *19*, 88. https://doi.org/10.1007/s41742-025-00754-2

Botle, A., Salgaonkar, S., Tiwari, R., & Barabde, G. (2024b). Unveiling heavy metal pollution dynamics in sediments of river Ulhas, Maharashtra, India: A comprehensive analysis of anthropogenic influence, pollution indices, and health risk assessment. *Environmental Geochemistry and Health, 46*, 419. https://doi.org/10.1007/s10653-024-02208-8

Botle, A., Salgaonkar, S., Tiwari, R., Barabde, G., & Herlekar, M. (2024c). Circular waste management system and greenhouse gases reduction. In V. Kumar, S. A. Bhat, S. Kumar & P. Verma (Eds.), *Environmental Engineering and waste management* (pp. 649–684). Springer Nature Switzerland, Cham. https://doi.org/10.1007/978-3-031-58441-1_22

Brauman, K. A., Daily, G. C., Duarte, T. K., & Mooney, H. A. (2007). The nature and value of ecosystem services: An overview highlighting hydrologic services. *Annual Review of Environment and Resources, 32*, 67–98. https://doi.org/10.1146/annurev.energy.32.031306.102758

Breitenmoser, L., Quesada, G. C., Bassi, N., Dkhar, N. B., Phukan, M., Kumar, S., Babu, A. N., Kierstein, A., Campling, P., & Hooijmans, C. M. (2022). Perceived drivers and barriers in the governance of wastewater treatment and reuse in India: Insights from a two-round Delphi study. *Resources, Conservation and Recycling, 182*, Article 106285.

Brugnach, M., & Ingram, H. (2011). Rethinking the role of humans in water management: Toward a new model of decision-making. In *Water, Cultural Diversity, and Global Environmental Change: Emerging Trends, Sustainable Futures?* (pp. 49–64). Springer Netherlands.

CAG. (2012). *Comptroller and auditor general of India, performance audit of water pollution in India*, CAG Report No. 21. of 2011–12. New Delhi.

Carpenter, S. R. (2008). Phosphorus control is critical to mitigating eutrophication. *Proceedings of the National Academy of Sciences, 105*, 11039–11040. https://doi.org/10.1073/pnas.0806112105

Carpenter, S. R. (2005). Eutrophication of aquatic ecosystems: Bistability and soil phosphorus. *Proceedings of the National Academy of Sciences, 102*, 10002–10005. https://doi.org/10.1073/pnas.0503959102

Chakraborty, P., Khuman, S. N., Selvaraj, S., Sampath, S., Devi, N. L., Bang, J. J., & Katsoyiannis, A. (2016). Polychlorinated biphenyls and organochlorine pesticides in River Brahmaputra from the outer Himalayan Range and River Hooghly emptying into the Bay of Bengal: Occurrence, sources and ecotoxicological risk assessment. *Environmental Pollution, 219*, 998–1006. https://doi.org/10.1016/j.envpol.2016.06.067

Chanapathi, T., & Thatikonda, S. (2020). Investigating the impact of climate and land-use land cover changes on hydrological predictions over the Krishna river basin under present and future scenarios. *Science of the Total Environment, 721*, Article 137736. https://doi.org/10.1016/j.scitotenv.2020.137736

Chessman, B. C. (2013). Do protected areas benefit freshwater species? A broad-scale assessment for fish in Australia's Murray-Darling Basin. *Journal of Applied Ecology, 50*, 969–976. https://doi.org/10.1111/1365-2664.12104

da Costa, J. A., de Souza, J. P., Teixeira, A. P., Nabout, J. C., & Carneiro, F. M. (2018). Eutrophication in aquatic ecosystems: A scientometric study. *Acta Limnologica Brasiliensia, 30*, Article e2.

Crossman, N. D., & Pollino, C. A. (2018). An ecosystem services and Bayesian modelling approach to assess the utility of water resource development in rangelands of north Australia. *Journal of Arid Environments, 159*, 34–44. https://doi.org/10.1016/j.jaridenv.2018.02.007

Cui, Z., Dou, Z., Chen, X., Ju, X., & Zhang, F. (2014). Managing agricultural nutrients for food security in China: Past, present, and future. *Agronomy Journal, 106*, 191–198. https://doi.org/10.2134/agronj2013.0381

Danegulu, A., Karki, S., Bhattarai, P. K., & Pandey, V. P. (2024). Characterizing urban flooding in the Kathmandu Valley, Nepal: The influence of urbanization and river encroachment. *Natural Hazards, 120*, 10923–10947. https://doi.org/10.1007/s11069-024-06650-w

Dar, S. A., Ganie, D. H., Teeli, J. I., & Bhat, S. U. (2023). A policy approach for sustainable governance of sand mining activities in NW Kashmir Himalayas. *Extractive Industries and Society, 13*, Article 101204. https://doi.org/10.1016/j.exis.2022.101204

Das, B. C. (2015). Socio-economic impact of a decaying river on fishermen: A case study of Taranipur Village, West Bengal. *International Research Journal of Management Science and Technology, 3*, 141–149.

Das, B. K., Debroy, P., Roy, A., Bayen, S.,& Chakraborty, H. (2022). Livelihood status and socio-economic condition of fishers of river Ganga. *Scientific Culture, 88*.

Das, M. K., Sharma, A. P., Vass, K. K., Tyagi, R. K., Suresh, V. R., Naskar, M., & Akolkar, A. B. (2013). Fish diversity, community structure and ecological integrity of the tropical River Ganges, India. *Aquatic Ecosystem Health & Management, 16*, 395–407.

Das, B. K., Debroy, P. I. Y. A. S. H. I., Roy, A., Bayen, S. U. P. R. I. T. I., & Chakraborty, H. (2022). Livelihood status and socio-economic condition of fishers of river Ganga. *Science Culture, 88*, 410–414.

Das, S., Singh, C. K., Sodhi, K. K., & Singh, V. K. (2023). Circular economy approaches for water reuse and emerging contaminant mitigation: Innovations in water treatment. *Environment, Development and Sustainability, 27*, 5753–5794. https://doi.org/10.1007/s10668-023-04183-z

Datry, T., Pella, H., Leigh, C., Bonada, N., & Hugueny, B. (2016). A landscape approach to advance intermittent river ecology. *Freshwater Biology, 61*, 1200–1213. https://doi.org/10.1111/fwb.12645

De Sousa, D. N. R., Mozeto, A. A., Carneiro, R. L., & Fadini, P. S. (2018). Spatio-temporal evaluation of emerging contaminants and their partitioning along a Brazilian watershed. *Environmental Science and Pollution Research, 25*, 4607–4620. https://doi.org/10.1007/s11356-017-0767-7

Debnath, J., Sahariah, D., Lahon, D., Nath, N., Chand, K., Meraj, G., Farooq, M., Kumar, P., Kanga, S., & Singh, S. K. (2022). Geospatial modeling to assess the past and future land use-land cover changes in the Brahmaputra Valley, NE India, for sustainable land resource management. *Environmental Science and Pollution Research, 30*, 106997–107020. https://doi.org/10.1007/s11356-022-24248-2

Deka, R. L., Mahanta, C., Pathak, H., Nath, K. K., & Das, S. (2013). Trends and fluctuations of rainfall regime in the Brahmaputra and Barak basins of Assam, India. *Theoretical and Applied Climatology, 114*, 61–71. https://doi.org/10.1007/s00704-012-0820-x

Dewan, A. M., & Yamaguchi, Y. (2009). Land use and land cover change in Greater Dhaka, Bangladesh: Using remote sensing to promote sustainable urbanization. *Applied Geography, 29*, 390–401.

Dey, S., Choudhary, S. K., Dey, S., Deshpande, K., & Kelkar, N. (2020). Identifying potential causes of fish declines through local ecological knowledge of fishers in the Ganga River, eastern Bihar, India. *Fisheries Management and Ecology, 27*, 140–154. https://doi.org/10.1111/fme.12390

Dhanakumar, S., Solaraj, G., & Mohanraj, R. (2015). Heavy metal partitioning in sediments and bioaccumulation in commercial fish species of three major reservoirs of river Cauvery delta region. *India. Ecotoxicol. Environ. Saf., 113*, 145–151. https://doi.org/10.1016/j.ecoenv.2014.11.032

Dhar, O. N., & Nandargi, S. (2003). Hydrometeorological aspects of floods in India. *Natural Hazards, 28*, 1–33. https://doi.org/10.1023/A:1021199714487

Dinesh Kumar, D. K., Malik, D. S., Sharma, A. K., & Varsha Gupta, V. G. (2019). Estimation of length-weight and length-length relationship of catfish, Rita rita inhabiting in the middle stretch of river Ganga.

Domingues, R. B., Barbosa, A. B., Sommer, U., & Galvão, H. M. (2012). Phytoplankton composition, growth and production in the Guadiana estuary (SW Iberia): Unraveling changes induced after dam construction. *Science of the Total Environment, 416*, 300–313.

Drobac, D., Tokodi, N., Lujić, J., Marinović, Z., Subakov-Simić, G., Dulić, T., Važić, T., Nybom, S., Meriluoto, J., & Codd, G. A. (2016). Cyanobacteria and cyanotoxins in fishponds and their effects on fish tissue. *Harmful Algae, 55*, 66–76.

DTE. (1999). *Coming back to life* [WWW Document]. Earth. URL https://www.downtoearth.org.in/environment/coming-back-to-life-19493 (accessed 4.12.25).

Dubey, M., Vellanki, B. P., & Kazmi, A. A. (2023). Removal of emerging contaminants in conventional and advanced biological wastewater treatment plants in India-a comparison of treatment technologies. *Environmental Research, 218*, Article 115012.

Dudgeon, D. (2011). Asian river fishes in the Anthropocene: Threats and conservation challenges in an era of rapid environmental change. *Journal of Fish Biology, 79*, 1487–1524. https://doi.org/10.1111/j.1095-8649.2011.03086.x

Dudgeon, D., Arthington, A. H., Gessner, M. O., Kawabata, Z.-I., Knowler, D. J., Lévêque, C., Naiman, R. J., Prieur-Richard, A.-H., Soto, D., & Stiassny, M. L. (2006). Freshwater biodiversity: Importance, threats, status and conservation challenges. *Biological Reviews, 81*, 163–182.

Dueñas, M.-A., Ruffhead, H. J., Wakefield, N. H., Roberts, P. D., Hemming, D. J., & Diaz-Soltero, H. (2018). The role played by invasive species in interactions with endangered and threatened species in the United States: A systematic review. *Biodiversity and Conservation, 27*, 3171–3183. https://doi.org/10.1007/s10531-018-1595-x

Dutta, V., Dubey, D., & Kumar, S. (2020). Cleaning the River Ganga: Impact of lockdown on water quality and future implications on river rejuvenation strategies. *Science of the Total Environment, 743*, Article 140756. https://doi.org/10.1016/j.scitotenv.2020.140756

Dwivedi, A. K. (2017). *Researchers in water pollution: A review*. https://doi.org/10.13140/RG.2.2.12094.08002

Dwivedi, S., Mishra, S., & Tripathi, R. D. (2018). Ganga water pollution: A potential health threat to inhabitants of Ganga basin. *Environment International, 117*, 327–338.

Dynesius, M., & Nilsson, C. (1994). Fragmentation and flow regulation of river systems in the Northern third of the world. *Science, 266*, 753–762. https://doi.org/10.1126/science.266.5186.753

Ekka, A., Keshav, S., Pande, S., Van Der Zaag, P., & Jiang, Y. (2022). Dam-induced hydrological alterations in the upper Cauvery river basin, India. *Journal of Hydrology: Regional Studies, 44*, Article 101231. https://doi.org/10.1016/j.ejrh.2022.101231

El-Sheekh, M., Abdel-Daim, M. M., Okba, M., Gharib, S., Soliman, A., & El-Kassas, H. (2021). Green technology for bioremediation of the eutrophication phenomenon in aquatic ecosystems: A review. *African Journal of Aquatic Science, 46*, 274–292. https://doi.org/10.2989/16085914.2020.1860892

Fuller, M. R., Doyle, M. W., & Strayer, D. L. (2015). Causes and consequences of habitat fragmentation in river networks. *Annals of the New York Academy of Sciences, 1355*, 31–51. https://doi.org/10.1111/nyas.12853

Gadgil, S., & Gadgil, S. (2006). The Indian monsoon, GDP and agriculture. *Economic and political weekly*, 4887-4895.

Gajbhiye, S., Meshram, C., Singh, S. K., Srivastava, P. K., & Islam, T. (2016). Precipitation trend analysis of Sindh River basin, India, from 102-year record (1901–2002). *Atmospheric Science Letters, 17*, 71–77. https://doi.org/10.1002/asl.602

García, J. (2021). Ecological engineering practice as a global strategy to prevent eutrophication and microalgae blooms. *Ecological Engineering, 161*, Article 106152.

Ghosh, K., & Muñoz-Arriola, F. (2023). Hysteresis and streamflow-sediment relations across the pre-to-post dam construction continuum in a highly regulated transboundary Himalayan River basin. *Journal of Hydrology, 624*, Article 129885. https://doi.org/10.1016/j.jhydrol.2023.129885

Ghosh, S., & Guchhait, S. K. (2014). Hydrogeomorphic variability due to dam constructions and emerging problems: A case study of Damodar River, West

Bengal, India. *Environment, Development and Sustainability, 16*, 769–796. https://doi.org/10.1007/s10668-013-9494-5

Gogoi, B., Kachari, A., Dutta, R., Darshan, A., & Narayan, D. (2015). Fishery based livelihood approaches and management of fishery resources in Assam, India. *Forest, 5*, 2–514.

Gopal, B. (2016). A conceptual framework for environmental flows assessment based on ecosystem services and their economic valuation. *Ecosystem Services, 21*, 53–58. https://doi.org/10.1016/j.ecoser.2016.07.013

Gopal, B. (2004). State of degradation and approaches to restoration of floodplain rivers in India. In *Symposium on the management of large rivers for fisheries* vol. 2, p. 79.

Gopal, B., & Chauhan, M. (2025). River sand and gravel mining: Striking a balance between ecology and economy. *International Journal of Ecology and Environmental Sciences, 51*, 171–179.

Gregory, K. J. (2006). The human role in changing river channels. *Geomorphology, 79*, 172–191.

Grill, G., Lehner, B., Thieme, M., Geenen, B., Tickner, D., Antonelli, F., Babu, S., Borrelli, P., Cheng, L., & Crochetiere, H. (2019). Mapping the world's free-flowing rivers. *Nature, 569*, 215–221.

Grizzetti, B., Lanzanova, D., Liquete, C., Reynaud, A., & Cardoso, A. C. (2016). Assessing water ecosystem services for water resource management. *Environmental Science & Policy, 61*, 194–203.

Grooten, M., & Almond, R. E. A. (Eds.) (2018). *Living planet report 2018: Aiming higher*. WWF--World Wide Fund for Nature, Gland, Switzerland.

Gruel, C. R., & Latrubesse, E. M. (2021). A monitoring system of sand mining in large rivers and its application to the Ayeyarwady (Irrawaddy) River. *Myanmar. Water, 13*, 2331. https://doi.org/10.3390/w13172331

Gupta, V., Kumar, D., Dwivedi, A., Vishwakarma, U., Malik, D. S., Paroha, S., Mohan, N., & Gupta, N. (2023). Heavy metal contamination in river water, sediment, groundwater and human blood, from Kanpur, Uttar Pradesh, India. *Environmental Geochemistry and Health, 45*, 1807–1818. https://doi.org/10.1007/s10653-022-01290-0

Guven, H., Ersahin, M. E., Ozgun, H., Ozturk, I., & Koyuncu, I. (2023). Energy and material refineries of future: Wastewater treatment plants. *Journal of Environmental Management, 329*, Article 117130.

Hackney, C. R., Darby, S. E., Parsons, D. R., Leyland, J., Best, J. L., Aalto, R., Nicholas, A. P., & Houseago, R. C. (2020). River bank instability from unsustainable sand mining in the lower Mekong River. *Nature Sustainability, 3*, 217–225. https://doi.org/10.1038/s41893-019-0455-3

Häder, D.-P., Banaszak, A. T., Villafañe, V. E., Narvarte, M. A., González, R. A., & Helbling, E. W. (2020). Anthropogenic pollution of aquatic ecosystems:

Emerging problems with global implications. *Science of the Total Environment, 713*, Article 136586.

Hermoso, V., Linke, S., Prenda, J., & Possingham, H. P. (2011). Addressing longitudinal connectivity in the systematic conservation planning of fresh waters. *Freshwater Biology, 56*, 57–70. https://doi.org/10.1111/j.1365-2427.2009.02390.x

Hussain, S. A., Irengbam, M., Barthwal, S., Dasgupta, N., & Badola, R. (2020). Conservation planning for the Ganga River: A policy conundrum. *Landscape Research, 45*, 984–999. https://doi.org/10.1080/01426397.2020.1808959

Immerzeel, W. W., Droogers, P., De Jong, S. M., & Bierkens, M. F. P. (2009). Large-scale monitoring of snow cover and runoff simulation in Himalayan river basins using remote sensing. *Remote Sensing of Environment, 113*, 40–49.

India WRIS. (2025). large_dams_in_india—INDIA WRIS WIKI [WWW Document]. URL https://indiawris.gov.in/wiki/doku.php?id=large_dams_in_india (accessed 4.16.25).

Ishaq, S., Jabeen, G., Arshad, M., Kanwal, Z., Un Nisa, F., Zahra, R., Shafiq, Z., Ali, H., Samreen, K. B., & Manzoor, F. (2023). Heavy metal toxicity arising from the industrial effluents repercussions on oxidative stress, liver enzymes and antioxidant activity in brain homogenates of Oreochromis niloticus. *Science and Reports, 13*, 19936. https://doi.org/10.1038/s41598-023-47366-4

Jain, V., Karnatak, N., Raj, A., Shekhar, S., Bajracharya, P., & Jain, S. (2022). Hydrogeomorphic advancements in river science for water security in India. *Water Secur., 16*, Article 100118.

Jain, V., Sonam, Singh, A., Sinha, R., & Tandon, S. K. (2020). Evolution of modern river systems: an assessment of 'landscape memory' in Indian river systems. *Episodes 43*, 535–551. https://doi.org/10.18814/epiiugs/2020/020035

Jena, J. K., & Gopalakrishnan, A. (2012). Aquatic Biodiversity Management in India. *Proceedings of the National Academy of Sciences, India Section B: Biological Sciences*. https://doi.org/10.1007/s40011-012-0108-z

Jenkins, M. (2003). Prospects for Biodiversity. *Science, 302*, 1175–1177. https://doi.org/10.1126/science.1088666

Kaiser, N. N., Feld, C. K., & Stoll, S. (2020). Does river restoration increase ecosystem services? *Ecosystem Services, 46*, Article 101206.

Kakade, A., Salama, E.-S., Han, H., Zheng, Y., Kulshrestha, S., Jalalah, M., Harraz, F. A., Alsareii, S. A., & Li, X. (2021). World eutrophic pollution of lake and river: Biotreatment potential and future perspectives. *Environmental Technology and Innovation, 23*, Article 101604. https://doi.org/10.1016/j.eti.2021.101604

Kamboj, V., Kamboj, N., & Sharma, S. (2017). Environmental impact of riverbed mining-a review. *International Journal of Scientific Research, 7*, 504–520.

Kantharajan, G., Yadav, A. K., Chandran, R., Singh, R. K., Mohindra, V., Krishnan, P., Kumar, K., Shukla, S. P., & Lal, K. K. (2022). Impact of terrestrial protected areas on the fish diversity and habitat quality: Evidence from tropical river Pranhita, India. *Journal for Nature Conservation, 68*, Article 126187.

Kaur, T., & Sinha, A. K. (2019). Pesticides in agricultural run offs affecting water resources: A study of Punjab (India). *Agricultural Science, 10*, 1381–1395. https://doi.org/10.4236/as.2019.1010101

Khalid, N. K., Devadasan, D., Aravind, U. K., & Aravindakumar, C. T. (2018). Screening and quantification of emerging contaminants in Periyar River, Kerala (India) by using high-resolution mass spectrometry (LC-Q-ToF-MS). *Environmental Monitoring and Assessment, 190*, 370. https://doi.org/10.1007/s10661-018-6745-9

Khan, A. S., Anavkar, A., Ali, A., Patel, N., & Alim, H. (2021). A review on current status of riverine pollution in India. *Biosciences Biotechnology Research Asia, 18*, 9–22. https://doi.org/10.13005/bbra/2893

Khurana, M. P. S., Kansal, B. D., & Setia, R. (2014). Long-term impact of irrigation with sewage water on cadmium concentration in soils and crops. *Agrochimica, 58*, 19–34.

Klingensmith, D. (2003). Building India's 'modern temples': Indians and Americans in the Damodar Valley Corporation 1945–60. Reg. Mod. Cult. Polit. Dev. India 122–42.

Konda, S., Nagendla, N. K., Williams, M., & Mudiam, M. K. R. (2024). Investigating pharmaceuticals and personal care products in Musi River: A exploratory study in Hyderabad, India. *Journal of Hazardous Materials Advances, 14*, Article 100420.

Koushlesh, S., Sajina, A. M., & Roshith, C. M. (2021). Ichthyofaunal diversity of the major Indian rivers: A review. *Journal of the Inland Fisheries Society of India, 53*, 22–35.

Kulkarni, R. (2023). *Interlinking of rivers could flood India's freshwater with invasive fish* [WWW Document]. Mongabay-India. URL https://india.mongabay.com/2023/12/interlinking-of-rivers-could-flood-indias-freshwater-with-invasive-fish/ (accessed 4.11.25).

Kumar, D., Malik, D. S., Kumar, N., Gupta, N., & Gupta, V. (2020a). Spatial changes in water and heavy metal contamination in water and sediment of river Ganga in the river belt Haridwar to Kanpur. *Environmental Geochemistry and Health, 42*, 2059–2079. https://doi.org/10.1007/s10653-019-00471-8

Kumar, K., Mishra, S., & Rao, K. (2011). Creating space for community-based conservation initiatives (CBCLs) in conventional academics. *International Journal of Peace and Development Studies, 2*, 26–36.

Kumar, N., Kumar Goyal, M., Kumar Gupta, A., Jha, S., Das, J., & Madramootoo, C. A. (2021). Joint behaviour of climate extremes across India: Past and future. *Journal of Hydrology, 597*, Article 126185. https://doi.org/10.1016/j.jhydrol.2021.126185

Kumar, R., Goyal, M. K., Surampalli, R. Y., & Zhang, T. C. (2024). River pollution in India: Exploring regulatory and remedial paths. *Clean Technologies and Environmental Policy, 26*, 2777–2799. https://doi.org/10.1007/s10098-024-02763-9

Kumar, R., Qureshi, M., Vishwakarma, D. K., Al-Ansari, N., Kuriqi, A., Elbeltagi, A., & Saraswat, A. (2022). A review on emerging water contaminants and the application of sustainable removal technologies. *Case Studies in Chemical and Environmental Engineering, 6*, Article 100219. https://doi.org/10.1016/j.cscee.2022.100219

Kumar, S., & Gill, G. S. (2015). Trend analysis of rainfall in Satluj River Basin, Himachal Pradesh, India. *World Scientific News*, 14.

Kumar, V., & Jain, S. K. (2011). Trends in rainfall amount and number of rainy days in river basins of India (1951–2004). *Hydrology Research, 42*, 290–306.

Kumar, V., Sharma, A., Kumar, R., Bhardwaj, R., Kumar Thukral, A., & Rodrigo-Comino, J. (2020b). Assessment of heavy-metal pollution in three different Indian water bodies by combination of multivariate analysis and water pollution indices. *Human and Ecological Risk Assessment: An International Journal, 26*, 1–16. https://doi.org/10.1080/10807039.2018.1497946

Kuo, P.-H., Shih, S.-S., & Otte, M. L. (2021). Restoration recommendations for mitigating habitat fragmentation of a river corridor. *Journal of Environmental Management, 296*, Article 113197.

Lee, J. H. (2013). An overview of phytoremediation as a potentially promising technology for environmental pollution control. *Biotechnology and Bioprocess Engineering, 18*, 431–439. https://doi.org/10.1007/s12257-013-0193-8

Lesage, E., Mundia, C., Rousseau, D. P. L., Van de Moortel, A. M. K., Du Laing, G., Meers, E., Tack, F. M. G., De Pauw, N., & Verloo, M. G. (2007). Sorption of Co, Cu, Ni and Zn from industrial effluents by the submerged aquatic macrophyte Myriophyllum spicatum L. *Ecological Engineering, 30*, 320–325.

Liang, Z., Xu, Y., Qiu, Q., Liu, Y., Lu, W., & Wagner, T. (2021). A framework to develop joint nutrient criteria for lake eutrophication management in eutrophic lakes. *Journal of Hydrology, 594*, Article 125883.

Lija, R. K., & Radhamany, P. M. (2023). Invasive plants in the Poovar region of Neyyar river. In C. T. Goerge, K. Satheeshkumar, K. V. Govindan, K. T. Chandramohanan, T. S. Swapna, G. K. Pramod & V. Balakrishan (Eds.), *Biological invasions issues biodiversity conservation management* (pp. 444–449). Kerala State Biodivers. Board Thiruvananthapuram.

Liu, X., Duan, L., Mo, J., Du, E., Shen, J., Lu, X., Zhang, Y., Zhou, X., He, C., & Zhang, F. (2011). Nitrogen deposition and its ecological impact in China: An overview. *Environmental Pollution, 159*, 2251–2264. https://doi.org/10.1016/j.envpol.2010.08.002

Lübbert, C., Baars, C., Dayakar, A., Lippmann, N., Rodloff, A. C., Kinzig, M., & Sörgel, F. (2017). Environmental pollution with antimicrobial agents from bulk drug manufacturing industries in Hyderabad, South India, is associated with dissemination of extended-spectrum beta-lactamase and carbapenemase-producing pathogens. *Infection, 45*, 479–491. https://doi.org/10.1007/s15010-017-1007-2

Mahajan, A. K., & Sharma, U. K. (2005). Deforestation and Geo-Environmental degradation in the Baner Khad Watershed, Himachal Pradesh, India. *Asian Journal of Water, Environment and Pollution, 2*, 39–46.

Mahapatra, B. K., & Mohanty, B. P. (2023). Invasive Alien Fish Species in India are threat to the Inland Aquatic Ecosystem. Advances in Fish Research.

Malik, D. S., Sharma, A. K., Kamboj, V., & Sharma, A. K. (2021). Length-weight relationships of coldwater fish species from the different stretches of upper Ganga basin in Garhwal region of Uttarakhand.

Manasi, S. (2013). Water pollution impacts on livelihoods: A case study of fishing communities in Tungabhadra Sub Basin. In S. Nautiyal, K. S. Rao, H. Kaechele, K. V. Raju & R. Schaldach (Eds.), *Knowledge systems of societies for adaptation and mitigation of impacts of climate change, environmental science and engineering* (pp. 347–365). Springer Berlin Heidelberg, Heidelberg. https://doi.org/10.1007/978-3-642-36143-2_21

Manasi, S., Latha, N., & Raju, K. V. (2009). *Fisheries and livelihoods in Tungabhadra Basin, India: Current status and future possibilities*. Bangalore.

Mantey, S. (2017). Galamsey, pollution destroying water bodies in Ghana—Water Company, March 22, 2017. Head Commun. Ghana Water Co.

Maroufi, N., & Hajilary, N. (2023). Nanofiltration membranes types and application in water treatment: A review. *Sustainable Water Resources Management, 9*, 142. https://doi.org/10.1007/s40899-023-00899-y

Mathew, M. M., Sreelash, K., Mathew, M., Arulbalaji, P., & Padmalal, D. (2021). Spatiotemporal variability of rainfall and its effect on hydrological regime in a tropical monsoon-dominated domain of Western Ghats, India. *Journal of Hydrology: Regional Studies, 36*, Article 100861.

Matthews, W. J., & Marsh-Matthews, E. (2007). Extirpation of red shiner in direct tributaries of Lake Texoma (Oklahoma-Texas): A cautionary case history from a fragmented river-reservoir system. *Transactions of the American Fisheries Society, 136*, 1041–1062.

Matin, S., & Behera, M. D. (2019). Studying evidence of land degradation in the Indian Ganga River Basin—a Geoinformatics approach. *Environmental Monitoring and Assessment, 191*(Suppl 3), 803.

Maurya, A. K., Swarnkar, S., & Prakash, S. (2024). Hydrological impacts of altered monsoon rain spells in the Indian Ganga basin: A century-long perspective. *Environ. Res. Clim., 3*, Article 015010.

Mawari, G., Kumar, N., Sarkar, S., Frank, A. L., Daga, M. K., Singh, M. M., Joshi, T. K., & Singh, I. (2022). Human health risk assessment due to heavy metals in ground and surface water and association of diseases with drinking water sources: A study from Maharashtra, India. *Environmental Health Insights, 16*. https://doi.org/10.1177/11786302221146020

Mazumder, M. K., Boro, F., Barbhuiya, B., & Singha, U. (2014). A study of the winter congregation sites of the Gangetic River Dolphin in southern Assam, India, with reference to conservation. *Global Ecology and Conservation, 2*, 359–366. https://doi.org/10.1016/j.gecco.2014.09.004

Menon, A. (2021). *When an army of women came together to revive a river* [WWW Document]. Vikalp Sangam. URL https://india.mongabay.com/2021/02/when-an-army-of-women-came-together-to-revive-a-river/ (accessed 4.12.25).

Meynell, P.-J., Metzger, M., & Stuart, N. (2021). Identifying ecosystem services for a framework of ecological importance for rivers in South East Asia. *Water, 13*, 1602. https://doi.org/10.3390/w13111602

Minhas, P. S., Saha, J. K., Dotaniya, M. L., Sarkar, A., & Saha, M. (2022). Wastewater irrigation in India: Current status, impacts and response options. *Science of the Total Environment, 808*, Article 152001.

Mirza, M. M. Q. (2004). The Ganges water diversion: Environmental effects and implications—an introduction. In *The Ganges water diversion: Environmental effects and implications* (pp. 1–12). Springer.

Mishra, R., Kumar, A., Singh, E., Kumar, S., Tripathi, V. K., Jha, S. K., & Shukla, S. K. (2022). Current status of available techniques for removal of heavy metal contamination in the river ecosystem. In *Ecological significance of river ecosystems* (pp. 217–234). Elsevier.

Mishra, V., Nanditha, J. S., Dangar, S., Chuphal, D. S., & Vegad, U. (2024). Drivers, changes, and impacts of hydrological extremes in India: A review. *Wires Water, 11*, Article e1742. https://doi.org/10.1002/wat2.1742

Mishra, V., Tiwari, A. D., & Kumar, R. (2022b). Warming climate and ENSO variability enhance the risk of sequential extremes in India. *One Earth, 5*, 1250–1259.

Mishra, V. K., & Tripathi, B. D. (2009). Accumulation of chromium and zinc from aqueous solutions using water hyacinth (Eichhornia crassipes). *Journal of Hazardous Materials, 164*, 1059–1063.

Mngeni, A., Musampa, C. M., & Nakin, M. D. V. (2017). The effects of sand mining on rural communities. *WIT Transactions on Ecology and the Environment, 210*, 443–453.

Montgomery, D. R., Collins, B. D., Buffington, J. M., & Abbe, T. B. (2003). Geomorphic effects of wood in rivers. In *American fisheries society symposium* (pp. 21–47).

Mooley, D. A., & Parthasarathy, B. (1983). Variability of the Indian summer monsoon and tropical circulation features. *Monthly Weather Review, 111*, 967–978.

Morais, P., Chícharo, M. A., & Chícharo, L. (2009). Changes in a temperate estuary during the filling of the biggest European dam. *Science of the Total Environment, 407*, 2245–2259.

Mujumdar, M., Bhaskar, P., Ramarao, M. V. S., Uppara, U., Goswami, M., Borgaonkar, H., Chakraborty, S., Ram, S., Mishra, V., Rajeevan, M., & Niyogi, D. (2020). Droughts and Floods. In R. Krishnan, J. Sanjay, C. Gnanaseelan, M. Mujumdar, A. Kulkarni & S. Chakraborty (Eds.), *Assessment of climate change over the Indian Region* (pp. 117–141). Springer Singapore. https://doi.org/10.1007/978-981-15-4327-2_6

Mukate, S. V., Panaskar, D. B., Wagh, V. M., & Baker, S. J. (2020). Understanding the influence of industrial and agricultural land uses on groundwater quality in semiarid region of Solapur, India. *Environment, Development and Sustainability, 22*, 3207–3238. https://doi.org/10.1007/s10668-019-00342-3

Mukherjee, A., Saha, D., Harvey, C. F., Taylor, R. G., Ahmed, K. M., & Bhanja, S. N. (2015). Groundwater systems of the Indian sub-continent. *Journal of Hydrology: Regional Studies, 4*, 1–14.

Naka, L. N., Werneck, F. P., Rosser, N., Pil, M. W., & Boubli, J. P. (2022). Editorial: The role of rivers in the origins, evolution, adaptation, and distribution of biodiversity. *Frontiers in Ecology and Evolution, 10*, 1035859. https://doi.org/10.3389/fevo.2022.1035859

Nanditha, J. S., & Mishra, V. (2022). Multiday precipitation is a prominent driver of floods in Indian river basins. *Water Resources Research, 58*, e2022WR032723. https://doi.org/10.1029/2022WR032723

Nanditha, J. S., Rajagopalan, B., & Mishra, V. (2022). Combined signatures of atmospheric drivers, soil moisture, and moisture source on floods in Narmada River basin, India. *Climate Dynamics, 59*, 2831–2851. https://doi.org/10.1007/s00382-022-06244-x

Napper, I. E., Baroth, A., Barrett, A. C., Bhola, S., Chowdhury, G. W., Davies, B. F., Duncan, E. M., Kumar, S., Nelms, S. E., & Niloy, M. N. H. (2021). The abundance and characteristics of microplastics in surface water in the transboundary Ganges River. *Environmental Pollution, 274*, Article 116348.

Nath, B., Ni-Meister, W., & Choudhury, R. (2021). Impact of urbanization on land use and land cover change in Guwahati city, India and its implication on declining groundwater level. *Groundwater for Sustainable Development, 12*, Article 100500. https://doi.org/10.1016/j.gsd.2020.100500

Neelavannan, K., & Sen, I. S. (2023). Microplastics in freshwater ecosystems of India: Current trends and future perspectives. *ACS Omega, 8*, 34235–34248. https://doi.org/10.1021/acsomega.3c01214

Nel, J. L., Roux, D. J., Abell, R., Ashton, P. J., Cowling, R. M., Higgins, J. V., Thieme, M., & Viesrs, J. H. (2009). Progress and challenges in freshwater conservation planning. *Aquatic Conservation: Marine and Freshwater Ecosystems, 19*, 474–485. https://doi.org/10.1002/aqc.1010

Nepal, S., & Shrestha, A. B. (2015). Impact of climate change on the hydrological regime of the Indus, Ganges and Brahmaputra river basins: A review of the literature. *International Journal of Water Resources Development, 31*, 201–218. https://doi.org/10.1080/07900627.2015.1030494

New, M., Todd, M., Hulme, M., & Jones, P. (2001). Precipitation measurements and trends in the twentieth century. *International Journal of Climatology, 21*, 1889–1922. https://doi.org/10.1002/joc.680

Ngo, T. H., Van, D.-A., Tran, H. L., Nakada, N., Tanaka, H., & Huynh, T. H. (2021). Occurrence of pharmaceutical and personal care products in Cau River, Vietnam. *Environmental Science and Pollution Research, 28*, 12082–12091. https://doi.org/10.1007/s11356-020-09195-0

Ngwane, M. (2024). Impact of deforestation on water quality in cameroon. *International Journal of Natural Sciences, 4*, 46–57. https://doi.org/10.47604/ijns.2661

Nielsen, M. C., & Jiang, S. C. (2020). Can cyanotoxins penetrate human skin during water recreation to cause negative health effects? *Harmful Algae, 98*, Article 101872.

Ning, C., Gao, P., Wang, B., Lin, W., Jiang, N., & Cai, K. (2017). Impacts of chemical fertilizer reduction and organic amendments supplementation on soil nutrient, enzyme activity and heavy metal content. *Journal of Integrative Agriculture, 16*, 1819–1831. https://doi.org/10.1016/S2095-3119(16)61476-4

NMCG (2017). Ganges River Dolphin conservation education programme educating schools and community for sustainable actions for conserving habitat of Ganges River Dolphin along Ganga River Basin (Phase II).

NMCG, N.M. for C.G., Ministry of Jal Shakti, Department of Water Resources, River Development &. Ganga Rejuvenation, Government of (2025). National Mission for Clean Ganga(NMCG) [WWW Document]. Natl. Mission Clean GangaNMCGMinistry Jal Shakti Dep. Water Resour. River Dev. Ganga Rejuvenation Gov. India. URL https://nmcg.nic.in/ (accessed 4.11.25).

Nobinraja, M., Aravind, N. A., & Ravikanth, G. (2023). Opening the floodgates for invasion—modelling the distribution dynamics of invasive alien fishes in India. *Environmental Monitoring and Assessment, 195*, 1411. https://doi.org/10.1007/s10661-023-12012-z

Pal, S. (2017). Impact of Tilpara barrage on backwater reach of Kushkarni River: A tributary of Mayurakshi River. *Environment, Development and Sustainability, 19*, 2115–2142. https://doi.org/10.1007/s10668-016-9833-4

Pal, S. (2016). Impact of Massanjore Dam on hydro-geomorphological modification of Mayurakshi River, Eastern India. *Environment, Development and Sustainability, 18*, 921–944. https://doi.org/10.1007/s10668-015-9679-1

Pan, B., Yuan, J., Zhang, X., Wang, Z., Chen, J., Lu, J., Yang, W., Li, Z., Zhao, N., & Xu, M. (2016). A review of ecological restoration techniques in fluvial rivers. *International Journal of Sediment Research, 31*, 110–119. https://doi.org/10.1016/j.ijsrc.2016.03.001

Panda, P. K., Panda, R. B., & Dash, P. K. (2018). The river water pollution in India and abroad: A critical review to study the relationship among different physico-chemical parameters. *American Journal of Water Resources, 6*, 25–38.

Panday, D. P., & Kumar, M. (2022). Climate indices and hydrological extremes: Deciphering the best fit model. *Environmental Research, 215*, Article 114301. https://doi.org/10.1016/j.envres.2022.114301

Pandey, D., Tiwari, A. D., & Mishra, V. (2022). On the occurrence of the observed worst flood in Mahanadi River basin under the warming climate. *Weather and Climate Extremes, 38*, Article 100520. https://doi.org/10.1016/j.wace.2022.100520

Pandey, L. K., Park, J., Son, D. H., Kim, W., Islam, M. S., Choi, S., Lee, H., & Han, T. (2019). Assessment of metal contamination in water and sediments from major rivers in South Korea from 2008 to 2015. *Science of the Total Environment, 651*, 323–333.

Pandit, A., Ekka, A., Das, B. K., Samanta, S., Chakraborty, L., & Raman, R. K. (2019). Fishers' livelihood diversification in Bhagirathi-Hooghly stretch of Ganga River in India. *Current Science, 116*, 1748–1752.

Parashar, V., Bara, S. K., Damde, D., Kumar, A., & Vyas, V. (2016). Assessment of the socioeconomic status of fishermen communities: A case study from a selected reach of River Narmada, India. *International Journal of Research in Fisheries and Aquaculture, 6*, 47–59.

Parua, P. K. (2010). *The Ganga: Water use in the Indian subcontinent*. Springer Science & Business Media.

Pathak, A. K., Kantharajan, G., Saini, V. P., Kumar, R., Dayal, R., Mohindra, V., & Lal, K. K. (2020). Fish community and habitat diversity profiling of Luni, an ephemeral saline river from Thar Desert of India for conservation and management. *Community Ecology, 21*, 303–316. https://doi.org/10.1007/s42974-020-00033-4

Pathak, H. (2023). Impact, adaptation, and mitigation of climate change in Indian agriculture. *Environmental Monitoring and Assessment, 195*, 52.

Patidar, K., Ambade, B., Verma, S. K., & Mohammad, F. (2023). Microplastic contamination in water and sediments of Mahanadi River, India: An assessment of ecological risk along rural-urban area. *Journal of Environmental Management, 348*, Article 119363.

Paul, P., & Chakraborty, S. (2016). Impact of inland fisheries on the socio-economic development: A focus on perspectives on development, Nadia District, West Bengal, India. *International Journal of Fisheries and Aquatic Studies, 6*, 59–76.

Pendergrass, A. G., Meehl, G. A., Pulwarty, R., Hobbins, M., Hoell, A., AghaKouchak, A., Bonfils, C. J., Gallant, A. J., Hoerling, M., & Hoffmann, D. (2020). Flash droughts present a new challenge for subseasonal-to-seasonal prediction. *Nature Climate Change, 10*, 191–199.

Poff, N. L., & Zimmerman, J. K. H. (2010). Ecological responses to altered flow regimes: A literature review to inform the science and management of environmental flows. *Freshwater Biology, 55*, 194–205. https://doi.org/10.1111/j.1365-2427.2009.02272.x

Pollock, M. M., Heim, M., & Werner, D. (2003). Hydrologic and geomorphic effects of beaver dams and their influence on fishes. In *American fisheries society symposium* (pp. 213–233).

Pownkumar, V., Ananthan, P. S., Ekka, A., & Qureshi, N. W. (2022). Fisheries as ecosystem services: A case study of the Cauvery river basin, India. *Frontiers in Environmental Science, 10*, Article 892012.

Prabha, S., Kumar, M., Kumar, A., Das, P., Al., & Ramanathan. (2013). Impact assessment of textile effluent on groundwater quality in the vicinity of Tirupur industrial area, southern India. *Environment and Earth Science, 70*, 3015–3022. https://doi.org/10.1007/s12665-013-2361-8

Pradhan, C., Chembolu, V., Bharti, R., & Dutta, S. (2023). Regulated rivers in India: Research progress and future directions. *ISH Journal of Hydraulic Engineering, 29*, 58–70. https://doi.org/10.1080/09715010.2021.1975319

Prasad, R., & Prasad, S. (2019). Algal Blooms and Phosphate Eutrophication of inland water ecosystems with special reference to India. *International Journal of Plant and Environment, 5*, 01–08. https://doi.org/10.18811/ijpen.v5i01.1

Prasad, R. R., & Nair, K. M. (2004). Integrated hydrogeological investigations and multipronged water conservation in selected watershed of Achankovil river basin. *Report of People's Research Organisation for grass root environmental science service (Progress), Hyderabad, 43.*

Prasood, S. P., Mukesh, M. V., Rani, V. R., Sajinkumar, K. S., & Thrivikramji, K. P. (2021). Urbanization and its effects on water resources: Scenario of a tropical river basin in South India. *Remote Sensing Applications: Society and*

Environment, 23, Article 100556. https://doi.org/10.1016/j.rsase.2021.100556

Praveen, B., Talukdar, S., Shahfahad, Mahato, S., Mondal, J., Sharma, P., Islam, A. R. M. T., & Rahman, A. (2020). Analyzing trend and forecasting of rainfall changes in India using non-parametrical and machine learning approaches. *Scientific Reports, 10*(1), 10342.

Proshad, R., Islam, S., Tusher, T. R., Zhang, D., Khadka, S., Gao, J., & Kundu, S. (2021). Appraisal of heavy metal toxicity in surface water with human health risk by a novel approach: A study on an urban river in vicinity to industrial areas of Bangladesh. *Toxin Reviews, 40*, 803–819. https://doi.org/10.1080/15569543.2020.1780615

Purnadurga, G., Lakshmi Kumar, T. V., Koteswara Rao, K., Rajasekhar, M., & Narayanan, M. S. (2018). Investigation of temperature changes over India in association with meteorological parameters in a warming climate. *International Journal of Climatology, 38*, 867–877.

Puth, L. M., & Post, D. M. (2005). Studying invasion: Have we missed the boat? *Ecology Letters, 8*, 715–721. https://doi.org/10.1111/j.1461-0248.2005.00774.x

Rajwant, M. A. K., & Sharma, U. K. (2005). Deforestation and geo-environmental degradation in the Baner Khad Watershed, Himachal Pradesh, India. *Asian Journal of Water, Environment and Pollution, 2*(1), 39–46.

Ramadoss, A., & Poyyamoli, G. (2011). Biodiversity conservation through environmental education for sustainable development-a case study from puducherry, India. *International Electronic Journal of Environmental Education, 1*(2)

RamyaPriya, R., & Elango, L. (2018). Evaluation of geogenic and anthropogenic impacts on spatio-temporal variation in quality of surface water and groundwater along Cauvery River, India. *Environmental Earth Sciences, 77*, 2. https://doi.org/10.1007/s12665-017-7176-6

Ranade, V. V., & Bhandari, V. M. (2014). *Industrial wastewater treatment, recycling and reuse*. Butterworth-Heinemann

Ranade, A., & Gurrapu, S. (2025). Long-term rainfall variability of Indian river basins in the context of global warming and climatic indices. *Journal of Water and Climate Change, 16*, 910–933.

Randhawa, K. S. (2025). A review on microplastic contamination in freshwater bodies, especially in Indian rivers with experimental insights on polyamide degradation. *Pigment Resin Technol*. https://doi.org/10.1108/PRT-10-2024-0110

Rao, K. N., & Jagannathan, P. (1963). Climatic changes in India. In *Proceedings of symposium on changes in climate* (pp. 49–66). UNESCO and WHO Rome.

Rashmi, I., Roy, T., Kartika, K. S., Pal, R., Coumar, V., Kala, S., & Shinoji, K. C. (2020). Organic and inorganic fertilizer contaminants in agriculture: Impact on soil and water resources. In M. Naeem, A. A. Ansari, & S. S. Gill (Eds.), *Contaminants in agriculture* (pp. 3–41). Springer International Publishing, Cham. https://doi.org/10.1007/978-3-030-41552-5_1

Rathore, S. S., Chandravanshi, P., Chandravanshi, A., & Jaiswal, K. (2016). Eutrophication: Impacts of excess nutrient inputs on aquatic ecosystem. *IOSR Journal of Agriculture and Veterinary Science, 9*, 89–96.

Ray, K., Pandey, P., Pandey, C., Dimri, A. P., & Kishore, K. (2019). On the recent floods in India. *Current Science, 117*, 204–218.

Renöfält, B. M., Jansson, R., & Nilsson, C. (2010). Effects of hydropower generation and opportunities for environmental flow management in Swedish riverine ecosystems. *Freshwater Biology, 55*, 49–67. https://doi.org/10.1111/j.1365-2427.2009.02241.x

Rentier, E. S., & Cammeraat, L. H. (2022). The environmental impacts of river sand mining. *Science of the Total Environment, 838*, Article 155877. https://doi.org/10.1016/j.scitotenv.2022.155877

Richter, B. D., & Thomas, G. A. (2007). Restoring environmental flows by modifying dam operations. *Ecology and Society, 12*, art12. https://doi.org/10.5751/ES-02014-120112

Rudra, K. (2006). Shifting of the Ganga and land erosion in West Bengal. *A socio-ecological viewpoint. Indian Institute of Management Calcutta*, 1-43.

Rumana, H. S., Jeeva, V., & Kumar, S. (2015). Impact of the low head dam/Barrage on fisheries—A case study of Giri River of Yamuna Basin (India). *Transylvanian Review of Systematical and Ecological Research, 17*, 119–138. https://doi.org/10.1515/trser-2015-0070

Sahana, M., Hong, H., & Sajjad, H. (2018). Analyzing urban spatial patterns and trend of urban growth using urban sprawl matrix: A study on Kolkata urban agglomeration, India. *Science of the Total Environment, 628*, 1557–1566.

Sahoo, M., & Kumar Yadav, R. (2022). The Interannual variability of rainfall over homogeneous regions of Indian summer monsoon. *Theoretical and Applied Climatology, 148*, 1303–1316. https://doi.org/10.1007/s00704-022-03978-w

Sajina, A. M., Sudheesan, D., Samanta, S., Paul, S. K., Bhowmick, S., Nag, S. K., Kumar, V., & Das, B. K. (2022). Development and validation of a fish-based index of biotic integrity for assessing the ecological health of Indian Rivers Mahanadi and Kathajodi-Devi. *Aquatic Ecosystem Health & Management, 25*(2), 25–35.

Salgaonkar, S., Botle, A., Barabde, G., & Herlekar, M. (2024). Micro-Nanoplastics from stormwater runoffs to water bodies: An in-depth investigation. In S. A. Bhat, V. Kumar, F. Li, & S. Kumar (Eds.), *Management of micro and nano-plastics in soil and biosolids* (pp. 95–117). Springer Nature Switzerland, Cham. https://doi.org/10.1007/978-3-031-51967-3_4

Samal, D. R., & Gedam, S. (2021). Assessing the impacts of land use and land cover change on water resources in the Upper Bhima river basin, India. *Environmental Challenges, 5*, Article 100251. https://doi.org/10.1016/j.envc.2021.100251

Sandilyan, S. (2016). Occurrence of ornamental fishes: A looming danger for Inland fish diversity of India. *Current Science*, 2099–2104.

Saravanan, P., Saravanan, V., Rajeshkannan, R., Arnica, G., Rajasimman, M., Baskar, G., & Pugazhendhi, A. (2024). Comprehensive review on toxic heavy metals in the aquatic system: Sources, identification, treatment strategies, and health risk assessment. *Environmental Research, 258*, Article 119440. https://doi.org/10.1016/j.envres.2024.119440

Saritha Vara, S. P. (2023). Agricultural activities and their environmental impact on surface waters: A review. *Tuijin Jishu Journal Propulsion Technology, 44*(2). https://doi.org/10.52783/tjjpt.v44.i2.216

Sarkar, B., & Islam, A. (2022). Assessing poverty and livelihood vulnerability of the fishing communities in the context of pollution of the Churni River, India. *Environmental Science and Pollution Research, 29*, 26575–26598. https://doi.org/10.1007/s11356-021-17719-5

Sarker, B., N. Keya, K. N., Mahir, F. I., Nahiun, K. M., Shahida, S., & Khan, R. A. (2021). Surface and ground water pollution: Causes and effects of urbanization and industrialization in South Asia. *Scientific Review*, 32–41. https://doi.org/10.32861/sr.73.32.41

Sathya, K., Nagarajan, K., Carlin Geor Malar, G., Rajalakshmi, S., & Raja Lakshmi, P. (2022). A comprehensive review on comparison among effluent treatment methods and modern methods of treatment of industrial wastewater effluent from different sources. *Applied Water Science, 12*, 70. https://doi.org/10.1007/s13201-022-01594-7

Savci, S. (2012). Investigation of effect of chemical fertilizers on environment. *APCBEE Procedia, 1*, 287–292. https://doi.org/10.1016/j.apcbee.2012.03.047

Setia, R., Dhaliwal, S. S., Kumar, V., Singh, R., Kukal, S. S., & Pateriya, B. (2020). Impact assessment of metal contamination in surface water of Sutlej River (India) on human health risks. *Environmental Pollution, 265*, Article 114907. https://doi.org/10.1016/j.envpol.2020.114907

Shameem, M. I. M., Momtaz, S., & Rauscher, R. (2014). Vulnerability of rural livelihoods to multiple stressors: A case study from the southwest coastal region of Bangladesh. *Ocean and Coastal Management, 102*, 79–87.

Sharma, B. M., Bečanová, J., Scheringer, M., Sharma, A., Bharat, G. K., Whitehead, P. G., Klánová, J., & Nizzetto, L. (2019). Health and ecological risk assessment of emerging contaminants (pharmaceuticals, personal care products, and artificial sweeteners) in surface and groundwater (drinking water)

in the Ganges River Basin, India. *Science of the Total Environment, 646*, 1459–1467.
Sharma, K., Rajan, S., & Nayak, S. K. (2024). Water pollution: Primary sources and associated human health hazards with special emphasis on rural areas. In *Water resources management for rural development* (pp. 3–14). Elsevier. https://doi.org/10.1016/B978-0-443-18778-0.00014-3
Sharma, P., Meher, P. K., Kumar, A., Gautam, Y. P., & Mishra, K. P. (2014). Changes in water quality index of Ganges river at different locations in Allahabad. *Sustainable Water Quality and Ecology, 3–4*, 67–76. https://doi.org/10.1016/j.swaqe.2014.10.002
Sharma, T., Vittal, H., Karmakar, S., & Ghosh, S. (2020). Increasing agricultural risk to hydro-climatic extremes in India. *Environmental Research Letters, 15*, Article 034010. https://doi.org/10.1088/1748-9326/ab63e1
Shukla, A. K., Ojha, C. S. P., Mijic, A., Buytaert, W., Pathak, S., Garg, R. D., & Shukla, S. (2018). Population growth, land use and land cover transformations, and water quality nexus in the Upper Ganga River basin. *Hydrology and Earth System Sciences, 22*, 4745–4770. https://doi.org/10.5194/hess-22-4745-2018
Shukla, S., & Gedam, S. (2018). Assessing the impacts of urbanization on hydrological processes in a semi-arid river basin of Maharashtra, India. *Modeling Earth Systems and Environment, 4*, 699–728. https://doi.org/10.1007/s40808-018-0446-9
Shymbin, B. W., & Nongbri, G. (2022). Impact of sand mining on the physical health of the river and the livelihood of the people: A case study of Umtyngngar River, Meghalaya. In G. S. Bhunia, U. Chatterjee, K. C. Lalmalsawmzauva, & P. K. Shit (Eds.), *Anthropogeomorphology, geography of the physical environment* (pp. 221–242). Springer International Publishing, Cham. https://doi.org/10.1007/978-3-030-77572-8_11
Simon, M., & Joshi, H. (2022). Story of the Ganga River: Its pollution and rejuvenation. In A. Mukherjee (Ed.), *Riverine systems* (pp. 21–55). Springer International Publishing, Cham. https://doi.org/10.1007/978-3-030-87067-6_2
Singh, A. K. (2025). Non-Native Invasive Fish Species (NIFS) in inland waters of India: Management issues, strategies and suggestions. *Fisheries Management and Ecology, 32*, Article e12747. https://doi.org/10.1111/fme.12747
Singh, A. K., & Lakra, W. S. (2011). Risk and benefit assessment of alien fish species of the aquaculture and aquarium trade into India. *Reviews in Aquaculture, 3*, 3–18. https://doi.org/10.1111/j.1753-5131.2010.01039.x
Singh, B. R., & Steinnes, E. (2020). Soil and water contamination by heavy metals. In *Soil processes and water quality* (pp. 233–271). CRC Press.
Singh, N. K., Gupta, G., Upadhyay, A. K., & Rai, U. N. (2019). Biological wastewater treatment for prevention of river water pollution and reuse:

Perspectives and challenges. In R. P. Singh, A. S. Kolok & S. L. Bartelt-Hunt (Eds.), *Water conservation, recycling and reuse: Issues and challenges* (pp. 81–93). Springer Singapore. https://doi.org/10.1007/978-981-13-3179-4_4

Singh, R. B., & Kumar, A. (2015). Climate variability and water resource scarcity in drylands of Rajasthan. *India. Geoenvironmental Disasters, 2*, 7. https://doi.org/10.1186/s40677-015-0018-5

Singh, S., Hariteja, N., Prasad, T. R., Raju, N. J., & Ramakrishna, C. (2020). Impact assessment of faecal sludge on groundwater and river water quality in Lucknow environs, Uttar Pradesh, India. *Groundwater for Sustainable Development, 11*, Article 100461.

Singh, S., Kumar, P., Parijat, R., Gonengcil, B., & Rai, A. (2024). Establishing the relationship between land use land cover, normalized difference vegetation index and land surface temperature: A case of Lower Son River Basin, India. *Geography and Sustainability, 5*, 265–275. https://doi.org/10.1016/j.geosus.2023.11.006

Sinha, A., & Sedai, A. K. (2025). National river conservation plan and water pollutants in India. *Economic Analysis and Policy, 85*, 2192–2206. https://doi.org/10.1016/j.eap.2025.02.031

Sinha, R., Mohanta, H., Jain, V., & Tandon, S. K. (2017). Geomorphic diversity as a river management tool and its application to the Ganga River, India. *River Research and Applications, 33*, 1156–1176. https://doi.org/10.1002/rra.3154

Sinha, R. K., & Eldho, T. I. (2018). Effects of historical and projected land use/cover change on runoff and sediment yield in the Netravati river basin, Western Ghats, India. *Environmental Earth Sciences, 77*, 111. https://doi.org/10.1007/s12665-018-7317-6

Sodhi, K. K., & Singh, C. K. (2023). A systematic review on the occurrence, fate, and remediation of SARS-CoV-2 in wastewater. *International Journal of Environmental Science and Technology, 20*, 8073–8086. https://doi.org/10.1007/s13762-022-04326-1

Sonu, T. S., Mohammed, F., & Bhagyanathan, A. (2022). The impact of upstream land use land cover change on downstream flooding: A case of Kuttanad and Meenachil River Basin, Kerala, India. *Urban Climate, 41*, 101089. https://doi.org/10.1016/j.uclim.2022.101089

Sreekanth, G. B., Mujawar, S., Lal, D. M., Mayekar, T., Stephen, J., Raghavan, R., Kumar, A. B., & Ingole, B. S. (2022). Modelling the mixed impacts of multiple invasive alien fish species in a closed freshwater ecosystem in India. *Environmental Science and Pollution Research, 29*, 58278–58296. https://doi.org/10.1007/s11356-022-19794-8

Srivastav, A. L. (2020). Chemical fertilizers and pesticides: role in groundwater contamination. In *Agrochemicals detection, treatment and remediation* (pp. 143–159). Elsevier. https://doi.org/10.1016/B978-0-08-103017-2.00006-4

Srivastava, H. N., Dewan, B. N., Dikshit, S. K., Rao, G. P., Singh, S. S., & Rao, K. R. (1992). Decadal trends in climate over India. *Mausam, 43*, 7–20.

Srivastava, P. K., Pradhan, R. K., Petropoulos, G. P., Pandey, V., Gupta, M., Yaduvanshi, A., Wan Jaafar, W. Z., Mall, R. K., & Sahai, A. K. (2021). Long-term trend analysis of precipitation and extreme events over Kosi River Basin in India. *Water, 13*, 1695.

Su, G., Logez, M., Xu, J., Tao, S., Villéger, S., & Brosse, S. (2021). Human impacts on global freshwater fish biodiversity. *Science, 371*, 835–838. https://doi.org/10.1126/science.abd3369

Sudarsan, J. S., Dogra, K., Kumar, R., Raval, N. P., Leifels, M., Mukherjee, S., Trivedi, M. H., Jain, M. S., Zang, J., Barceló, D., Mahlknecht, J., & Kumar, M. (2024). Tricks and tracks of prevalence, occurrences, treatment technologies, and challenges of mixtures of emerging contaminants in the environment: With special emphasis on microplastic. *Journal of Contaminant Hydrology, 265*, Article 104389. https://doi.org/10.1016/j.jconhyd.2024.104389

Sugunan, V. V. (1995). Reservoir fisheries of India. FAO, Rome, Italy. Fish. Tech. Pap. 345.

Sun, J., Du, W., Lucas, M. C., Ding, C., Chen, J., Tao, J., & He, D. (2023). River fragmentation and barrier impacts on fishes have been greatly underestimated in the upper Mekong River. *Journal of Environmental Management, 327*, Article 116817.

Szlag, D. C., Sinclair, J. L., Southwell, B., & Westrick, J. A. (2015). Cyanobacteria and cyanotoxins occurrence and removal from five high-risk conventional treatment drinking water plants. *Toxins, 7*, 2198–2220.

Tariq, A., & Mushtaq, A. (2023). Untreated wastewater reasons and causes: A review of most affected areas and cities. *International Journal of Chemical and Biochemical Science, 23*, 121–143.

Taxak, A. K., Murumkar, A. R., & Arya, D. S. (2014). Long term spatial and temporal rainfall trends and homogeneity analysis in Wainganga basin, Central India. *Weather and Climate Extremes, 4*, 50–61.

Thakur, J. (2022). Mapping of deforestation and forest degradation associated with responsible drivers in the man river basin, central India. *Geodesy and Cartography, 48*, 218–223. https://doi.org/10.3846/gac.2022.14951

Thapliyal, V., & Kulshrestha, S. M. (1991). Climate changes and trends over India. *Mausam, 42*, 333–338.

Thomsen, M. S., Byers, J. E., Schiel, D. R., Bruno, J. F., Olden, J. D., Wernberg, T., & Silliman, B. R. (2014). Impacts of marine invaders on biodiversity depend on trophic position and functional similarity. *Marine Ecology Progress Series, 495*, 39–47.

Tierney, P. A., Caffrey, J. M., Vogel, S., Matthews, S. M., Costantini, E., & Holland, C. V. (2020). Invasive freshwater fish (Leuciscus leuciscus) acts as a sink for a parasite of native brown trout Salmo trutta. *Biological Invasions, 22*, 2235–2250. https://doi.org/10.1007/s10530-020-02253-1

TIES, (2017). My river my life a school based river conservation project For Chalakudy Puzha, Kerala, India Preparation of Community Biodiversity Register.

Tong, L. (2023). The importance of river water for environment and society. *Journal of Hydrogeology & Hydrologic Engineering, 12*. https://doi.org/10.4172/2325-9647.1000257

Tripathi, S., Kumar, R., Mondal, P., Thakur, A. K., Singh, S., Kaur, G., Sharma, M., & Kumar, M. (2023). Urban fingerprinting on the River Ganges: A perspective of emerging contaminants status. *Total Environment Research Themes, 5*, Article 100028. https://doi.org/10.1016/j.totert.2023.100028

Upadhyay, S. K., Rajput, V. D., Kumari, A., Espinosa-Saiz, D., Menendez, E., Minkina, T., Dwivedi, P., & Mandzhieva, S. (2023). Plant growth-promoting rhizobacteria: A potential bio-asset for restoration of degraded soil and crop productivity with sustainable emerging techniques. *Environmental Geochemistry and Health, 45*, 9321–9344. https://doi.org/10.1007/s10653-022-01433-3

Valentim, H. I. L., Feio, M. J., & Almeida, S. F. P. (2024). Fluvial protected areas as a strategy to preserve riverine ecosystems—a review. *Biodiversity and Conservation, 33*, 439–462. https://doi.org/10.1007/s10531-023-02774-w

Van Looy, K., Tormos, T., & Souchon, Y. (2014). Disentangling dam impacts in river networks. *Ecological Indicators, 37*, 10–20.

Varma, K., & Jha, P. K. (2024). Phosphorus distribution in the water and sediments of the Ganga and Yamuna Rivers, Uttar Pradesh, India: Insights into pollution sources, bioavailability, and eutrophication implications. *Environmental Monitoring and Assessment, 196*, 336. https://doi.org/10.1007/s10661-024-12499-0

Varma, V. G., Jha, S., Raju, L. H. K., Kishore, R. L., & Ranjith, V. (2022). A review on decentralized wastewater treatment systems in India. *Chemosphere, 300*, Article 134462.

Vashisht, D., Kumar, A., Mehta, S. K., & Ibhadon, A. (2020). Analysis of emerging contaminants: A case study of the underground and drinking water samples in Chandigarh, India. *Environmental Advances, 1*, Article 100002.

Vass, K., Das, M., Tyagi, R., Katiha, P., Samanta, S., Srivastava, N., Bhattacharjya, B., Suresh, V., Pathak, V., & Chandra, G. (2011). Strategies for sustainable fisheries in the Indian part of the Ganga-Brahmaputra river basins. *International Journal of Ecology and Environmental Sciences, 37*, 157–218.

Vasseghian, Y., Nadagouda, M. M., & Aminabhavi, T. M. (2024). Biochar-enhanced bioremediation of eutrophic waters impacted by algal blooms. *Journal of Environmental Management, 367*, Article 122044.

Vazhayil, A. M., & Wantzen, K. M. (2023). 34 community-based conservation initiatives in India: A participative movement driven by river culture. *River Culture–Life as a dance to the rhythm of the waters. UNESCO Publishing, Paris, 10*, 241–260.

Venkatachalapathy, R., & Karthikeyan, P. (2015). Diatom indices and water quality index of the Cauvery river, India: Implications on the suitability of bio-indicators for environmental impact assessment. In Mu, Ramkumar, K. Kumaraswamy & R. Mohanraj (Eds.), *Environmental management of river basin ecosystems* (pp. 707–727). Springer Earth System Sciences. Springer International Publishing, Cham. https://doi.org/10.1007/978-3-319-13425-3_31

Verma, D., Meena, R. H., Sukhwal, A., Jat, G., Meena, S. C., Upadhyay, S. K., & Jain, D. (2023). Effect of ZSB with graded levels of zinc fertilizer on yield and zinc uptake under maize cultivation. *Proceedings of the National Academy of Sciences, India Section B, 93*, 379–385. https://doi.org/10.1007/s40011-022-01433-4

Verma, N. P. (2016). Water pollution in India: An overview. *Indian Journal of Law and Justice, 7*, 22.

Wang, Y., Yang, Z., Shen, Z., Tang, Z., Niu, J., & Gao, F. (2011). Assessment of heavy metals in sediments from a typical catchment of the Yangtze River, China. *Environmental Monitoring and Assessment, 172*, 407–417. https://doi.org/10.1007/s10661-010-1343-5

Wasson, R. J. (2018). Riverine flood hazard: Disaster risk reduction in India. *Proceedings of the Indian National Science Academy, 99*. https://doi.org/10.16943/ptinsa/2018/49502

Wear, S. L., Acuña, V., McDonald, R., & Font, C. (2021). Sewage pollution, declining ecosystem health, and cross-sector collaboration. *Biological Conservation, 255*, Article 109010.

Welcomme, R. L. (1995). Relationships between fisheries and the integrity of river systems. *Regulated Rivers: Research and Management, 11*, 121–136. https://doi.org/10.1002/rrr.3450110110

Wilkinson, J. L., Boxall, A. B. A., Kolpin, D. W., Leung, K. M. Y., Lai, R. W. S., Galbán-Malagón, C., Adell, A. D., Mondon, J., Metian, M., Marchant, R. A., Bouzas-Monroy, A., Cuni-Sanchez, A., Coors, A., Carriquiriborde, P., Rojo, M., Gordon, C., Cara, M., Moermond, M., Luarte, T., & Teta, C. et al. (2022). Pharmaceutical pollution of the world's rivers. *Proceedings of the National Academy of Sciences, 119*, Article e2113947119. https://doi.org/10.1073/pnas.2113947119

Williams, M., Kookana, R. S., Mehta, A., Yadav, S. K., Tailor, B. L., & Maheshwari, B. (2019). Emerging contaminants in a river receiving untreated wastewater from an Indian urban centre. *Science of the Total Environment, 647,* 1256–1265.

WWF, (2022). *Building a nature-positive society, Living planet report*. CBS Library, Frederiksberg.

Xi, H., Li, T., Yuan, Y., Chen, Q., & Wen, Z. (2023). River ecosystem health assessment based on fuzzy logic and harmony degree evaluation in a human-dominated river basin. *Ecosystem Health* and *Sustainability, 9*.

Xia, Y., Zhang, M., Tsang, D. C. W., Geng, N., Lu, D., Zhu, L., Igalavithana, A. D., Dissanayake, P. D., Rinklebe, J., Yang, X., & Ok, Y. S. (2020). Recent advances in control technologies for non-point source pollution with nitrogen and phosphorous from agricultural runoff: Current practices and future prospects. *Applied Biological Chemistry, 63*, 8. https://doi.org/10.1186/s13765-020-0493-6

Yadav, I. C., Devi, N. L., Syed, J. H., Cheng, Z., Li, J., Zhang, G., & Jones, K. C. (2015). Current status of persistent organic pesticides residues in air, water, and soil, and their possible effect on neighboring countries: A comprehensive review of India. *Science of the Total Environment, 511*, 123–137.

Yang, S., Twiss, M. R., Fernando, S., Grimberg, S. J., & Yang, Y. (2022). Mitigation of Cyanobacterial Harmful Algal Blooms (cHABs) and cyanotoxins by electrochemical oxidation: From a bench-scale study to field application. *ACS ES&T Engineering, 2*, 1160–1168. https://doi.org/10.1021/acsestengg.1c00344

Zahoor, I., & Mushtaq, A. (2023). Water pollution from agricultural activities: A critical global review. *International Journal of Chemical and Biochemical Science, 23*, 164–176.

Zhang, H. (2022). Lake management and eutrophication mitigation: Coming down to Earth—In *situ monitoring, scientific management and well-organized collaboration are still crucial*. Water.

Zhang, X., Obringer, R., Wei, C., Chen, N., & Niyogi, D. (2017). Droughts in India from 1981 to 2013 and implications to wheat production. *Science and Reports, 7*, 44552. https://doi.org/10.1038/srep44552

Zhang, Y., Whalen, J. K., Cai, C., Shan, K., & Zhou, H. (2023). Harmful cyanobacteria-diatom/dinoflagellate blooms and their cyanotoxins in freshwaters: A nonnegligible chronic health and ecological hazard. *Water Research, 233*, Article 119807.

Zou, W., Tolonen, K. T., Zhu, G., Qin, B., Zhang, Y., Cao, Z., Peng, K., Cai, Y., & Gong, Z. (2019). Catastrophic effects of sand mining on macroinvertebrates in a large shallow lake with implications for management. *Science of the Total Environment, 695*, Article 133706. https://doi.org/10.1016/j.scitotenv.2019.133706

CHAPTER 3

Transformations in India's River Systems: The Impact of Climate Change on Ecosystem Dynamics and Water Quality

Akshay Chavan, Akshay Botle, Neha Mourya, and Umesh Kakde

3.1 Introduction

Rivers play an important role in human society and the environment. They are crucial since they supply water for agriculture, industries, and household purposes (Botle et al., 2023; Grizzetti et al., 2019; Tickner et al., 2017). They also play a significant role in maintaining rich biodiversity and ecosystem integrity since they support many aquatic

A. Chavan · N. Mourya · U. Kakde (✉)
Department of Botany, The Institute of Science, Dr. Homi Bhabha State University, Mumbai, India
e-mail: drumeshkakde@iscm.ac.in

A. Chavan
e-mail: 22phdbot.akshay@iscm.ac.in

N. Mourya
e-mail: resbot.neha@iscm.ac.in

S. Ghosh (ed.), *Reimagining Indian Rivers for Sustainability*, Palgrave Studies in Environmental Sustainability,
https://doi.org/10.1007/978-3-032-04754-0_3

and terrestrial flora and fauna (Koehler, 2008; Lynch et al., 2023). Different rivers offer notable opportunities for recreational and cultural services, including aesthetic experience, development tourism, and relaxation activities (Botle et al., 2025; Harrison et al., 2010). These significant river system services are under threat from climate change, and therefore, climate change is among the most complicated global problems (Přívara & Přívarová, 2019). Increased atmospheric temperature, temperature of water, melting of glacial and snow cover, changed precipitation regimes (increased frequency and intensity), drought, and modification of the hydrological cycle are the key effects of climate change (Ougahi et al., 2022).

Due to these climate-altering occurrences, there is an alteration in India's river basins, river ecosystems, impacting various functional, structural, resilience, and ecological components (Grimm et al., 2013; Hossain et al., 2017; Li et al., 2018). These phenomena affect frequency and timing of peak discharge, flood and drought frequency, river base flow, and river water quantity and quality (Fig. 3.1) (Pittock & Finlayson, 2011; Schneider et al., 2013; Singh et al., 2014). Moreover, the solubility of oxygen decreases because of climate change, causing lower DO levels, which can negatively affect the health, dietary pattern, swimming habits, and breeding capacity of aquatic animals (Ali et al., 2022).

Apart from that, climate change influences the physical properties and morphology of rivers due to high sediment load, channel morphology change, and increased river bank erosion (Mittal et al., 2016; Thapa et al., 2024). For example, Monsoonal rainfall is enhanced due to climate change processes, resulting in increased peak flow, channel morphology change, frequency of floods increased, and river hydrology is changed in the Godavari and Ganga river basins (Hengade et al., 2018; Khan et al., 2018). These changes are due to natural and anthropogenic processes and occur in other rivers of the world (Li et al., 2024).

Flood hazard, flood displacement, drought, water poverty, reduced agricultural production, and declining water resources are the main concerns that are faced by human societies due to the altered flow

A. Botle
Department of Environmental Science, The Institute of Science, Dr. Homi Bhabha State University, Mumbai, India
e-mail: 23res.evs.akshay@iscm.ac.in

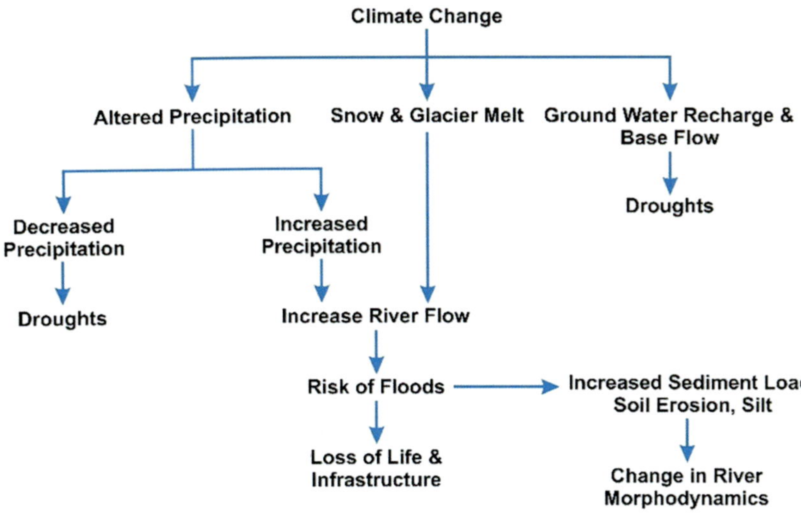

Fig. 3.1 Impact of climate change on Rivers

and river shape (Ashmore & Church, 2001). The Ganga-Brahmaputra-Meghana river basin flood is the classic example in which around 1200 mortalities and significant crop and infrastructure loss were recorded in 2017 (Uhe et al., 2019).

Despite much research highlighting these issues, most of the available literature is either scattered or concentrates on short-term fluctuations. There is limited research that concentrates on long-term, cumulative, and cascading climate change impacts on rivers. This chapter aims to analyze the adverse effects of climate change on the biological, chemical, and physical components of the river ecosystem. This research also evaluates case studies and considers larger societal and ecological repercussions. Through the focus on the deficit in scientific knowledge and its influence, this chapter contributes to the development of policies, climate adaptation plans, and river basin management policies. Lastly, this data is necessary for ensuring the river systems' integrity during an evolving climate era.

3.2 Climate Change in the Indian Sub-Continent

The Indian subcontinent experiences severe issues from climate change, including rising temperatures, changes in precipitation patterns, and increasing intensities of harsh weather (Varghese et al., 2024). These changes have widespread impacts on agriculture and river basins (Jha et al., 2021). India experienced a substantial warming during the past century, with surface temperature increases of approximately 1 C during winter and 1.1 C during the post-monsoon season (Dash et al., 2007).

The Indian Meteorological Department (IMD) indicated that this warming has been rising since 2000 (Purnadurga et al., 2018). It's interesting to note that the pattern is not seasonal. Temperature seasonal anomalies indicate that the post-monsoon and summer monsoon seasons differ by roughly 0.8 °C, which could change patterns of atmospheric circulation (Dash et al., 2007).

River flow and discharge in India majorly depend on the monsoon rainfall. Due to the rising temperatures, monsoons have become unpredictable, including changes such as duration and intensity. According to the IPCC Sixth Assessment Report (2021), India is susceptible to hydrological disturbance and is one of the most climate-sensitive nations (Maharjan et al., 2021; Poncelet et al., 2010). These disruptions are now exposed as catastrophic climatic events. For example, the Uttarakhand flash floods in 2013, the Himachal Pradesh cloudbursts in 2023, and the Kerala floods in 2018. These events have been connected to rising temperatures, glacial melt, and abnormal rainfall (Chevuturi & Dimri, 2016; Hunt & Menon, 2020; Nandargi et al., 2016).

India's important basins, including the Brahmaputra, Ganges, Indus, Cauvery, and Godavari, are facing different climate-related impacts. The most flood-sensitive area, the Brahmaputra basin, is facing riverbank erosion, recurring flash floods that displace individuals annually (Pradhan et al., 2021). As well as, the Indo-Gangetic plain is facing extreme floods due to glacial retreat from the Himalayas and intense monsoonal precipitation, which is impacting millions (Hengade et al., 2018; Nepal & Shrestha, 2015).

For the development of a successful water resource policy in peninsular India, it is necessary to show high rainfall in the Godavari River Basin by the Coupled Model Intercomparison Project Phase 5 (CMIP5) climate models and the Variable Infiltration Capacity (VIC) hydrology model (Hengade et al., 2018). Even though there are water shortages because of decreasing rainfall in southern Indian river basins such as the Krishna and Cauvery (Mishra et al., 2021).

There is an urgent need for climate-resilient management strategies in river basins due to these interconnected crises. The important data to display on these upcoming issues is obtained from the most useful datasets of the India Water Resources Information System (WRIS) of the MoWR and Plans on Climate Change (SAPCCs) of the MoEFCC (Ferrant et al., 2014; Kumar et al., 2024; Raihan & Tuspekova, 2022; Saryal, 2018).

3.3 Climatic, Geophysical, and Hydrological Changes in Rivers

3.3.1 Altered Rainfall Pattern

India's water cycle relies significantly on the monsoon. The country experiences approximately 80% of the nation's yearly precipitation between June and September (Jena et al., 2015). Climate change affects the dynamics of monsoons, rainfall pattern, and their predictability and intensity in India. All these changes have significant impacts on the hydrology of rivers (Katzenberger et al., 2021).

3.3.2 Enhance Rainfall Frequency and Intensity

The most apparent impacts of climate change on precipitation patterns are the heightened intensity of rainfall (Giorgi et al., 2019; Myhre et al., 2019; Sabade et al., 2011). The primary reasons for the Indian summer monsoon are Sea-land differential heating and the northward shift of the intertropical convergence zone (ITCZ). The southwest ocean brings moisture toward land, which leads to rain (Gadgil, 2018). Evaporation from the Central and Southern Indian Ocean is the primary source of moisture (Dey & Döös, 2021). The western equatorial belt and Arabian Sea are warming at a fast rate as a result of climate change, leading to enhanced evaporation and atmospheric water content, no doubt contributing to the rise in Indian summer monsoon rainfall (Sharma et al., 2023; Skliris et al., 2022). India is experiencing an increase in more frequent and more severe short-duration rain, particularly during the monsoon season.

This trend is characterized by an increase in very heavy and heavy precipitation, and a reduction in moderate and light precipitation (Kumari et al., 2022). The river basins are highly affected by the changes, and the potential risk of flash flooding and associated hazards exists. This trend is most clearly evident in the Himalayan region (Bhardwaj et al., 2021). The frequency of extreme precipitation events is witnessed in different parts of India, for instance, in the North-eastern part, the Himalayan foothills, and the Western Ghats. It is known, from reports, that the intensity and frequency of country monsoon precipitation have increased to a large degree since the 1950s (Priya et al., 2017). They caused catastrophic flooding that was estimated to cost the country approximately 3 billion dollars every year (Roxy et al., 2017). Heavy precipitation extensively influences catchment hydrology, resulting in enhanced runoff and sediment in river systems, which impacts the river's water quality (Qiu et al., 2021). Rivers such as the Narmada and other Indian rivers repeatedly experience flooding that impacts society and geomorphology (Kale, 2002).

For instance, frequent flooding occurred in the Narmada River in 1970, 1973, 1984, 1990, 1994, and 2013 (Mangukiya et al., 2022). Consequently, many Indian peninsular rivers experienced severe flooding with maximum discharge (10,000–80,000 m^3/s) (Kale et al., 1997). Unprecedentedly heavy precipitation led to cataclysmic flooding in Kerala (August 2018). Severe precipitation of 56% more and runoff conditions leading to the flooding had no precedence in the 66-year history between 1951 and 2017 (Mishra & Nagaraju, 2019; Mishra & Shah, 2018).

These changes have two main effects: first, they overwhelm flood control measures, resulting in destruction; second, they increase the risk of flash floods, which directly affect riverine infrastructure and communities living near riverbanks. Comprehensive flood risk assessment and management strategies are needed to manage this kind of disastrous event.

3.3.3 Reduced Rainfall in Arid and Semi-Arid Areas

Inadequate monsoon rainfall dramatically lowers the water levels in the river systems that rely on surface runoff from rain-fed catchments. Reduced precipitation affects the relationship between river flow and the recharge of groundwater, primarily in monsoon-dependent precipitation regions. For example, the Indian peninsular region possesses many river basins that are largely supplied by rainfall. This is a multifaceted relationship that shifts in different times and places (Sreelash et al., 2018).

Minor river basins, which are more sensitive to climate change, receive less emphasis (Lutz et al., 2016).

Since less rainfall and irregular distribution of rain, states such as Rajasthan and Gujarat, located in arid and semi-arid areas, are experiencing critical problems resulting in extended droughts and critical water scarcity (Bandyopadhyay et al., 2016; Mehta & Yadav, 2021). These situations are exhibiting adverse effects on the rivers of Rajasthan and Gujarat, namely the Luni and Tapi rivers (Das et al., 2020).

3.3.4 Snow and Glacier Melt

1.3 billion humans from eight countries are relying on the Himalayan Mountain range for freshwater. It is the third-largest ice and snow deposit on the planet (Miller et al., 2012). The essential Indian subcontinental rivers, such as Brahmaputra, Ganga, and Indus, are sourced from here (Ahluwalia et al., 2013). A billion individuals rely on these rivers for their drinking water, catchments, agricultural irrigation, and to sustain a varied ecosystem (Sud et al., 2015). These are crucial cryosphere regions impacted by climate change (Kulkarni et al., 2011).

3.3.4.1 Himalayan Glacier Retreat

In the last several decades, Himalayan glaciers have been shrinking and retreating at extremely high rates (Kumar & Sharma, 2023; Sigdel et al., 2020). Since the 1980s, with climate warming, the rate of glacial retreat has increased (Sigdel et al., 2020). Central Himalaya's Gangapurna glacier has retreated at a rate of 24.4 m/year from 1976–2016. Similarly, Annapurna III retreated at the rate of 7.54 m/year during 1950–2016 (Sigdel et al., 2020). Moreover, Sikkim glaciers retreated at the rate of 3.9–63.9 m/year during 1988–2018 (Kumar & Sharma, 2023). Glacier retreat increases meltwater in rivers, influencing river flow.

As the glaciers retreat, river hydrology goes through two phases. First, glacier retreat results in an increased contribution of meltwater to river discharge. For instance, summer flows are projected to increase by 5.6% to 19.8% as temperature rises by 0.7 to 2.6 °C from 2039 to 2070 (Latif et al., 2020). Also, mean discharge rose by 7.9×106 m^3/year in the Shule River catchment in the Tibetan plateau between 1957 and 2010, with glacier melt accounting for about 187.99×108 m^3 or 33.4% of total discharge (Xu et al., 2019). Nevertheless, this initial rise in river runoff is succeeded by a second decline in water resources as glaciers get smaller.

This tipping point, or "peak water," has already occurred in roughly half of the 56 large-scale glacierized drainage basins that have been studied globally (Huss & Hock, 2018). River runoff slowly declines after peak water, particularly in late summer. By 2100, one month of the melting season can witness more than a 10% decline in runoff in one-third of these basins, with Central Asia being most affected (Huss & Hock, 2018). Table 3.1 provides a list of India's significant glaciers impacted due to climate change.

Table 3.1 India's major glaciers are associated with river systems

Glacier	*Associated rivers*	*River basin*	*Region*	*Climate change impacts*	*Reference*
Gangotri glacier	Ganga	Ganga basin	Uttarakhand	Rapid retreat, decreased snow accumulation, and early melting	(Singh et al., 2023a, 2023b)
Yamunotri glacier	Yamuna	Yamuna basin	Uttarakhand	Shrinking mass balance, Lower discharge in pre-monsoon, and glacier fragmentation	(Kumar et al., 2020)
Zemu glacier	Teesta	Brahmaputra basin	Sikkim	Significant retreat, glacial lake formation, GLOF risk increasing	(Bajracharya & Mool, 2009; Islam & Patel, 2022)
Baltoro glacier	Shyok (Indus tributary)	Indus Basin	Karakoram Region	Shows mixed trends: stable or slow growth due to Karakoram Anomaly (localized cooling)	(Usman & Furuya, 2018)

(continued)

Table 3.1 (continued)

Glacier	*Associated rivers*	*River basin*	*Region*	*Climate change impacts*	*Reference*
Rimo glacier	Shyok (Indus tributary)	Indus basin	Eastern Karakoram, Ladakh	Glacier retreating, debris cover affecting the melt rate	(Muhammad et al., 2019)x
Milam glacier	Goriganga (Kali, tributary of Ghaghara/ Sarda)	Ganga basin	Uttarakhand	Loss in glacier volume, disrupted downstream flow, hazardous glacial lakes	(Mal et al., 2019)
Chorabari glacier	Mandakini (Alak-nanda, Ganga tributary)	Ganga basin	Uttarakhand (Kedar-nath)	Sudden lake burst caused 2013 Kedarnath floods, high GLOF risk	(Bhambri et al., 2016)
Dokriani glacier	Bhagirathi (Ganga)	Ganga basin	Uttarakhand	Monitored by Wadia Institute, retreat of ~ 18 m/year, snowline rising	(Hasnain & Thayyen, 1999)
Durung Drung glacier	Stod (Zanskarm Indus tributary)	Indus basin	Zanskar (Jammu and Kashmir)	Retreating rapidly, streamflow alterations, and alpine biodiversity at risk	(Rashid & Majeed, 2018)

The extent and timing of these hydrological changes vary across the Himalayan regions. For the western Himalayas, with limited precipitation, the impact of retreating glaciers on river flow will be stronger. On the other hand, the eastern Himalayas may experience a later peak in river discharge and reduced glacial retreat rate owing to summer snow (Rees & Collins, 2006).

3.3.5 Waning Snow Cover and Inter-Seasonal Variability of Flow

Seasonal snow contributes significantly to river flows in hilly areas. Snow falls in the winter months and then melts over the spring and summer seasons, which ensures streamflow of rivers (Jenicek et al., 2016). Snowmelt provides around 60% annual flow in the Liddar River of the western Himalaya, while glacier ice provides as low as 2% (Jeelani et al., 2012).

The increase in temperature is causing a decrease in both the proportion of snowfall and snow depth, consequently resulting in increased winter runoff and reduced late spring due to the advanced snowmelt (Blahušiaková et al., 2020). Decreased summer flow, reduced summer groundwater storage, and longer and more intense summer low flow periods are all the results of this shift toward more intense and frequent temperature-driven snow dearth (Dierauer et al., 2021). Rivers such as Ravi, Chenab, Jhelum, and Beas, which are sourced from the Western Himalayas, exhibit these effects (Sharma et al., 2012; Thind et al., 2019). The annual flow of these rivers is supplied by snow and glacier melt. For example, hydrological modelling and isotope study of the Beas River at Manali reveals that glacier melt contributes 50–52% to the overall river discharge (Ahluwalia et al., 2013). Likewise, about 49% of the Chenab River at Akhnoor (Singh et al., 1997).

3.3.6 Formation of Glacial Lakes and Glacial Lake Outburst Floods (GLOFs) Hazard

The development of moraine-dammed lakes through the sudden melting of glaciers has heightened the risk of GLOFs (Bajracharya & Mool, 2009), which can lead to disastrous impacts on downstream populations and infrastructure (Zhang et al., 2021). 329 glacial lakes greater than 0.05 km^2, 23 of which are very high risk and 50 high risk, in the Indian Himalayas (Dubey & Goyal, 2020). Due to chronic glacier melting induced by climate change, over 5000 moraine-dammed lakes are created and enlarged in high mountain ecosystems, particularly within the Hindu Kush-Karakoram-Himalaya-Nyaingentanglha (HKKHN) region and the Himalayas (Furian et al., 2021; Veh et al., 2020). In the same vein, the north-western Himalayas of Uttarakhand are facing quick glacier melting and rising incidents of GLOFs and other hazards (Islam & Patel, 2022; Singh & Pandey, 2024). The June 2013 Kedarnath disaster was a calamitous GLOF occurrence that claimed more than 6000 lives and left large-scale destruction in Uttarakhand. The event was caused by intense monsoon rainfalls, with 390 mm of rainfall reported between June 10 and 17 after unusually quick snow cover loss in the previous weeks (Allen et al., 2016).

The Teesta and Alaknanda river basins of the Himalayas are highly vulnerable to GLOFs. 203 new glacial lakes were formed in the Teesta River basin between 2000 and 2018, and 82 of them were recently formed in 2011–2018, reflecting the quick glacial retreat and lake area expansion (Islam & Patel, 2022). A list of 143 glacial lakes was prepared for the Teesta River basin, and 18 of them were listed as potentially dangerous (Aggarwal et al., 2016). In downstream societies, early warning and disaster risk reduction depend heavily on the ongoing monitoring and evaluation of hazards.

3.3.7 Impacts on Base Flow and Groundwater Recharge

Increased atmospheric warming and changed evapotranspiration mechanisms as a result of global warming, which cause measurable changes in soil water regimes, river base flow, and groundwater recharge (Bell et al., 2010; Yuan et al., 2008). One component of a river's discharge from groundwater release is base flow. It is vital in maintaining streamflow, supporting human use of water, and the aquatic ecosystem, particularly during drought (Barclay et al., 2020). It plays an important role in contributing to overall streamflow, at 36%-80% in various river systems (Fan et al., 2023).

In India, base flow has a very important role to play in the health of rivers, especially in dry and semi-dry regions of seasonal rainfall, e.g., Shivganga river in Western Maharashtra (Kadam et al., 2022). The surface and ground water systems are actually disturbed by climate change and anthropogenic, having significant effects on water resources. Climate change is also intensifying these effects. Ground water replenishment, ground water flow into streams, and hydrological cycles are being changed due to the increase in temperature and rising rate of evapotranspiration (Earman & Dettinger, 2011; Van Roosmalen et al., 2007). Ground water recharge is reduced due to the increase in intensity and variability of precipitation patterns, leading to more runoff of water than percolation (Tashie et al., 2016). Groundwater depletion is having a negative effect on the Tungabhadra and Bhima rivers (Meenu et al., 2013).

Human activities, and more specifically, groundwater exploitation, are also responsible for the lower levels of groundwater (Li et al., 2014). India has led to the alarming reduction of groundwater, impacting food security and rural livelihoods. These unsustainable groundwater overexploits are dominated by irrigation, sustaining food production for about 170 million people (Zaveri et al., 2016). Every year, India uses over 230 km^3 of groundwater, of which over 60% is used for irrigated agriculture (Chinnasamy et al., 2013). These impacts of groundwater overextraction are not restricted to regional depletion. Unsustainable pumping can lead to the inversion of the hydraulic gradient, with a resulting loss of water to aquifers instead of being recharged by groundwater (Grogan et al., 2017). Groundwater excessive use in excess of natural replenishment in most regions is of special concern, as it leads to the diminishment of river flows and impacts freshwater ecosystems (De Graaf et al., 2024).

3.4 Geomorphological and Sediment Transport Modifications

Rivers play a vital role in landscape evolution through continuous processes of river soil erosion, increased sediment load deposition, and conveyance (Legout et al., 2018). Climate change processes, such as the intensification of monsoonal rainfall, exacerbate such geomorphological processes extensively (Khan et al., 2018). Such changes are highly effective due to the high population load on river basins and diverse physiography.

3.4.1 Sediment Load and Soil Erosion Increased

Altered precipitation, like more frequent and more intense rain, leading to flash flooding, intensifying hillslope soil erosion, and riverbank collapse (Brandolini et al., 2018; Kayitesi et al., 2022). Deforestation, removal of vegetation, land clearing for agriculture, and urbanization are major causes of sediment transport and soil erosion. These processes enhance the exposure of the soil surface and reduce soil stability (Kumar et al., 2022). Sedimentation and soil erosion have significant impacts on water storage, reservoir capacity, and hydropower output (Bravo-Linares et al., 2024). Extinction, reduction, and degradation of species are caused by excess sedimentation (Tagliabue et al., 2023). Uncontrolled runoff influences aquatic organisms and the river water environment with soil nutrients, pesticides, fertilizers, and other pollutants, leading to water quality deterioration (Rashmi et al., 2022).

3.4.2 *River Morpho-Dynamics*

River flow regimes, sediment transport, and channel boundaries regulate the river channels and are referred to as river morpho-dynamics (Church & Ferguson, 2015). Climate change is altering these dynamics significantly (Praskievicz, 2015). In Indian Himalayan Rivers, morpho-dynamics are being significantly impacted by climate change. For instance, in the Alaknanda River over the past 20 years, flash floods and cloudburst disasters increased the frequency in the valley due to more intense precipitation (Joshi & Kumar, 2006). This disaster brought about a change in sediment movement, which frequently produced channel blocks, debris flow, and landslides. As also in the Kosi River, also referred to as "Sorrow of Bihar," showed excessive avulsion and channel migration due to excessive sediment loads (Slingerland & Smith, 2004).

3.5 Water Quality Issues

River water quality is seriously threatened by the effects of increasing temperature, changing precipitation, and global warming (Jiang & Zhang, 2015; Sahoo et al., 2024; Todd et al., 2012). As many chemical and biological processes take place at an increased rate at high temperatures, the rate of growth of phytoplankton, aquatic vegetation, macrophytes, and other living organisms will be influenced by increasing river water temperature (Yang et al., 2019). River water quality worsens, and the process of eutrophication becomes intensified.

The multifaceted interaction between biochemical oxygen demand (BOD), water quality, and water temperature is influenced by different factors. Increased biochemical reactions due to higher temperatures in arid areas significantly increase BOD (Chapra et al., 2021). Climate change influences different water quality parameters, i.e., dissolved organic matter (DOM), DO, pH, chemical oxygen demand (COD), water temperature, etc. (Fig. 3.2). Warming of river water decreases the efficiency of river waters to hold DO, further worsening water quality (Chapra et al., 2021).

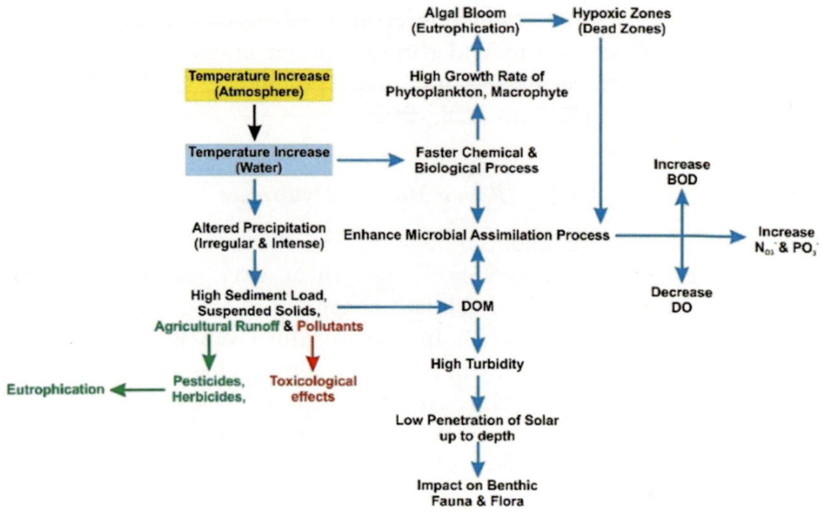

Fig. 3.2 Climate change & Water quality relation

Eutrophication and subsequent decomposition may lead to the formation of a hypoxic zone, causing devastating impacts on the aquatic ecosystem. As microorganisms use the oxygen available in water through the decomposition of algae, leading to very low DO levels (King et al., 2012; Prathumratana et al., 2008). DOM has significant effects on light penetration in aquatic ecosystems, impacting the growth of benthic flora and fauna. Increased DOM is inversely correlated with water transparency, which reduces solar penetration (Pace et al., 2012).

3.5.1 Concentration of Contaminants

Contaminants such as heavy metals, microorganisms, chemicals, and nutrients are deteriorating the river water (Botle et al., 2024). Higher concentrations of contaminants in rivers are present during droughts because of low flow and low water volumes. Climate change exacerbated this with more intense and frequent droughts in most regions (Chattopadhyay et al., 2017; Whitehead et al., 2009). With increased rainfall variability and longer lean flow durations, rivers experience lower discharge, reducing their natural ability to dilute contaminants. This reduced ability to dilute leads to the accumulation and persistence of contaminants in the water column from agricultural runoff, domestic sewage, industrial effluent, and urban drainage at higher concentrations (Qin et al., 2020). These contaminants increase the risk of toxic algal blooms and have negative effects on the aquatic ecosystem (Whitehead et al., 2009).

Excess pollution due to the dumping of untreated and poorly treated water in the Yamuna River is a major issue, especially in Delhi. During dry seasons, these pollutants and toxic substances pile up due to low river discharge (Harada & Karn, 2001; Misra, 2010). In studies, this is indicated by increased BOD and COD (Joshi et al., 2022; Sharma & Kansal, 2011). Jaiswal et al. (2019) indicated that the upper Yamuna zone has "excellent-to-good" water quality, but the middle zone, especially influenced by Delhi's large drains, has "poor-to-marginal" quality. Surprisingly, the quality of the river's water is improved in the downstream zone of Delhi because of the dilution and depuration effects of tributaries such as the Betwa and Chambal rivers (Jaiswal et al., 2019).

The impacts of such a quantity of pollutants are numerous. The pollutants are physically detrimental, oxygen-depleting, and chemically toxic to induce oxidative stress, DNA damage, and other toxic effects (Benson et al., 2022). Not only does it destroy the ecological health of the river, but it also affects its quality for drinking, bathing, and irrigation. For human communities, the concerns are increased water treatment expenses, increased incidence of waterborne illnesses, and increased water-based livelihoods like fishing and agriculture.

3.6 River Temperature and Ecosystem Effects

Climate change has a strongly influential impact on riverine diversity on all spatial scales, ranging from species to biomes. With increasing atmospheric temperature, river water temperature increases, with huge implications for freshwater communities (Bellard et al., 2012; Trimmel et al., 2018).

3.6.1 Rise in Temperature of River Water and Aquatic Life

Global warming has been reported to affect the relationship between river water temperature and the concentration of DO in rivers (Danladi

Bello et al., 2017). Increasing trends in river water temperatures, especially during low flow, are a cause of concern in rivers like the Ganga, Mahanadi, and Yamuna. These, along with increasing pollution and urbanization, are exerting tremendous pressure on water quality and ecosystem integrity (Bowes et al., 2020). Reduced DO concentration due to a rise in water temperature has become one of the emerging concerns. This can be extremely harmful to fish and invertebrates, particularly to sensitive species. Fish and invertebrates can be affected by decreased growth, stress, impaired reproduction, and even death when exposed to hypoxia (Saari et al., 2018). For example, salmonids and other cold-water fishes need greater DO levels than warm-water fishes, and thus, they are more vulnerable to DO depletion (Chapra et al., 2021). Low DO levels have species-specific effects (male or female, adult or juvenile, etc.) and also on factors like exposure, duration, tolerance physiology, and temperature. For example, in the case of fish species, *Reinhardtius hippoglossoides* (Greenland halibut) juveniles were less tolerant of hypoxia compared to adults, and with larger critical oxygen thresholds (Dupont-Prinet et al., 2013). Similarly, the adult females of amphipod *Gammarus pseudolimnaeus*, were more sensitive to anoxia and hypoxia than males and juveniles (Hoback & Barnhart, 1996).

Ravindra et al. (2003) report that during summer season, DO values were very low in the Yamuna River in Delhi. Low DO values are toxic to aquatic organisms and can cause the death of fish (Ravindra et al., 2003). Moreover, Lokhande and Tare (2021) report that the DO and ammoniacal nitrogen values of Delhi to Agra are not appropriate for indigenous aquatic life due to unfavorable conditions (Lokhande & Tare, 2021). *Platanista gangetica* (Ganges river dolphin) is also affected by climate change as well as human activities. They are highly dependent upon certain habitat characteristics, including water temperature and depth. They prefer cold, deep pools of water and are sensitive to the fluctuation of DO levels, which are crucial for their echolocation and survival (Choudhary et al., 2012).

Endangered species, including *Nilssonia gangetica* (Ganges softshell turtle), *Tor putitora* (Himalayan Mahaseer), *Gavialis gangeticus* (Gharial) (Critically endangered), etc., are already under severe threat due to climate change. With each passing year, the chances of extinction of these species are increasing. Temperature is a key factor in fish reproduction through various mechanisms. It acts differently on spring- and autumn-spawning fishes. Decreasing temperatures induce reproduction in autumn-spawning fishes, whereas rising temperatures induce reproductive development in spring-spawning fishes (Pankhurst & Munday, 2011). In a few instances, rising temperatures can abbreviate spring spawning and delay autumn spawning, and the effect could be complete reproductive inhibition (Pankhurst & Munday, 2011). The migratory fish species *Tor putitora*, migrates seasonally in pre-monsoon and monsoon months for breeding (Bhatt & Pandit, 2016). Rising and more variable conditions can result in reproductive failure or disruption of migration timing (Joshi et al., 2018). Temperature rise and CO2 rise strongly suppressed native species, whereas non-native species (thermally tolerant) exhibited more tolerant responses to environmental change and change in community composition (Sorte et al., 2013). Rivers, which support a rich assemblage of endangered and endemic species, are most susceptible. These ecological alterations have ripple effects, affecting not only species but also human societies that are dependent on rivers for food, economic livelihood, and cultural values.

3.6.2 Wetland Ecosystems Under Stress

Riverine wetlands are source ecosystems that preserve biodiversity and support the hydrologic cycle. Over 6% of the Earth's surface and most of the global biodiversity is covered by wetlands (Junk et al., 2013). Climate change, river dynamic oscillations, and human activities are placing tremendous pressure on riverine wetlands by altering water temperature, sea levels, and hydrology (Hudon et al., 2018; Sileshi et al., 2020). These alterations can be detrimental to the delicate balance of wetland ecosystems, undermining their capacity to fulfil essential functions, including flood control, water filtration, and supporting many species (Junk et al., 2013).

The internationally recognized Ramsar Convention wetland, Harike Wetland, is situated at the meeting point of the Beas and Sutlej rivers in Punjab (Chopra et al., 2001). As an international migratory bird habitat, it is exposed to a range of environmental stresses, including desiccation, seasonally caused eutrophication, and heavy metal contamination. These, as well as changed upstream flow and sedimentation, are all contributing significantly to its ecology and structure.

3.7 Socioeconomic Impacts

Climate-induced change in river systems has extensive socioeconomic consequences, particularly in India, whose rivers provide the backbone for rural, urban, and indigenous livelihoods. River flow regimes deteriorate, water quality deteriorates, and variability in the quantity of water available are seriously affecting India's food security, economic growth, and social cohesion. India's national river and crores of people's lifeline, the Ganga River, is faced with unprecedented pressure from mounting population growth, climate change, and water pollution (Matta et al., 2020, 2022).

3.7.1 *Livelihood and Agriculture*

Indian agriculture is majorly dependent upon freshwater resources and rice and wheat cultivation accounts for more than 80% of agricultural water utilization in India (Singh et al., 2023a, 2023b). Indian agriculture is therefore very sensitive to shifts in water availability and river flows due to such extensive water dependence. Climate change, altering food patterns, and increasing non-agricultural demands further stress water resources, constituting a massive challenge for irrigated agriculture (Rosegrant et al., 2009).

Rain deficit has also emerged as the main cause of crop loss in dry areas such as Marathwada, Maharashtra, due to rising temperatures. Agrarian crisis is fueled by the gradual transition from water-saving crops such as pearl millet and sorghum to water-unsustainable crops such as sugarcane (Zachariah et al., 2020). Yield decline, indebtedness, and suicides among farmers have emerged as major issues in monsoon-based agricultural areas, in particular for marginal and small-scale farmers without irrigation (Pande & Savenije, 2016; Urfels et al., 2020).

3.7.2 *Urban Water Supply*

Urban areas are actually facing increasing water deficits, particularly those that rely on the surface water of rivers. This is due to the fact that the rapid rate of urbanization, population, and economic development put pressure on water resources (Van Den Brandeler et al., 2019). Climate change will also increase these, and even decrease the supply of water in some areas (Padowski & Gorelick, 2014).

Chennai and Bengaluru, located near the river basins, are not immune to water crises due to a myriad of reasons either. Chennai is experiencing increasing supply–demand deficits for water, with unsatisfied demand increasing in normal and deficit rainfall years (RaziSadath et al., 2023). Southern Indian Cauvery and Krishna river basins are faced with serious challenges due to climate change and increasing water demand, which are exacerbating inter-state water sharing conflicts further (Bhave et al., 2018; Raje et al., 2014). In Hyderabad, the Musi River, which is one of the main providers of freshwater, has now become highly polluted and hydrologically degraded (Sonkar et al., 2024).

3.7.3 River-Dependent Indigenous Communities

Riverine and indigenous communities residing in river basins like the Brahmaputra, Narmada, and Chambal are quite vulnerable to climate change in river systems. Fishing communities are highly reliant on rivers for livelihood and cultural purposes, yet are exposed to increasing pressures from environmental change and anthropogenic activities. Rising water temperature, habitat degradation, and pollution deplete fish stocks, and these directly affect food security and the livelihood of these communities (Arantes et al., 2022; Sarkar & Islam, 2022). Riverbank erosion displacement has a number of adverse effects on the displaced community. Displaced people suffer social, economic, and cultural stigma in the host communities (Barua et al., 2019).

The socioeconomic base and livelihoods of the Mishing people of Majuli, the largest river island in the Brahmaputra basin (Das, 2016), Bangladesh's Char land people, living on riverine islands, have been negatively impacted by regular floods, erosion, siltation, and other climate disasters, making them extremely climate-vulnerable (Al Mamun et al., 2023; Sarker et al., 2019). The Narmada basin, one of the highly regulated catchments in central India, is home to over 16 million people and is negatively impacted by disrupted flow regimes that interfere with tribal communities' ritual water practices and the agricultural cycle (Rickards et al., 2020).

3.8 Indian River Case Studies

India's extensive river basins present strong examples of climate change impacts on ecology, hydrology, and governance. From Brahmaputra's monsoon-fed basins to glaciers of the Ganga headwaters, and inter-state conflicts around the Cauvery (Fig. 3.3), each basins present challenging and worthwhile lessons in climate change adaptation.

3.8.1 The Ganga Basin

The Ganga River has its source at about 4100 meters above mean sea level in India's Uttarakhand state in the Uttarkashi district from the Gaumukh of the Gangotri Glacier (Fig. 3.4) (Jain & Singh, 2020). It is the second-largest basin in India and the fourth-largest in the world, and the most economically valuable and populous. It is a member of the Ganga-Bramhaputra-Meghna, draining into Bangladesh (4%), China (4%), Nepal

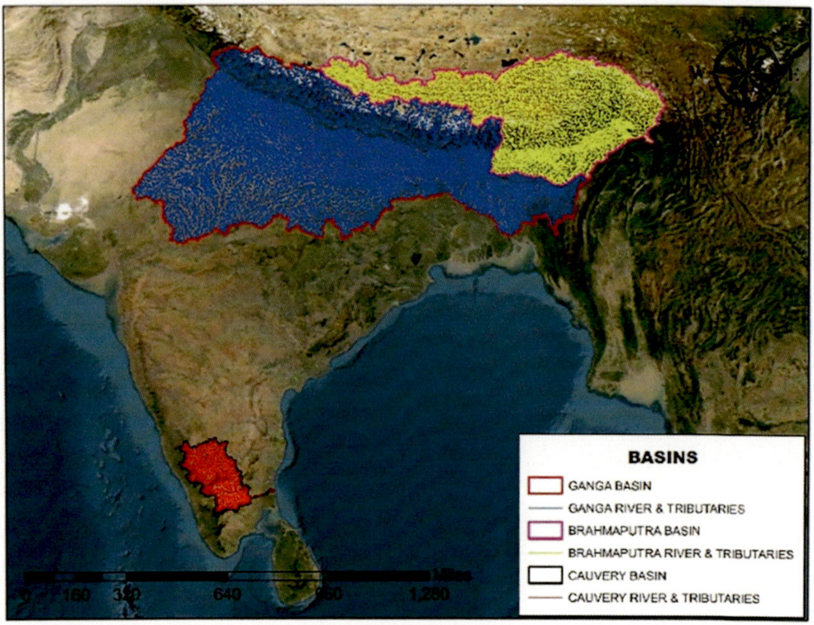

Fig. 3.3 Map of India's major river basins (Brahmaputra, Ganga, and Cauvery)

(13%), and India (79%) (Nepal & Shrestha, 2015). More than 520 million people in 11 Indian states, including Uttarakhand, Himachal Pradesh, Rajasthan, Uttar Pradesh, Bihar, Madhya Pradesh, Jharkhand, Delhi, Haryana, and West Bengal, depend on the Ganga River. It accounts for 40% of the gross domestic product of India, and thus its importance in the economy of the country cannot be overemphasized (Dangar & Mishra, 2021).

Climate change is a major threat to the headwaters of Himalayan glaciers supplying the river are melting fast due to warming and black carbon accumulation (Li et al., 2016; Sigdel et al., 2020). It has also caused concern regarding seasonal flow variability, with the possible short-term rise in discharge but long-term decline in available water.

According to the projections of regional climate models, the average annual temperature of the Ganga basin will increase by 1-4°C between the years 2010 and 2050 (Moors et al., 2011). The winter rainfall may be converted from snowfall to rain due to this warming (Moors et al.,

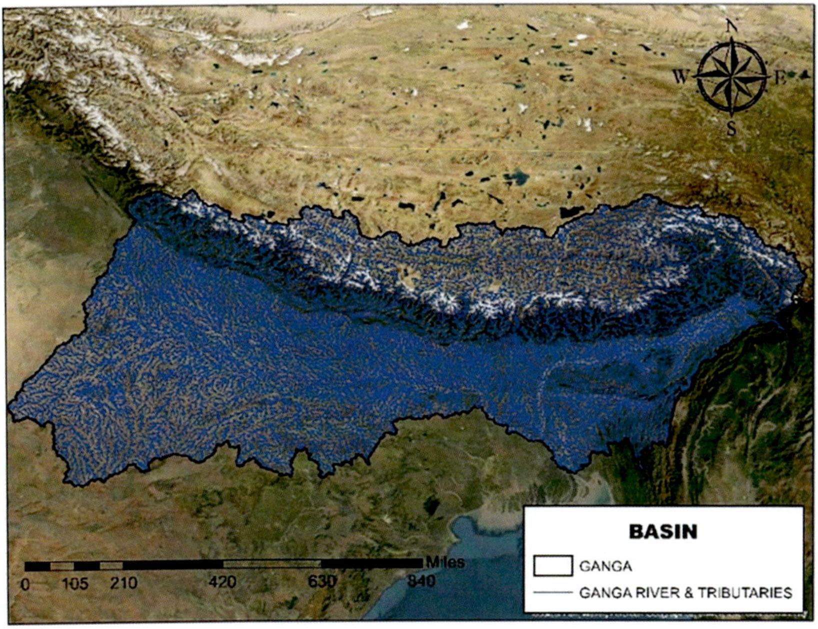

Fig. 3.4 Map of the Ganga river basin

2011). This, in turn, can change the river flow (Kundzewicz et al., 2008). The river temperature, heat budget, and hydrologic cycle also get influenced by other parameters such as heat flow between the river and the surrounding air, river bank, and land; evaporation and condensation processes; resistance to heat flow provided by the river banks to the river bed; and the sun's insolation absorbed and reflected, etc. (Webb et al., 2008).

The abrupt melting of glaciers in mountains due to an increase in temperature causes variability in the discharge regimes of rivers (Bliss et al., 2014; Milner et al., 2009), leading to floods and other catastrophes (Lutz et al., 2014). It has been found that the mean rate of retreat of the Gangotri glacier is 19 m/year. It has retreated over 850 m in the latter 25 years of the twentieth century (Jain & Singh, 2020). This will have serious implications for the flow dynamics of the Ganga (Bolch et al., 2012; Immerzeel et al., 2012). It has been found from research studies that there is a decreasing trend in glacial mass balance between 2017 and 2021, and there have been recurring GLOF events (Mondal et al., 2023).

Climate warming is anticipated to decrease Himalayan glaciers and affect downstream river flows. Early enhancement of river flow is conceivable through enhanced melting, but long-term loss of the glaciers would lead to less flow, particularly in small and heavily glacierized basins (Rees & Collins, 2006). This is most concerning in the case of the Ganga River, where glacial ice-melt contribution toward stream discharge is up to 40% for summer and monsoon months (Maurya et al., 2011). Plains and rivers have been discovered to have higher minimum water temperatures, affecting the spawning habits of fish and the distribution of species (Vass et al., 2009).

Adaptation and restoration—The Namami Gange Programme is an integrated approach to respond to the multi-faceted problems confronting the Ganges River ecosystem. The multi-dimensional program is applicable in the context of the range of threats to the Ganges, such as pollution caused by sewage and industrial effluent, disposal of dead bodies, forest clearance, over-use of fertilizers and pesticides (Rai et al., 2024; Rao, 2001). Namami Gange Programme's emphasis on promoting ecological flow is applicable since research has demonstrated that the preservation of ecological flows has the potential to reduce the impact of noise on threatened species such as the *Platanista gangetica* (Dey et al., 2019). In addition, as the projected increases in temperature and precipitation have the potential to influence levels of contamination and flood levels, the program must consider the potential impacts of climate change on river hydrology and water quality (Whitehead et al., 2015).

3.8.2 *The Brahmaputra: Climate Refugees, Flooding, and Erosion*

The Brahmaputra River originates at an altitude of approximately 5150 meters in south-west Tibet and flows for over 2900 kilometers in Tibet (China), India, and Bangladesh (Fig. 3.5) (Sarma et al., 2015). The river carries an extremely high sediment load and silt in its unsmooth flow from the Tibet high plateau to the Assam valley, and thus it is highly unstable in the valley. This slit and sediment load in river caused by deforestation caused by shifting cultivation which increases soil erosion. The river has over 100 tributaries, of which 15 large tributaries originate in the north due to the melting of Eastern Himalayan snow (Khan et al., 2016; Sarma et al., 2015).

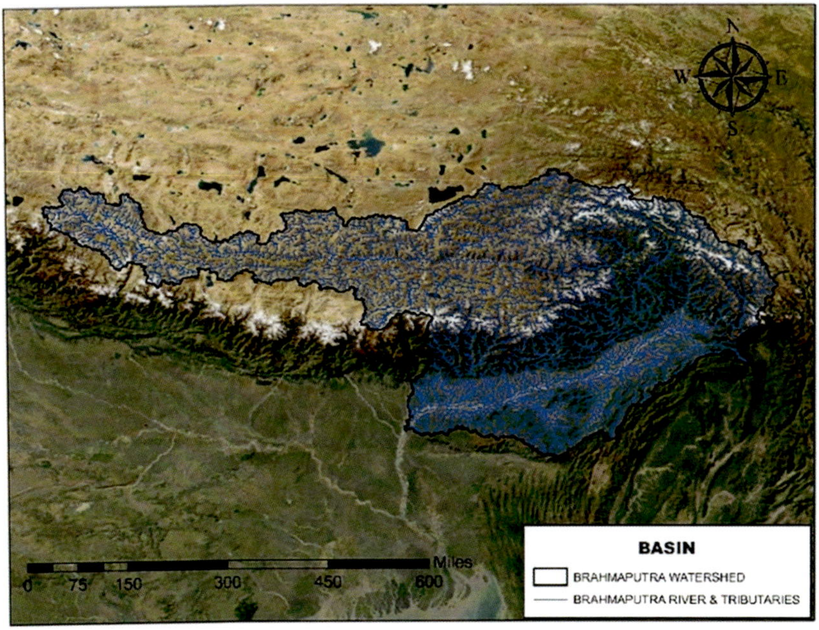

Fig. 3.5 Map of the Brahmaputra River basin

The climatic and geographical conditions make the Brahmaputra River extremely prone to landslides, devastating floods, and other calamities. The region witnesses extremely heavy monsoon rainfall, with over 80% of the total annual rainfall (Panthi et al., 2015). The Brahmaputra basin experiences landslides and flash floods due to seismic activity, human-induced activities, and high rains.

Climate change will increase the intensity and frequency of extreme weather events, which will enhance these risks (Oo et al., 2023). During the summer monsoon season (June to September), Assam is suffering severe flooding, causing widespread damage and relocation (Dhar & Nandargi, 2000). River width has also increased over time, from a mean of 5949 m during 1912–1928 to 9012 m during 2009, in addition to the gigantic sediment load because of the 1950 earthquake in Assam and large-scale deforestation in the Himalayas (Sarma & Acharjee, 2018).The blocking of river courses by landslides and subsequent flooding elsewhere is a serious issue because it can cause enormous damage to infrastructure, agriculture, and human settlements. To reduce such hazards to the least possible level, full flood and landslide monitoring systems have to be installed, land-use planning improved, and community resilience models constructed for hazardous areas (Wadhawan et al., 2020).

Adaptation and restoration—The Brahmaputra River basin is seriously under threat from environmental stresses like floods, riverbank erosion, and climate change impacts, and influences livelihood and resilience of the people (Ayeb-Karlsson et al., 2016; Pradhan et al., 2021). The Brahmaputra basin requires effective adaptation and restoration with joint management by basin countries, with high priority to equitable benefit sharing, disaster risk management, and resilience building (Pradhan et al., 2021). Adaptation action in the basin includes changing agricultural practices, adopting alternative livelihoods, and migration as adaptation measures (Ayeb-Karlsson et al., 2016).

Natural Flood Management (NFM) practices, including woodland cover restoration and wetland restoration, can be used for peak flow mitigation and provision of co-benefits in ecosystems. These measures do come with trade-offs, like impacts on food production, that need to be addressed on a case-by-case basis (Iacob et al., 2014).

Development of eco-tourism and indigenous knowledge application in flood management is proposed as a way to enhance social resilience for the Mishing community (Das, 2016). Policy and adaptation planning must learn from individuals already subject to environmental stress and shock in the future. Because of the underestimation of flood hazard risk in recent observations (Rao et al., 2020). There is a need to develop robust baseline estimates of natural climate variability to inform adaptation measures.

3.8.3 Cauvery River: Transboundary Water Management Issue Under Climate Change

The Cauvery River originates in the Brahmagiri range in the Kodagu region of Karnataka. The river empties into the Bay of Bengal after traversing more than 800 km through Tamil Nadu and Karnataka (Fig. 3.6) (Singh & Rajamani, 2001). It has come to be an ecology, culture, and economy of the region of South India. Being a vital river system in South India, it subsidizes agriculture, drinking water, and ecosystems at various stages of its course (Solaraj et al., 2010). The climate change-induced transboundary river impacts commonly include changes in the periodic flow regime, an increase in variability, and a potential decrease in total water supply.

Agriculture, water, livelihood, and infrastructure are facing climate change challenges in the Cauvery basin. The shifting rainfall pattern and temperature increase are making drought and water scarcity worse and more pronounced in the basin (Halder et al., 2024). The dispute between southern Indian Karnataka (formerly Mysore) and Tamil Nadu over the waters of the Cauvery is another instance of enhanced sophistication under the pressure of climate change in transboundary water management. The dispute has a very rich and complicated history, dating back to the British colonial period. The Madras Presidency and the Mysore State signed an agreement in 1924 to share waters, which was supposed to reduce tensions but instead created the framework for more disputes (Gately, 1974). Post-Independence till today, the problem is nurtured by the dam projects put in place by the two states that originated the conflict, giving Tamil Nadu an occasion to approach the Central government for relief. Consequently, the Cauvery Water Disputes Tribunal (CWDT) came into being to resolve the issue. Interim orders about water sharing between the states were passed in 1998 by the Cauvery River Authority (CRA). Then followed the 2013 award of the CWDT, which allotted specific quantities of water to the states (Bhave et al., 2018). The award necessitated the discharge of 177.25 TMC of water by Karnataka to Tamil Nadu in a normal year. However, monsoon years are troublesome because

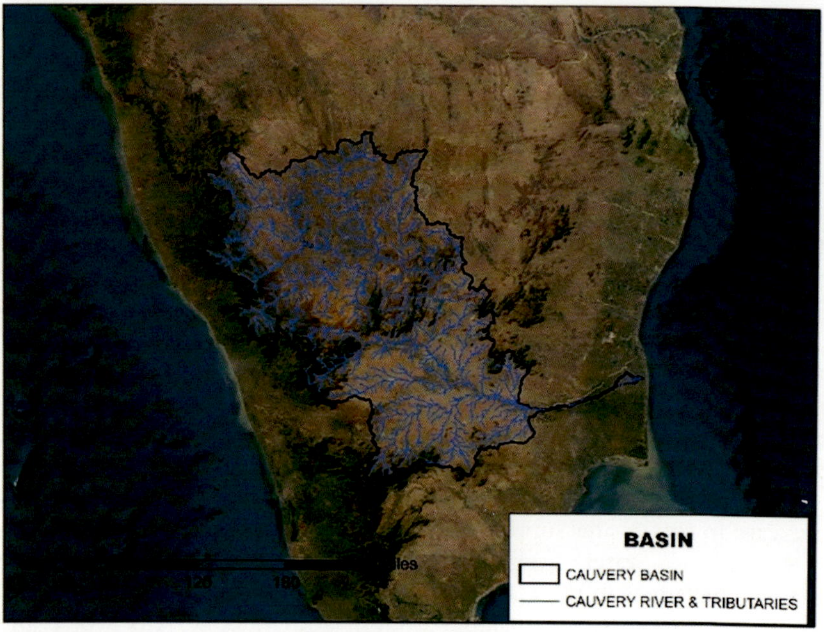

Fig. 3.6 Map of the Cauvery River basin

of uneven rainfall, which affects the quantity of available water (Anand et al., 2020).

The controversy brings to light the intricate equilibrium between demand and availability of water in the Cauvery basin. The basin is extremely susceptible to future uncertainties and changes, most importantly, of climate and socioeconomic factors. This vulnerability emphasizes the strong decision-making strategies under uncertainty (DMUU) to effectively manage water resources (Bhave et al., 2018). In 2018, the Supreme Court designated Cauvery as a national asset and applied the equality principle to the distribution of river water. Therefore, this judgment, along with Article 262 of the Constitution that empowers parliament to resolve inter-state river disputes and the Seventh Schedule in which water resources have been placed for legislative purposes, is the pool of legislation under which the Cauvery issue is being resolved (Bhave et al., 2018). These incidents suggested the need for adaptive management options with the involvement of stakeholders for designing water allocation policies in the Cauvery basin. The dispute presents the issues of the management of water resources and certain inter-state coordination in India.

3.9 Adaptation Strategies

River basin climate change adaptation policies need to be integrated and context-based because of differences in ecological and socioeconomic contexts. Adaptation mechanisms should work in India with scientific anticipation, indigenous knowledge, institutional transformation, and civic-led governance (Saikanth et al., 2023). Adaptation policies must address the problems with regard to local contexts.

3.9.1 Climate-Resilient River Basin Planning

India's conventional water management has indeed been sectoral and reactive, yet more and more now there is a responsive awareness for more integrated, basin-level policies with regard to climate change impact. Climate change is reorganizing hydrological regimes in India with predictions of 1.5 to 4.3 °C warming and increased precipitation of 6 to 14% in the 2080s relative to 1961 to 1990, affecting rivers and increasing the risk of floods, droughts, and other climatic hazards (Sikka et al., 2018; Wu et al., 2012).

Since it views the entire river as a single, cohesive entity, Integrated River Basin Management (IRBM) provides a framework for coordinated management of water, land, and ecosystems. The approach is especially relevant to climate change adaptation and sustainable water management (Benson & Lorenzoni, 2017). IRBM focuses on the importance of comprehending upstream–downstream interactions in hydrological processes for efficient water resource planning. Figure 3.7 depicts the key elements of IRBM. These intricate interactions are conditioned by land-use modification, erosion, snow storage, and global climate change effects on flood regime, water availability, and sedimentation (Nepal et al., 2014). The approach also takes into consideration balancing various requirements of water demand in the basin, including domestic consumption, irrigation, environmental requirements, and hydropower (Arfanuzzaman & Abu Syed, 2018).

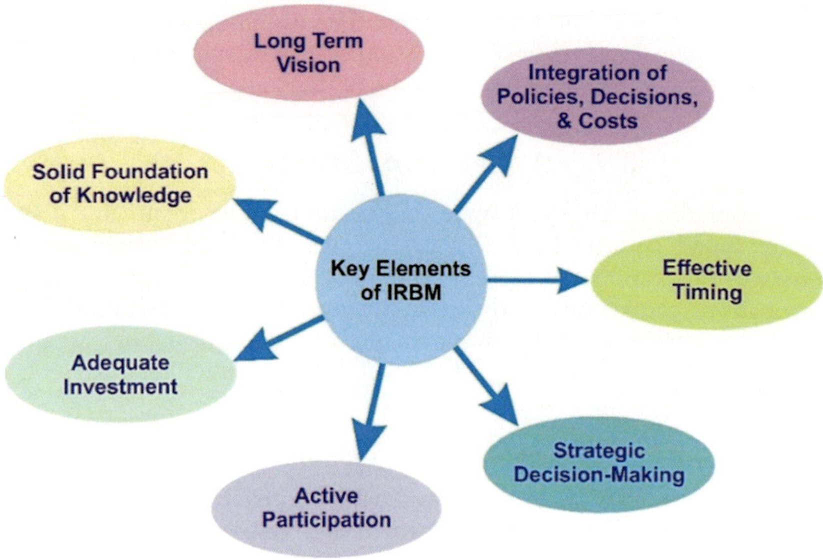

Fig. 3.7 IRBM key elements

The Ken-Betwa River link project (KBLP) is enabled by a River Basin Organization (RBO), a flood control and water shortage program under India's overall interlinking rivers plan. This river connecting project intends to irrigate the drought-prone Bundelkhand region by transferring excess water from the Ken River in Madhya Pradesh (MP) to the Betwa in Uttar Pradesh (UP) (Patel & Anuragi, 2024). The interlinking plan has been sanctioned by the center as India's first interlinking plan, and presents scope and challenge for water management and nature conservation (Lakra et al., 2011). However, while the plan aims at flood regulation and water shortage, it is a concern for the potential impact on aquatic life and biodiversity.

Digital technologies like river basin simulation models, remote sensing, and Geographical Information System (GIS)-based vulnerability mapping are actually becoming essential for climate-resilient decision-making and design of water allocation compacts among states. Remote sensing and GIS have proven useful in mapping the potential areas of groundwater in hard rock basaltic basins (Pradhan, 2009). The technologies integrate thematic layers like topography, geology, and hydrogeomorphology for estimating groundwater potential indices. Similarly, spatial modelling through the Analytic Hierarchy Process (AHP) and GIS has been applied for mapping waterlogged hazard and risk areas in river basins (Sar et al., 2015). The technique integrates topographical information, satellite information, and socioeconomic information for identifying risky zones.

While satellite photos are immediately available in digital form, the process of gathering and storing data in a GIS starts with the digitization of point data and thematic maps (Chang, 2017). Results from the analysis of satellite images can be utilized in GIS, and interpolation is frequently needed to fill in the gaps in point data (Pradhan, 2009). GIS software is used to combine all data, regardless of its source, and align it with a single reference grid. Final outputs like maps and statistical data can be produced by reprocessing intermediate results (Fig. 3.8) (Chang, 2017).

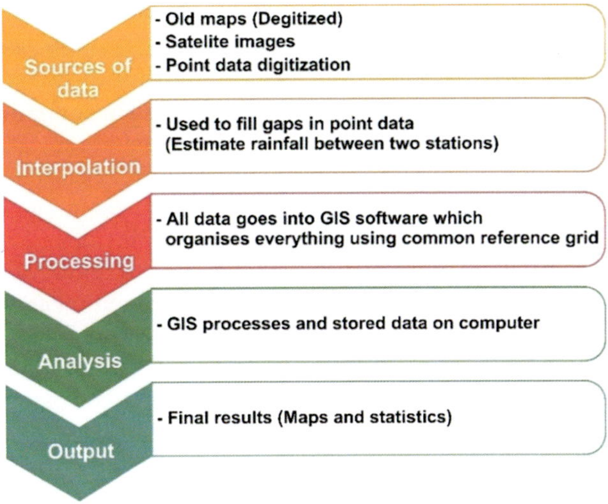

Fig. 3.8 Data acquisition procedure and storing in a GIS

3.9.2 *Nature and Traditional Solutions*

For renewal of the river system in order to restore the ecosystem and tackle climate resilience, we can approach nature-based Solutions (NbS), which are floodplain reconnection, wetland restoration, and riparian buffer restoration. Amelioration, conservation, green infrastructure, and appropriate practices are the different interventions that can be utilized by NbS in the river system (Simelton et al., 2021). Goals such as improving water quality, conserving biodiversity, and flood regulation can be achieved by the intermediaries collectively (Alejo et al., 2025; Turner et al., 2022). Tank systems, stepwells, sacred forests, and floodplain agriculture are Traditional ecological knowledge (TEK) and community conservation practices that serve to conserve India's rivers and associated ecosystems. For water management and climate adaptations in a sustainable manner, these generation-old ways are extremely crucial and beneficial.

A range of TEK practices has evolved among the North-eastern Indian community. These include biodiversity conservation, sacred forest, the cultural belief system for supporting watershed management, traditional agriculture, and irrigation practice (Das et al., 2023). Such as woodland and sacred grove conservation, along with plant diversity and other ecological utilization, is practiced by the Byans Valley community (Negi et al., 2018). Proper implementations and local conditions are the main reasons behind the success of afforestation, which is beneficial for river catchment hydrology. The careful implementation, planning, and monitoring are very crucial to be effective in the Namami Gange Programme. This program stabilizes the riverbank, infiltration, and flood risk reduction through afforestation. This clearly indicates the significance of cultural beliefs for environmental conservation. Surprisingly, these kinds of practices are also observed all over the world. For instance, the acequia irrigation system of the southwestern United States assists in crop irrigation, riparian vegetation, groundwater recharge, and a healthy floodplain agroecosystem (Fernald et al., 2007). This clearly indicates that traditional knowledge is very significant in ecological conservation besides its traditional functions. Modern ecological sciences need to come together with TEK for cost-effective and sustainable solutions for climate change.

3.9.3 *Community Co-Governance and Participation*

The measures taken by the Indian government, i.e., the Atal Bhujal Yojana, to address the issue of the groundwater in regions such as Solapur (Mustaq Shaikh & Farjana Birajdar, 2024) and Namami Ganga, to clean the Ganga (Matta et al., 2020). The Jal Shakti Abhiyan announcement in 2019 encourages water conservation by the people, particularly in water-scarce districts (Table 3.2). Desilting of tanks, rainwater harvesting, and planting of trees are a few of the activities that are periodically carried out by the women and youth groups of the area. It is necessary for the optimum utilization of the people with the help of the government for the success of the scheme. The main objective of this Yojana is to bring all the existing schemes for the utilization of sustainable groundwater with the active participation of the local people and stakeholders.

Table 3.2 Important policies and river protection programs in India

Policy/ Program	*Target river(s) / Basin(s)*	*Climate-related Objectives*	*Implementing agency*	*References*
Namami Ganges Mission	Ganga River and tributaries	restore wetlands, improve river flow and resilience, afforestation, and reduce pollution	Ministry of Jal Shakti (NMCG) (2014)	(Bowes et al., 2020)
National River Conservation Plan (NRCP)	Ganga, Yamuna, Krishna, Godavari, etc	Reduce untreated sewage, improve water quality, enhance health of ecosystem	Ministry of Environment, Forest and Climate Change (MoEFCC) (1995) (revised)	(Bowes et al., 2020)
Ganga River Basin Management Plan	Ganga basin	Long-term basin-wide strategy, climate resilience, sustainable flow management	IIT Consortium (led by IIT Kanpur) under Ministry of Environment and Forests (MoEF) (2015)	(Dangar & Mishra, 2021)

(continued)

Table 3.2 (continued)

Policy/ Program	*Target river(s) / Basin(s)*	*Climate-related Objectives*	*Implementing agency*	*References*
Yamuna Action Plan (Phases I–III)	Yamuna river	Urban wastewater control, ecological rejuvenation	MoEFCC with JICA Support (1993)	(Sharma & Kansal, 2011)
Integrated Ganga Conservation Mission	Ganga basin	Flow regulation, addressing climate - induced flow variability	Ministry of Jal Shakti (2015)	(Fulazzaky, 2014; Vass et al., 2009)
Indus Water Treaty (with adaptive reviews)	Indus river system	Flow assurance amid glacial retreat, transboundary water sharing,	India-Pakistan Indus Water Commission (1960) (adaptive reviews ongoing)	(Haines, 2017)
Integrated Coastal Zone Management (ICZM)	Mahanadi, Godavari, Krishna deltas	Protect delta ecosystems, manage climate-induced salinity, erosion, cyclones,	MoEFCC with World Bank funding (2007)	(Pramanik et al., 2016)
Atal Bhujal Yojana	Groundwater-dependent river basins	Climate-resilient aquifer management, recharge support to stressed rivers,	Ministry of Jal Shakti, World Bank (2020)	(Mustaq Shaikh & Farjana Birajdar, 2024)
National Action Plan on Climate Change (NAPCC)—Water Mission	All major river basins	Enhance recharge and storage, climate-resilient planning, promote water use efficiency,	NITI Aayog, MoJS (2008)	(Pandve, 2009)
CWC's Flood Forecasting System	Brahmaputra, Ganga, Mahanadi, etc	Protect riverine infrastructure and life forecast climate-induced floods,	Central Water Commission (CWC) (Modernized in 2018)	(Ganguli et al., 2020; Pandey et al., 2022)
State Riverfront Development Projects	Sabarmati, Gomti, Mithi, others	Climate-responsive urban planning, river rejuvenation through green infrastructure	State Urban Departments with Central Assistance Varies (2000s–2020s)	(Khan et al., 2021; Simons et al., 2023)

3.10 Conclusion

Climate change transformed India's river systems with lasting effects on water security, livelihood, and ecology. Rising temperature, disturbed precipitation patterns, and snow/glacier melting are inducing droughts, poor water quality, and enhanced flooding, which are influencing hydrological regimes. Due to change, an increased pressure from pollution, exploitation of water resources, and urbanization is generated. The effect is felt on human societies, aquatic richness, and river form, which are reliant on such water systems. Rising erosion, sediment load, and river channel dynamics are remolding the landscape and promoting flood hazards. Due to altered flow regimes, breeding cycle and migration cycle are disturbed, as well as due to warming water and low DO, aquatic animals are threatened. Deaths, displacements, and cultural disruption are felt by riverine and indigenous societies. For such, a holistic and responsive approach toward river basin management is required. Science knowledge combined with TEK provides NbS for climate-resilient strategies and emphasizes community participation. We can rely on the IRBM, which applies the newest modelling and remote sensing technology, to enhance flood control and water management. Increased research, monitoring, and capacity building can assist India in lessening the impacts of climate change on its rivers. Climate-resilient infrastructure, institutional capacity, and water use efficiency require investment to manage transboundary waters. At last, we can preserve India's rivers from climate change by applying creativity, determination, and cooperation.

References

Aggarwal, A., Jain, S. K., Lohani, A. K., & Jain, N. (2016). Glacial lake outburst flood risk assessment using combined approaches of remote sensing, GIS and dam break modelling. *Geomatics, Natural Hazards and Risk, 7*, 18–36. https://doi.org/10.1080/19475705.2013.862573

Ahluwalia, R. S., Rai, S. P., Jain, S. K., Kumar, B., & Dobhal, D. P. (2013). Assessment of snowmelt runoff modelling and isotope analysis: A case study from the western Himalaya, India. *Annals of Glaciology, 54*, 299–304. https://doi.org/10.3189/2013AoG62A133

Al Mamun, A., Islam, A. RMd. T., Alam, G. M. M., Sarker, M. N. I., Erdiaw-Kwasie, M. O., Bhandari, H., & Mallick, J. (2023). Livelihood vulnerability of char land communities to climate change and natural hazards in Bangladesh:

An application of livelihood vulnerability index. *Natural Hazards, 115*, 1411–1437. https://doi.org/10.1007/s11069-022-05599-y

Alejo, C., Luers, A., Ventimiglia, A., & Matthews, H. D. (2025). *Maximizing nature-based Solutions using Artificial Intelligence to align global biodiversity, climate, and water targets*. https://doi.org/10.1101/2025.01.06.631540

Ali, B., & Anushka, Mishra, A. (2022). Effects of dissolved oxygen concentration on freshwater fish: A review. *International Journal of Fisheries and Aquatic Studies, 10*, 113–127. https://doi.org/10.22271/fish.2022.v10.i4b.2693

Allen, S. K., Rastner, P., Arora, M., Huggel, C., & Stoffel, M. (2016). Lake outburst and debris flow disaster at Kedarnath, June 2013: Hydrometeorological triggering and topographic predisposition. *Landslides, 13*, 1479–1491. https://doi.org/10.1007/s10346-015-0584-3

Anand, B., Karunanidhi, D., Subramani, T., Srinivasamoorthy, K., & Suresh, M. (2020). Long-term trend detection and spatiotemporal analysis of groundwater levels using GIS techniques in Lower Bhavani River basin, Tamil Nadu, India. *Environment, Development and Sustainability, 22*, 2779–2800. https://doi.org/10.1007/s10668-019-00318-3

Arantes, C. C., Laufer, J., Pinto, M. D. D. S., Moran, E. F., Lopez, M. C., Dutka-Gianelli, J., Pinto, D. M., Chaudhari, S., Pokhrel, Y., & Doria, C. R. C. (2022). Functional responses of fisheries to hydropower dams in the Amazonian Floodplain of the Madeira River. *Journal of Applied Ecology, 59*, 680–692. https://doi.org/10.1111/1365-2664.14082

Arfanuzzaman, M., & Abu Syed, M. (2018). Water demand and ecosystem nexus in the transboundary river basin: A zero-sum game. *Environment, Development and Sustainability, 20*, 963–974. https://doi.org/10.1007/s10668-017-9915-y

Ashmore, P., & Church, M. (2001). *The impact of climate change on rivers and river processes in Canada (No. 555)*. https://doi.org/10.4095/211891

Ayeb-Karlsson, S., Van Der Geest, K., Ahmed, I., Huq, S., & Warner, K. (2016). A people-centred perspective on climate change, environmental stress, and livelihood resilience in Bangladesh. *Sustainability Science, 11*, 679–694. https://doi.org/10.1007/s11625-016-0379-z

Bajracharya, S. R., & Mool, P. (2009). Glaciers, glacial lakes and glacial lake outburst floods in the Mount Everest region, Nepal. *Annals of Glaciology, 50*, 81–86. https://doi.org/10.3189/172756410790595895

Bandyopadhyay, N., Bhuiyan, C., & Saha, A. K. (2016). Heat waves, temperature extremes and their impacts on monsoon rainfall and meteorological drought in Gujarat, India. *Natural Hazards, 82*, 367–388. https://doi.org/10.1007/s11069-016-2205-4

Barclay, J. R., Starn, J. J., Briggs, M. A., & Helton, A. M. (2020). Improved prediction of management-relevant groundwater discharge characteristics throughout river networks. *Water Resources Research, 56*. https://doi.org/10.1029/2020WR028027

Barua, P., Rahman, S. H., & Molla, M. H. (2019). Impact of river erosion on livelihood and coping strategies of displaced people in South-Eastern Bangladesh. *IJMRM, 2*, 34. https://doi.org/10.1504/IJMRM.2019.103275

Bell, J. E., Sherry, R., & Luo, Y. (2010). Changes in soil water dynamics due to variation in precipitation and temperature: An ecohydrological analysis in a tallgrass prairie. *Water Resources Research, 46*. https://doi.org/10.1029/2009WR007908

Bellard, C., Bertelsmeier, C., Leadley, P., Thuiller, W., & Courchamp, F. (2012). Impacts of climate change on the future of biodiversity. *Ecology Letters, 15*, 365–377. https://doi.org/10.1111/j.1461-0248.2011.01736.x

Benson, D., & Lorenzoni, I. (2017). Climate change adaptation, flood risks and policy coherence in integrated water resources management in England. *Regional Environmental Change, 17*, 1921–1932. https://doi.org/10.1007/s10113-016-0959-6

Benson, N. U., Agboola, O. D., Fred-Ahmadu, O. H., De-la-Torre, G. E., Oluwalana, A., & Williams, A. B. (2022). Micro(nano)plastics prevalence, food web interactions, and toxicity assessment in aquatic organisms: A review. *Frontiers in Marine Science, 9*, Article 851281. https://doi.org/10.3389/fmars.2022.851281

Bhambri, R., Mehta, M., Dobhal, D. P., Gupta, A. K., Pratap, B., Kesarwani, K., & Verma, A. (2016). Devastation in the Kedarnath (Mandakini) Valley, Garhwal Himalaya, during 16–17 June 2013: A remote sensing and ground-based assessment. *Natural Hazards, 80*, 1801–1822. https://doi.org/10.1007/s11069-015-2033-y

Bhardwaj, A., Wasson, R. J., Chow, W. T. L., & Ziegler, A. D. (2021). High-intensity monsoon rainfall variability and its attributes: A case study for Upper Ganges Catchment in the Indian Himalaya during 1901–2013. *Natural Hazards, 105*, 2907–2936. https://doi.org/10.1007/s11069-020-04431-9

Bhatt, J. P., & Pandit, M. K. (2016). Endangered Golden mahseer Tor putitora Hamilton: A review of natural history. *Reviews in Fish Biology and Fisheries, 26*, 25–38. https://doi.org/10.1007/s11160-015-9409-7

Bhave, A. G., Conway, D., Dessai, S., & Stainforth, D. A. (2018). Water resource planning under future climate and socioeconomic uncertainty in the Cauvery river basin in Karnataka, India. *Water Resources Research, 54*, 708–728. https://doi.org/10.1002/2017WR020970

Blahušiaková, A., Matoušková, M., Jenicek, M., Ledvinka, O., Kliment, Z., Podolinská, J., & Snopková, Z. (2020). Snow and climate trends and their impact on seasonal runoff and hydrological drought types in selected mountain catchments in Central Europe. *Hydrological Sciences Journal, 65*, 2083–2096. https://doi.org/10.1080/02626667.2020.1784900

Bliss, A., Hock, R., & Radić, V. (2014). Global response of glacier runoff to twenty-first century climate change. *JGR Earth Surface, 119*, 717–730. https://doi.org/10.1002/2013JF002931

Bolch, T., Kulkarni, A., Kääb, A., Huggel, C., Paul, F., Cogley, J. G., Frey, H., Kargel, J. S., Fujita, K., Scheel, M., Bajracharya, S., & Stoffel, M. (2012). The state and fate of Himalayan Glaciers. *Science, 336*, 310–314. https://doi.org/10.1126/science.1215828

Botle, A., Salgaonkar, S., Tiwari, R., Ambadekar, S., & Barabde, G. R. (2023). Brief status of contamination in surface water of rivers of India by heavy metals: A review with pollution indices and health risk assessment. *Environmental Geochemistry and Health, 45*, 2779–2801. https://doi.org/10.1007/s10653-022-01463-x

Botle, A., Salgaonkar, S., Tiwari, R., & Barabde, G. (2025). Heavy metal pollution in river Ulhas, Maharashtra, India: Unraveling contamination dynamics through pollution indices, multivariate analysis, and health risk assessment. *nternational Journal of Environmental Research, 19*, 88. https://doi.org/10.1007/s41742-025-00754-2

Botle, A., Salgaonkar, S., Tiwari, R., & Barabde, G. (2024). Unveiling heavy metal pollution dynamics in sediments of river Ulhas, Maharashtra, India: A comprehensive analysis of anthropogenic influence, pollution indices, and health risk assessment. *Environmental Geochemistry and Health, 46*, 419. https://doi.org/10.1007/s10653-024-02208-8

Bowes, M. J., Read, D. S., Joshi, H., Sinha, R., Ansari, A., Hazra, M., Simon, M., Vishwakarma, R., Armstrong, L. K., Nicholls, D. J. E., Wickham, H. D., Ward, J., Carvalho, L. R., & Rees, H. G. (2020). Nutrient and microbial water quality of the upper Ganga River, India: Identification of pollution sources. *Environmental Monitoring and Assessment, 192*, 533. https://doi.org/10.1007/s10661-020-08456-2

Brandolini, P., Pepe, G., Capolongo, D., Cappadonia, C., Cevasco, A., Conoscenti, C., Marsico, A., Vergari, F., & Del Monte, M. (2018). Hillslope degradation in representative Italian areas: Just soil erosion risk or opportunity for development? *Land Degradation and Development, 29*, 3050–3068. https://doi.org/10.1002/ldr.2999

Bravo-Linares, C., Ovando-Fuentealba, L., Muñoz-Arcos, E., Kitch, J. L., Millward, G. E., López-Gajardo, R., Cañoles-Zambrano, M., Del Valle, A., Kelly, C., & Blake, W. H. (2024). Basin scale sources of siltation in a contaminated hydropower reservoir. *Science of the Total Environment, 914*, Article 169952. https://doi.org/10.1016/j.scitotenv.2024.169952

Chang, K. (2017). Geographic information system. In D. Richardson, N. Castree, M. F. Goodchild, A. Kobayashi, W. Liu, & R. A. Marston (Eds.), *International encyclopedia of geography* (pp. 1–9). Wiley. https://doi.org/10.1002/9781118786352.wbieg0152

Chapra, S. C., Camacho, L. A., & McBride, G. B. (2021). Impact of global warming on dissolved oxygen and BOD assimilative capacity of the world's rivers: Modeling analysis. *Water, 13*, 2408. https://doi.org/10.3390/w13172408

Chattopadhyay, S., Edwards, D. R., Yu, Y., & Hamidisepehr, A. (2017). An assessment of climate change impacts on future water availability and droughts in the Kentucky river basin. *Environmental Processes, 4*, 477–507. https://doi.org/10.1007/s40710-017-0259-2

Chevuturi, A., & Dimri, A. P. (2016). Investigation of Uttarakhand (India) disaster-2013 using weather research and forecasting model. *Natural Hazards, 82*, 1703–1726. https://doi.org/10.1007/s11069-016-2264-6

Chinnasamy, P., Hubbart, J. A., & Agoramoorthy, G. (2013). Using remote sensing data to improve groundwater supply estimations in Gujarat, India. *Earth Interactions, 17*, 1–17. https://doi.org/10.1175/2012EI000456.1

Chopra, R., Verma, V. K., & Sharma, P. K. (2001). Mapping, monitoring and conservation of Harike wetland ecosystem, Punjab, India, through remote sensing. *International Journal of Remote Sensing, 22*, 89–98. https://doi.org/10.1080/014311601750038866

Choudhary, S., Dey, S., Dey, S., Sagar, V., Nair, T., & Kelkar, N. (2012). River dolphin distribution in regulated river systems: Implications for dry-season flow regimes in the Gangetic basin. *Aquatic Conservation, 22*, 11–25. https://doi.org/10.1002/aqc.1240

Church, M., & Ferguson, R. I. (2015). Morphodynamics: Rivers beyond steady state. *Water Resources Research, 51*, 1883–1897. https://doi.org/10.1002/2014WR016862

Dangar, S., & Mishra, V. (2021). Natural and anthropogenic drivers of the lost groundwater from the Ganga River basin. *Environmental Research Letters, 16*, Article 114009. https://doi.org/10.1088/1748-9326/ac2ceb

Danladi Bello, A.-A., Hashim, N., & Mohd Haniffah, M. (2017). Predicting impact of climate change on water temperature and dissolved oxygen in tropical rivers. *Climate, 5*, 58. https://doi.org/10.3390/cli5030058

Das, A., Gujre, N., Devi, R. J., & Mitra, S. (2023). A review on traditional ecological knowledge and its role in natural resources management: North East India, a cultural paradise. *Environmental Management, 72*, 113–134. https://doi.org/10.1007/s00267-021-01554-y

Das, D. (2016). Changing climate and its impacts on Assam, Northeast India. *Bandung: Journal of the Global South, 2*, 1–13. https://doi.org/10.1186/s40728-015-0028-4

Das, J., Gayen, A., Saha, P., & Bhattacharya, S. K. (2020). Meteorological drought analysis using standardized precipitation index over Luni River Basin in Rajasthan, India. *SN Applied Sciences, 2*, 1530. https://doi.org/10.1007/s42452-020-03321-w

Dash, S. K., Jenamani, R. K., Kalsi, S. R., & Panda, S. K. (2007). Some evidence of climate change in twentieth-century India. *Climatic Change, 85*, 299–321. https://doi.org/10.1007/s10584-007-9305-9

De Graaf, I. E. M., Marinelli, B., & Liu, S. (2024). Global analysis of groundwater pumping from increased river capture. *Environmental Research Letters, 19*, Article 044064. https://doi.org/10.1088/1748-9326/ad383d

Dey, D., & Döös, K. (2021). Tracing the origin of the South Asian summer monsoon precipitation and its variability using a novel lagrangian framework. *Journal of Climate, 34*, 8655–8668. https://doi.org/10.1175/JCLI-D-20-0967.1

Dey, M., Krishnaswamy, J., Morisaka, T., & Kelkar, N. (2019). Interacting effects of vessel noise and shallow river depth elevate metabolic stress in Ganges river dolphins. *Science and Reports, 9*, 15426. https://doi.org/10.1038/s41598-019-51664-1

Dhar, O. N., & Nandargi, S. (2000). A study of floods in the Brahmaputra basin in India. *International Journal of Climatology, 20*, 771–781. https://doi.org/10.1002/1097-0088(20000615)20:7%3c771::AID-JOC518%3e3.0.CO;2-Z

Dierauer, J. R., Allen, D. M., & Whitfield, P. H. (2021). Climate change impacts on snow and streamflow drought regimes in four ecoregions of British Columbia. *Canadian Water Resources Journal / Revue Canadienne des Ressources Hydriques, 46*, 168–193. https://doi.org/10.1080/07011784.2021.1960894

Dubey, S., & Goyal, M. K. (2020). Glacial lake outburst flood hazard, downstream impact, and risk over the Indian Himalayas. *Water Resources Research, 56*, e2019WR026533. https://doi.org/10.1029/2019WR026533

Dupont-Prinet, A., Vagner, M., Chabot, D., & Audet, C. (2013). Impact of hypoxia on the metabolism of Greenland halibut (*Reinhardtius hippoglossoides*). *Canadian Journal of Fisheries and Aquatic Sciences, 70*, 461–469. https://doi.org/10.1139/cjfas-2012-0327

Earman, S., & Dettinger, M. (2011). Potential impacts of climate change on groundwater resources—a global review. *Journal of Water and Climate Change, 2*, 213–229. https://doi.org/10.2166/wcc.2011.034

Fan, L., Kuang, X., Or, D., & Zheng, C. (2023). Streamflow composition and water "imbalance" in the Northern Himalayas. *Water Resources Research, 59*, e2022WR034243. https://doi.org/10.1029/2022WR034243

Fernald, A. G., Baker, T. T., & Guldan, S. J. (2007). Hydrologic, riparian, and agroecosystem functions of traditional *Acequia* irrigation systems. *Journal of Sustainable Agriculture, 30*, 147–171. https://doi.org/10.1300/J064v30n02_13

Ferrant, S., Caballero, Y., Perrin, J., Gascoin, S., Dewandel, B., Aulong, S., Dazin, F., Ahmed, S., & Maréchal, J.-C. (2014). Projected impacts of climate change on farmers' extraction of groundwater from crystalline aquifers in

South India. *Science and Reports, 4*, 3697. https://doi.org/10.1038/srep03697

Fulazzaky, M. (2014). Challenges of integrated water resources management in Indonesia. *Water, 6*, 2000–2020. https://doi.org/10.3390/w6072000

Furian, W., Loibl, D., & Schneider, C. (2021). Future glacial lakes in High Mountain Asia: An inventory and assessment of hazard potential from surrounding slopes. *Journal of Glaciology, 67*, 653–670. https://doi.org/10.1017/jog.2021.18

Gadgil, S. (2018). The monsoon system: Land–sea breeze or the ITCZ? *Journal of Earth System Science, 127*, 1. https://doi.org/10.1007/s12040-017-0916-x

Ganguli, P., Nandamuri, Y. R., & Chatterjee, C. (2020). Analysis of persistence in the flood timing and the role of catchment wetness on flood generation in a large river basin in India. *Theoretical and Applied Climatology, 139*, 373–388. https://doi.org/10.1007/s00704-019-02964-z

Gately, D. (1974). Sharing the gains from regional cooperation: A game theoretic application to planning investment in electric power. *International Economic Review, 15*, 195. https://doi.org/10.2307/2526099

Giorgi, F., Raffaele, F., & Coppola, E. (2019). The response of precipitation characteristics to global warming from climate projections. *Earth System Dynamics, 10*, 73–89. https://doi.org/10.5194/esd-10-73-2019

Grimm, N. B., Chapin, F. S., Bierwagen, B., Gonzalez, P., Groffman, P. M., Luo, Y., Melton, F., Nadelhoffer, K., Pairis, A., Raymond, P. A., Schimel, J., & Williamson, C. E. (2013). The impacts of climate change on ecosystem structure and function. *Frontiers in Ecology & Environment, 11*, 474–482. https://doi.org/10.1890/120282

Grizzetti, B., Liquete, C., Pistocchi, A., Vigiak, O., Zulian, G., Bouraoui, F., De Roo, A., & Cardoso, A. C. (2019). Relationship between ecological condition and ecosystem services in European rivers, lakes and coastal waters. *Science of the Total Environment, 671*, 452–465. https://doi.org/10.1016/j.scitotenv.2019.03.155

Grogan, D. S., Wisser, D., Prusevich, A., Lammers, R. B., & Frolking, S. (2017). The use and re-use of unsustainable groundwater for irrigation: A global budget. *Environmental Research Letters, 12*, Article 034017. https://doi.org/10.1088/1748-9326/aa5fb2

Haines, D. (2017). *Rivers divided*. Oxford University Press. https://doi.org/10.1093/acprof:oso/9780190648664.001.0001

Halder, B., Rana, B., Juneng, L., Pande, C. B., Alshehery, S., Elsahabi, M., Yadav, K. K., Sammen, S. S., & Naganna, S. R. (2024). Cloud computing-based estimation of Peninsular India's long-term climate change impacts on rainfall, surface temperature, and geospatial indices. *Geomatics, Natural Hazards and Risk, 15*, 2381635. https://doi.org/10.1080/19475705.2024.2381635

Harada, H., & Karn, S. K. (2001). Surface water pollution in three urban territories of Nepal, India, and Bangladesh. *Environmental Management, 28*, 483–496. https://doi.org/10.1007/s002670010238

Harrison, P. A., Vandewalle, M., Sykes, M. T., Berry, P. M., Bugter, R., De Bello, F., Feld, C. K., Grandin, U., Harrington, R., Haslett, J. R., Jongman, R. H. G., Luck, G. W., Da Silva, P. M., Moora, M., Settele, J., Sousa, J. P., & Zobel, M. (2010). Identifying and prioritising services in European terrestrial and freshwater ecosystems. *Biodiversity and Conservation, 19*, 2791–2821. https://doi.org/10.1007/s10531-010-9789-x

Hasnain, S. I., & Thayyen, Renoj. J. (1999). Controls on the major-ion chemistry of the Dokriani glacier meltwaters, Ganga basin, Garhwal Himalaya, India. *Journal of Glaciology, 45*, 87–92. https://doi.org/10.3189/S0022143000003063

Hengade, N., Eldho, T. I., & Ghosh, S. (2018). Climate change impact assessment of a river basin using CMIP5 climate models and the VIC hydrological model. *Hydrological Sciences Journal, 63*, 596–614. https://doi.org/10.1080/02626667.2018.1441531

Hoback, W. W., & Barnhart, M. C. (1996). Lethal limits and Sublethal effects of hypoxia on the amphipod Gammarus Pseudolimnaeus. *Journal of the North American Benthological Society, 15*, 117–126. https://doi.org/10.2307/1467437

Hossain, K., Yadav, S., Quaik, S., Pant, G., Maruthi, A. Y., & Ismail, N. (2017). Vulnerabilities of macrophytes distribution due to climate change. *Theoretical and Applied Climatology, 129*, 1123–1132. https://doi.org/10.1007/s00704-016-1837-3

Hudon, C., Jean, M., & Létourneau, G. (2018). Temporal (1970–2016) changes in human pressures and wetland response in the St. Lawrence River (Québec, Canada). *Science of the Total Environment, 643*, 1137–1151. https://doi.org/10.1016/j.scitotenv.2018.06.080

Hunt, K. M. R., & Menon, A. (2020). The 2018 Kerala floods: A climate change perspective. *Climate Dynamics, 54*, 2433–2446. https://doi.org/10.1007/s00382-020-05123-7

Huss, M., & Hock, R. (2018). Global-scale hydrological response to future glacier mass loss. *Nature Climate Change, 8*, 135–140. https://doi.org/10.1038/s41558-017-0049-x

Iacob, O., Rowan, J. S., Brown, I., & Ellis, C. (2014). Evaluating wider benefits of natural flood management strategies: An ecosystem-based adaptation perspective. *Hydrology Research, 45*, 774–787. https://doi.org/10.2166/nh.2014.184

Immerzeel, W. W., Van Beek, L. P. H., Konz, M., Shrestha, A. B., & Bierkens, M. F. P. (2012). Hydrological response to climate change in a glacierized catchment in the Himalayas. *Climatic Change, 110*, 721–736. https://doi.org/10.1007/s10584-011-0143-4

Islam, N., & Patel, P. P. (2022). Inventory and GLOF hazard assessment of glacial lakes in the Sikkim Himalayas, India. *Geocarto International, 37*, 3840–3876. https://doi.org/10.1080/10106049.2020.1869332

Jain, C. K., & Singh, S. (2020). Impact of climate change on the hydrological dynamics of River Ganga, India. *Journal of Water and Climate Change, 11*, 274–290. https://doi.org/10.2166/wcc.2018.029

Jaiswal, M., Hussain, J., Gupta, S. K., Nasr, M., & Nema, A. K. (2019). Comprehensive evaluation of water quality status for entire stretch of Yamuna River, India. *Environmental Monitoring and Assessment, 191*, 208. https://doi.org/10.1007/s10661-019-7312-8

Jeelani, G., Feddema, J. J., Van Der Veen, C. J., & Stearns, L. (2012). Role of snow and glacier melt in controlling river hydrology in Liddar watershed (western Himalaya) under current and future climate. *Water Resources Research, 48*, 2011WR011590. https://doi.org/10.1029/2011WR011590

Jena, P., Azad, S., & Rajeevan, M. (2015). Statistical selection of the optimum models in the CMIP5 dataset for climate change projections of Indian monsoon rainfall. *Climate, 3*, 858–875. https://doi.org/10.3390/cli3040858

Jenicek, M., Seibert, J., Zappa, M., Staudinger, M., & Jonas, T. (2016). Importance of maximum snow accumulation for summer low flows in humid catchments. *Hydrology and Earth System Sciences, 20*, 859–874. https://doi.org/10.5194/hess-20-859-2016

Jha, S., Das, J., & Goyal, M. K. (2021). Low frequency global-scale modes and its influence on rainfall extremes over India: Nonstationary and uncertainty analysis. *International Journal of Climatology, 41*, 1873–1888. https://doi.org/10.1002/joc.6935

Jiang, C., & Zhang, L. (2015). Climate change and its impact on the eco-environment of the three-rivers headwater region on the Tibetan Plateau, China. *IJERPH, 12*, 12057–12081. https://doi.org/10.3390/ijerph121012057

Joshi, K. D., Das, S. C. S., Pathak, R. K., Khan, A., Sarkar, U. K., & Roy, K. (2018). Pattern of reproductive biology of the endangered golden mahseer *Tor putitora* (Hamilton 1822) with special reference to regional climate change implications on breeding phenology from lesser Himalayan region, India. *Journal of Applied Animal Research, 46*, 1289–1295. https://doi.org/10.1080/09712119.2018.1497493

Joshi, P., Chauhan, A., Dua, P., Malik, S., & Liou, Y.-A. (2022). Physicochemical and biological analysis of river Yamuna at Palla station from 2009 to 2019. *Science and Reports, 12*, 2870. https://doi.org/10.1038/s41598-022-06900-6

Joshi, V., & Kumar, K. (2006). Extreme rainfall events and associated natural hazards in Alaknanda valley, Indian Himalayan region. *Journal of Mountain Science, 3*, 228–236. https://doi.org/10.1007/s11629-006-0228-0

Junk, W. J., An, S., Finlayson, C. M., Gopal, B., Květ, J., Mitchell, S. A., Mitsch, W. J., & Robarts, R. D. (2013). Current state of knowledge regarding the world's wetlands and their future under global climate change: A synthesis. *Aquatic Sciences, 75*, 151–167. https://doi.org/10.1007/s00027-012-0278-z

Kadam, A., Wagh, V., Jacobs, J., Patil, S., Pawar, N., Umrikar, B., Sankhua, R., & Kumar, S. (2022). Integrated approach for the evaluation of groundwater quality through hydro geochemistry and human health risk from Shivganga river basin, Pune, Maharashtra, India. *Environmental Science and Pollution Research, 29*, 4311–4333. https://doi.org/10.1007/s11356-021-15554-2

Kale, V. S. (2002). Fluvial geomorphology of Indian rivers: An overview. *Progress in Physical Geography: Earth and Environment, 26*, 400–433. https://doi.org/10.1191/0309133302pp343ra

Kale, V. S., Hire, P., & Baker, V. R. (1997). Flood hydrology and geomorphology of Monsoon-dominated Rivers: The Indian Peninsula. *Water International, 22*, 259–265. https://doi.org/10.1080/02508069708686717

Katzenberger, A., Schewe, J., Pongratz, J., & Levermann, A. (2021). Robust increase of Indian monsoon rainfall and its variability under future warming in CMIP6 models. *Earth System Dynamics, 12*, 367–386. https://doi.org/10.5194/esd-12-367-2021

Kayitesi, N. M., Guzha, A. C., & Mariethoz, G. (2022). Impacts of land use land cover change and climate change on river hydro-morphology- a review of research studies in tropical regions. *Journal of Hydrology, 615*, Article 128702. https://doi.org/10.1016/j.jhydrol.2022.128702

Khan, A. S., Anavkar, A., Ali, A., Patel, N., & Alim, H. (2021). A review on current status of Riverine pollution in India. *Biosciences Biotechnology Research Asia, 18*, 9–22. https://doi.org/10.13005/bbra/2893

Khan, M. Y. A., Daityari, S., & Chakrapani, G. J. (2016). Factors responsible for temporal and spatial variations in water and sediment discharge in Ramganga River, Ganga Basin, India. *Environmental Earth Sciences, 75*, 283. https://doi.org/10.1007/s12665-015-5148-2

Khan, S., Sinha, R., Whitehead, P., Sarkar, S., Jin, L., & Futter, M. N. (2018). Flows and sediment dynamics in the Ganga River under present and future climate scenarios. *Hydrological Sciences Journal, 63*, 763–782. https://doi.org/10.1080/02626667.2018.1447113

King, A. J., Tonkin, Z., & Lieshcke, J. (2012). Short-term effects of a prolonged blackwater event on aquatic fauna in the Murray River, Australia: Considerations for future events. *Marine and Freshwater Research, 63*, 576. https://doi.org/10.1071/MF11275

Koehler, A. (2008). Water use in LCA: Managing the planet's freshwater resources. *International Journal of Life Cycle Assessment, 13*, 451–455. https://doi.org/10.1007/s11367-008-0028-6

Kulkarni, A. V., Rathore, B. P., Singh, S. K., & Bahuguna, I. M. (2011). Understanding changes in the Himalayan cryosphere using remote sensing techniques. *International Journal of Remote Sensing, 32*, 601–615. https://doi.org/10.1080/01431161.2010.517802

Kumar, C. S., Vani, D. P. D., Damodar, K., Manoharan, G., & Veerapaga, N. (2024). Climate change mitigation through AI solutions. In S. Bhartiya, V. Bhatt, & A. E. Jimenez Ruiz (Eds.), *Advances in hospitality, tourism, and the services industry* (pp. 37–59). IGI Global. https://doi.org/10.4018/979-8-3693-4135-3.ch003

Kumar, M., Sharif, M., & Ahmed, S. (2020). Impact of urbanization on the river Yamuna basin. *International Journal of River Basin Management, 18*, 461–475. https://doi.org/10.1080/15715124.2019.1613412

Kumar, P., & Sharma, M. C. (2023). Frontal changes in medium-sized glaciers in Sikkim, India during 1988–2018: Insights for glacier-climate synthesis over the Himalaya. *iScience, 26*, 107789. https://doi.org/10.1016/j.isci.2023.107789

Kumar, R., Kumar, A., & Saikia, P. (2022). Deforestation and forests degradation impacts on the environment. In V. P. Singh, S. Yadav, K. K. Yadav, & R. N. Yadava (Eds.), *Environmental degradation: Challenges and strategies for mitigation, water science and technology library* (pp. 19–46). Springer International Publishing, Cham. https://doi.org/10.1007/978-3-030-95542-7_2

Kumari, A., Kumar, P., Dubey, A. K., Mishra, A. K., & Saharwardi, Md. S. (2022). Dynamical and thermodynamical aspects of precipitation events over India. *International Journal of Climatology, 42*, 3094–3106. https://doi.org/10.1002/joc.7409

Kundzewicz, Z. W., Mata, L. J., Arnell, N. W., Döll, P., Jimenez, B., Miller, K., Oki, T., Şen, Z., & Shiklomanov, I. (2008). The implications of projected climate change for freshwater resources and their management. *Hydrological Sciences Journal, 53*, 3–10. https://doi.org/10.1623/hysj.53.1.3

Lakra, W. S., Sarkar, U. K., Dubey, V. K., Sani, R., & Pandey, A. (2011). River inter linking in India: Status, issues, prospects and implications on aquatic ecosystems and freshwater fish diversity. *Reviews in Fish Biology and Fisheries, 21*, 463–479. https://doi.org/10.1007/s11160-011-9199-5

Latif, Y., Ma, Y., Ma, W., Muhammad, S., Adnan, M., Yaseen, M., & Fealy, R. (2020). Differentiating snow and glacier melt contribution to runoff in the Gilgit river basin via degree-day modelling approach. *Atmosphere, 11*, 1023. https://doi.org/10.3390/atmos11101023

Legout, C., Droppo, I. G., Coutaz, J., Bel, C., & Jodeau, M. (2018). Assessment of erosion and settling properties of fine sediments stored in cobble bed rivers: The Arc and Isère alpine rivers before and after reservoir flushing. *Earth Surface Processes and Landforms, 43*, 1295–1309. https://doi.org/10.1002/esp.4314

Li, D., Wu, S., Liu, L., Zhang, Y., & Li, S. (2018). Vulnerability of the global terrestrial ecosystems to climate change. *Global Change Biology, 24*, 4095–4106. https://doi.org/10.1111/gcb.14327

Li, H., Xu, C.-Y., Beldring, S., Tallaksen, L. M., & Jain, S. K. (2016). Water resources under climate change in Himalayan basins. *Water Resources Management, 30*, 843–859. https://doi.org/10.1007/s11269-015-1194-5

Li, J., Wang, G., Song, C., Sun, S., Ma, J., Wang, Y., Guo, L., & Li, D. (2024). Recent intensified erosion and massive sediment deposition in Tibetan Plateau rivers. *Nature Communications, 15*, 722. https://doi.org/10.1038/s41467-024-44982-0

Li, X., Li, G., & Zhang, Y. (2014). Identifying major factors affecting groundwater change in the North China plain with grey relational analysis. *Water, 6*, 1581–1600. https://doi.org/10.3390/w6061581

Lokhande, S., & Tare, V. (2021). Spatio-temporal trends in the flow and water quality: Response of river Yamuna to urbanization. *Environmental Monitoring and Assessment, 193*, 117. https://doi.org/10.1007/s10661-021-08873-x

Lutz, A. F., Immerzeel, W. W., Shrestha, A. B., & Bierkens, M. F. P. (2014). Consistent increase in High Asia's runoff due to increasing glacier melt and precipitation. *Nature Clim Change, 4*, 587–592. https://doi.org/10.1038/nclimate2237

Lutz, S. R., Mallucci, S., Diamantini, E., Majone, B., Bellin, A., & Merz, R. (2016). Hydroclimatic and water quality trends across three Mediterranean river basins. *Science of the Total Environment, 571*, 1392–1406. https://doi.org/10.1016/j.scitotenv.2016.07.102

Lynch, A. J., Cooke, S. J., Arthington, A. H., Baigun, C., Bossenbroek, L., Dickens, C., Harrison, I., Kimirei, I., Langhans, S. D., Murchie, K. J., Olden, J. D., Ormerod, S. J., Owuor, M., Raghavan, R., Samways, M. J., Schinegger, R., Sharma, S., Tachamo-Shah, R., Tickner, D., ... Jähnig, S. C. (2023). People need freshwater biodiversity. *Wires Water, 10*, Article e1633. https://doi.org/10.1002/wat2.1633

Maharjan, A., Tuladhar, S., Hussain, A., Mishra, A., Bhadwal, S., Ishaq, S., Saeed, B. A., Sachdeva, I., Ahmad, B., Ferdous, J., & Hassan, S. M. T. (2021). Can labour migration help households adapt to climate change? Evidence from four river basins in South Asia. *Climate and Development, 13*, 879–894. https://doi.org/10.1080/17565529.2020.1867044

Mal, S., Mehta, M., Singh, R. B., Schickhoff, U., & Bisht, M. P. S. (2019). Recession and morphological changes of the debris-covered Milam glacier in Gori Ganga Valley, Central Himalaya, India, derived from satellite data. *Frontiers in Environmental Science, 7*, 42. https://doi.org/10.3389/fenvs.2019.00042

Mangukiya, N. K., Mehta, D. J., & Jariwala, R. (2022). Flood frequency analysis and inundation mapping for lower Narmada basin, India. *Water Practice and Technology, 17*, 612–622. https://doi.org/10.2166/wpt.2022.009

Matta, G., Kumar, A., Nayak, A., & Kumar, P. (2022). Appraisal of spatial–temporal variation and pollution source estimation of Ganga River system through pollution indices and environmetrics in Upper Ganga basin. *Applied Water Science, 12*, 33. https://doi.org/10.1007/s13201-021-01552-9

Matta, G., Nayak, A., Kumar, A., & Kumar, P. (2020). Water quality assessment using NSFWQI, OIP and multivariate techniques of Ganga River system, Uttarakhand. *India. Appl Water Sci, 10*, 206. https://doi.org/10.1007/s13201-020-01288-y

Maurya, A. S., Shah, M., Deshpande, R. D., Bhardwaj, R. M., Prasad, A., & Gupta, S. K. (2011). Hydrograph separation and precipitation source identification using stable water isotopes and conductivity: River Ganga at Himalayan foothills. *Hydrological Processes, 25*, 1521–1530. https://doi.org/10.1002/hyp.7912

Meenu, R., Rehana, S., & Mujumdar, P. P. (2013). Assessment of hydrologic impacts of climate change in Tunga-Bhadra river basin, India with HEC-HMS and SDSM. *Hydrological Processes, 27*, 1572–1589. https://doi.org/10.1002/hyp.9220

Mehta, D., & Yadav, S. M. (2021). An analysis of rainfall variability and drought over Barmer District of Rajasthan, Northwest India. *Water Supply, 21*, 2505–2517. https://doi.org/10.2166/ws.2021.053

Miller, J. D., Immerzeel, W. W., & Rees, G. (2012). Climate change impacts on glacier hydrology and river discharge in the Hindu Kush–Himalayas: A synthesis of the scientific basis. *Mountain Research and Development, 32*, 461–467. https://doi.org/10.1659/MRD-JOURNAL-D-12-00027.1

Milner, A. M., Brown, L. E., & Hannah, D. M. (2009). Hydroecological response of river systems to shrinking glaciers. *Hydrological Processes, 23*, 62–77. https://doi.org/10.1002/hyp.7197

Mishra, A. K., & Nagaraju, V. (2019). Space-based monitoring of severe flooding of a southern state in India during south-west monsoon season of 2018. *Natural Hazards, 97*, 949–953. https://doi.org/10.1007/s11069-019-03673-6

Mishra, V., & Shah, H. L. (2018). Hydroclimatological Perspective of the Kerala Flood of 2018. *Journal of the Geological Society of India, 92*, 645–650. https://doi.org/10.1007/s12594-018-1079-3

Mishra, V., Thirumalai, K., Jain, S., & Aadhar, S. (2021). Unprecedented drought in South India and recent water scarcity. *Environmental Research Letters, 16*, Article 054007. https://doi.org/10.1088/1748-9326/abf289

Misra, A. K. (2010). A river about to die: Yamuna. *JWARP, 02*, 489–500. https://doi.org/10.4236/jwarp.2010.25056

Mittal, N., Bhave, A. G., Mishra, A., & Singh, R. (2016). Impact of human intervention and climate change on natural flow regime. *Water Resources Management, 30*, 685–699. https://doi.org/10.1007/s11269-015-1185-6

Mondal, S. K., Patel, V. D., Bharti, R., & Singh, R. P. (2023). Causes and effects of Shisper glacial lake outburst flood event in Karakoram in 2022. *Geomatics, Natural Hazards and Risk, 14*, 2264460. https://doi.org/10.1080/19475705.2023.2264460

Moors, E. J., Groot, A., Biemans, H., Van Scheltinga, C. T., Siderius, C., Stoffel, M., Huggel, C., Wiltshire, A., Mathison, C., Ridley, J., Jacob, D., Kumar, P., Bhadwal, S., Gosain, A., & Collins, D. N. (2011). Adaptation to changing water resources in the Ganges basin, northern India. *Environmental Science & Policy, 14*, 758–769. https://doi.org/10.1016/j.envsci.2011.03.005

Muhammad, S., Tian, L., & Khan, A. (2019). Early twenty-first century glacier mass losses in the Indus Basin constrained by density assumptions. *Journal of Hydrology, 574*, 467–475. https://doi.org/10.1016/j.jhydrol.2019.04.057

Mustaq, S., Farjana, B. (2024). Harmony in Hydroinformatics: Integrating AI and IEC for sustainable groundwater conservation in Solapur. *Int. J. Sci. Res. Arch. 11*, 2163–2175. https://doi.org/10.30574/ijsra.2024.11.1.0294

Myhre, G., Alterskjær, K., Stjern, C. W., Hodnebrog, Ø., Marelle, L., Samset, B. H., Sillmann, J., Schaller, N., Fischer, E., Schulz, M., & Stohl, A. (2019). Frequency of extreme precipitation increases extensively with event rareness under global warming. *Science and Reports, 9*, 16063. https://doi.org/10.1038/s41598-019-52277-4

Nandargi, S., Gaur, A., & Mulye, S. S. (2016). Hydrological analysis of extreme rainfall events and severe rainstorms over Uttarakhand, India. *Hydrological Sciences Journal, 61*, 2145–2163. https://doi.org/10.1080/02626667.2015.1085990

Negi, V. S., Pathak, R., Sekar, K. C., Rawal, R. S., Bhatt, I. D., Nandi, S. K., & Dhyani, P. P. (2018). Traditional knowledge and biodiversity conservation: A case study from Byans Valley in Kailash Sacred Landscape, India. *Journal of Environmental Planning and Management, 61*, 1722–1743. https://doi.org/10.1080/09640568.2017.1371006

Nepal, S., Flügel, W.-A., & Shrestha, A. B. (2014). Upstream-downstream linkages of hydrological processes in the Himalayan region. *Ecological Processes, 3*, 19. https://doi.org/10.1186/s13717-014-0019-4

Nepal, S., & Shrestha, A. B. (2015). Impact of climate change on the hydrological regime of the Indus, Ganges and Brahmaputra river basins: A review of the literature. *International Journal of Water Resources Development, 31*, 201–218. https://doi.org/10.1080/07900627.2015.1030494

Oo, K. T., Haishan, C., & Jonah, K. (2023). Climate change impact on the trigger of natural disasters over South-Eastern Himalayas foothill region of Myanmar: Extreme rainfall analysis. *International Journal of Geophysics, 2023*, 1–20. https://doi.org/10.1155/2023/2186857

Ougahi, J. H., Cutler, M. E. J., & Cook, S. J. (2022). Modelling climate change impact on water resources of the Upper Indus Basin. *Journal of Water and Climate Change, 13*, 482–504. https://doi.org/10.2166/wcc.2021.233

Pace, M. L., Reche, I., Cole, J. J., Fernández-Barbero, A., Mazuecos, I. P., & Prairie, Y. T. (2012). PH change induces shifts in the size and light absorption of dissolved organic matter. *Biogeochemistry, 108*, 109–118. https://doi.org/10.1007/s10533-011-9576-0

Padowski, J. C., & Gorelick, S. M. (2014). Global analysis of urban surface water supply vulnerability. *Environmental Research Letters, 9*, Article 104004. https://doi.org/10.1088/1748-9326/9/10/104004

Pande, S., & Savenije, H. H. G. (2016). A sociohydrological model for smallholder farmers in Maharashtra, India. *Water Resources Research, 52*, 1923–1947. https://doi.org/10.1002/2015WR017841

Pandey, A. C., Kaushik, K., & Parida, B. R. (2022). Google Earth engine for large-scale flood mapping using SAR data and impact assessment on agriculture and population of Ganga-Brahmaputra Basin. *Sustainability, 14*, 4210. https://doi.org/10.3390/su14074210

Pandve, H. (2009). India′s national action plan on climate change. *Indian Journal of Occupational and Environmental Medicine, 13*, 17. https://doi.org/10.4103/0019-5278.50718

Pankhurst, N. W., & Munday, P. L. (2011). Effects of climate change on fish reproduction and early life history stages. *Marine and Freshwater Research, 62*, 1015. https://doi.org/10.1071/MF10269

Panthi, J., Dahal, P., Shrestha, M., Aryal, S., Krakauer, N., Pradhanang, S., Lakhankar, T., Jha, A., Sharma, M., & Karki, R. (2015). Spatial and temporal variability of rainfall in the Gandaki river basin of Nepal Himalaya. *Climate, 3*, 210–226. https://doi.org/10.3390/cli3010210

Patel, J. K., & Anuragi, R. B. (2024). Identifying the ecotourism potential locations using AHP and weighted overlay analysis: A case of Bundelkhand region in Central India. *Papers in Applied Geography, 10*, 17–33. https://doi.org/10.1080/23754931.2023.2254786

Pittock, J., & Finlayson, C. M. (2011). Australia's Murray - Darling Basin: Freshwater ecosystem conservation options in an era of climate change. *Mar. Freshwater Res., 62*, 232. https://doi.org/10.1071/MF09319

Poncelet, A., Gemenne, F., Martiniello, M., & Bousetta, H. (2010). A country made for disasters: environmental vulnerability and forced migration in

Bangladesh. In T. Afifi, & J. Jäger (Eds.), *Environment, forced migration and social vulnerability* (pp. 211–222). Springer Berlin Heidelberg. https://doi.org/10.1007/978-3-642-12416-7_16

Pradhan, B. (2009). Groundwater potential zonation for basaltic watersheds using satellite remote sensing data and GIS techniques. *Open Geosciences, 1*. https://doi.org/10.2478/v10085-009-0008-5

Pradhan, N. S., Das, P. J., Gupta, N., & Shrestha, A. B. (2021). Sustainable management options for healthy rivers in South Asia: The case of Brahmaputra. *Sustainability, 13*, 1087. https://doi.org/10.3390/su1303 1087

Pramanik, M. K., Biswas, S. S., Mondal, B., & Pal, R. (2016). Coastal vulnerability assessment of the predicted sea level rise in the coastal zone of Krishna-Godavari delta region, Andhra Pradesh, east coast of India. *Environment, Development and Sustainability, 18*, 1635–1655. https://doi.org/10.1007/s10668-015-9708-0

Praskievicz, S. (2015). A coupled hierarchical modeling approach to simulating the geomorphic response of river systems to anthropogenic climate change. *Earth Surf Processes Landf, 40*, 1616–1630. https://doi.org/10.1002/esp.3740

Prathumratana, L., Sthiannopkao, S., & Kim, K. W. (2008). The relationship of climatic and hydrological parameters to surface water quality in the lower Mekong River. *Environment International, 34*, 860–866. https://doi.org/10.1016/j.envint.2007.10.011

Přívara, A., & Přívarová, M. (2019). Nexus between climate change. *Displacement and Conflict: Afghanistan Case. Sustainability, 11*, 5586. https://doi.org/10.3390/su11205586

Priya, P., Krishnan, R., Mujumdar, M., & Houze, R. A. (2017). Changing monsoon and midlatitude circulation interactions over the Western Himalayas and possible links to occurrences of extreme precipitation. *Climate Dynamics, 49*, 2351–2364. https://doi.org/10.1007/s00382-016-3458-z

Purnadurga, G., Lakshmi Kumar, T. V., Koteswara Rao, K., Rajasekhar, M., & Narayanan, M. S. (2018). Investigation of temperature changes over India in association with meteorological parameters in a warming climate. *International Journal of Climatology, 38*, 867–877. https://doi.org/10.1002/joc.5216

Qin, G., Liu, J., Xu, S., & Wang, T. (2020). Water quality assessment and pollution source apportionment in a highly regulated river of Northeast China. *Environmental Monitoring and Assessment, 192*, 446. https://doi.org/10.1007/s10661-020-08404-0

Qiu, J., Shen, Z., Leng, G., & Wei, G. (2021). Synergistic effect of drought and rainfall events of different patterns on watershed systems. *Science and Reports, 11*, 18957. https://doi.org/10.1038/s41598-021-97574-z

Rai, P., Kumar, J. S. Y., & Sen, A. (2024). Ganga, GAP, and lockdown: Potential threats to the biodiversity of the river. *Journal of Water and Climate Change, 15*, 5482–5500. https://doi.org/10.2166/wcc.2024.322

Raihan, A., & Tuspekova, A. (2022). Nexus between emission reduction factors and anthropogenic carbon emissions in India. *Anthropocene Science, 1*, 295–310. https://doi.org/10.1007/s44177-022-00028-y

Raje, D., Priya, P., & Krishnan, R. (2014). Macroscale hydrological modelling approach for study of large scale hydrologic impacts under climate change in Indian river basins. *Hydrological Processes, 28*, 1874–1889. https://doi.org/10.1002/hyp.9731

Rao, M. P., Cook, E. R., Cook, B. I., D'Arrigo, R. D., Palmer, J. G., Lall, U., Woodhouse, C. A., Buckley, B. M., Uriarte, M., Bishop, D. A., Jian, J., & Webster, P. J. (2020). Seven centuries of reconstructed Brahmaputra River discharge demonstrate underestimated high discharge and flood hazard frequency. *Nature Communications, 11*, 6017. https://doi.org/10.1038/s41467-020-19795-6

Rao, R. J. (2001). Biological resources of the Ganga River, India. *Hydrobiologia, 458*, 159–168. https://doi.org/10.1023/A:1013173204091

Rashid, I., & Majeed, U. (2018). Recent recession and potential future lake formation on Drang Drung glacier, Zanskar Himalaya, as assessed with earth observation data and glacier modelling. *Environment and Earth Science, 77*, 429. https://doi.org/10.1007/s12665-018-7601-5

Rashmi, I., Karthika, K. S., Roy, T., Shinoji, K. C., Kumawat, A., Kala, S., & Pal, R. (2022). Soil erosion and sediments: A source of contamination and impact on agriculture productivity. In M. Naeem, J. F. J. Bremont, A. A. Ansari & S. S. Gill (Eds.), *Agrochemicals in soil and environment* (pp. 313–345). Springer Nature Singapore. https://doi.org/10.1007/978-981-16-9310-6_14

Ravindra, K., Ameena, M., Monika, R., & Kaushik, A. (2003). Seasonal variations in physico-chemical characteristics of River Yamuna in Haryana and its ecological best-designated use. *Environmental Monitoring and Assessment , 5*, 419. https://doi.org/10.1039/b301723k

RaziSadath, P. V., RinishaKartheeshwari, M., & Elango, L. (2023). WEAP model based evaluation of future scenarios and strategies for sustainable water management in the Chennai Basin, India. AQUA—Water infrastructure. *Ecosystems and Society, 72*, 2062–2080. https://doi.org/10.2166/aqua.2023.144

Rees, H. G., & Collins, D. N. (2006). Regional differences in response of flow in glacier-fed Himalayan rivers to climatic warming. *Hydrological Processes, 20*, 2157–2169. https://doi.org/10.1002/hyp.6209

Rickards, N., Thomas, T., Kaelin, A., Houghton-Carr, H., Jain, S. K., Mishra, P. K., Nema, M. K., Dixon, H., Rahman, M. M., Horan, R., Jenkins, A., & Rees, G. (2020). Understanding future water challenges in a highly regulated indian river basin—Modelling the impact of climate change on the hydrology of the Upper Narmada. *Water, 12*, 1762. https://doi.org/10.3390/w12061762

Rosegrant, M. W., Ringler, C., & Zhu, T. (2009). Water for agriculture: Maintaining food security under growing scarcity. *Annual Review of Environment and Resources, 34*, 205–222. https://doi.org/10.1146/annurev.environ.030308.090351

Roxy, M. K., Ghosh, S., Pathak, A., Athulya, R., Mujumdar, M., Murtugudde, R., Terray, P., & Rajeevan, M. (2017). A threefold rise in widespread extreme rain events over central India. *Nature Communications, 8*, 708. https://doi.org/10.1038/s41467-017-00744-9

Saari, G. N., Wang, Z., & Brooks, B. W. (2018). Revisiting inland hypoxia: Diverse exceedances of dissolved oxygen thresholds for freshwater aquatic life. *Environmental Science and Pollution Research, 25*, 3139–3150. https://doi.org/10.1007/s11356-017-8908-6

Sabade, S. S., Kulkarni, A., & Kripalani, R. H. (2011). Projected changes in South Asian summer monsoon by multi-model global warming experiments. *Theoretical and Applied Climatology, 103*, 543–565. https://doi.org/10.1007/s00704-010-0296-5

Sahoo, P. K., Datta, R., Rahman, M. M., & Sarkar, D. (2024). Sustainable environmental technologies: Recent development, opportunities, and key challenges. *Applied Sciences, 14*, 10956. https://doi.org/10.3390/app142310956

Saikanth, D. R. K., Kumar, S., Rani, M., Sharma, A., Srivastava, S., Vyas, D., Singh, G. A., & Kumar, S. (2023). A comprehensive review on climate change adaptation strategies and challenges in agriculture. *IJECC, 13*, 10–19. https://doi.org/10.9734/ijecc/2023/v13i113138

Sar, N., Chatterjee, S., & Das Adhikari, M. (2015). Integrated remote sensing and GIS based spatial modelling through analytical hierarchy process (AHP) for water logging hazard, vulnerability and risk assessment in Keleghai river basin, India. *Modeling Earth Systems and Environment, 1*, 31. https://doi.org/10.1007/s40808-015-0039-9

Sarkar, B., & Islam, A. (2022). Assessing poverty and livelihood vulnerability of the fishing communities in the context of pollution of the Churni River, India. *Environmental Science and Pollution Research, 29*, 26575–26598. https://doi.org/10.1007/s11356-021-17719-5

Sarker, M. N. I., Wu, M., Alam, G. M. M., & Shouse, R. C. (2019). Livelihood vulnerability of Riverine-Island dwellers in the face of natural disasters in Bangladesh. *Sustainability, 11*, 1623. https://doi.org/10.3390/su11061623

Sarma, J. N., & Acharjee, S. (2018). A study on variation in channel width and braiding intensity of the Brahmaputra River in Assam, India. *Geosciences, 8*, 343. https://doi.org/10.3390/geosciences8090343

Sarma, J. N., Acharjee, S., & Murgante, B. (2015). Morphotectonic study of the Brahmaputra basin using geoinformatics. *Journal of the Geological Society of India, 86*, 324–330. https://doi.org/10.1007/s12594-015-0318-0

Saryal, R. (2018). Climate change policy of India: Modifying the environment. *South Asia Research, 38*, 1–19. https://doi.org/10.1177/0262728017745385

Schneider, C., Laizé, C. L. R., Acreman, M. C., & Flörke, M. (2013). How will climate change modify river flow regimes in Europe? *Hydrology and Earth System Sciences, 17*, 325–339. https://doi.org/10.5194/hess-17-325-2013

Sharma, D., & Kansal, A. (2011). Water quality analysis of River Yamuna using water quality index in the national capital territory, India (2000–2009). *Applied Water Science, 1*, 147–157. https://doi.org/10.1007/s13201-011-0011-4

Sharma, S., Ha, K.-J., Yamaguchi, R., Rodgers, K. B., Timmermann, A., & Chung, E.-S. (2023). Future Indian Ocean warming patterns. *Nature Communications, 14*, 1789. https://doi.org/10.1038/s41467-023-37435-7

Sharma, V., Mishra, V. D., & Joshi, P. K. (2012). Snow cover variation and streamflow simulation in a snow-fed river basin of the Northwest Himalaya. *Journal of Mountain Science, 9*, 853–868. https://doi.org/10.1007/s11629-012-2419-1

Sigdel, S. R., Zhang, H., Zhu, H., Muhammad, S., & Liang, E. (2020). Retreating glacier and advancing forest over the past 200 years in the Central Himalayas. *JGR Biogeosciences, 125*, e2020JG005751. https://doi.org/10.1029/2020JG005751

Sikka, A. K., Islam, A., & Rao, K. V. (2018). Climate-smart land and water management for sustainable agriculture. *Irrigation and Drainage, 67*, 72–81. https://doi.org/10.1002/ird.2162

Sileshi, A., Awoke, A., Beyene, A., Stiers, I., & Triest, L. (2020). Water purifying capacity of natural riverine wetlands in relation to their ecological quality. *Frontiers in Environmental Science, 8*, 39. https://doi.org/10.3389/fenvs.2020.00039

Simelton, E., Carew-Reid, J., Coulier, M., Damen, B., Howell, J., Pottinger-Glass, C., Tran, H. V., & Van Der Meiren, M. (2021). NBS framework for agricultural landscapes. *Frontiers in Environmental Science, 9*, Article 678367. https://doi.org/10.3389/fenvs.2021.678367

Simons, S., Kinjawadekar, A., & Kinjawadekar, T. A. (2023). Assessing the impacts of ecological framework of Indian riverfront revitalization projects. *Environment, Development and Sustainability, 26*, 27553–27583. https://doi.org/10.1007/s10668-023-03771-3

Singh, D., Tsiang, M., Rajaratnam, B., & Diffenbaugh, N. S. (2014). Observed changes in extreme wet and dry spells during the South Asian summer monsoon season. *Nature Clim Change, 4*, 456–461. https://doi.org/10.1038/nclimate2208

Singh, G., & Pandey, A. (2024). Climate change induced disasters and highly vulnerable infrastructure in Uttarakhand, India: Current status and way forward towards resilience and long-term sustainability. *Sustainable and Resilient Infrastructure, 9*, 145–167. https://doi.org/10.1080/23789689.2023.2253409

Singh, J., Singh, V., Ojha, C. S. P., & Arora, M. K. (2023a). Assessment of recent changes (2011–2020) in glacier thickness and runoff variability in Gangotri glacier, India. *Hydrological Sciences Journal, 68*, 2223–2242. https://doi.org/10.1080/02626667.2023.2258861

Singh, P., & Rajamani, V. (2001). Geochemistry of the floodplain sediments of the Kaveri River, Southern India. *Journal of Sedimentary Research, 71*, 50–60. https://doi.org/10.1306/042800710050

Singh, P., Jain, S. K., & Kumar, N. (1997). Estimation of snow and glacier-melt contribution to the Chenab River, Western Himalaya. *Mountain Research and Development, 17*(1), 49. https://doi.org/10.2307/3673913

Singh, V., Stevanović, M., Jha, C. K., Beier, F., Ghosh, R. K., Lotze-Campen, H., & Popp, A. (2023b). Assessing policy options for sustainable water use in India's cereal production system. *Environmental Research Letters, 18*, Article 094073. https://doi.org/10.1088/1748-9326/acf9b6

Skliris, N., Marsh, R., Haigh, I. D., Wood, M., Hirschi, J., Darby, S., Quynh, N. P., & Hung, N. N. (2022). Drivers of rainfall trends in and around Mainland Southeast Asia. *Frontiers in Climate, 4*, Article 926568. https://doi.org/10.3389/fclim.2022.926568

Slingerland, R., & Smith, N. D. (2004). River avulsions and their deposits. *Annual Review of Earth and Planetary Sciences, 32*, 257–285. https://doi.org/10.1146/annurev.earth.32.101802.120201

Solaraj, G., Dhanakumar, S., Rutharvel Murthy, K., & Mohanraj, R. (2010). Water quality in select regions of Cauvery Delta River basin, southern India, with emphasis on monsoonal variation. *Environmental Monitoring and Assessment, 166*, 435–444. https://doi.org/10.1007/s10661-009-1013-7

Sonkar, V., Kashyap, A., Pallarés-Vega, R., Sasidharan, S. S., Modi, A., Uluseker, C., Jambu, S. C., Mohapatra, P. K., Larsen, J., Graham, D., Thatikonda, S., & Kreft, J.-U. (2024). Field Monitoring and Hydraulic Modelling Quantify Untreated Wastewater as Dominant Source of AMR in a Small River Running through a Big City. *BioRxiv*. https://doi.org/10.1101/2024.12.21.629897

Sorte, C. J. B., Ibáñez, I., Blumenthal, D. M., Molinari, N. A., Miller, L. P., Grosholz, E. D., Diez, J. M., D'Antonio, C. M., Olden, J. D., Jones, S. J., & Dukes, J. S. (2013). Poised to prosper? A cross-system comparison of climate change effects on native and non-native species performance. *Ecology Letters, 16*, 261–270. https://doi.org/10.1111/ele.12017

Sreelash, K., Sharma, R. K., Gayathri, J. A., Upendra, B., Maya, K., & Padmalal, D. (2018). Impact of rainfall variability on river hydrology: A case study of Southern Western Ghats, India. *Journal of the Geological Society of India, 92*, 548–554. https://doi.org/10.1007/s12594-018-1065-9

Sud, R., Mishra, A., Varma, N., & Bhadwal, S. (2015). Adaptation policy and practice in densely populated glacier-fed river basins of South Asia: A systematic review. *Regional Environmental Change, 15*, 825–836. https://doi.org/10.1007/s10113-014-0711-z

Tagliabue, A., Matterson, K. O., Ponti, M., Turicchia, E., Abbiati, M., & Costantini, F. (2023). Sediment and bottom water eDNA metabarcoding to support coastal management. *Ocean & Coastal Management, 244*, Article 106785. https://doi.org/10.1016/j.ocecoaman.2023.106785

Tashie, A. M., Mirus, B. B., & Pavelsky, T. M. (2016). Identifying long-term empirical relationships between storm characteristics and episodic groundwater recharge. *Water Resources Research, 52*, 21–35. https://doi.org/10.1002/2015WR017876

Thapa, S., Sinclair, H. D., Creed, M. J., Mudd, S. M., Attal, M., Borthwick, A. G. L., Ghimire, B. N., & Watson, C. S. (2024). The impact of sediment flux and calibre on flood risk in the Kathmandu Valley, Nepal. *Earth Surface Processes and Landforms, 49*, 706–727. https://doi.org/10.1002/esp.5731

Thind, P. S., Chandel, K. K., Sharma, S. K., Mandal, T. K., & John, S. (2019). Light-absorbing impurities in snow of the Indian Western Himalayas: Impact on snow albedo, radiative forcing, and enhanced melting. *Environmental Science and Pollution Research, 26*, 7566–7578. https://doi.org/10.1007/s11356-019-04183-5

Tickner, D., Parker, H., Moncrieff, C. R., Oates, N. E. M., Ludi, E., & Acreman, M. (2017). Managing rivers for multiple benefits–A coherent approach to research, policy and planning. *Frontiers in Environmental Science, 5*. https://doi.org/10.3389/fenvs.2017.00004

Todd, A. S., Manning, A. H., Verplanck, P. L., Crouch, C., McKnight, D. M., & Dunham, R. (2012). Climate-change-driven deterioration of water quality in a mineralized watershed. *Environmental Science and Technology, 46*, 9324–9332. https://doi.org/10.1021/es3020056

Trimmel, H., Weihs, P., Leidinger, D., Formayer, H., Kalny, G., & Melcher, A. (2018). Can riparian vegetation shade mitigate the expected rise in stream temperatures due to climate change during heat waves in a human-impacted pre-alpine river? *Hydrology and Earth System Sciences, 22*, 437–461. https://doi.org/10.5194/hess-22-437-2018

Turner, B., Devisscher, T., Chabaneix, N., Woroniecki, S., Messier, C., & Seddon, N. (2022). The role of nature-based solutions in supporting social-ecological resilience for climate change adaptation. *Annual Review of Environment and Resources, 47*, 123–148. https://doi.org/10.1146/annurev-environ-012220-010017

Uhe, P. F., Mitchell, D. M., Bates, P. D., Sampson, C. C., Smith, A. M., & Islam, A. S. (2019). Enhanced flood risk with 1.5 °C global warming in the Ganges–Brahmaputra–Meghna basin. *Environmental Research Letters, 14*, 074031. https://doi.org/10.1088/1748-9326/ab10ee

Urfels, A., McDonald, A. J., Krupnik, T. J., & Van Oel, P. R. (2020). Drivers of groundwater utilization in water-limited rice production systems in Nepal. *Water International, 45*, 39–59. https://doi.org/10.1080/02508060.2019.1708172

Usman, M., & Furuya, M. (2018). Interannual modulation of seasonal glacial velocity variations in the Eastern Karakoram detected by ALOS-1/2 data. *Journal of Glaciology, 64*, 465–476. https://doi.org/10.1017/jog.2018.39

Van Den Brandeler, F., Gupta, J., & Hordijk, M. (2019). Megacities and rivers: Scalar mismatches between urban water management and river basin management. *Journal of Hydrology, 573*, 1067–1074. https://doi.org/10.1016/j.jhydrol.2018.01.001

Van Roosmalen, L., Christensen, B. S. B., & Sonnenborg, T. O. (2007). Regional differences in climate change impacts on groundwater and stream discharge in Denmark. *Vadose Zone Journal, 6*, 554–571. https://doi.org/10.2136/vzj2006.0093

Varghese, S. J., Surendran, S., Rajendran, K., Ghosh, S., Kitoh, A., & Ashok, K., (2024). Precipitation scaling in extreme rainfall events and the implications for future Indian monsoon: Analysis of high-resolution global climate model simulations. *Geophysical Research Letters, 51*, e2023GL105680. https://doi.org/10.1029/2023GL105680

Vass, K. K., Das, M. K., Srivastava, P. K., & Dey, S. (2009). Assessing the impact of climate change on inland fisheries in River Ganga and its plains in India. *Aquatic Ecosystem Health & Management, 12*, 138–151. https://doi.org/10.1080/14634980902908746

Veh, G., Korup, O., & Walz, A. (2020). Hazard from Himalayan glacier lake outburst floods. *Proceedings of the National Academy of Sciences of the United States of America, 117*, 907–912. https://doi.org/10.1073/pnas.1914898117

Wadhawan, S. K., Singh, B., & Ramesh, M. V. (2020). Causative factors of landslides 2019: Case study in Malappuram and Wayanad districts of Kerala, India. *Landslides, 17*, 2689–2697. https://doi.org/10.1007/s10346-020-01520-5

Webb, B. W., Hannah, D. M., Moore, R. D., Brown, L. E., & Nobilis, F. (2008). Recent advances in stream and river temperature research. *Hydrological Processes, 22*, 902–918. https://doi.org/10.1002/hyp.6994

Whitehead, P. G., Sarkar, S., Jin, L., Futter, M. N., Caesar, J., Barbour, E., Butterfield, D., Sinha, R., Nicholls, R., Hutton, C., & Leckie, H. D. (2015). Dynamic modeling of the Ganga river system: impacts of future climate and socio-economic change on flows and nitrogen fluxes in India and Bangladesh. *Environmental Science: Processes & Impacts, 17*, 1082–1097. https://doi.org/10.1039/C4EM00616J

Whitehead, P. G., Wilby, R. L., Battarbee, R. W., Kernan, M., & Wade, A. J. (2009). A review of the potential impacts of climate change on surface water quality. *Hydrological Sciences Journal, 54*, 101–123. https://doi.org/10.1623/hysj.54.1.101

Wu, Y., Liu, S., & Abdul-Aziz, O. I. (2012). Hydrological effects of the increased CO2 and climate change in the Upper Mississippi River Basin using a modified SWAT. *Climatic Change, 110*, 977–1003. https://doi.org/10.1007/s10584-011-0087-8

Xu, M., Kang, S., Wang, X., Pepin, N., & Wu, H. (2019). Understanding changes in the water budget driven by climate change in cryospheric-dominated watershed of the northeast Tibetan Plateau, China. *Hydrological Processes, 33*, 1040–1058. https://doi.org/10.1002/hyp.13383

Yang, J., Wang, F., Lv, J., Liu, Q., Nan, F., Liu, X., Xu, L., Xie, S., & Feng, J. (2019). Process of eutrophication is. *Environmental Monitoring and Assessment, 191*, 688. https://doi.org/10.1007/s10661-019-7847-8

Yuan, X., Xie, Z., Zheng, J., Tian, X., & Yang, Z. (2008). Effects of water table dynamics on regional climate: A case study over east Asian monsoon area. *Journal of Geophysical Research, 113*, 2008JD010180. https://doi.org/10.1029/2008JD010180

Zachariah, M., Mondal, A., Das, M., AchutaRao, K. M., & Ghosh, S. (2020). On the role of rainfall deficits and cropping choices in loss of agricultural yield in Marathwada. *Environmental Research Letters, 15*, Article 094029. https://doi.org/10.1088/1748-9326/ab93fc

Zaveri, E., Grogan, D. S., Fisher-Vanden, K., Frolking, S., Lammers, R. B., Wrenn, D. H., Prusevich, A., & Nicholas, R. E. (2016). Invisible water, visible impact: Groundwater use and Indian agriculture under climate change. *Environmental Research Letters, 11*, Article 084005. https://doi.org/10.1088/1748-9326/11/8/084005

Zhang, T., Wang, W., Gao, T., & An, B. (2021). Simulation and assessment of future glacial lake outburst floods in the Poiqu river basin. *Central Himalayas. Water, 13*, 1376. https://doi.org/10.3390/w13101376

CHAPTER 4

Navigating River Rehabilitation: A Holistic Perspective on Integrated Watershed Management for Environmental Sustainability

Sayli Salgaonkar, Akshay Botle, and Gayatri Barabde

4.1 Introduction

4.1.1 River Restoration: Definition and Initiatives in India

The progress of human civilization has been significantly influenced by rivers. Alongside huge rivers, numerous significant civilizations, including Mesopotamia, Egypt, China, and the Indus Valley, developed. Rivers are

S. Salgaonkar (✉) · A. Botle · G. Barabde
Department of Environmental Science, The Institute of Science, Dr. Homi Bhabha State University, Mumbai 400032, India
e-mail: 23res.evs.sayli@iscm.ac.in

G. Barabde
Department of Analytical Chemistry, The Institute of Science, Dr. Homi Bhabha State University, Mumbai 400032, India

S. Ghosh (ed.), *Reimagining Indian Rivers for Sustainability*, Palgrave Studies in Environmental Sustainability,
https://doi.org/10.1007/978-3-032-04754-0_4

important for many reasons, such as practical, ecological, artistic, and spiritual (Postel & Richter, 2012). Human activities around the world have become the cause of impairment of aquatic environments (Gleick, 2003). Through several interventions of varying degrees, the human subsystem has impacted the river or fluvial processes, disrupting the river system beyond the point of resilience. In contrast to urban-dominated areas, where there are issues of sewage discharge, municipal drainage, industrial effluents, waste water flow, and infrastructure constraints on river channels that have resulted in the addition of toxic contaminants, modification of flow of river, and system instability, river ecosystems in agrarian settings are experiencing amplified nutrient and sediment loads, flow alteration, and ruining of habitat (Botle et al., 2024). In addition, there is an increase in water extraction for a variety of uses and extensive sand mining from the riverbed. The riverine environment suffered irreversible harm as a result of all these activities. The current state of affairs is so concerning that river restoration is seen as a critical task that must be completed immediately and carefully (Chattopadhyay & Harilal, 2017).

River restoration is the activity meant to "protect and rehabilitate the physical and biotic processes of a river in a way that is conducive to the progression of ecosystems towards their natural state" (Koehn et al., 1997). The idea of river management has evolved over time (Table 4.1). In the 1950s, when water was plentiful and only seen as a commodity, the focus was on analyzing the quality of river water. By 2000, however, there had been a slow shift in perspective, with rivers now being viewed as an ecosystem component and efforts focused on enhancing the human-river relationship within the broader context of the human–environment relationship.

Table 4.1 Evolving view of river restoration

Year	*Activity*
1950	Regular assessment of the quality of the water
1960	Constructing water input stations and a sewer system
1980	Enhancing urban landscaping, riverfront amenities, and accessibility between rivers and cities
1990	River and environment restoration project
2000	Enhancement of human-, community-, and nature-river relationships

Although river management for controlling floods and developing irrigation has a long history in India, initiatives to restore rivers at the government level began with "Clean Ganga project" in the 1980s. To lead the efforts, the "National Ganga River Basin Authority (NGRBA)" was founded with main goals to create a management plan for the Ganga basin; regulate the measures related to river ecosystem and management in the Ganga basin States; and prevent, control, and mitigate pollution in the Ganga to preserve its water quality. Another initiative involving the State and Central Governments is the restoration of the Yamuna River. In addition, government agencies and non-governmental organizations have made a number of other endeavors. Seven Rajasthani streams have been revitalized by the Tarun Bharat Sanga. This is one of the best illustrations of how grassroots community involvement can revitalize a river. Following the formulation of a thorough restoration plan, the Nagpur Municipal Corporation began restoring the Nagpur rivers in 2013. The Government of Tamil Nadu's Chennai River Restoration Trust has started the ecorestoration of the Cooum River, Adyar Creek, and Adyar Creek Estuary (Puranik & Kulkarni, 2014).

4.1.2 Watershed Management – An Overview

A watershed is defined by topography and is the whole region above a point on a stream or river that drains past that point. It can be a good socioeconomic-political unit for management planning and execution in addition to being a helpful unit for physical studies. Watersheds can range in size from a small area drained by a freshwater body to thousands of square kilometers (Wang et al., 2016).

Watershed management (WSM) is defined as "the study of the relevant characteristics of a watershed aimed at the sustainable distribution of its resources and the process of creating and implementing plans, programs, and projects to sustain and enhance watershed function that affect the plant, animal, and human communities within a watershed boundary" (Wang et al., 2016). Because it works well in a variety of climates, watershed management is one of the most extensively used technologies. Watershed technology can also be used to improve and preserve the fertility of the soil, which is declining drastically as a result of intensifying agriculture. More significantly, it attracted policy interest in certain nations because of its ability to preserve water and soil in arid areas (Ratna Reddy et al., 2017).

The idea of integrated watershed management has gained popularity as a result of the development of watershed management. By combining different social, technical, and institutional aspects with conservation, social, and economic goals, integrated watershed management expands on the fundamentals of watershed management (German et al., 2007). Examining more than just the hydrology of watershed resources is crucial, as demonstrated by the integrated watershed management method. It aims to preserve ecosystem services and biodiversity while balancing the requirements of humans and the environment (Bakker, 2012). The rapid advancement of computer science and geospatial technology is largely responsible for the improvements made to this integrative approach in recent decades. Explicit interfaces for communication between decision makers, communities, public interest groups, and other stakeholders have been made possible by the integration of remote sensing imagery, geographical information systems, global positioning systems, metaanalysis techniques, computer simulation models, and access to extensive databases (Wang et al., 2016).

A resource region's development is planned as part of the WSM concept. Both the common and private property regimes are included in the resource region. The best and most sustainable use of the resources is the main objective of planning. Therefore, treatment of watersheds should be a comprehensive approach that incorporates several biophysical, socioeconomic, and hydrological factors. As a technology based on land, WSM might assist prevent soil erosion, preserve and enhance in situ soil moisture, and enhance water resources, particularly groundwater in rainfed areas. It basically implies using both technical and non-technical measures to improve watershed or rainfall catchment region management. To adapt the biophysical and hydrological circumstances to the demands of the local communities, technical interventions are necessary. To increase the effectiveness of the technical interventions, non-technical (socioeconomic) tools are also needed (Ratna Reddy et al., 2017).

4.1.3 River Restoration Techniques

Prioritizing restoration actions, defining the restoration aim, and clarifying the restoration planning process are all important. In order to prioritize restoration goals and design restoration solutions, a restoration program must first determine the damaged functions of the affected rivers, the extent of the damage, and the limits of the river systems.

Fig. 4.1 Techniques of river restoration

The primary issues that damaged rivers deal with are either deteriorating water quality due to pollution discharges, changed hydrological processes brought on by the installation of hydraulic facilities, or both. The diagnosis of river risks can be used to inform the implementation of effective restoration techniques (Pan et al., 2016). The ways in which restoration can be characterized are discussed in Fig. 4.1.

4.1.4 Catchment Management

By better managing a river's catchment, watershed management encompasses actions intended to enhance several facets of river, such as quality of water and flow pattern. Catchment managing encompasses both direct interventions, like re-vegetation, to improve the catchment's state and legislative measures, such land use regulations.

A river's terrain, geology, land cover, location within the river network, and climate can all affect surface water runoff and the kind and makeup of flows that enter the river (Poff & Zimmerman, 2010). Notably, the supply of water resources can be impacted by watershed deterioration (Le Maitre et al., 2007). Addressing catchment characteristics that have altered the amount, makeup, and timing of sediment, water, and other elements entering a river system is the goal of catchment management and restoration.

Improved grazing land management to reduce fine sediment yields in the catchment; flood risk management measures in the upper and middle areas of the catchment to retain water and reduce rapid runoff into rivers; and direct interventions to improve the condition of existing land, including land and water conservation strategies, such as re-vegetating the catchment or rehabilitating gullies and erosion-prone zones, are all examples of catchment restoration (Glendell & Brazier, 2014).

A lot of catchment restoration strategies include direct farm-level intervention to lessen diffuse agricultural pollution, like sediment loss and nutrient runoff. For instance, ditches and lime-sand filters can be used to reduce nutrient runoff (Kirkkala et al., 2012); buffer strips between fields and rivers can be designated (Catchment Restoration Fund, 2014); and erosion in gullies, along drains, and along creeks can be controlled by fencing and restoring eroded banks (Alt et al., 2009).

4.1.5 Flow Modification

The goal of flow modification is to return a river to its natural flow pattern. This category mainly focuses on legislative and regulatory actions intended to modify river flows, such as those that alter the "volume, frequency, duration, or timing of flows." One of the key factors influencing the health of rivers is flow regime (Bunn & Arthington, 2002). The physical habitat of streams and rivers is largely determined by flow, river species have changed their life cycles in response to natural flow patterns, flows preserve lateral (such as a river-floodplain connection) and longitudinal (such as upstream–downstream) natural patterns, and unwanted pest species spread more easily when flow regimes change.

Various components of the flow regime carry out distinct ecological tasks, including channel form maintenance, habitat maintenance, spawning, and wetland flooding (Speed et al., 2013). The capacity of rivers to carry out their tasks may be affected by changes in their flows brought about by instream management (like building dams) and diversions (like using water for irrigation). Restoring ecologically significant hydrological variables is usually the goal of restoration efforts that entail altering the flow regime.

Restoring environmental flows can be accomplished in one of the following ways: 1. Regulations requiring dam operators and other instream infrastructure operators to manage their operations in a way that improves environmentally significant flows. 2. Modifications to the water allocation regime may restrict the quantity and timing of water consumption by water users granting abstraction licenses. For instance, setting restrictions that limit or prohibit water abstractions during significant flow events, or lowering the overall volume of water that can be taken under water licenses. 3. In order to achieve restoration results, obtaining water licenses aids in keeping the water associated with those licenses in the river channel. To reduce the amount of water being extracted from the river, for instance, a government agency, community organization, or environmental NGO could buy water licenses with the intention of essentially retiring the licenses. Such actions depend on the presence of a moderately complex water allocation system (Speed et al., 2013).

4.1.6 Dam Removal

Removing dams or altering or adapting already-existing infrastructure to lessen its detrimental ecological effects is referred to as dam removal or retrofitting. Longitudinal hydrological connectedness is hampered by dam building, and reservoir ecological processes can somewhat change the temperature and flow regime. (Mladenovic et al., 2013; Pan et al., 2016). The river may be redirected or pumped around the location if a dam is completely removed "in the dry." As an alternative, dams could be gradually reduced over several years in order to remove them in phases and restore the original impoundment more gradually (American Rivers and Trout Unlimited, 2002).

River health can be significantly impacted by dams and other instream infrastructure, including changes to the pattern of flow, sediment and nutrient entrapment behind a barrier, and aquatic life migratory hindrance. Particularly, storage dams and hydroelectric peaking facilities have the potential to upset the flow regime and cause significant daily and seasonal changes that deviate significantly from the levels of natural flow. Degradation of a river's channel, flood zone, and coastal delta, as well as the loss of biologically significant habitat, can arise from a decrease in sediment migrating downstream. The survival of populations of many riverine species depends on maintaining natural patterns of longitudinal and lateral connection. One of the main causes of habitat fragmentation is dams, which limit species' ability to migrate up and down the river corridor (World Commission on Dams, 2000). By removing obstacles to upstream and downstream migration, reestablishing natural seasonal flow variations, preventing siltation of breeding and feeding habitat above the dam, permitting debris, small rocks, and nutrients to pass below the dam, eliminating unnatural fluctuations in temperature below the dam, and removing turbines that kill fish, dam removal benefits aquatic organisms, especially fish (American Rivers and Trout Unlimited, 2002).

4.1.7 Riparian Management

The re-vegetation of the riparian zone is one aspect of riparian management, whereas controlling or eliminating invasive species, like cattle or weeds, is another aspect of it. Between the river and the activity in the nearby watershed, the riparian zones serve as a buffer. The "food web" and other crucial connections between the terrestrial and water habitats can be preserved, sediment runoff can be captured, and nutrients can be filtered by the riparian zone. In addition to having a high diversity of flora and fauna, riparian zones can be significant habitats in and of themselves. Because riparian vegetation provide shade, it affects the temperature of the water, especially in smaller rivers and streams.

River health may suffer from a variety of consequences when riparian buffers are removed, such as enhanced stream sedimentation and turbidity due to weakened banks (Barling & Moore, 1994), reduced capacity to trap and remove contaminants (Lowrance et al., 1997) and nutrients from runoff (Vought et al., 1994), increased water temperature and light penetration, which leads to increased primary production, decreased leaf litter and terrestrial invertebrate inputs, which decrease production (Pusey & Arthington, 2003), decreased stream width and reduced woody debris inputs (Sweeney & Czapka, 2004), and decreased aquatic habitat are some of the negative effects of riparian buffer removal on river health (Baxter et al., 2005).

The management of riparian vegetation, including land use and planning controls to limit the removal of riparian vegetation as well as active interventions to maintain the health of riparian vegetation; management of livestock, such as banning them from the riparian zone through fencing or creating crossings and watering points to reduce the need for stock to access the river channel; management of invasive weeds; and re-vegetation of the riparian zone, either actively (i.e., planting vegetation) or passively (removing disturbances and allowing for natural recovery) (Speed et al., 2016).

4.1.8 Instream Habitat Improvement

In natural settings, a river's intricate flow patterns play a key role in propelling the geomorphic processes that mold the instream habitat and channel elements. River channels have frequently been altered by humans through dredging, relocation, containment, and straightening. These changes frequently result in the river becoming disconnected from the floodplain and the loss of instream habitat.

Large woody debris is being reintroduced to increase physical habitat and offset stream incision. Large woody debris can be reintroduced in more complicated ways, such as artificial logjams with multiple large trees anchored into the riverbed or bank, or simply by planting fast-growing tree species on the riverbank that will be recruited over time by the natural process of bank erosion (Mott, 2010). By lowering flow velocity, groynes, retards, and log sills lessen the erosion and encourage the accumulation of silt on the outer banks. This may lead to the development of habitat (Yossef & De Vriend, 2011).

4.1.9 Bank Stabilization

The goal of bank stabilization is to minimize or completely stop bank material from eroding or sinking into the river channel (Florsheim et al., 2008). High turbidity and sediment deposition downstream are caused by erosion of bed and bank material, both of which can be harmful to biota. Excessive bank erosion rates are caused by underlying processes that are intended to be addressed by bank stabilization techniques. Where changes in river ecosystems, such as a changed flow regime, land development, or the loss of riparian vegetation, have caused faster erosion, stabilization may be necessary (Speed et al., 2016).

In addition to preventing the loss of vegetation and improving habitat for biota, stabilization of banks assists in reducing sediment loads and improving recreational opportunities by the riverbank by permitting the creation of new recreational regions.

4.1.10 Chanel Reconfiguration

Around world, channel-based or "hydromorphological" restoration initiatives are very prevalent. Where disturbances or previous management practices have simplified the channel form—for instance, when rivers have been straightened or otherwise channelized—or where instream habitat has been lost—for instance, as a result of mining or dredging—these strategies seek to restore the diversity of physical habitat (Bernhardt & Palmer, 2007). Projects to reconfigure channels are typically a reaction to disruptions that are so severe that it is doubtful that the channel can return to its ideal state in a reasonable amount of time by removing current pressures and stressors and altering the flow regime. Improved water conveyance in flood-prone areas—usually by slowing floodwater flow and lowering flood peaks—as well as the mitigation of unstable river beds and banks, increased sediment transport, riparian habitat enhancement, and enhanced possibilities for recreation are common justifications for channel reconfiguration (Roni & Beechie, 2012).

4.1.11 Water Quality Management

Restoring water quality is a major objective of "river restoration" programs and initiatives as declining quality of water is one of the biggest hazards to river. The several metrics that are used to assess the quality of water, such as "temperature, turbidity, conductivity, pH, nutrient levels, and the presence of heavy metals," can all be improved by management actions.

River water pollution mostly consists of organic pollutants, heavy metal pollutants, nutrients (such as nitrogen, phosphorus, and others), and so forth. Point source pollution and non-point source pollution are the two categories of external contaminants that affect the river. Because of its magnitude and broad range of pollutants, non-point source pollution is more challenging to control than point source pollution (Pan et al., 2016). To lower the quantity and kind of waste that enters the river ecosystem, point source pollution loads may be reduced by lowering the quantity and kind of pollutants generated, for instance, by bettering operational procedures and enhancing the processing of pollutant waste. Water treatment techniques that lower the risk of water contamination and the quantity of water that needs to be treated include surface water diversion and interception, ore processing water recycling, drainage water collection and storage in tailing dams, evaporation ponds, and the installation of liners and covers for waste rock and ore piles (Lottermoser, 2010). Reverse osmosis, ion exchangers, membrane filters, and settling ponds are examples of active treatment methods that remove impurities from the water column. The use of microorganisms to reduce metal precipitation, plant uptake of contaminants (bioremediation), and filtering through soil and artificial wetlands are examples of passive treatment techniques (Wolkersdorfer, 2008). Similar to wastewater treatment plants, inline filtration systems are frequently used to manage industrial pollutant loads; however, biological filtration, which uses bio-retention basins and artificial wetlands, is frequently a more economical method of treating effluents.

In certain cases, the contaminants that are already present in a river ecosystem may be of a kind or concentration that makes it impossible to attain the necessary river health outcomes in a reasonable amount of time by merely lowering additional pollution inputs and enhancing flows. In some situations, more direct interventions might be needed, such as physical techniques like dredging to remove contaminated sediments or isolation and coverage, which include covering a contaminated riverbed with layers of fine sand or mud to isolate the contamination section; enhanced flows to dilute pollutants; aeration to artificially raise oxygen levels, especially to aid in the breakdown of organic contaminants (Speed et al., 2016). Algal blooms can be addressed chemically, for instance, by using algaecides. Mechanical methods (such as the gas flotation method, Plocher system, ultrasonic, fixed mechanical facilities, or movable vessels) can be used to successfully remove blue-green algae from areas where they are abundant (Pan et al., 2016). Large expanses of algal blooms can be removed using mechanical techniques, and aquatic ecosystems is unlikely to be adversely affected. Biological techniques include the usage of plants to absorb and eliminate pollutants or the introduction of microorganisms. Microorganisms have a great role in the degradation of contaminants in water. Dosing microorganisms into water can accelerate the breakdown of contaminants when there are no microorganisms present and the water is severely contaminated (Eduardo, 2008).

4.1.12 *Approaches for Watershed Management*

To maximize the development of land, water, and plant resources and sustainably support the basic needs of humans and animals, the integrated watershed management strategy recommends integrating technology within the natural bounds of a drainage area. By providing all the facilities needed for maximum output, this strategy seeks to raise the earning potential of the average person and raise their level of living. To accomplish its goal, integrated watershed management recommends using improved seed varieties, integrated nutrient management, integrated pest management, water harvesting in ponds, groundwater recharge to increase the potential for water resources, stress on crop diversification, and other measures (Singh, 2000).

The consortium model places a strong emphasis on community involvement and group efforts, involving “key stakeholders,” “governmental” and “non-governmental” groups, and other institutions. Multidisciplinary abilities and skills are necessary for watershed management. The remarkable effects in the watershed that have been documented are largely due to easy access and prompt counsel for farmers. These result in increased farmers’ knowledge and their capacity to seek advice from the appropriate individuals when issues emerge. Engineering, social science, farming, animal husbandry, forestry, marketing, and economics are among the transdisciplinary skills needed. Having all the necessary support and expertise in one place isn’t always feasible. In order to increase the efficacy of the many watershed programs and interventions, the consortium method combines the knowledge of several fields (Wani & Garg, 2009).

4.1.13 *Technologies for Watershed Management*

From flooding and landslides to water depletion and soil degradation, watershed can have a range of physical changes. Therefore, integrating a wide range of spatial and temporal data is necessary for the planning and evaluation of WSM.

4.1.14 *Remote Sensing (RS)*

A watershed’s land use/cover, physiography, soil distribution, drainage characteristics, lake temperature, water level, and algal dynamics can all be surveyed, identified, classified, and monitored using this precise and up-to-date data source (Hampton, 2013; Pandey et al., 2007). RS can supply spatial input data for watershed management models either by itself or in conjunction with "geographic information systems (GIS)." This is particularly helpful for isolated, difficult-to-reach locations inside a watershed. Melting of seasonal snow, the development of WSM plans for conservation planning, flood damage and drought assessments, environmental impact evaluations of water resource projects, and many other types of information can also be produced in real time by using RS as model input data. By creating alternate management scenarios, this technology can also help politicians create watershed management plans more effectively (Wang et al., 2016). All things considered, utilizing data that is remotely sensed—like satellite imagery—allows for analysis on a larger spatial scale than would be possible with just in situ data. However, as RS cannot currently offer the required specific species data and more research is needed to fully understand the link between satellite-derived and in situ data, it is not a substitute for in situ observations (Hampton, 2013).

4.1.15 Geographic Information System

"GIS" is an essential tool for managing watersheds because it can be used to illustrate the effects of various management scenarios and to model the effects of human activity on watershed conditions. Understanding the dynamics and processes that influence the biological, chemical, and physical environments of watersheds is made possible by these modeling and visualization tools. Combinations of slope, aspect, and hydrological-response units can be analyzed in hydrological simulation models thanks to the use of GIS, which produces more spatial detail than the majority of other hydrological models (Brooks et al., 2012). The constraints of computer-based models with regard to data preparation and visualization have been addressed by GIS in conjunction with the Internet's better access to data and information (Choi et al., 2005). GIS is a helpful tool for giving decision-makers management information because of its high-quality outputs, simple updating features, and ability to test management solutions (Wang et al., 2016).

4.1.16 Big Data

The term "big data," which refers to the exponential expansion and accessibility of both structured and unstructured data, includes five fundamental features: complexity, volume, velocity, diversity, and variability (Davenport & Dyché, 2013). Large amounts of data can yield more precise analyses that could result in more assured and knowledgeable decision-making, which would therefore enable increased operational effectiveness, lower costs, and lower risk. In order to develop and test different watershed management scenarios, time should be spent gathering comprehensive data. The management of watersheds responsibly and to the best of our ability is crucial not only locally for the watershed of concern but also worldwide (Gijzen, 2013). In order to sort through massive data repositories and find new connections, patterns, and trends—as well as categorize and group information into groups—data mining combines statistical and mathematical methods with pattern recognition technologies (Larose, 2004).

4.1.17 Global Positioning System

A freely available space-based worldwide navigational system, the "Global Positioning System (GPS)" tells users their exact location and time for practically any place on Earth (Brooks et al., 2012). By employing "GPS" to pinpoint the precise location of gauges, monitoring equipment, study plots, etc., data precision can be improved. This data can be added to GIS or database management systems to increase the accuracy and quality of the data (Brooks et al., 2012). Tracking the movement of fish, birds, and mammals within watersheds using GPS and radio-tracking has shown to be extremely beneficial, allowing for in-depth research on the location and timing of specific habitat utilization.

4.1.18 Meta analysis

A statistical method for combining quantitative results from multiple research and condensing them into general conclusions is called meta-analysis. By using this method, a statistically convincing synthesis can be produced by combining research that may not be decisive in and of themselves. Despite being an uncommon method for assessing hydrology, a usefulness analysis shows that it is a potential method for combining findings from multiple research (Locatelli & Vignola, 2009). Policymakers and decision-makers who are working with a wealth of data, frequently conducted as separate studies and reviews, some of which may be contradictory, may find this thorough compilation of material helpful. There must be enough data and research available to synthesize for metaanalysis to be successful, though, which could be a constraint when applying this method to specific watershed components.

4.1.19 Simulation Modeling

A helpful method for achieving optimal management strategies that balance many advantages of land and water resources in a watershed is watershed simulation modeling, also known as hydrologic simulation. In order to do this, watershed processes and their linkages are analyzed, and management scenarios are developed and evaluated that concurrently take into account upstream soil conservation, midstream land use, and downstream reservoir level sediment control (Heathcote, 2009; Lee et al., 2013; Qi & Altinakar, 2013). For past, present, and future streamflow regimes, these models produce water balance data (amount and related hydraulic parameters, source and channel, residence time, etc.) by simulating the dynamic behavior of flow and storage processes (Brooks et al., 2012). Because of this, they are more helpful to managers than model results from earlier studies that offer knowledge and insight into the dynamics of watersheds but might not be applicable outside of the watershed where those studies were conducted (Westervelt, 2001). For the purpose of developing input parameters and analyzing and visualizing simulation results, these models frequently require integration with GIS or remote sensing technology (Heathcote, 2009).

The impact of land-use change, pollution, sustainable resource development (Mocanu et al., 2013), water supply (Chung et al., 2008), soil conservation (Rejani et al., 2016), and stakeholder analysis in integrated watershed management (Mutekanga et al., 2013) are just a few of the many recent studies on watershed decision-making support systems. In order to enable watershed managers and planners create, comprehend, and assess different WSM strategies while taking into consideration the objectives and interests of multiple stakeholders, user-friendly "decision support systems (DSS)" are required. The DSS should incorporate computer programs with elements of GIS, database management systems

(DBMS), decision and simulation models, and user interfaces that are simple to comprehend (Wang et al., 2016).

4.1.20 River Restoration in India

Since they supply many people with energy, affordable transportation, irrigation, drinkable water, and a means of subsistence, rivers play a vital part in Indian culture. The combined catchment area of India's 12 main river basins is 25.3 lakh square kilometers. With a catchment area of 11 lakh square kilometers, the "Ganga, Brahmaputra, and Meghna" rivers combine to form the largest river basin. With catchment areas above one lakh square kilometers, there are four further significant river basins: the "Indus, Mahanadi, Godavari, and Krishna." Eighty-one percent of India's land area is made up of river basins (Bedarkar et al., 2018).

However, the majority of Indian rivers have been reduced to mere sewage-carrying drains due to pollution. Since just 37% of sewage generated is cleaned, untreated sewage dumped into waterways is the main source of pollution. Indian rivers are susceptible to a variety of climate change-related effects, in addition to declining water quality and quantitative stress (Table 4.1). Gosain et al. (2006) quantify how climate change affects twelve Indian rivers. According to the report, the country would face extreme floods in certain areas and severe droughts in others under the GHG scenario. Across India, a number of non-governmental organizations are working to restore rivers. Through a number of programs, including the Thames-Ganges Twinning Programme from 2010 to 2014, the Ganga basins: Dolphin conservation program, the Wetlands Conservation program, policy initiatives, and the Education and Awareness Campaign, WWF India has made a significant contribution to the revitalization of the Ganga River (Bedarkar et al., 2018).

For the purpose of improving water quality, the "Central Pollution Control Board (CPCB)" in India has identified 302 contaminated river segments among the country's 275 rivers. The combined length of the contaminated river segments is 12,363 kilometers. Since 1985, the Indian government has been working on a large-scale river restoration project. In 1985, the centrally supported "Ganga Action Plan (GAP)" was introduced, marking the beginning of river restoration in India. It was expanded to encompass additional rivers in 1995 under the name "National River Conservation (NRCP)." By implementing pollution abatement programs in designated polluted river segments, the NRCP aimed to enhance the water quality of rivers.

Three important case studies on Adyar Poonga, Mula-Mutha, and Sabarmati river restoration. In order to improve the environment, build social infrastructure, and promote sustainable development, the "Sabarmati riverfront development project" has been hailed as one of India's first river restoration initiatives. It has served as an inspiration for other significant river projects, such as the Yamuna Action Plan, Gomti Action Plan, and others (Bedarkar et al., 2018). The pollution abatement project of river Mula-Mutha, which has received the highest cost allocation under "NRCP," comprises building structures for enhancing the treatment capacity of present sewage, usage of technology for infrastructural facilities monitoring and generating public awareness. The rehabilitation of the Adyar Poonga river is the subject of the third case study. By repairing the deteriorated region and recovering the coastal ecology of Adyar Creek, which was caused by the exploitation of the wetland due to urban sprawl, the ecological park was established with the primary goal of beautifying the area surrounding the Adyar estuary and creek.

4.1.21 Watershed Management Programs in India

With full national support, the watershed development plan for shifting agriculture regions was introduced in 1987–1988 and covered all seven northeastern states as well as the states of "Andhra Pradesh" and "Orissa." Through suitable soil conservation and watershed management methods in impacted areas, the program sought to assist 25,000 Jhumia families. The impacted areas have stabilized thanks to these actions. This program covers 0.59 million hectares of land until 2011–2012. Based on the dual ideas of integrated WSM and sustainable farming systems, the "National Watershed Development Project for Rainfed Areas (NWDPRA)" program was introduced in "1990–1991" across twenty-five states and two union territories. From 2000 to 2001, the NWDPRA program was incorporated into the "Scheme for Macro Management of Agriculture (MMA)." Currently, twenty-eight states and two union territories are implementing this plan. An area of 268.35 acres was developed up till 2011–2012. "The Desert Development Program (DDP)," "the Drought Prone Area Program (DPAP)," and the Food for Work Program were started in 1972 and 1973. In 1987, these programs embraced the watershed concept. From the beginning to 2011–2012, an area of 15.2 million hectares under

"DPAP" and 222.39 acres under "DDP" were covered. Watershed-based wasteland development was another goal of the "National Wasteland Development Board's 1989 Integrated Wasteland Development Projects Scheme (IWDP)." IWDP covered 252 acres between its establishment and 2011–2012. Various watershed programs have been implemented through external financial institutions like the "World Bank, Swiss Development Cooperation, Danish International Development Agency, and Department for International Development" in addition to national watershed programs (Bhan, 2013). Under Indian policy, WSM has transformed in social, economical, and ecological ways. Also, experience of India in WSM is considered pivotal in understanding the rural livelihoods and resource degradation issues in rainfed regions. Most South Asian nations, including Nepal, Sri Lanka, Pakistan, and others, have now embraced the participatory and livelihoods approach to WSM, while some of these nations have not yet advanced institutionally (Ratna Reddy et al., 2017).

4.1.22 Anthropogenic Activities and Environmental Crisis in the Rivers

Rivers offer a unique set of products and services that the general public values much and that are closely related to the dynamics of their flow and how that flow interacts with the environment. However, the majority of rivers are found in watersheds that are somewhat strained by human endeavors such as construction, dams, or extraction (Fig. 4.2). River channel pattern and floodplain sedimentation are mostly determined by land cover and agricultural practices, which are also important contributors in river sediment supply (Macklin & Lewin, 2019). With reach-scale confinement, urban river changes over the centuries have mostly occurred inside growing built-up areas. However, they can also have downstream effects, particularly through the dissemination of contaminants, the alteration of natural sediment dynamics, and more ostentatious and higher runoff produced by paved surfaces and lined streams (Chin, 2006). Water and energy needs were met in part by the construction of dams and the storage of water in reservoirs. Although dams have a significant positive impact on local or regional economics, the natural health of rivers is severely harmed by them. Dams obstruct the upstream and downstream migration of aquatic animals that are mobile, like anadromous or catadromous fish. Additionally, dams significantly impact river ecosystems by modifying the natural flow regime, water temperature, sediment movement, and vegetation communities in floodplains (Palmer et al., 2009). Rivers have been significantly impacted by the conversion from natural land cover to anthropogenic purposes, especially because it has diminished the natural capacity of surrounding riverine ecosystems and landscapes to absorb and filter water flows. Significant contaminants, such as excessive sediment, fertilizers, and pesticides, have been introduced into freshwater environments by farming (Silk & Ciruna, 2013). Industrial and municipal pollutants were disposed into tiny rivers where pollutants were collected since waterpower was the catalyst for the "Industrial Revolution" (Lewin, 2013).

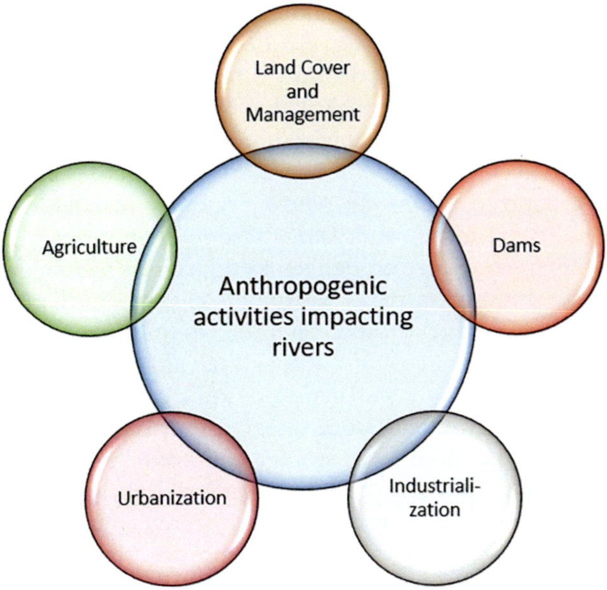

Fig. 4.2 Anthropogenic activities affecting rivers

4.1.23 *Impact of Climate Change in Achieving Environmental Sustainability and River Resilience*

Because they migrate laterally across the terrain and alter the channels, rivers are naturally active systems that are always acclimating to variations in sediment and water inputs. A healthy river adapts to changes in the environment and climatic regime by going through these changes. However, because many rivers are impacted by development, dams, and water extraction, their capacity to adapt to changes in the flow of water and material may be compromised. Additionally, the new temperature and precipitation regimes predicted as a result of climate change will occur much more quickly than historical climate shifts (Palmer et al., 2005). Managers might have to react to natural disasters like floods, droughts, erosion, and species extinction on the spot if proper planning is not done beforehand. Severe flow events can cause significant river bank erosion, endanger delicate riparian ecosystems and perhaps lowering downstream water quality due to increased suspended sediment loads. To prevent future environmental deterioration or a species' extinction, reactive management measures can be required. Rivers in extensively urbanized basins or whose water is managed for various uses will require the most costly and severe reactive interventions. River restoration efforts to restore eroded banks or projects to rebuild in-stream habitat may be required in rivers with notable surges in discharge. More stormwater infrastructure should be installed to stop similar incidents in the future. Other strategies to absorb high flow energy and preserve water quality include the construction of off-channel storage basins or wetland areas. Removing silt from reservoir bottoms may be a temporary fix to enable additional water storage and maybe prevent catastrophic dam breaches; but, this may become logistically challenging if rates of sediment accumulation in reservoirs rise in future climates (Palmer et al., 2009). Reactive projects could include replanting native vegetation by the river with drought-resistant vegetation, removing unwanted non-native species that establish themselves, or creating passage for fishes to allow to move between isolated river reaches during drought times in areas with higher temperatures and less precipitation. If dams are in place, summer flow releases could be utilized to support wildlife and plants in drying-up downstream river reaches. Other environmental advantages, such as improved water quality and regenerated fish populations, may result from proactive steps to restore rivers' natural ability to mitigate the effects of climate change. Levee-setbacks to remove infrastructure from the floodplain, absorb floods, and permit the recovery of riparian vegetation are a few examples of such interventions, as are stormwater management in developed basins and land purchase along the river (Palmer et al., 2008). Establishing drought-tolerant plant species may aid in preventing erosion of the riparian corridor along rivers in areas that are predicted to endure hot, dry spells. The adaptive potential of species may be enhanced by concentrating on boosting genetic diversity and population size through plantings or fish stocking. Increased physical habitat variability in the channel may be beneficial to aquatic fauna, and river fauna may be protected in a number of ways by replanting or expanding any deteriorated riparian buffers, such as by offering additional shade and preserving sources of allochthonous input (Brown, 2003).

4.2 Conclusion

India needs to manage its rivers in a methodical and scientific manner. Research confirms that Indian rivers are susceptible to the "frequency and intensity" of climate change impacts. Therefore, we must manage droughts and floods in an adaptive manner. Indian river restoration initiatives ought to take a scientific approach, relying on engineering design, scientific monitoring, and environmental modeling methodologies. Additionally, there must to be more community interest and participation. River restoration should be viewed as an ongoing, multi-year endeavor, with a restoration strategy that periodically adapts to the demands and difficulties of economic expansion. More significantly, each restoration plan should include specific, quantifiable goals that can be connected to effect analysis.

It is clear that WSM has evolved to a more comprehensive method to resource management, using management techniques—integrated and adaptive—to take into consideration the landscape's components. This comprehensive strategy has been greatly aided by technological developments. Sharing information within and between disciplines is now easier, data collection methods and model development are more accurate, and integrated, multilevel analysis that produces more comprehensive information about the watershed system from an ecological and social perspective is now feasible. Decision-makers may now evaluate a watershed's condition and enact the required rules with the use of more thorough and multifaceted information thanks to advancements in technology and watershed management. Social problems, environmental degradation, and the health and management of a watershed can all be addressed by the application of integrated management strategies, sophisticated data collection and analysis techniques, and the consideration of both ecological and socioeconomic concerns.

References

Alt, S., Jenkins, A., & Lines-Kelly, R. (2009). *Saving soil: A landholder's guide to preventing and repairing soil erosion.*

American Rivers and Trout Unlimited. (2002). *Exploring Dam Removal: A decision-Making Guide.* http://www.americanrivers.org/assets/pdfs/dam-removal-docs/Exploring_Dam_Removal-A_Decision-Making_Guide6fdc.pdf?477ed5

Bakker, K. (2012). Water security: Research challenges and opportunities. *Science, 337*(6097), 914–915. https://doi.org/10.1126/science.1226337

Barling, R. D., & Moore, I. D. (1994). Role of buffer strips in management of waterway pollution: A review. *Environmental Management, 18*(4), 543–558. https://doi.org/10.1007/BF02400858

Baxter, C. V., Fausch, K. D., & Carl Saunders, W. (2005). Tangled webs: Reciprocal flows of invertebrate prey link streams and riparian zones. *Freshwater Biology, 50*(2), 201–220. https://doi.org/10.1111/j.1365-2427.2004.01328.x

Bedarkar, M., Dhingra, U. G., Mishra, M., & Gopalkrishnan, S. (2018). Analysis of global river restoration experiences: Learnings and policy measures in the Indian context. *Asian Journal of Water, Environment and Pollution, 15*(2), 203–215. https://doi.org/10.3233/AJW-180033

Bernhardt, E. S., & Palmer, M. A. (2007). Restoring streams in an urbanizing world. *Freshwater Biology, 52*(4), 738–751. https://doi.org/10.1111/j.1365-2427.2006.01718.x

Bhan, S. (2013). Land degradation and integrated watershed management in India. *International Soil and Water Conservation Research, 1*(1), 49–57. https://doi.org/10.1016/S2095-6339(15)30049-6

Botle, A., Salgaonkar, S., Tiwari, R., & Barabde, G. (2024). Unveiling heavy metal pollution dynamics in sediments of river Ulhas, Maharashtra, India: A comprehensive analysis of anthropogenic influence, pollution indices, and health risk assessment. *Environmental Geochemistry and Health, 46*(10), 419. https://doi.org/10.1007/s10653-024-02208-8

Brooks, K. N., Ffolliott, P. F., & Magner, J. A. (2012). *Hydrology and the management of watersheds*. Wiley. https://books.google.com/books?hl=en&lr=&id=tVWX47zODhIC&oi=fnd&pg=PR13&dq=Brooks+KN,+Ffolliott+PF,+Magner+JA+(2013)+Hydrology+and+the+management+of+watersheds,+4th+edn.+Wily-Blackwell,+Ames,+p+552&ots=XL8dm4oLpB&sig=Hq7Q8h-2TxcOrRyOcyavU91cKdI

Brown, B. L. (2003). Spatial heterogeneity reduces temporal variability in stream insect communities. *Ecology Letters, 6*(4), 316–325. https://doi.org/10.1046/j.1461-0248.2003.00431.x

Bunn, S. E., & Arthington, A. H. (2002). Basic Principles and Ecological Consequences of Altered Flow Regimes for Aquatic Biodiversity. *Environmental Management, 30*(4), 492–507. https://doi.org/10.1007/s00267-002-2737-0

Catchment Restoration Fund. (2014). *Environment agency summary report. Department for Environment Food and Rural Affairs*. https://assets.publishing.service.gov.uk/government/uploads/system/uploads/attachment_data/file/330464/pb14179-catchment-restoration-fund-report.pdf

Chattopadhyay, S., & Harilal, K. N. (2017). River restoration in Kerala: Developing a co-evolutionary Framework and river restoration action plan for Thiruvananthapuram city. *Centre for Development Studies Thiruvananthapuram, India*. https://cds.edu/wp-content/uploads/2021/02/River-Restroration_Chattopadhyay-report.pdf

Chin, A. (2006). Urban transformation of river landscapes in a global context. *Geomorphology, 79*(3–4), 460–487.

Choi, J.-Y., Engel, B. A., & Farnsworth, R. L. (2005). Web-based GIS and spatial decision support system for watershed management. *Journal of Hydroinformatics, 7*(3), 165–174.

Chung, G., Lansey, K., Blowers, P., Brooks, P., Ela, W., Stewart, S., & Wilson, P. (2008). A general water supply planning model: Evaluation of decentralized treatment. *Environmental Modelling & Software, 23*(7), 893–905.

Davenport, T. H., & Dyché, J. (2013). Big data in big companies. *International Institute for Analytics, 3*(1–31). https://www.academia.edu/download/32021923/Big-Data-in-Big-Companies.pdf

Eduardo, D. (2008). *Microbial biodegradation: Genomics and molecular biology*. Caister Academic Press.

Florsheim, J. L., Mount, J. F., & Chin, A. (2008). Bank erosion as a desirable attribute of rivers. *BioScience, 58*(6), 519–529.

German, L., Mansoor, H., Alemu, G., Mazengia, W., Amede, T., & Stroud, A. (2007). Participatory integrated watershed management: Evolution of concepts and methods in an ecoregional program of the eastern African highlands. *Agricultural Systems, 94*(2), 189–204.

Gijzen, H. (2013). Big data for a sustainable future. *Nature, 502*(7469), 38–38.

Gleick, P. H. (2003). Global Freshwater Resources: Soft-Path Solutions for the 21st Century. *Science, 302*(5650), 1524–1528. https://doi.org/10.1126/science.1089967

Glendell, M., & Brazier, R. (2014). *Evaluating the benefits of catchment management for multiple ecosystem services. Environment Agency, UK*. www.gov.uk/government/uploads/system/uploads/attachment_data/file/366175/Evaluating_the_water_quality_benefits_of_land_management_change_-_report.pdf

Gosain, A. K., Rao, S., & Basuray, D. (2006). Climate change impact assessment on hydrology of Indian river basins. *Current Science, 1*, 346–353.

Hampton, S. E. (2013). Understanding Lakes Near and Far. *Science, 342*(6160), 815–816. https://doi.org/10.1126/science.1244732

Heathcote, I. W. (2009). *Integrated watershed management: Principles and practice*. Wiley.

Kirkkala, T., Ventelä, A.-M., & Tarvainen, M. (2012). Long-Term Field-Scale Experiment on Using Lime Filters in an Agricultural Catchment. *Journal of Environmental Quality, 41*(2), 410–419. https://doi.org/10.2134/jeq2010.0429

Koehn, J. D., Nicol, S. J., & Lucas, A. M. (1997). In R.J. Banens, P. Crabb (Eds.), *River restoration through integrated management* (pp. 87–94).

Larose, D. T. (2004). *Discovering knowledge in data: An introduction to data mining* (1st Ed.). Wiley. https://doi.org/10.1002/0471687545

Le Maitre, D. C., Milton, S. J., Jarmain, C., Colvin, C. A., Saayman, I., & Vlok, J. H. (2007). Linking ecosystem services and water resources: Landscape-scale hydrology of the Little Karoo. *Frontiers in Ecology and the Environment, 5*(5), 261–270. https://doi.org/10.1890/1540-9295(2007)5[261:LESAWR]2.0.CO;2

Lee, Y., Yoon, T., & Shah, F. A. (2013). Optimal Watershed Management for Reservoir Sustainability: Economic Appraisal. *Journal of Water Resources Planning and Management, 139*(2), 129–138. https://doi.org/10.1061/(ASCE)WR.1943-5452.0000232

Lewin, J. (2013). Enlightenment and the GM floodplain. *Earth Surface Processes and Landforms, 38*(1), 17–29. https://doi.org/10.1002/esp.3230

Locatelli, B., & Vignola, R. (2009). Managing watershed services of tropical forests and plantations: Can meta-analyses help? *Forest Ecology and Management, 258*(9), 1864–1870.

Lottermoser, B. (2010). Mine Wastes. *Springer, Berlin Heidelberg*. https://doi.org/10.1007/978-3-642-12419-8

Lowrance, R., Altier, L. S., Newbold, J. D., Schnabel, R. R., Groffman, P. M., Denver, J. M., Correll, D. L., Gilliam, J. W., & Robinson, J. L. (1997). Water quality functions of riparian forest buffers in Chesapeake Bay watersheds.

Environmental Management. https://repository.si.edu/bitstream/handle/10088/17707/serc_Lowrance_etal_1997_EnvironManage_21_687_712.pdf

Macklin, M. G., & Lewin, J. (2019). River stresses in anthropogenic times: Large-scale global patterns and extended environmental timelines. *Progress in Physical Geography: Earth and Environment, 43*(1), 3–23. https://doi.org/10.1177/0309133318803013

Mladenovic, M. B., Kolarov, V., & Damjanovic, V. (2013). Sediment regime of the Danube River in Serbia. *International Journal of Sediment Research, 28*(4), 470–485.

Mocanu, M., Vacariu, L., Drobot, R., & Muste, M. (2013). Information-centric systems for supporting decision-making in watershed resource development. In *2013 19th International Conference on Control Systems and Computer Science* (pp. 611–616). https://ieeexplore.ieee.org/abstract/document/6569327/

Mott, N. (2010). *Fish live in trees too: Rive rehabilitation and large woody debris*. Staffordshire Wildlife Trust.

Mutekanga, F. P., Kessler, A., Leber, K., & Visser, S. (2013). The use of stakeholder analysis in integrated watershed management. *Mountain Research and Development, 33*(2), 122–131.

Palmer, M. A., Bernhardt, E. S., Allan, J. D., Lake, P. S., Alexander, G., Brooks, S., Carr, J., Clayton, S., Dahm, C. N., Follstad Shah, J., Galat, D. L., Loss, S. G., Goodwin, P., Hart, D. D., Hassett, B., Jenkinson, R., Kondolf, G. M., Lave, R., Meyer, J. L., & Sudduth, E. (2005). Standards for ecologically successful river restoration. *Journal of Applied Ecology, 42*(2), 208–217. https://doi.org/10.1111/j.1365-2664.2005.01004.x

Palmer, M. A., Lettenmaier, D. P., Poff, N. L., Postel, S. L., Richter, B., & Warner, R. (2009). Climate Change and River Ecosystems: Protection and Adaptation Options. *Environmental Management, 44*(6), 1053–1068. https://doi.org/10.1007/s00267-009-9329-1

Palmer, M. A., Reidy Liermann, C. A., Nilsson, C., Flörke, M., Alcamo, J., Lake, P. S., & Bond, N. (2008). Climate change and the world's river basins: Anticipating management options. *Frontiers in Ecology and the Environment, 6*(2), 81–89. https://doi.org/10.1890/060148

Pan, B., Yuan, J., Zhang, X., Wang, Z., Chen, J., Lu, J., Yang, W., Li, Z., Zhao, N., & Xu, M. (2016). A review of ecological restoration techniques in fluvial rivers. *International Journal of Sediment Research, 31*(2), 110–119. https://doi.org/10.1016/j.ijsrc.2016.03.001

Pandey, A., Chowdary, V. M., & Mal, B. C. (2007). Identification of critical erosion prone areas in the small agricultural watershed using USLE, GIS and remote sensing. *Water Resources Management, 21*(4), 729–746. https://doi.org/10.1007/s11269-006-9061-z

Poff, N. L., & Zimmerman, J. K. H. (2010). Ecological responses to altered flow regimes: A literature review to inform the science and management of environmental flows. *Freshwater Biology, 55*(1), 194–205. https://doi.org/10.1111/j.1365-2427.2009.02272.x

Postel, S., & Richter, B. (2012). *Rivers for life: Managing water for people and nature*. Island Press. https://books.google.com/books?hl=en&lr=&id=jKgthwMcvQEC&oi=fnd&pg=PR3&dq=Postel,+S.+and+B.+Richter+(2012).+Rivers+for+Life:+Managing+Water+for+People+and+Nature.+Island+Press.&ots=Y2_B15wc4e&sig=havEYarBEefCegG9jLAa1GTF-vQ

Puranik, V., & Kulkarni, A. (2014). Study of rejuvenation and restoration aspects of Nag river. *International Journal of Applied Research and Studies, 3*(9), 1–7.

Pusey, B. J., & Arthington, A. H. (2003). Importance of the riparian zone to the conservation and management of freshwater fish: A review. *Marine and Freshwater Research, 54*(1), 1–16.

Qi, H., & Altinakar, M. S. (2013). Integrated Watershed Management with Multiobjective Land-Use Optimizations under Uncertainty. *Journal of Irrigation and Drainage Engineering, 139*(3), 239–245. https://doi.org/10.1061/(ASCE)IR.1943-4774.0000537

Ratna Reddy, V., Saharawat, Y. S., & George, B. (2017). Watershed management in South Asia: A synoptic review. *Journal of Hydrology, 551*, 4–13. https://doi.org/10.1016/j.jhydrol.2017.05.043

Rejani, R., Rao, K. V., Osman, M., Srinivasa Rao, Ch., Reddy, K. S., Chary, G. R., Pushpanjali, & Samuel, J. (2016). Spatial and temporal estimation of soil loss for the sustainable management of a wet semi-arid watershed cluster. *Environmental Monitoring and Assessment*, *188*(3), 143. https://doi.org/10.1007/s10661-016-5143-4

Roni, P., & Beechie, T. (2012). *Stream and watershed restoration: A guide to restoring riverine processes and habitats*. Wiley. https://books.google.com/books?hl=en&lr=&id=mvXiu-4oEz8C&oi=fnd&pg=PR3&dq=Roni,+P.+and+Beechie,+T.+2013.+Stream+and+Watershed+Restoration.+A+Guide+to+Restoring+Riverine+Processes+and+Habitats.+John+Wiley+%26+Sons,+Ltd.+Hoboken,+New+Jersey&ots=lTXL29rtxB&sig=JwkoblgMElF-sALf8eFPtqStKQg

Silk, N., & Ciruna, K. (2013). *A practitioner's guide to freshwater biodiversity conservation*. Island Press. https://books.google.com/books?hl=en&lr=&id=5qAUpnCBKNwC&oi=fnd&pg=PR2&dq=Silk+N,+Ciruna+K+(2005)+A+practitioner%E2%80%99s+guide+to+freshwater+biodiversity+conservation.+Island+Press,+Washington,+DC,+p+409&ots=MyscGPGcxD&sig=soCr-v2NfkjiyYFuTZIjc6DDVY0

Singh, R. V. (2000). *Watershed planning and management*. Yash Publishing House.

Speed, R., Tickner, D., Naiman, R., Gang, L., Sayers, P., Yu, W., Yuanyuan, L., Houjian, H., Jianting, C., Lili, Y., & Zhongnan, Z. (2016). *River restoration: A strategic approach to planning and management*. UNESCO Publishing.

Speed, R., Yuanyuan, L., Zhiwei, Z., Le Quesne, T., & Pegram, G. (2013). *Basin water allocation planning: Principles, procedures and approaches for basin allocation planning*. Asian Development Bank, GIWP, UNESCO, and WWF-UK. https://core.ac.uk/download/pdf/161801145.pdf

Sweeney, B. W., & Czapka, S. J. (2004). Riparian forest restoration: Why each site needs an ecological prescription. *Forest Ecology and Management, 192*(2–3), 361–373.

Vought, L.B.-M., Dahl, J., Pedersen, C. L., & Lacoursiere, J. O. (1994). Nutrient retention in riparian ecotones. *Ambio, 1*, 342–348.

Wang, G., Mang, S., Cai, H., Liu, S., Zhang, Z., Wang, L., & Innes, J. L. (2016). Integrated watershed management: Evolution, development and emerging trends. *Journal of Forestry Research, 27*(5), 967–994. https://doi.org/10.1007/s11676-016-0293-3

Wani, S. P., & Garg, K. K. (2009). *Watershed management concept and principles*. https://oar.icrisat.org/id/eprint/3914

Westervelt, J. (2001). *Simulation modeling for watershed management*. Springer. https://books.google.com/books?hl=en&lr=&id=1u86gEblVUwC&oi=fnd&pg=PR5&dq=Westervelt+J+(2001)+Challenges+in+management+and+modeling.+Simulation+modeling+for+watershed+management.+Springer,+New+York,+pp+9%E2%80%9326&ots=SfL7uNcffU&sig=sD6AY-PFHoH1IYvguGBLGxRb04g

Wolkersdorfer, C. (2008). *Water management at abandoned flooded underground mines: Fundamentals, tracer tests, modelling, water treatment*. Springer. https://books.google.com/books?hl=en&lr=&id=tHuu7QtUurwC&oi=fnd&pg=PR9&dq=Wolkersdorfer,+C.+2008.+Water+Management+at+Abandoned+Flooded+Underground+Mines:+Fundamentals,+Tracer+Tests,+Modelling,+Water+Treatment,+A.+Springer:+Berlin.+p+465.&ots=_3L3HnTE8U&sig=8ViJyCU8Pd_84JnxM1-MzN5rbcI

World Commission on Dams. (2000). *Dams and development: A new framework for decision-making*.

Yossef, M. F. M., & De Vriend, H. J. (2011). Flow Details near River Groynes: Experimental Investigation. *Journal of Hydraulic Engineering, 137*(5), 504–516. https://doi.org/10.1061/(ASCE)HY.1943-7900.0000326

CHAPTER 5

Evaluating the Link Between River Contamination and Public Health: Investigating Heavy Metals, Pesticides, and Their Socioeconomic Impacts

Advait G. Burande, Pratibha S. Agrawal, and Komal R. Chandankhede

5.1 Introduction

Statistically (Fig. 5.1), our planet is composed of about 79% water, of which 97.5% is saline. Only 2.5% of this remaining water is fresh, and just 0.3% of this freshwater is found on the surface—primarily in rivers—while 30.2% exists as groundwater and 69.5% is locked in ice and glaciers. Of this 0.3%, only 0.264% is present in rivers(Kabeto et al., 2022). Rivers, therefore, play a critical role by providing nearly half of the global population's domestic water needs and supplying around 50% of the water used

A. G. Burande (✉) · P. S. Agrawal · K. R. Chandankhede
Department of Applied Chemistry, Laxminarayan Innovation Technological University, Nagpur, India
e-mail: advait.burande369@gmail.com

S. Ghosh (ed.), *Reimagining Indian Rivers for Sustainability*, Palgrave Studies in Environmental Sustainability,
https://doi.org/10.1007/978-3-032-04754-0_5

for irrigation, making them essential not only for drinking but also for food production.

Every substance deposited on Earth has the potential to contaminate groundwater and river water, particularly water-soluble substances that can easily pass through soil layers. Atmospheric gases may also interact with river water, altering its composition. Even minor variations can introduce toxins and pathogens, leading to health issues such

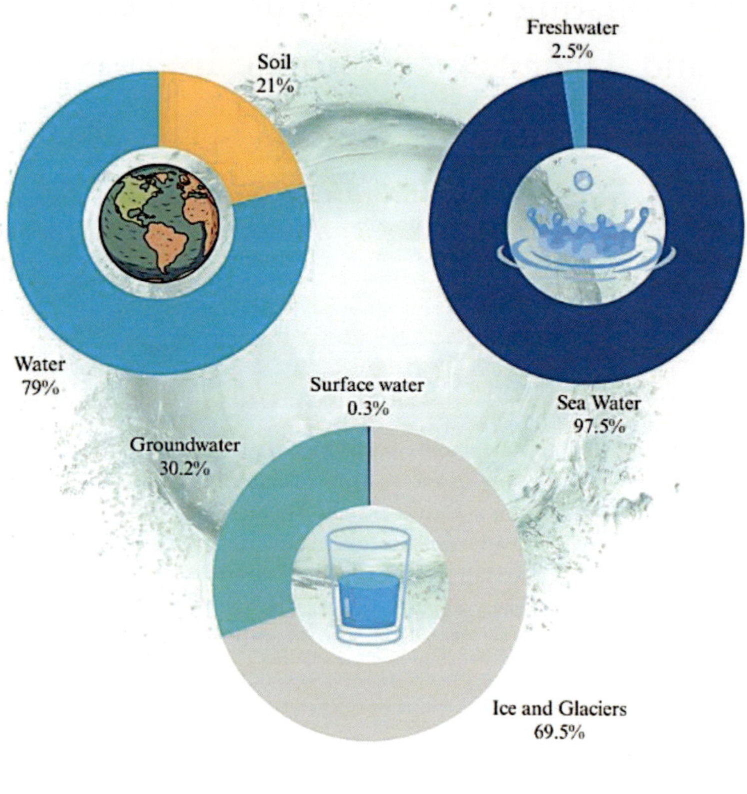

Fig. 5.1 Statistical analysis of water on Earth: (a) Percentage of water on Earth, (b) Composition of water on Earth, (c) Composition of freshwater on Earth

as poisoning, cholera, and cancer. Contaminated water adversely impacts aquatic organisms, disrupts food chains, and can pollute broader ecosystems. Polluted irrigation water can poison crops, affecting food security and the livelihoods of communities. Once contaminated, water is difficult and expensive to clean, requiring complex remediation and ongoing monitoring (Wu et al., 2020).

River water pollution arises from multiple sources, including seeping rainfall, soil components, fertilizers, spills, industrial and municipal discharges, mining, and construction activities (Fig. 5.2). Minerals or contaminants trapped in soil can migrate to groundwater or infiltrate rivers. Toxins such as heavy metals are particularly concerning due to their ability to bioaccumulate and enter food chains, posing serious long-term risks. (Roy & Zahid, 2021).

Heavy metals like chromium, copper, mercury, and lead are especially worrisome because of their toxicity and persistence in the environment. These metals can originate from both industrial and natural sources and often accumulate in aquatic organisms, ultimately impacting human health. Anthropogenic activities in aquatic systems can be detrimental to both flora and fauna, with lead contamination being one such example (Siddiqui et al., 2020; Xu & Li, 2020).

Eutrophication, caused by an excess of nutrients (often from fertilizers and sewage), leads to algal blooms and oxygen depletion, endangering aquatic life and reducing water quality. Microplastic contamination is an

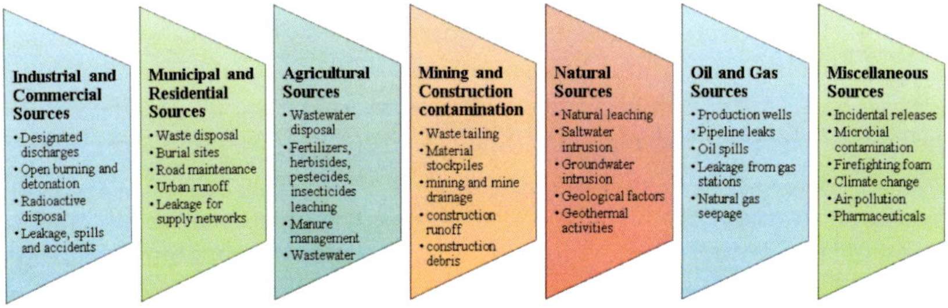

Fig. 5.2 Sources of River water pollution: **(a)** Industrial and Commercial Sources, **(b)** Municipal and Residential Sources, **(c)** Agricultural Sources, **(d)** Mining and Construction Contamination, **(e)** Natural Sources, **(f)** Oil and Gas Sources, **(g)** Miscellaneous Sources

emerging concern, with microplastics now being detected in both bottled and tap water, and their health impacts are still being understood.

Other forms of contamination include domestic and industrial effluents, improper disposal of electronic and toxic waste, and increased salinity, all of which can render water unsafe for consumption and agriculture. These pollutants also threaten endemic species, as seen in rivers like the Tamiraparani, were endangered fish species face risks from industrial and domestic pollution. A significant challenge is the synergistic impact of mixed pollutants, which can amplify environmental and health issues. While water quality analysis often focuses on physical and chemical factors, biological indicators such as plankton diversity are also crucial for assessing river health. Seasonal and geographic variations in water quality further complicate monitoring and management (2012).

Advances in monitoring and treatment technologies, such as high-precision instrumentation and innovative biosorption methods, offer new avenues for pollution control. However, many water sources still exhibit poor quality, with heavy metal concentrations in fish tissue often exceeding safe limits. Insufficient water availability in some regions exacerbates pollutant buildup and water degradation. India has witnessed several major water contamination incidents with severe impacts on human health, wildlife, and the environment. The global impact of river pollution and public health impact is illustrated in Fig. 5.3. Table 5.1 summarizes some of these past tragedies, highlighting the urgent need for effective conservation and pollution prevention measures.

Given the vast scope of river water contamination as depicted in Fig. 5.4 Major Rivers in India, this chapter focuses on hydro-geo-chemical analysis, health issues, and environmental hazards (see Fig. 5.5 Objectives of the article: (a) Hydro-geo-chemical Analysis, (b) Health Issues, and (c) Environmental Hazard). Special attention is given to the persistent and future environmental crises in India's river basins, driven by increasing human modifications and the growing influence of climate change. The chapter also explores the importance of integrated management and restoration strategies for achieving environmental sustainability and river resilience.

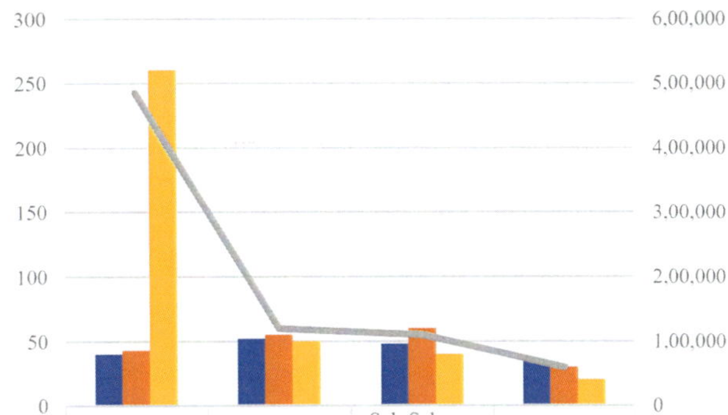

	Global	South Asia	Sub-Saharan Africa	Latin America
% Rivers Severely Polluted	40	52	48	34
% Population at Health Risk	43	55	60	30
Economic Losses (USD Billion)	260	50	40	20
Annual Deaths Linked to Water Pollution	4,85,000	1,20,000	1,10,000	60,000

Fig. 5.3 River pollution and public health impact

5.2 Hydro-Geo-Chemical Analysis

Hydro-geo-chemical analysis is a significant tool in defining water make-up in terms of its chemicals and is a significant tool in determination and estimation of a variety of chemical contents in river water. By analyzing these factors, scientists can have a deeper understanding of the role of humans and naturally developed factors in water systems, and in return, can inform effective strategies for successful water resource management and conservation of environment. Complexity in river water make-up, however, is a critical challenge, in that it is prone to numerous fluctuations and variances, and therefore, cannot possibly have an absolute level of accuracy in defining its make-up. These complex dynamics have a variety of factors, including geologic, atmospheric, and anthropogenetically derived processes, that can act and react with each other in complex forms. In this study, four significant parameters of hydro-geo-chemical analysis will be discussed, refer Fig. 5.6. All of these key parameters will have a full discussion, with minor ones discussed in a concise form, in an attempt to have a full understanding of the issue and its role in water resource management and conservation of environment (Singh et al., 2012). For more details on specific pollutant, refer Table 5.4.

Table 5.1 Major past incidents caused due to water contamination

Year	*Location/Incident*	*Contaminant(s) Involved*	*Impact/Outcome*
1984	Bhopal, India	Methyl isocyanate (MIC)	Thousands killed, long-term health effects, environmental devastation
1956	Minamata, Japan	Mercury	Minamata disease, neurological disorders, deaths
1976	Seveso, Italy	Dioxin (TCDD)	Health disorders, animal deaths, environmental contamination
1986	Chernobyl, Ukraine	Radioactive isotopes	Widespread radioactive contamination, health impacts, ecological damage
2000	Baia Mare, Romania	Cyanide and heavy metals	Major fish kill, water supply contamination, ecological disaster
2014	Flint, Michigan, USA	Lead	Lead poisoning, public health crisis
2010	Hungary (Ajka)	Red mud (heavy metals, alkaline waste)	River and soil contamination, fish kill, health, and environmental hazards
2017	Bellandur Lake, Bangalore, India	Industrial effluents, sewage	Toxic foam, fire incidents, water unfit for use
1999	Walkerton, Canada	E. coli, Campylobacter	Waterborne disease outbreak, deaths, illness
2018	Cape Town, South Africa	Severe drought, water scarcity	Water rationing, “Day Zero” crisis, public health, and economic impacts

5.2.1 Physical Parameters

See Fig. 5.6.

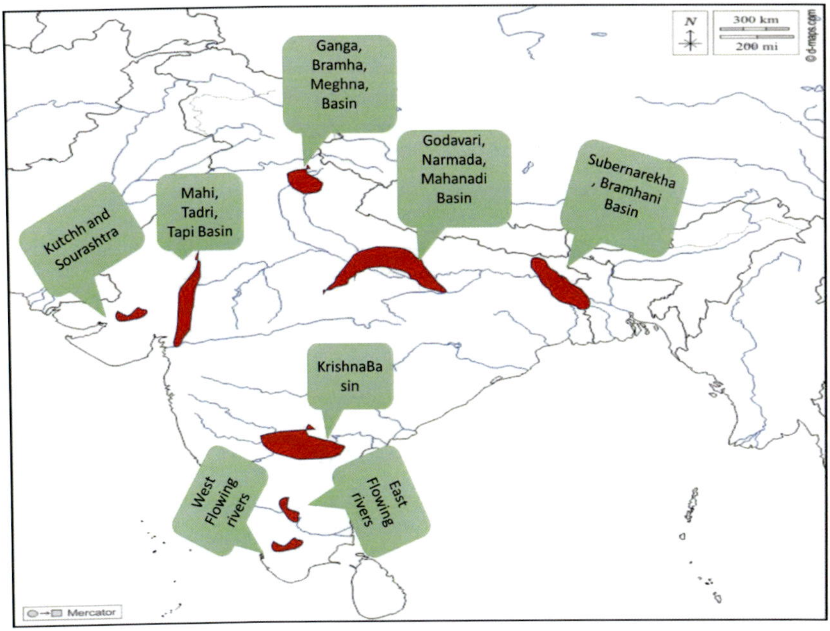

Fig. 5.4 Major Rivers in India

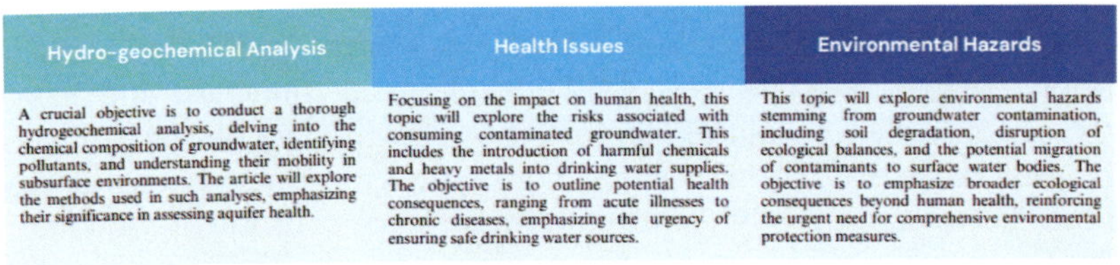

Hydro-geochemical Analysis	Health Issues	Environmental Hazards
A crucial objective is to conduct a thorough hydrogeochemical analysis, delving into the chemical composition of groundwater, identifying pollutants, and understanding their mobility in subsurface environments. The article will explore the methods used in such analyses, emphasizing their significance in assessing aquifer health.	Focusing on the impact on human health, this topic will explore the risks associated with consuming contaminated groundwater. This includes the introduction of harmful chemicals and heavy metals into drinking water supplies. The objective is to outline potential health consequences, ranging from acute illnesses to chronic diseases, emphasizing the urgency of ensuring safe drinking water sources.	This topic will explore environmental hazards stemming from groundwater contamination, including soil degradation, disruption of ecological balances, and the potential migration of contaminants to surface water bodies. The objective is to emphasize broader ecological consequences beyond human health, reinforcing the urgent need for comprehensive environmental protection measures.

Fig. 5.5 Objectives of the article: (a) Hydro-geo-chemical Analysis, (b) Health Issues, and (c) Environmental Hazard

5.2.1.1 *Temperature (T)*

Temperature is a critical water quality variable with significant implications for riverine chemical, physical, and biological processes. Temperature variation affects water density and viscosity, and subsequently, flow regimes and pollutant dispersal. For example, temperature gradients can induce

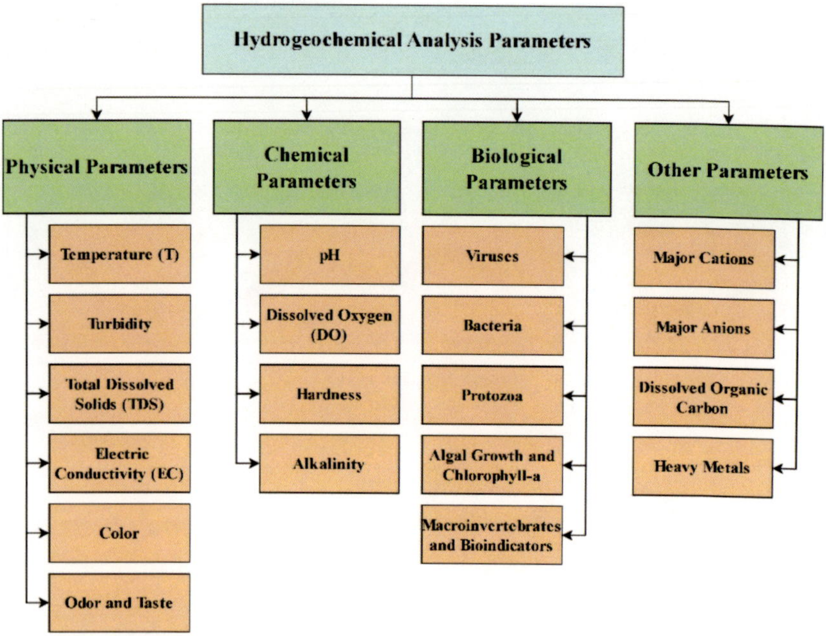

Fig. 5.6 Hydro-geo-chemical analysis parameters

thermal stratification, with an impact on solutes and solutes distribution. Temperature controls solubility of metal, minerals, and gases, and, in consequence, solutes concentrations, including concentrations of dissolved oxygen and other solutes. For example, warmer water will have reduced concentrations of dissolved oxygen, a key constituent for aquatic life. Temperature accelerates reaction rates and can, therefore, quicken organic matter decomposition and cycling of nutrients. Temperature is important in aquatic organism's metabolism, growth, and reproduction. All species have an optimal temperature range, and temperature variation in water can alter species distribution and abundance, with a potential impact on sensitive species. High temperatures can, in addition, promote pathogen development and disease transmission (Diamantini et al., 2018; Jiang et al., 2014; Kedra, 2020; Toffolon & Piccolroaz, 2015).

The temperature of water can have many effects both on its quantity and its quality. High temperatures can make some minerals and metal

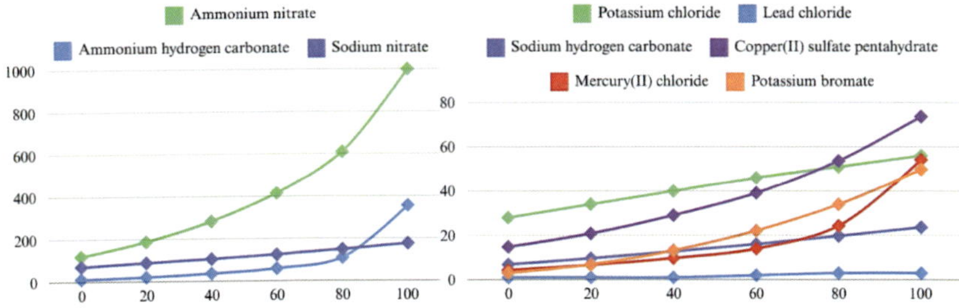

Fig. 5.7 Impact of temperature on solubility of certain chemicals

become more soluble (as demonstrated in Fig. 5.7), but can cause a loss in solubility in gasses, such as oxygen. High temperatures can cause an increased microbial activity and degradation of organic contaminants (Brooks & Steele-Macinnis, 2019). High temperatures can cause a loss in density and viscosity in water, and affect flow and transportation. Low temperatures can have an opposite effect in these factors.

Global warming brings about an increase in temperature in the atmosphere, and river water temperature is impacted directly by it. Long-term observations have been examined in statistics, and in most cases, river water temperature variation and temperature variation in the atmosphere with global warming have a less than one proportion, between 0.5°C and 0.85°C. To a certain extent, it increases evaporation and water-to-air heat transfer with an increase in temperature in the atmosphere (Jiang et al., 2014; Kedra, 2020; Ozaki et al., 2003).

Impervious surfaces in urban areas trap and store heat and, in consequence, generate warmer runoffs and warmer temperatures in streams. Empirical studies have confirmed that urbanization, in terms of changing land use, can generate a significant rise in temperatures in stream water, with amplitude and duration of temperature rise changing with urban density (McBean et al., 2022). Dam releases can change and decide river temperatures in the downstream, with warmer water in winter and cooler water in summer through releases at alternate reservoir depths. For example, the Three Gorges Dam manipulates outflow thermal regime with an apparent winter rise in river water temperature and an interannual temperature pattern (thermal regime) lag, in relation to inflow (Shi et al., 2021; Toffolon & Piccolroaz, 2015). Solar radiation is a strong source of heat, particularly during summer. Solar radiation intensity and duration contribute a lot towards water temperature. There is a high correlation between river water temperature variation and solar radiation intensity, and most prominently in summer (Ozaki et al., 2003).

River discharge determines water velocity and depth, and both can vary with atmospheric heat exchange, changing in concert with it. Higher discharges will, in general, stabilize water temperature variation. In regulated and snow-fed, high-elevation reservoirs' case, discharge is a key river water temperature controlling variable (Toffolon & Piccolroaz, 2015; Zhu & Piotrowski, 2020).

Direct measurements with sensors produce dependable information for thermal regimes analysis. Thermal regimes serve a basis for temperature trends and model calibration with such a measurement. Thermal satellite remote sensing, utilizing satellite observations such as Landsat, provides temperature information with high extendibility in terms of space. Thermal satellite remote sensing can apply in areas with no data and in thermal variation studies in terms of extendibility in space. However, satellite remote sensing processing for river temperatures with accuracy entails overcoming boundary effects and compensating for systematic errors (Shi et al., 2021).

Models such as artificial neural networks (ANN), adaptive neuro-fuzzy inference systems (ANFIS), and Gaussian process regression, make prediction of water temperature in relation to environment factors including temperature, discharge, and solar radiation. Unlike process-based models, such models require less detail information in terms of input but can have limitations in extrapolating prediction out of range with measured values (Shi et al., 2021; Toffolon & Piccolroaz, 2015; Zhu & Piotrowski, 2020).

Process-based deterministic models, such as MIKE HYDRO, MIKE SHE, and ECO Lab, can simulate river processes governing heat exchange in a river. They require a complete suite of river, meteorologic, and land use information for model input. By contrast, such models can produce a high level of detail in thermal dynamics but high complexity and high demand for information, and hence, can have limited use (McBean et al., 2022; Toffolon & Piccolroaz, 2015).

5.2.1.2 Turbidity

Turbidity measures scattered and absorbed light when it travels through water, not when it is transmitted. Turbidity is a consequence of scattering, and it happens with the presence of particles in water in a state of suspension (see Fig. 5.8). Turbidity can be measured with a nephelometer, an apparatus that estimates scattered intensity at a specific angle (most commonly 90 degrees) of a source of light. It is most measured in Nephelometric Turbidity Units (NTU) and Formazin Nephelometric Units (FNU). Turbidity can be measured in situ in a river with sensors, and these allow continuous observation for us. There are sensors utilizing techniques of scattering of light in an optoelectronic manner, and these can provide real-time information. Turbidity can range between low values (clear water) and high values (very high turbid water). For example, in Xitiaoxi River, it can range between 2 and 298 NTU, and in Ramganga River, it can range between 2.3 and 112 NTU (Allam et al., 2020; Bright & Mager, 2020; Göransson et al., 2013; Lawler et al., 2006; Wu et al., 2014; Yao et al., 2016; Zhang et al., 2024).

Suspended sediment is one of the most important sources of turbidity. Silt, clay, and fine particles fall in this group. Channel run-off and channel eroding, and rain and surface run-off, are important sources of riverine suspended sediment. Microbes, decaying organic matter, and bacteria can cause increased turbidity. Mineral and particulate, fine, and insoluble matter cause increased turbidity. Mining, shipping, and agricultural run-off can contribute a lot towards increased turbidity. Mining and ore washing can introduce a lot of sediment in a river. Shipments can disturb

Fig. 5.8 Turbid water

river bottoms' sediments. Agrarian run-off can cause increased turbidity through increased run-off through eroding. Storms and excessive rain can cause increased run-off and, subsequently, increased turbidity through run-off. Wind velocity, run-off, and vegetative cover variation can cause variation in turbidity in a season (Allam et al., 2020; Göransson et al., 2013; Lawler et al., 2006; Wu et al., 2014; Zhang et al., 2024).

Increased turbidity can hinder the intensity of penetrating through the water column. It affects aquatic flora, algal and other producers, for photosynthesis, and general ecosystem integrity. Turbidity can modify water temperature through less penetrating sunlight. It therefore affects absorption and distribution of solar radiation. High turbidity can hinder aquatic life through degradation of breeding habitats, reduced availability of foods, and impairment of gill function. It can hinder interstitial spaces in river bottoms, providing habitats for aquatic species. Heavy metal and organic toxins have a high attraction for suspended particles. Turbidity can therefore be associated with carriage of such toxins (Allam et al., 2020; Bright & Mager, 2020; Lannergård et al., 2019; Toffolon & Piccolroaz, 2015; Wu et al., 2014; Zhang et al., 2024).

Turbidity tends to increase in a river system in a direction towards down reaches in a river system. Rivers' upper reaches (part near origin) have lesser turbidity in consideration of forests and lesser shipping activity. Turbidity can increase in middle reaches with lesser cover of vegetative cover, and in down reaches, agricultural and construction activity can contribute towards increased non-point source pollution. Turbidity can even vary with a change in the season. For example, in a river, in spring, minimum and in winter, maximum turbidity can occur. Turbidites vary with a change in vegetative cover, with rain, and with a change in the velocity of wind. In a river such as a river such as the Ramganga, during a period of pre-monsoon, an increase in turbidity over post-monsoon values can occur. Storms can cause a sharp rise in turbidity through run-off and erosion of the land surface. Meandering of a river, Reynolds number of flow, velocity gradient of flow, and density of particles can cause a variation in turbidity even in a cross-sectional plane of a river (Allam et al., 2020; Göransson et al., 2013; Zhang et al., 2024).

Turbidity is at times taken as a proxy for concentrations of suspended solids because both factors have a high correlation with one another. Nevertheless, its relation can vary with both particle composition and particle size. Turbidity can, in specific cases, be taken in use in predicting concentrations of total phosphorus. It has been confirmed to have a relation with concentrations of specific metal in surface water (Bright & Mager, 2020; Göransson et al., 2013; Lannergård et al., 2019; Lawler et al., 2006; Yao et al., 2016).

5.2.1.3 *Total Dissolved Solids (TDS)*

Total dissolved solids, such as organic matter (in minor concentrations) and inorganic salts (calcium, magnesium, potassium, sodium, bicarbonates, chlorides, and sulfates), are present in a given water sample. TDS can be used to represent water quality, salinity, and water hardness, and to detect contaminating matter and pollutant matter. TDS can even affect chemical and biological water characteristics, such as solubility of metal and nutrient, toxicity of contaminating matter, and viability of aquatic organism.

The TDS of groundwater can vary naturally with character and amount of rock and soil minerals, residence period of water in an aquifer, and chemical state of water. Several minerals, such as sodium, potassium, calcium, magnesium, chloride, sulfate, carbonate, and nitrate, can contribute towards water TDS in a variety of fashions. How long water stays in contact with minerals, the larger, the larger will its ability to dissolve them and its TDS will become. Chemical state of water, such as temperature, oxidation–reduction potential, and dissolved oxygen, can affect water TDS through changing chemical reaction taking (Saikrishna et al., 2020).

The TDS of water can impact its usability for consumption for humans and animals, and for use in domestic, agricultural, and industries. However, TDS of water will not have a direct consequence in terms of its impact on the health of water drinkers, for its value will rely on the character and level of the dissolved solids, and not its totality. Thus, no guideline value for TDS for health purpose, Fig. 5.9 demonstrates desirability of TDS in water. Nonetheless, TDS is a significant parameter that must be monitored and regulated to maintain river water quality (Raj et al., 2023; Weber-Scannell & Duffy, 2007).

Some hotspots with high values of TDS are coastal areas, arid areas, volcanic areas, and areas with industries, in which river water can face seawater intrusion, evaporation, geothermal activity, and discharge of effluents. Some of the hotspots with low values of TDS are high-elevation areas, high-rainfall areas, and areas with forests, in which river water can face rain, melting of snow, and organic matter.

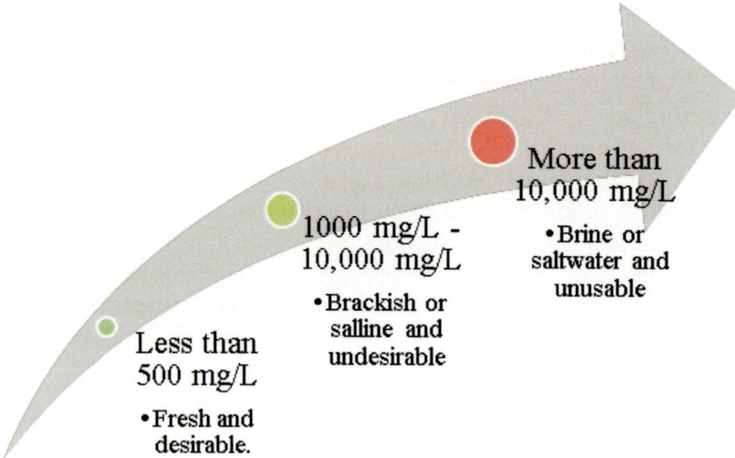

Fig. 5.9 Total dissolver solids in water

5.2.1.4 Electric Conductivity (EC)

Electrical conductivity (EC), a presentation of water's capacity for conduction of an electric current. It tends to depend on presence and type of compounds in water, such as salt, metal, and minerals. EC is measured through ion concentration, temperature, and ionic intensity. EC increase can detect presence of contaminating factors such as metal, salt, and agricultural and industrial effluents. EC decrease can detect excessive dilution and presence of non-ion species such as organic matter (Ebrahimi et al., 2022; TERZIĆ et al., 2021). EC can affect water contamination in a range of ways. For example, high EC can detect intrusion of seawater in freshwater sources, and in such a scenario, availability and source water quality for consumption and irrigation can become compromised. High EC can detect salt and nutritional salt and nutrient leaching in agricultural lands, and in such a scenario, salinization and nutritional contamination can occur in lands and in groundwater. Low EC can detect presence of organic contaminating factors such as herbicides, pesticides, and petroproducts, and in such a scenario, humans and wildlife can suffer consequences, refer Fig. 5.10 for more details (Atangana & Oberholster, 2021; Qu et al., 2022; Tabassum et al., 2019).

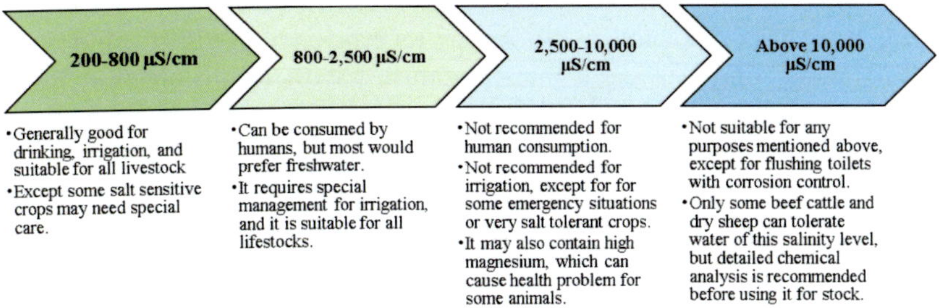

Fig. 5.10 Water usability based on electrical conductivity

The hotspots for high EC are the same as TDS, as high ions increase EC while high organic matter decreases EC. Water salinity is measured by electrical conductivity (EC) in micro-siemens per centimeter (µS/cm) (Gueddari et al., 2022).

5.2.2 *Chemical Parameters*

5.2.2.1 *Acidity (pH)*

pH is a minus base 10 logarithm of H_2 ion concentration and can naturally vary with types of rock and soil minerals passing through contact with river and with river discharge composition, etc. Chemical and biological character of water varies with its pH, such as solubility of metal and nutrients, toxicity of pollutant, and aquatic organism health, for example, rainwater, having a faint acid pH, can solubilize rock and soil minerals (e.g., Calcium Carbonate) when it percolates in groundwater, river water, etc. Different rocks, for example, basalt, sandstone, and limestone, have a mix of variety of minerals and can change river water pH in a variety of manner. How long water is in contact with minerals, the more it will react with them and change its pH, etc. The chemical form of water, its temperature, oxidation–reduction potential, organic matter, dissolved heavy metal and dissolved oxygen, can change its pH through its effect on chemical reaction, etc. (Turyanti & Chaerunnisa, 2017).

The acidity and alkalinity of water, measured through its pH, can affect its corrosivity to metal and its efficiency in eradicating germs. Water with a high level of acidity and high level of alkalinity can make pipes and plumbing equipment's become rusted, and can make dangerous metal enter water. It can affect the odor, and even the looks of water. Water with a high level of alkalinity can hinder its efficiency in disinfecting water and in eradicating pathogen microorganisms with its disinfectant, chlorine. Thus, it is best for its level of pH to fall between 6.5 and 8.5, about its source and its level of quality. Because its impact is not direct towards its drinkers' health, no guideline value for its level is calculated in terms of its impact towards its drinkers' health. In order for its level of pH to rise, lime and soda ash addition and caustic soda can be utilized; and for its level of pH to fall, its use of carbonic acid injection and organic and inorganic acids can be utilized.

pH is one of the most critical factors to monitor and manage regarding river water quality. Limestone areas, alkaline lakes, and urban areas, for example, have high-pH hotspots, whose water can become affected through carbonate dissolution, evaporation, and discharge of cement, respectively. Mining areas, acid rain areas, and peat areas, for example, have low-pH hotspots, whose water can become affected through sulfide oxidation, atmospheric deposition, and organic acid production, respectively.

5.2.2.2 Dissolved Oxygen (DO)

Dissolved oxygen (DO) is an important river water quality criterion for a variety of factors, including its requirements for aquatic life survival. DO is free, uncombine oxygen in water, and its value is an important river health gauge. All aquatic life requires a plentiful DO source, and low DO values will result in an unbalanced environment with dead aquatic animals, odors, and aesthetic factors. As DO values fall, aquatic animals will have to adapt to changing respiration, and in extreme scenarios, even suspend activity, with a detrimental impact on development and reproduction. ([CSL STYLE ERROR: reference with no printed form.]; Cox, 2003; Majeed et al., 2022; Zounemat-Kermani et al., 2019).

DO levels in a river rely on a high level of factors; these include: temperature of water, wind activity, and aquatic and algal presence and aquatic flora. Solubility in water decreases with an increase in temperature, according to Henry's Law. DO concentrations will fall during summer when DO consumption through respiration is at its high level. DO is added to water through photosynthesis in aquatic flora and algal presence, but dead flora and algal decay consume DO for consumption. Overgrowth of algal presence can produce widespread bacterial growth, consuming DO in water. DO is consumed through aquatic organism respiration and organic matter decomposition (Cox, 2003; Fitri et al., 2021; Liu et al., 2020; Zhi et al., 2021; Zhong et al., 2021).

DO levels temporally and spatially vary in a river system. In shallow and actively moving streams, mixing is effectively accomplished, and photosynthesis is facilitated with rich lights. DO concentrations in such a habitat, therefore, are high in relation to deeper and sluggish habitats. Figure 5.11 shows the stagnant water indicating low to moderate DO levels. Rich organic matter habitats, responsible for heightened respiration, utilize DO. DO in water column is reduced through direct discharge of DO-poor sewage effluents. Water infrastructure development, such as reservoirs and dams, can alter flow regimes and affect DO dynamics (Cox, 2003; Li et al., 2020; Zhi et al., 2021; Zhong et al., 2021).

Low DO level, i.e., hypoxia (for DO < 2 mg/L), is a critical threat to aquatic environments. Fish kill and mortalities of benthic animals can result in hypoxia. Production of hypoxia is a result of an imbalance between DO consumption and DO availability and between and between DO concentrations in water masses and between them through lateral

Fig. 5.11 Low dissolved oxygen in water

exchange. (Li et al., 2020; Liu et al., 2020; Zhi et al., 2021; Zhong et al., 2021).

Various studies have examined DO's spatio-temporal distributions in riverine environments. For example, in Pearl River Estuary (PRE), DO concentrations overall in urban reaches are less and with distance downstream, but then fall off in the open sea again. In urban reaches in the PRE, hypoxia has been measured and in intermittent periods in upper reaches, with increased sewage discharge being a cause, beginning particularly after 2000. In southeastern parts of the PRE, DO stratification is larger during summertime, and hypoxia at the bottom has been measured, particularly in the Hong Kong region. In addition, DO concentration is inversely correlated with particulate organic carbon and ammonia. (Li et al., 2020; Liu et al., 2020).

Survey data in a variety of studies unveil DO's importance as a water quality variable. In Kelantan River Basin, Malaysia, DO between 6.8 and 8.4 mg/L in 2005 in a period of drought, with a minor dip in a period of flood, DO concentrations in 2010 were lowered with augmented land use and sewage disposal. (Fitri et al., 2021). In India, in the Henwal River, a composite pollution index (CPI) with DO one of its constituent factors determined that river water was poorly contaminated (Siddiqui et al., 2020). In America, mean DO at the surface is in use in determination of river and lake water condition classes for the US EPA. For them, > 5 ppm is in a satisfactory state, 3 to 5 ppm fair, and 3 ppm and less poor. In US, 7% of its lakes showed poor DO at the surface. ([CSL STYLE ERROR: reference with no printed form.]). In St. Johns River, Florida, DO and a variety of chemical factors such as temperature, pH and salinity have been in use in creation of prediction model for water quality (Zounemat-Kermani et al., 2019). In a study in a cascade river, Yongjiang, China, DO increased in a direction river head towards river mouth, with reaeration rate larger than consumption. (Zhong et al., 2021).

5.2.2.3 Hardness

Hardness measures general ion concentrations of multivalent ions, specifically calcium (Ca^{2+}) and magnesium (Mg^{2+}), in water. Ions naturally occur in compounds and affect the suitability of water for use in a range of applications. Ions naturally occur through rock weathering, specifically in sedimentary rocks, including limestone and dolomite rocks. There are two forms of hardness: carbonate (temporary) and non-carbonate (permanent). Carbonate hardness occurs through calcium and magnesium bicarbonates and is removed via boiling. On the other hand, non-carbonate hardness occurs through other salts and cannot be removed via boiling. Total hardness is a summation of ion concentrations responsible for hardness. It is measured in terms of $CaCO_3$ in terms of mg/L, and is used in distinguishing between soft, moderately hard, hard, and tough water (Akkanen & Ukkonen, 2001; Islam et al., 2020; Thompson et al., 2012). Figure 5.12 demonstrates levels of water hardness.

High hardness can include such defects as impairing soap efficiency, pipe scales, and having a harmful effect on aquatic life. Caused almost exclusively by Ca^{2+} and Mg^{2+}, other metal ions such as iron, aluminium, barium, and strontium can cause hardness as well. It is important to note that hardness is not equivalent to Total Dissolved Solids (TDS), a general measurement of overall ion concentrations of all such species in solution. It is not necessarily contributing to hardness (Akkanen & Ukkonen, 2001; Wimalawansa 2016).

Evidence highlights that geographical location and other factors can cause a variation in hardness. For example, in India's Henwal River, a variation range of 115.0 ± 0.10 mg/L and 246.0 ± 0.11 mg/L with a

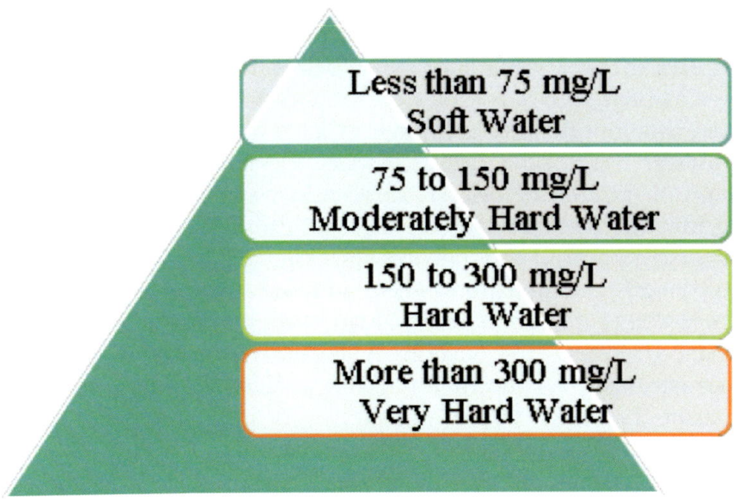

Fig. 5.12 Hardness levels in water

high value during summer at site Jajal, respectively, have been encountered. In Chittagong, Bangladesh, water was saline, hard, and alkaline with high Na^+, Ca^{2+}, Cl^-, and HCO_3^- concentrations. In addition, a variation range of 70 to 495 mg/L in terms of hardness was encountered in a study in India's Kosi River in Uttarakhand. All such variances imply that geology and environment at a local level vary with hardness. (Islam et al., 2020).

Hardness also regulates other factors of water quality, including pH and alkalinity. Calcium ions, in particular, act as a pH stabilizer through their buffers. There is a sophisticated relationship between alkalinity and hardness; in one kind of water, both can have a similar level in terms of having been extracted from similar minerals. In some, one can have a larger value in relation to the other one. That can mean that water will have travelled through a water softener, or chloride, nitrate, and sulfate have a high level in it.

5.2.3 *Biological Parameters*

5.2.3.1 *Viruses*

Viruses in river water can cause waterborne diseases. They can survive in water for long periods, and cause diseases such as diarrhoea, gastroenteritis, and pneumonia. (Altintas et al., 2015) (Rusiñol & Girones, 2019). Thus, viruses are also important factors in water quality analysis. Generally, viruses in river water indicate faecal contamination. Traditionally, bacterial markers such as E. Coli were used to evaluate viral contamination in water, though often not accurately (Gall et al., 2015).

Viruses are responsible factors in river water quality analysis through causation of waterborne disease, survival in water for long periods, and having low infectious dosages. Viruses can cause disease, including diarrhoea, gastroenteritis, and pneumonia. Viral pathogens in river water indicate faecal contamination. The traditional use of bacterial markers such as E. coli cannot accurately evaluate the level of viral contamination.

Specific viruses such as hepatitis A, hepatitis E, rotavirus, adenovirus, and norovirus are a concern for having a relation with severe disease (Altintas et al., 2015). Viruses can survive for long periods in water and have low infectious dosages (Rusiñol & Girones, 2019). For instance, contact with a single particle of rotavirus can produce a 31% infection chance (Gall et al., 2015). Besides, viruses such as polyomaviruses and cytomegalovirus, shed in urine, can transmit via water. Viruses, with a size range of 15 to 90 nm, have a small size and cannot pass through a filter, and can resist disinfectants (Altintas et al., 2015). Besides, enteric viruses have been recovered in potable water in concentrations of 1–20 PFU per 1000 liters (Carter, 2005).

Several factors contribute to viral pollution in surface water, including: distance between the discharge of wastewater, epidemic period and initial virus concentration, survival and transportation character of the virus, and environment such as temperature, flow velocity, salinity, and p.h (Altintas et al., 2015; Hamza et al., 2011; Saylan et al., 2019; Skraber et al., 2004). Somatic coliphages have a strong relation with the presence of a virus genome, in contrast with thermotolerant coliforms. (Carter, 2005; Kiulia et al., 2010; Rusiñol & Girones, 2019). Viral genomes can be detected with molecular techniques such as PCR, but the presence of a genome does not confirm presence of infectious particles. Virus detection is changing with changing seasons; noroviruses in winter, and enteroviruses in summer. Efficient techniques for detection and removal become a necessity, for a virus can have a very low infectious dose. (Altintas et al., 2015; Hamza et al., 2011; Skraber et al., 2004). Refer Table 5.2 for more details on specific river water viruses.

5.2.3.2 Bacteria

Bacteria are most frequently utilized for river water contaminating indications use. There are certain types, such as faecal indicator bacteria, and these make up the first preference for microbiological risk evaluation. Examples include *E. coli* and enterococcal strains, and these can occur in humans and animals' intestines. In water, presence signifies faecal contaminating presence. It is an indication of a likely presence of pathogenical pathogens, including viruses. Traditional microbiological methods, such as through cultures, are most frequently utilized for numbering such types of bacteria. Nevertheless, even when such methods can be reliable, they can become cumbersome and long (Saylan et al., 2019).

Enterococcus faecalis is a microbe utilized for faecal water pollution with an indicative organism role. It carries virulence factors responsible for biofilm establishment. Thermotolerant coliforms and enterococci have been used for bacterial indications of faecal pollution. *E. coli* is utilized worldwide for an indication of probable faecal contaminations (Saylan et al., 2019; Skraber et al., 2004).

However, there are limitations in using bacterial indicators alone. Not all bacterial indicators will have a direct correlation with viruses.

Table 5.2 Viruses in river water, short analysis, associated diseases, and transmission routes

Virus	*Family*	*Genome*	*Size (nm)*	*Associated diseases*	*Transmission*
Enteroviruses	Picornaviridae	ssRNA	20–30	Gastroenteritis, Meningitis, Respiratory	Faecal-oral route, contaminated water
Norovirus	Caliciviridae	ssRNA	27–38	Gastroenteritis	Faecal-oral route, contaminated water and surfaces
Adenovirus	Adenoviridae	dsDNA	70–90	Respiratory, Gastroenteritis, Ocular	Faecal-oral route, contaminated water
Rotavirus	Reoviridae	dsRNA	70–77	Gastroenteritis	Faecal-oral route, contaminated water
Hepatitis A Virus	Picornaviridae	ssRNA	27	Hepatitis	Faecal-oral route, contaminated water
Hepatitis E Virus	Hepeviridae	ssRNA	27–34	Hepatitis	Faecal-oral route, contaminated water
Polyomaviruses	Polyomaviridae	dsDNA	40–50	Often Asymptomatic, potential for disease in immunocompromised	Urine, contaminated water
Cytomegalovirus	Herpesviridae	dsDNA	150–200	Often Asymptomatic, potential for immunocompromised diseases	Urine, contaminated water

Some bacterial indicators will not depend on viral population and will underreport for viral pollution. Hence, bacterial indicators can be useful for detecting faecal contamination, but not for an overall picture of water quality, specifically for viral contamination (Fang et al., 2017; Gómez-Gómez et al., 2020; Gunasekaran et al., 2022; Haramoto, 2018).

Bacteria such as *Acinetobacter spp.*, and *Ralstonia pickettii*, produce waterborne infection. Blood infection can arise through contaminated water sources, and it occurs specifically in an immunocompromised individual. In a similar manner, *Acinetobacter spp.*, can occur in rural potable water sources and can cause infection in a medical environment. All such microorganisms unveil the role of microbial contaminating in a medical environment. Certain microorganisms such as Legionella, Enterobacteriaceae, and Pseudomonas aeruginosa, produce waterborne infection, as well. Refer Table 5.3 for more details on specific waterbone bacteria.

5.2.3.3 Protozoa

Certain protozoa, including *Cryptosporidium spp.*, *Giardia lamblia*, *Toxoplasma gondii*, and *Entamoeba histolytica*, act as faecal water contaminating matter indicators in river water. Transmission of parasites occurs through the faecal-oral route. Cryptosporidium is a zoonotic species, and therefore, can be transmitted between animals and humans. This increases the potential for water contamination from both sources. Protozoa have resistant forms (resistant stages including cysts and oocysts) that allow them to survive in the environment, and can survive in water even in cases with partially conducted processes of water treatment. Protozoa are important waterborne pathogens that cause a variety of disease in humans. Giardia lamblia is a common cause of diarrhoeal disease transmissible through contaminated water. These pathogens have been added to lists of microbial contaminating matter of concern for waterborne transmission according to the United States Environmental Protection Agency (EPA). Some of the protozoa bioaccumulate contaminating matter in the column of water. They bioaccumulate contaminating matter and transmit them to other living forms in a food web (Sara et al., 2023; DİNÇ 2021; Boughattas et al., 2025; Jiang et al., 2024).

The presence of river water with protozoa can be owed to discharge of sewage effluents. Research validates that effluents in wastewater have protozoa including *Cryptosporidium spp.* and *Giardia duodenalis*, and therefore river water can become contaminated with sewage works effluents. Even treated effluents can contain protozoa, indicating flaws in conventional treatment processes (Sara et al., 2023; Boughattas et al., 2025; Haramoto, 2018).

Table 5.3 Bacteria in water, transmission, potentials and associated diseases

Bacteria	*Transmission*	*Potentials*	*Associated diseases*
Escherichia coli	Faecal-oral route, contaminated water, food	Used worldwide as indicator of faecal contamination	Gastroenteritis, diarrhoea
Enterococcus faecalis	Faecal contamination of water, can also adhere to bile drain materials	Can form biofilms	Bloodstream infections, other infections
Enterobacter spp.	Contaminated water, hospital water systems	Can grow in nonsterile water, prolonged hospital stays	Bacteraemia, bloodstream infections
Pseudomonas aeruginosa	Contaminated water sources (tap water, therapy pools, tubs), biofilms in hospital water systems	Can form biofilms, well adapted to survive in low-nutrient conditions, common in hospital water systems	Bloodstream infections, pneumonia, infections in immunocompromised patients
Legionella	Inhalation of contaminated water aerosols (showers, faucets)	Can grow in nonsterile water, in hospital water systems, especially when water temperatures are suitable	Bloodstream infections, pneumonia, Legionnaires' disease
Serratia marcescens	Moist environments, intravenous solutions, catheters, soaps, disinfectants	Thrives in moist environments, including intravenous solutions	Bloodstream infections, outbreaks related to total parenteral nutrition, infections in different clinical specimens

(continued)

Table 5.3 (continued)

Bacteria	*Transmission*	*Potentials*	*Associated diseases*
Ochrobactrum spp.	Contaminated water, soil, biological products, poor hand hygiene, catheter flushing solution contamination	Can be isolated from water	Bloodstream infections, especially in immunocompromised patients, chemotherapy, or organ transplant recipients
Ralstonia pickettii	Contaminated water and water-related devices, intravenous solutions, dialysis water treatment facilities	Can grow in PVC piping, biofilm producer	Bloodstream infections, particularly in immunocompromised patients, infections related to infusion of contaminated water for injection

5.2.3.4 Chlorophyll-A and Algae

Chlorophyll-a (Chl-a) is yet another key parameter for river water quality evaluation, specifically in terms of algal growth and eutrophication (Björn et al., 2009). Chl-a can act as an algal abundance proxy, and can be determined through a spectrophotometer reading of extracted algal chlorophyll at certain wavelengths (Jones & Lee, 1982).

Measuring Chl-a level is an efficient method for estimating algal abundance and biomass, and a less tedious and less reliable alternative for direct algal count (Jones & Lee, 1982). High Chl-a concentrations in general mean increased algal growth, and in most cases, increased algal growth can mean a sign of algal growth stimulated through an abundance of nutrients (nutrient enrichment, eutrophication, and possibly water quality impairments). High Chl-a concentrations can mean algal blooms, and algal blooms can mean extreme raw water impairments in quality.

Chlorophyll levels have a direct correlation with water concentrations of nutrients, and particularly with phosphorus. OECD (Organization for Economic Cooperation and Development)'s model for eutrophication identifies a relation between loading of nutrients, in this case, phosphorus (P), and water factors for eutrophication, using Chl-a as a significant marker. Higher phosphorus concentrations have a propensity to cause an algal growth rise, and consequently, in Chl-a concentrations. Nevertheless, Chl-a and total phosphorus relation can vary with regards to other factors in the environment, including availability of light (Oliver et al.).

Chl-a measurements can be used for predicting algal water quality-related planktonic algal problems in domestic water sources. High concentrations of Chl-a can denote times of high raw water quality issues, such as off-tastes and odors, high trihalomethane (THM) concentrations, reduced filter runs, and high demand for chlorine. Besides, a Chl-a level/threshold intensity for odors can be constructed for the following off-tastes and odors. Besides, an increase in Chl-a concentrations can denote an increase in requirements for additional treatment. For example, off-tastes and odors in the upper Ohio River exhibited a high level of association with Chl-a, and algal origin, and not industrial disposal, could therefore be determined.

Several factors will have an impact on Chl-a concentrations in water, including: Availability of nutrients, including availability of phosphorus and nitrogen, have a direct impact on algal growth (Oliver et al.; Halliday et al., 2016). Availability of penetrating light in the water column will have an impact on algal photosynthesis and growth. Turbidity and shading, in contrast, will inhibit penetrating light and restrict algal growth. For example, riparian shading has been shown to manage algal growth through restriction of penetrating light in the river (Halliday et al., 2016; Oliver et al.). Rate of river flow will have an impact on algal and nutrient transportation, and temperature stratification, and both will have an impact on algal growth and Chl-a concentrations (Jones & Lee, 1982; Wille, 2000). In impoundments, reduced flow rates can increase residence times for water and stimulate algal blooms. Temperature for algal growth will have an impact on algal growth, with warmer temperatures tending to promote increased growth rates. Salinity will have an impact on algal species composition in a river, and algal biomass in general. In deeper water, Chl-a concentrations can vary with depth, particularly when thermal stratification occurs. Algae will congregate in upper regions of the water column in regions with increased availability of penetrating light (Ballschmiter & Katz, 1969; Huang et al., 2020; Zhang et al., 2023).

5.2.3.5 *Macroinvertebrates and Bioindicators*

Macroinvertebrates are universally recognized to serve as useful bioindicators for river water quality assessments through susceptibility to changing environments, relatively long-life cycles, and restricted locomotory capabilities (Fierro et al., 2017; López-López & Sedeño-Díaz, 2015). Macroinvertebrates are animals, including insects, crustaceans, and worms, that can be seen with the naked eye. They make an important constituent of aquatic life and serve in contributing towards the food web, providing a mechanism for the transmittal of energy and toxins to upper trophic levels (López-López & Sedeño-Díaz, 2015) (Kripa et al., 2013; López-López & Sedeño-Díaz, 2015).

Community diversity reflects ecosystem well-being. Several macroinvertebrates react with variable sensitivities towards toxins, and thus, can perceive organic and metal contaminations (López-López & Sedeño-Díaz, 2015; Shimba & Jonah, 2016). Most macroinvertebrates have poor powers of locomotory escape and have a completed life cycle in aquatic habitats. Having poor powers of locomotory escape, macroinvertebrates become susceptible to cumulative pollutant loads at a location over a period of time, offering a temporally integrated record of water quality (Fierro et al., 2017; López-López & Sedeño-Díaz, 2015). Macroinvertebrates are relatively easy to sample using inexpensive equipment such as kick nets and D-nets. Sampling can be done with minimal human resources, making it a cost-effective method for routine monitoring programs (Kripa et al., 2013; López-López & Sedeño-Díaz, 2015; Shimba & Jonah, 2016). Identification at family level can occur for macroinvertebrates, and in most cases, bioassessment can use family level, not necessarily species level, and therefore, species level can be avoided (Fierro et al., 2017). Several approaches have been proposed for employing macroinvertebrates in water quality assessments such as biotic indices, multimetric indices, and multivariate approaches (Fierro et al., 2017).

Biotic indices, such as Hilsenhoff Biotic Index (HBI), apply numerical values in assigning values to a range of taxa in terms of organic tolerance (Fierro et al., 2017; López-López & Sedeño-Díaz, 2015). In general, a numerical value portrays water quality, with smaller values signifying healthy and larger values signifying degraded ones. Biological Monitoring Working Party (BMWP) and Family Biotic Index (FBI) are also applied in aquatic ecosystem health evaluation through numerical values for single bioindicators at a level of a specific taxonomy. All such indices apply a principle that healthy streams will have most types of species sensitive to pollution present in them. Some indices, such as the Tanzania River

Scoring System (TARISS), have been developed specifically for use in a specific region.

Multimetric indices, such as Index of Biotic Integrity (IBI) and Benthic Index of Biotic Integrity (B-IBI), use a suite of several of these such as taxa richness, composition, trophic morphology, and tolerance to pollution in an endeavor to make a fuller evaluation of water quality. IBI values at locations of testing are compared with values achieved at native habitats in the region. IBI brings together information regarding a range of community characters in one overall value for representing ecologic integrity of a water body (López-López & Sedeño-Díaz, 2015).

Multivariate methods use models to transpose physicochemical river property onto actual organism, comparing predicted organism with organism in ideal, unimpacted state. Multivariate techniques use such models to identify potentially degraded locations through deviation from ideal state. River Invertebrate Prediction and Classification System (RIVPACS) is a common multivariate index predicting predicted macroinvertebrate community composition through use of environment information (Fierro et al., 2017).

Functional approaches analyse the functional feeding groups (FFG) of macroinvertebrates, such as shredders, grazers, collectors, and predators, to assess the ecological status of the river. The distribution of these groups along a river continuum reflects the energy source of the ecosystem. For example, shredders are associated with coarse particulate organic matter, while collectors are more prevalent in downstream reaches where fine particulate organic matter is abundant.

Functional approaches dissect functional feeding groups (FFG) such as shredders, grazers, collectors, and predators in an analysis of a river's state of affairs ecologically. Where such groups occur in a river continuum, it reflects a source of energy in an ecosystem. For example, shredders have a relation with coarse particulate organic matter, and collectors in augmented abundance occur in high availability sections in a river with a high level of fine particulate organic matter.

Several factors can impact macroinvertebrate community diversity and structure, and in consequence, bioassessment: Agriculture, Factors in water quality, Hydrological connectivity, Flows in a river, Alteration in habitats, Temporal variation. Agricultural activity, such as transformation of native forests for agricultural purpose, can have a significant impact on macroinvertebrate communities through run-off of nutrients, sedimentation, and use of pesticides. Alteration in factors in water such as temperature, pH, dissolved oxygen, conductivity, and turbidity have a direct impact on abundance, diversity, and community structure of macroinvertebrates. Hydrological connectivity between a river and its nearby water bodies can impact diversity and distribution of macroinvertebrates. Velocity of a river is directly proportional to its dispersal capacity and its impact in distribution of macroinvertebrates. Physical manipulation of a river channel, such as channelization and construction, can impact habitats of macroinvertebrates and cause a loss in diversity. Assemblage of macroinvertebrates can vary immensely with changing times of a year.

Despite their utility, a range of problems in using macroinvertebrates for bioindicator purpose can occur. Identification at family level will in most instances be enough, but species level will at times become a necessity, and such identifications require expertise in taxonomy and can become labor-involving. There can sometimes occur a lack of full life history information for a lot of species of macroinvertebrate, and for that reason, it can become uninterpretable to respond in them to an environment change. Macroinvertebrates can become less well-known in general to the layman in comparison with fish, and for that reason, communicating bioassessment information with them can become unfeasible. It can require a high density of samples and a range of metrics in a river to make its biological state, and in more complex systems, even more.

Water Lettuce (*Pistia stratiotes*) is an aquatic plant, commonly grown in rivers and lakes. It acts as a bioindicator, as its presence indicate high nutrient level, eutrophication, and poor water quality in river water. Its presence highlights nutrients in water and organic contamination. Figure 5.13 shows the Google Map image of growth of water lettuce near Mula Mutha river of Maharashtra, India. The sudden overgrowth (bloom) indicates the imbalance in ecosystem of river.

5.2.4 *Other Parameters*

5.2.4.1 *Major Cations*

Calcium (Ca_2^+), magnesium (Mg_2^+), sodium (Na^+), and potassium (K^+) are the major cations found in river water or any other water body. Calcium and magnesium are contaminants causing water hardness, which can affect soap lathering and pipe scaling. Sodium and chloride ions cause salinity. This can affect aquatic life adversely and damage irrigation soil. Potassium is a macronutrient for plants. Its presence in river water can

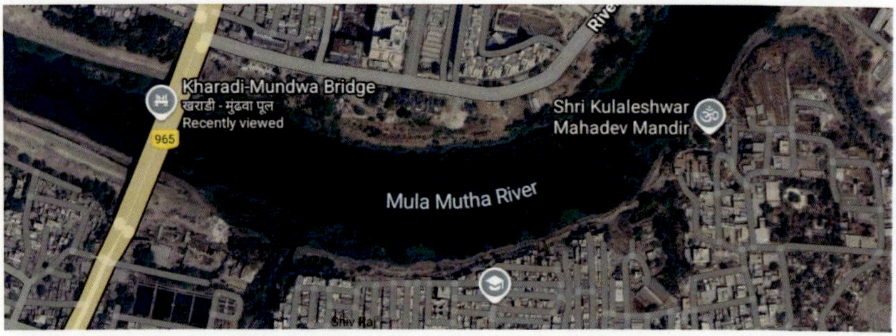

Fig. 5.13 Water Lettuce in river

cause aquatic and algal growth and development in aquatic plants. High values for sodium and chloride can represent an effect of saltwater intrusion, salt, and possibly saltwater intrusion, and can affect aquatic life and suitability for use in irrigation. Balance between river water cations is important for aquatic ecosystem integrity, with species having a range of tolerance to a range of ions. Cations in water, overall, most commonly represent a result of mineral dissolution, including calcite, dolomite, halite, and feldspar. Limestone areas, coastal areas, and arid areas, with water possibly affected by carbonate dissolution, seawater intrusion, and evaporation, represent a selection of hotspots for high values for these cations. Granite areas, mountainous areas, and wet areas, with water possibly affected by silicate weathering, rain, and melting of snow, represent a selection of hotspots for low values for these cations (Whelton & He, 2014).

5.2.4.2 *Calcium (Ca^{2+})*

Calcium ions in river water have both positive and negative impacts, on aquatic life, water quality, and in humans (Meride & Ayenew, 2016a). Calcium and magnesium together highly contribute to water hardness. Calcium can counteract toxicity in specific metals through competition with them for channels in fish gills for absorption. Higher concentrations of calcium can inhibit chloride paracellular permeability in fish gills and skin, reducing chloride toxicity. In the presence of magnesium, flocculation and sedimentation of fine particles can occur with calcium, clarifying the column of water. In hard water, toxicity in fluoride can be counteracted with calcium through CaF_2 precipitation. Calcium ions can alter microenvironments in fish gills, and through them, sodium bicarbonate ($NaHCO_3$) impacts in aquatic life can become an issue. Higher concentrations of calcium can seal aquatic organism gills and skin cellular junctions, reducing epidermal permeability (1987).

There is no documented report of any detrimental impact on health specifically produced by calcium in potable water. Drinking water can act as a complementary source of intake of calcium. Adult daily intake (RDI) of calcium is about 1000 mg. Water can contribute to it, with a part of tap water providing a maximum of 13% of the RDI and a part of mineral water providing over 40%. Different regions have a totally different level of tap water with a high level of calcium. Drinking water can contribute towards a daily intake of calcium, with Canadian tap water having an average of 18.2 to 21.8 mg/L and as high as 112.8 mg/L at times. In America, tap water can range between 8.3 and 131 mg/L, with an average of 50.6 mg/L. In Canada, Spring water contains an average of 21.8 mg/L of calcium, and mineral water has calcium an average of 208 mg/L. Some specific types of mineral water can have over 40% of a daily intake of calcium. Purified water will have a negligible level of calcium (Morr et al., 2006).

High concentrations in drinking water can inhibit absorption of osteoporosis drugs, including bisphosphonate drugs, and, therefore, such drugs must not be consumed with mineral water, and in a few cases, with high-calcium tap water. Drinking water with high concentrations of calcium can inhibit absorption of heavy metal, and a low-calcium diet sensitizes rats to lead and reduces plasma level of cholesterol. Calcium can be removed from water in softening processes such as ion exchange and chemical precipitation. Water softeners, such as Brita, can remove a considerable proportion of calcium in water, even 89% (1987).

5.2.4.3 *Magnesium (Mg2+)*

Magnesium (2 +) is an important mineral for many processes in living organisms, including generating energy, supporting enzyme activity, DNA synthesis, and relaxing muscles. Magnesium can react with other chemicals and minerals in water, such as carbonate, sulfate, and chloride, and form a range of compounds and precipitates. Magnesium can also affect water hardness and pH (with an estimation of 10.3 for magnesium oxide and 10.43 for magnesium hydroxide), and have an impact on water quality and its treatment. The criterion for magnesium in ambient water is 35 mg/L. It is considered to be sufficient to cover nutritional demand with no deleterious consequences, but can denote unnatural contamination in case of its presence in potable water. The nutritional allowance for magnesium for a single day varies between 50 and 450 mg/day. Magnesium is not a general cause for concern for its impact in case of consumption in ordinary food quantities (1984). Additional magnesium in the organism is eliminated in general via urine. Despite, high magnesium supplements can cause nausea, and in extreme cases, severe complications, and even more in case of kidney disease in humans. Higher concentrations can cause pipeline scaling, corrosion, and pipe lining impairment (Rapant et al., 2017). Methods such as ion exchange, distillation, RO, chemical perception (using lime or soda ash), and nanofiltration can be used to remove excess magnesium from water.

5.2.4.4 Sodium (Na^+)

Sodium (+1) is a vital element for all living organisms, including humans. It supports biological processes, including fluid balance, nerve transmission, and muscle contraction. It can also interact with other minerals and chemicals in water, such as chloride, bicarbonate, and fluoride, to form various compounds and solutions. Sodium can also affect the salinity and pH of water. This can have implications for water quality and treatment Sodium in drinking water can arise through mineral weathering, such as in rocks and sediments, or through contamination, such as in sewage and agricultural chemicals. Home softeners can add sodium to drinking water through ion exchange, exchanging sodium for hardness minerals, adding approximately 8 mg/L of sodium for each grain per gallon (17.118 ppm) of removed hardness.

High sodium intake can result in hypertension and heart disease; sodium-concerned individuals have to monitor overall intake. As per the EPA, a level of 20 mg/L for sodium is recommended. Suggestions include a restriction of overall intake to 2,300 mg a day. Average Americans normally consume 4,000 to 6,000 mg of sodium daily. About 77% of this dietary sodium comes from processed foods (2014a). Sodium intake globally is 4310 mg/day, which is equivalent to 10.78 g/day salt. Double the level recommended by the WHO, less than 2000 mg/day for adults (equivalent to < 5 g/day salt). About 1.89 million deaths a year occur with excessive sodium intake (2023). Taste impairment and corrosion occur at high sodium concentrations.

5.2.4.5 *Potassium (K^+)*

Potassium (1 +) ion is an important mineral for many processes in living, including nervous transmission, contraction in a muscle, and heart function. Potassium can affect salinity and pH in water (13.5 to 14 for KOH at 25°C), and such an effect can have an impact for water treatment and quality. It is recommended to have its level less than 12 mg/L (Wang et al., 2020). Total potassium in a human organism is 110 to 140 g. Deficit in potassium is in a negligible but can result in heart arrhythmia, depression, weakness in a muscle etc.(Meride & Ayenew, 2016b).

5.2.4.6 *Ammonium (NH_4^+)*

Ammonium is a form of nitrogen in water through both anthropogenic and natural processes. Ammonium is a beneficial nutrient for animals and plants, but a problem for both environment and human health when in high concentrations. Ammonium can react with chemicals in water, such as chlorine, oxygen, and nitrate, to form a variety of compounds and solutions (Liang et al., 2022). 'Ammonia' is a name for two species of ammonia chemicals, in equilibrium in water: un-ionized ammonia (NH_3) and ionized ammonium ion (NH_4^+). Species proportion is a function of physicochemical character of water, namely its temperature and its pH. Ammonia is toxic to aquatic life, but ammonium ion is not. Ammonium is not a matter of concern for health, but an operational guideline value of 1.5 mg/L measured ammonium (NH_4^+) for potable water is recommended, in case high concentrations produce off-tastes and odors and hinder disinfection processes (Zhang et al., 2019). Removal processes such as RO, ion exchange with selective resins (e.g., chelating and thiol group resin) and adsorption can remove excessive ammonium in water. High concentrations of NH_4 occur in agricultural, urban, and sewage effluents, and in effluents in wastewater, in which water can be subjected to fertilizer use, sewage discharge, and microbial nitrification. Predicted nitrogen level in river water from 2010 to 2050 is illustrated in Fig. 5.14.

5.2.4.7 *Major Anions*

Major anions in river water have a source predominantly in reduced compounds, organic matter oxidation, and mineral dissolution. Common examples include bicarbonate (carbonate), chloride (halite), and sulfate (gypsum or sulfide oxidation). High river water concentrations can occur in agricultural regions (fertilizer application in agricultural lands), coastal regions (intrusion of seawater), and mining regions (sulfide oxidation

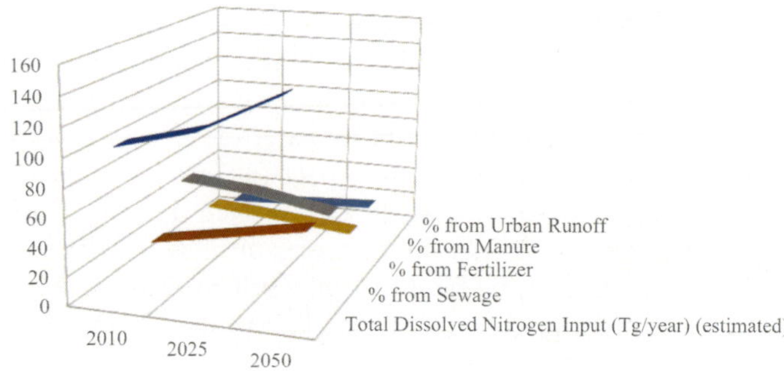

	2010	2025	2050
Total Dissolved Nitrogen Input (Tg/year) (estimated)	106	120	147
% from Sewage	25	35	45
% from Fertilizer	55	50	40
% from Manure	20	15	10
% from Urban Runoff	10	12	15

Fig. 5.14 Sources of Nitrogen Pollution in Rivers (Global, 2010–2050, Worst-case Scenario)

in mining areas). On the other hand, low concentrations can occur in regions with high humidity, hills, and forests (in regions with high rainfall, melting of snow, and contribution which contributes to organic matter). Removal of anions in water can be a difficult process. It can involve steps like boiling, ion exchange, reverse osmosis, distillation, activated alumina, and bone char.

5.2.4.8 Fluoride (F^-)

Fluoride (F-) is a necessary mineral for us. But if it is consumed in excessive amounts, it can cause harm. It can lead to dental fluorosis, skeletal fluorosis, and a range of disorders in general. Fluoride can react with magnesium, iron, and calcium, and with other chemicals and minerals present in water, and form compounds and precipitates with them. Fluoride can even affect water's hardness and its pH, and such an effect can have an impact on water's quality and its treatment.

The established a guideline for a value of 1.5 mg/L for fluoride (F) in potable water, for it is at a level that creates an ideal balance between the positive and negative consequences of fluoride. Nevertheless, it is prudent that a level of fluoride in potable water is altered in relation to the environment, nutrition, and consumption of supplementary sources of fluoride, such as toothpaste, mouthwash, and supplements. In case potable water contains a level of fluoride over 1.5 mg/L, a use of a device for defluorination, or a use of packaged water, is prudent (Ling et al., 2022).

Some hotspots with high values of F include volcanic, arid, and industrial regions, in which geothermal activity, evaporation, and use and production of fluoride can affect the water. Some of the hotspots with low values of F include high-elevation regions, high-rainfall regions, and high-forest regions, in which water can be affected by rain, melting of snow, and organic matter.

5.2.4.9 *Chloride (Cl^-)*

Chloride (Cl-) is one of the most important blood electrolytes involved in acid–base and body fluid balancing. It can occur in numerous sources, such as fertilizers, salted highways, effluents, and sewage, and is added in water for disinfection. Cl^- can affect odor, taste, and color in water and pipe and fixture corrosion (Roy, 2019; Sonone et al., 2021). Chloride can have a destructive impact in aquatic life and flora, robbing them of water and nutrition, and, in consequence, leading them to wither and die. WHO has established a 5-mg/L allowable upper level for potable water for chlorine (2008).

5.2.4.10 *Nitrate (NO_3^-)*

Nitrate (NO3-) can both contribute and cause both water and human disease, but contribute to high volumes for a high yield in plants. Nitrate can move out of fertilized lands, and out of animal manures, and out of septic systems, and even out of other sources, into groundwater. Nitrate can move off lands and into surface water, such as river, lake, and stream, and its determined allowable level of contaminating, for nitrates, is 10 mg/L (Abascal et al., 2022).

5.2.4.11 *Bicarbonate (HCO_3^-)*

Bicarbonate is responsible for regulating water's pH level. It keeps its level of pH in a balanced state. Bicarbonate will react with chemicals and minerals including iron, magnesium, and calcium in water. It forms new compounds and solutions. Bicarbonate can affect water alkalinity and hardness. It can affect water treatment and its quality (Anawar et al.,

2004). Bicarbonate is not harmful to one's health (Benaabidate et al., 2021).

5.2.4.12 *Sulfate (SO_4^-)*

Sulfate is responsible for maintaining water's pH level. Sulfate prevents extreme shifts in pH (Eastoe et al., 2022). Sulfate also mixes with chemicals and minerals such as iron, magnesium, and calcium in water. Sulfate forms various compounds and solutions in such a reaction. Sulfate can even impact water hardness and alkalinity. Sulfate can affect water quality and its treatment in such a case. Sulfate in water must not exceed 250 mg/L for safe consumption (Benaabidate et al., 2021). Sulfate in high concentrations can generate off-flavors and cause rusting (Sharma & Kumar, 2020).

5.2.4.13 *Dissolved Organic Carbon*

Dissolved Organic Carbon (DOC) is organic matter in water that is important in terms of aquatic life and water quality. It is a gauge of organic matter small enough to pass through an incredibly thin filter (0.22 to 0.7 micrometers). DOC is an important part of the planet's carbon budget and in a direct role in water quality and its life forms. DOC can be derived from nature, such as plants, animals, and soil, or through human actions, such as in wastewater. It can react with other compounds in water, such as metal and minerals, and with chlorine. DOC can modify aspects of water such as its pH, its hardness, its alkalinity, and its color. While DOC is not detrimental to your health, it can cause water to become a problem with its odor and taste, rust, or form unsightly byproducts when water is disinfected. There are specific requirements for concentrations of DOC in certain countries or regions (Chapelle, 2022).

High DOC values can act as an indicator of presence of organic matter, both natural and anthropogenic, in water, and low values can act as an indicator of presence of water with high oxidation and mineralization. High value hotspots include wetland, forests, and urban areas, in which water can have an impact of sewage water, peat, and plant debris. Arid lands, lakes, and groundwater have low value hotspots, in which water can have an impact of evaporation, biodegradation, and photo degradation (Regan et al., 2017; van der Jagt, 2013). Methods such as bio-filtration, RO, coagulation, and activated carbon adsorption can remove water's DOC.

5.2.5 *Heavy Metals*

5.2.5.1 *Soluble Iron (Fe)*

Soluble iron is iron that can be dissolved in water, most commonly in the form of the ferrous ion (Fe2 +). Soluble iron, ferric ion, iron bacteria, and organic ion, in addition to iron, can be present in water. Soluble iron can be present in groundwater, and most specifically in wells, and can occur naturally in rocks and in soils. Soluble iron can enter river water through groundwater, such as through rocks and soils, and through such actions of humans such as mining, use of fertilizers, and processes in industries. Soluble iron can have positive and negative effects, both in terms of water and for humans' overall general well-being and health. Iron in water, when in a state of being in a dissolved state, is not a concern for health. Nevertheless, an aesthetic or operational guideline value of 0.3 mg/L measured iron (Fe) for potable water is recommended, with high concentrations having a negative impact (Choudhury et al., 2022; Ghosh et al., 2020). Oxidation and filtering, ion exchange, and water softeners remove iron most effectively from water.

Some hotspots with high values of iron (Fe) include volcanic, mining, or industrial regions, whose water can be affected by geothermal activity, sulfide oxidation, or iron consumption and production. Some of the hotspots with low values of iron include regions with limestone, forests, or groundwater, whose water can be affected by carbonate dissolution, organic matter, or natural filtering (Sukmilin & Sangsirimongkolying, 2021).

5.2.5.2 *Soluble Arsenic (As)*

Soluble arsenic is a form of arsenic that can be solvated in water, predominantly in trivalent arsenate oxyanions (ASI) and pentavalent arsenate oxyanions (As(V)). Arsenic is a naturally occurring metalloid that can poison drinking water and food. Long-term exposure to arsenic can cause extreme medical conditions, including cancer, skin lesions, and cardiovascular disease. Long-term exposure to arsenic in a river can have disastrous impacts both for the environment and for humans consuming water, irrigating, and for any other use in a river.

Arsenic in river water can be absorbed by animals and plants and then move through the food web, and in the long term, affect humans. The guideline value for arsenic, according to its effect on humans, is 10 μg/L for potable water. Nevertheless, an estimated 140 million persons in at least 70 countries have consumed water with concentrations of arsenic over the guideline value of the WHO (2022a). It is prudent that, apart from its level in potable water, its level in potable water must differ with its surrounding environment, such as its climate, its diet, and its intake of arsenic through sources such as food, earth, and atmosphere (Shaji et al., 2021). Some high-value hotspots include volcanic areas, arid areas, groundwater areas, and mining areas, in which geothermal activity, evaporation, and sulfide oxidation can result in a change in water. Removal of arsenic in water can be done by anion exchange, electrocoagulation, distillation, and adsorptive membranes.

5.2.5.3 Soluble Manganese (Mn)

Soluble manganese is a form of manganese that can dissolve in water, most often in its divalent state (Mn2 +). Soluble manganese in groundwater, including in wells, can occur naturally in rocks and soils, but also through human activity such as mining, use in fertilizers, and in industries and make its way to rivers. Soluble manganese is a neurotoxin, with neurological symptoms (thinking and cognition impairments) a cause for concern, most particularly for babies and children. Babies who have bottle feeds with water that contains manganese are most at risk. The guideline value for potable water for manganese is 80 μg/L for precaution in consideration of uncertainties regarding its consequences for health. Soluble manganese naturally occurs in sources of water, and concentrations can become augmented through human activity, most particularly in reduced or acid groundwater (Desta et al., 2022).

Effective treatment methods are available to reduce manganese levels in drinking water to below 0.05 mg/L, including oxidation/filtration and reverse osmosis. Manganese can cause aesthetic problems like discoloration and staining in water at levels above 0.02 mg/l, even below health guideline values (2022b). Manganese contamination in drinking water is a global issue affecting both developed and developing countries. Infants have higher manganese absorption and retention compared to adults due to immature metabolic processes, increasing their vulnerability. Some of the hotspots of low Mn values are limestone areas, forested areas, or groundwater, where the water may be influenced by carbonate dissolution, organic matter, or natural filtration (Athirah et al., 2019; Li et al., 2021).

5.2.5.4 Soluble Cadmium (Cd)

Cadmium is a high molecular weight toxic metal and occurs in minute concentrations in natural water. It is an important river water analysis parameter in consideration of its toxicity and bioaccumulation in living organisms. In general, in natural water, the concentration of cadmium is below 1 μg/l, but in industrially contaminated and mining areas, it can increase. In water, cadmium can present in a range of forms including simple ions, ion pairs, organic and inorganic compounds, and metal ions sorbed onto particulates and colloids. All these forms have a variable degree of liability and affect their toxicity, bioavailability, bioaccumulation, and transport (2014b, 2022b).

Cadmium is one of the most toxic chemicals for animals and humans. Exposure to cadmium, even at low concentrations, decimates bone density and composition. Children have high susceptibility with developing bones, and hence, increased susceptibility to its effects. Cadmium accumulates in kidneys, and in older persons, it can cause kidney disease. Cadmium is a carcinoma when inhaled, but little is established for its role in producing carcinoma when taken orally. Cadmium can inhibit proper function of the endocrine in aquatic animals and can even affect aquatic animals' behavior. There is a guideline value for 3 ug/L (3 ppb) for cadmium in drinking water (2011).

Cadmium enters both naturally and anthropogenically in the environment. Volcanic eruptions and burning of forests can naturally discharge cadmium. Anthropogenics include mining, processing, disposal, use of phosphate fertilizers, and burning of fossil fuels, and cause cadmium contamination. Drinking water can become contaminated with cadmium through use of cadmium-plated solders, galvanized pipes, or through effluents in industries. Tobacco and foods such as leafy foods, sunflower seeds, peanuts, potatoes, organ meat, and shellfish can have cadmium in them.

Analytical methods can be used to evaluate concentrations of cadmium in water. There is a general technique, atomic absorption spectrometry (AAS), that can be used to evaluate concentrations of cadmium, with a 0.1 μg/l detectability when a furnace technique is adopted. There is a specific technique, graphite furnace atomic absorption spectrometry (GFAAS), that has been used to evaluate concentrations of cadmium in river water, following a kinetic solvent extraction technique with a chelating reagent, dithizone. The current advancements include—A new kinetic technique for speciating river water cadmium, in which extraction time can be adjusted with dithizone; a technique for speciating river water cadmium species, in which species can be partitioned into "highly labile," "moderately labile," and "slowly labile" species, with free cadmium ion extracted into chloroform in a dithizonato complex in one minute; and a new anodic stripping voltammetry (ASV) technique, in addition, can evaluate concentrations of electroactive cadmium (Tokuo SHIMIZU & Yoshinori HIRAOKA, n.d.).

Cadmium speciation is an important consideration when river water is examined. Cadmium can occur in a range of forms, and these forms can be characterized in terms of lability. "Very labile" metal complexed forms have high rates of metal complex dissociation, but "inert" complexed forms of metal are resistant to change. Some methods differentiate between species of cadmium and classify them "highly labile," "moderately labile," "slowly labile," and "inert." Variability in lability between species of cadmium can be determined with kinetic methods, such as a dithizone extraction in a solvent extraction method. In one analysis, in a river water, 10.1, 3.2, and 2.2 ng/l of "highly labile," "moderately labile," and "slowly labile" species, respectively, were measured. "Inert" species can be extracted with acid treatment. Cadmium in river water can become complexed with high molecular weight organic compounds that could have a strong association with humic compounds. Over a period, distribution between species of cadmium can vary, with a loss in "highly labile" and an increase in "moderately labile" species (Duinker & Kramer, 1977).

Cadmium levels in river water vary with location and sources of pollution. In a study of Missouri's Old Lead Belt's Big River and Flat River Creek, sediments ranged high in concentrations at all locations in and surrounding industries, sometimes even in values for screening. Maximum values ranged near sewage outfalls and near heaps of tailings, with values of 403 μg g-1. Concentrations in whole small sunfish bodies were greater in relation to concentrations in muscle tissue, a fact corroborated with cadmium's propensity for accretions in kidneys and in the liver (Gale et al., 2004).

5.2.5.5 Zinc (Zn)

Zinc can migrate into water in regions with high metal concentrations in the ground, including in locations with disposal of hazardous wastes. Zinc can occur in minor concentrations in practically all igneous rocks. Sulfides, including wurtzite and sphalerite, produce most zinc ores. Contamination of river water with metal comes about through runoffs at the surface contaminating the ground and filtering through groundwater aquifers. Most sources of releases of zinc into water and the environment occur through processes in metal production, burning of coal for industries, burning of trash, and tire wear and tear in cars. 1–300 mg/kg of zinc naturally occur in soils, estimates an estimate.

The concentration of zinc in groundwater and in surface water is less than 10 μg/liter in general; and in groundwater, between 10 and 40 μg/liter. However, through zinc leaching out of pipe and pipe fittings, zinc concentrations in tap water can become many times larger in value. Zinc concentrations in groundwater and in surface water in general don't become larger than 0.01 and 0.05 mg/l (Krok et al., 2022), respectively, but through zinc solubility in pipes, tap water can have concentrations many times larger in value. Most zinc in soil doesn't enter a water solution and is immobile, but under specific circumstances, zinc enters a water solution. RO, chelation (with amino carboxylic acids, thiol-containing compounds, and ionophores), ion exchange and chemical precipitation can remove zinc and purify water.

5.2.5.6 Soluble Uranium (U)

Soluble uranium is a uranium compound that can be dissolved in water, most frequently in ion form (UO_2^{2+}). Phosphate fertilizers—in agricultural lands, leached, and in industries, if leaked—contribute a considerable source of U-contamination in groundwater. In agricultural regions, high cases of discovering uranium concentrations in values over 0.5 micrograms per liter have been discovered. Drinking well water directly is the dominant way in which humans become contaminated with uranium (U), contributing over 99% of the toxic and radioactivity dose to consuming humans.

Soluble uranium can have both beneficial and detrimental effects both in the ecosystem and in humans' health. Long-term contact even with low concentrations of U isotope can cause cancer, and at high concentrations, kidney disease as well. It is a catalyst for a reaction between H_2O_2 and DNA, too. Measured uranium (U) in potable water guideline value, calculated through a basis of health, is 0.030 mg/L, for it is at such a level that one can have an ideal proportion between danger and benefit. However, it is preferable that uranium in potable water is lowered in proportion with the surrounding environment, such as climate, nutrition, and intake through foods, soil, and atmosphere, respectively. In case uranium concentrations in water fall in a range over 0.030 mg/L, one can utilize a uranium filtering device, and in extreme cases, one can utilize bottled water (Liesch et al., 2015).

The ingestion dose resulting from uranium in drinking water varies across age groups, ranging from as low as 0.03 to as high as 28.3 microsieverts per year. The lifetime cancer risk associated, ranges from 4.3×10^8 to 1.7×10^5, with an average of 4.8×10^6. Despite these variations, the average cancer risk remains notably lower than the recommended value for radiation exposure, which is 10^3. RO, coagulation, and Uranex process (ion exchange designed for uranium) can be used for removal of uranium from water.

5.3 Health Issues

Contaminated river water is a major issue as it is widespread and long-term. Confirmatory case studies worldwide validate its severity. In 1972, a disastrous episode in Minamata, Japan, involved methylmercury in a chemical plant spilling into water sources, contaminating fish and shellfish and, subsequently, causing mercury poisoning in humans (Yu et al., 2022). There followed 2,265 cases, including 460 deaths, with widespread neurological impairment, birth defects, and high fatality rates. Figure 5.15 demonstrates some of its tragic images. In Bangladesh (1980s to date), naturally occurring arsenic in groundwater contaminated water sources, and widespread chronic arsenic poisoning afflicted millions and an estimated thousand deaths followed. Health consequences include dermal lesions, cancers, cardiovascular disease, and impairments.

These issues vary from increased susceptibility to cancer, respiratory disease, and development concerns. Rural areas in all parts of the world face such concerns as high agricultural fertilizers contaminating groundwater and water sources with nitrates. There have not been any fatalities, but such ailments such as blue baby syndrome in newborns and increased susceptibility to cancer have been reported. India, in its instance, grappled with groundwater fluoride contamination (since the 2010s), affecting millions with skeletal fluorosis, producing aches in bones, stiffness in joints, and tooth decay. In spite of a decline in pesticides in Ganga basin since last decade, inorganic, carcinogen-related, and microbial contaminations have increased, putting several lives at risk (Dwivedi et al., 2018). All such case studies present a complex and widespread impact of contaminated water over humans, and therefore, immediate

Fig. 5.15 Minamata Japan disaster images

and overall interventions have to be taken in an attempt to reverse its impact and secure public welfare. For more details on specific pollutant, refer Table 5.4.

5.3.1 Respiratory System

The respiratory system is through which our bodies obtain oxygen and rid ourselves of carbon dioxide. It is composed of organs and tissue working together, starting with the nose in which air is cleaned, conditioned, and moistened. Air then comes in through voice box (larynx) and throat (pharynx) and then down through the windpipe (trachea). It then comes in through tubes (bronchi) that pass into the lungs.

Several chemicals can hurt our respiratory system when inhaled, when taken in through an injection, or when contaminated water enters via them (Cao et al., 2021). For example, a common water pollutant, chlorine, can cause irritation in our lungs. Cough, wheeze, and shortness of breath can occur through such irritation. Acute bronchitis, a disease characterized by a cough, shortness of breath, and a constricting chest, can occur with chlorine as well. Widespread inhalation of a lot of chlorine can make such a problem even worse.

Ammonia, another compound, can cause problems, too. High levels can cause lung irritation and develop into a case of pneumonia, specifically in susceptible individuals. Ammonia can cause hypersensitivity pneumonitis, in such a case, the lungs become inflamed and produce allergic symptoms. Nitrates in water have a similar problem with breathing, too. Long-term use of high levels can destroy the airways and lungs, and possibly develop into a case of chronic obstructive pulmonary disease (COPD). Nitrates can convert to nitrites, too, and disturb the way in which oxygen is transported in the body and cause pulmonary edema, a collection of fluid in the lungs (Kumar et al., 2022).

Heavy metals, including arsenic, when consumed in drinking water, can even result in difficulty in breathing when taken orally. Long-term consumption of arsenic is responsible for increased susceptibility towards lung cancer. Volatile organic compounds (VOCs) represent a group of chemicals that can cause irritation in the respiratory system and contribute towards asthma and similar disease. In a Bangladesh study, it was noticed that subjects taking water with arsenic developed susceptibility towards respiratory disease such as chronic bronchitis and a persistent cough. In a comparative study with and without arsenic, a high correlation between intake of arsenic-contaminated water and difficulty in breathing was noticed. With increased concentrations of arsenic, susceptibility towards such disease increased, and it can cause irreparable damage to the respiratory system (Milton & Rahman, 2002).

5.3.2 *Digestive System*

Digestive system is responsible for processing food, absorbing nutrients, and waste removal from body. Mouth, esophagus, stomach, intestines, pancreas, and other related organs and glands together form digestive system (S. and S. 2017). Minerals such as magnesium, calcium, potassium, and sodium play an important for health and function of digestive system. Magnesium supports muscle function in the digestive tract, calcium regulates muscle contractions, potassium helps maintain fluid balance, and sodium in necessary nutrient absorption.

Digestive system directly comes in contact with water and food. This makes it highly sensitive as it has to eliminate contaminates and ensure absorption of necessary nutrients. Contaminated water can directly introduce microbes or toxins to digestive system, making it venerable. Exposure to arsenic, mercury, lead, cadmium, chromium, pesticides and industrial solvents can cause inflammation and irritation to stomach. Resulting symptoms may range from nausea, vomiting to abdominal pain.

Nitrates, sulfates, and heavy metals like copper can influence Ulcerative colitis and Crohn's disease, both forms of inflammatory bowel disease (IBD). Though fluoride intake from water is necessary for dental health, excessive fluoride can cause dental fluorosis (Sharma et al., 2023). Certain contaminants can even kill helpful gut microbes, affecting absorption of nutrients and increased chances of infections.

5.3.3 *Reproductive System*

The reproductive system consists of organs for the creation of new life. Organs for continuity of species, and one of the basic characteristics that we are alive. Female organs include vagina, uterus, double ovary, and double fallopian tube, while males include testes, vas deferens, prostate gland, and penis. These organs constitute the major components for reproduction. Upon maturity, these organs make gametes. Proceeding to facilitate fertilization, pregnancy, and giving birth. And the cycle goes on.

Water contaminants including Bisphenol A (BPA), phthalates, and pesticides including diclofenac and chlorpyrifos cause 'Endometriosis.' It is a disorder when lining of uterus comes out of it. It bursts during each period, like any lining in uterus, and cause a lot of pain, discharge during micturition, and increased menses flow. There have been indications that these chemicals disturb hormonal and metabolite processes. It creates an abnormality in body cycles. Inflammation can occur and in turn contribute towards developing endometriosis. Polycystic Ovary Syndrome (PCOS) can develop through hormonal imbalance and metabolite disturbances, developed through exposure of Bisphenol A (BPA), phthalates, and Persistent Organic Pollutants (POPs) (Hama Aziz et al., 2023). Not only that but in males, it decreases sperm count, decreases motion of sperm, and abnormalities in tails.

Ovarian cancer is also associated with arsenic, nitrate/nitrite, and trichloroethylene exposure. Despite not knowing the mechanism, it is suspected that arsenic and nitrates cause oxidative stress. This inflicts damage to cells and increases the susceptibility to ovarian cancer. Not only ovarian but even testicular cancer is associated with consumption of nitrate/nitrite, heavy metal, arsenic and cadmium, and herbicide (atrazine) in drinking water. It is seen in studies that atrazine disrupts hormonal signaling and heavy metal is involved in oxidative stress. Together, it increases susceptibility to testicular and ovarian cancer.

Most of male infertility cases following exposure to lead, arsenic, phthalate, and pesticides including chlorpyrifos and dichlorodiphenyldichloroethane (DDT). Long-term and continuous exposure to such toxins can cause impairment in sperm development, loss in motility, and impairment in function, and therefore male infertility can result subsequently.

Nitrates is also one of the most important risks to human reproductive health, as per several studies. (Bukowski et al., 2001) described a complicated association of nitrate exposure and impact on reproductive health. The study comprised 210 intrauterine growth restriction (IUGR) cases, 336 premature deliveries, and 4,098 controls. Results showed a dose–response relationship. High nitrate levels were directly associated with risks of IUGR and prematurity. Maternal complications, including infections, hypertension, and bleeding, are found to be linked to methemoglobin. Methemoglobin levels in complicated pregnancies range from 0.1% to 11.2%, and in normal pregnancies from 0.4% to 2.8% (Tabacova et al., 1997). These findings indicate adverse effects of nitrate exposure in drinking water. While the study does not declare a definite relationship, the statistics suggest the need of preventive measures.

5.3.4 *Nervous System*

The most complex and least understood part of our body, nervous system is a web of complexly interconnected neurons and supporting cells and tissue. Spread out in coordinate form and joined together at brain and spinal cord connecting all the systems in whole body to conduct operations and respond towards stimuli (internal and external). It is composed of brain, spinal cord, and peripheral nerves, and together work to interpret sensations, initiate, and coordinate motion, and regulate bodily processes such as heart beat, breathing, thinking and digestions.

Minerals like magnesium, potassium, and calcium enable a nerve cell to convey and pass a message. Vitamin D, vitamin B12, and omega-3 fatty acids, in addition, preserve neuronal function and protect us from neurodegenerative disease (Yu et al., 2022).

Lead, arsenic, mercury, and organophosphates can cause neurodevelopmental disorders autism spectrum disorder (ASD) and attention-deficit hyperactivity disorder (ADHD). Autism spectrum disorder (ASD) is a lifetime mental condition, affecting behavior of a person. Rather than a specific disorder it is a spectrum of conditions, where an individual faces constant communication and behavioral problems. Chemicals disturb brain development and neuronal signaling, causing increased risk of neurodevelopmental disorders. Similarly, arsenic, lead, mercury, and fluoride have many times associated with cognitive decline and dementia. These chemicals can accumulate in the brain over time. Overtime they damage neuronal structures, disrupt function, and accelerates cognitive decline. Pesticides (organochlorines, pyrethroids) and manganese have connections to Parkinson's disease. It is a progressive neurodegenerative disorder, causing motor and cognitive impairments. Contaminants can induce oxidative stress and neurodegeneration, leading to the development of Parkinson's disease in the future (Preisner et al., 2021).

Arsenic, lead, mercury, and cyanide exposure can also lead to peripheral neuropathy. It is a condition determined by damaged peripheral nerves. This results in symptoms such as pain, numbness, and weakness in the extremities. Cyanide, lead, and methylmercury is often linked with motor neuron diseases like amyotrophic lateral sclerosis (ALS). ALS is rare condition in which motor neurons degrade over time, leading to muscle weakness, paralysis, and eventual death.

5.3.5 Skin

Skin is the greatest organ in human body, which comes directly in contact with outside. It is a protective envelope, and composed of several layers—epidermis, dermis, and subcutaneous tissue. Its major functions are thermoregulation, sensation, and immunity.

Contaminated water can directly come in contact with skin during bathing, causing skin diseases. Chlorine, generally used in disinfection of swimming pools, trihalomethanes and haloacetic acids can cause contact dermatitis if came in contact with skin (Bhattarai & Bhasin, 2022). This can cause irritation, producing burning and itching sensations. Eczema is a very frequent inflammatory disease generally characterized by dry, itchy, and inflamed skin. Sulfates and nitrates dissolved in water can damage skin and worsen eczema symptoms.

Arsenic, lead, nickel, and chromium, if present in water, can cause irritation skin leading to psoriasis and skin irritations. Psoriasis is an autoimmune condition leading to excessive cellular proliferation of the epidermal layers resulting in scale formation, epidermal thickening, erythema. Chronic exposure to this type condition can aggravate psoriasis and can stimulate more inflammation and irritation (Bhattarai & Bhasin, 2022).

5.3.6 Immune System

The immune system is a protective mechanism for the body. It protects us against pathogens, toxins, and foreign invaders. The immune system consists of white blood cells, antibodies, lymph nodes, thymus, bone marrow, and spleen. It is a collection of cells, tissue, and organs working together to detect and destroy danger. Defense forces maintain harmony in the body, as well.

Though contaminants destroy immune function, immune system has certain ingredients that it must have in its function. These chemicals include Zinc for immune cell development and function, vitamin C that act as an antioxidant and helps in function of numerous immune cells, vitamin D for immune cell function, iron for blood development in red blood cells and for immunity, and then the selenium for antioxidant and supporting immune function (Ma et al., 2023).

Water polluted with arsenic, mercury, lead, and Persistent Organic Pollutants (POPs) can lead to chronic inflammation and breakdown in immune tolerance. This has the potential to trigger autoimmune responses. Nitrates, pesticides (organophosphates), and chlorinated byproducts in water confuse immune development. This makes us more vulnerable to allergies and allergic diseases, such as asthma and eczema.

Not only that but, toxins in heavy metal such as arsenic, lead, mercury, and cadmium can pressurize immune cell function. That reduces our immunity and sensitizes us to pathogens and viruses. Chemicals including fluoride, nitrate, and chlorinated byproducts can penetrate mucous membranes in body, evading immune responses. That sensitizes us to bacterial, viral, and fungal infection. Exposure to toxins including arsenic, nitrates, and chlorinated byproducts can produce long-term inflammation. That triggers inflammatory ailments including Crohn's disease and ulcerative colitis.

5.3.7 Cardiovascular System

Cardiovascular system comprises of heart, blood vessels and blood flowing through them, transporting oxygen, nutrients, hormones, and waste products. It also maintains homeostasis, a state of balance in body. The heart works as a pump, providing essentials through blood.

Minerals play an important role in its functions. Potassium present in body, regulates blood pressure and heart functions; calcium is necessary for muscles, including muscles of heart; magnesium regulates blood pressure and controls heartbeat pattern; sodium regulates blood pressure and fluid balance. Chemicals like oxygen, CO2, and hormones work to regulate heart rate and blood vessel constriction.

Lead, arsenic, mercury, cadmium, chromium, and beryllium, if entered in body, can increase the risk of atherosclerosis and buildup plaque in arteries. This can later on lead to coronary heart disease (CHD). Arsenic, lead, cadmium, cyanide, and nitrates in contaminated water can cause hypertension. This increases the risk of heart disease, stroke, and kidney problems. Arsenic, lead, cyanide, and pesticides dissolved in river water trigger arrhythmias, a condition where heartbeat pattern beat very fast or very slow or even fluctuate. Arsenic, lead, cadmium, and selenium can also cause cardiomyopathy, a condition where heart muscle weakens. This can result in heart failure, stroke, or other complications (Teasdale & Panegyres, 2023).

Lead and mercury can cause pericarditis and inflammation of sac around heart. Chest pain, difficulty breathing, and fluid buildup around heart are some of its symptoms. Pesticides in water, such organochlorides, arsenic, lead, and cadmium, can cause cognitive heart failure (CHF), hindering heart's ability to pump blood and cause fatigue, breadth shortness and some other symptoms.

5.3.8 Renal System

The renal system, including kidneys and urinary tract, plays a role in general wellness in humans. The kidneys filter out waste and excess chemicals in our blood, and the tract executes disposal of urine through kidneys to the bladder for storing and eventual expulsion. To function effectively, kidneys require intake of certain chemicals such as potassium, sodium, calcium, and magnesium, involved in regulating fluid, balancing acid and base, and nervous and muscle function (Corona et al., 2022). All such significant electrolytes are present in water and foods, and proper intake is significant for effective kidney function.

Contaminated river water can bring toxic chemicals in our body, that can damage our renal system. Chemicals likes heavy metals, for example, arsenic, lead, mercury, and cadmium, can cause serious medical complications when present in river water. It can not only cause a reversible disease, such as Acute Kidney Injury (AKI), but can even develop into a state, such as of chronic kidney disease, with long-term impairment of kidneys (Corona et al., 2022). Chronic Kidney Disease of Unknown Etiology (CKDu) is a complex disease with a tendency to affect rural communities which has less access to pure water and sanitation. Arsenic, fluoride, cadmium, heavy metals, and agrochemicals in sources of groundwater and rivers are suspected to cause CKDu after consumption, though they are speculations only with no strong evidence. Exposure to specific chemicals in water can even have a variety of kidney disease. Like for instance, Glomerular Nephritis (an inflammatory glomeruli disease) can be caused through consumption of water contamination. These contaminants include arsenic, mercury, lead, cadmium, and organic solvents. It can cause inflammation, which can impair kidney function. This can become a cause for future development of renal disease. Diseases such as hypertension, a heart disease that causes stroke, can become worse through contact with arsenic, lead, and cadmium in drinking water. Interstitial Nephritis, Metabolic Acidosis, and nephrolithiasis (kidney stones) can develop through drinking contaminated heavy metals, nitrates, and variations in concentrations of calcium, magnesium, fluoride, and silica in water (Bekkouche et al., 2023; Perin et al., 2023).

Prolonged exposure to certain water impurities can even result in a heightened chance of cancer risks in the kidneys and urinary tracts. For example, Urothelial Carcinoma, a cancer risk in the bladder, can develop with long-term contact with arsenic, chlorinated metabolites, and trichloroethylene in drinking water. With a prolonged duration and high level of contact with such impurities, urothelial carcinoma is at a high risk, and thus, a healthy river water environment is significant to evade such cancer risks. Overall, access to safe and clean drinking water must be assured to protect our kidney system and prevent kidney-related complications.

5.3.9 Hepatic System

The hepatic system (liver) is responsible for metabolic, detoxification, and regulatory functions essential to living. It plays a central role in processing nutrients, filtering toxins from the blood, producing bile for digestion, and regulating various biochemical processes in the body.

Several chemicals found in water sources can impact the hepatic system and contribute to various liver disorders. Acute chemical-induced hepatitis can result from exposure to arsenic, mercury, lead, heavy metals like cadmium and chromium, pesticides such as organochlorines and glyphosate, and industrial solvents like trichloroethylene and carbon tetrachloride (Melebary, 2023). These chemicals can directly damage liver cells, disrupt metabolic pathways, and induce oxidative stress, leading to acute liver inflammation and dysfunction.

5.3.9.1 Genetical Impact

The relationship between water pollution and its genetic impact on human health is an emerging and complex field of study. While direct health impacts of certain chemicals have been well characterized, transgenerational consequences through genetic modifications are in development. Some of the processes involved in mediating chemical water pollutant's genetic impact include direct DNA damage, epigenetics, and selective pressures.

Direct DNA damage may occur when chemicals like arsenic and chromium induce mutations, deletions, and chromosomal translocations. Such alterations are heritable to offspring when they happen in germ cells (sperm or egg). Epigenetic alterations, on the other hand, alter gene expression but not the DNA sequence. Some chemicals in polluted river water can disrupt such epigenetic marks, potentially changing gene expression in future offspring. Environmental pollution can, through selective pressure, act in a way that favors survival in individuals with variations in their genome that render them resistant to chemicals. In the long run, such resistant variations become predominant in a population (Landrigan et al., 2023).

Several chemicals in contaminated riverwater have been associated with potential genetic impact. Exposure to arsenic, for example, can induce DNA damage and enhance susceptibility to certain cancers with a genetic etiology. Exposure to lead has been associated with aberrant processes in DNA methylation, with potential for impact during cognitive development through epigenetic changes. Exposure to mercury can induce DNA damage and enhance susceptibility to neurodevelopmental disease.

Nitrates in riverwater can form nitrites in the body, causing DNA damage and a cancer risk. Certain pesticides have been associated with epigenetics and birth defects and with inducing childhood cancers. Solvents in industry cause DNA damage and a reproductive impairment.

Uranium decay products like radon emit alpha radiation, cause DNA damage, and a cancer risk. Tritium, a radioisotope of hydrogen, can impact DNA stability and cell replication when ingested by the body. Cesium-137 and strontium-90 radioisotopes, which are created in nuclear accidents and in weapons testing, can find their way into riverwater, and become deposited in bones, with long-term radiation and genetic effect.

5.4 Environmental Issues

5.4.1 Impact on Ecosystems

Contaminated rivers (as shown in Fig. 5.16) is a big concern and can have serious long-term and large-scale impacts, especially for wildlife ecosystems dependent on it. Even a small variation in river water composition can change the balance of ecosystem, introduce invasive species, or erase entire ecosystem. For more details on specific pollutant, refer Table 5.4.

5.4.1.1 Impact on Flora

Flora is a collective term for all the plant species within an ecosystem. River contamination poses a great threat to the flora. Especially, nitrates

Fig. 5.16 Water pollution

and heavy metals are the major pollutants that can critically damage flora. Many research findings show the alarming rate of elevated nitrate levels in plants, hindering root growth and nutrient uptake. Nitrate concentration higher than 10 mg/L in water can stunt root development and reduce water uptake (Ren et al., 2022). Furthermore, empirical data indicate that heavy metals, such as lead (Pb) and arsenic (As), accumulate in plant tissues, causing physiological stress and impairing growth.

River water can introduce contaminants in soil as well. A study revealed that concentration of lead exceeding 0.5 mg/kg in soil can be hazardous for local fauna. Lead can bioaccumulate, leading to inhibition of seed germination, root elongation, and chlorophyll synthesis, resulting in reduced biomass production (Srikanth et al., 2022). Arsenic has also potential in bioaccumulation in vegetables in contaminated soil. Some studies even revealed the increased arsenic levels, beyond permissible levels, in edible plants at certain areas (Kazemi et al., 2022).

Heavy metals accumulated in plants can disrupt soil–plant interactions and affect nutrient cycling. Heavy metals such as manganese are often correlated with physiochemical properties of soils (Kapitalchuk et al., 2018). Empirical data demonstrates that heavy metals can influence soil ph, organic carbon content, and exchangeable cations. These findings highlight the complexity of relations between river water, contamination, soil, and fauna. Research conducted in contaminated wetland areas, resulting in accumulation of pollutants in plants, soil, and water, has shown decline in plant diversity and changes in community structure (Ghezzi et al., 2023). Additionally, loss of keystone plant species, such as ones involved in phytoremediation, due to contaminated water, can damage ecosystems even further.

5.4.1.2 Impact on Fauna

Fauna refers to all animal life within a particular region collectively. It can encompass variety of organisms, ranging from bacteria to elephants. Pollution significantly impacts benthic macroinvertebrate communities. In the Carpathian rivers of southern Poland, pollution led to a massive increase in Oligochaeta density, while diversity remained stable in less polluted mountain rivers. However, in heavily polluted areas, sensitive taxa such as Ephemeroptera and Trichoptera were extinct, indicating severe ecological degradation. Metal contamination affects different biological groups in rivers. Studies show that macroinvertebrates are more sensitive to metal contamination than fish, with significant reductions in mayfly abundance observed in contaminated sites. This suggests that macroinvertebrates are effective indicators of metal pollution in aquatic environments (Namba et al., 2020, 2021).

Contaminants in river water can have cytotoxic and mutagenic effects on fauna. In the Marrecas River, Brazil, bullfrog tadpoles exhibited cellular changes due to agrochemical contamination (dos Santos et al., 2021). Urban pollutants, like xenobiotic contaminants, can damage river food webs. In South Wales, UK, urban wastewater discharges led reduced diversity in benthic invertebrates. This affected the entire food web by reducing prey, reducing apex predators (Windsor et al., 2019). Microplastic pollution is also one of the persistently growing treats to riverine ecosystems. In UK, microplastics accumulated in river beds affected benthic fauna unbalancing the entire ecosystem (Woodward et al., 2021).

River damming and subsequent water quality changes can severely affect aquatic fauna. In the Sabor River dam in Portugal, dam construction led to increased water temperature, electric conductivity, and nutrient accumulation, which in turn caused a decline in native fish species and an invasion of exotic species. The ecological status of benthic macroinvertebrate assemblages also deteriorated, indicating a shift from good-fair to fair-poor conditions (Santos et al., 2017).

Moreover, soil contamination with heavy metals and petroleum derivatives disrupts soil ecosystems, affecting edaphic fauna and potentially impacting terrestrial organism dependent on these habitats (de Souza et al., 2013). Additionally, Malard and Hervant (Hervant & Malard, 1999) discussed the adaptations of groundwater animals, including crustaceans, to low oxygen concentrations in hypogean environments. While these adaptations enable survival in subtoxic conditions, pollutants can exacerbate physiological stress, affecting the health and viability of hypogean fauna.

5.5 Impact on Agriculture

Agriculture is in significant threat due to river water contamination. Agriculture itself is a source contaminant such pesticides, fertilizers, and organic matter. Contaminated water can cause numerous issues such as low crop yield, plant damage, low stress tolerance, lowering nutrient value and most importantly accumulation of toxic contaminants in food. Not only that, but in long-term, constant accumulation of heavy metals, ions, and biological contaminants can damage soil fertility, pH, texture, and its water absorption capability. For sustainable healthy agriculture, correct water source is a must. Water source compatibility with agriculture is a multidimensional topic, including many aspects such as crop type, soil quality, solar irradiance, water accumulation on agricultural land, etc. But in general sense low levels of nitrates in water can support agriculture, though high levels may be harmful. Many regions across the globe are affected by arsenic, nitrates, pesticides, and fluoride. For example, studies conducted in Shahpur block, Bhojpur dist. Bihar, India revealed alarming levels of arsenic contamination in groundwater. The maximum concentrations reached 598 ppb. The sample water was obtained by a hand pump and the maximum arsenic content in drinking water is 10 ppb. It is already far beyond consumption limit, but if this water is used for irrigation, it will not only damage crops, but also harm the people consuming the agricultural products (Singh & Ghosh 2014).

Water pollution in Triffa Plain in Morocco is due to intensive agriculture practices. The contaminants are nitrates, salinity, and pesticides. Deterioration of water quality over years is due to the over use of water and improper use of agrochemicals. Which is detrimental to us, at the same time to agricultural productivity and environment health (Kadaoui et al., 2019). A survey in Giao Thuy District, Nam Dinh Province, Vietnam, also highlighted the risk of nitrate contamination in shallow groundwater due to agricultural activities. Same can be speculated for the nearby rivers, all the contaminants are naturally washed away, adding to river contamination (Huyen et al., 2020).

In the Salinas River, California, agricultural drain water entering rivers, introduced organophosphate pesticides like chlorpyrifos and diazinon in it. This has cause acute toxicity in aquatic organisms. Such situations can disrupt the natural balance. Not only that, but it can kill beneficial organisms from agricultural land. This can affect agricultural productivity and sustainable agriculture ecosystem (Anderson et al., 2003). In the Guarani Aquifer region, nitrogen fertilizer increased nitrate levels in water bodies (LIMA et al., 2023). In south-west Victoria, Australia, dairy farming has high levels of nitrogen and coliforms in river water. This degrades water quality and introduce toxins to agricultural land. It also promotes the growth of harmful microorganisms, infecting both water and food (via agriculture) (Aburto-Medina et al., 2019).

The agricultural run-off water can directly impact river water quality. It allows harmful pesticides, herbicides, fertilizers, and nitrates to infiltrate river. This can affect downstream aquatic ecosystem; thought it is variable due to multiple factors involved. Different types of cultivation can alter the hydrochemistry of streams. Like in central Poland, concentrations of nitrates, with some other chemicals, in river water increased due to agricultural land run-off (Stępniewski et al., 2023). Huangshai river is another example of pollution due to agriculture and animal husbandry. It was contaminated by pesticides and fertilizers which introduced hazardous metals in river. This has led to several health concerns to both aquatic life and human's dependent on river water (Zhang et al., 2022). Same in India where farmers have shown their concerns about river water contamination (Jatav et al., 2023). Agricultural water intrusion in river water can damage other agricultural land, affect fish ecosystem, and poison human lives.

There are modeling applications such as MODFLOW and MT3DMS, which are powerful tools for simulating water flow. They can also help understand the contaminant transport in agricultural regions. This can aid in risk assessment and management of contamination of river water. These models incorporate complex parameters to simulate contaminant behavior in water systems. It can even enable us to study the potential impacts of groundwater and river water pollution on agriculture and ecosystem. Additionally, modeling studies help identify best remediation strategies for contamination removal (Elbeltagi et al., 2023).

5.6 Persistent and Future Environmental Crisis in Indian River Basins

India's river basins are increasingly threatened by persistent and emerging environmental crises, primarily driven by escalating human modifications. Rapid urbanization, industrialization, dam construction, and land-use changes have significantly altered natural flow regimes, reduced groundwater recharge, and increased pollutant loads. These modifications disrupt the ecological balance, leading to habitat fragmentation, loss of biodiversity, and the decline of endemic aquatic species.

Furthermore, over-extraction of water for agriculture and industry, coupled with climate change-induced variability in rainfall patterns, exacerbates water scarcity, and degrades water quality. The cumulative impacts of these stressors make river ecosystems more vulnerable to extreme events such as floods and droughts. If current trends continue, the resilience of India's rivers to recover from disturbances will be severely compromised, posing long-term risks to human health, food security, and overall environmental sustainability.

5.6.1 *Impact of Climate Change on River Sustainability and Resilience*

Climate change is emerging as a major driver of riverine environmental change in India. Rising temperatures, altered precipitation patterns, and increased frequency of extreme weather events are already affecting river flows, water quality, and ecosystem health. Reduced rainfall and higher evaporation rates lead to lower river discharge, concentrating pollutants and intensifying water quality issues. Conversely, intense rainfall events can cause flooding, mobilizing contaminants and overwhelming water treatment infrastructure.

These changes threaten the resilience of river systems—their ability to absorb disturbances and recover. Climate-induced shifts in river hydrology can disrupt fish migration, alter sediment transport, and affect the timing and extent of floodplain inundation, all of which are critical for maintaining ecological integrity. To achieve environmental sustainability, it is essential to integrate climate adaptation strategies into river basin management, such as restoring wetlands, enhancing riparian buffers, and promoting water-efficient agricultural practices.

Appendix

See Table 5.4.

Table 5.4 Water pollutants, Sources, Zones and impacts on human health and environment

Contaminant	*Manmade sources*	*Zones of occurrence*	*Impact on human life*	*Impact on environment*
Lead	Activities include mining, industry, plumbing, coal, gasoline, and water additives	galena, anglesite, minum, and cerussite	Affects critical areas of the brain, and neurological disorders In Children- memory loss, behavioral problems, anemia, schizophrenia, Alzheimer's disease, Parkinson's disease, and other neurodegenerative disorders (Fu & Xi, 2020)	Accumulation in fish results in hypocalcemia by impairing monocytes' basolateral transport mechanisms, which disrupts the electrochemical gradient and ion regulation
Chromium	human activities like mining, burning coal, and disposing of industrial waste in water receivers	Because of processes involving the interaction of ultramafic rocks and water	multi-organ poisoning in humans, cancer of the respiratory tract, asthma, allergies, and renal damage, abdominal tumors (Smith & Steinmaus, 2009)	Tumor formation, infertility, birth defects, and respiratory issues Causes anatomical alterations like hypertrophy and paraplegia at the gill epithelium
Mercury	chemical spills, or improper disposal of mercury-containing materials, mining, tanning, Electroplating industries	Found typically at low levels	pharyngitis, dysphagia, abdominal pain, nausea, vomiting, bloody diarrhea, and shock Later on, hepatitis, nephritis, anuria, stomatitis, and loosening of the teeth happen (Stolzenburg et al., 1986)	These impacts may impair genes, proteins, tissues, cells, and ultimately the ability of marine fish to survive, grow, and behave

(continued)

Table 5.4 (continued)

Contaminant	*Manmade sources*	*Zones of occurrence*	*Impact on human life*	*Impact on environment*
Cadmium	Ni–Cd batteries, the color of plastic, use of Cd-containing fertilizers	Organic-rich sediments, like black shale. phosphorites and carbonates contain cadmium, which leads to higher geogenic Cd concentra-tions in associated rock types	Kidney stone formation, hypercalciuria, lung cancer and prostate cancer, renal failure, and implications for hypertension	System, antioxidant defense system, reproductive regulation system, and tissue integrity. Cause a variety of epigenetic modifications (Genchi et al., 2020b)
Zinc	Production of metals, the industrial burning of coal, the incineration of waste, and the wear and tear on automobile tires	Igneous rocks	Production of metals, the industrial burning of coal, the incineration of waste, and the wear and tear on automobile tires (Rehman et al., 2018). Hypovolemic, and hypertension Anxiety, sadness, somnolence, dragging, fatigue, dizziness, and shock	It most likely kills fish by destroying their gill tissues at acutely toxic concentrations
Nickel	Septic Systems, Urban Run-off, Extracting and Processing Ores	Natural Weathering of Rocks Containing Nickel	Lung fibrosis, kidney, cardiovascular, and respiratory tract cancer. Skin irritation, diarrhea, vomiting, and stomach pain	Adverse effects on growth, development, and reproduction. Abnormalities in eggs, larvae, or juvenile stages (Genchi et al., 2020a)

(continued)

Table 5.4 (continued)

Contaminant	*Manmade sources*	*Zones of occurrence*	*Impact on human life*	*Impact on environment*
Manganese	mining, waste disposal, and manufacturing	mafic and ultramafic rocks	Tremors, balance issues, behavioral abnormalities, mood swings, IQ decline, and developmental delays speech impairment and motor dysfunction (Kondakis et al., 1989; Usepa, 2004)	Swimming behavior, stunted growth, and increased mortality
Iron	Corrosion of iron pipes, faucets, plumbing fixtures	subterranean rock	gastrointestinal problems like diarrhea, vomiting, and stomach pain Teeth and Clothes Staining	Excessive iron concentrations encourage the growth of algae, which can interfere with other plants' ability to receive sunlight and alter their habitats and feeding habits. A large number of algae reduces the freshness of the water (Isibor et al., 2020)

(continued)

Table 5.4 (continued)

Contaminant	*Manmade sources*	*Zones of occurrence*	*Impact on human life*	*Impact on environment*
Copper	pollution, domestic wastewater, and mining wastewater	regions of tectonic activity, such as subduction zones, where magma and hydrothermal fluids can be generated and transported to the Earth's surface	nausea, vomiting, and diarrhea as well as irritation in the mucosal lining Wilson's disease Menkes disease	Copper can interfere with this olfactory function in fish leading to toxicity, stress, and disorientation. gill damage, liver necrosis (Izah et al., 2016)
Cobalt	volcanic eruptions, forest fires, mining, and industrial activities	Cobalt and cobalt ores are found in two zones: weathered oxide and unweather sulfide deposits	hyperthyroidism or hypothyroidism. anemia or polycythemias hearing loss, tinnitus, nerve problems, and vision impairments	disruptions in their growth pigment production and antioxidant responses DNA damage, genotoxicity
Molybdenum	Spills or waste disposal	Rocky Mountains of North America, the Ural Mountains, Andes of South America	joint pain and symptoms resembling gout (Chappell et al., 1979)	disrupt the nitrogen fixation process (Gallo et al., 2023)
Arsenic	Mining Farming Industrial process	The region with Geothermal activity, Agricultural soil	Kidney and bladder damage Heart failure Lung cancer Reproductive problems Intestinal bleeding Skin cancer (Hong et al., 2017)	Apoptosis and necrosis in fish cells

(continued)

Table 5.4 (continued)

Contaminant	*Manmade sources*	*Zones of occurrence*	*Impact on human life*	*Impact on environment*
Halogens	Spills, leaks, infiltration, run-off, waste disposal	estuarine and marine environments	Cancer Blood disorder Lack of coordination Fetus development, miscarriage Dermatological issues	Disruption of the endocrine system in aquatic animals
Selenium	Sewage effluent, agricultural run-off, and industrial wastewater	Sedimentary rocks, shales, coal and phosphate deposits, and soils Arid regions	Keshan disease, Kashin Beck disease, Nerve impairment, Reduce the production of thyroid hormones (Werkneh et al., 2023)	deformities or mortality in fish Affecting species diversity, nutrient cycling
Nitrates	Animal Urine, Run-off from fertilizers, and Food processing waste	breakdown of organic matter and nitrogen fixation Groundwater is shallow, unconfined, and overlain by permeable soils	Methemoglobinemia, Thyroid problems, Cancer of the stomach and intestines (Ramalingam et al., 2022; Ward et al., 2018)	overstimulate the growth of algae and aquatic plants, hypoxic, anoxic
Radon	Mining and milling of uranium ore, nuclear power plant, hydraulic fracturing (Ademola & Ojeniran, 2017)	Rocks in the terrestrial crust	Lung cancer, leukemia, malignant melanoma, prostate cancer, kidney cancer (Ganguly, 2023)	illness and genetic flaws in aquatic life

(continued)

Table 5.4 (continued)

Contaminant	*Manmade sources*	*Zones of occurrence*	*Impact on human life*	*Impact on environment*
Radium	Coal ash, oil and gas production, nuclear waste	Granite bedrock	depression of the immune system, anemia, cataracts, fractured teeth, bone, liver, and breast cancer (Canu et al., 2011)	Mutations decreased fertility in bones and shells (Chowdhury et al., 2004)
Uranium	disposal of spent fuel from nuclear power facilities Excess nitrate in agricultural practices, phosphate fertilizer uses in oil and gas mining	minerals such as carnotite, uraninite, and autunite in rocks and ores processing phosphate rocks	brain, reproductive system, xenobiotic metabolism, antioxidant system, and intestinal inflammatory pathways	Disruption of aquatic organisms' normal immune, respiratory, digestive, and nervous systems (Corlin et al., 2016)

Conclusion

River water contamination is a major problem, having long-term and large-scale impact. Contaminants can enter food chains easily, through drinking water and food sources as well. Heavy metals can bioaccumulate in plants and animals, and enter higher levels of food chain, toxication every level. Major parameters of river water should be kept in check, to preserve local health and environment. Contaminated water risk human health, and can cause range of issues including physical, mental, and genetic. Not only human health but it can impact local ecosystem, damaging flora, and fauna. Prevention is necessary as remediation is costly and challenging.

References

Abascal, E., Gómez-Coma, L., Ortiz, I., & Ortiz, A. (2022). Global diagnosis of nitrate pollution in groundwater and review of removal technologies. *Science of the Total Environment, 810.*

Aburto-Medina, A., Shahsavari, E., & Salzman, S. A., et al. (2019). Elucidation of the microbial diversity in rivers in south-west Victoria, Australia impacted by rural agricultural contamination (dairy farming). *Ecotoxicology and Environmental Safety, 172*, 356–363. https://doi.org/10.1016/j.ecoenv.2019.01.112

Ademola, J. A., & Ojeniran, O. R. (2017). Radon-222 from different sources of water and the assessment of health hazard. *Journal of Water and Health, 15*(1).

Akkanen, J., & Ukkonen, J. V. K. K. (2001). Effects of water hardness and dissolved organic material on bioavailability of selected organic chemicals. *Environmental Toxicology and Chemistry, 20*(10).

Allam, M., Khan, M. Y. A., & Meng, Q. (2020). Retrieval of turbidity on a spatio-temporal scale using landsat 8 SR: A case study of the Ramganga River in the Ganges Basin, India. *Applied Sciences (Switzerland), 10*. https://doi.org/10.3390/app10113702

Altintas, Z., Gittens, M., Pocock, J., & Tothill, I. E. (2015). Biosensors for waterborne viruses: Detection and removal. *Biochimie, 115*, 144–154.

Anawar, H. M., Akai, J., & Sakugawa, H. (2004). Mobilization of arsenic from subsurface sediments by effect of bicarbonate ions in groundwater. *Chemosphere*. https://doi.org/10.1016/j.chemosphere.2003.08.030

Anderson, B. S., Hunt, J. W., & Phillips, B. M., et al. (2003). Ecotoxicologic impacts of agricultural drain water in the Salinas River, California, USA. *Environmental Toxicology and Chemistry, 22*, 2375–2384. https://doi.org/10.1897/02-427

Atangana, E., & Oberholster, P. J. (2021). Using heavy metal pollution indices to assess water quality of surface and groundwater on catchment levels in South Africa. *Journal of African Earth Sciences*. https://doi.org/10.1016/j.jafrearsci.2021.104254

Athirah, A. A., Saad, N. A., Akhir, M. F. M., & Zakaria, N. A. (2019). Manganese removal in groundwater treatment using marble. *International Journal of Integrated Engineering, 11*(2).https://doi.org/10.30880/ijie.2019.11.02.006

Aziz, K. H. H., Mustafa, F. S., Omer, K. M., Hama, S., Hamarawf, R. F., & Rahman, K. O. (2023). Heavy metal pollution in the aquatic environment: Efficient and low-cost removal approaches to eliminate their toxicity: A review. *RSC Advances, 13*(26), 17595–17610.

Ballschmiter, K., & Katz, J. J. (1969). An infrared study of Chlorophyll-Chlorophyll and Chlorophyll-Water interactions. *Journal of the American Chemical Society, 91*, 2661–2677. https://doi.org/10.1021/ja01038a044

Bekkouche, I., Shishonin, A. Y., & Vetcher, A. A. (2023). Recent development in biomedical applications of oligonucleotides with triplex-forming ability. *Polymers, 15* (Basel).

Benaabidate, L., Zian, A., & Sadki, O. (2021). Hydrochemical characteristics and quality assessment of water from different sources in Northern Morocco. In *Global groundwater* (pp. 261–274). Elsevier.

Bhattarai, K. R., & Bhasin, M. (2022). Water pollution is the main problem for human health. *Journal of Algebraic Statistics, 13*(2).

Björn, L. O., Papageorgiou, G. C., Blankenship, R. E., & Govindjee, A. (2009). A viewpoint: Why chlorophyll a? *Photosynth Res 99*, 85–98.

Boughattas, S., Ben Ayed, L., Mirjalali, H., et al. (2025). Molecular identification and characterization of waterborne protozoa among stranded Dugong dugon. *Marine Pollution Bulletin, 211*. https://doi.org/10.1016/j.marpolbul.2024.117454

Bright, C. E., & Mager, S. M. (2020). A national-scale study of spatial variability in the relationship between turbidity and suspended sediment concentration and sediment properties. *River Research and Applications, 36*, 1449–1459. https://doi.org/10.1002/rra.3679

Brooks, H. L., & Steele-Macinnis, M. (2019). A model for the solubility of minerals in saline aqueous fluids in the crust and upper mantle. *American Journal of Science*. https://doi.org/10.2475/09.2019.02

Bukowski, J., Somers, G., & Bryanton, J. (2001). Agricultural contamination of groundwater as a possible risk factor for growth restriction or prematurity. *Journal of Occupational and Environmental Medicine*. https://doi.org/10.1097/00043764-200104000-00016

Canu, I. G., Laurent, O., & Pires, N., et al. (2011). Health effects of naturally radioactive water ingestion: The need for enhanced studies. *Environmental Health Perspectives*. https://doi.org/10.1289/ehp.1003224

Cao, X., Coyle, J. P., & Xiong, R. et al. (2021). Invited review: human air-liquid-interface organotypic airway tissue models derived from primary tracheobronchial epithelial cells—overview and perspectives. *In Vitro Cellular & Developmental Biology-Animal, 57*(2).

Carter, M. J. (2005). Enterically infecting viruses: Pathogenicity, transmission and significance for food and waterborne infection. *Journal of Applied Microbiology, 98,* 1354–1380. Blackwell Publishing Ltd.

Chapelle, F. (2022). Dissolved organic carbon in groundwater systems. https://doi.org/10.21083/978-1-77470-016-7, ISBN: 9781774700167.

Chappell, W. R., Meglen, R. R., & Moure-Eraso, R. et al. (1979). United States Environmental Protection Agency, National Service Center for Environmental Publications (NSCEP). https://nepis.epa.gov/Exe/ZyPURL.cgi?Dockey=2000Z0FV.TXT

Choudhury, N., Siddiqua, T. J., & Ahmed, S. M. T., et al. (2022). Iron content of drinking water is associated with anaemia status among children in high groundwater iron areas in Bangladesh. *Tropical Medicine and International Health*. https://doi.org/10.1111/tmi.13710

Chowdhury, S., Husain, T., & Veitch, B., et al. (2004). Human health risk assessment of naturally occurring radioactive materials in produced water—A case study. *Human and Ecological Risk Assessment*. https://doi.org/10.1080/10807030490887203

Corlin, L., Rock, T., & Cordova, J., et al. (2016). Health effects and environmental justice concerns of exposure to uranium in drinking water. *Current Environmental Health Reports*. https://doi.org/10.1007/s40572-016-0114-z

Corona, R., Ordaz, B., & Robles-Osorio, L. et al. (2022). Neuroimmunoendocrine link between chronic kidney disease and olfactory deficits. *Frontiers in Integrative Neuroscience, 16*.

Cox, B. A. (2003). A review of currently available in-stream water-quality models and their applicability for simulating dissolved oxygen in lowland rivers. In *Science of the total environment* (pp. 335–377). Elsevier.

de Souza, R. B., Guilherme, T., & Aparecida, C. et al. (2013). Soil contamination with heavy metals and petroleum derivates: Impact on edaphic fauna and remediation strategies. In *Soil processes and current trends in quality assessment*, 175.

Desta, W. M., Lemma, D. B., & Tessema, T. A. (2022). Removal of Iron and Manganese from groundwater by using aeration and natural sand filtration techniques. *Chemistry Africa*. https://doi.org/10.1007/s42250-022-00486-2

Diamantini, E., Lutz, S. R., & Mallucci, S., et al. (2018). Driver detection of water quality trends in three large European river basins. *Science of the Total Environment, 612*, 49–62. https://doi.org/10.1016/j.scitotenv.2017.08.172

DİNÇ, G. (2021). Some foodborne and waterborne Protozoa. *Journal of Istanbul Veterinary Sciences, 5*, 107–112. https://doi.org/10.30704/http-www-jivs-net.948361

dos Santos, F. I., Mizobata, A. A., & Suyama, G. A., et al. (2021). Cytotoxicity and mutagenicity of the waters of the Marrecas River (Paraná, Brazil) to bullfrogs (Lithobates catesbeianus). *Environmental Science and Pollution Research, 28*, 21742–21753. https://doi.org/10.1007/s11356-020-12026-x

Duinker, J. C., & Kramer, C. J. M. (1977). An experimental study on the speciation of dissolved zinc, cadmium, lead and copper in river rhine and north sea water, by differential pulsed anodic stripping voltammetry. *Marine Chemistry, 5*(3).

Dwivedi, S., Mishra, S., & Tripathi, R. D. (2018). Ganga water pollution: A potential health threat to inhabitants of Ganga basin. *Environment International, 117*, 327–338. https://doi.org/10.1016/j.envint.2018.05.015

Eastoe, C., Hibbs, B., Merino, M., & Dadakis, A. (2022). Origins of sulfate in groundwater and surface water of the rio grande floodplain, Texas, USA

and Chihuahua, Mexico. *Hydrology, 9*(6). https://doi.org/10.3390/hydrology9060095
Ebrahimi, P., Guarino, A., & Allocca, V., et al. (2022). Hierarchical clustering and compositional data analysis for interpreting groundwater hydrogeochemistry: The application to Campi Flegrei volcanic aquifer (south Italy). *Journal of Geochemical Exploration*. https://doi.org/10.1016/j.gexplo.2021.106922
Elbeltagi, A., Srivastava, A., & Kushwaha, N. L., et al. (2023). Meteorological data fusion approach for modeling crop water productivity based on ensemble machine learning. *Water (Switzerland)*. https://doi.org/10.3390/w15010030
Fang, Y., Hu, Z., & Zou, Y., et al. (2017). Improving nitrogen utilization efficiency of aquaponics by introducing algal-bacterial consortia. *Bioresource Technology, 245*, 358–364. https://doi.org/10.1016/j.biortech.2017.08.116
Fierro, P., Valdovinos, C., & Vargas-Chacoff, L. et al. (2017). Macroinvertebrates and fishes as bioindicators of stream water pollution. In *Water quality*. InTech.
Fitri, A., Nizam, K., & Maulud, A. et al. (2021). Spatial and temporal distribution of dissolved oxygen and suspended sediment in Kelantan River Basin. https://doi.org/10.2991/aer.k.210204.011
Fu, Z., & Xi, S. (2020). The effects of heavy metals on human metabolism. *Toxicology mechanisms and methods, 30.*
Gale, N. L., Adams, C. D., & Wixson, B. G. et al. (2004). Lead, zinc, copper, and cadmium in fish and sediments from the Big River and Flat River Creek of Missouri's Old Lead Belt. *Environmental Geochemistry and Health, 26*(1), 37–49.
Gall, A. M., Mariñas, B. J., Lu, Y., & Shisler, J. L. (2015). Waterborne viruses: A barrier to safe drinking water. *PLoS Pathog, 11*(6).
Gallo, S., Nania, G., & Caruso, V., et al. (2023). Bioaccumulation of trace elements in the muscle of the Blackmouth Catshark Galeus melastomus from Mediterranean waters. *Biology (Basel)*. https://doi.org/10.3390/biology12070951
Ganguly, P. (2023). Medical geology related to different trace elements deficiency and toxicity diseases. *International Journal for Research in Applied Science and Engineering Technology, 11.* https://doi.org/10.22214/ijraset.2023.55616
Genchi, G., Carocci, A., & Lauria, G. et al. (2020a). Nickel: Human health and environmental toxicology. *International Journal of Environmental Research and Public Health, 17.*
Genchi, G., Sinicropi, M. S., & Lauria, G. et al. (2020b). The effects of cadmium toxicity. *International Journal of Environmental Research and Public Health, 17.*

Ghezzi, L., Arrighi, S., & Petrini, R., et al. (2023). Arsenic contamination in groundwater, soil and the food-chain: Risk management in a densely populated area (Versilia Plain, Italy). *Applied Sciences (Switzerland)*. https://doi.org/10.3390/app13095446

Ghosh, G. C., Khan, Md. J. H., & Chakraborty, T. K., et al. (2020). Human health risk assessment of elevated and variable iron and manganese intake with arsenic-safe groundwater in Jashore. *Bangladesh. Sci Rep, 10*, 5206. https://doi.org/10.1038/s41598-020-62187-5

Gómez-Gómez, B., Volkow-Fernández, P., & Cornejo-Juárez, P. (2020). Bloodstream infections caused by waterborne bacteria. *Current Treatment Options in Infectious Diseases, 12*, 332–348. https://doi.org/10.1007/s40506-020-00234-5

Göransson, G., Larson, M., & Bendz, D. (2013). Variation in turbidity with precipitation and flow in a regulated river system-river Göta Älv, SW Sweden. *Hydrology and Earth System Sciences, 17*, 2529–2542. https://doi.org/10.5194/hess-17-2529-2013

Gueddari, H., Akodad, M., & Baghour, M., et al. (2022). The salinity origin and hydrogeochemical evolution of groundwater in the Oued Kert basin, northeastern of Morocco. *Sci Afr*. https://doi.org/10.1016/j.sciaf.2022.e01226

Gunasekaran, D., Gerchman, Y., & Vernick, S. (2022). Electrochemical detection of waterborne bacteria using bi-functional magnetic nanoparticle conjugates. *Biosensors (Basel), 12*. https://doi.org/10.3390/bios12010036

Halliday, S. J., Skeffington, R. A., & Wade, A. J., et al. (2016). Riparian shading controls instream spring phytoplankton and benthic algal growth. *Environmental Science. Processes & Impacts, 18*, 677–689. https://doi.org/10.1039/c6em00179c

Hamza, I. A., Jurzik, L., Überla, K., & Wilhelm, M. (2011). Evaluation of pepper mild mottle virus, human picobirnavirus and Torque teno virus as indicators of fecal contamination in river water. *Water Research, 45*, 1358–1368. https://doi.org/10.1016/j.watres.2010.10.021

Haramoto, E. (2018). Detection of waterborne protozoa, viruses, and bacteria in groundwater and other water samples in the Kathmandu Valley, Nepal. In *IOP Conference series: Earth and environmental science*. Institute of Physics Publishing

Hervant, F., & Malard, F. (1999).Oxygen supply and the adaptations of animals in groundwater. *Freshwater Biology, 41.*

Hong, Y. S., Ye, B. J., & Kim, Y. M., et al. (2017). Investigation of health effects according to the exposure of low concentration arsenic contaminated ground water. *International Journal of Environmental Research and Public Health.* https://doi.org/10.3390/ijerph14121461

Huang, B., Yan, D., & Ouyang, C., et al. (2020). Chloropicrin fumigation alters the soil phosphorus and the composition of the encoding alkaline phosphatase PhoD gene microbial community. *Science of the Total Environment, 711*. https://doi.org/10.1016/j.scitotenv.2019.135080

Huyen, D. T., Da, L. N., & Toan, V. D. et al. (2020). Assessment of water quality in coastal zone of the red river estuary (Nam dinh province, vietnam). In *APAC 2019—Proceedings of the 10th International Conference on Asian and Pacific Coasts* (pp. 837–843).

Isibor, P. O., Imoobe, T. O. T., & Dedeke, G. A., et al. (2020). Health risk indices and zooplankton-based assessment of a tropical rainforest river contaminated with iron, lead, cadmium, and chromium. *Science and Reports*. https://doi.org/10.1038/s41598-020-72526-1

Islam, M. S., Hoque Majumder, S. M. M., & Author, A. (2020). Alkalinity and hardness of natural waters in Chittagong city of Bangladesh Article in. *International Journal of Science and Business, 4*(1). https://doi.org/10.5281/zenodo.3606945

Izah, S. C., Chakrabarty, N., & Srivastav, A. L. (2016). A review on heavy metal concentration in potable water sources in Nigeria: Human health effects and mitigating measures. *Exposure and health, 8*(2).

Jafarzadeh-Ghoushchi, S., & Dorosti, S. (2017). Effects of exposure to a variety of waste on human health—A review. *Journal of the Liaquat University of Medical and Health Sciences, 16*(1).

Jatav, S. S., Bharati, N., & Nayak, S. (2023). Farmer's perception of river water contamination in India: A case study of Rohini River in Maharajganj, India. *Agricultural Science Digest—A Research Journal*. https://doi.org/10.18805/ag.D-5738

Jiang, J., Sharma, A., Sivakumar, B., & Wang, P. (2014). A global assessment of climate-water quality relationships in large rivers: An elasticity perspective. *Science of the Total Environment, 468–469*, 877–891. https://doi.org/10.1016/j.scitotenv.2013.09.002

Jiang, Y., Yuan, Z., & Wang, Y., et al. (2024). Wastewater-based intestinal protozoa monitoring in Shanghai, hina. *Microbiology Spectrum, 12*. https://doi.org/10.1128/spectrum.04032-23

Jones, R. A., & Lee, G. F. (1982). Chlorophyll—A raw water quality parameter. In *Journal / American Water Works Association* (pp. 490–494).

Kabeto, J., Adeba, D., Regasa, M. S., & Leta, M. K. (2022). Groundwater potential assessment using GIS and remote sensing techniques: Case study of west Arsi zone. *Ethiopia. Water (Switzerland)*. https://doi.org/10.3390/w14121838

Kadaoui, M., Bouali, A., & Arabi, M. (2019). Assessment of physicochemical and bacteriological groundwater quality in irrigated Triffa Plain, North-East of Morocco. *Journal of Water and Land Development*. https://doi.org/10.2478/jwld-2019-0050

Kapitalchuk, I. P., Sheshnitsan, T. L., Sheshnitsan, S. S., & Kapitalchuk, M. V. (2018). Migration of manganese, zinc, copper and molybden in landscape-geochemical catena of the lower dnester valley. *South of Russia: Ecology, Development*. https://doi.org/10.18470/1992-1098-2018-2-96-112

Kazemi, A., Esmaeilbeigi, M., & Ansari, A., et al. (2022). Alterations and health risk assessment of the environmental concentration of heavy metals in the edible tissue of marine fish (Thunnus tonggol) consumed by different cooking methods. *Reg Stud Mar Sci*. https://doi.org/10.1016/j.rsma.2022.102361

Kedra, M. (2020). Regional response to global warming: Water temperature trends in semi-natural mountain river systems. *Water (Switzerland), 12*. https://doi.org/10.3390/w12010283

Kiulia, N. M., Netshikweta, R., & Page, N. A., et al. (2010). The detection of enteric viruses in selected urban and rural river water and sewage in Kenya, with special reference to rotaviruses. *Journal of Applied Microbiology, 109*, 818–828. https://doi.org/10.1111/j.1365-2672.2010.04710.x

Kondakis, X. G., Makris, N., & Leotsinidis, M., et al. (1989). Possible health effects of high manganese concentration in drinking water. *Archives of Environmental Health*. https://doi.org/10.1080/00039896.1989.9935883

Kownacki, A., & Szarek-Gwiazda, E. (2022). The impact of pollution on diversity and density of benthic macroinvertebrates in mountain and Upland Rivers. *Water (Basel), 14*, 1349. https://doi.org/10.3390/w14091349

Kripa, P. K., Prasanth, K. M., Sreejesh, K. K., & Thomas, T. P. (2013). *Aquatic macroinvertebrates as bioindicators of stream water quality-a case study in Koratty*. Kerala.

Krok, B., Mohammadian, S., & Noll, H. M., et al. (2022). Remediation of zinc-contaminated groundwater by iron oxide in situ adsorption barriers—From lab to the field. *Science of the Total Environment*. https://doi.org/10.1016/j.scitotenv.2021.151066

Kumar, R., Manna, C., & Padha, S., et al. (2022). Micro(nano)plastics pollution and human health: How plastics can induce carcinogenesis to humans? *Chemosphere*. https://doi.org/10.1016/j.chemosphere.2022.134267

Landrigan, P. J., Raps, H., & Cropper, M., et al. (2023). The Minderoo-Monaco commission on plastics and human health. *Annals of Global Health*. https://doi.org/10.5334/aogh.4056

Lannergård, E. E., Ledesma, J. L. J., Fölster, J., & Futter, M. N. (2019). An evaluation of high frequency turbidity as a proxy for riverine total phosphorus concentrations. *Science of the Total Environment, 651*, 103–113. https://doi.org/10.1016/j.scitotenv.2018.09.127

Lawler, D. M., Petts, G. E., Foster, I. D. L., & Harper, S. (2006). Turbidity dynamics during spring storm events in an urban headwater river system: The

Upper Tame, West Midlands, UK. *Science of the Total Environment, 360*, 109–126. https://doi.org/10.1016/j.scitotenv.2005.08.032

Li, R., Zhang, Y., & Chu, W. et al. (2021). Natural mineral for remediation of iron-and manganese-contaminated groundwaters. *Polish Journal of Environmental Studies*. https://doi.org/10.15244/pjoes/98996

Li, X., Lu, C., & Zhang, Y., et al. (2020). Low dissolved oxygen in the Pearl River estuary in summer: Long-term spatio-temporal patterns, trends, and regulating factors. *Marine Pollution Bulletin, 151*. https://doi.org/10.1016/j.marpolbul.2019.110814

Liang, Y., Ma, R., & Nghiem, A., et al. (2022). Sources of ammonium enriched in groundwater in the central Yangtze River Basin: Anthropogenic or geogenic? *Environmental Pollution*. https://doi.org/10.1016/j.envpol.2022.119463

Liesch, T., Hinrichsen, S., & Goldscheider, N. (2015). Uranium in groundwater—Fertilizers versus geogenic sources. *Science of the Total Environment*. https://doi.org/10.1016/j.scitotenv.2015.05.133

Lima, C. D., Batista, L. V., Garpelli, L. N., Santos, V. D., Quaggio, C. S., Gastmans, D., & Garpelli, L. N. et al. (2023).Assessment of river-aquifer interaction and nitrogen contamination in an agricultural zone located in the guarani aquifer system outcrop area. *Anais da Academia Brasileira de Ciências, 95*. https://doi.org/10.1590/0001-3765202320220609

Ling, Y., Podgorski, J., & Sadiq, M., et al. (2022). Monitoring and prediction of high fluoride concentrations in groundwater in Pakistan. *Science of the Total Environment*. https://doi.org/10.1016/j.scitotenv.2022.156058

Liu, G., He, W., & Cai, S. (2020). Seasonal variation of dissolved oxygen in the southeast of the Pearl River Estuary. *Water (Switzerland), 12*. https://doi.org/10.3390/w12092475

López-López, E., & Sedeño-Díaz, J. E. (2015). Biological indicators of water quality: The role of fish and macroinvertebrates as indicators of water quality. *Environmental indicators* (pp. 643–661). Springer.

Ma, Y. B., Xie, Z. Y., & Hamid, N., et al. (2023). Recent advances in micro (nano) plastics in the environment: Distribution, health risks, challenges and future prospects. *Aquatic Toxicology*. https://doi.org/10.1016/j.aquatox.2023.106597

Majeed, O. S, Nashaat, M. R., & Al-Azawi, A. J. M. (2022). Physicochemical parameters of river water and their relation to zooplankton: A review. In *IOP Conference Series: Earth and Environmental Science. Institute of Physics* (Vol. 1120, No. 1, p. 012040).

McBean, E., Bhatti, M., & Singh, A., et al. (2022). Temperature modeling, a key to assessing impact on rivers due to urbanization and climate change. *Water (Switzerland), 14*. https://doi.org/10.3390/w14131994

Melebary, S. J. (2023). Heavy metal toxicity and remediation in human and agricultural systems: An updated review. *Advances in Animal and Veterinary Sciences, 11.*

Meride, Y., & Ayenew, B. (2016a). Drinking water quality assessment and its effects on residents health in Wondo genet campus. *Ethiopia. Environmental Systems Research, 5.* https://doi.org/10.1186/s40068-016-0053-6

Meride, Y., & Ayenew, B. (2016b). Drinking water quality assessment and its effects on residents health in Wondo genet campus. *Ethiopia. Environmental Systems Research, 5*, 1. https://doi.org/10.1186/s40068-016-0053-6

Milton, A. H., & Rahman, M. (2002). Respiratory effects and arsenic contaminated well water in Bangladesh. *International Journal of Environmental Health Research, 12*, 175–179. https://doi.org/10.1080/09603120220129346

Morr, S., Cuartas, E., Alwattar, B., & Lane, J. M. (2006). How much calcium is in your drinking water? A survey of calcium concentrations in bottled and tap water and their significance for medical treatment and drug administration. *HSS Journal, 2*, 130–135.

Namba, H., Iwasaki, Y., Heino, J., & Matsuda, H. (2020). What to survey? A systematic review of the choice of biological groups in assessing ecological impacts of metals in running waters. *Environmental Toxicology and Chemistry, 39*, 1964–1972. https://doi.org/10.1002/etc.4810

Namba, H., Iwasaki, Y., & Morita, K., et al. (2021). Comparing impacts of metal contamination on macroinvertebrate and fish assemblages in a northern Japanese river. *PeerJ, 9*, Article e10808. https://doi.org/10.7717/peerj.10808

Oliver, R. L., Hart, B. T., & Olley, J. et al. The darling river: algal growth and the cycling and sources of nutrients Executive summary. https://www.researchgate.net/publication/285863799_The_Darling_River_Algal_Growth_and_the_Cycling_and_Sources_of_Nutrients

Ozaki, N., Fukushima, T., & Harasawa, H., et al. (2003). Statistical analyses on the effects of air temperature fluctuations on river water qualities. *Hydrological Processes, 17*, 2837–2853. https://doi.org/10.1002/hyp.1437

Perin, M., Chinigò, G., & Genova, T. et al. (2023).The impact of plasma membrane ion channels on bone remodeling in response to mechanical stress, oxidative imbalance, and acidosis. *Antioxidants, 12.*

Preisner, M., Smol, M., & Szołdrowska, D. (2021). A toxic-free environment ambition in the light of the polish baltic sea coastal zone pollution by heavy metals. *Desalination Water Treat*. https://doi.org/10.5004/dwt.2021.27325

Qu, S., Duan, L., & Shi, Z. et al. (2022).Hydrochemical assessments and driving forces of groundwater quality and potential health risks of sulfate in a coalfield, northern Ordos Basin, China. *Science of the Total Environment*. https://doi.org/10.1016/j.scitotenv.2022.155519

Raj, A., Gupta, A., Gupta, N., & Bhagyawant, S. S. (2023). Effect of water TDS, on the growth of plant (Phaseolus vulgaris). *International Journal of Plant & Soil Science*. https://doi.org/10.9734/ijpss/2023/v35i122977

Ramalingam, S., Panneerselvam, B., & Kaliappan, S. P. (2022). Effect of high nitrate contamination of groundwater on human health and water quality index in semi-arid region, South India. *Arabian Journal of Geosciences*. https://doi.org/10.1007/s12517-022-09553-x

Reddy, S., Sunitha, V., Suvarna, B., Reddy, M. R. (2022). Groundwater quality and its potential health impacts nearby inactive mines using PIG and geospatial technology, Southwestern part of Cuddapah basin, Andhra Pradesh, South India. *Groundwater for Sustainable Development, 17*. https://doi.org/10.1016/j.gsd.2022.100742

Regan, S., Hynds, P., & Flynn, R. (2017). An overview of dissolved organic carbon in groundwater and implications for drinking water safety. *Hydrogeology Journal*. https://doi.org/10.1007/s10040-017-1583-3

Rehman, K., Fatima, F., Waheed, I., & Akash, M. S. H. (2018). Prevalence of exposure of heavy metals and their impact on health consequences. *Journal of Cellular Biochemistry*. https://doi.org/10.1002/jcb.26234

Ren, H., Tian, L., & Zhu, Y., et al. (2022). Nitrogen and water addition alter nitrogen uptake preferences of two dominant plant species in a typical Inner Mongolian steppe. *Kexue Tongbao/chinese Science Bulletin*. https://doi.org/10.1360/TB-2021-1202

Roy, J. W. (2019). Endobenthic organisms exposed to chronically high chloride from groundwater discharging along freshwater urban streams and lakeshores. *Environmental Science and Technology*. https://doi.org/10.1021/acs.est.9b02288

Roy, S. K., & Zahid, A. (2021). Assessment of river water–groundwater–seawater interactions in the coastal delta of Bangladesh based on hydrochemistry and salinity distribution. *SN Appl Sci*. https://doi.org/10.1007/s42452-021-04389-8

Rusiñol, M., & Girones, R. (2019). Summary of excreted and waterborne viruses. In *Water and sanitation for the 21st Century: Health and microbiological aspects of excreta and wastewater management* (Global Water Pathogen Project). Michigan State University.

Saikrishna, K., Purushotham, D., & Sunitha, V. et al. (2020). Data for the evaluation of groundwater quality using water quality index and regression analysis in parts of Nalgonda district, Telangana, Southern India. *Data Brief*. https://doi.org/10.1016/j.dib.2020.106235

Santos, R. M. B., Sanches Fernandes, L. F., & Cortes, R. M. V., et al. (2017). Integrative assessment of river damming impacts on aquatic fauna in a

Portuguese reservoir. *Science of the Total Environment, 601–602*, 1108–1118. https://doi.org/10.1016/j.scitotenv.2017.05.255

Sara, N., Farzaneh, S., Hamed M., & Kobra, O. (2023). Toward waterborne protozoa setection using sensing technologies. *Frontiers in Microbiology*. https://doi.org/10.3389/fmicb.2023.1118164

Saylan, Y., Erdem, Ö., Cihangir, N., & Denizli, A. (2019). Detecting fingerprints of waterborne bacteria on a sensor. *Chemosensors, 7*. https://doi.org/10.3390/CHEMOSENSORS7030033

Shaji, E., Santosh, M., & Sarath, K. V., et al. (2021). Arsenic contamination of groundwater: A global synopsis with focus on the Indian Peninsula. *Geoscience Frontiers*. https://doi.org/10.1016/j.gsf.2020.08.015

Sharma, M. K., & Kumar, M. (2020).Sulphate contamination in groundwater and its remediation: An overview. *Environmental monitoring and assessment, 192*.

Sharma, S., Ibkar, A., & Javid, D. et al. (2023). Bio remedial process: A review on removal of fluoride from the waste water. *Ecology, Environment and Conservation*. https://doi.org/10.53550/eec.2023.v29i03s.073

Shi, X., Sun, J., & Xiao, Z. (2021). Investigation on river thermal regime under dam influence by integrating remote sensing and water temperature model. *Water (Switzerland), 13*. https://doi.org/10.3390/w13020133

Shimba, M. J., & Jonah, F. E. (2016). Macroinvertebrates as bioindicators of water quality in the Mkondoa River, Tanzania, in an agricultural area. *African Journal of Aquatic Science, 41*, 453–461. https://doi.org/10.2989/16085914.2016.1230536

Shimizu, T., Hiraoka, Y., Kikuchi, M., & Uehara, N. (2005). A novel kinetic method for the sepciation of cadmium in river water. *Analytical Sciences, 21*(2).

Siddiqui, N. A., Tauseef, S. M., & Dobhal, R. (Eds.). (2020). *Advances in water pollution monitoring and control*. Springer Singapore.

Singh, A., & Ghosh, A. K. (2014). Groundwater arsenic contamination and its implications: A case study of Shahpur block of Bhojpur District, Bihar. *International Journal of Modern Engineering Research, 4*(8).

Singh, A. K., Mondal, G. C., & Singh, T. B. et al. (2012). Hydrogeochemical processes and quality assessment of groundwater in Dumka and Jamtara districts, Jharkhand, India. *Environmental Earth Sciences*. https://doi.org/10.1007/s12665-012-1658-3

Skraber, S., Gassilloud, B., & Gantzer, C. (2004). Comparison of coliforms and coliphages as tools for assessment of viral contamination in river water. *Applied and Environment Microbiology, 70*, 3644–3649. https://doi.org/10.1128/AEM.70.6.3644-3649.2004

Smith, A. H., & Steinmaus, C. M. (2009).Health effects of arsenic and chromium in drinking water: Recent human findings. *Annual review of public health, 30*.

Sonone, S. S., Jadhav, S., Sankhla, M. S., & Kumar. R. (2021). Water contamination by heavy metals and their toxic effect on aquaculture and human health through food chain. *Letters in Applied NanoBioScience, 10*(2).

Srikanth, H., Kumari, K., & Meena, R. (2022).Contamination of groundwater with heavy metals and its impact on soil and plants—A review. *Ecology, Environment and Conservation*. https://doi.org/10.53550/eec.2022.v28i07s.054

Stępniewski, K., Karger, M., & Łaszewski, M. (2023). The impact of various types of cultivation on stream water quality in central Poland. *Water (Basel), 16*, 50. https://doi.org/10.3390/w16010050

Stolzenburg, T. R., Stanforth, R. R., & Nichols, D. G. (1986). Potential health effects of mercury in water supply wells. *Journal American Water Works Association*. https://doi.org/10.1002/j.1551-8833.1986.tb05676.x

Sukmilin, A., & Sangsirimongkolying, R. (2021). Removal of iron from groundwater by ozonation: The response surface methodology for parameter optimization. *Environment and Natural Resources Research*. https://doi.org/10.32526/ennrj/19/2020286

Tabacova, S., Balabaeva, L., & Little, R. E. (1997). Maternal exposure to exogenous nitrogen compounds and complications of pregnancy. *Archives of Environmental Health*. https://doi.org/10.1080/00039899709602209

Tabassum, R. A., Shahid, M., & Dumat, C., et al. (2019). Health risk assessment of drinking arsenic-containing groundwater in Hasilpur, Pakistan: Effect of sampling area, depth, and source. *Environmental Science and Pollution Research*. https://doi.org/10.1007/s11356-018-1276-z

Teasdale, N., & Panegyres, P. K. (2023).Climate change in Western Australia and its impact on human health. *Journal of Climate Change and Health, 12*.

Terzić, J., Filipović, M., & Boljat, I., et al. (2021). Groundwater level and electrical conductivity datasets acquired within pumping tests on Ilovik Island in Croatia. *Data in Brief*. https://doi.org/10.1016/j.dib.2021.107180

Thompson, M. Y., Brandes, D., & Kney, A. D. (2012). Using electronic conductivity and hardness data for rapid assessment of stream water quality. *Journal of Environmental Management, 104*, 152–157. https://doi.org/10.1016/j.jenvman.2012.03.025

Toffolon, M., & Piccolroaz, S. (2015). A hybrid model for river water temperature as a function of air temperature and discharge. *Environmental Research Letters, 10*. https://doi.org/10.1088/1748-9326/10/11/114011

Turyanti, A., & Chaerunnisa, C. (2017). The estimation of rainwater acidity level based on the ambient air pollutants concentration (Case study: DKI Jakarta). *Agromet, 31*(2). https://doi.org/10.29244/j.agromet.31.2.71-79

Usepa. (2004). *Drinking water health advisory for manganese*. US Environmental Protection Agency Office of Water. Washington, DC EPA-822-R-04-003. https://doi.org/EPA-822-R-04-003

van der Jagt H. (2013). Water Analysis | Overview. https://doi.org/10.1016/B978-0-12-409547-2.00567-9

Wang, H., Feng, C., & Deng, Y. (2020). Effect of potassium on nitrate removal from groundwater in agricultural waste-based heterotrophic denitrification system. *Science of the Total Environment*. https://doi.org/10.1016/j.scitotenv.2019.134830

Ward, M. H., Jones, R. R., & Brender, J. D. et al. (2018). Drinking water nitrate and human health: An updated review. *International Journal of Environmental Research and Public Health, 15*(7)

Weber-Scannell, P. K., & Duffy, L. K. (2007). Effects of total dissolved solids on aquatic organisms: A review of literature and recommendation for salmonid species. *American Journal of Environmental Sciences, 3*(1).

Werkneh, A. A., Gebretsadik, G. G., & Gebru, S. B. (2023). Review on environmental selenium: Occurrence, public health implications and biological treatment strategies. *Environmental Challenges, 11.*

Whelton, P. K., & He, J. (2014). Health effects of sodium and potassium in humans. *Current Opinion in Lipidology, 25*(1).

Wille, S. (2000). *Influence of light on algal growth in the lower Willamette River*. Portland, OR.

Wimalawansa, J. S. (2016). Effect of water hardness on non-communicable diseases, including chronic kidney disease of multifactorial origin (CKDmfo/CKDuo). *Journal of Environment and Health Science, 2*, 1–11. https://doi.org/10.15436/2378-6841.16.029

Windsor, F. M., Pereira, M. G., Tyler, C. R., & Ormerod, S. J. (2019). Persistent contaminants as potential constraints on the recovery of urban river food webs from gross pollution. *Water Research, 163*, Article 114858. https://doi.org/10.1016/j.watres.2019.114858

Woodward, J., Li, J., Rothwell, J., & Hurley, R. (2021). Acute riverine microplastic contamination due to avoidable releases of untreated wastewater. *Nature Sustainability, 4*, 793–802. https://doi.org/10.1038/s41893-021-00718-2

Wu, C., Fang, C., Wu, X., & Zhu, G. (2020). Health-risk assessment of arsenic and groundwater quality classification using random forest in the Yanchi Region of Northwest China. *Expo Health*. https://doi.org/10.1007/s12403-019-00335-7

Wu, J. L., Ho, C. R., & Huang, C. C., et al. (2014). Hyperspectral sensing for turbid water quality monitoring in freshwater rivers: Empirical relationship between reflectance and turbidity and total solids. *Sensors (Switzerland), 14*, 22670–22688. https://doi.org/10.3390/s141222670

Xu, M., & Li, C. (2020). *Application of the water footprint: Water stress analysis and allocation*. Springer Singapore.

Yao, H., Zhuang, W., & Qian, Y., et al. (2016). Estimating and predicting metal concentration using online turbidity values and water quality models in two rivers of the Taihu Basin, Eastern China. *Plos One, 11*. https://doi.org/10.1371/journal.pone.0152491

Yu, H., Chen, Q., & Qiu, W., et al. (2022). Concurrent water- and food-borne exposure to microplastics leads to differential microplastic ingestion and neurotoxic effects in zebrafish. *Water Research*. https://doi.org/10.1016/j.watres.2022.118582

Zhang, J., Liu, Q., & Liu, M., et al. (2023). Algae in a drinking water reservoir: Development of an Air–Water–Algal Growth Model (AWAM) for long-term prediction. *Water (Switzerland), 15*. https://doi.org/10.3390/w15203567

Zhang, L., Tan, X., & Chen, H., et al. (2022). Effects of agriculture and animal husbandry on heavy metal contamination in the aquatic environment and human health in Huangshui River Basin. *Water (Basel), 14*, 549. https://doi.org/10.3390/w14040549

Zhang, M., Yan, R., & Gao, J., et al. (2024). A framework for characterizing spatio-temporal variation of turbidity and drivers in the navigable and turbid river: A case study of Xitiaoxi River. *Water (Switzerland), 16*. https://doi.org/10.3390/w16172503

Zhang, W., Zhang, R., & Yang, Y., et al. (2019). Removal of high concentrations of ammonium from groundwater in a pilot-scale system through aeration at the bottom layer of a chemical catalytic oxidation filter. *International Journal of Environmental Research and Public Health*. https://doi.org/10.3390/ijerph16203989

Zhi, W., Feng, D., & Tsai, W. P., et al. (2021). From hydrometeorology to river water quality: Can a deep learning model predict dissolved oxygen at the continental scale? *Environmental Science and Technology, 55*, 2357–2368. https://doi.org/10.1021/acs.est.0c06783

Zhong, M., Liu, S., & Li, K., et al. (2021). Modeling spatial patterns of dissolved oxygen and the impact mechanisms in a Cascade River. *Frontiers in Environmental Science, 9*. https://doi.org/10.3389/fenvs.2021.781646

Zhu, S., & Piotrowski, A. P. (2020). River/stream water temperature forecasting using artificial intelligence models: A systematic review. *Acta Geophysica, 68*, 1433–1442.

Zounemat-Kermani, M., Seo, Y., & Kim, S. et al. (2019). Can decomposition approaches always enhance soft computing models? Predicting the dissolved oxygen concentration in the St. Johns River, Florida. *Applied Sciences (Switzerland), 9*. https://doi.org/10.3390/app9122534

(2012). Key indicators of drinking water, sanitation, hygiene and housing condition in India.

National Lakes Assessment Report. In: United States environmental protection agency. https://nationallakesassessment.epa.gov/webreport/. Accessed 25 Jan 2025a.

National Rivers and Streams Assessment: The Third Collaborative Survey. In United States environmental protection agency.

(1987). Guidelines for Canadian drinking water quality: Guideline technical document—Calcium.

(1984). US environmental protection agency: Ambient surface water quality standards documentation. In US EPA. https://www.epa.gov/sites/default/files/2015-06/documents/ny_hh_396_w_12201984.pdf. Accessed 2 Feb 2025.

(2014a). Sodium in drinking water.

(2014b). Cadmium and drinking water information sheet.

(2023). Sodium reduction.

(2008). Drinking water chlorination.

(2022a). Arsenic. In: World Health Organization. https://www.who.int/newsroom/fact-sheets/detail/arsenic. Accessed 2 Feb 2025.

(2022b). Guidelines for drinking water quality: Fourth edition incorporating the first and second addenda. In World Health Organization. https://www.who.int/publications/i/item/9789240045064. Accessed 2 Feb 2025.

(2011). Cadmium in drinking-water background document for development of WHO guidelines for drinking-water quality.

CHAPTER 6

Water Quality and Ecological Health of a Himalayan River Modified by Hydroelectric Projects Using Benthic Macroinvertebrate Bioindicators

Sandeep Kumar and Prakash Nautiyal

6.1 Introduction

Human civilization has historically occurred along the banks of rivers or near water bodies, underscoring the essential services the aquatic ecosystems provide to the mankind (Gleick, 2000). The conservation of these resources would be one of the earliest attempts of natural resource management in human history. Despite this, the continued exploitation

S. Kumar (✉)
Department of Zoology, Doctor Harisingh Gour Vishwavidyalaya (A Central University), Sagar, India
e-mail: sandeepkgu@gmail.com

S. Kumar · P. Nautiyal
Aquatic Biodiversity Unit, Department of Zoology, Hemvati Nandan Bahuguna Garhwal University (A Central University), Srinagar, India

S. Ghosh (ed.), *Reimagining Indian Rivers for Sustainability*, Palgrave Studies in Environmental Sustainability,
https://doi.org/10.1007/978-3-032-04754-0_6

of riverine resources for economic gains has resulted in significant degradation of biological diversity and the disruption of ecosystem structure and functioning (Dudgeon et al., 2006; Haase et al., 2025). Among the anthropogenic stressors, the construction of hydroelectric projects (HEPs) is recognized as the major threat to the riverine ecosystems (He et al., 2024). The science of the regulated river science, particularly advanced in European, North American, and Australian rivers contexts, has highlighted the ecological alterations induced by HEPs, including flow regime modification, habitat fragmentation, and decline in aquatic biodiversity (Dewson et al., 2007; Lillehammer & Saltveit, 1984; Walker, 1985; Wards & Stanford 1979, 1983, 1995). This has led to a persistent dilemma: while hydroelectric power is widely promoted as a renewable energy source with relatively low carbon emissions, its ecological impacts are often underestimated in favor of perceived economic and developmental gains. The release of environmental flows identified as sole solution for ecological sustenance is either not released or if released do not fulfil ecological needs and rarely subjected to systematic monitoring (Lin et al., 2024; Weisberg et al., 1990). As a result, the environmental costs associated with hydroelectric projects are frequently overlooked at different levels; during planning, decision-making and management.

For the sustainable management and conservation of riverine ecosystems and other aquatic resources, the concept of river health has emerged as a comprehensive framework (Karr & Chu, 1999). This framework emphasizes the maintenance of biological integrity (Palmer et al., 2005). The European Union Water Framework Directive (WFD, 2000) mandates the assessment of river health based on multiple biological quality elements, benthic invertebrates, fish, periphyton, and macrophytes. Analogous to human health, just as clinical indicators reflect human health status, biological indicators serve as diagnostic tools to assess the ecological condition of rivers, providing a reference point for management and restoration efforts. Among the biological quality elements, benthic macroinvertebrates are widely recognized as reliable bioindicators of water quality due to following reasons: These occupy key positions in aquatic food webs, links the grazing and detritus food chains; mediate energy flow and nutrient cycling; are relatively easy to sample and identify. Furthermore, these exhibit varying degrees of tolerance to the organic pollution, habitat degradation, altered flow regimes, and substrate composition, hydropower alterations (Nautiyal et al., 2019;

Rosenberg & Resh, 1993; Wright et al., 2000) makes them effective indicators of ecological integrity. The relatively long-life cycles tied to the natural hydrology in the riverine ecosystems enable the detection of both current and historical environmental conditions. Since the macroinvertebrates are widely studied group and hence has resulted in the development of number of biotic indices of wide acceptance in developed and under developing countries including India. Thus, the community is often preferred to determine the water quality and level of organic pollution in the water bodies.

Precise evaluation of water body degradation resulting from anthropogenic pressures is essential for the effective management and conservation of aquatic resources (Cox et al., 2022). In the current study, the Bhagirathi River, which is the source tributary of the Ganga River system, is examined for its water quality and overall ecosystem health. The river is serially fragmented by four hydroelectric projects (HEPs): Maneri Bhali Stage-I (MB-I), Maneri Bhali Stage-II (MB-II), Tehri HEP, and Koteshwar HEP, along its lower ~121 km stretch. As a consequence, approximately 27% of this river section has been transformed into lentic habitats (reservoirs), while the remaining riverine segments either receive very low flows downstream (d/s) of dams and barrages or experience regulated discharges downstream of powerhouse outlets. These hydropower-impacted stretches have previously been studied in terms of benthic macroinvertebrates, benthic diatoms, fishery with community structure and composition shown to be adversely affected (Kumar & Nautiyal, 2019a, b; Nautiyal et al., 2019). In the present study, benthic macroinvertebrate-based indices—including the Biological Monitoring Working Party (BMWP) (Armitage et al., 1983), Functional Feeding Group (FFG) (Merritt & Cummins, 2005), and Lotic Invertebrate Flow Evaluation (LIFE) (Extence et al., 1990) were applied with the objective of evaluating the impacts of hydropower modifications on water quality (specifically organic pollution levels), functional ecological status, flow conditions, and consequently, the ecological health of the riverine sections of the Bhagirathi River. This study contributes to the Government of India's initiative, under the Ministry of Jal Shakti, to identify polluted river sections for improved and sustainable management of aquatic resources, and supports the National River Conservation Directorate (NRCD) in its aim of preserving aquatic ecosystems, considered as part of the VIKSIT BHARAT 2047 vision. Furthermore, determining the degree of degradation across river segments subjected to varied flow regimes will assist policymakers in identifying critical stretches and in formulating management strategies for sustainably conserving hydropower-impacted sections of the Himalayan River.

6.2 Methodology

6.2.1 Study Area

The Bhagirathi River originates from the Gangotri glacier, specifically from the Goumukh snout located in the Indian Himalayan region, and flows for approximately 191 km before its confluence with the Alaknanda River at Devprayag, where the Ganga River is formed. The lower ~121 km stretch of the Bhagirathi is fragmented by a sequence of four hydroelectric projects (HEPs). The MB-I project diverts the river through a 9 km long tunnel, leaving around 10 km of the river channel devoid of natural flow. As the diverted flow is returned to the main channel via the Tiloth powerhouse at Uttarkashi (an urbanized area), the river is again diverted downstream (d/s) by MB-II through a 25 km long tunnel, leaving a further 25 km section with low flow. D/s of the MB-II powerhouse at Dhrasu, the tailrace of the Tehri HEP complex continues for another 35 km. Further d/s ~19 km, the river is impounded by the Koteshwar Dam, which is part of the larger Tehri-Koteshwar Dam Complex (TKDC). The final ~20 km section upstream (u/s) of the Devprayag confluence receives regulated discharge as per the peaking requirements of the hydropower plants.

For the present study (Fig. 6.1), six sampling stations were selected. Station S1 is located 2.5 km upstream of the MB-I dam and represents a near-pristine, free-flowing section of the river; it is considered the reference site. Stations S2, S4, and S5 are located downstream of various hydraulic structures: specifically, 5 km d/s of the MB-I dam/barrage, 1 km d/s of the MB-II barrage, and 25 km d/s of S4, respectively. These river sections continually receive low flows due to upstream diversion structures. Stations S3 and S6 are located d/s of the Tiloth powerhouse (MB-I) and approximately 20 km d/s of the Koteshwar powerhouse, respectively. These sites experience regulated flows released according to the peaking operations of the respective HEPs. Each station thus differs in terms of flow condition and its proximity to dams, barrages, or powerhouses. Furthermore, the selection of these stations departs from traditional impact assessment approaches, which typically focus on immediate upstream and downstream segments of HEP structures where community disturbance is obvious.

6.2.2 *Benthic Macroinvertebrate Collection and Analysis*

Benthic macroinvertebrate sampling was conducted monthly from October 2017 to September 2018 using a standard D-frame kick net sampler (mesh size: 500 µm). At each sampling station, ten quadrats were laid to sample multiple habitat types. The collected samples were preserved in 5% formaldehyde in labeled plastic tubes and transported to the laboratory for further analysis. Taxonomic identification of benthic organisms was performed up to the family level using a dissection microscope, following standard taxonomic literature (Dudgeon, 1999; Edington & Hildrew, 1995; Edmondson, 1959) and the online portal www.macroinvertebrates.org.

Taxa inventories were compiled for the computation of biological indices. The Biological Monitoring Working Party (BMWP) score (Armitage et al., 1983) was calculated using tolerance values to determine water quality classes and associated levels of organic pollution. Functional Feeding Groups (FFGs) were assigned following Cummins et al. (2005) and categorized as follows: Scrapers (graze periphyton, non-filamentous algae, associated detritus, microflora, fauna, and fecal material of size 0.01–1 mm), Filter Collectors (suspension feeders filtering particles from the water column, 0.01–1 mm), Gatherer-Collectors (deposit feeders ingesting sediment or collecting particles from deposited areas, 0.05–1 mm), Predators (capture and consume prey or tissues >0.05 mm), and Shredders (consume conditioned or live vascular plant tissue and coarse particulate organic matter >1 mm). The Lotic Invertebrate Flow Evaluation (LIFE) index (Extence et al., 1990) was also employed to assess flow condition. The LIFE index is particularly valuable for detecting ecological shifts associated with flow modifications.

Physicochemical parameters and BMWP scores were first assessed for normality, followed by appropriate significance tests to evaluate variations among sampling stations. Pearson's correlation analysis was conducted to assess relationships between BMWP scores and physicochemical parameters showing significant variation across stations. All statistical analyses were performed using PAST software version 4.17.

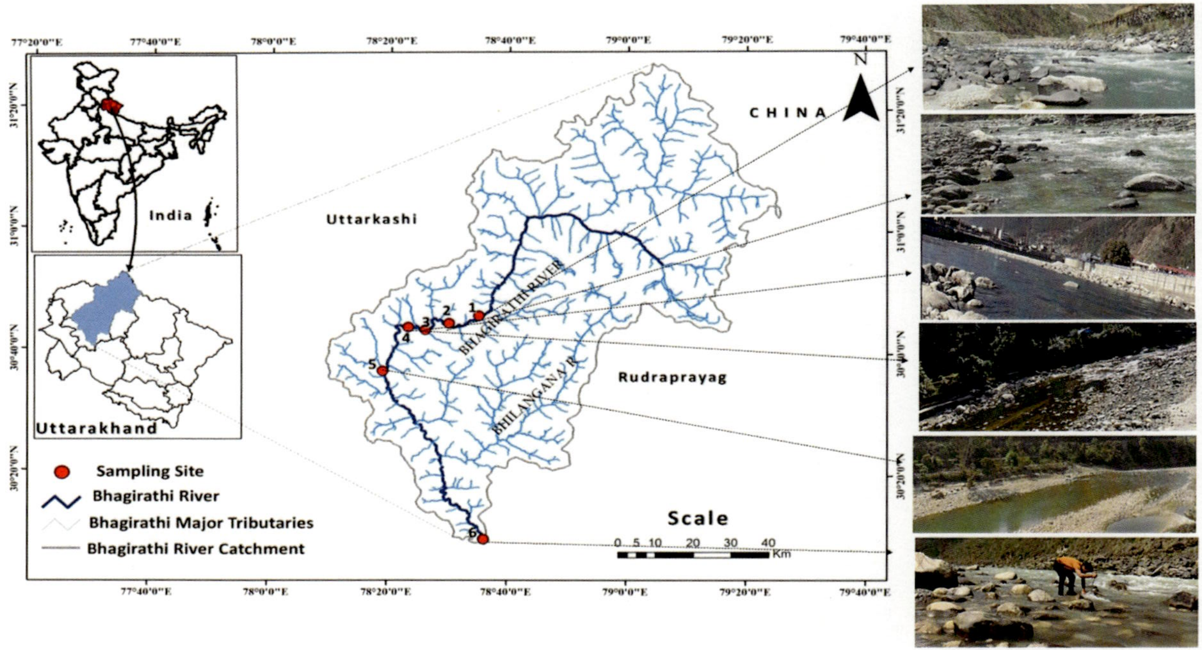

Fig. 6.1 Location map showing sampling stations in the Bhagirathi river. The images of the sampling location in the right reflects the different flow conditions at each station. S2, S4 and S5 shows low flows while S3 and S6 show regulated discharge in contrast to the natural flow at S1. The original map is taken FROM Kumar and Nautiyal (2019b)

6.3 Results and Discussion

6.3.1 Physicochemical Environment

The mean water temperature (WT) was 10.7 ± 1.08 at S1 rises sharply to 13.6 ± 0.9 at S2, 12.5 ± 1.04 at S3, 13.6 ± 0.8 at S4, 15 ± 0.8 at S5 and 15.5 ± 1.08 at S6, hence lack a characteristic longitudinal trend (Table 6.1). The 2.9 °C rise in WT from S1 to S2 was an exceptional within a short distance and has never been recorded earlier in the Himalayan rivers which is believed to rise by 1 °C at 25 km distance (Nautiyal, 2010; Nautiyal et al., 2004; Pathak et al., 2015). The WT falls at S3 is attributable to the discharge coming out of the 9 km long tunnel. This significant ($p < 0.05$) rise in the WT in the low-flow sections is attributable to the low discharge of the water along with the increased retention time, enhancing the rate of heating in these sections (Kumar & Nautiyal, 2019a, b; Poff & Zimmerman, 2010; Sharma & Nautiyal, 2023). The increase in the WT can be assumed to affect the endemic cold water fish fauna in these sections. The water velocity (WV) exhibited sharp decline from S1 (89 ± 3.1 m/s) to the low-flow sections S2 (65 ± 8.6 m/s), S4 (66 ± 6.6 m/s) and S5 (66 ± 6.7 m/s) depicting the low flow conditions d/s the barrage/dam. However, the higher WV at S3 (85 ± 6.5) and S6 (75 ± 6.8) shows the discharge released back to the river section. The disruptive pattern of WV is the foremost consequence of any hydroelectric projects.

The dissolved oxygen (DO) exhibits the significant low level at S2 (7 ± 0.14), S3 (6.7 ± 0.16), S4 (7 ± 0.1) and S5 (6.8 ± 0.11) compared to S1 (8 ± 0.07). The low DO at S2, S4 and S5 can be attributed to the inverse relation with WT and WV. The low oxygenation at S3 is possibly since the released water from the powerhouse comes through the long tunnel preventing atmospheric mixing. Nitrate tends to increase S1 to S2 and attains a sharp rise at S3, decline at S4 and further increases downstream. The sharp increase at S3 is possibly due to the non-point source of pollution from the urbanized banks along the S3 increases bacterial degradation. Total alkalinity, total hardness, Phosphate and Silicate exhibited a longitudinal increasing trend; however, the increase was not significant. The significant differences in the WT, WV, pH and DO among the stations indicates these parameters are the most effected parameters in the fragmented sections of the Bhagirathi River Singh and Singh et al., 1994).

Table 6.1 The Physicochemical characteristic showing longitudinal trends from S1 to S6 in the Bhagirathi river. Water temperature, water velocity, pH, Dissolved oxygen, and Nitrate show significant difference among the stations. Synonyms: r-spearman correlation coefficient

	S1	*S2*	*S3*	*S4*	*S5*	*S6*	*ANOVA* (p *value*)	*r*
Water temperature (°C)	**10.7** ± 1.08	13.6 ± 0.9	12.5 ± 1.04	13.6 ± 0.8	15 ± 0.8	**15.5** ± 1.08	**0.01**	**-0.48**
Water velocity (cm/sec)	**89** ± 3.1	**65** ± 8.6	85 ± 6.5	66 ± 6.6	66 ± 6.7	75 ± 6.8	**0.02**	**0.02**
pH	8.4 ± 0.1	8.5 ± 0.1	8.4 ± 0.1	8.6 ± 0.1	**8.7** ± 0.1	**8.7** ± 0.1	**0.01**	**-0.26**
Dissolved Oxygen (mg/l)	**8** ± 0.07	7 ± 0.14	**6.7** ± 0.16	7 ± 0.1	6.8 ± 0.11	7.8 ± 0.06	**0.002**	**0.4**
Free Carbon dioxide (mg/l)	**1** ± 0.2	**1.3** ± 0.2	1.1 ± 0.1	0.8 ± 0.2	1.1 ± 0.2	0.9 ± 0.2	0.14	
Total Alkalinity (mg/l)	**28** ± 1.7	29 ± 2	32 ± 1.9	33 ± 2	36 ± 2.3	**43** ± 3.3	**0.33**	
Total hardness (mg/l)	**41** ± 2.5	**42** ± 2.5	43 ± 1.6	44 ± 2	**51** ± 1.3	49 ± 3.1	0.5	
Nitrate (ug/l)	**16.2** ± 2	17.6 ± 2.3	22.2 ± 3.7	19 ± 2.7	32.6 ± 3.3	**36.1** ± 3.8	**0.0004**	**-0.4**
Phosphate (ug/l)	5.3 ± 0.9	5.6 ± 1.0	5.6 ± 0.9	5.9 ± 1.1	7.2 ± 0.9	6.7 ± 0.7	0.5	
Silicate (mg/l)	**0.72** ± 0.1	0.76 ± 0.1	0.86 ± 0.1	0.78 ± 0	**0.86** ± 0.1	0.85 ± 0.1	0.61	

The mean water temperature (WT) at Station S1 was 10.7 ± 1.08 °C and increased sharply to 13.6 ± 0.9 °C at S2, followed by 12.5 ± 1.04 °C at S3, 13.6 ± 0.8 °C at S4, 15 ± 0.8 °C at S5, and 15.5 ± 1.08 °C at S6, indicating a lack of a characteristic longitudinal trend (Table 6.1). The 2.9 °C rise in WT from S1 to S2 over a relatively short distance is exceptional and unprecedented in Himalayan rivers, which typically show a 1 °C increase over approximately 25 km (Nautiyal, 2010; Nautiyal et al., 2004; Pathak et al., 2015). The decline in WT at S3 is attributed to the release of water from a 9 km long tunnel. The significant increase ($p < 0.05$) in WT in low-flow sections can be explained by decreased discharge and increased retention time, both of which enhance the rate of heating (Kumar & Nautiyal, 2019a, b; Poff & Zimmerman, 2010; Sharma & Nautiyal, 2023). This rise in temperature may negatively impact endemic cold water fish species in these areas. Water velocity (WV) showed a marked decline from S1 (89 ± 3.1 m/s) to the low-flow sections S2 (65 ± 8.6 m/s), S4 (66 ± 6.6 m/s), and S5 (66 ± 6.7 m/s), reflecting flow reduction d/s of barrages or dams. In contrast, elevated WV values at S3 (85 ± 6.5 m/s) and S6 (75 ± 6.8 m/s) indicate restored flow due to water release from powerhouses. This disrupted WV pattern is a primary hydrological alteration associated with hydroelectric development. Dissolved oxygen (DO) concentrations were significantly lower at S2 (7 ± 0.14 mg/L), S3 (6.7 ± 0.16 mg/L), S4 (7 ± 0.1 mg/L), and S5 (6.8 ± 0.11 mg/L), compared to S1 (8 ± 0.07 mg/L). Reduced DO levels at S2, S4, and S5 can be attributed to their inverse relationship with elevated WT and reduced WV. At S3, low DO may result from flow releases through long tunnels, which hinder atmospheric oxygenation. Nitrate levels showed an increasing trend from S1 to S2, followed by a sharp rise at S3, a decline at S4, and a gradual increase downstream. The pronounced elevation at S3 is likely due to non-point source pollution from urbanized banks, leading to enhanced bacterial decomposition. Total alkalinity, total hardness, phosphate, and silicate concentrations displayed a general downstream increase; however, these changes were not statistically significant. Significant differences in WT, WV, pH, and DO across the stations indicate that these parameters are the most impacted by hydrological fragmentation along the Bhagirathi River (Singh & Singh, 1994).

6.3.2 *Benthic Macroinvertebrate Community-Based Assessment*

The water quality assessment based on BMWP score showed variation at temporal and spatial scale across the river sections (Table 6.2. At the reference station (S1), the water quality corresponds to Water quality class III (BMWP score: 41–70) during the months of November to April, class IV (Score: 11–40) during October, May, June and deteriorated to class V during August and September. At S2, water quality remained withing class IV from February to June and improved to class II during November to Jan and September. At S3 and S4, water quality was predominantly within Class IV, except Class V in May at S3. At S5, Class III water quality was observed from November to February and again from April to May, while Class II was recorded in March, and Class IV prevailed during October and from June to September. Station S6 exhibited consistently poor water quality (Class V) throughout the year, with the exception class IV during April.

BMWP indicates water quality class III suggesting moderate level of pollution November to April, class IV (Score: 11–40) during October, May, June and deteriorated to class V during August and September. In contrast, quality class III (BMWP score: 41–70) during the months of at S2, water quality remained withing class IV from February to June and improved to class II during November to Jan and September. At S3 and S4, water quality was predominantly within Class IV, except Class V in May at S3. At S5, Class III water quality was observed from November to February and again from April to May, while Class II was recorded in March, and Class IV prevailed during October and from June to September. Station S6 exhibited consistently poor water quality (Class V) throughout the year, with the exception class IV during April. The annual BMWP score indicates class II water quality indicating slightly impacted

Table 6.2 Variations in the BMWP scores reflecting different water quality classes at each station in the Bhagirathi river

	Oct	Nov	Dec	Jan	Feb	Mar	Apr	May	Jun	Jul	Aug	Sep	BMWP Score	WQ Class	Color code
S1	25	45	43	70	62	62	43	28	16		4	4	>100	I	
S2	NS	58	51	48	35	21	38	38	14			53	71-100	II	
S3	NS	28	31	21	23	23	16	6	14				41-70	III	
S4	31	31	29	31	31	35	35	23	14		14	14	11-40	IV	
S5	19	50	54	60	54	73	56	53	19		15	19	0-10	V	
S6	4	4			4	4	12				4				

water quality with low organic pollution; class IV. BMWP score show significant ($p < 0.05$) differences among the sampling stations (Fig. 6.2). The box whisker plots indicate much variations in the BMWP scores at S1 and S5. The BMWP score show narrow range at S6 followed by S3, S4 and S2. The Mann-Whittney pair wise test indicates significant differences in BMWP between S1–S6, S2–S3, S2–S4, S2–S6, S3–S5, S3–S6, S4–S5, S4–S6 and S5–S6.

Differences among and between stations clearly reflect modifications in water quality across the river continuum. The BMWP index has been demonstrated as one of the most effective tools for assessing organic pollution and associated degradation in freshwater systems, particularly in the UK and European rivers (Armitage et al., 1983). Given the morphological and topographical similarities, its applicability has been extended to Himalayan rivers as well (Nautiyal et al., 2019). At the reference station (S1), BMWP scores generally classified water quality as Class III during most months, indicating moderate levels of organic loading. In contrast, persistent Class IV at S2, and predominantly at S3 and S4, suggest increased organic pollution in these sections. The elevated pollution levels and impaired water quality at S2 and S4 can largely be attributed to reduced flows downstream of hydroelectric project (HEP) diversions. Reduced discharge increases water retention time, leading to enhanced sedimentation, and is likely exacerbated by increased water temperature and decreased dissolved oxygen levels (Poff & Zimmerman,

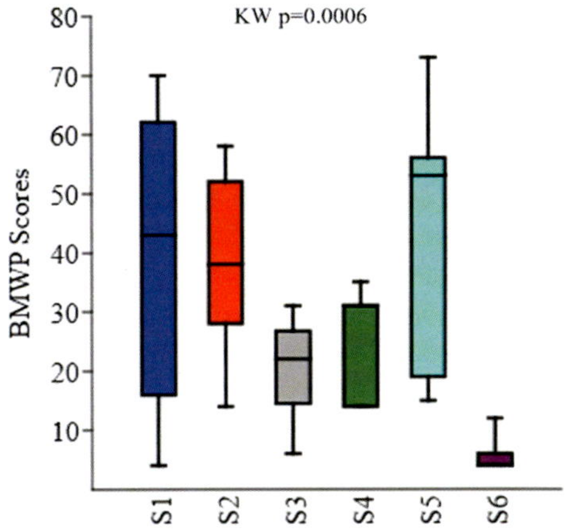

Fig. 6.2 Box Whisker plot showing variations in the BMWP scores at different stations in the Bhagirathi River

2010; Sharma & Nautiyal, 2023). At S3, although flow is regulated, the degraded water quality is likely influenced by non-point pollution sources, particularly from the urbanized region of Uttarkashi, and a cumulative flushing effect that limits macroinvertebrate colonization. Due to the sampling of multiple microhabitats, the riffle zones at S3 frequently flushed showed minimal colonization. Conversely, pool habitats, less impacted by flushing, supported taxa such as chironomid red bloodworms, which are classic indicators of organic pollution. The poor BMWP scores at S3, therefore, can be attributed to both habitat disturbance and sewage input. These observations are supported by Singh and Nautiyal (1990) and Kumar and Nautiyal (2019b), who reported the dominance of chironomids at S3. Water quality at S5 showed relative improvement with increasing distance d/s MB-II barrage. This can be linked to greater macroinvertebrate richness, a factor known to positively influence BMWP scores (Bonada et al., 2006). However, despite better scores, benthic assemblages at S5 were predominantly composed of pollution-tolerant taxa (Kumar & Nautiyal, 2019a, b), revealing a limitation of the BMWP method, which considers presence-absence but not abundance. This limitation highlights the need for further calibration and adaptation of biotic indices for Himalayan River systems. A notable finding of the study was the decline in BMWP scores during flood months at all stations. Seasonal fluctuations in benthic community structure, particularly during high flow conditions, are known to influence biotic index performance (Growns et al., 1995). During such months, taxonomic richness is typically low or negligible, leading to low index scores. Station S6, situated further d/s in a highly regulated stretch, recorded the lowest taxa diversity and consistently poor water quality (Class V), reflecting diminished macroinvertebrate colonization due to altered flow and possible cumulative pollution effects. Thus, the reliability of the BMWP index in Himalayan rivers, particularly under conditions like those at S3 and S6 is somewhat limited. The presence of tolerant taxa at these sites (Kumar & Nautiyal, 2019b) further substantiates their degraded ecological condition. Previous findings of increased eutrophication at S2, S3, S4, and S5 using diatom-based indices (Kumar & Nautiyal, 2024), except at S6, suggest that a multi-biotic approach is crucial for a comprehensive assessment of riverine ecosystem health. Persistent degradation at S3 and S6, followed by S4, S2, and S5, indicates the greatest impacts occur in highly regulated sections and areas closest to diversion structures. Conversely, improved water quality observed at S5 reflects partial riverine resilience, likely aided by contributions from downstream tributaries that help restore flow and dilution capacity.

Although originally developed for temperate rivers in the UK, BMWP is now widely adopted and often modified through additions such as abundance weighting, seasonal consideration, and regional calibration (Walley & Hawkes, 1997). Wright et al. (1984) have emphasized that BMWP should be used in conjunction with complementary biotic indices and physicochemical parameters for accurate ecological assessments.

In the current study, Spearman's correlation was performed using only those physicochemical variables that exhibited significant spatial variation. BMWP scores showed a moderate negative correlation with water temperature and nitrate levels, a weak negative correlation with pH, and moderate positive correlations with dissolved oxygen and water velocity. These relationships suggest that water quality degradation is closely linked to reduced flow conditions in the impacted sections.

6.3.3 *Functional Analysis*

The Functional Feeding Group (FFG) analysis revealed a dominance of Gatherer-Collectors (GC, 44%), followed by Scrapers (Sc, 36%), Filter Collectors (FC, 12%), Shredders (Sh, 5%), and Predators (Pr, 3%) (Fig. 6.3). At Station S2, a notable shift in functional composition was observed, with a 10% decline in GC and a 16% reduction in Sc (to 20%), accompanied by an increase in FC (17%) and slight increases in Sh (7%) and Pr (4%) compared to the reference site (S1). This shift was marked by a 13% rise in FC and a 9% decline in Sc, indicating a transition toward collector-based feeding strategies. The community composition at S5 closely resembled that of S2, reflecting similar trophic conditions.

The functional structure at S3 and S6 was markedly simplified, with GC (68%) and Sc (32%) at S3, and GC (63%) and Shredders (37%) at S6, demonstrating a loss of functional diversity in these highly regulated sections. Functional organization of benthic macroinvertebrates has increasingly been recognized as a key indicator for evaluating the ecological impacts of hydroelectric projects (Kumar & Nautiyal, 2019a, 2019b; Wang et al., 2016; Xiaocheng et al., 2008). The observed decline in Sc and GC, along with increases in FC and Sh at S2, S4, and S5—located approximately 7 km downstream of the reference station, further suggests shifts in functional composition likely driven by hydrological alterations.

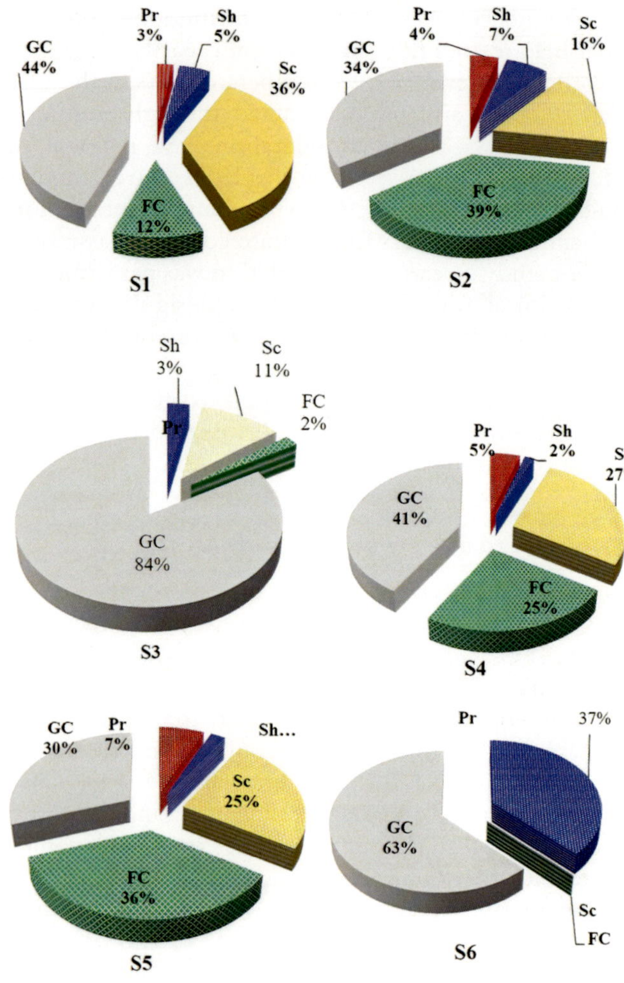

Fig. 6.3 Relative abundance of macroinvertebrate functional feeding groups at different stations in the Bhagirathi river. Acronyms: Sh-Shredders; Sc–Scrappers; GC–Gatherer-Collectors; Pr–Predators. FFGs Lack definite longitudinal trend; GC and Sc share declined (S1–S2); increased (S3) and alternately increased and decreased (S3–S6). FC share increased (S1–S2); declined (S3, S4) and exhibits further increase at S5

The elevated proportion of FC at S2 and S5 is attributed to the abundance of fine particulate organic matter (FPOM), while the reduction of FFG diversity in the flow-depleted (F-D) stretch of MB-II may be due to the interruption of downstream nutrient transport by the MB-II barrage. These findings are consistent with previous observations by Bredenhand and Samways (2009), who noted a greater abundance of filter feeders downstream from barrages compared to immediately downstream sites, in the Eerste River, South Africa. Similarly, Lin et al., (2024) reported an increased proportion of Gatherer-Collectors downstream of small hydropower projects on the Oujiang River, China.

The absence of a consistent longitudinal trend in FFG composition across the Bhagirathi River underscores the disruption of the natural river continuum caused by multiple HEP installations and indicates ecological degradation in ecosystem functioning.

6.3.4 *Lotic Invertebrate Flow Evaluation (LIFE) Index (Extence et al., 1999)*

LIFE scores were calculated as 8.5, 7.0, 9.5, 7.5, 7.6, and 7.6 at stations S1 through S6, respectively (Fig. 6.4). The highest score at S3 reflects conditions of rapid to fast-flowing water, while the reference station S1 exhibited moderate to fast flow. In contrast, the flow regimes at S2, S4, S5, and S6 were categorized as slow to moderate.

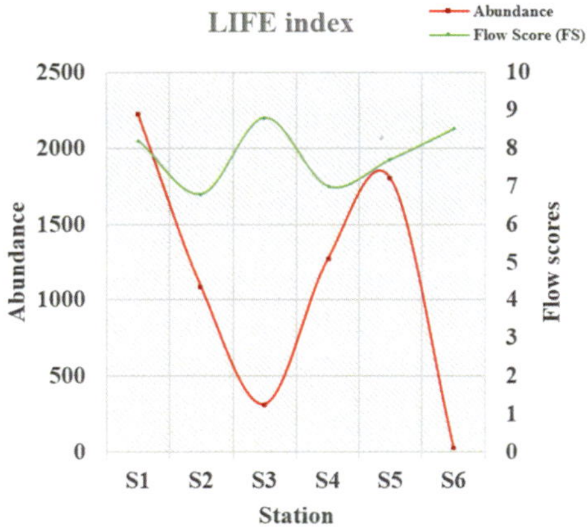

Fig. 6.4 Variations in the LIFE score and abundance of the taxa at each station

LIFE analysis indicated that stations S2, S4, and S5, despite occasional fluctuations, experienced modified flow conditions, which in turn influenced macroinvertebrate community structure and composition. These findings support the utility of the LIFE index as an effective tool for assessing hydropower-induced flow modifications in Himalayan River systems.

6.3.5 Comparison With the Water Quality Evaluation Through Diatom-Based Indices (Kumar & Nautiyal, 2024) at the Same Stations

Kumar and Nautiyal (2024) reported water quality in the studied sampling station S1-S6 through diatom-based indices and reported slightly impacted water quality at stations S2, S3, and S4, while good water quality at S1, S5, and S6 through Pollution Sensitivity Index (IPS) and the Index of Saprobity/Eutrophication (IDE/S). When compared with benthic macroinvertebrate-based indices, only stations S1, S2, and S3 showed a consistent pattern. Notably, pollution appeared minimal at S2, S3, and S4, though the indices classified these under different water quality classes.

An improvement in water quality at S5 was indicated by both the BMWP score and diatom-based indices. However, functional metrics, particularly the Lotic Invertebrate Flow Evaluation (LIFE) index respectively revealed a predominance of collectors FFG and low flow conditions. At station S5, constant low flows combined with tributary inflows d/s of the MB-II barrage support the resilience of both diatom and macroinvertebrate communities. However, the dominance of collector taxa indicates an increased dependence on fine particulate organic matter (FPOM), present in both suspended and transportable forms. Meanwhile, a reduced presence of grazer taxa indicates a decline in the grazing food chain. The increased availability of FPOM is further confirmed through low LIFE flow scores, reflecting higher accumulation of particulate matter.

At S6, diatom-based indices indicated good water quality, which may be valid given that macroinvertebrate colonization at this site was likely affected by flow regulation and periodic flushing events that prevent nutrient accumulation. Additionally, the diatom taxa recorded at S6 were sensitive to organic pollution and largely composed of polyoxybiontic species (Kumar & Nautiyal, 2023), suggesting colonization despite low overall abundance. If condition like S6 persist, benthic diatoms may serve as more reliable indicators of water quality or pollution while macroinvertebrate for habitat deterioration. At S3, organic pollution levels were found to be lower according to diatom-based indices than macroinvertebrate-based assessments. These discrepancies between diatom and macroinvertebrate indices are attributed to the differential sensitivities of these communities to environmental stressors.

6.4 Conclusion

The cascade of hydroelectric projects along the Bhagirathi River has significantly impacted riverine ecosystem health, as evidenced by deterioration in water quality classes, disruption of functional community organization, and notable shifts in macroinvertebrate assemblages in response to altered flow regimes. These impacts were most pronounced in flow-regulated sections (S3 and S6), where hydrological modifications were severe. Among the low-flow stations (S2, S4, and S5), ecological degradation was greater at sites closer to diversion structures (e.g., S4), whereas reduced impacts at more distant locations (S2 and S5) suggest partial ecological resilience and recovery of benthic communities with increasing distance from hydropower infrastructure.

Downstream impacts persisted well beyond the powerhouse outlets, as evidenced at S6, where regulated discharges limited ecological restoration. The disturbances observed in this study highlight the urgent need to maintain ecologically appropriate environmental flows to support long-term ecosystem sustainability. This study underscores the importance of continuous biomonitoring of hydropower-impacted stretches in the Himalayan region to inform adaptive river management and promote sustainable conservation of aquatic biodiversity and hence riverine ecosystems of Himalaya.

Acknowledgements Head, Department of Zoology, HNB Garhwal University is acknowledged for lab facilities. First author acknowledges CSIR for CSIR-NET SRF fellowship.

References

Armitage, P. D., Moss, D., Wright, J. F., & Furse, M. T. (1983). The performance of a new biological water quality score system based on macroinvertebrates over a wide range of unpolluted running-water sites. *Water Research, 17*(3), 333–347. https://doi.org/10.1016/0043-1354(83)90188-4

Bonada, N., Prat, N., Resh, V. H., & Statzner, B. (2006). Developments in aquatic insect biomonitoring: A comparative analysis of recent approaches. *Annual Review of Entomology, 51*, 495–523.

Bredenhand, E., & Samways, M. J. (2009). Impact of a dam on benthic macroinvertebrates in a small river in a biodiversity hotspot: Cape Floristic Region, South Africa. *Journal of Insect Conservation, 13*, 297–307.

Cox, B., Oeding, S., & Taffs, K. (2022). A comparison of macroinvertebrate-based indices for biological assessment of river health: A case example from the sub-tropical Richmond River Catchment in northeast New South Wales, Australia. *Ecological Indicators, 106*, Article 105479.

Cummins, K. W., Merritt, R. W., & Andrade, P. C. N. (2005). The use of invertebrate functional groups to characterize ecosystem attributes in selected streams and rivers in South Brazil. *Studies on Neotropical Fauna and Environment, 40*(1), 69–89. https://doi.org/10.1080/01650520400025720

Dewson, Z. S., James, A. B., & Death, R. G. (2007). A review of the consequences of decreased flow for instream habitat and macroinvertebrates. *Journal of the North American Benthological Society, 26*(3), 401–415.

Dudgeon, D. (1999). *Tropical Asian Streams: Zoobenthos, Ecology and Conservation*. Hong Kong University Press.

Dudgeon, D., Arthington, A. H., Gessner, M. O., Kawabata, Z. I., Knowler, D. J., Lévêque, C., Naiman, R. J., Prieur-Richard, A. H., Soto, D., Stiassny, M. L., & Sullivan, C. A. (2006). Freshwater biodiversity: Importance, threats, status and conservation challenges. *Biological Reviews*, *81*(2), 163–182.

Edington, J. M., & Hildrew, A. G. (1995). *A revised key to the caseless caddis larvae of the British Isles with notes on their ecology.* Freshwater Biological Association Scientific Publication No. 53, Freshwater Biological Association, Ambleside, UK.

Edmondson, W. T. (1959). *Fresh-water biology* (2nd ed.). John Wiley & Sons.

European Parliament, & Council of the European Union. (2000). Directive 2000/60/EC establishing a framework for Community action in the field of water policy. *Official Journal of the European Communities*, L 327, 1–73. Retrieved from https://eur-lex.europa.eu/legal-content/EN/TXT/?uri=CELEX:32000L0060

Extence, C. A., Balbi, D. M., & Chadd, R. P. (1990). Biological monitoring of river water quality using macroinvertebrate community analysis: A practical assessment. *Environmental Pollution, 60*(1–2), 77–90. https://doi.org/10.1016/0269-7491(90)90109-Q

Gleick, P. H. (2000). A look at twenty-first century water resources development. *Water international, 25*(1), 127–138.

Growns, J. E., Chessman, B. C., McEvoy, P. K., & Wright, I. A. (1995). Rapid assessment of rivers using macroinvertebrates: Case studies in the Nepean River and Blue Mountains. *NSW Australian Journal of Ecology, 20*, 130–141.

Haase, P., Cortés-Guzmán, D., He, F., Jupke, J. F., Mangadze, T., Pelicice, F. M., Palmer, M. A., Rolls, R. J., Schäfer, R. B., Welti, E. A., & Sinclair, J. S. (2025). Successes and failures of conservation actions to halt global river biodiversity loss. *Nature Reviews Biodiversity*, 1–15.

He, F., Zarfl, C., Tockner, K., Olden, J. D., Campos, Z., Muniz, F., Svenning, J. C., & Jähnig, S. C. (2024). Hydropower impacts on riverine biodiversity. *Nature Reviews Earth & Environment*, *5*(11), 755–772.

Karr, J. R., & Chu, E. W. (1999). *Restoring life in running waters: Better biological monitoring*. Island Press.

Kumar, S., & Nautiyal, P. (2019a). Community features of benthic macroinvertebrates and bioassessment of the stretch of Bhagirathi River impacted by Maneri Bali Stage I Hydroelectric Project. *International Journal of Ecology and Environmental Sciences, 45*(3), 311–315.

Kumar, S., & Nautiyal, P. (2019b). State of benthic macroinvertebrate community in Bhagirathi river ecosystem impacted by hydroelectric projects. *Indian Journal of Ecology, 46*(4), 760–767.

Kumar, S., & Nautiyal, P. (2023). Impact assessment: Diatom flora of free-flowing and fragmented stretches of serially impounded bhagirathi river (Indian Himalayan region). In *Insights into the world of diatoms: From essentials to applications* (pp. 21–35). Springer Nature Singapore.

Kumar, S., & Nautiyal, P. (2024). Assessment of ecological state of serially impounded Himalayan River using diatoms-based matrices: Identification of critical stretches for management. *Ecological Frontiers, 44*(2), 321–326.

Lillehammer, A., & Saltveit, S. J. (Eds.). (1984). *Regulated rivers*. Universitetsforlaget AS.

Lin, Z., Qi, X., Li, M., Duan, Y., Gao, H., Liu, G., Khan, S., Mu, H., Cai, Q., Messyasz, B., & Wu, N. (2024). Differential impacts of small hydropower plants on macroinvertebrate communities upstream and downstream under ecological flow. *Journal of Environmental Management*, *370*, 123070.

Nautiyal, P. (2010). The River Ganga: A case of conflicting demands. *Aquatic Ecosystem Health & Management, 13*(1), 104–112.

Nautiyal, P. Kumar, S. Sharma, N. Chauhan, K. & Kala, K. (2019). Cold-water capture fishery in the Ganga headwaters having operational hydropower projects: A failing co-existence. *Journal of the Inland Fisheries Society of India, 5*(1), 6–23.

Nautiyal, P., Mansi, Sharma, N., Kumar, P., & Singh, D. (2018). Biological monitoring of water quality before and after completion of the hydroelectric project on the river Alaknanda at Srinagar. In N. Siddiqui, S. Tauseef, & K. Bansal (Eds.), *Advances in health and environment safety: Springer transactions in civil and environmental engineering* (pp. 219–227). Springer Nature Singapore Pte Ltd. https://doi.org/10.1007/978-981-10-7122-5_22

Nautiyal, P., Mishra, A. S., Singh, H. R., & Dobriyal, A. K. (2004). Longitudinal variation in the community structure of epilithic macroinvertebrates in the cold water substratum of the Ganga River in the Himalayas, India. *Limnology, 5*(1), 15–22.

P. H. (2000). A look at twenty-first century water resources development. *Water International*, *25*(1), 127–138.

Palmer, M. A., Bernhardt, E. S., Allan, J. D., Lake, P. S., Alexander, G., Brooks, S., Carr, J., Clayton, S., Dahm, C. N., Follstad Shah, J., Galat, D. L., Loss, S. G., Goodwin, P., Hart, D. D., Hassett, B., Jenkinson, R., Kondolf, G. M., Lave, R., Meyer, J. L., ... Sudduth, E. (2005). Standards for ecologically successful river restoration. *Journal of Applied Ecology, 42*(2), 208–217. https://doi.org/10.1111/j.1365-2664.2005.01004.x

Pathak, S. K., Prasad, S., & Pathak, T. (2015). Determination of water quality index river Bhagirathi in Uttarkashi, Uttarakhand, India. *International Journal of Research Granthaalayah, 3*(9), 1–7.

Poff, N. L., & Zimmerman, J. K. (2010). Ecological responses to altered flow regimes: A literature review to inform the science and management of environmental flows. *Freshwater biology, 55*(1), 194–205.

Rosenberg, D. M., & Resh, V. H. (Eds.). (1993). *Freshwater biomonitoring and benthic macroinvertebrates*. Chapman & Hall.

Sharma, N., & Nautiyal, P. (2023).Hydropower development impacts on the benthic macroinvertebrates in the Indian Himalayan region. *International Journal of Environmental Studies, 80*(5), 1235–1247.

Sharma, P., Sharma, N., Kumar, S., Kavita, & Kala, K. (2019). Coldwater capture fishery in the Ganga headwaters having operational hydropower projects: A failing co-existence. *Journal of the Inland Fisheries Society of India, 6*.

Singh, H. R., & Nautiyal, P. (1990). Altitudinal changes and the impact of municipal sewage on the community structure of macrobenthic insects in the torrential reaches of the River Ganges in the Garhwal Himalaya (India). *Acta Hydrobiologica, 32*(3–4), 407–421.

Singh, H. R., Nautiyal, P., Dobriyal, A.K., Pokhriyal, R. C. Negi, M., Baduni, V., ... Gautam, A. (1994). Water quality of the river Ganga (Garhwal Himalayas). *Aceta Hydrobiologia 36*(1): 3–11.

Walker., K. F. (1985). A review of the ecological effects of river regulation in Australia. *Hydrobiologia* 125: 111–129.

Walley, W. J., & Hawkes, H. A. (1997). A computer-based development of the Biological Monitoring Working Party Score System incorporating abundance rating, site type, and indicator value. *Water Research, 31*(2), 201–210.

Wang, H., Chen, Y., Liu, Z., & Zhu, D. (2016). Effects of the "Run-of-River" hydro scheme on macroinvertebrate communities and habitat conditions in a Mountain River of Northeastern China. *Water, 8*(1), 31.

Ward, J. V., & Stanford, J. A. (1979b). *The ecology of regulated streams* (pp. 398). Plenum Press.

Ward, J. V., & Stanford, J. A. (1983). The serial discontinuity concept of lotic ecosystems. In T. D. Fontaine & S. M. Bartell (Eds.), *Dynamics of lotic ecosystems* (pp. 29–42). Ann. Arbor Sciences, Publishers Inc.

Ward, J. V., & Stanford, J. A. (1995). Ecological connectivity in alluvial river ecosystems and its disruption by flow regulation. *Regulated Rivers: Research & Management, 11*(1), 105–119.

Weisberg, S. B., Janicki, A. J., Gerritsen, J., & Wilson, H. T. (1990). Enhancement of benthic macroinvertebrates by minimum flow from a hydroelectric dam. *Regulated Rivers: Research & Management, 5*(3), 265–277.

Wright, J. F., Moss, D., Armitage, P. D., & Furse, M. T. (1984). A preliminary classification of running-water sites in Great Britain based on macro-invertebrate species and the prediction of community type using environmental data. *Freshwater Biology, 14*(3), 221–256.

Wright, J. F., Sutcliffe, D. W., & Furse, M. T. (Eds.). (2000). *Assessing the biological quality of freshwaters: RIVPACS and other techniques*. Freshwater Biological Association.

Xiaocheng, F., Tao, T., Wanxiang, J., Fengqing, L., Naicheng, W., Shuchan, Z., & Qinghua, C. (2008). Impacts of small hydropower plants on macroinvertebrate communities. *Acta Ecologica Sinica, 28*(1), 45–52.

CHAPTER 7

River Ecosystem Services: A Systematic Assessment of the Berai River, with Special Emphasis on River Aggradation, River Degradation, Impact of Riverine Badlands, and Their Transformation

Shreya Basak and Arindam Sarkar

7.1 Introduction

7.1.1 Review of Riverine Ecosystem Services

Riverine ecosystems, which are interconnected systems related to rivers and their surrounding floodplains, are among the region's most vibrant and essential ecological systems. These river-associated ecosystems provide various services vital for environmental sustainability, human well-being, and socio-economic development. Rivers serve as critical natural

S. Basak (✉)
Department of Geography, University of Calcutta, Kolkata, India
e-mail: shreyabasak0607@gmail.com

S. Ghosh (ed.), *Reimagining Indian Rivers for Sustainability*, Palgrave Studies in Environmental Sustainability,
https://doi.org/10.1007/978-3-032-04754-0_7

corridors, linking diverse landscapes, maintaining water balance, and supporting a wide range of biodiversity. Ecosystem services can generally be grouped into four main categories (Millennium Ecosystem Assessment, 2005). In the river system of the Chhotanagpur plateau fringe, ecosystem services are particularly significant. The river, which originates from the plateau fringe area of Chhotanagpur, provides provisional ecosystem services (Water supply: For drinking, agriculture, and industry; Fisheries and aquatic food resources: Capture and culture of fish and invertebrates; Material resources: Sand, gravel, and other raw materials), Regulating Services (Flood regulation: Natural absorption and gradual release of floodwaters; Water purification: Filtration by riparian plants and sedimentation processes; Climate regulation: Carbon sequestration and maintenance of microclimate; Erosion control: Stabilization by vegetation along riverbanks), Supporting Services (Nutrient cycling: Transport and transformation of nutrients; Soil formation and maintenance: Through the movement and deposition of sediment; Habitat provision: For aquatic and riparian biodiversity, including migratory species), and Cultural Services (Spiritual and religious significance; Recreation; Aesthetic values: Scenic landscapes and cultural identity).

7.1.2 Significance of the Berai River

A tributary situated in the lateritic region of West Bengal holds significant ecological, socio-economic, and cultural importance within its catchment area (Fig. 7.18). Although relatively small compared to major Himalayan rivers, the Berai is crucial in delivering ecosystem services that sustain rural livelihoods and biodiversity in a fragile environment. Despite its modest size, the Berai River is a vital ecological resource for the lateritic plains of West Bengal. Its varied ecosystem services maintain environmental and hydrological balance, as well as social and cultural integrity, in the area. Protecting the Berai requires a collaborative strategy that combines scientific knowledge, policy advocacy, and active community engagement.

A. Sarkar
Department of Geography, Purash Kanpur Haridas Nandi College, University of Calcutta, Kanpur Puras, India

7.1.3 River Aggradation

River aggradation refers to the elevation of a riverbed caused by sediment deposition, typically resulting from a decrease in flow velocity or an increase in sediment load (Fig. 7.1). This process frequently occurs in regions with severe erosion in the upper catchment and insufficient transport capacity downstream (Knighton, 1998). In the Chhotanagpur Plateau area, rivers such as the Damodar and Subarnarekha demonstrate significant aggradation due to intense soil erosion from lateritic uplands, deforestation, and unregulated mining. The Damodar River, historically labeled the "Sorrow of Bengal," gathers considerable amounts of silt and sand in its middle and lower reaches due to high sediment inflow from its plateau source, resulting in frequent aggradation and subsequent flooding in the plains below (Sinha & Friend, 1994). The accumulation of coarse sediments alters the riverbed shape, affecting water conveyance and ecological processes (Fig. 7.19).

Fig. 7.1 Schematic conceptual diagram of river aggradation, degradation, and transition zone

7.1.4 River Degradation

River degradation involves lowering riverbeds caused by erosion, incision, and sediment removal, which can destabilize riverbanks and change aquatic habitats (Fig. 7.1). This issue is often exacerbated by human activities such as dam construction, channelization, and sand mining (Surian & Rinaldi, 2003). In rivers originating from the Chhotanagpur Plateau, particularly the Subarnarekha and Barakar, degradation is evident in areas affected by excessive sand extraction and regulated upstream flow. For example, stretches of the Subarnarekha River near Ranchi have faced vertical erosion due to sediment deficiency and increased Stream Energy, resulting in channel incision and bedrock exposure (Sarkar, 2020). This degradation disrupts geomorphic balance and endangers groundwater recharge and riparian ecosystems (Fig. 7.9).

7.1.5 Formation and Transformation of Riverine Badlands

The Riverine Badlands represent eroded, desolate landscapes formed through vigorous fluvial incision and soil erosion within semi-arid lateritic regions. The Berai River Basin has undergone significant badland formation in recent decades, particularly near Shinhati and Jam Doba (Fig. 7.3). These Badlands are formed by seasonal rainfall, deforestation, and the vulnerability of lateritic soil structures (Singh & Kale, 2009). Previously regarded as degraded areas, contemporary research indicates that the Badlands, when managed sustainably, can support unique microhabitats and may serve as potential ecotourism attractions (Rudra, 2010). Their transformation is contingent upon implementing effective land-use policies, community adaptation strategies, and restoration initiatives, including afforestation and the construction of check dams.

7.1.6 Studying These Processes Offers Insights Into Environmental Sustainability and Landscape Transformation

Studying these processes provides critical insights into environmental sustainability. Examining the dynamics of landscape transformation reveals key insights into the mechanisms that support environmental sustainability and landscape changes. This investigation enhances our understanding of the challenges related to environmental sustainability and the pathways of landscape evolution.

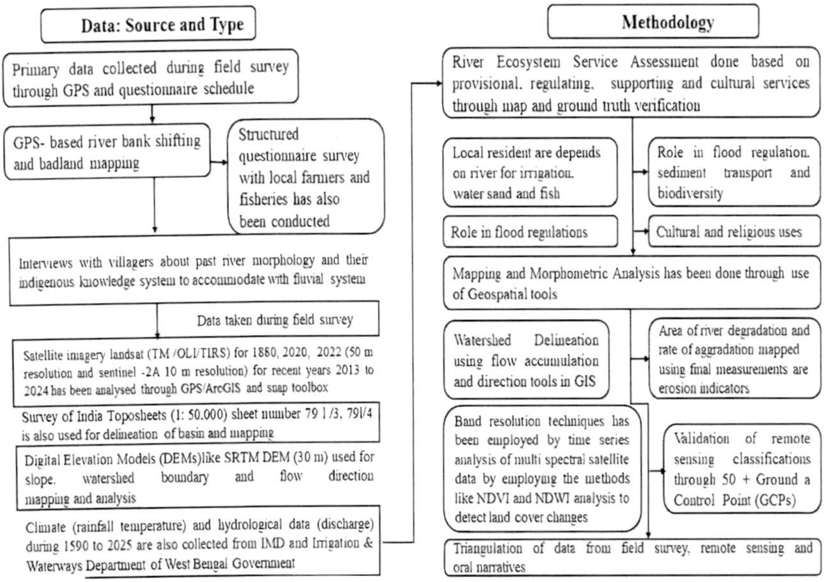

Fig. 7.2 Methodological Flowchart of the research

7.2 Literature Survey

Rivers offer numerous ecosystem services, including provisioning services (like water, fish, and sand), regulating services (such as flood control and sediment trapping), cultural services (for example, religious significance and aesthetics), and supporting services (which encompass nutrient cycling and habitats for biodiversity) (MEA, 2005). Smaller rivers, such as the Berai River, which originates from the Chhotanagpur Plateau fringe and traverses the lateritic terrains of West Bengal, play a significant yet often overlooked role in regional sustainability. Understanding the ecosystem services of the Berai River is essential for effective integrated river basin management. Aggradation within river systems accumulates sediment, resulting in shallower channels, expanded floodplains, and reduced flow capacity (Knighton, 1998). For the Berai River, this issue is exacerbated by the erosion of lateritic soil from nearby slopes and the seasonal deposition of coarse sediment by flash floods. Consequently, this aggradation alters the riverbed structure, lowers groundwater recharge, and disrupts aquatic ecosystems (Sinha & Friend, 1994). These changes frequently result in decreased water retention, increased vulnerability to flooding, and deterioration of provisioning services, including irrigation and drinking water supply (Fig. 7.15c).

In contrast, river degradation caused by bed incision and bank erosion can lower the water table and sever the connection between the river and its floodplain (Surian & Rinaldi, 2003). In the lower stretches of the Berai River, activities like unregulated sand extraction, deforestation, and gullying triggered by monsoons contribute to this degradation. This deterioration threatens riparian vegetation, harms agricultural land, and diminishes the diversity of in-stream habitats. The observed narrowing and deepening of the river channel negatively affect fish populations and aquatic plant communities, thus reducing the river's supporting and regulating ecosystem services (Kondolf, 1997). Integrating ecosystem service frameworks with geomorphic processes, such as aggradation, degradation, and badland transformation, provides a comprehensive perspective on river health and utility (Brierley & Fryirs, 2005). For the Berai River, systematic mapping using geospatial technologies, such as Landsat and Sentinel imagery, in conjunction with participatory rural assessments, can pinpoint areas of high value for ecosystem services and geomorphic risks. Additionally, incorporating local communities' indigenous knowledge, particularly about seasonal sediment flows and land-use practices, can significantly enhance management strategies. A dynamic river health card combining biophysical, ecological, and socio-economic indicators could effectively guide long-term conservation initiatives in the Berai catchment.

7.3 Formulation of Research Gap

While significant research exists on river ecosystem services and fluvial geomorphology in larger river basins, smaller rivers such as the Berai River, which emerges from the edges of the Chhotanagpur Plateau, are considerably under-studied. The existing literature lacks a coherent integration of ecosystem service evaluations with physical geomorphic processes, such as aggradation, degradation, and the creation of riverine Badlands. Additionally, there is scarce documentation on how morphodynamics, primarily driven by human activities and natural erosion, affects the provisioning, regulating, and cultural services of small lateritic rivers.

This study aims to address this gap through a combination of field observations, geomorphic analyses, and evaluations of ecosystem services in the context of a vulnerable yet ecologically important river system (Fig. 7.19).

7.4 Research Questions

Research questions formulated through the literature survey offer insightful deliberation. Key questions arise regarding the Berai River's ecosystem services and how these services are spatially and temporally distributed. Understanding the dominant processes and drivers of river aggradation and degradation in the Berai River basin is essential. The most significant inquiry is how riverine Badlands have developed along the Berai River and their impact on surrounding land use and river health. Researchers may have questions about how aggradation, degradation, and badland transformation influence the ecosystem services of the Berai River. Questions about strategies to sustain and restore river ecosystem services in the Berai River basin are raised.

7.5 Research Objectives

First, systematically identify and categorize the different ecosystem services linked to the Berai River, emphasizing four main types: provisioning services (including water supply, fishery resources, and agricultural irrigation), regulating services (such as flood control, water purification, and carbon sequestration), cultural services (encompassing recreational opportunities, aesthetic appeal, and cultural heritage), and supporting services (like nutrient cycling, habitat provision, and soil formation).

Second, thoroughly investigate the geomorphic processes influencing the Berai River, focusing on river aggradation and degradation by analyzing high-resolution satellite imagery, conducting detailed field surveys, and examining sediment samples. This will include assessing changes in river morphology over time, identifying sediment deposition patterns, and correlating these changes with hydrological dynamics and human activities.

Third, map and assess the spatial extent, formation mechanisms, and evolutionary processes of riverine Badlands within the Berai River catchment. This involves using geographic information systems (GIS) and remote sensing technologies to outline the Badlands and conducting field observations to understand the factors affecting their development, including rainfall patterns, soil composition, and land-use changes.

Fourth, devise comprehensive management strategies for conserving, rehabilitating, and sustainably utilizing the Berai River's ecosystem services. These strategies should engage stakeholders, promote community education, establish protected areas, and explore innovative methods to balance ecological health with the socio-economic needs of local populations.

7.6 Significance of the Study

This study holds significant importance for multiple reasons. First, it presents original research on a small river system in eastern India that has been largely overlooked, underscoring its ecological and socio-economic importance (Figs. 7.18 and 7.19). Second, by linking geomorphic changes to ecosystem services, the study provides an interdisciplinary viewpoint essential for effective river basin planning and management. Third, this research can serve as a foundational reference for regional policymakers, environmental planners, and local communities, aiding in the implementation of targeted conservation efforts. Lastly, it enhances the academic discussion surrounding riverine Badlands, a crucial yet poorly understood aspect of the lateritic belts in West Bengal.

7.7 Limitations of the Study

Several constraints have been noted regarding the study. Researching the Berai River basin is challenging for multiple reasons. Political and other restrictions make accessibility throughout the basin area unfeasible. Accessing remote badlands during monsoon seasons can hinder field data collection. Various human activities have altered the landscape in the basin. The area's owner is unwilling to allow a survey. Shinhati Badland has been affected by the administrator's actions, making a proper survey in that area impossible (Fig. 7.14a). The availability of temporal data for satellite imagery and sediment records may limit the analysis of long-term changes. Additionally, no database captures the socio-economic status related to ecosystem service valuation.

7.8 Location and Landscape of the Study Area

The Berai River Basin is located in the western Bankura district of West Bengal (Fig. 7.3). It is a small, non-perennial river basin with ecological and geomorphological significance. The Berai basin is part of the Rarh Bengal region, the easternmost area of the Chhotanapur plateau. The landscape features gently sloping ridges and dissected valleys. The soil is predominantly lateritic, sandy loam, and gravelly in the wasteland areas, with patches of red soil. The river experiences peak flow during the monsoon season, and water remains until the post-monsoon period. The Berai is a tributary of larger streams, including the Dwarkeshwar. At Pratapur (Latitude: 22°40′18.88″N, Longitude: 87°46′41.39″E), the Dwarkeshwar and Shilabati converge and continue southwest as the Rupnarayan River. The riverbed of the Berai is primarily sandy, with evidence of seasonal aggradation and degradation. Channel shifting and dry river stretches are common along the main course of the Berai River. The basin experiences a Tropical Monsoon Climate, with annual rainfall ranging from 1100 to 1400mm. This area receives the most rainfall between June and September, with temperatures ranging from 10 °C to 45 °C. Evapotranspiration is high due to dry summer conditions.

Agricultural activities in the basin include the cultivation of seasonal paddy, mustard, pulses, and various vegetables (Figs. 7.11b and 7.11c). The area primarily focuses on dual- and monocropping systems. Significant patches of mixed dry deciduous species have been found in the upper and middle courses of the Berai River. Riverine badlands, like Jamdoba near Singhargoriya (Latitude: 23°5′57.11″N, Longitude: 87°12′57.03″E) and Shinhati (Latitude: 23°5′25.48″N, Longitude: 87°16′21.54″E), have been identified along the left bank of the Berai River (Fig. 7.3). Some parts of the basin are experiencing erosion due to anthropogenic activities and deforestation, which contribute to land degradation (Fig. 7.14b). The Berai River Basin faces numerous challenges, including land degradation caused by runoff, erosion, deforestation, the removal of bed materials from the river, unscientific agricultural practices, and unsustainable land use. The area is characterized by scattered rural settlements, where livelihoods depend on agriculture, forest products, fishing during the monsoon, and small-scale livestock farming. Water management is primarily traditional, often utilizing check dams, ponds, and dug wells.

7.9 Data and Methods

7.9.1 Data: Source and Type

Primary data were collected in the field using GPS and a questionnaire schedule. GPS-based riverbank shifting and badland mapping were conducted. Structured questionnaire surveys were conducted with local farmers and fishers during the fieldwork. Interviews with village elders gathered information about past river morphology and their indigenous knowledge systems related to the fluvial environment. Ground photographs were also taken during the field survey. Satellite imagery from Landsat (TM, OLI/TIRS) for the years 1990, 2000, 2010, 2020, and 2024 (30 m resolution) and Sentinel-2A (10 m resolution) for recent years (2019–2024) were analyzed using QGIS/ArcGIS and the SNAP Toolbox—survey of India toposheets (1:50,000)—Sheet Nos. 73I/3 and 73I/4 were also utilized to delineate the basin and map it. Digital Elevation Models (DEMs), such as the SRTM DEM (30 m) for slope (Fig. 7.5), watershed boundary (Fig. 7.4), flow direction, and the ASTER GDEM for cross-verification, were used for mapping and analysis. Climate data (rainfall and temperature) and hydrological data (discharge) from 1990 to 2025 were collected from the India Meteorological Department (IMD) and the Irrigation & Waterways Department of the Government of West Bengal. Datasets from the National Remote Sensing Centre (NRSC) and supervised classification from satellite imagery were also employed.

7.9.2 Methodology and Validation

A River Ecosystem Service Assessment was conducted based on provisioning, regulating, supporting, and cultural services through map verification and ground-truthing (Fig. 7.9). The results were evaluated based on the local dependence on the river for irrigation, water, sand, and fish, as well as its role in flood regulation, sediment transport, biodiversity, and cultural and religious uses. Mapping and morphometric analysis were conducted using geospatial tools. Watershed delineation was performed using flow accumulation and direction tools in QGIS (Figs. 7.2 and 7.4). Band ratio techniques were employed through time-series analysis of multispectral satellite data, utilizing methods such as NDVI (Figs. 7.6a and 7.6b) and NDWI (Figs. 7.7a and 7.7b) analysis to detect changes in land cover. Cross-sectional changes in the channel profile were compared over time using GPS data and satellite-derived channel width. Sediment deposition zones were identified using false color composites (FCC) and slope analysis. Areas of river degradation and the rate of aggradation were mapped using field measurements and erosion indicators. Badland mapping and transformation were manually digitized from high-resolution images (Google Earth Pro) for gully and badland features (Figs. 7.14a and 7.15a). Transformation was tracked using multi-temporal imagery and field-based erosion pins. Areas of vegetation recovery or anthropogenic reclamation were identified with NDVI thresholds. Community perception and participatory mapping were important targets during field surveys. Focus group discussions (FGDs) and participatory sketch mapping with local stakeholders were conducted to gauge perceptions of changes in ecosystem services resulting from changes in the Badlands or the river, including both positive and negative impacts. Data triangulation was performed using information from field surveys, remote sensing, and oral narratives. The remote sensing classifications were validated using over 50 ground control points.

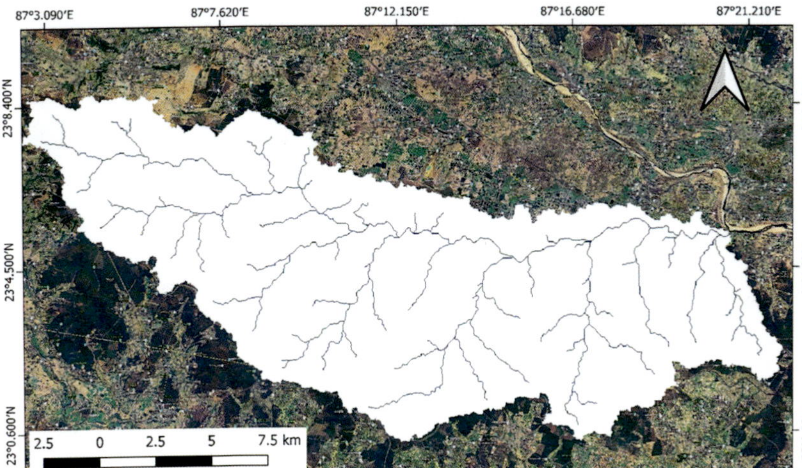

Fig. 7.3 Location of the Berai River Basin. This river basin is probably located at 23°13′28″N and 23°30′25″N Latitude and 86°53′13″E and 87°07′30″E Longitude. The Berai basin spans across the Bankura district of West Bengal

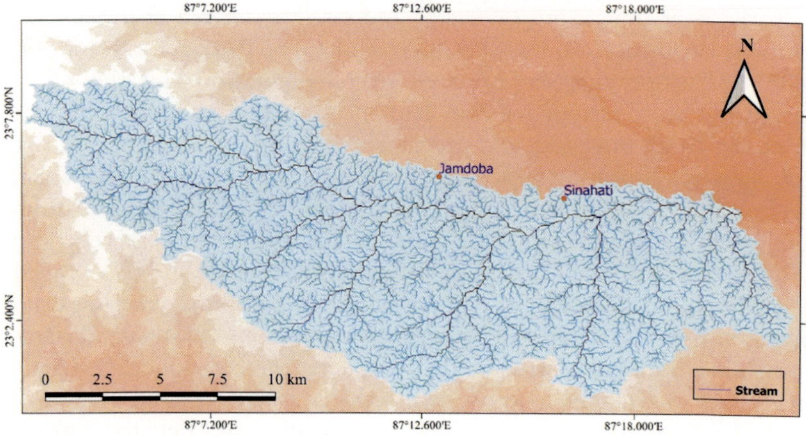

Fig. 7.4 The Berai River Basin stretches between 23°13′28″N and 23°30′25″N Latitude and 86°53′13″E and 87°07′30″E Longitude. It serves as a significant sub-basin in the Dwarakeswar River System

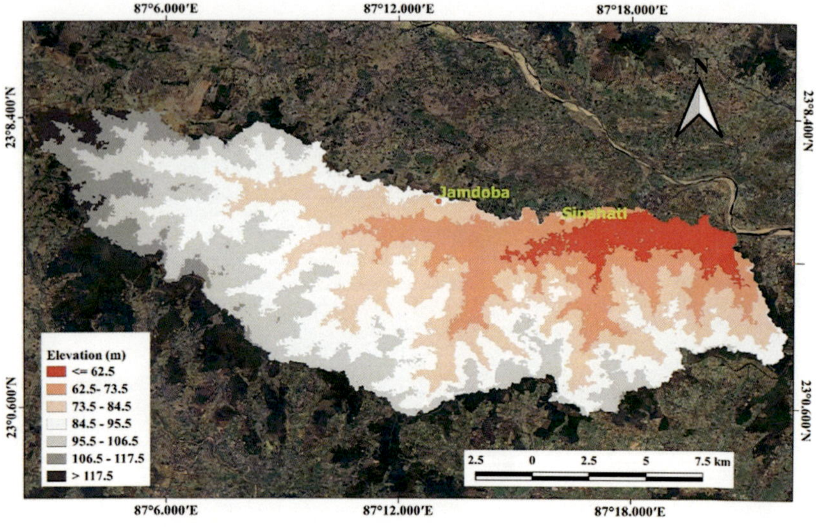

Fig. 7.5 Physiography map of the Berai River Basin. The basin elevation varies from 120 to 60 meters above sea level. Jamdoba Badland is located in the middle course, and Shinhati Badland is located in the lower catchment region. Both badlands are highly transformed by the river flow

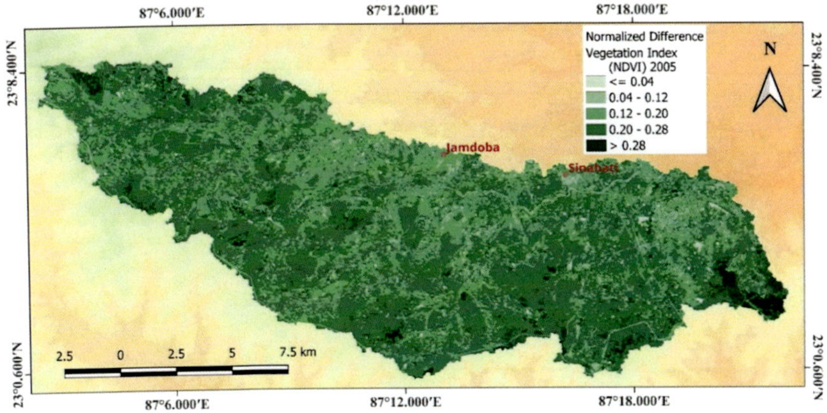

Fig. 7.6a Normalized Difference Vegetation Index (NDVI) of the Berai River Basin for 2005. This map illustrates the NDVI for the Berai River Basin in 2005, showing the density and vitality of vegetation in the region. An NDVI range on the map of <0.04 denotes barren land, water bodies, or settlements; 0.04–0.12 signifies sparse vegetation; 0.12–0.20 represents light vegetation; 0.20–0.28 indicates moderate vegetation, and >0.28 reflects dense and healthy vegetation in the basin

7.10 Observations and Explanations

7.10.1 Fluvial Geomorphological Configuration of Berai River

The Berai River originates from the plateau region of Chhotanagpur. This plateau region is the source of many rivers, characterized by a dissected peneplain, and features Precambrian rocks such as gneiss and granite, a diverse landscape, numerous plentiful resources, and specific climate characteristics. The upper stream of this river has a massive amount of vertical erosion, resulting in a V-shaped channel. During the monsoon season, the river flows rapidly due to heavy rainfall, resulting in flash floods. This river transports clastic sediment material such as gravel, boulders, and cobble with its flow in the upper stream area. For this, the erosion process is very active, creating steep slopes. The erosion process primarily occurs along the bed and banks of the river, creating a straight and deepest channel. In the downstream floodplains area, the course river shifts into a broader, braided region, producing many oxbow lakes and point bars due to slope reduction (Fig. 7.8). In this downstream, the river carries silts, clay, and

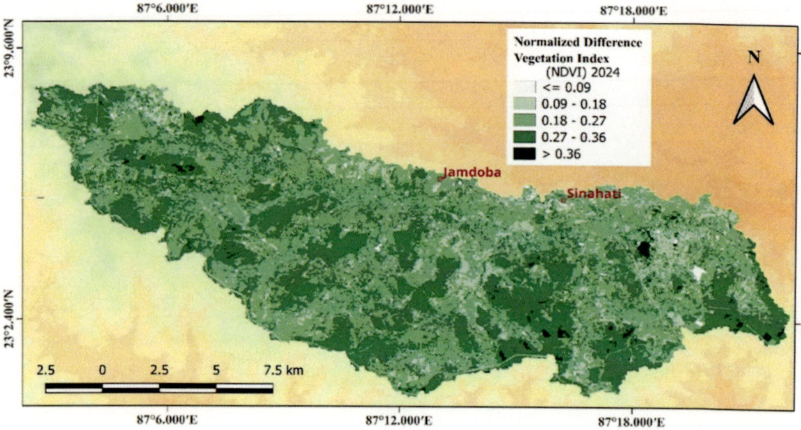

Fig. 7.6b Normalized Difference Vegetation Index (NDVI) of the Berai River Basin for 2024. Dark green regions (>0.36) indicate likely dense forests/healthy vegetation. In contrast, non-vegetated surfaces (−0.09); stressed vegetation/dry grassland/fallow fields (−0.09 to 0.18); moderate vegetation/shrubland/agricultural fields (0.18 to 0.27); healthy vegetation/croplands/mixed forests between 0.27 and 0.36 ranges. The Basin displays a notably lower NDVI, especially in the left bank of the river, where vegetation is sparse due to lateritic soil or quarrying activities

fine sands with its flow, and as a result, the velocity of the river becomes slower. This flow process varies depending on the intensity of rainfall. During the monsoon season, the bankside faces overflooding, but in the dry season, the low flow velocity affects the river's discharge rate.

Berai is a non-perennial river located on the western plateau fringe of West Bengal. The river basin has dynamic characteristics. The geological composition of this river basin is mainly dominated by laterite, granite-gneiss, and weathering schist. The erosion process is very high in the upper and lower catchment areas, which helps to create the steep gradients (Fig. 7.10). The basin on this river is classified into three categories-

a. The upper basin is known as the source zone (Figs. 7.20a, 7.20b, and 7.20c). This portion of the basin was formed due to intensive runoff, bedrock erosion, and the dissection of a plateau region with very steep slopes, resulting in a narrow valley and a V-shaped channel. Due to the hard lateritic crust, infiltration is inferior in this basin area.

b. The middle portion of the basin is the tributary junction of small alluvial fans. In this transition zone, seasonal bars are formed with sand and gravel.
c. The lower portion of the river basin, also known as the floodplain zone, where a gentle slope is present, and the channel is morphologically transformed from meandering to sinuous (Fig 7.16). The morphology of the Berai River Basin is evident in episodic hydrological events where dynamic fluvial processes are active, such as high erosional processes in the upper basin. In contrast, depositional processes dominate the lower basin.

The Berai River is an example of a rain-fed river system. This river's hydrological behaviors depend on seasonal rainfall in monsoonal months, soil properties, characteristics of terrain, and the pattern of land use. The water level rapidly increases from June to September due to heavy rainfall, so flood events occur quickly. Conversely, during dry months, the riverbeds filled with sand become almost dry and continue with a minimal flow. The river's catchment is characterized by undulating terrain and lateritic soils. The lateritic and red loamy soils have low water retention capabilities (Fig. 7.15c). These categories of soil contribute to rapid surface runoff during the monsoon season. In the catchment area, sparse vegetation covers that control the infiltration rate. The presence of lateritic soil and sparse vegetation causes the intensive rate of soil erosion (Fig. 7.10). The region receives an intensive annual rainfall range of 1100–1400 mm due to the southwestern monsoon. The vast amount of rainwater leads to excess runoff, influencing the river's flow regime. In the dry season, the channel faces a dry bed or isolated pools due to the poor aquifer connectivity in the lateritic soil profile. Their runoff rates are fast in the plateau upland where lateritic crust is available, but the infiltration rate is low. Their infiltration rate and runoff are moderate in the mid-lope zone, where red loamy soil is available. However, their infiltration rate and sediment-laden runoff are very high in an alluvial fringe zone with a gentle slope.

The drainage morphology of Berai River refers to the physical configuration of Berai Course, networks such as drainage density, drainage pattern, drainage slope, and relief (Fig. 7.4). The Berai River drainage pattern is dendritic in the upper catchment area due to the uniform challenging rock terrain in the plateau region (Fig. 7.4). However, in the alluvial zone, the pattern looks like semi-parallel types. The pattern indicated the mature stage of the river's drainage evolution. The drainage density of Berai defines the channel spacing within this river basin (Fig. 7.4). The Berai River in West Bengal is the transition zone between the Chhotanagpur plateau and the Rarh Plains. The basin of Berai is dominated by semi-arid red loamy soil and lateritic soil, with challenging rock terrain present in massive quantities. During monsoonal months, runoff rates increase due to intense rainfall, while infiltration rates are limited. The first and second order streams are developed in this river (Fig. 7.4). These morphometries indicate the moderate to high drainage density in the river basin (Fig. 7.4). The density range varies between 2.5 and 4.0 km/km^2, which means a well-integrated and dissected drainage network of the Berai River (Fig. 7.4). Due to this high drainage density, the surrounding area of this river is rapidly eroded in a short time due to rainfall, and erosion rates are becoming high due to concentrated flow.

The episodic flow regime pattern is the dynamic character of the Berai River, where the interaction of hydrological forces completes different morphological processes. These two processes help to contribute to the continuous shape of the river's channel and floodplain development. In the case of the Berai River, the morpho dynamics mainly focuses on different morphological processes, including erosion, sediment deposition, channel migration, etc. The channel shape of the Berai River is different in other zones. Due to the intensive vertical erosional process, the river channel in the plateau origin area looks like a V-shaped valley. This erosion process is continued for a steep gradient and low-discharge conditions in the dry season. In this upper stream area, the river carries many bedload sediment materials with its flow, including cobbles and gravels during high discharges in monsoonal months (mainly June to September). During these months, the river faces flash floods due to the excessive discharge rate exceeding the bank's capacity. These floors and the banks of the upper valley frequently experience high erosion.

i. In the transition zone of the river, the flow shifts towards the braided channels (Fig. 7.1). This shifting process is dependent on the flow condition and total sediment load. Sediments are accumulated on the point bars during the low-flow periods. The flow velocity decreases in this area due to the gradient flattening. In that case, the outer meander bends are transformed into highly eroded prone areas, and the inner bends face a sediment depositional condition (Fig. 7.16).

ii. But in the lower area of the river, the channel transforms into a sinuous pattern where floodplains are becoming wider like natural levees due to outer bank deposition. In this meandering channel, the water flows faster on the outer bank while the water moves slowly on the inner bank (Fig. 7.16). The fine sediments are continuously deposited during flood events, forming terraces or alluvial fans.

iii. During monsoon months, the discharge rate far exceeds that of the dry season, which leads to overbank flooding. In the dry season, sometimes the river faces sporadic discharge, leaving disconnected pools within the riverbed. Flow is interrupted in the transition and lower portion of the channel due to resistance, such as grasses and shrubs vegetation, sediment depositional landforms like gravel bars and points, etc. The river flow velocities are reduced during the monsoon season, when the floodplain deposition process continues with silt and clay, whereas channel silting and sedimentation occur during the dry season. Some bed landforms, including dunes and ripples, originate in the lower catchment area. These bed landforms are formed in the riverbed when water flows over sediments during high flow conditions.

7.10.2 Ecosystem Services: Classification and Quantification of Benefits From the River

Fluvial ecosystem services are defined as the benefits that river ecological systems provide to human society. The Berai River ecosystem services are directly linked with seasonal hydrological cycles and geomorphological landscape dynamics (Fig 7.9).

7.10.3 Provision Ecosystem Services

The Berai River serves as a vital water source for agriculture in the surrounding basin. It facilitates seasonal farming, primarily of rice, mustard, and vegetables, through the use of shallow tube wells and seasonal channels. The river also plays a crucial role in supporting fisheries and the livelihoods of local communities. Traditional fishing techniques and seasonal fish captures are essential for providing subsistence and dietary protein, particularly for Scheduled Tribe (ST) and Scheduled Caste (SC) communities. Additionally, the sediments from the Berai River are a key source of raw materials, with clay and fine silt being utilized for brick production and pottery, especially in the Kumartuli-style idol-making practices of nearby regions.

Berai is a non-perennial river, meaning it only flows when the local precipitation rate is exceptionally high. This limited water accessibility creates a vital situation for the surrounding communities. Rainwater falls into the river channel through surface runoff, which supplies the necessary water for the local traditional irrigation process. The kharif crops are predominantly supported by adequate water, such as paddy cultivation. Some livestock of surrounding rural households depend on the river's seasonal flow, such as cattle and goats. This river conveys sediment from the Chhota Nagpur plateau and local uplands, depositing it in the lower stream to form beautiful alluvial floodplains. These highly nutritious soils boost soil fertility around agricultural regions and reduce the need for artificial fertilizers. This fluvio-natural mechanism is essential for achieving long-term agrarian productivity.

7.10.4 Regulating Ecosystem Services

Flood Mitigation and Groundwater recharge are essential regulatory services provided by the Berai River in the basin area. The seasonal increase in water levels during the monsoon season helps replenish aquifers, restore soil moisture, and mitigate flash floods, thereby protecting agricultural lands and rural infrastructure. Additionally, the Berai River significantly contributes to Soil Retention and Erosion Control within the basin. The presence of riparian vegetation and the river's natural flow pattern help stabilize riverbanks and curtail lateritic soil erosion, a significant issue in the Birbhum-Bankura transition zone. Moreover, the Berai River plays a crucial role in Water Purification in this area. Its natural sedimentation processes and vegetative buffer zones function as effective filters, absorbing excess nutrients and minimizing contamination levels from adjacent settlements.

In the middle and lower stretches of Berai, the transported sediments are slowly deposited and accumulated in the riverbed and on the floodplain. Due to this aggradation process, the land surface is built up, and the floodplains become stable, forming many point bars and natural levees that support vegetation growth. However, in the upper stretches, the river exhibits its erosive behavior, characterized by channel incision, gully formation, and bank erosion (Fig 7.10 and 7.14d). In this zone, agricultural productivity also reduces, and due to an increase in downstream sediment load, the aquatic habitats and channel morphology have become disrupted. The riparian vegetation lining the banks of the Berai River plays a vital role in absorbing atmospheric CO_2. Grasses, shrubs, and trees store organic carbon within the soil and their biomass. When floodwaters recede, they leave behind organic debris that enriches the soil with carbon, which becomes trapped in the floodplain sediments, contributing to local climate regulation.

7.10.5 Supporting Ecosystem Services

Sometimes, the badlands replenish the topsoil layers through erosion, which is deposited during floods. Nevertheless, these all-depositional materials, such as clay and silt, help maintain productive agricultural land, prevent soil exhaustion from continuous farming, and support terrace farming in steep-slope areas. Many organic materials, such as leaf litter and eroded topsoil from plateau regions, are carried downstream and redistributed during floods. This nutrient cycle process helps enrich the soil surface downstream, promotes microbial activity, and ensures nutrient availability for crop and vegetation growth.

7.10.6 Cultural Ecosystem Services

The Berai River plays a vital role in cultural and aesthetic services, as well as in religious practices and folk traditions. The river holds ritual significance during festivals, where local water deities are honored at temporary ghats. Additionally, the Berai River basin possesses considerable educational and research value, contributing to the development of moral values. It serves as a living laboratory for the study of small river hydrology, seasonal flow dynamics, and human-river interactions in lateritic areas. Local communities closely link the river to cultural identity, seasonal cycles, and place-based knowledge systems, such as traditional weather forecasting and indicators of soil fertility.

The Berai River primarily supports the maintenance and restoration of habitat and the regulation of biodiversity. It offers a habitat for amphibians, aquatic insects, crustaceans, and native fish, which enhances local trophic interactions and pollination for nearby farmland. Additionally, the Berai River is crucial for nutrient cycling and sediment transport. During its flow season, the river promotes the movement of organic and inorganic nutrients between upland and lowland zones, enriching the soils of the floodplain. The Berai River's seasonal differentiation helps tie the local communities' cultural identity. The local people can celebrate their post-monsoon harvesting festivals after harvesting paddy crops, which depend on the availability of water from the river. Some bathing rituals are performed during specific months, coinciding with the river's cycle.

7.10.7 *Anthropogenic Pressures and Service Degradation*

The Berai River has faced challenges from sand mining, riverbed disturbances, encroachment, agricultural expansion, and pollution from domestic waste over the past few decades. Unregulated extraction of sand and lateritic materials jeopardizes habitat integrity and alters flow patterns. Farming in riparian areas diminishes the buffer capacity, leading to increased erosion and sedimentation. Without a structured waste management system, both biodegradable and non-biodegradable waste flow into the river, disrupting aquatic ecosystems (Fig. 7.14c).

7.11 Changes in the Fluvial System

The Berai River, located in the Bankura district of West Bengal, is a small river system with significant ecological and cultural importance. It forms part of the broader semi-arid lateritic area of Rarh Bengal and has experienced considerable alterations due to natural and human influences.

7.11.1 *Changes in River Morphodynamics*

The morphodynamics of the Berai River encompasses the shifts and changes in its shape, structure, and sediment transport over time. Mapping and ground surveys have revealed that the river's channel exhibits significant lateral shifts, particularly around confluence areas and agricultural plains. During monsoon seasons, small-scale braiding occurs mainly due to sediment overload and variable flow patterns. Erosion is prominent on the outer bends, whereas deposition occurs on the concave banks. These alterations intensify during peak flow periods, specifically from June to September. Upstream, particularly in the foothills, signs of degradation, such as channel incision caused by headward erosion and deforestation, are evident. The riverbed is becoming more aggraded in the downstream areas near Bishnupur due to substantial silt and sand accumulation, partly from upland erosion and human activities like sand mining. A comparison of historical topographic maps and recent satellite digital elevation models (such as SRTM and Carto DEM) reveals a decrease in the bank's full width and depth in several sections, attributed to sediment clogging and vegetation overgrowth (Fig. 7.5).

7.11.2 Hydrological Changes in the Basin

The hydrological changes in the Berai River illustrate modifications in its water volume, timing, and flow variability. Notable seasonal variability has been detected, with significant discharge during the monsoon (flood peaks) and near-dry conditions in the summer, which has expanded the annual amplitude. The base flow of the river has seen a striking reduction. The decrease in groundwater recharge in the basin stems from reduced infiltration, increased surface runoff, and the overuse of land and water resources, as well as excessive groundwater extraction for agricultural and household purposes, all of which contribute to a lower base flow during the dry season (December to May). Additionally, transforming the badland regions within the Berai basin has compromised the river's hydrological configuration and sediment budget. Over recent decades, the Berai River has rapidly lost its perennial status, with several stretches that were once perennial now becoming ephemeral or seasonal, due to climatic variability and water extraction.

7.11.3 Lulc and Its Transformation

Changes in land cover and land use (LULC) patterns within the Berai River basin have directly and indirectly impacted the fluvial system. The upstream and midstream areas have experienced a significant loss of forest cover, particularly in *sal* and mixed deciduous forests, due to rapid deforestation and land conversion for various human activities. This has increased both the potential and rate of soil erosion and the discharge of additional sediment into the main channel of the Berai

River. Converting lateritic uplands to fish breeding and paddy fields has further exacerbated soil erosion. In some areas along the river, irrigation water diversion has reduced surface runoff. The construction of check dams and bridges along the main channel has negatively affected the river's flow (Fig. 7.11a). Human encroachment for agriculture, settlements, and brick kilns has occupied several areas of the Berai River basin floodplain. Unregulated sand mining near roads and ghats, particularly during the dry season, has disrupted the sediment balance and stability of the riverbed. After the monsoon season, much of the floodplain remains uncultivated or is overtaken by invasive species, which alter the riparian vegetation and destabilize bank profiles. The fluvial system of the Berai River is undergoing a multi-dimensional transformation: morphodynamically, the river is becoming shallower and more irregular in shape, while hydrologically, it is exhibiting more seasonal and unpredictable flow patterns. Regarding land use and land cover (LULC), human activities have significantly disrupted the river's natural equilibrium through deforestation, sand mining, and land conversion(Adnan & Atkinson, 2020). Such changes necessitate immediate attention to river basin management, focusing on riparian buffer restoration, erosion control, and sustainable land-use practices.

7.11.4 *Human-Induced Transformation*

Human involvement transformed the fragile geomorphic system, especially in the badland region. Shinhati and Jamdoba are highly erodible regions on a lateritic and red loamy soil profile (Figs. 7.14d and 7.15a, 7.15b). Nevertheless, massive forest destruction for daily activities such as fuelwood collection, fodder production, and agricultural expansion has affected these regions. This event involved slope destabilization through the control of erosion in the gully channel and slope retreat. In the Shinhati badlands, the unscientific construction of many check dams is designed to control erosion by reducing the sediment load. Check dam construction enhances floodplain stability and modifies natural flow regimes. All gully channels are related to the Berai River, a sediment supplier in the badland area, primarily during the monsoons. Vegetation can compact the soil structure. However, overgrazing has affected the topsoil layer in this area by reducing the regeneration of vegetation. As a result, the bare surface is more susceptible to runoff and gully expansion, especially during the monsoon season. In the floodplain area, illegal sand extraction has disturbed the natural equilibrium. This mining activity led to a natural aggradation process near the downstream stretches, resulting in the shallowing of the Berai channel, the formation of point bars, natural levees, and the instability of the floodplain. During the dry season, isolated pools and disconnected flow forms occur due to sediment blockage, which affects ecosystem resilience and reduces aquifer connectivity. Surrounding the badlands area, agricultural productivity is declining due to the poor soil conditions. This viable, unsustainable condition affected the farming and livelihoods of the local communities.

7.12 Discussion

7.12.1 River Aggradation: Evidence, Factors, and Mechanisms

River gradation is a noticeable geomorphic process in the middle and lower stretches of the Berai Basin. In these two stretches, sediments are deposited due to monsoonal flow, catchment degradation, and human interference (Fig. 7.23).

7.12.2 Evidence of River Aggradation

a. The riverbed of Berai became broader and shallower due to the deposition of fine silt, clay, and sand in the middle and lower stretches. This channel infilling process is active, especially near the Shinhati and Jamdoba Badlands. The eroded materials fall from the badlands, reducing the channel depth and forming a lateral bar.
b. In the contiguous floodplain, alluvial deposits are highly concentrated in the soil, which indicates overbank sedimentation, especially during flash flood events in the monsoonal season. Sometimes, sediment layers increase the frequency of crop burial during the post-monsoon.
c. During the dry season, when the discharge rate is low, specific geomorphic fluvial landforms form, including mid-channel bars, point bars, and sand-deposited islands (Fig. 7.8). These landforms indicate that the sediment load has exceeded its transport capacity.

7.12.3 Factors of River Aggradation

a. The Berai River originates from the degraded uplands of the Chhotanagpur Plateau, which are characterized by lateritic soil, deforestation, and overgrazing, resulting in destabilized slopes. From the nearest badland, soil erosion produces tremendous sediment in the river, mainly in monsoonal runoff.
b. In the monsoon season, the high-volume discharge carries coarse sediment in the downstream zone, which settles rapidly in flatter form and reduces flow velocity.
c. Berai's channel gradient sharply decreases from its incised to depositional zone. As a result, when the river enters the alluvial floodplain zone, its energy drops due to sediment deposition.
d. In the Shinhati badland, construct many check dams to prevent the erosion rate and the extension of the gully channel. However, this informal embankment on check dam construction restricts the natural flow, like trapping sediment in the upstream zone. In the upper stretches, sand mining is a cause of riverbed degradation; however, it also contributes to the aggradation process in the downstream zone. During monsoon, agricultural runoff also increases the nutrient-rich sediment supply in the riverbed.

7.12.4 Mechanisms of River Aggradation

Overbank flood depositions work as a mechanism of Berai Aggradation. During monsoons, when the discharge rate is very high, the bankside of Berai ruptures, and river water spreads across the floodplains. During this overbank flood, coarse materials drop near the channel, and fine silt materials are more towards the outer parts. This event leads to the formation of natural levees and vertical accretion. River degradation refers to the lowering of the riverbed elevation due to excessive erosion. This geomorphic activity continues in Berai's upstream and transitional zone.

7.12.4.1 Evidence of River Degradation

a. In the upper stream of Berai, the riverbed became lower (up to 1.5 m), especially near the Jamdoba-degraded badlands. In post-monsoon, the dependent channels are visible, indicating the high-velocity flow of the Scouring riverbed of Berai.

b. In the Downstream stretches of Berai, the bankside is vertically cut down through erosion, creating steep and wide spear-shaped margins. In this area, soil collapses during the monsoon period, indicating channel widening and lateral instability.
c. Sometimes the reduction of riparian green vegetation indirectly indicates river degradation and destabilization. The roots of the vegetation help stabilize the banks. However, due to deforestation activity or loss of vegetation, the margins of Berai progressively retreat.
d. Many gully channels are directly connected to this main river. Berai contributes sediment to the badland areas and is also a deterioration agent.

7.12.4.2 *Factors of River Degradation*

a. The Berai River originates in rapidly erosion-prone lateritic uplands, characterized by very low infiltration rates. Due to steep slopes, the incision and headward erosion rate are very high. These lithological settings contribute to river degradation.
b. Intensive removal of coarse material from the riverbed is initiating headward degradation in Berai.
c. Deforestation activity in badlands and plateau fringes has increased overland flow through destabilizing the slopes.

7.12.4.3 *Mechanisms of River Degradation*

a. All gully channels extend their upslope into badland areas, which connect with the main Berai channel and depend on the riverbed.
b. In the downstream area, sediment supply is reduced due to the trapping of upstream sediment by constructing check dams. These events cause the river to erode its bed and margins.

7.12.5 *Impact of Riverine Badland Formation on Configuration and Function of the Basin*

7.12.5.1 *Impact on River Basin Configuration*

Shinhati and Jamdoba, both badlands, consist of laterite soil. These two badlands originated due to loose alluvial deposition and a highly erodible area. Especially during the monsoon months, bank cutting, sheet erosion, and the formation of gully channels lead to erosion due to the torrents of seasonal rainfall. These events contribute to lateral migration and irregularity, depending on the Berai channel. This disruption activity simulates the unpredictable channel behavior and coherence of the Berai River. Due to the breakdown of soil structure in the riverine badland area, numerous gully and rill networks, as well as intermittent channels, are developing across the surrounding landscape. As a result, the terrain transforms into a highly fragmented and irregular land mosaic, increasing drainage density. This expanded drainage pattern leads to an improvement in the hydrological stress of the Berai basin. Shinhati and Jamdoba badlands are associated with bare soil and infrequent green vegetation (Fig. 7.4). During rainfall months, bursts of sediment flow into the river channel, which disrupts the gradual sediment movement pattern. Due to the formation of a local aggradation zone, the longitudinal channel stability was lost, and the erosional and depositional cycle was again stimulated.

7.12.5.2 Impact on River Basin Function

During the peak flow months, the riverine badland area serves as a primary sediment source for Berai, deriving fine and coarse sediments from the gully channel and steep gully walls. This aggradation process decreases the discharge capacity of the river channel and increases the risk of overbank flooding, particularly in the flatter stretch of the Berai Basin. On the barren surface of the badlands, the infiltration capacity decreases, resulting in intense surface runoff. These surface flows disrupt the formation of the floodplain. Due to the expansion of the badland area, the total amount of vegetation cover and soil fertility has decreased. These uncertain situations can impact water, fertile land, habitat diversity, the nutrient cycle, water flow moderation, and sedimentation in the river ecosystem. This morphodynamics condition also disrupts channel avulsion, habitat fragmentation, the river's ability, and the braiding pattern.

7.12.6 Transformation of Riverine Badland and Catchment Process

7.12.6.1 Nature of Morphology and Formation of Badland

Initially, badland forms where the lithological structure is combined with erodible laterite and red loamy soil. Surrounding the Berai basin, the Shinhati and Jamdoba badlands feature steep slopes and experience high-intensity episodic rainfall, leading to surface erosion and the formation of gullies and rills in these areas. Headward gully erosion, slope retreat, and bank collapse lead to the expansion of badlands, reshaping the surrounding catchment's morphological structure in Berai.

7.12.6.2 *Catchment Process Driving Transformation*

During the monsoon season, approximately 1100–1400 mm of annual rainfall occurs in the Berai River basin, resulting in intensive surface runoff over the laterite soil surface. These climatic events generate sheet flow over the surface and concentrated runoff in the channelized formation. Due to the absence of adequate vegetation, this sheet flow event directly contributes to deepening the gully channel and sediment transport into the leading Berai Channel. This sediment component is mixed with clay, fine sand, silt, and other materials, carried by the water flow in the downstream area. It contributes to river aggradation in the middle and lower stretches of the Berai River. These events alter the Berai channel's morphological structure, reduce the river's flow capacity, and transform the floodplain dynamics into point bars and natural levees. In a badland area, the ground infiltration connectivity is inferior due to limited water retention and baseflow generation; as a result, the river experiences disconnected flow during the dry season (Fig. 7.15c).

7.12.7 *Hotspot River Aggradation and Degradation*

Based on current understanding and ongoing fieldwork in the Berai River basin, several locations have been identified as hotspots for river aggradation and degradation. The processes and indicators of aggradation in the Berai River have been recognized downstream of Purandarpur–Sona Bandh, characterized by broad, shallow, sandy stretches with significant sedimentation, along with channel splitting due to bar formation. The deposition of fine silt during the post-monsoon period, along with sand mining and low-flow conditions, exacerbates aggradation at the confluence near Bishnupur. Natural levee development occurs with fine alluvial deposits in the floodplain area, especially during peak surface flow. Degradation of the Berai River has been observed in the upstream and upland badland transition zone, attributed to rapid surface runoff, active gullying, and headward erosion. Active bank cutting and channel incision, along with agricultural runoff and deforestation, contribute to increased erosion in various areas of the Berai River basin. The deepened channel bed and steep banks indicate degradation.

7.12.8 Estimation of Loss in Ecosystem Services

In the non-perennial Berai River ecosystem, degradation and depletion occur through the expansion of riverine badlands, channel incision, erosion of the floodplain, and sediment overload, leading to geomorphic and hydrological disturbances.

7.12.9 Quantification of Deterioration in Provisioning Ecosystem Services

In laterite soil, the surface retention and baseflow connectivity became reduced due to channel deepening and disconnection of the aquifer. During dry months, the reduction in irrigation supply levels is typically around 20–40%, which can impact Kharif cultivation and increase the cost of alternative water sources, such as tubewells and tankers. During monsoonal months, sediments are deposited near the floodplain area. However, upstream erosion shifted the sediment from fine, nutrient-rich silt to coarse sands and gravels. This degradation factor decreases the organic carbon from the topsoil profile and reduces crop productivity. Due to this loss, farmers rely heavily on chemical fertilizers to maintain crop yields, and soil fertility is compromised. Erosion also affected vegetation cover near the riparian zone. The considerable amount of standing biomass loss resulted in a 40% reduction in local fuelwood collection, leading to increased fuel costs.

7.12.10 Quantification of Deterioration in Regulating Ecosystem Services

Gully expansion has considerably slashed the sediment holding capacity in the upland area of the Berai riverbed and increased suspended load movement towards the downstream portion (Fig. 7.17). This degradational movement destabilized the floodplain, reducing its natural capacity to buffer floods, particularly during peak monsoon season. Consequently, the risk of flash flood hazards increased in the downstream villages.

7.12.11 Indicators of Badland Transformation

Degraded badland areas, such as Shinhati Badland and Jamdoba Badland, are characterized by steep slopes, dissected terrain, and low vegetation density. These uncertain conditions can be stabilized from erodible terrain to a persistent landform. Identifying the necessary indicators can help restore the landscape by improving surface stability, soil structure and fertility, plant succession, and hydrological buffering capacity, as well as the construction of check dams, among other measures.

- In the soil profile, organic matter is represented by soil organic carbon, which helps enhance water retention and the nutrient cycle and contributes to the microbial activity of the badland area (Fig. 7.15c). Fertility restoration can accumulate flood-deposited materials such as silt and organic matter in the floodplain area. These organic inputs gradually increase the topsoil's Nitrogen, Phosphorus, and Potassium (N, P, K) levels.
- Plantation can significantly resist gully expansion, stabilize soil, and enhance ecosystem resilience. Plant succession can transform low-biodiversity colonizers into highly diverse communities. Plantation can lead to soil stability by controlling the erosion rate in the badland area. Grass cover might be a perfect option for restricting soil loss.
- Check dams are the most effective structural invention in badland topography. In the Shinhati and Jamdoba badlands, small check dams and bunds constructed at the source area of the gully channels help concentrate rills.
- In the southern section of the gully channel of Shinhati badland near the Berai River, the initiation of fish hatchery ponds at the end of gullies can control soil loss (Figs. 7.14d and 7.14e). Even during the monsoon season, when the discharge rate and overland flow become high, these ponds also serve as sediment catchments.

7.13 Policy Framing, Implications, and Sustainability

The Berai River basin is characterized by monsoonal hydrology. The transformation of the riverine badland is due to the aggradation and degradational cycle of the Berai River. Some strategic policy framing and broader implications are necessary to prevent the transformation.

7.13.1 Integrated River Basin Management

Due to the episodic flow regime, integrated river basin management is crucial for this non-perennial river. This management involves many coordinates about water, land, and sediment behavior along the river's stretches. Integrated River Basin Management mainly focuses on erosion control through synchronizing upland stretches with sediment management in the downstream area; optimizing sediment retention through the construction of a check dam without resisting the flow riverine connectivity; managing gully erosion in the riverine badland area and stabilizing the floodplain area downstream of the river. These effective management practices can monitor the sediment load, retention of soil moisture, and maintain the drainage morphometry of the Berai River during monsoon and dry season through involving the local government, the district board of water, and community-based organizations.

7.13.2 Balancing Ecosystem Preservation With Human Needs

To maintain the ecological integrity and local livelihoods surrounding the river, where people are most probably dependent on the seasonal flow of the catchment (Fig. 7.18). The local farmers practice their irrigation process on a nutrient-rich floodplain area based on the seasonal flow of Berai. In this sediment depositional part, kharif crops are cultivated, which are the main economic backbone of the local communities. Promoting eco-agricultural activity by practicing traditional water harvesting and low-cost irrigation technology should enhance resilience policies, especially in dry months. The locals sometimes practice cultural integration, such as ritual baths in the Berai River. For this, it should be necessary to preserve the flow regimes of the river and maintain the water quality, which can affect the ecosystem cycle of the river.

7.13.3 River Restoration

The river must be managed in the upper reaches portion to restore the degraded part of Berai, mainly through physical activity and ecological rehabilitation. This stretch area, primarily a headward erosion zone, should be restored through bioengineering solutions, re-contouring, and step pool systems to stabilize the channel morphology. The surrounding riparian zone can be restored by planting grass and shrub vegetation to create a buffer to prevent intensive runoff. Check dam construction in the badland area can restrict the flow and trap sediment from the upstream area. This scientific implication should control the erosion and flow continuity.

7.13.4 Soil Conservation

In the Berai Basin, highly erodible lateritic and red loamy soil must be conserved to maintain the low water retention. Due to steep slopes, terracing, mulching, and crop covering processes can reduce the runoff intensity and increase the groundwater infiltration. In the badland area, interventions should include a micro-catchment bunds system, gully plugs, and vegetation cover to prevent the extension of the gully channel. During monsoon, the organic-rich sediment is transported downstream. This physical depositional process must encourage the build-up of organic carbon in soil particles, which enhances the soil characteristics and structure and also benefits crop productivity.

7.13.5 Afforestation

In a semi-arid lateritic region, afforestation must be implemented near the riverbanks and the degraded upstream area. Adaptation of native plants led to ecological rehabilitation and livelihood enhancement. Fast-growing plants such as grasses and shrubs should be promoted to manage soil erosion in the badland area. Development can be made possible by involving local communities in ecosystem restoration. Plantation along the riverine area can increase carbon storage by trapping organic carbon in the floodplain area.

7.13.6 Sustainable Sediment Management

Effective sediment dynamics are needed to manage the entire river profile of Berai. To prevent the riverbed from deepening, the extraction of sediment materials such as sand should be regulated, which also benefits Berai's baseflow condition. Scientifically, the re-distribution of sediment trapped in check dams is essential to avoid maximum mass flow and extension of aggradation in the upstream, which humans control. To mitigate the impact of overbank flood hazardous conditions, seasonal desolation should be implemented in the floodplain area. These sustainable practices can be made possible by providing subsidies, Payment for Ecosystem Services (PES), and applying different schemes to reduce environmental degradation.

7.14 Community Concerns: Promotion of Traditional Knowledge for Sustainable River use

In the context of the Berai River, the Traditional Knowledge System (TKS) plays a vital role in sustainable river use among the local communities. This system informs the communities about traditional water harvesting strategies, such as creating an earthen check dam and field channels, following the alignment of seasonal cropping patterns during monsoonal discharge, and practices of soil conservation processes like mulching. These traditional practices help to improve water use, preserve soil fertility, and maintain the ecological balance. Local communities along the river mainly depend on the seasonal flow related to their agriculture. However, high erosion, sedimentation, and badland transformation directly affect their livelihoods and ecosystem depletion (Fig 7.18). For this, they must be concerned with irrigation cost, native vegetation, agricultural productivity, and cultural practices related to their livelihood (Fig 7.12 and 7.18)). Residents must practice bamboo fencing, construction of small bunds, crop diversification, and the rotational cropping pattern to reduce soil loss, exhaustion, and overland flow. To address all these concerns, some community organizations, panchayats, and government boards like water boards must collaborate with them. These organizations all provide knowledge and training to the surrounding residents about these practices' causes, effects, results, and benefits. Blending traditional wisdom with science can also make the ecosystem of the Berai River resilient.

7.15 Major Findings

River aggradation patterns are observed and acknowledged along the main course of the Berai River. Notable aggradation zones were identified in midstream and downstream sections, where sediment deposits have created sandbars, point bars, and braided channels. Seasonal flooding and sediment influx from degraded uplands, driven by monsoons, have heightened the aggradation processes. The encroachment of agricultural lands into these depositional areas has modified the natural sediment flow, thereby decreasing the capacity of the main channel of the Berai River. The Berai River's main course recognizes river degradation dynamics. The upper catchment and transitional areas of the Berai River are facing serious degradation issues. We see problems like channel incision, bank erosion, and headward cutting (Fig. 7.10). Much of this is due to human activities like deforestation, overgrazing, and pulling too much groundwater out of the ground. These actions have destabilized the riverbanks and made the degradation even worse. Historical analysis of satellite images has revealed a notable decline in riverbed elevation across several sections over the past two decades. The considerable impact of riverine Badlands on the Berai River system has also been recognized. These Badlands, particularly those close to the lateritic uplands, have been crucial sediment sources, contributing to aggradation downstream and causing channel instability. The Badlands are experiencing transformations, with increased gullying, surface runoff, and vegetation loss leading to sediment overload.

Initial signs of stabilization have been observed in certain areas, especially where community initiatives such as afforestation and soil bunding have been implemented. Transformation of the badland area into different human-use landscapes plays a crucial role in the morphodynamics and hydrological behaviors of the Berai River. It has also been written about and documented how the riverine landscape has changed throughout the Berai basin (Fig. 7.8). The river's course, width, and morphology change dynamically, especially in the regions where aggradation and degradation occur. Land use and land cover (LULC) analysis indicates a gradual decline in riparian vegetation and a conversion from wetland to agricultural land along the riverbanks. Human interventions, including embankments, check dams, and sand mining, have altered natural fluvial processes, often worsening sedimentation and erosion (Figs. 7.13a, 7.13b, and 7.13c). The present study has assessed the ecosystem services provided by the Berai River. These services include provisioning services (water for irrigation, fisheries, and clay for pottery) (Figs. 7.12 and 7.14e), regulating services (flood control in aggraded areas), and cultural services (ritual and religious functions near ghats). Nevertheless, numerous services face growing risks from morphodynamic instability, pollution, and shifts in land use.

7.16 Conclusion

The Berai River system showcases the complex river dynamics in a semi-arid, lateritic environment, where natural geomorphic processes and human influences are closely connected (Fig. 7.9). The delicate balance between aggradation and degradation has led to significant changes in the landscape and the ecosystem, impacting the amount and quality of these ecosystems' services. Riverine badlands serve as a source of sediment and a factor of instability, but they also offer unique opportunities for restoration if approached with a mindful ecological perspective (Fig. 7.19). Every day, the river's evolving features underscore the need for a basin-wide management policy that encompasses scientific assessments, local assessments, and civic participation. Community efforts to curb further anthropogenic impacts, restore riparian habitats, and rehabilitate the Badlands would significantly enhance the sustainability of the Berai River ecosystem and preserve its resources for the region's residents (Fig. 7.22).

The current research has led to several recommendations for Conservation and Management, considering the study's findings and limitations. To address the ongoing challenges related to the Berai River, Community-Based River Management practices should be adopted by reinforcing local institutions and self-help groups to effectively monitor and maintain river health. Maintaining the Ecological Flow of the Berai River is essential for ensuring seasonal flow continuity, which can be achieved through check dams and vegetative channels (Figs. 7.20c and 7.21). Eco-restoration Programs should focus on riparian reforestation and gully plugging to reduce sedimentation and enhance water retention. Additionally, it is important to incorporate Traditional Knowledge systems by documenting and integrating local knowledge and practices for climate adaptation and water conservation in the Berai River basin. Although the Berai River is relatively small, it serves as a vital ecological artery for the lateritic plains of West Bengal (Fig. 7.19). Its various ecosystem services play a critical role in sustaining the environmental and hydrological balance, as well as the socio-cultural framework of the region. Ensuring the protection of the Berai requires a collaborative approach that harmonizes scientific knowledge, policy support, and community involvement.

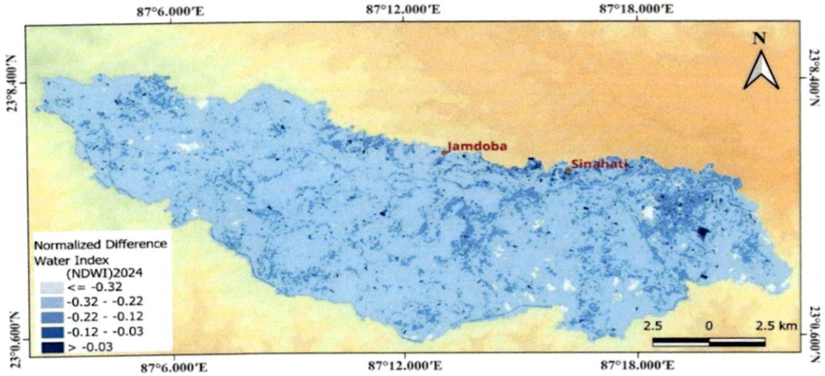

Fig. 7.7a Normalized Difference Water Index of the Berai River Basin for 2005. This map displays the Normalized Difference Water Index (NDWI) for the Kopai River Basin in 2005, highlighting the availability of surface water and wetness conditions throughout the region. The NDWI ranges depicted on the map are as follows: <−0.25 (Dehydrated/vegetated/barren surfaces/very low moisture), −0.25 to −0.18 (slightly moist soil/possible degraded or fallow land), −0.18 to −0.10 (Moderate dryness), −0.10 to −0.03 (Moist soils or potential waterlogged areas/minimal water reflection), and >−0.03 (Surface water body)

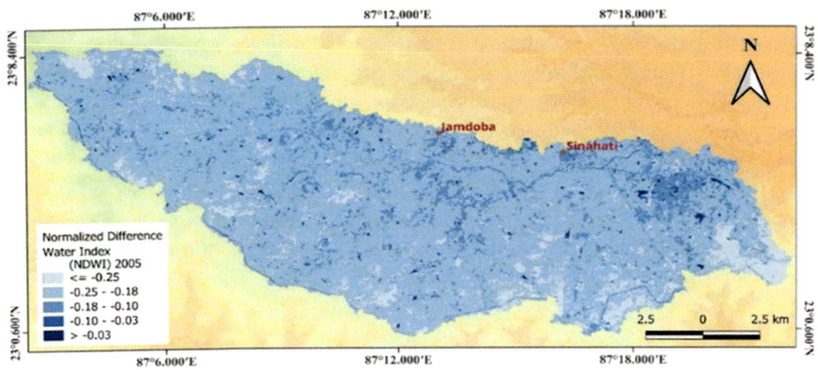

Fig. 7.7b Normalized Difference Water Index of the Berai River Basin for the year 2024. Darker blue regions signify high surface water (>−0.03) while lighter blue to white shades indicate minimal or absent surface water (<−0.32). Areas surrounding the river that appear light blue to white are linked to seasonal wetlands or floodplains that are moist but not saturated, as well as rain-fed zones that dry rapidly and agricultural areas with varying soil moisture levels

Fig. 7.8 This image showcases a semi-arid riverine landscape at Chakdaha-Madhuban, Bankura (Latitude: 23° 5′39.39″ N, Longitude: 87°20′32.27″E), where the river Berai meets the Dwarkeshwar. It illustrates a seasonal stream highlighting essential river aggradation and ecological stress characteristics. A narrow, shallow stream winds through the scene, clearly showing reduced water flow typical of non-perennial rivers during post-monsoon or dry seasons. The stream's sinuous shape and exposed mid-channel bars reveal ongoing sediment deposition (aggradation). A dark area along the stream may indicate bank erosion, but the prevalence of sandy and silty bars points to sediment overload from upstream sources. The image features standard sandbars and point bars in aggradation zones where the sediment load surpasses the river's carrying capacity. These deposits are sparsely vegetated, indicating that aggradation has occurred and that the sites are nearing stability. The vast floodplain is devoid of vegetation or has scant grass, indicating periodic flooding, erosion, or human activity such as grazing or farming. On the left bank, grasses and reeds dominate the riparian area, which helps to stabilize the bank and indicates damaged riparian vegetation in transition. The far tree cover suggests a move towards more stable, less disturbed upland zones. The striking contrast between the lush uplands and sandy lowlands highlights the continuous fluvial changes. Aggradation Prevails: This landscape exemplifies aggradation dominance, where sediment deposition outpaces removal. Significant implications of the features depicted in this image include reduced water-holding capacity of the river, increased flood risk during monsoons due to choked channels, and a decline in aquatic biodiversity and ecosystem services

Fig. 7.9 This image displays the Berai riverine environment, highlighting elements that demonstrate seasonal hydrology and riparian vegetation dynamics, most likely within a degraded or aggradation fluvial situation. It has a shallow, softly flowing channel with brownish water, indicating a moderate to low discharge, probably after recent rains or during the post-monsoon season. The presence of floating aquatic plants, likely Eichhornia or other invasive species, suggests nutrient-rich stagnant zones, indicative of channel aggradation and reduced flow velocity. The expansive, flat area between the riverbanks and the islands, as well as the lateral bars, is covered with grass and aquatic weeds, suggesting sediment deposition. The vegetation-covered depositional zones confirm that aggradation has stabilized, indicating this river segment has been accumulating and retaining sediment over various seasons. On the right bank, a diverse community of tall grasses and shrubs typically shows early successional species establishing on degraded or aggraded banks. In contrast, the left bank has high tree cover, which could indicate a semi-stable natural levee or a largely undisturbed area. The right-side floodplain appears to be more open, possibly used for grazing or subjected to repeated human pressure. The dense vegetation in the channel, particularly floating weeds, suggests a diminished flushing capacity, often a result of reduced water discharge and sediment overload, indicating ongoing transformations in the riverine ecosystem driven by both natural deposition and human-altered catchment processes

Fig. 7.10 This image illustrates a riverbank slope that showcases distinct sedimentary and erosion features. The layered profile consists of a dark brown upper horizon, rich in organic matter, which is typical of topsoil. Beneath it lies a yellowish to reddish sandy-silt layer, which is looser and poorly compacted, typical of fluvial deposits. These layers likely indicate episodic sediment deposition, each stratum formed during high-flow flood events, shaping the river's aggradation history. The steep, eroded face displays crumbling blocks and loose clods, indicating active bank failure due to lateral erosion during high discharge, slumping, and sheet erosion, particularly following rainfall. The presence of angular clumps also suggests physical weathering, possibly caused by dry–wet or freeze–thaw cycles. Sparse grasses and pioneer vegetation struggle to establish themselves, especially at the slope's top and base. The absence of deep-rooted plants suggests an unstable substrate and ongoing soil erosion or human-induced clearance (e.g., grazing, land-use change). This exposure exemplifies classic riverbank degradation, characterized by the loss of cohesive vegetation cover, oversteepening of the bank (likely spurred by channel migration or monsoonal flash floods), and poor soil binding capacity in the alluvial material, resulting in accelerated slumping. Such degraded banks significantly contribute to increased sediment load downstream, the formation of mid-channel bars through redeposition (i.e., aggradation elsewhere), and potential badland expansion if left unchecked, particularly in lateritic and semi-arid regions

Fig. 7.11a Mandy Pole Bridge (Berai River Bridge) (Latitude: 23° 5′31.12″ N, Longitude: 87°11′55.77″E) is an important bridge that spans the Berai River in Bankura district, West Bengal. It contributes significantly to regional infrastructure by improving connectivity across the river. This bridge, constructed with a robust, balanced cantilever design on a solid base, exhibits a dedication to longevity and stability, especially in areas prone to fluctuating water levels and seasonal floods. Local names frequently have cultural or historical significance, which may not necessarily be reflected in official documents

Fig. 7.11b Mandy Pole (Berai River Bridge) (Latitude: 23° 5′31.12″ N, Longitude: 87°11′55.77″E), road connected between Arbani and Singer garia village of Bankura district. A paddy field was constructed in the cross-sectional area by modification of the natural landscape

Fig. 7.11c Photograph is taken from the Mandy Pole (Berai River Bridge) (Latitude: 23° 5′31.12″ N, Longitude: 87°11′55.77″E). This is the upstream of the Berai River. The cross-sectional region contains a paddy field, the outcome of natural landscape transformation. It decreases the cross-sectional area of the Berai River

Fig. 7.12 The Berai Irrigation Scheme (BIS) was developed and launched by the Irrigation & Waterways Department of the West Bengal government. This medium-scale irrigation project is located in the Onda block of Bankura district, West Bengal, and has an irrigation potential of 4,100 hectares. It intends to greatly increase regional agricultural productivity through interconnected canals, especially during the dry winter cropping season. The project's budget is ₹3.71 crore, and it is expected to be completed by January 3, 2025. As part of the broader West Bengal Accelerated Development of Minor Irrigation Project (WBADMIP), BIS is implemented by the Water Resources Investigation & Development Department (WRIDD) with support from the World Bank. This project aims to improve water availability and strengthen Water User Associations (WUAs) to enhance irrigated agriculture throughout the state. Phase II, which spans from December 2023 to June 2029, will include 22 districts, 234 blocks, and 2,522 villages, with a total budget of $211 million (World Bank, WBADMIP). Incorporating the Berai Irrigation Scheme into WBADMIP demonstrates the state's commitment to improving irrigation infrastructure and encouraging sustainable farming practices. (WBADMIP)

Fig. 7.13a Kamar Kata Dam (Latitude: 23° 5′34.10″N, Longitude: 87°10′35.10″E) is located within the Berai Irrigation Project. This dam is part of the Berai Irrigation Scheme, a medium-scale irrigation initiative overseen by the Irrigation & Waterways Department of the Government of West Bengal. The Berai Irrigation Scheme aims to increase agricultural output by providing a regular water supply for irrigation. Although particular information about Kamar Kata Dam's capacity and the approximate area it irrigates is not easily available, the Berai project makes an important contribution to the region's broader irrigation network

Fig. 7.13b Kamar Kata Dam: The Berai Irrigation Project. The Kamar Kata Dam is an important component of West Bengal's Berai Irrigation Scheme, a medium-sized project located in the Onda block of Bankura district. This initiative proposes to supply irrigation for around 4,100 hectares of agriculture, considerably benefiting local farming communities, particularly during the dry period. Both the Kamar Kata Dam and the Berai Irrigation Scheme are essential components of West Bengal's initiative to boost agricultural growth and manage water resources in the region

Fig. 7.13c This photograph was captured at Kamar Kata Dam (Latitude: 23° 5′34.10″N, Longitude: 87°10′35.10″E), situated within the Berai Irrigation Project, downstream of the Berai River. In the center, a small remnant of an old river bridge appears, which was once the sole means of communication in the area. However, following the construction of the Kamar Kata Dam, the bridges were washed away due to the dam's flow regulation, especially during peak flows. The image also highlights a distinct signature of the exposed right bank's depositional profile and sediment stratigraphy, which is a result solely of the flow regulation of the Berai River at the dam

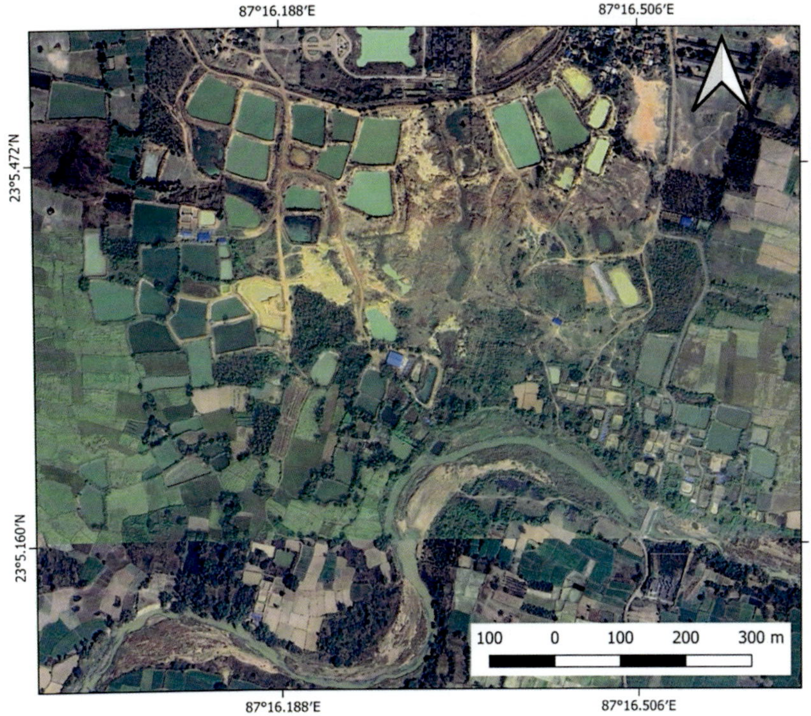

Fig. 7.14a The transformation of the Shinhati badland area into a hatchery zone in the Ramsagar region (Bishnupur subdivision, Bankura district) signifies a major shift in land use, influenced by ecological and economic factors. Previously part of the Shinhati Badland, this region was marked by lateritic soil, rugged terrain, severe gully erosion, and limited vegetation, making it unsuitable for agriculture. However, through deliberate human efforts, particularly after implementing water harvesting and minor soil conservation projects, select areas have been reclaimed. Over the last twenty years, bolstered by local expertise and state-backed fisheries programs, these degraded lands have been transformed into seasonal ponds and developed into fish hatcheries. The practice of pisciculture has emerged as a viable livelihood option for local communities. Key contributions to this change include the construction of embankments, the excavation of shallow basins, and enhancements in water retention capacity

Fig. 7.14b Shinhati badland, is tucked along the left bank (Latitude: 23° 5′14.41″N, Longitude: 87°16′26.28″E) of the Berai River, has narrow and winding channel which can be observed near the right bank. It provide hinds about the river's potential flow area within this unique landscape. This channel has become noticeably constricted over time. As a result, diminished sediment transport from the surrounding Badlands. This area characterized by its rugged terrain. This area experiencing a remarkable morphometric transformation into a serene fish culture pond. This transformation reflects the delicate balance between nature's processes and human influence

Fig. 7.14c The left bank of the Berai River (Latitude: 23° 5′14.41″N, Longitude: 87°16′26.28″E) at Shinhati is experiencing significant collapse mainly due to severe erosion. This erosion is exacerbated by seasonal high-water discharge and uncontrolled sand mining. The river's meandering path combined with loose alluvial soil and inadequate embankments accelerates lateral erosion. Also, human activities such as agriculture close to the bank and deforestation further weaken the riverbank's stability. Monsoon floods exacerbate the situation, resulting in land loss, property damage, and the displacement of local communities. Immediate action through riverbank protection efforts and sustainable land-use practices is essential to prevent further degradation and to protect nearby settlements

Fig. 7.14d A Photograph was taken from the Shinhati badland (Latitude: 23° 5′21.50″ N, Longitude: 87°16′19.78″E), currently modified to a fish culture pond area. In this photograph, the gully sidewall is exposed due to erosion, with a significant fluted structure and signature of deep weathering and leaching. A light brown portion of the photographs represents the pallid zone formed due to erosion and leaching. A consecutive thin layer of hard pan is also present in this photograph

Fig. 7.14e The fish breeding and culture ponds (Latitude: 23° 5′21.50" N, Longitude: 87°16′19.78″E) are part of the Ram Sagar Hatchery Project, situated in the Shinhati gully-eroded Badlands. This innovative program aims to transform a damaged terrain into a thriving aquaculture hub. The project uses naturally occurring gully channels and water-retaining depressions to produce fish-rearing ponds, promoting long-term livelihood options. Minor earthworks and bunding change the geomorphology of the Badlands, preventing erosion and retaining water. By combining scientific hatchery procedures with local participation, the project improves fish output, boosts food security, and helps to restore ecological balance. Furthermore, it exemplifies how marginal lands can be leveraged for economic gains while addressing environmental degradation

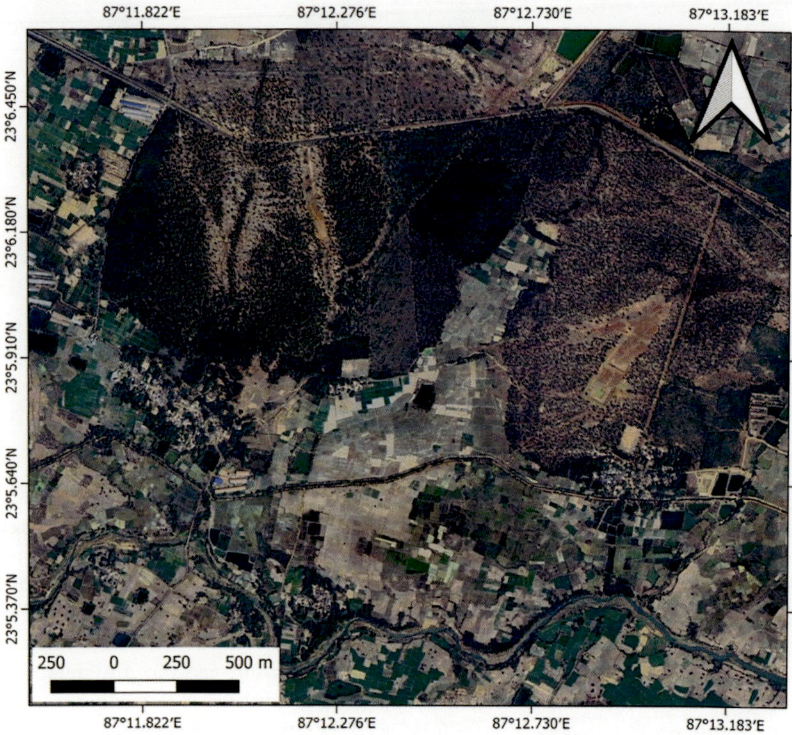

Fig. 7.15a Jamdoba forest area located near Teliberia (Latitude: 23° 5′44.66″N, Longitude: 87°13′34.46″E). In the Jamdoba forest area, several gullies have been formed. The forest authority constructed sedimentation ponds to catch sediment and prevent gully erosion

Fig. 7.15b The Jamdoba Badland (Latitude: 23° 5′56.38″N, Longitude: 87°12′57.61″E) is situated in the lateritic landscape of the Rarh region (Onda block, Bankura district, West Bengal) and exhibits characteristic signs of surface erosion due to water movement. Two primary erosion processes are evident here: rill erosion and sheet erosion, which significantly impact the topography of the Badlands. The image shows the Sheet Erosion Zone, indicating that the area's upper slopes and interfluves tend to be smoother and more gently rolling. The development of the badland landscape involves the removal of a consistent layer of topsoil through surface runoff during heavy rainfall. This area is characterized by sparse vegetation, exposed lateritic crusts, hardened ferricrete, the movement of fine sediments during the monsoon, slight discoloration, and notable textural differences in satellite images, which range from reddish-brown to light tan due to soil erosion. These elements contribute to the initial stages of badland formation, leading to soil nutrient depletion and the creation of microchannels. In the Rill Erosion Zone, particularly on the middle to lower slopes where sheet flow converges, water accumulates in narrow pathways, carving into the soil to form small channels known as rills. If not managed, rills can evolve into gullies, resulting in significant land degradation and increased sediment transport downstream

Fig. 7.15c Water retention as well as a sedimentation pond has been created by the Forest department at Jamdoba Badland (Latitude: 23° 5′56.38″N, Longitude: 87°12′57.61″E) to reduce gully erosion. An embankment borders this sedimentation pond. This strategically located pond intercepts surface runoff, allowing sediments to settle before water disperses further downstream. Encircled by a reinforced earthen embankment, the pond reduces the erosive energy of flowing water, stabilizing the surrounding badland slopes. It also helps in groundwater recharge and supports micro-ecosystems during dry spells. By trapping sediments, the structure prevents the expansion of rills into deep gullies, contributing to sustainable land restoration and erosion control in this severely degraded landscape

Fig. 7.16 The photograph shows a stunning view from the Madhuban Manakali Mandir, which sits gently near the mysterious confluence of the Berai and Dwarkeshwar rivers. A small, meandering route has been created by carefully diverting some of the river's flow, specifically during the dry season when the waters recede. Previously, this gorgeous area was graced by a solid concrete bridge, which regrettably succumbed to the violent currents of the Berai during high water, illustrating the force and instability of nature at this spectacular river confluence

Fig. 7.17 This photograph showcases a temporary culvert-like structure built across a shallow section of the Berai River near Chakdaha village, where it meets the Dwarkeshwar River. Residents built a sandbag bridge and culvert to use during the dry season, with a ferry service accessible during the monsoon. The construction is mostly made of sandbags with circular pipes or culverts inserted into the base to allow water to flow beneath. This temporary causeway allows pedestrians and light vehicles to pass while reducing the impact on river flow. It is a low-cost, temporary option that is commonly used in regions where permanent bridges are not feasible owing to economic constraints or seasonal circumstances. The photographs provide a clear depiction of river morphology. The water body resembles a small, meandering stream or distributary, characterized by visible sandbars and an exposed riverbed. The river exhibits evidence of seasonal flow, as indicated by the width of its dry banks, which suggests higher water levels during the monsoons. Erosion and deposition patterns are apparent, highlighting cut banks and point bars typical of small, dynamic channels. The photographs also convey spatial information about the surrounding vegetation and terrain. This area features sparse vegetation, grassy clumps, and semi-arid scrubland, indicative of seasonal rainfall and potentially containing lateritic or alluvial soils. The right side reveals a slightly elevated bank covered in tall grass, while the left side displays exposed pipes and altered soil, possibly indicating past modifications to the water flow or construction efforts. The photographs powerfully illustrate human intervention in the Berai river system. The construction of the culvert-like barrier and visible pipes on the left bank indicates active river management or rural development initiatives, likely aimed at irrigation, local connectivity, or flood control. In the background, a brick kiln chimney, a tall vertical structure, suggests nearby industrial activity, which could impact river health and sediment dynamics

Fig. 7.18 Confluence of the Beari River with the Dwarkeshwar River. This map illustrates a detailed area of the Dwarkeshwar River basin and confluence of Berai River near Bishnupur in Bankura district, West Bengal, India, showcasing both natural and man-made features. The Dwarkeshwar River flows from north to south in this stretch, winding through the region and serving as a significant hydrological element. It features village names, temples, schools, bus stops, playgrounds, and healthcare centers, highlighting the socio-cultural fabric. The presence of bridges, roads, and dense clusters of rural settlements indicates strong connectivity and a relatively developed rural infrastructure

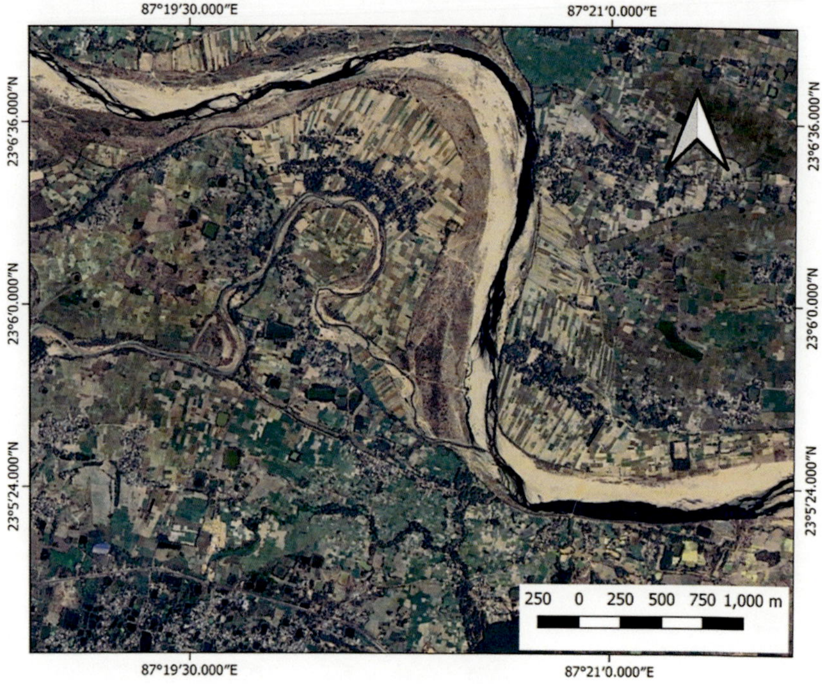

Fig. 7.19 The image shows the confluence of the Beari and Dwarkeshwar rivers during the pre-monsoon season. The graphic depicts the merging of fluvial channels along the lowland floodplain, highlighting changes in sediment load, flow dynamics, and channel form. This junction is significant in defining the local hydrological regime and sediment deposition patterns. Vegetation along the banks and land use around the confluence zone indicate significant human–environment interactions. The Beari River, as a minor tributary, contributes seasonal runoff and sediment flux to the Dwarkeshwar River, influencing downstream hydrodynamics and ecological conditions

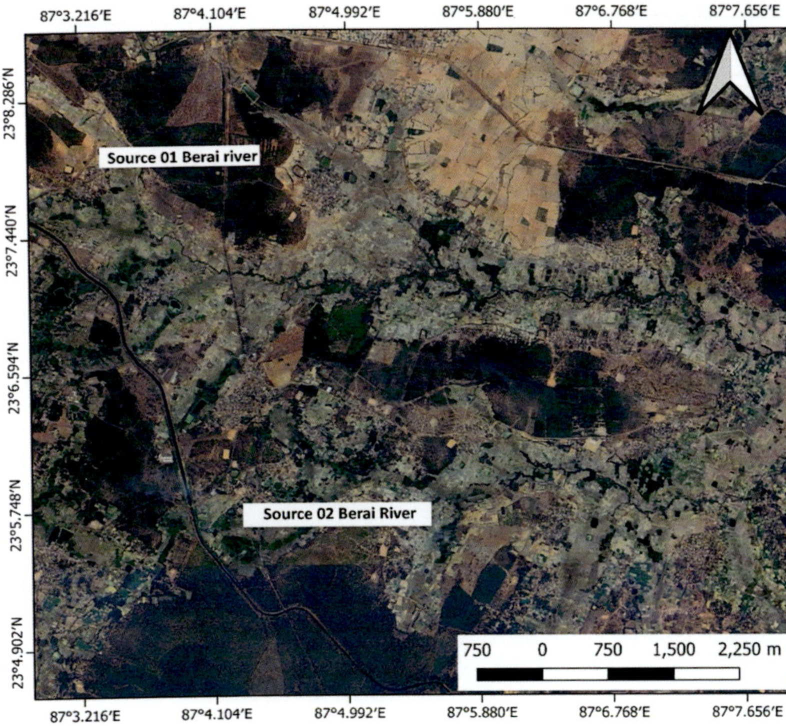

Fig. 7.20a The source region of the Berai River lies in the eastern fringes of the Chhotanagpur Plateau in West Bengal. This upland terrain, composed mainly of Precambrian rocks, gives rise to numerous seasonal streams, one of which forms the Berai River. The area experiences moderate to heavy rainfall throughout the monsoon season, which contributes to the river's formation via surface runoff and groundwater emergence. The region is characterized by dense foliage, forested slopes, and lateritic soil, which sustain a varied range of flora and wildlife. The terrain's gradient and lithological features influence the river's initial flow, channel formation, and sediment transport processes

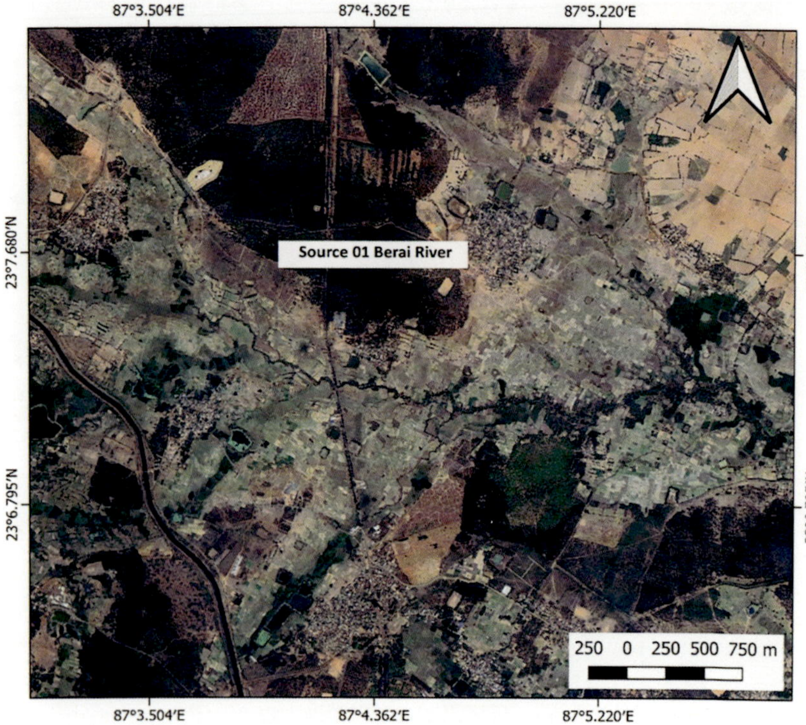

Fig. 7.20b The Berai River has two subsequent sources and both has its critical flow direction and characteristics. In contrast, they don't have a reliable source region pertaining to rivers which begins in plateau terrains. The source area of this river from a plateau can fluctuate significantly due to various factors, especially rainfall patterns and other external influences that impact water flow and source sustainability. Using Google Earth Pro, a comprehensive analysis of satellite imagery has pinpointed two primary sources of the Berai River. The first source is situated in a geographic cluster that includes several localities: Mandarbani, located at Latitude 23° 7′ 46.15″ N and Longitude 87° 4′ 12.56″ E; Dubrakun at Latitude 23° 8′ 44.59″ N and Longitude 87° 4′ 11.44″ E; Hirapur, which can be found at Latitude 23° 8′ 47.47″ N and Longitude 87° 3′ 20.07″ E; Digshuli located at Latitude 23° 8′ 35.01″ N and Longitude 87° 2′ 37.90″ E; and finally, Haryaki placed at Latitude 23° 7′ 43.72″ N and Longitude 87° 2′ 43.43″ E. This concentration of locations emphasizes the importance of these areas which consider as critical hydrological sources for the Berai River, highlighting how geographical factors significantly influence water sources in plateau regions

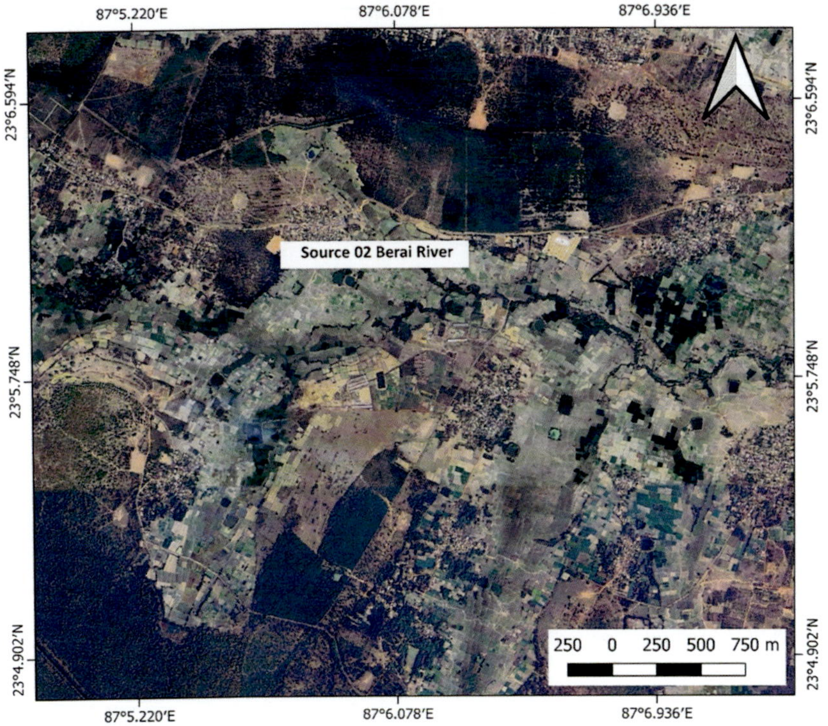

Fig. 7.20c The Berai River's second source originates in the plateau fringe zone. This area, an extension of the Chhotanagpur Plateau, features undulating landscapes and scattered hillocks composed of rugged rock formations, including granite and gneiss. Seasonal rains significantly aid in stream formation through sheet flow and rill erosion. Numerous temporary streams from the highlands converge, gradually developing the Berai River. This secondary source boosts the river's flow during monsoons and is crucial for supplying water to agricultural fields and downstream settlements

Fig. 7.21 Small bridge constructed on the Berai River. This satellite image shows a culvert-based river crossing over a shallow, winding stream called Berai, located at approximately 23°46′46.752" N and 87°20′16.771" E. It serve as a rural access like road which linking the two banks of the Berai river. In the crossing visible vehicles tracks which indicating regular uses of the access. The culvert construction runs diagonally across the channel and looks to be made up of multiple embedded culverts that allow for restricted water movement. The sparse vegetation and broken topography on both banks of the river indicate a semi-arid or lateritic environment, potentially prone to erosion and sedimentation. Multiple braided channels and exposed sandy beds reflect seasonal variations in water output. The map scale and north arrow make spatial orientation and measurements easier, with a 40-meter crossing span displayed. While these constructions are useful, they may restrict aquatic communication and alter the river's natural geomorphology

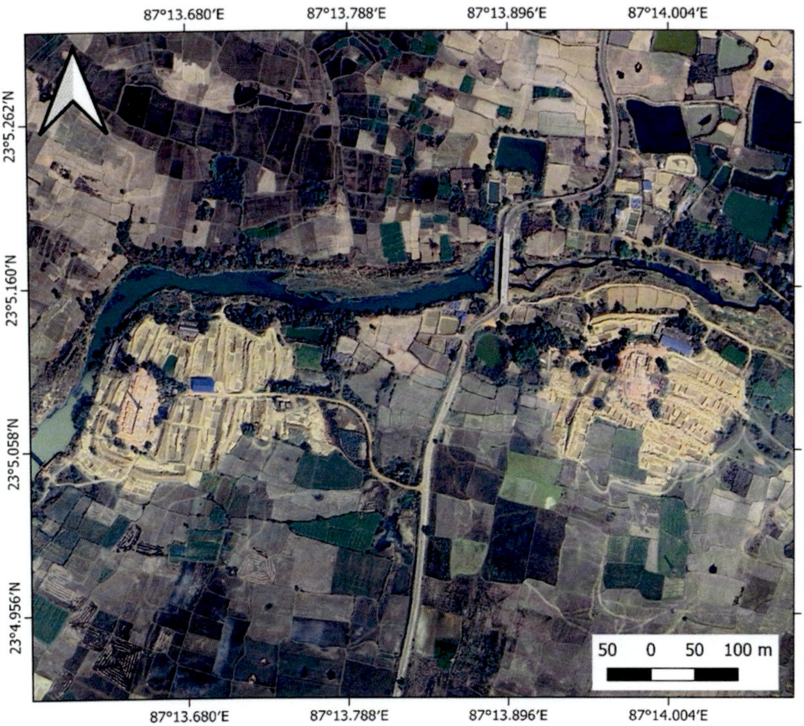

Fig. 7.22 It is a significant instances of artificial interference in Berai fluvial system. This concrete bridge are a remarkable evidenced which supported by a sturdy central pillar which entends over the river. Adjoining the bridge two expansive brick kilns are stand on the right bank of the river. These kilns are highlight the industrial expansion. These brick kilns with their billowing smoke and rhythmic clamor, not only reshape the landscape but also illustrate the intricate relationship between nature and human enterprise along this vital waterway

Fig. 7.23 In this image shows a crucial railway bridge under the part of south-western railway network which covers the Berai River near Bishnupur town in Bankura District, West Bengal. This major infrastructure enhances regional rail connectivity, supporting both passenger and freight transport, and plays a crucial role in linking rural areas with larger urban and industrial centers. In the satellite image or field photograph, a significant land-use change is apparent on the right (southern) bank of the Berai River, especially along the concave curve of the meandering waterway. This site is geomorphologically important due to active erosion and sediment transport processes. A substantial brick kiln has been noted here, raising concerns about soil degradation, riverbank instability, and human pressure on the fluvial ecosystem. The contrast between the railway infrastructure and industrial activities, such as the brick kiln, highlights a transforming riverine landscape influenced by both development efforts and extractive land-use practices. This development underscores a broader trend of infrastructure-driven changes in peri-urban river regions, frequently compromising ecological balance and hydrological integrity

References

Abdi, H. (2007). The Bonferroni and Šidák corrections for multiple comparisons. In N. J. Salkind (Ed.), *Encyclopedia of measurement and statistics* (pp. 103–107). Sage. https://doi.org/10.4135/9781412952644

Adhikari, R. N., & Mondal, M. K. (2015). Tectonogeomorphic analysis of the upper Kangsabati River basin, West Bengal, India: Implications for drainage development. *Arabian Journal of Geosciences, 8*(12), 11069–11083. https://doi.org/10.1007/s12517-015-1917-7

Adimalla, N., & Li, P. (2019). Occurrence, health risks, and geochemical mechanisms of fluoride and nitrate in groundwater of the rock-dominant semi-arid region, Telangana State, South India. *Human and Ecological Risk Assessment: An International Journal, 25*(1–2), 81–103. https://doi.org/10.1080/10807039.2018.1522002

Adnan, M. S. G., & Atkinson, P. M. (2020). Exploring the impact of climate and land use/cover changes on streamflow in a transboundary river basin: A case study in the Ganges–Brahmaputra–Meghna River Basin. *Science of the Total Environment, 704*, Article 135360. https://doi.org/10.1016/j.scitotenv.2019.135360

Ahmad, M. J., Ahamed, I., & Alam, M. A. (2022). Applying hydromorphological indices and RS-GIS-based multi-criteria decision analysis to identify flood susceptibility in Bangladesh. *Environmental Challenges, 8*, Article 100564. https://doi.org/10.1016/j.envc.2022.100564

Al-Hussein, A. I. A., Al-Saffawi, A. Y. T., Al-Shaker, Y. M. S., & Al-Hussein, A. (2024). *Possible human health risks of nitrates in drinking water: A case study of groundwater in Wana district, Northern Iraq*. https://doi.org/10.31413/nat.v11i2.15742

Barman, D., & Nandy, S. (2021). An appraisal of groundwater quality for drinking and irrigation purposes in Birbhum District, West Bengal, India. *Arabian Journal of Geosciences, 14*(14), 1310. https://doi.org/10.1007/s12517-021-07256-4

Balamurugan, P., Shankar, K., Nagavinothini, R., Vijayasurya, K., & Kumar, P. (2020). *Non-carcinogenic risk assessment of groundwater in the southern part of Salem district in Tamilnadu, India*. https://doi.org/10.4067/s0717-97072020000104697

Barua, S., & Sinha, R. (2022). River aggradation and degradation: Indicators, causes, and implications. In R. Sinha & S. Barua (Eds.), *Fluvial geomorphology and river engineering* (pp. 61–79). Springer. https://doi.org/10.1007/978-3-030-82014-5_4

Bhattacharya, R., Mukhopadhyay, A., & Ghosh, S. (2012). Impact of rainfall variability on river regime: Case study of the Subarnarekha River basin, eastern India. *Current Science, 102*(7), 1034–1041.

Brierley, G., & Fryirs, K. (2005). *Geomorphology and river management: Applications of the river styles framework*. Blackwell Publishing.

Chatterjee, M., & Prasad, N. (2020). Riverine morphology and socio-economic environment: A review. *International Journal of Current Research, 12*(12), 356–362.

Das, S., & Pal, S. (2020). GIS-based morphometric analysis and its implications: A study of Chandan River Basin in Eastern India. *Modeling Earth Systems and Environment, 6*(2), 1061–1077. https://doi.org/10.1007/s40808-019-00709-7

Das, G. K. (2024). *River systems of West Bengal: Water and environments*. Springer. https://doi.org/10.1007/978-3-031-53480-5

Das, U., Bajpai, R., & Chakrabortty, D. (2021). River regulation and associated geo-environmental problems: A case study of lower reaches of Shilabati River Basin, West Bengal, India. *International Journal of Current Research, 13*(12), 198–205.

Erencin, Z. (2000). *C-Factor mapping using Remote Sensing and GIS: A case study of Lom Sak/Lom Kao, Thailand*. https://core.ac.uk/download/56347214.pdf

Ghosh, S., & Dandapat, S. (2021). Spatiotemporal changes in river morphology and sediment dynamics using remote sensing and GIS: A case study of the Kangsabati River, India. *Sustainable Water Resources Management, 7*(3), 1–17. https://doi.org/10.1007/s40899-021-00528-8

Ghosh, D., Banerjee, M., Pal, S., & Mandal, M. (2022). Spatio-temporal variation of channel migration and vulnerability assessment: A case study of Bhagirathi River within Barddhaman District, West Bengal, India. In P. K. Shit, B. Bera, A. Islam, S. Ghosh, & G. S. Bhunia (Eds.), *Drainage basin dynamics* (pp. 331–344). Springer. https://doi.org/10.1007/978-3-030-79634-1_14

Ghosh, S., & Das, B. C. (2021). *Anthropogeomorphology of Bhagirathi-Hooghly river system in India*. Routledge.

Grabowski, R. C., & Gurnell, A. M. (2016). Hydrogeomorphology-ecology interactions in river systems. *River Research and Applications*. https://doi.org/10.1002/rra.2974

Irrigation & Waterways Department, Government of West Bengal. (n.d.). *Medium irrigation projects*. Retrieved from https://www.wbiwd.gov.in/index.php/applications/medium_irrigation(wbiwd.gov.in)

Jahan, C. S., & Pal, S. (2020). Riverine ecosystem services and land use dynamics in the lower Gangetic floodplain: A socio-ecological assessment. *Ecological Indicators, 114*, Article 106321. https://doi.org/10.1016/j.ecolind.2020.106321

Kaur, R., & Singh, S. (2019). Assessment of river water quality using water quality index (WQI): A case study of Sutlej River, Punjab. *Applied Water Science, 9*, 1–10. https://doi.org/10.1007/s13201-019-1001-0

Kale, V. S. (2002). Fluvial geomorphology of Indian rivers: An overview. *Progress in Physical Geography, 26*(3), 400–433. https://doi.org/10.1191/0309133302pp343ra

Knighton, D. (1998). *Fluvial forms and processes: A new perspective*. Arnold.

Kondolf, G. M. (1997). Hungry water: Effects of dams and gravel mining on river channels. *Environmental Management, 21*(4), 533–551.

Mandal, A. (2020). GIS-based morphometric analysis for sustainable watershed management: A case study from the Ayodhya Hills, India. *Environmental Earth Sciences, 79*, 1–13. https://doi.org/10.1007/s12665-020-8870-y

Mandal, S., & Pal, S. (2021). Understanding land degradation and its impact on livelihoods in the lateritic Birbhum District, West Bengal. *Environmental Management, 68*(1), 59–76. https://doi.org/10.1007/s00267-021-01527-3

Mondal, M. K., & Satpati, L. N. (2017). Quantitative assessment of the morphometric characteristics and micro-watershed prioritization using RS and GIS in the Sarujore River basin, Bankura, West Bengal. *Remote Sensing Applications: Society and Environment, 8*, 43–60. https://doi.org/10.1016/j.rsase.2017.08.001

Mitra, R., & Mandal, D. K. (2023). Assessment of livelihood vulnerability in the riparian region of the Tista River, West Bengal, India. *GeoJournal, 88*, 811–839. https://doi.org/10.1007/s10708-022-10645-0

Millennium Ecosystem Assessment (MEA). (2005). *Ecosystems and human well-being: Synthesis*. Island Press.

Pal, S., & Saha, S. (2014). A study of the channel characteristics of the Dwarakeswar River in the lateritic terrain of Eastern India. *Geography Journal, 2014*, Article ID 147498. https://doi.org/10.1155/2014/147498

Pareta, K., & Pareta, U. (2017). Geomorphological analysis and hydrological potential zone of Baira River watershed, Churah in Chamba District of Himachal Pradesh, India. *Indonesian Journal of Science and Technology, 2*(1), 31–47. https://doi.org/10.17509/ijost.v2i1.5987

Roy, A., & Bera, S. (2021). Evaluating morphometric characteristics for groundwater potential using GIS and remote sensing in the Chandan-Dwarakeswar River basin, India. *Modeling Earth Systems and Environment, 7*(4), 2421–2433. https://doi.org/10.1007/s40808-020-00975-5

Roy, P., & Pal, S. (2022). Spatio-temporal variability of channel migration and its impact on the local environment: A study of the Kunur River, West Bengal, India. *Geojournal, 87*(3), 2119–2136. https://doi.org/10.1007/s10708-020-10277-6

Rudra, K. (2010). *River morphology and flood management in West Bengal*. State Water Investigation Directorate, Government of West Bengal.

Saha, S., & Pal, S. (2021). Understanding the influence of geomorphology on flood risk: A study of the upper Dwarakeswar River Basin, West Bengal, India. *Natural Hazards, 109*(2), 1391–1414. https://doi.org/10.1007/s11069-021-04835-5

Sinha, R., Jain, V., & Tandon, S. K. (2021). Fluvial system response to climate and tectonic changes in the Himalayan foreland: A review. *Geological Society, London, Special Publications, 489*(1), 1–18. https://doi.org/10.1144/SP489-2018-195

Sinha, R., & Friend, P. F. (1994). River systems and their sediment flux in the alluvial plains of northern India. *Sedimentology, 41*(2), 317–335.

Singh, S., & Kale, V. S. (2009). *Introduction to geomorphology*. Prayag Pustak Bhawan, Allahabad.

Surian, N., & Rinaldi, M. (2003). Morphological response to river engineering and management in alluvial channels in Italy. *Geomorphology, 50*(4), 307–326.

Sarkar, A. (2020). Impacts of anthropogenic activities on riverbed morphology: A case study of the Subarnarekha River near Ranchi, Jharkhand. *Journal of Geomorphological Studies in Eastern India, 12*(1), 45–53.

Srivastava, R. K., & Bhattacharya, A. R. (2022). Integrated remote sensing and GIS approach for morphometric analysis of a river basin in eastern India. *Applied Geomatics, 14*(1), 55–68. https://doi.org/10.1007/s12518-021-00382-5

Thakur, J. K., & Pandey, A. C. (2021). Application of geo-spatial technology in hydrological studies of Indian river basins: A review. *Journal of the Geological Society of India, 97*(1), 11–25. https://doi.org/10.1007/s12594-021-1712-4

TendersOnTime. (2024). *Improvement of Irrigation Potential by Resuscitation of Berai Irrigation Project Block Onda Dist Bankura*. Retrieved from https://www.tendersontime.com/india/details/improvement-irrigation-potential-resuscitation-berai-irrigation-project-block-onda-dist-bankura-6a747f2/(TendersOnTime)

West Bengal Accelerated Development of Minor Irrigation Project (WBADMIP). (n.d.). *Project Overview*. Retrieved from https://wbadmip.wb.gov.in/(wbadmip.wb.gov.in)

World Bank. (2023, June 9). *World Bank approves $148 Million financing for minor irrigation project to support water availability, increase food security for 960,000 People in West Bengal*. Retrieved from https://www.worldbank.org/en/news/press-release/2023/06/09/world-bank-approves-148-million-financing-for-minor-irrigation-project-to-support-water-availability-increase-food-secur(WorldBank)

Zhang, Q., Xu, P., & Qian, H. (2019). Assessment of Groundwater Quality and Human Health Risk (HHR) Evaluation of Nitrate in the Central-Western

Guanzhong Basin, China. *International Journal of Environmental Research and Public Health, 16*(21), n/a.

CHAPTER 8

Restoring the Environmental Balance: Addressing Human Interventions and Their Effects on the Sustainability of the Silabati River in West Bengal, India

Ananta Gope and *Shambhu Nath Sing Mura*

8.1 Introduction

For men may come and men may go,
But I go on forever.

—*The Brook*, (Alfred Lord Tennyson, 1842)

A. Gope (✉) · S. N. S. Mura
Department of Geography, Vivekananda Mahavidyalaya, Sripally, Burdwan, India
e-mail: anantagopegeo@gmail.com

S. Ghosh (ed.), *Reimagining Indian Rivers for Sustainability*, Palgrave Studies in Environmental Sustainability,
https://doi.org/10.1007/978-3-032-04754-0_8

> Coming to the countryside, I do not see human beings as separate entities. Just as a river flows over the land, the currents of human beings, with uproar, move in a zigzag path through jungles, villages, towns, and cities forever—as if it has no end.
>
> —*Chinnapatra,* (Rabindranath Tagore, 1968, p. 94; Letter No.38)

These contrasting views of the river illustrate its deep connection to human, society, and civilization. Alfred Tennyson beautifully contrasted the river's unceasing flow with the transient nature of human existence in his poem "The Brook." In a more complex depiction, Rabindranath Tagore presented the river as a metaphor for the dynamic movement of human populations. Despite their different tones, both Tennyson and Tagore emphasize the deep, inseparable connection between the river and humanity—one that lies in the eternal flow of the river and the ongoing progress of civilization.

From these reflections arises a paradox concerning the interrelationship between rivers and human life. While a river may seem to have a simple name, its significance transcends its literal form, serving as the very lifeblood of human society. Every point on Earth's surface lies within the reach of a river basin, illustrating the river's omnipresence. The health of a river and the well-being of society are intricately linked. Since ancient times, rivers have played a crucial role in sustaining agriculture, animal husbandry, transportation, industry, waste management, and human habitation through forming floodplains, estuaries, and deltas. Rivers are indispensable—whether in the context of ancient, rudimentary societies or in the complex, technology-driven societies of the Anthropocene. However, in the Anthropocene era, driven by excessive greed, humans have exploited resources rapaciously from river without fully understanding the mechanisms that sustain the health and vitality of river ecosystems.

As global populations continue to grow, the demand for water increases, driven by the needs of agriculture, industry, commerce, and urbanization. Consequently, efforts to harness and control the flow of rivers have intensified, aiming to secure water resources for the future. Societies increasingly rely on technological interventions—dams, barrages, embankments, and channelization—to regulate and manage rivers, while the unchecked extraction of stones, gravel, and sand from riverbeds exacerbates these challenges. Moreover, the encroachment of human activity is steadily reducing the size of river valleys. Agriculture—particularly on riverbeds and dried-up reservoir beds—alongside the construction of brick kilns, settlements along riverbanks, and the proliferation of road and rail bridges across rivers, further diminishes these vital ecosystems. In addition, pollution from agricultural runoff and urban settlements has polluted the water of the rivers and thereby tilted the ecological equilibrium. Unfortunately, the Silabati River is not exempt from these devastating challenges.

Extensive research has been conducted on the Silabati River, focusing on a range of environmental, hydrological, and geomorphological aspects of the basin. Dolui and Ghosh (2013) examined the causes and impacts of flooding in the Ghatal Block, identifying both natural and anthropogenic factors contributing to frequent inundation. Subsequently, Shit et al. (2013) investigated the variability of gully erosion within a micro-catchment area of the Silabati Basin. Their findings revealed a strong positive correlation between slope gradient, rainfall intensity, and the size of the runoff-contributing area with the severity of gully erosion in the Garbeta Badland region of Paschim Medinipur district. Bhandari and Roy (2018) used the Analytic Hierarchy Process (AHP) to assess flood vulnerability across the Silabati Basin. Their study aimed to develop a reliable flood forecasting model through the integration of remote sensing data. To delineate flood risk zones, they employed a multi-criteria decision-making framework alongside GIS-based spatial analysis. Das et al. (2020) focused on the spatial dimensions of river regulation in the lower reaches of the Silabati River. Utilizing remote sensing and GIS techniques, they analysed the impacts of human interventions on channel morphology and the ecological settings of adjacent floodplains. Halder et al. (2020) examined trends in groundwater levels and drought across the basin using the Standard Groundwater Level Index (SGLI), offering valuable insights into the temporal variability of subsurface water resources. Mahala et al. (2020) identified water erosion, vegetation loss, and soil degradation as the principal land degradation processes within the basin. The middle reaches, characterized by lateritic soil cover, were found to be more susceptible to erosion than other areas. In a related study, Patel et al. (2021) emphasized the role of anthropogenic activities in altering the geomorphic diversity of the Silabati Basin. Intense human interventions, particularly in the lower basin, have led to significant channel modifications due to sand mining. Additionally, adjacent terraces and floodplains have experienced vegetation loss and degradation resulting from deforestation and the quarrying of lateritic formations. Some areas have also been developed as tourist destinations, accelerating ecological degradation. Das (2022) undertook a comprehensive study of the Silabati River's history, mythology, and geographical context, also evaluating the physicochemical properties and overall quality of the river water. Most recently, Mandal and Kayal (2023) applied geospatial techniques to identify and analyse the hydro-geomorphic characteristics of the Silabati River, contributing to a better understanding of its fluvial behaviour.

To date, no research has sufficiently addressed the direct impact of human activities on the life cycle of the Silabati River, nor has it explored alternative management strategies to restore its stability—gaps that this study seeks to fill. This research aims to provide a comprehensive analysis of the landform characteristics shaped by the interplay of endogenetic and exogenetic processes, further modified by human interventions across different sections of the river basin. In addition to interpreting the geomorphic processes and landforms of the Silabati River, the study also highlights the detrimental effects of human activity on the river's health. It proposes remedial measures designed to restore and sustain the ecological and geomorphic integrity of the basin.

8.2 The Study Area

The Silabati River basin extends from 86°38′37″ E to 87°43′35″ E longitudes and from 22°28′31″ N to 23°14′21″ N latitudes (Fig. 8.1). The total area of the basin is approximately 3,897 km^2, with a perimeter of 578.5 km. It mainly flows through the *Rarh* Bengal region, which forms part of the greater Bengal Basin. The river basin is bounded to the north by the Dwarakeswar River basin and to the south by the Kangsabati River basin. The climate of the region is characterized by hot summers and cool winters, with summer temperatures exceeding 45 °C and winter temperatures dropping below 6 °C. The area receives an average annual rainfall of approximately 150 cm. The climatic conditions of the basin area are characterized by an oppressively hot summer, high humidity throughout the year, and well-distributed rainfall during the monsoon season. The year can be broadly divided into four distinct seasons:

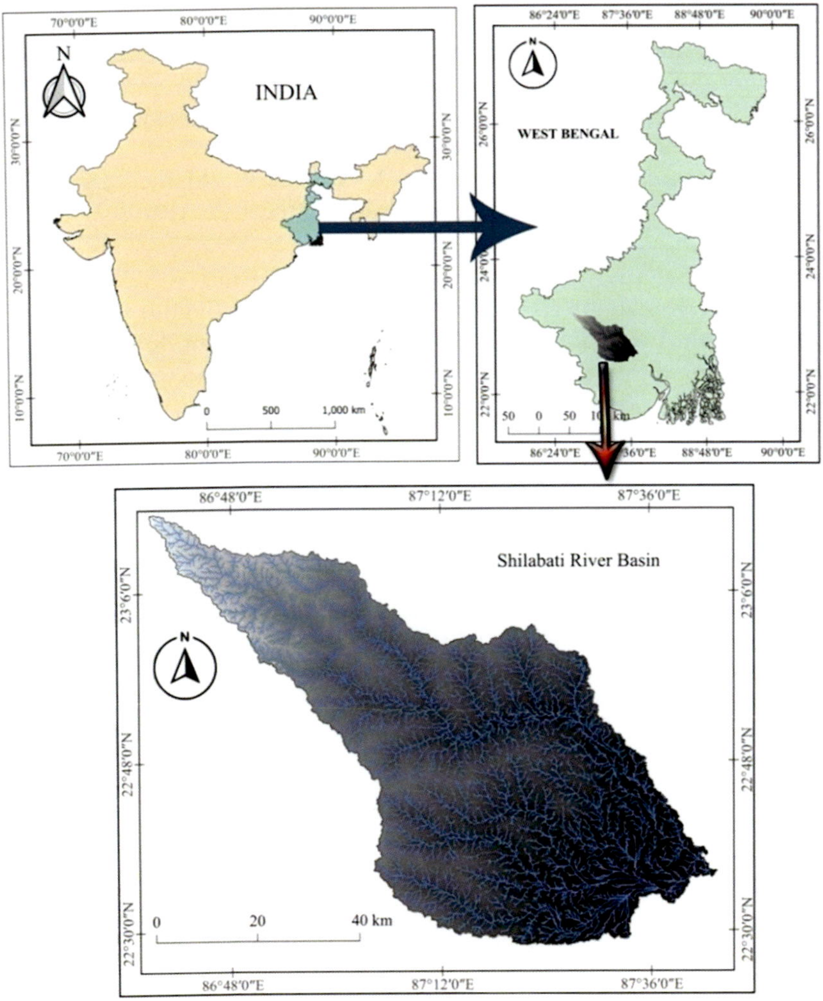

Fig. 8.1 Location of the study area

i. Summer Season (March to May): This period is marked by extremely high temperatures, with May being the hottest month. The oppressive heat is intensified by the high moisture content in the air.

ii. Southwest Monsoon Season (June to September): This season brings the bulk of the annual rainfall. July is typically the rainiest month.
iii. Post-Monsoon Season (October to mid-November): This short transitional period follows the monsoon.
iv. Winter Season (mid-November to February): Marked by cooler temperatures, January is the coldest month of the year.

The average annual rainfall in the district of Purulia is 1,217.7 mm, Bankura receives about 1,155.2 mm, and Paschim Medinipur sees around 1,421.1 mm. A significant portion of this rainfall—about 83% in Purulia, 80% in Bankura, and 77% in Paschim Medinipur—occurs during the southwest monsoon months (June to September) (IMD, 2008).

8.2.1 Extreme Weather Records

i. Heaviest 24-hour Rainfall:

   a. Purulia: 438.7 mm on 18 June 1898
   b. Bankura: 599.2 mm at Taldangra on 10 August 1950
   c. Paschim Medinipur: 400.0 mm at Dantan on 9 June 1994

ii. Highest Maximum Temperatures Recorded:

   a. Purulia: 46.3 °C on 20 May 1972
   b. Bankura: 47.4 °C on 20 May 1972
   c. Paschim Medinipur: 47.2 °C on 12 June 1942

iii. Lowest Minimum Temperatures Recorded:

   a. Purulia: 3.8 °C on 13 January 1989
   b. Bankura: 5.1 °C on 31 January and 2 February 1967
   c. Paschim Medinipur: 5.0 °C on 3 January 1971

8.2.2 Rainfall Data from Key Observatories in the Basin Area

i. Manbazar (Purulia District):

   a. Annual Rainfall: 1,227.3 mm
   b. Heaviest 24-hour Rainfall: 242.3 mm on 11 July 1942

ii. Indpur (Bankura District):

a. Annual Rainfall: 1,005.2 mm
b. Heaviest 24-hour Rainfall: 307.6 mm on 3 August 1922

iii. Taldangra (Bankura District):

a. Annual Rainfall: 1,197.2 mm
b. Heaviest 24-hour Rainfall: 599.2 mm on 10 August 1950

iv. Simlapal (Bankura District):

a. Annual Rainfall: 1,276.3 mm
b. Heaviest 24-hour Rainfall: 403.1 mm on 7 August 1969

v. Chandrakona (Paschim Medinipur District):

a. Annual Rainfall: 939.0 mm
b. Heaviest 24-hour Rainfall: 221.5 mm on 25 June 1907

vi. Ghatal (Paschim Medinipur District):

a. Annual Rainfall: 1,292.4 mm
b. Heaviest 24-hour Rainfall: 256.0 mm on 22 September 1878

The rainfall pattern in the region is erratic and shows considerable year-to-year variability (Table 8.1 and Fig. 8.2). A small amount of rainfall occurs during the winter months, occasionally influenced by Western Disturbances. Additionally, storms and depressions originating from the Bay of Bengal sometimes affect the area during the pre- and post-monsoon seasons, resulting in widespread heavy rainfall. *Kalbaisakhi* (local thunderstorms) are a common phenomenon during the hot weather season. These afternoon storms are often violent and bring significant rainfall. While they can cause damage to property, they also provide a welcome drop in temperature and help the farming community by moistening the soil, thereby facilitating seed sowing (India Meteorological Department, 2008, 2024; District Gazetteers: Bankura, 1968). The average annual runoff of the Silabati River during the monsoon months, recorded between 1948 and 1951 at Baharamuri village within the Khatra police station jurisdiction, was 13,000 acre-feet (Banerji, 1968). The topography of the upper and middle reaches of the basin is predominantly undulating, marked by small hills rising to about 150 metres. The general slope of the land is directed toward the southeast. The soils in the upper and middle parts of the basin are primarily red and lateritic, while the lower part is covered with alluvial soil, formed by recent silt deposition.

Table 8.1 Month-Wise Rainfall (mm) Statistics of Bankura (2012–2024)

Year	*January*	*February*	*March*	*April*	*May*	*June*	*July*	*August*	*September*	*October*	*November*	*December*	*Total (mm)*
2012	43.9	16.4	7.9	85.9	44.6	173.2	301.4	345.8	242.7	35.9	38	15	1350.7
2013	0.9	15	22.9	72.6	342.4	369.7	289.8	368.4	260.8	398	0	0	2140.5
2014	0.8	38.3	8	3.3	84.7	85.7	313.7	323.4	164.1	53.2	0	0.3	1075.5
2015	17.4	1.5	7.1	85.6	55.9	152.2	467.5	230.6	96.5	12.7	0	0.2	1127.2
2016	6.1	10.2	15.6	0.8	101.3	175.2	264.8	445.5	268.9	46.9	0.7	0	1336
2017	0	0	16.9	27.9	76.3	228.8	634.2	330.4	186.1	249.1	25.3	5.2	1780.2
2018	0	0.1	9.1	149.6	76.5	235	328.9	315.2	151	27.6	0.5	40.6	1334.1
2019	0	33.8	23.2	69.5	175.9	113.8	226	319.5	267.3	173.1	11.6	12.3	1426
2020	44.4	2.9	61.6	58.2	153.5	266.6	278.8	315.8	197.7	58.3	3.6	0.7	1442.1
2021	0	2	9.3	53.2	254.7	440.7	481	151.2	466	195	38.1	54.8	2146
2022	30	62.8	0	6.6	120.3	136.7	148.1	403.8	193.9	97.6	0	0	1199.8
2023	0	0	53.6	29.3	128.3	102.9	175.2	337.9	198.9	209.1	0.3	84.4	1319.9
2024	5.1	19.4	45.7	4	168.6	80.9	5.1	5.1	5.1	5.1	5.1	5.1	354.3
Total	148.6	202.4	280.9	646.5	1783	2561.4	3914.5	3892.6	2699	1561.6	123.2	218.6	18,032.3
Monthly Average Rainfall (mm)	11.43	15.56	21.60	49.73	137.15	197.03	301.11	299.43	207.61	120.12	9.47	16.81	1387.1

Source IMD (2024)

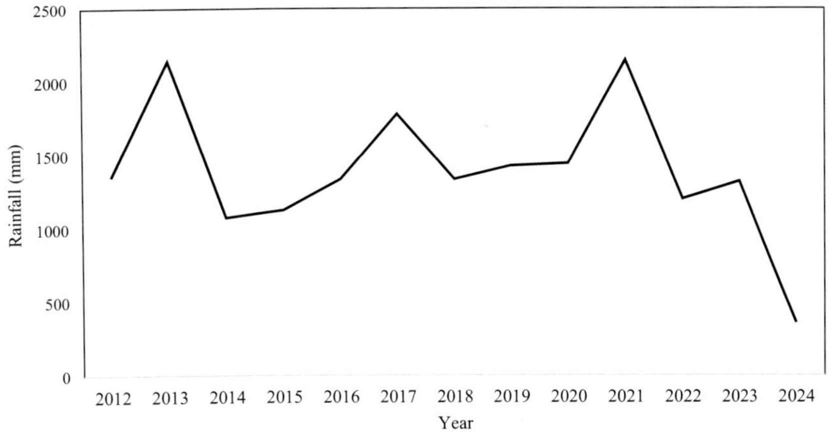

Fig. 8.2 Year-wise fluctuation of rainfall (mm) of Bankura (2012–2024)

At any point along the river, the shape of the channel is influenced primarily by drainage and sediment supply. Typically, the area of the drainage basin expands downstream as tributaries join the master stream, leading to an increase in discharge. The channel bed of the Silabati River is divided into two distinct sections: a bedrock channel in the upstream region and an alluvial channel in the downstream region. The upstream part of the Silabati River channel is strongly influenced by the resistant nature of bedrock formations. Structural controls—such as underlying geological strata, joints, fractures, and lineaments—significantly affect the river's morphology. These channels are often marked by gravel clusters and transverse ribs. In this section, channel morphology is primarily shaped by erosional processes, with block excavation, abrasion, and plucking being the three dominant forms of erosion. Sedimentation occurs in various parts of the middle and lower reaches of the river valleys, particularly when the flow capacity and retention capacity of the river are reduced. Common causes of sedimentation in the Silabati River channel include reduced discharge in the lean season, decreased slope, increased cross-sectional area in the middle reaches, enhanced boundary resistance, flow segregation, and the presence of obstacles within the flow in the lower reaches.

8.3 Methodology

The present study integrates intensive field investigations with geospatial technology. The objectives clearly indicate that river is a natural entity that carves its landforms depending on the existing geotectonic, lithological, geomorphic, and geo-hydrological characteristics of its basin. Simultaneously, these created landforms and the overall sustainability of the river are also significantly influenced by human activities. In the present study, the authors aim to assess the human-induced impacts on the Silabati River and propose remedial measures to restore its sustainability. Accordingly, the study focuses on the physical and geomorphological characteristics of the basin area as altered by human interventions. To achieve the stated objectives, various maps and datasets have been utilized, including the Topographic Maps, SRTM data, LANDSAT data 2011 and 2025. Using geospatial software such as QGIS 2.18, MapInfo 10.0, and ArcGis 10.4, a range of maps and charts have been generated. These outputs have been analysed and interpreted to draw meaningful conclusions.

8.4 The River and Its Basin

Rivers are like mothers for the country … These water-filled streams fill the earth with the power of procreation and energy. It is these streams flowing on the landmass which act as the blood-pumping arteries in the body.
—*Kalidasa in Raghuvansham* (Radhakant Bharati, 2009, p. 1)

I love the river very much, and I love the sky also. There is an excellent union between the river and the sky - colours in colours, lights in shadows - just like the echo of the sky. The sky does not get its echo in the earth except on the water.
—*Bhanusingher Patrabali,* (Rabindranath Tagore, 2016, p. 105)

Great poets have long recorded the profound driving force of rivers in shaping human civilization and nurturing life. Kalidasa celebrated the productive power and essential presence of rivers in the natural world, while Tagore reflected on the river's poetic dialogue with the sky, capturing its spiritual and aesthetic essence.

The name Silabati (also known as Silai) is derived from the Sanskrit word "*Sila*," meaning stone or rock, reflecting the rocky nature of much of the riverbed (Fig. 8.3). The river originates at Baragram village (23°15′N & 86°39′E) under the Puncha police station of Purulia district, West Bengal, India (District Gazetteer: Puruliya, 1985). It flows predominantly southeastward, following the regional slope, through the districts of Purulia, Bankura, and Paschim Medinipur. In Purulia district, the Silabati River originates and flows southeastward, confined to the Puncha CD Block. In Bankura district, the river passes through the Indpur, Hirbandh, Khatra, Taldangra, and Simlapal CD Blocks, flowing in a northwest to southeast direction. After crossing Simlapal, the river enters in Paschim Medinipur district and continues through Garbeta II and Garbeta I CD Blocks. In CD Blocks—Chandrakona I and II, it shifts to a southerly direction. From Keshpur CD Block, it turns north-northeast, passing through Debra, Daspur I, and Ghatal CD Blocks. In Purulia, the river runs for about 20 km, receiving several small tributaries known locally as *Jhore*. The catchment area here is mostly wooded, and the river is non-perennial. The riverbed in this section is marked by exposed rocks. Silabati enters Bankura district near Salanpur in the Indpur police station and continues southeastward through Taldangra and Simlapal, exiting the district near Dhuliapur. The river's length in Bankura is approximately 56 km (Banerji, 1968). The principal left-bank tributary of the Silabati is the Jaypanda River, which is 43.2 km long and originates near Chaitandihi village in the Indpur police station. Other tributaries in Bankura include the Purandar (12 km) and Champakhal (10.4 km), both of which merge with Silabati in Paschim Medinipur. Several smaller rivulets (*Jhore*) also join at acute angles, following the regional slope. This stretch of the river features several small, picturesque waterfalls (Fig. 8.4).

Upon entering Paschim Medinipur, the river's course becomes winding. Initially flowing eastward, it turns southeast and then southward through the Ghatal subdivision. Near Narajol, it sharply turns northward, eventually meeting the Dwarakeswar River at Bandar. The combined flow forms the Rupnarayan River, which merges with the Hooghly River and ultimately flows into the Bay of Bengal. In Paschim Medinipur, main tributaries include Kulbai (or Kubai), Tamal, and Parang. Kulbai is the primary right-bank tributary, originating near Tangasol, crossing the Midnapore-Bankura railway line, and joining the Silabati after receiving the Tamal River near Mugbasan (Bandopadhyay, 2007). Tamal originates near the Ramgarh-Kantapahari area. Their

Fig. 8.3 Rocky riverbed of Silabati River

combined flow meets the Silabati near Narajol (Basu, 2002; Das, 2022). Both tributaries are seasonal and mainly active during the monsoon. The Parang River, another right-bank tributary, originates in Lalgarh, flows through Godapiyasal, and joins the Silabati near Keshpur. The Khetia River or Khetia Khal, is a left-bank tributary, flows parallel to the Silabati from Kharkushma to Krishnapur, and then again joins with the principal river at a right angle near Daspur.

The estimated annual discharge of the river is 2,644 million cubic metres—2,093 million during the monsoon and 551 million during the non-monsoon period. The estimated sediment load is 0.22 million

Fig. 8.4 Picturesque small waterfall in Silabati River near Harhmashra village

tonnes (Rudra, 2016). In its lower reaches, affected by tidal fluctuations, a small stretch of the Silabati is navigable year-round. However, the river frequently causes flooding, particularly in Banka, Khirpai, and Ghatal regions. Geomorphologically, the upper catchment is characterized by erosion, while the lower part is depositional. Deforestation due to agricultural expansion and settlement growth has led to increased sedimentation, causing silting in the river's lower course. A pickup barrage was constructed on the Silabati at Kadamdeuli (22°45′54.30″ N, 86°49′29.82″ E) in Indpur police station as part of the Kangsabati River Valley Project, completed in 1975. Malfunctions in the barrage could jeopardize the broader project, necessitating rehabilitation and repair (Fig. 8.5).

8.5 Surface Elevation and Landforms

The area under study showcases a range of unique physical features distinguishing it from other regions of West Bengal. The solid and crystalline rock consists of Precambrian and Archaean formations, while the rocks on the surface are primarily intrusive in origin—mainly granite, gneiss, and quartzite. This geological makeup of rocks, combined with climatic weathering, has given rise to distinct geomorphic features. In the upper

Fig. 8.5 Kadamdeuli pick-up barrage on Silabati River

and middle parts of the basin, laterite soils have developed on such landscapes.

Structurally, the region lies along the outer boundary of the Ranchi peneplain. This lower peneplain landscape features a gently sloping, eroded platform, with elevations ranging from below 50 to above 200 metres above mean sea level (Fig. 8.6). The area experiences active degradation, resulting in dissected badland patches (locally called *Khowai*), and residual hills known as *Dungri*. These typically rounded hills result from exfoliation weathering, particularly on gneissic rocks, forming dome-shaped features called tors. These low hills are remnants of the ancient Chotanagpur Plateau. The overall landscape exhibits alternating series of ridges-and-valleys, dotted with knolls and isolated hillocks, adding to the region's topographic diversity. Figure 8.6 categorizes the area into five elevation zones:

i. Below 50 m: Covers 35.60%, primarily in the eastern basin.
ii. 50 m–100 m: Covers the largest portion—48.33%, mainly in the central basin.
iii. 100 m–150 m: Covers 3.49%, mainly in western Bankura.
iv. 150 m–200 m: Covers 12.42%, mainly in eastern Purulia, and
v. Above 200 m: Makes up the remaining 0.18%, mainly near the river's source.

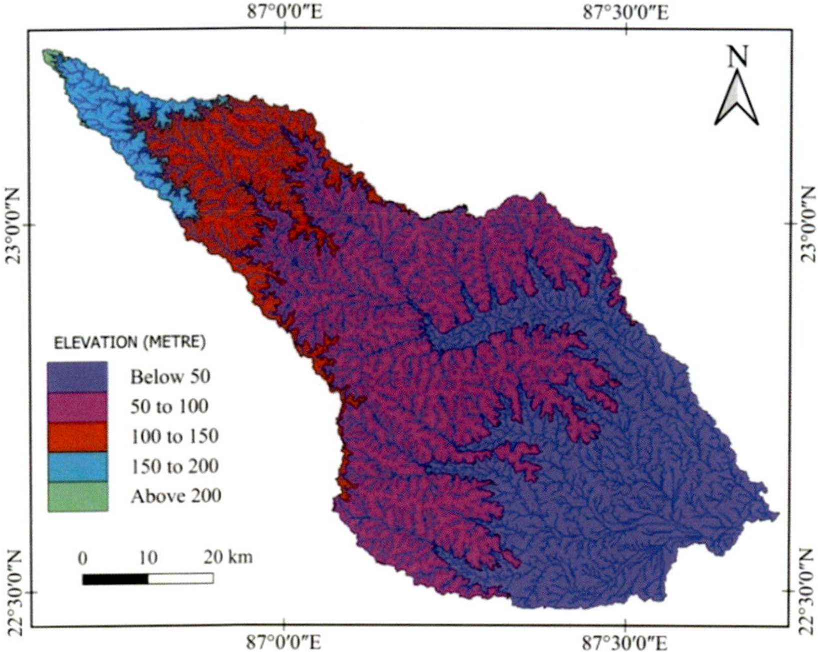

Fig. 8.6 Elevation zones of Silabati River basin

The Silabati River, with its many tributaries, follows a northwest to southeast course, dissecting the gently undulating plains into a series of parallel strips. The southeastern portions of the basin form a segment of the Gangetic Delta. In this part, active delta-building and floodplain formation continue. This area lies at elevations below 20 metres. The river and its right bank tributaries have extensive sedimentary features, such as extensive floodplains, backwaters, and deltas. River flow patterns such as meandering, braiding, and anastomosing are clearly visible in certain sections of the river. These give rise to a variety of micro-geomorphic features including: channel bars, natural levees, meander scars, scour routes, neck cut-offs, chute cut-offs, sand plugs, clay plugs, and lakes of different shapes (Fig. 8.7a). Many of these features have been altered by human interventions like road and railway construction, artificial levees, culverts, wooden bridges, and agricultural practices on the floodplain (Fig. 8.7b).

8.6 Geological Characteristics

The area under study and its surrounding region encompasses the south-eastern part of the Chotanagpur Plateau (O'Malley, 1995). This region is largely characterized by a rolling surface that merges with plains in the east and rugged uplands in the west. The erosional hills (locally known as *dungri*) stand as remnants on the ancient plateau surface, which has since been reshaped into the present landscape through active processes of planation.

Geologically, the entire basin area is divided into three distinct sectors:

i. The Western Sector—dominated by crystalline rocks of Proterozoic age.
ii. The Middle Sector—comprised of laterite formations of Pleistocene age.
iii. The Eastern Sector—composed of older and newer alluvial deposits from the Holocene period.

For ease of analysis, based on lithological characteristics, the basin is further subdivided into four distinct topographic terrain units:

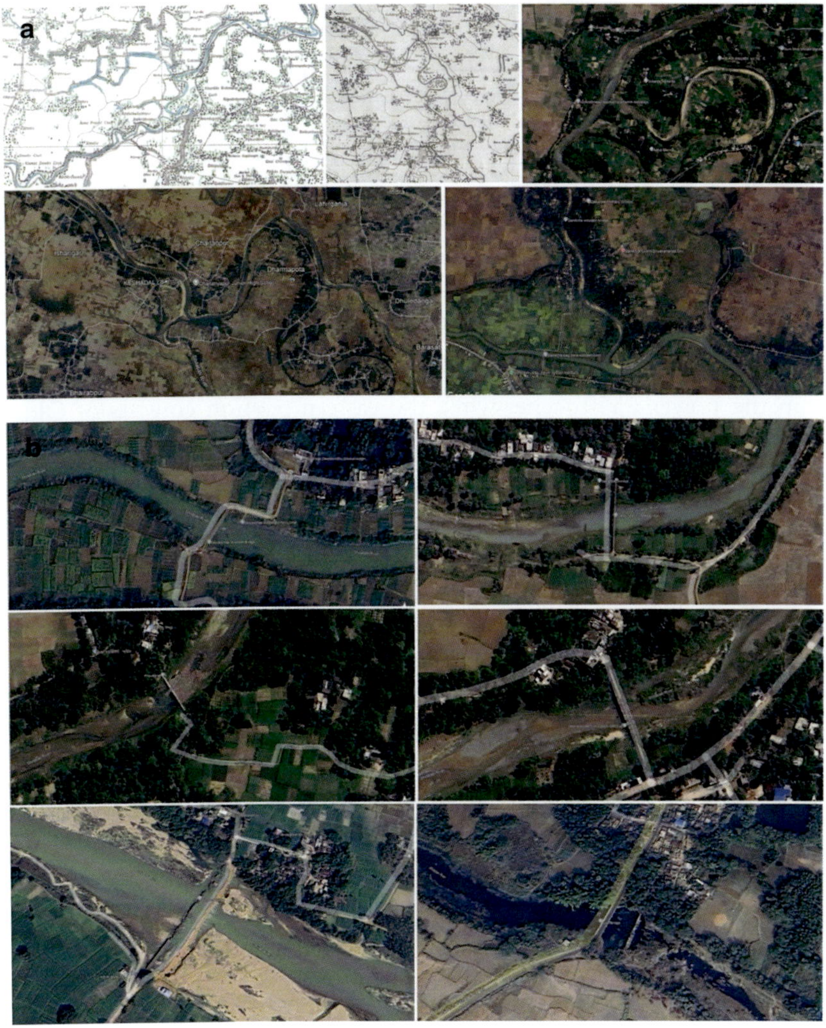

Fig. 8.7 (**a**) Micro-geomorphic features on the Silabati floodplain. (**b**) Cross bridges on the Silabati River

i. Hard Crystalline Terrain—located in the upstream part of the basin, mainly within the districts of Purulia and western Bankura.
ii. Lateritic Terrain—found in the districts of eastern Bankura and western Paschim Medinipur.
iii. Older Alluvial Terrain—present in parts of Paschim Medinipur district.
iv. Recent Alluvial Terrain—located in the southeastern part of the Paschim Medinipur district.

In summary, the hilly and undulating region of the northwest is composed of large granite-gneiss formations from the Middle and Upper Proterozoic era. The central part is dominated by extensive red laterite formations from the Lower Pleistocene, while the eastern portion consists of Older and Newer Alluvium from the Upper Holocene (Fig. 8.8 and Table 8.2).

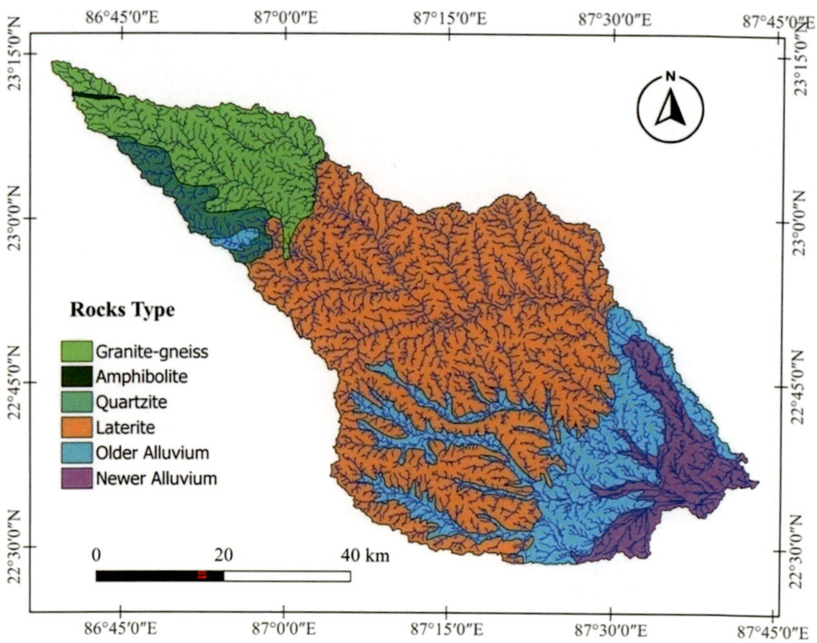

Fig. 8.8 Surface geology of the Silabati River basin

Table 8.2 Geological characteristics of the Basin Area

Serial number	*Rocks*	*Area covered (sq. km)*	*Area (%)*	*Geology formation*
1	Granite-gneiss	498.3	12.80	Chhotanagpur granite-gneiss (Middle-Upper Proterozoic)
2	Amphibolite	3.9	0.10	DalmaVolcanics (upper Protozoic)
3	Quartzite	140.7	3.59	Unclassified sedimentaries (Lower-Middle Proterozoic)
4	Laterite	2223	57.04	Lalgarh Formation (Lower Pleistocene)
5	Older Alluvium	714.02	18.33	Panskura Formation (Upper Holocene)
6	Newer Alluvium	317.3	8.14	Present Day Flood Plain Deposition (Upper Holocene)

The upstream, western part of the basin is underlain by crystalline rocks of Upper and Middle Proterozoic age. These include granite, gneiss, amphibolites, and quartzite—typically weathered to varying depths and characterized by joints, fissures, and fractures. In the central portion, a thick mantle of laterite deposits from the Pleistocene age overlays the crystalline basement. In the western part of this region, a thin primary laterite cap exists above the crystalline rocks. This layer gradually increases in thickness toward the east and south until it is eventually covered by recent alluvium. The eastern areas feature a thick, detrital laterite capping, succeeded by substantial deposits of yellow and grey sticky clay, silt, sandy clay, and sand—ranging in texture from fine to very coarse—along with gravels. River valley zones are predominantly underlain by alluvial sediments of varying grades, including clay, silt, sand, and gravel. The rolling upper and middle sections of the basin, which constitute a significant portion of the total area, exhibit sloping, often unbounded terrain. These areas are highly susceptible to surface runoff and erosion. The soils here are generally shallow, acidic, low in fertility, and possess limited water-holding capacity.

8.7 Human-Induced impacts on the Silabati River

The Cenozoic Era consists of two periods: the Tertiary and the Quaternary. The Quaternary period is further divided into three epochs: the Pleistocene, the Holocene, and the Anthropocene. The term *Anthropocene* is derived from the Greek words *anthropos*, meaning "humans," and *kainos*, meaning "new." Thus, the Anthropocene refers to the "new epoch of humans" (Schwagerl, 2014, p. 9). The Anthropocene idea was born through a short article published in the newsletter of the International Geosphere-Biosphere Programme (IGBP) written jointly by Crutzen and Stoermer (2000). Two years later, Crutzen attempted to rename the Holocene to the Anthropocene in an article entitled "The Geology of Humankind" published in the scientific journal *Nature*. He noted that, "Unless there is a global catastrophe — a meteorite impact, a world war or a pandemic—mankind will remain a major environmental force for many millennia" (2002, p. 23). In his 1864 book *Man and Nature*, Marsh was one of the first to document human influences on the environment. Italian geologist Stoppani in 1873 recognized the activities of human being as a new telluric force, and spoke of the Anthropozoic era. Other notable works in this field include Fischer's 1915 journal article *Man as a Geological Factor*, Sherwood's 1922 work *Man as a Geological Agent*, Vernadsky's *1926* book *Geochemistry and the Biosphere* and Fels's 1935 book *Man as a Transformer of Earth* (Schwagerl, 2014, p. 56). These authors established that humans have become significant geological and geomorphologic agents in the Anthropocene (Price et al., 2011). Since the Industrial Revolution (beginning around 1760), technological and scientific advancements have enabled humans to become powerful geomorphic agents, capable of accelerating or decelerating geomorphic processes in various ways. Nearly all geomorphic processes—including fluvial, glacial, periglacial, coastal, and aeolian processes—have been profoundly affected and modified by human activities, both intentional and unintentional, across different geomorphic environments (Singh, 2019). It is indeed a real fact that, in the era of Anthropocene, humans play a key role in transforming the riverscape.

Rivers are key agents in shaping terrestrial landscapes. Fluvial morphology—the study of landforms created by river action—is primarily a descriptive science (Grade, 2006). It involves careful observation and interpretation of landforms produced by running water. Fluvial geomorphology, which examines the interactions between river channel forms and processes across different spatial and temporal scales, provides important insights into this (Charlton, 2008). More broadly, it includes both channelized and unchannelized flow, which moves down slope due to gravity (Gabler et al., 2009). Likewise, Schumm pointed out that a fluvial system is basically a physical system with a distinct history (Knighton, 1998). Fluvial geomorphologists focus on both the functional characteristics of rivers over short time spans and their evolutionary tendencies over longer periods.

Human role has been exercised in changing river channels for more than 4000 years (Gregory, 2006). A key question is how humans alter drainage basins, channel processes, and landforms. Human impact is so widespread and intense that it significantly influences river processes and forms. Lewin classifies channel changes into two broad categories: (i) *autogenic changes*, and (ii) *allogenic changes* (Grade, 2006). Autogenic changes are inherent to the river system and are related to natural river regimes. These include processes such as avulsion, channel migration, cutoffs, and crevassing. In contrast, allogenic changes occur as a result of external factors, such as climatic fluctuations, changes in sediment load, and alterations in discharge due to human activities.

Human-induced changes in fluvial processes can be classified as direct or indirect (Knighton, 1998). Direct (or channel-phase) changes directly modify the river channel itself. These include river regulation and channel modifications. River regulation involves activities like constructing dams, barrages, weirs, and canal networks, as well as building rail and road bridges across rivers at specific locations. Channel modifications include actions such as stream gravel extraction, bank stabilization through artificial embankments, channel straightening, and the widening and deepening of river channels. These engineering works are typically aimed at mitigating the effects of flooding, erosion, and deposition.

Indirect (or land-phase) changes, on the other hand, occur outside the river network but still influence the river's discharge and sediment load. These include land use and land cover changes, as well as alterations to land drainage systems. Wohl (2014) similarly pointed out that human impacts on rivers can be either direct or indirect. Indirect impacts arise from activities that do not directly alter channel form or processes but still affect the river system. For example, changes in atmospheric chemistry, as well as shifts in land use and land cover, can have indirect effects. Modifications in vegetation are often accompanied by changes in topography and soil characteristics. Direct impacts, however, involve activities within the river network that directly alter the channel's form or processes. Examples include large-scale flow regulation via dam construction and smaller-scale alterations, such as embankments, dredging, channelization, bank stabilization, and in-stream mining of placer metals and sediment for construction aggregates.

Human activities have dramatically altered sediment yields in rivers around the world. Increases in sediment yield are mainly driven by land use and land cover changes, accelerated soil erosion, mining, and urbanization. However, sediment yield in many rivers has decreased due to the construction of large dams and barrages, which trap sediment in reservoirs. As a result, human interventions have fundamentally changed the transfer and storage of sediment within most global drainage basins, leading to significant effects on channel morphology and behaviour. The global annual sediment shift caused by human activities through mineral extraction is 57,000 million tonnes (mt). In comparison, the annual transport of sediment to the oceans by rivers is about 22,000 mt. This means that human activities contribute to sediment transport nearly three times more than rivers do (Douglas & Lawson, 2000).

Like many rivers in eastern India, the Silabati River faces significant human-induced pressures that have disrupted its basin ecosystem, leading to environmental degradation. These impacts are primarily driven by increasing human activities within its basin, such as deforestation, gravel and boulder extraction from the riverbed, agricultural expansion in the upper catchment area, sand extraction in the middle reaches, and encroachment of the river valley in its lower course.

In the upper reaches of the Silabati River's catchment area, deforestation has been substantial due to the expansion of agriculture. Over time, the population in these areas has grown, leading to the extension of agricultural plots upwards from the valley. This has resulted in significant soil erosion, reduced water quality, and an increased sediment load in the river. The loss of forest cover disrupts the natural hydrological cycle, causing more rapid runoff and reduced groundwater recharge. Deforestation and unsustainable agricultural practices have accelerated soil erosion in the upper and middle catchment areas, leading to the development of badland topography in specific parts of the Jaypanada and Silabati rivers (Fig. 8.9). As a result, the increased sediment load has caused siltation in the lower reaches of the rivers. These silt deposits have clogged the riverbeds, impairing both navigability and water quality.

This, in turn, affects aquatic ecosystems and reduces agricultural productivity. Additionally, domestic waste disposal and agricultural runoff have contributed to pollution in the river. Chemical fertilizers and pesticides used in farming flow into the river during the monsoon, contaminating the water with harmful substances. Domestic and small-scale industrial waste further deteriorates water quality, making the river unsuitable for drinking, fishing, and irrigation in certain areas (Table 8.3).

The Silabati River, part of the Bhagirathi-Hooghly system, is influenced by a tropical monsoon climate. Its hydrology is characterized by a plateau stream that is strongly affected by the southwestern monsoon, exhibiting a leptokurtic hydrograph (Tables 8.1 and 8.3; Figs. 8.10 and 8.11). During the rainy season, the river's water level rises rapidly and overflows its banks downstream. In the upstream areas, the Silabati

Fig. 8.9 Soil erosion and ravine formation in the Silabati River basin

Table 8.3 Season-Wise Rainfall (mm) Pattern of Bankura (2012–2024)

Year	*Winter*	*Pre-Monsoon*	*SW Monsoon*	*Post-Monsoon*	*Annual (mm)*
2012	60.3	138.4	1063.1	88.9	1350.7
2013	15.9	437.9	1288.7	398	2140.5
2014	39.1	96	886.9	53.5	1075.5
2015	19.1	148.6	946.8	12.7	1127.2
2016	16.3	117.7	1154.4	47.6	1336
2017	0	121.1	1379.5	279.6	1780.2
2018	0.1	235.1	1030.3	68.6	1334.1
2019	33.8	268.6	926.6	197	1426
2020	48	273.3	1058.9	61.9	1442.1
2021	56.8	317.2	1538.9	233.1	2146
2022	92.8	126.9	882.5	97.6	1199.8
2023	84.4	211.2	814.9	209.4	1319.9
2024	29.6	218.3	96.2	10.2	354.3
Total (mm)	496.2	2710.3	13067.7	1758.1	18032.3
Average (mm)	38.17	208.48	1005.21	135.24	1387.10

Source IMD (2024)

is prone to heavy floods, particularly during the monsoon months, locally known as *Hurpah ban*, although it is generally fordable at other times. The river is narrow in the upper reaches, with steep and stable banks, which allows sudden discharges to move downstream (Fig. 8.12). However, in the lower reaches, the riverbed consists of low sand and clay banks that fail to contain excess water, leading to flooding. In the lower course, these floods are different in nature—they cause water stagnation for extended periods, resulting in significant property and life losses. To control flooding in the lower reaches, embankments have been built along both sides of the river (Fig. 8.13). During floods, water overflows these embankments, spreading to surrounding areas. After the floodwaters recede, the embankments prevent the stagnant water from returning to the river, which increases the frequency of floods. Furthermore, these embankments prevent the river's silt from spreading to surrounding agricultural lands. Over the years, large volumes of mud carried by the river during floods have settled on the riverbed, raising its level. Consequently, even small amounts of rainfall can now cause floods.

These issues were noted long ago in the works of Willcocks (1930), *Lectures on the Ancient System of Irrigation in Bengal and Its Application*

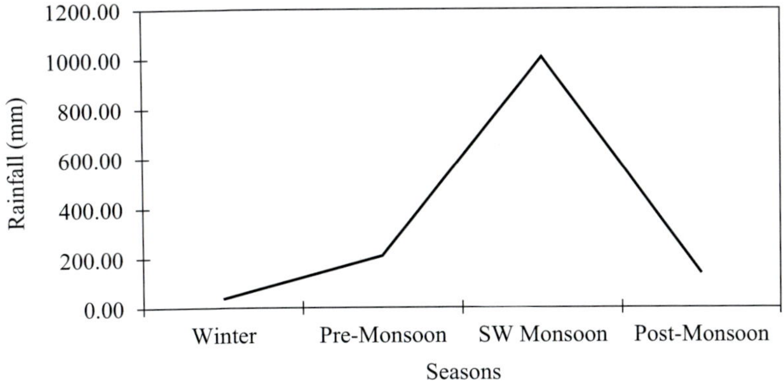

Fig. 8.10 Season-wise distribution of rainfall (mm) of Bankura (2012–2024) (*Source*: IMD, 2024)

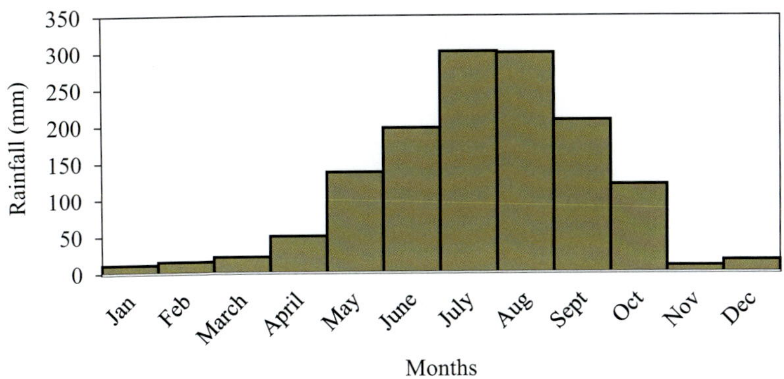

Fig. 8.11 Month-wise distribution of rainfall (mm) of Bankura (2012–2024) (*Source* IMD, 2024)

to Modern Problems; Mukherjee (1938), *The Changing Face of Bengal: A Study in Riverine Economy*; Majumdar (1942), *Rivers of The Bengal Delta*; and Bagchi (1944), *The Ganges Delta*. These books explore the interrelationship between rivers, water, silt, and agriculture. Willcocks's analysis of the ancient irrigation system of Bengal emphasized the importance of overflow irrigation—where farmers would cut canals to channel

Fig. 8.12 Narrow river course of Silabati in the upstream area

swollen river water into agricultural lands, which would deposit fine silt. Majumdar (1942) discussed the deteriorating condition of Bengal's rivers due to human interference in the past and the lack of effective action in the present. Mukherjee examined river management and the agricultural economy of colonial Bengal. He argued that the overflow irrigation system and the agricultural practices of ancient Bengal were in harmony with nature. In Bengal, which depends on monsoon winds, rainfall occurs for four months during the monsoon season, leaving the remaining 8 months dry. To address this, large ponds were constructed to store monsoon water. However, due to human negligence, many of these ponds have since dried up.

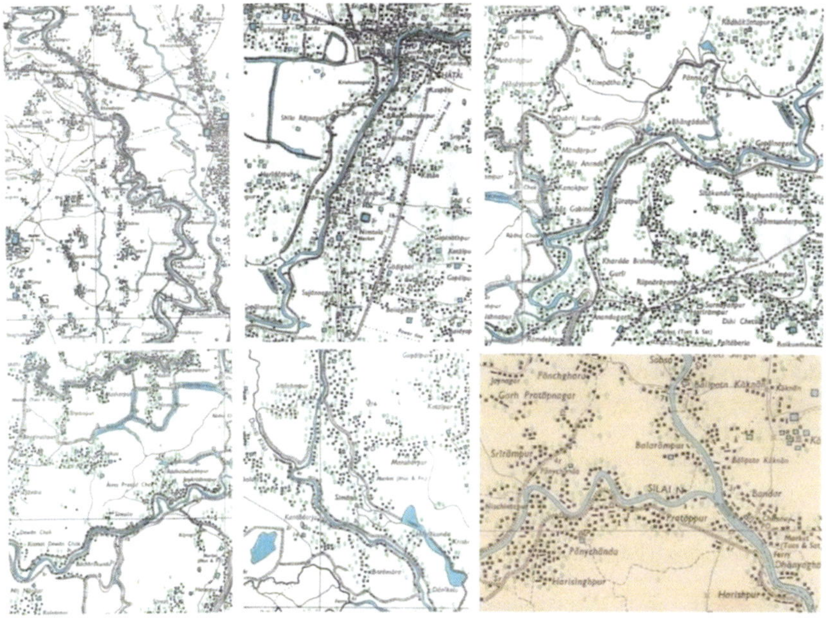

Fig. 8.13 Artificial embankment of Silabati River in the downstream area

During the late eighteenth century, Zamindars and British engineers began constructing embankments along the riverbanks. Willcocks described these embankments as "Satanic chains." Majumdar (1942, p. 19) argued that "constructing flood embankments as a flood control measure would be like mortgaging future generations to derive temporary benefits for the present generation." The river frequently causes flooding in its lower courses, especially in the Banka, Khirpai, and Ghatal regions. The construction of numerous weirs and barrages, such as the Kadamdeuli barrage in the upper reaches, has altered the river's natural flow. These interventions can cause localized flooding both upstream and downstream, disrupting ecosystems and communities. Additionally, poorly designed infrastructure exacerbates water logging in certain pockets of the basin and obstructs the natural flow of the river. The increasing demand for water—due to agriculture, industry, and domestic use—has led to over-extraction, particularly from the river's lower reaches. This has reduced water flow, especially during the non-monsoon period, disrupting the river's ecological balance and limiting water availability for various uses.

8.7.1 Impact of Dams

Rivers, as described by sixth-century BC Taoist engineer Chia Jang, are like the mouths of infants: "if one tried to stop them up they only yelled the louder or were suffocated" (Joseph Needham, *Science and Civilization of China*, 1971, quoted in McCully, 1998, p. 213). This metaphor encapsulates the natural resistance of rivers to being confined. In a similar vein, Tagore (2009, p. 200) reflects on the broader societal impact of such interventions, poignantly stating:

> *"Why did the river die? I made a dam*
> *to catch her,*
> *To have her for all the times.*
> *For this, the river died."*
> *Durakankha*, (Rabindranath Tagore, 2009, p. 200)

Both Needham and Tagore highlight the consequences of human attempts to control rivers. Needham suggests that rivers inevitably react when obstructed, while Tagore offers a deeper reflection, linking the act of damming a river to the broader consequences for human society. He symbolizes the river's death as the result of humanity's desire to control nature, an act that disrupts not only the river's flow but also the balance of the surrounding environment.

One of the most profound modifications to fluvial landscapes in the Anthropocene has been the widespread construction of dams, barrages, and canal networks. Dams fundamentally alter the use of rivers and natural resources, often resulting in the redistribution of benefits from local river users to new groups of beneficiaries at the regional or national level (WCD Report, 2000). By significantly altering natural flow patterns and impeding the movement of fish and other aquatic organisms, large dams can severely disrupt natural river production systems—particularly fisheries, flood-recession agriculture, and dry-season grazing (Richter et al., 2010).

Dams have far-reaching effects on flow regimes, channel morphology, sediment transport, and ecological processes. The construction of dams, locks, and weirs has severely disrupted the longitudinal connectivity of rivers. These anthropogenic changes have significantly altered river channel morphology (Surian, 1999). Reservoirs behind dams trap sediments, reduce flood peaks, and modify the seasonal distribution of flows. When sediment-free water—often referred to as "hungry water"—is released from a dam, the absence of sediment leads to an energy surplus. If this excess energy exceeds the resistance of the riverbed, it can result in bed erosion and channel incision (Kondolf, 1997). This process is most pronounced in areas with fine-grained substrates and where flood peak reductions are relatively minor.

The downstream impacts of both large and small dams have been extensively documented by scholars such as Petts (1979), Williams and Wolman (1984), Brandt (2000), Grant et al. (2003), Batalla et al. (2004), Magilligan and Nislow (2005), Graf (2005, 2006), Petts and Gurnell (2005), Schmidt and Wilcock (2008), Işik et al. (2008), Hupp et al. (2009), Sun et al. (2012), Dai and Liu (2013), Rubin et al. (2015), Han et al. (2020), among others. Similarly, the impacts on the upstream sections of rivers have been well documented by Evans et al. (2007) and Liro (2014). In addition, river channelization processes—including straightening, channel enlargement, erosion control, and embankment construction—further alter channel dynamics.

In the upstream parts of the Silabati River, a large number of earthen-boulder cross-dams have been constructed along the main channel by local peasants to regulate water for irrigation during the dry season. These structures, while beneficial for irrigation, reduce the navigability and longitudinal connectivity of the river. Moreover, they contribute to flooding during the rainy months (Fig. 8.14). In addition to these small-scale dams, there is a small pick-up barrage constructed across the river at Kadamdeuli in Bankura district, as part of the Kangsabati Reservoir Project. Completed in 1975, this barrage has significantly altered the river's flow. The catchment area of the river is 358 km^2 at the dam site, located in the Purulia and Bankura districts. The river descends from an elevation of about 198 m to about 83 m at the dam site, with an average slope of 1.75 m/km. The catchment area, lacking significant forest cover, experiences rapid runoff, leading to flash floods. During the monsoon, the river's discharge is poor between two floods. The absence of irrigation facilities has also hindered agricultural development in the area. The total submerged area of the Kadamdeuli dam and reservoir is 1,865.27 ha. (4,609.08 acres), located in the Indpur police station of Bankura district. A total of 5,422 individuals were displaced due to the construction of the

Silabati reservoir. The principal features of Kadamdeuli dam and reservoir are represented in Table 8.4.

The Silabati reservoir at Kadamdeuli benefits from inter-basin water transfer, realized through the Left Bank Feeder Canal of the Kangsabati River Valley Project. The Supur Main Canal, which has a water transfer capacity of 118.85 cumecs, is connected to the Kadamdeuli reservoir. Additionally, a canal from the Kadamdeuli reservoir extends into the Indpur Main Canal, which also has a capacity of 118.85 cumecs. The canal is bifurcated into the Simlapal Branch Canal (capacity: 11.54 cumecs) and the Bishnupur Branch Canal (capacity: 106.50 cumecs), which further divides into two smaller branches, the Ghatal Branch Canal (capacity: 38.14 cumecs) and the Kotulpur Branch Canal (capacity: 34.27 cumecs). However, seepage loss through these canals poses a significant challenge, reducing irrigation efficiency. Seepage rates vary depending on the area, with the western rocky area experiencing a loss of 2.94 cumecs/M.Sq.m, the laterite area losing 19.40, and the eastern alluvial area losing 2.44 (Government of West Bengal, 2007). Another constraint in the command area is the lack of proper water courses and field channels, which hampers the efficient delivery of water to the tail ends. Additionally, inadequate

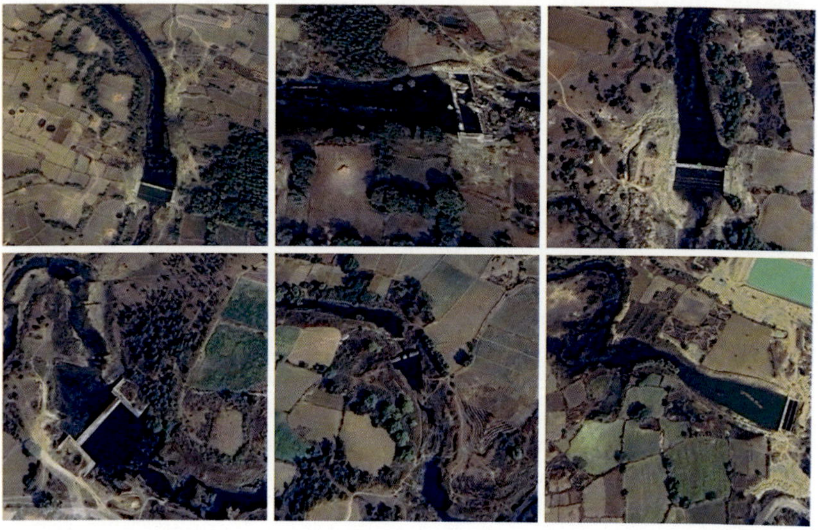

Fig. 8.14 Earthen-boulder cross dams in the upstream course of Silabati River

Table 8.4 Salient features of the Silabati Dam and Reservoir at Kadamdeuli

Feature	*Details*
Name of the river	Silabati
Spillway crest level	22.00 m G.T.S
Live storage	6,415.21 h.m
Maximum height of the dam	22.54 m G.T.S
Dead storage	2,165.04 h.m
Gross storage	8,580.25 h.m
Reservoir submergence at FRL	1,865.27 ha
Maximum flood discharge	5,072 cumecs
Deepest bed level	83.14 m G.T.S
F.R.L	102.00 m G.T.S
D.S.L	96.01 m G.T.S
Top of the dam	106.00 m G.T.S
Total length of earthen portion	3,765 m
Number of gates	8
Size of gates	11.5 m × 10.25 m
Length of link canal (lined)	16.10 km

drainage arrangements lead to water stagnation in low-lying paddy fields during the monsoon, damaging crops. In the *Rabi* season, the high porosity of the soil and water loss through seepage prevents water from reaching the tail-end fields.

8.7.2 *Impact of Gravel and Sand Extraction*

After freshwater, sand is the most consumed natural resource globally (Gavriletea, 2017; Rentier & Cammeraat, 2022; Villioth, 2014). Sand and gravel are essential for infrastructure development and construction. Global annual sand consumption is estimated at 15 billion tonnes, with a trade value of approximately 70 billion dollars (Villioth, 2014). This growing demand is driven by industrialization, urbanization, and population growth, particularly in developing countries such as China, India, and several regions in Africa. Sand is commonly extracted from active river channels and floodplain terraces using five primary methods: in-stream mining, wet pit mining, dry pit mining, bar excavation, and bar skimming (Kondolf, 1994; Rentier & Cammeraat, 2022). The ecological, infrastructural, and social impacts of sand and gravel extraction directly affect the livelihoods of nearly 3 billion people living along rivers (Bendixen et al., 2019). Large-scale river sand mining—estimated at around 50 billion tonnes annually—has severe consequences, including morphological alterations, water quality degradation, destruction of aquatic habitats, and depletion of groundwater (Tripathy et al., 2025).

According to Collins and Dunne (1990), gravel extraction from riverbeds can lead to bed degradation and altered river morphology. This degradation may reduce flood levels by lowering riverbed elevations, thereby limiting the delivery of overbank sediments to adjacent floodplains. Rapid degradation can also result in bank collapse and reduced groundwater levels, potentially destroying riparian vegetation. While there are some perceived benefits—such as reduced flood damage, improved navigation, and increased water availability for urban areas—these are often outweighed by negative consequences. These include increased irregularities in the riverbed slope, enhanced bank erosion, disrupted navigation during dry seasons, and saltwater intrusion in downstream regions (Luo et al., 2007). The scale and severity of these impacts depend on various factors, including extraction location and method, sediment type and availability, fluvial dynamics, flood risk, mining rate and intensity, and the sensitivity of local biodiversity and river-dependent communities (Harrison et al., 2005). Unregulated sand mining significantly alters the geomorphic character of river channels and valleys. In many regions, sand and gravel are being removed faster than natural replenishment rates (Bendixen et al., 2019). In-stream gravel mining disrupts the balance between sediment supply and the river's transport capacity, altering channel geometry and bed elevation (Kondolf, 1994). Mining pits in riverbeds extend degradation and aggradation processes both upstream and downstream. As water flows into these pits, the increased gradient intensifies stream power, triggering upstream erosion through head-cutting or knick point migration. Downstream, sediment-starved "hungry water" aggressively erodes riverbeds and banks (Kondolf, 1994, 1997), lowering channels, complicating drinking water extraction, and accelerating scouring that threatens infrastructure such as bridges and embankments. Channel incision contributes to overall instability, causing bank erosion, channel migration, and declining groundwater levels. This instability leads to the loss of riparian vegetation, disrupted ecosystems, degraded wildlife habitats, and increased water turbidity due to sediment suspension during mining activities. Sand plays a vital role as a permeable layer facilitating water infiltration and groundwater recharge. Its removal impedes vertical and lateral water movement, causing riverbeds to dry and groundwater levels to fall. Deep mining pits can significantly lower water tables, while reduced stream flow and groundwater availability collectively lead to declining vegetation and increased bank erosion, further promoting geomorphic and ecological degradation and biodiversity loss. Despite these environmental challenges, sand mining provides employment and contributes to local economies. However, it also displaces communities, harms fisheries and tourism, and frequently involves illegal and unsustainable practices.

In southern Bengal, particularly in rivers originating from the Chotanagpur Plateau, illegal sand mining is rampant due to high demand and profit potential. Many operators bypass regulations, resulting in considerable environmental degradation and governance issues. Gravel and sand extraction is a widespread practice in the Silabati River Basin. Gravel is primarily extracted from the basin's upper reaches, while sand mining dominates the middle reaches. Key sand mining zones include the CD Blocks of Garbeta-I, Garbeta-II, Chandrakona-I, Chandrakona-II, and Daspur-I (Table 8.5). These anthropogenic activities have significant long-term physical and social impacts. In the Silabati River, sand is often excavated from one side of the riverbank, compromising land stability (Fig. 8.15), leading to bank collapses and occasional landslides. The excavation and transport of sand also generate substantial dust pollution, posing health risks to workers and local communities. To mitigate environmental damage, sand should preferably be extracted from abandoned stream channels or inactive floodplains rather than from active river channels or floodplains. River diversions for mining purposes should be avoided. Furthermore, extraction should not occur near erosion-prone areas such as concave banks or within 200–500 metres of critical hydraulic infrastructure, including pumping stations, water intakes, and bridges. Mining activities must maintain the river's natural flow capacity, and excavation depths should not exceed three metres.

8.7.3 *Impact of Land Use and Land Cover Change (LU and LCC)*

Land is a product of nature and the foundation for the origin and spread of human civilization. It serves as the carrier and supplier of essential resources for human beings. In a broader context, land use refers to how land is utilized for various purposes to meet human needs (Mather, 1986). Land use (LU) and land cover change (LCC) are dynamic and ongoing processes, primarily driven by human activities. In recent decades, tropical regions have experienced some of the most

Table 8.5 Existing mining leases in the Silabati River Basin (Govt. of West Bengal, 2022)

CD Block	*Mouza*	*J.L. No*
CHANDRAKONA-I	Jara	152
	Kuldaha	284
	Manikkundu	151
CHANDRAKONA-II	Shirsa	73
	Dharmapota	74
	Chhotakna	122
	Ishnagar	45
	Panchami	39
	Dhaikhanda	132
GARHBETA-I	Bhattagram	399
	Kalikapur	900
	Raghunathpur	906
	Uchhaladanga	507
	Kankrasol	593
	Bara Panchrol	494
	Lakshyatapal	937
	Barai	905
	Gangani	485
	Bhalukmura	795
GARHBETA-II	Suniakon	406
	Kenja	164
	Joypur	203
	Kantore	204
	Kankdaha	209
	Bhagardanga	366
DASPUR-I	Kismatnarajol	16

rapid LU and LCC, with climate change (CC) further influencing hydrological and geomorphological processes in river systems (Kayitesi et al., 2022). As the human population continues to grow, the demand for land also increases. As a result, a predominant trend has been the conversion of forested areas into agricultural land and human settlements. These changes have significantly altered the components of the water balance, leading to variations in evaporation rates, peak flow intensities, flood frequencies, and river morphology. Human activities disrupt the natural rainfall-runoff patterns, altering the hydrological regime of watersheds. Additionally, they contribute to changes in river channel morphology through processes such as accretion and degradation. LU and LCC can substantially affect the hydro-geomorphologic characteristics of watersheds, influencing flood regimes, soil erosion rates, and sediment yields.

Fig. 8.15 Sand excavating parts of Silabati River

These impacts may lead to the formation of channels and gullies, the expansion of headwater streams, disruption of longitudinal connectivity, changes in flood levels, shifts in sediment transport, and periodic erosion events. Such geomorphologic changes further modify the physical characteristics of drainage basins. Moreover, land use changes impact the downstream transport of water, sediments, nutrients, and other materials. These changes, in turn, affect water quality, aquatic habitats, and the morphology of river channels and floodplains over both short and medium time scales (James & Lecce, 2013). Despite these effects, accurately quantifying the extent to which human activities have transformed river systems due to land use change remains a significant challenge.

The study area was once part of the *Jungle Mahal* region (Banerji, 1968; District Gazetteer, Puruliya, 1985). Historical information, data, and documents reveal a specific pattern of land utilization, river water management, and sedimented pockets from river systems that were once suitable for agriculture, forestry, and other economic activities. However, this ecological setup has been forced to change in later periods. In the Silabati Basin, land is used at both the macro and micro levels for various purposes such as cultivation, settlement, forestry, seasonal grazing, roads, and water bodies. At a more localized level, land is utilized for activities such as the extraction of stones, gravel, and sand from riverbeds, as well as for establishing brick kilns near river valleys

(Fig. 8.16), compost pits, threshing crops (*Khamar*), schools, community places (*Atchala*), temples, mosques, burial grounds, crematoriums, and playgrounds. In the upper and middle reaches of the basin, the topography is hilly, interspersed with hillocks and dissected by low ridges and valleys. Consequently, agricultural activities are confined to the lowland areas, settlements are located in midland areas, and arboreal forests, trees, and shrubs dominate the upland areas (Table 8.6). In contrast, downstream areas have developed built-up areas, particularly settlements and transportation lines, in the upland regions.

A comparative analysis of the Land Use and Land Cover (LU & LC) statistics of the Silabati River Basin is presented in Tables 8.7 and

Fig. 8.16 Brick kilns near Silabati River

Table 8.6 Interrelation between Land, Soil, and Vegetation in the Upstream Area

Serial number	*Land type*	*Textural grades of soil*	*Vegetation types*
1	High land (*Tanr*)	Light (*Kankars* mixed with rock fragments)	Covered by scrub jungle merging into the forests. Patches of grasses and shrubs grow in Rainy season
2	High land (*Baid*)	Light (sandy and sandy loams)	Isolated stands of *Tal* (Borassus flabellifera), *Khejur* (Phoenix sylevestris), *Ata* (Annonaceae), *Arjun* (Terminalia arjuna), *Siris* (Dalbergia lanceolaria), *Tentul* (Tamarindus indica)
3	Low land (*Sole/Bahal*)	Heavy (clay loam and clay soil)	Sparse vegetations of *babla* (Acacia arabica), *Neem* (Azadirachta indica), *Jam* (Syzygium cumini) and *Akanda* (Calotropis giganteum)
4	Transition land (*Kanali*)	Medium (loamy soil)	Covered with Savannah like widely spaced low tree of A*am* (Mangifera indica), *Jam* (Eugenia jambolana), *Kanthal* (Artocarpus integrifolia), *Bot* (Ficus benghalensis), *Aswatha* (Ficus religiosa), *Shimul* (Bombax malabaricum)

8.8, using data from 2011 and 2025 (Figs. 8.17 and 8.18). The analysis reveals a 6.65% decrease (equivalent to 259.29 sq. km) in vegetative area over the 14-year period. Water bodies in the basin remain minimal, accounting for only 1.23% in 2011 and declining to 0.32% in 2025. This represents a reduction of 2.38% (93.24 sq. km). Additionally, the area of riverside bare sand has decreased by 0.51% (19.89 sq. km). Conversely, there has been a 9.55% increase (72.42 sq. km) in current fallow land, including built-up areas. Historically, a substantial portion of the river basin—particularly the upper basin—was covered by forests. However, the expansion of agricultural activities and settlement development over time has significantly reduced both forest cover and water bodies. The growing population has contributed to the expansion of built-up areas. The loss of forest cover has also led to increased sediment deposition in the riverbeds, thereby reducing the rivers' carrying capacity. This diminished capacity often results in overflows during periods of high discharge, leading to more frequent flood events.

Table 8.7 Land use land cover status of Silabati River Basin, 2011

Land use/land cover	*Area (sq.km.)*	*% of Area*
Vegetation	1586.63	40.71
Water bodies	105.38	2.70
Riverside bare sand	48.1	1.23
Current fallow land (including built-up area)	2156.89	55.35

Table 8.8 Land use land cover status of Silabati River Basin, 2025

Land use/land cover	*Area (sq.km.)*	*% of Area*
Vegetation	1327.34	34.06
Water bodies	12.14	0.32
Riverside bare sand	28.21	0.72
Current fallow land (including built-up area)	2529.31	64.90

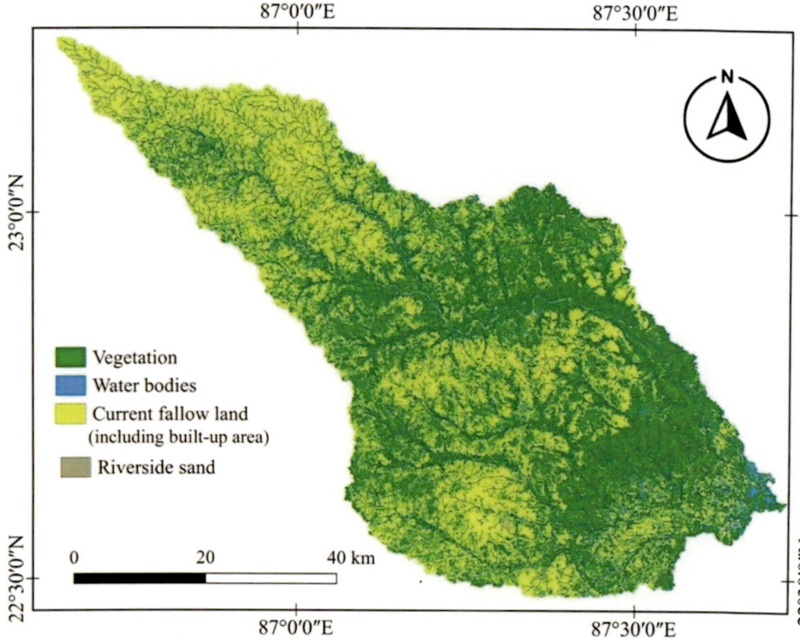

Fig. 8.17 Land use land cover of Silabati River Basin, 2011 (*Source* Landsat-5 (07 May 2011))

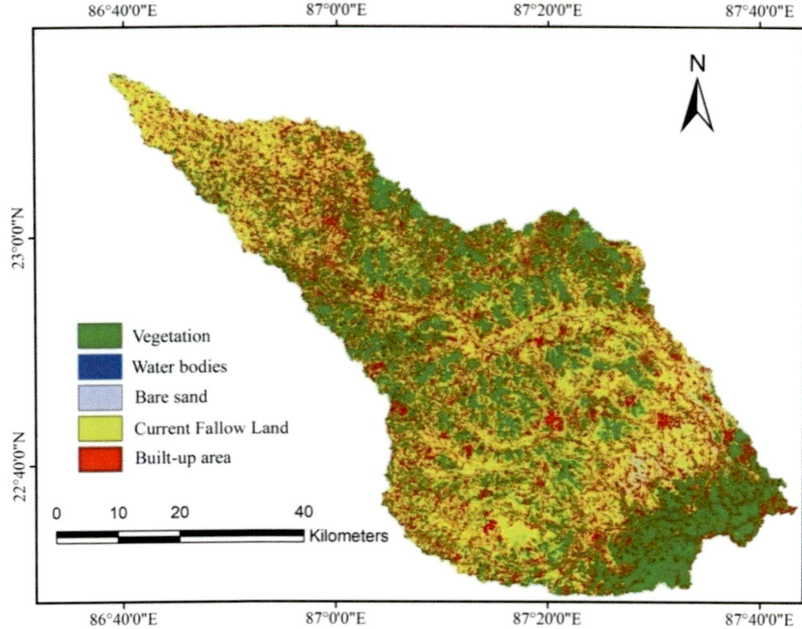

Fig. 8.18 Land use land cover of Silabati River basin, 2025 (*Source* LANDSAT 8 (26 March 2025))

8.8 Restoring the Sustainability of the Silabati River

Man, if he wishes to control nature, must to a great extent follow her, for she has her own wisdom. If he seriously disturbs the balance and rhythm in which nature delights, her vengeance often follows quickly, and man has no escape.

—(Radhakamal Mukherjee, 2009, p. 12)

Supersession of nature may be tolerable to a certain extent, thereafter comes the phase of destruction ... it is a difficult problem of modern civilization that how human can make a consistency in the livelihood process in his self-composed gigantic complex artificial system by ignoring the simple sanitation and curable system that exist within the natural system

—*Upekshita Pally,* (Rabindranath Tagore, 1961, pp. 528–529)

The above quotes depict a complex ambivalent relationship between nature and humanity. The underlying message of these quotes is that humans must respect and abide by the laws of nature in all their developmental endeavors. If humans consistently pursue a development process that violates natural limits, then humans will inevitably face nature's revenge, which will ultimately be harmful to humanity. In the first quote, Mukherjee suggests that the balance and rhythm of nature are fundamental to our survival. Any disruption of this balance brings consequences that cannot be avoided. This resonates with the larger philosophical belief that human activities should be in harmony with the environment in order to maintain long-term prosperity. Tagore's second quote deepens the idea that while temporary advances in technological or artificial systems may seem tolerable, they ultimately lead to destruction when disconnected from the underlying order of nature. Essentially, these two perspectives highlight the need for a balanced approach to human development - one that is consistent with the well-being of people and the environment without ignoring the inherent wisdom of nature.

Restoring the sustainability of the Silabati River requires a multifaceted approach, combining regulatory measures, community engagement, and innovative restoration techniques. Key actions to mitigate deforestation and soil erosion include promoting afforestation and reforestation in the middle and upper catchment areas of the river basin and along both banks of the rivers and canal networks.

The Silabati River flows slowly from its source to its mouth, leading to water shortages in the upstream areas and water abundance in the downstream regions. Therefore, irrigation is needed where there is water scarcity, while a drainage system is required in areas with water surplus. In the upstream areas of the river, irrigation systems are almost non-existent or very limited. During the rainy season, the river's flow increases rapidly, causing it to overflow its banks in the downstream areas. Due to the river's natural flow and heavy rainfall, there is an excess of water in the lower reaches, but no effective drainage system is in place. The only flood control system currently available is the embankment.

The topography of the Silabati River basin in the upstream section is predominantly wavy, with lowlands on both sides of the river, which are ideal for agriculture. These lowlands are known locally as *sole* or *bahal* land. Over time, agricultural activities have expanded upslope, creating terraced lands, referred to as *Kanali* and *Baid* lands. *Tanr* lands, located at higher elevations, cover extensive areas. The basin showcases a symbiotic relationship between topography and vegetation, resulting in three distinct vegetation landscapes:

1. Cultivated lands on the more watered plains.
2. Savannah-like landscapes with widely spaced trees, grasses, and shrubs on undulating terrain.
3. Scrub jungles that gradually transition into forests in the uplands.

Although the original forests have been cleared, secondary dry deciduous forests are regenerating in the uplands. Many species from the original forests persist in the form of remnants or isolated stands (Fig. 8.19). The existing forest cover has largely been replaced by planted *Eucalyptus* (Eucalyptus globules) and *Akasmoni* (Acacia auriculiformis) trees (Fig. 8.20). Other important plant species, such as *Bel* (Aegle marmelos), *Dumur* (Ficus hispida), *Pipal* (Ficus religiosa), and *Banyan* (Ficus benghalensis), are found in scattered patches, while *Kadam* (Anthocephalus cadamba), *Jackfruit* (Artocarpus integrifolia), and *Arjun* (Terminalia arjuna) are common along the riverbanks. The scrub jungle of the uplands includes species such as *Kurchi* (Holarrhena antidysenterica), *Asan* (Terminalia tomentosa), and a variety of medicinal plants like *Tulsi* (Ocimum tenuiflorum), *Pudina* (Mentha spicata), and *Lavender* (Lavandula). Other plants, such as *Bans* (Dendrocalamus strictus), *Date Palm* (Phoenix sylvestris), and *Tal* (Borassus flabellifera), grow in *tanr* lands (Fig. 8.21).

The combination of alternating dry and wet seasons, shallow soil depth, red and lateritic soil, and the presence of *kankars* (hard limestone nodules) in the soil presents significant challenges for effective forest management. Encouraging the planting of native tree species, such as *Babla* (Acacia nilotica), *Neem* (Azadirachta indica), *Aam* (Mangifera indica), *Jam* (Syzygium cumini), *Kanthal* (Artocarpus heterophyllus), *Tentul* (Tamarindus indica), *Sal* (Shorea robusta), *Shegun* (Tectona grandis), *Sishu* (Dalbergia sissoo), *Mahua* (Madhuca longifolia), and *Siris* (Albizia lebbeck), will help stabilize the soil, reduce runoff, and improve water retention.

A healthy river ecosystem relies heavily on the health of its watershed. Restoring the watershed with forests, wetlands, and healthy soils minimizes the occurrence of devastating floods, reduces the risk of droughts,

Fig. 8.19 Isolated stands of original forest

mitigates soil erosion, and reduces the sediment entering the riverbed. It also enhances the river's ability to filter pollutants and provides diverse wildlife habitats (McCully, 1998). Forest cover and soils act as sponges, absorbing rainfall and gradually releasing water into the river or allowing it to percolate into aquifers. Forests slow the flow of surface water and can even increase regional rainfall.

Fig. 8.20 Planted trees of eucalyptus and *Akasmoni*

Common Property Resources (CPRs) such as land, water, and forests provide sustainable livelihoods for many people in the Silabati River Basin. However, due to over-exploitation and poor management, these resources have rapidly degraded. Integrated watershed management practices are crucial for the sustainable management of land, water, and vegetation in the basin. These practices include ecological, economic, and social components (Roy, 2005). By implementing soil and water conservation measures, promoting forestry development, and enhancing agriculture, livestock, and horticulture—surface runoff, soil erosion, and sedimentation of the riverbed can be reduced.

Fig. 8.21 Important plant species in Silabati basin

The Silabati River is primarily rain-fed, and its flow is mostly seasonal. The agro-climatic conditions in the upstream part of the basin are well-suited for plantations and horticultures. Productions of *mango, guava, cashew nut, jackfruit, banana, mosambi, bel, papaya, citrus fruits,* as well as medicinal and aromatic plants are well suited in such agro-climatic condition. Commercial cultivation of medicinal plants like *Sarpgandha, Vasaka, Kalmegh, Safed Musli, Stevia,* and *Nayantara* may be explored. Minor fruits such as *amla, custard apple, karonda, sapota, tamarind, bel, jam,* and *drumstick* also have significant potential under these agro-climatic conditions.

The dry climate of the basin is also conducive to animal husbandry and seed production. Certain parts of the basin have land use patterns and climatic conditions suitable for *mulberry* and A*rjun* plantations, supporting silkworm rearing. Both mulberry-based sericulture and traditional *tasar* culture hold considerable scope for further development in the upper and middle part of the basin. The open-field cultivation of flowers such as *marigold*, *tuberose*, and *jasmine* can significantly enhance the local farming landscape. Activities based on *date palm* and *palmyra palm*, aimed at sustainable employment through value addition and product diversification, offer promising opportunities. Additionally, the upstream parts are rich in *Polash* (Butea monosperma) and *Kul* (Ziziphus mauritiana) trees. *Polash* and *Kul* trees are especially suitable for *lac* cultivation. The local villagers possess traditional skills in *lac* cultivation. Formation of Self-Help Groups (SHGs) could facilitate the promotion and scaling up of this activity.

To reduce soil erosion, methods such as contour farming and the construction of check dams should be implemented. During the monsoon, large volumes of loose soil enter the river due to rill and gully erosion. Constructing check dams at the confluence points of first-order and second-order streams can prevent excessive soil loss, increase irrigation potential, and promote groundwater recharge. In the upstream, several dams have been constructed in the main channel of the Silabati River. As a result, the connectivity of the main river has been affected. The main flow of the river should be kept active without dams.

Afforestation and reforestation with indigenous tree species above these check dams will further mitigate soil erosion. Key species to be planted include *Sal* (Shorea robusta), *Polash* (Butea monosperma), *Kendu* (Diospyros melanoxylon), *Shegun* (Tectona grandis), *Sishu* (Dalbergia sissoo), *Mahua* (Madhuca indica), *Gamhar* (Gmelina arborea), *Haritaki* (Terminalia chebula), *Bahera* (Terminalia bellirica) and *Bhalai* (Semecarpus anacardium) on the *tanr* land. *Neem* (Azadirachta indica), *Arjun* (Terminalia arjuna), *Shimul* (Bombax ceiba), *Babla* (Acacia nilotica), and *Siris* (Albizia lebbeck) would be planted on the walls of agricultural plots. Additionally, farmers should be encouraged to adopt sustainable farming practices such as agroforestry and organic farming, which reduce reliance on chemical fertilizers and pesticides. By increasing cropping intensity, green fodder production, and milk yield, the economic sustainability of communities in the upper and middle reaches of the Silabati basin will be improved, and at the same time, the health status of the catchment area will also be healthy.

Rainwater harvesting techniques, such as pond excavation and recharge wells, can alleviate pressure on the river during dry seasons, helping maintain a consistent flow throughout the year. The proper maintenance of infrastructure, such as the Kadamdeuli barrage, is also critical. Regular repairs and the installation of effective water regulation mechanisms will help mitigate downstream flooding and maintain the river's ecological health. The success of the integrated watershed management program depends on local people's participation through a participatory approach, ensuring equitable resource distribution and empowering marginalized groups, including women.

While there are no major industrial hubs or large towns along the Silabati River, small towns such as Molian, Lalbazar, Simlapal, Garbeta, and Ghatal, along with tourist spots like Gangani (famous for ravine topography), are located near the riverbanks. These areas face water quality issues, particularly from industrial discharge, sewage, and agricultural runoff. Stringent regulations should be enforced to ensure wastewater treatment in industrial and urban areas. Establishing sewage and effluent treatment plants, along with improved waste management systems, will significantly enhance water quality in the river.

The existence of swamps, marshes, and other water bodies, such as paleo-channels, lakes, and cut-off channels in the lower courses of the river, helps maintain its health by naturally filtering pollutants from sewage and agricultural runoff. These natural water bodies also provide habitats for fish, aquatic animals, and riverine organisms, reducing the intensity and severity of flooding. Certain open lands must be reserved for flooding. Restoring the natural floodplains in the lower basin and improving flood management systems will help reduce flood frequency and severity. Creating flood retention basins and implementing sustainable land-use planning to avoid construction in flood-prone areas will mitigate the adverse impacts of flooding on local communities. The restoration of paleo-channels and wetlands in the river's lower course will help absorb excess floodwater, filter pollutants, and prevent flood damage.

In the downstream section, the Silabati River naturally flows through the floodplain formed by its flow over many years, moving at a slow pace. This alluvial soil is fertile, and water is easily retained in it, allowing for cultivation. Water accumulates beneath the soil, seeping from above. In this area, canals can be easily cut for water distribution. Even in the event of reduced rainfall, the irrigation system is less likely to be disrupted. Flooding is a natural occurrence of the river's cycle, not a disaster (Biswas, 2002). Floodplains are formed by the river itself over many years, as it floods and then empties. If left uncontrolled, the river will reach a senile stage, but it will not become stagnant. However, if the main river channel is artificially controlled with dams and embankments, as an imprudent person might do, it could become stagnant, and its plains could become barren land.

The Silabati River basin lies in the monsoon climate zone, where rainfall is often uncertain, sudden, and intense. When the upstream areas are washed away rapidly, the sediment load in the river increases, and large amounts of sediment accumulate in the downstream valleys, causing floods. Erosion and deposition are closely connected. Soil without trees cannot retain water, leading to erosion, which intensifies the water's force downstream. Stopping abnormal erosion will prevent excessive sediment accumulation and flooding. Therefore, afforestation and proper forest conservation upstream of the river are crucial. Additionally, preventing riverbank erosion by planting trees, maintaining adequate water flow during normal rainfall, and allowing natural flooding will help maintain a balanced ecosystem.

A comprehensive monitoring system is needed to track the river's health and its catchment area. Regular data collection on water quality, sediment load, and biodiversity will inform future management decisions. Establishing a robust ecological and hydrological database will provide critical insights for restoration and management strategies.

8.9 Conclusion

To ensure the long-term sustainability of the Silabati River, a combination of afforestation, sustainable agricultural practices, erosion control, regulated sand extraction, pollution management, floodplain restoration, and continuous scientific monitoring must be implemented. These strategies will not only enhance the well-being of local communities but also restore the river's ecological balance, securing its role as a vital resource for future generations. However, the success of these initiatives depends on the collective sense of responsibility and involvement of the people.

Acknowledgements The authors are grateful to the editor of the book and to the anonymous reviewer for helpful and critical comments in revision of this essay. The writers are also grateful to Dr. Uttam Kumar Hazra, Librarian, Vivekananda Mahavidyalaya, who provided technical support.

References

Bagchi, K. (1944). *The Ganges Delta*. University of Calcutta.

Bandopadhyay, D. K. (2007). *Banglar Nadnadi*. Dey's Publishing.

Banerji, A. K. (1968). *West Bengal District Gazetteers: Bankura*. Govt. of West Bengal.

Basu, A. K. (2002). *Paschimbanger Nad Nadi*. Mitra and Ghosh Publishers.

Batalla, R. J., Gomez, C. M., & Kondolf, G. M. (2004). Reservoir-induced hydrological changes in the Ebro River basin (NE Spain). *Journal of Hydrology, 290*, 117–136.

Bendixen, M., Best, J., Hackney, C., & Iversen, L. L. (2019). Time is running out for sand. *Nature, 571*, 29–31.

Bhandari, U., & Roy, M. (2018). Assessment of GIS and remote sensing- based flood vulnerability of Shilabati River Basin using AHP method. *International Journal of Creative Research Thoughts, 6*(2), 311–327.

Bharati, R. (2009) (Translated by Sandeep Sinha). *Rivers of India*. New Delhi: National Book Trust, India.

Biswas, A. (2002).Banyake Banchiye Rakha: Banya Niye Benche Thaka (in Bengali). In R. Roy (Ed.). *Pratiti-Arabinda Biswas Smaranika* (pp.48–57). Unique Phototype.

Brandt, S. A. (2000). Classification of geomorphological effects downstream of dams. *CATENA, 40*, 375–401.

Charlton, R. (2008). *Fundamentals of fluvial geomorphology*. Routledge.

Collins, B., & Dunne, T. (1990). *Fluvial geomorphology and river-gravel mining: A guide for planners, case studies included*. California Department of Conservation-Division of mines and Geology. https://ia600203.us.archive.org/5/items/fluvialgeomorpho98coll/fluvialgeomorpho98coll.pdf

Crutzen, P. J. (2002). Geology of mankind. *Nature, 415*(23).

Crutzen, P. J., & Stoermer, E. F. (2000, May). The "Anthropocene". *IGBP Newsletter 41*.

Dai, Z., & Liu, J. T. (2013). Impacts of large dams on downstream fluvial sedimentation: An example of the Three Gorges Dam [TGD] on the Changjiang [Yangtze River]. *Journal of Hydrology, 480*, 10–18.

Das, G. K. (2022). Shilabati river: Its environment. *Indian Science Cruiser, 36*(5), 10–15.

Das, U., Bajpai, R., & Chakrabortty, D. (2020). River regulation and associated geo-environmental problems: A case study of lower reaches of shilabati river basin, West Bengal, India. *International Journal of Ecology and Environmental Sciences, 2*(4), 233–240.

Dolui, G., & Ghosh, S. (2013). Flood and its effects: A case study of Ghatal Block, Paschim Medinipur, West Bengal. *International Journal of Scientific Research, 2*(11), 248–252.

Douglas, I., & Lawson, N. (2000). The Human Dimensions of Geomorphological work in Britain. *Journal of Industrial Ecology, 4*(2), 9–33. https://doi.org/10.1162/108819800569771

Evans, J. E., Huxley, J. M., & Vincent R. K. (2007). upstream channel changes following dam construction and removal using a GIS/ remote sensing approach. *Journal of the American Water Resources Association, (JAWRA) 43*(3), 1–15. https://doi.org/10.1111/j.1752-1688.2007.00055.x

Gabler, R. E., Petersen, J. F., Trapasso, L. M., & Sack, D. (2009). *Physical geography*. Brooks/Cole.

Gavriletea, M. D. (2017). Environmental impacts of sand explitation. Analysis of sand market. *Sustainability, 9*, [Article 1118], 1–26.

Govt. of India: India Meteorological Department. (2008).Climate of West Bengal. New Delhi: Controller of Publications. Retrieved May26, 2025, https://www.imdpune.gov.in/library/public/Climate%20of%20WestBengal.pdf

Govt. of India: India Meteorological Department. (2024). Rainfall Statistics of India (2012–2025). New Delhi: Ministry of Earth Sciences. Retrieved on May26, 2025, https://hydro.imd.gov.in/hydrometweb/(S(ib0yuoyvc4euk145ncm3yhyf))/landing.aspx

Government of West Bengal. (1985). *West Bengal District Gazetteers: Puruliya.*

Government of West Bengal. (2007). *Kangsabati Command Area Development Authority* [Annual Report; 2004–05].

Govt. of West Bengal. (2022). *District Survey Report, Paschim Medinipur District (For mining of minor minerals).* Retrieved May 2, 2025, http://www.dmm.gov.in/pdfs/DSR/DSR_of_Paschim_Medinipur.pdf

Grade, R. J. (2006). *River morphology.* New Age International [p] Ltd.

Graf, W. L. (2005). Geomorphology and American dams: The scientific, social, and economic context. *Geomorphology, 71*, 3–26.

Graf, W. L. (2006). Downstream hydrologic and geomorphic effects of large dams on American rivers. *Geomorphology, 79*, 336–360.

Grant, G. E., Schmidt, J. C., & Lewis, S. H. (2003). A geological framework for interpreting downstream effects of dams on rivers. *Americian Geophysical Union*, 209–225.

Gregory, K. J. (2006). The human role in changing river channels. *Geomorphology, 79*, 172–191.

Halder, S., Biswas Roy, M., & Roy, P. K. (2020). Analysis of groundwater level trend and groundwater drought using Standard Groundwater Level Index: A case study of an eastern river basin of West Bengal, India. *SN Applied Sciences [Springer Nature Switzerland AG], 2*(507), 1–24. Retrieved April 28,2025,https://link.springer.com/content/pdf/https://doi.org/10.1007/s42452-020-2302-6.pdf

Han, M., Brierley, G., Li, B., Li, Z., & Li, X. (2020). Impacts of flow regulation on geomorphic adjustment and riparian vegetation succession along an anabranching reach of the Upper Yellow River. *CATENA, 190*, 1–12.

Harrison, D. J., Fidgett, S., Scott, P. M., MacFarlane, M., Mitchell, P., Eyre, J. M., & Weeks, J. M. (2005). Sustainable River mining of aggregates in developing countries. In Marker et al. (Eds.). *Sustainable minerals operations in the developing world*. Geological Society, London, Special Publications, 250, 35–46.

Hupp, C. R., Pierce, A. R., & Noe, G. B. (2009). Floodplain geomorphic processes and environmental impacts of human alteration along coastal plain rivers, USA. *Wetlands, 29*(2), 413–429.

Isik, S., Dogan, E., Kalin, L., Sasal, M., & Agiralioglu, N. (2008). Effects of anthropogenic activities on the Lower Sakarya River. *CATENA, 75*, 172–181.

James, L. A., & Lecce, S. A. (2013). Impacts of land-use and land-cover change on river systems. In J. F. Shroder (Ed.), *Treatise on geomorphology* (Vol. 9, pp. 768–793). Academic Press.

Kayitesi, N. M., Guzha, A. C., & Mariethoz, G. (2022). Impacts of land use land cover change and climate change on river hydro-morphology—A review of research studies in tropical regions. *Journal of Hydrology, 615*, 1–14.

Knighton, D. (1998). *Fluvial forms and processes: A New Perspective*. Arnold.

Kondolf, G. M. (1994). Geomorphic and environmental effects of instream gravel mining. *Landscape and Urban Planning, 28*, 225–243.

Kondolf, G. M. (1997). Hungry water: Effects of dams and gravel mining on river channels. *Environmental Management, 21*(4), 533–551. https://people.wou.edu/~taylors/g407/kondolf_97.pdf

Liro, M. (2014). Conceptual model for assessing the channel changes upstream from Dam reservoir. *Quaestiones Geographicae, 33*(1), 61–74.

Luo, X.-L., Zeng, E. Y., Ji, R.-Y., & Wang, C.-P. (2007). Effects of in-channel sand excavation on the hydrology of the Pearl River Delta, China. *Journal of Hydrology, 343*, 230–239.

Magilligan, F. J., & Nislow, K. H. (2005). Changes in Hydrologic regime by dams. *Geomorphology, 71*, 61–78.

Mahala, A., et al. (2020). Land degradation processes of Silabati River Basin, West Bengal, India: A physical perspective. In P. K. Shit (Ed.), *Gully Erosion studies from India and surrounding regions, advances in science, technology & innovation* (pp. 265–278). Springer Nature Switzerland AG.

Majumdar, S. C. (1942). *Rivers of the Bengal Delta*. University of Calcutta.

Mandal, D., & Kayal, P. (2023).A hydrogeometric assessment of the Silabati River system in West Bengal, India: A remote sensing and GIS-based approach. *Indian Journal of Spatial Science, 14*(4), 71–82.

Marsh, G. P. (1864). *Man and nature*. The Belknap Press.

Mather, A. S. (1986). *Land use*. Longman.

McCully, P. (1998). *Silenced rivers: The ecology and politics of large dams*. Orient Longman Ltd.
Mukherjee, R. K. (1938). *The changing face of Bengal: A study in riverine economy*. University of Calcutta.
O'Malley, L. S. S. (1995) [Reprint]. Bengal district Gazetteers: Bankura. Calcutta: The Bengal Secretariat Book Depot.
Patel, P. R., Mondal, S., & Prasad, R.(2021). Modifications of the geomorphic diversity by anthropogenic interventions in the Silabati River Basin. In B. C. Das, et al. (Eds.), *Anthropogeomorphology of Bhagirathi-Hooghly river system in India* (Chapter-10, pp. 331–356). Taylor &Francis Group, LLC.
Petts, G. E. (1979). Complex response of river channel morphology subsequent to reservoir construction. *Progress in Physical Geography: Earth and Environment, 3*(3), 329–362. https://doi.org/10.1177/030913337900300302
Petts, G. E., & Gurnell, A. M. (2005). Dams and geomorphology: Research progress and future directions. *Geomorphology, 71*, 27–47.
Price, S. J., Ford, J. R., Cooper, A. H., & Neal, C. (2011). Humans as major geological and geomorphological agents in the Anthropocene: The significance of artificial ground in Great Britain. *Philosophical Transactions of the Royal Society A, 369*, 1056–1084.
Rentier, E. S., & Cammeraat, L. H. (2022). The environmental impacts of river sand mining. *Science of the Total Environment, 838*(1), [Article 155877], 1–8.
Richter, B. D., Postel, S., Revenga, C., Scudder, T., Lehner, B., Churchill, A., & Chow, M. (2010). Lost in development's shadow: The downstream human consequences of dams. *Water Alternatives, 3*(2), 14–42.
Roy, U. N. (2005). *People's participation in watershed management*. Kanishka Publishers.
Rubin, Z. K., Kondolf, G. M., & Carling, A. P. (2015). Anticipated geomorphic impacts from Mekong basin dam construction. *International Journal of River Basin Management, 13*(1), 105–121.
Rudra, K. (2016). *State of India's Rivers for India Rivers Week, 2016, West Bengal*. Retrieved April 28, 2025, https://sandrp.in/wp-content/uploads/2017/03/west-bengal.pdf
Schmidt, J., & Wilcock, P. R. (2008). Metrics for assessing the downstream effects of dams. *Water Resources Research, 44*, 1–19.
Schwagerl, C. (2014). *The Anthropocene: The human era and how it shapes our planet*. Synergetic Press.
Singh, S. (2019). Changing perspectives in geomorphology. *Journal of Indian Geomorphology, 7*, 1–11.
Shit, P. K., Bhunia, G. S., & Maiti, R. (2013). Assessment of factors affecting ephemeral gully development in Badland Topography: A case study at Garbheta Badland (Paschim Medinipur, West Bengal, India). *International Journal of Geosciences, 4*, 461–470.

Stoppani, A. (1873).*Corso di Geologia*. Milano.

Sun, Z., Huang, Q., Opp, C., Hennig, T., & Marold, U. (2012). Impacts and implications of major changes caused by the Three Gorges Dam in the Middle Reaches of the Yangtze River, China. *Water Resources Management*. Retrieved March 27, 2025, https://doi.org/10.1007/s11269-012-0076-3

Surian, N. (1999). Channel changes due to River regulation: The case of the Piave River, Italy. *Earth Surface Processes and Landforms, 24*, 1135–1151.

Tagore, R. (2009). *Rabindra Rachanaboli*, Vol-II. Visva-Bharati Granthan Bivag.

Tagore, R. (1961). *Rabindra Rachanaboli*, Vol-XIII. Visva-Bharati Granthan Bivag.

Tagore, R. (1968). *Chhinapatra*. Biswabharati Granthan Bivag.

Tagore, R. (2016). *Bhanu Singher Patraboli*. Visva-Bharati Granthan Bivag.

Tennyson, A. L. (1842). *The Brook*. Retrieved April 28, 2025 https://tennysoncountry.com/2020/04/21/the-brook/

Tripathy, I. M., Mahto, S. S., Bhagat, C., Modi, A., Jain, V., & Mohapatra, P. K. (2025). A review of river sand mining: Methods, impacts, and implications. *Next Research, 2*, 1–20.

Villioth, J. (2014). *Building an economy on quicksand*. Retrieved April 22,2 025, http://www.ejolt.org/2014/08/building-an-economy-on-quicksand/

Willcocks, W. (1930). *Ancient system of irrigation in Bengal*. University of Calcutta.

Williams, G. P., & Wolman, M. G. (1984). *Downstream effects of dams on alluvial rivers*. U. S. Geological Survey Professional Paper, vol. 1286. U.S. Government Printing Office, Washington, DC.

Wohl, E. (2014). *Rivers in the landscape: Science and management*. John Wiley & Sons. Ltd.

World Commission on Dams. (2000). *Dams and development: A new framework for decision-making*. Earthscan Publications Ltd.

CHAPTER 9

Anthropogenic Impacts on the Banka River: Lead Bioaccumulation in *Filopaludina bengalensis* and Associated Health Risks Assessment

Ayan Mondal, Surajit Ghosh, Ahamadul Hoque Mandal, Santu Ghosh, Kishore Dhara, Nimai Chandra Saha, and Shubhajit Saha

9.1 Introduction

All life forms depend on water for survival, but anthropogenic activities have significantly impacted freshwater ecosystems through industrial discharge and heavy metal contamination (Oladimeji et al., 2024). Lead

A. Mondal
Department of Zoology, Government General Degree College, Mohanpur, India

S. Ghosh · A. H. Mandal · S. Saha (✉)
Ecotoxicology Laboratory, Department of Zoology, The University of Burdwan,

S. Ghosh (ed.), *Reimagining Indian Rivers for Sustainability*, Palgrave Studies in Environmental Sustainability,
https://doi.org/10.1007/978-3-032-04754-0_9

(Pb) pollution has emerged as a critical environmental concern, particularly in regions with extensive industrialization (Levin et al., 2021; Zhang et al., 2024). The Banka River in Purba Bardhaman district, India, serves as an intricate part for local low-income communities, supporting both biodiversity (Mondal et al., 2020) and human livelihoods (Ghosh et al., 2022; Sharma et al., 2021). However, the discharge of industrial effluents has led to increased levels of lead in the aquatic environment, raising concerns about bioaccumulation in freshwater organisms and subsequent human exposure (Dey et al., 2025; Fulke et al., 2024). *Filopaludina bengalensis*, a gastropod species commonly found in the Banka River, is widely consumed due to its high nutritional value (Dhara et al., 2022b). As a benthic organism, it is highly susceptible to heavy metal accumulation from sediment and water, making it an effective bioindicator of lead contamination (Dhara et al., 2022b). Heavy metal pollution in freshwater systems can lead to toxic effects in aquatic organisms, with potential consequences for entire ecosystems (Dhara et al., 2022a). The pathways and rates of lead uptake in freshwater gastropods remain inadequately understood, yet their bioaccumulation potential poses a significant risk to human health through dietary exposure. Prolonged exposure to lead can cause neurological, renal, and cardiovascular disorders in humans, necessitating a thorough investigation into its accumulation and potential health risks (Collin et al., 2022a).

The present study aims to assess the extent of lead bioaccumulation in both external and internal organs of *F. bengalensis* and its associated health risks for human consumers. Field samples of gastropod tissue, sediment, and water were collected for laboratory analysis, while environmental parameters such as pH, dissolved oxygen, and temperature were

---

Bardhaman, India
e-mail: ssaha@zoo.buruniv.ac.in

S. Ghosh
Publication Unit, Registrar Section, The University of Burdwan, Bardhaman, India

K. Dhara
Office of the Dy. Director of Fisheries, Gour Banga Zone, Malda, India

N. C. Saha
Department of Zoology, Bidhannagar College, Kolkata, India

measured in situ. Additionally, a survey was conducted among local populations reliant on the Banka River to determine the potential health effects of lead exposure (Ghosh et al., 2020). The findings suggest that waterborne lead plays a significant role in the bioaccumulation process, with individuals above 40 years of age and those from lower-income groups facing the highest health risks. A high hazard quotient (HQ) value was observed among frequent consumers, indicating a potential risk of chronic lead toxicity. To mitigate the adverse effects of lead pollution, comprehensive risk assessment models are required, integrating biokinetic studies and dietary exposure evaluations. Understanding the bioaccumulation dynamics of lead in freshwater organisms, along with its potential for biomagnification, is essential for developing pollution control strategies and safeguarding public health. Various technologies such as adsorption, chemical precipitation, and bioremediation may be proposed to remove heavy metals from wastewater, but their effectiveness in real-world scenarios needs further evaluation (Ghosh et al., 2020; Niloy et al., 2024). The role of policy measures in regulating industrial discharge and improving wastewater treatment infrastructure is crucial in addressing this growing environmental threat (Ghosh et al., 2024, 2025a, b; Singh et al., 2023).

This study underscores the urgent need for environmental monitoring, regulatory measures, and community awareness to address lead contamination in freshwater ecosystems. By evaluating the pathways of lead accumulation and its effects on both aquatic life and human populations, this research contributes to the broader understanding of heavy metal pollution and its long-term implications (Raj & Das, 2023). Future studies should focus on developing sustainable remediation techniques and enhancing public health strategies to reduce lead exposure in vulnerable communities.

9.2 Material and Methodology

Filopaludina bengalensis specimens were collected biweekly from nine different sampling sites of the Banka River, which carries overflow from these ponds, and is significantly affected by urban runoff, accumulating large amounts of loose soil, plastic waste, household refuse, industrial byproducts from rice mills, and even pathological waste. Slum-dwelling communities in nearby wards (19, 21, and 35) depend on this river for daily activities, making pollution a major concern (Ghosh & Maji, 2011).

Water samples were collected in 1-liter high-density polyethylene (HDPE) bottles, pre-washed with 10% hydrochloric acid (HCl), and deionized water to ensure sample integrity. Since mussels are bottom feeders, water samples were taken from just above the benthic sediment layer (0.6–1.4 meters depth) using a Kemmerer-type water sampler. Samples were immediately stored at 4°C and transported to the laboratory, where they were filtered through syringe filters (Sigma-Aldrich, USA) to remove particulates.

Sediment samples were collected from the upper benthic layer (0–0.1 meters depth) using a benthic grab sampler, as this is the primary habitat range of *Filopaludina bengalensis (*Keeley et al., 2021*)*. These sediment samples were stored in sterilized polyethylene zip pouches at 4 °C before further analysis. In the laboratory, the samples were dried at 65 °C for 24 hours in a hot air oven, ground into a fine powder, and sieved to remove large particles (Sheng & Wang, 2021). Live specimens of *Filopaludina bengalensis* were also collected from the river and transported in polyethylene fish carrier bags for controlled experimental studies (Dhara et al., 2022c).

Physico-chemical parameters of the water were measured on-site. Water temperature was recorded using LUTRON-PTM-816, pH was measured with EUTECH-pHTest-10, and dissolved oxygen (DO) was assessed using LUTRON-PDO-520 (Mondal et al., 2022b). Total hardness (TH) and alkalinity (AL) were determined through the EDTA method and titrimetric method, respectively (Dhara et al., 2022c).

Lead concentrations in water (PbW), sediment (PbS), and gastropod tissues (PbT) were measured using an Atomic Absorption Spectrophotometer (GBC Avanta, Victoria, Australia) with a resolution of 0.0044 absorbance per 0.25 pg change in characteristic mass (Wu et al., 2023). Acid digestion was performed to dissolve the samples for metal analysis. The tissue samples were first dried at 110°C, weighed, and then digested with a 3:1 mixture of nitric acid and hydrochloric acid. The digested samples were heated on a hot plate for 15 minutes, filtered into volumetric flasks, and diluted to a final volume of 50 mL. The lead bioaccumulation concentrations in gastropod tissues were expressed in mg/kg, with tissue weights standardized to 1 mg per fixed volume (Gawad, 2018).

An artificial neural network (ANN) model was constructed to analyze the relationship between environmental factors and lead bioaccumulation in *Filopaludina bengalensis*. The physico-chemical parameters of water were used as input variables (covariates), while the lead concentrations in gastropod tissues, sediment, and water served as output (dependent) variables.

A multi-layer perceptron neural network was developed, using a hyperbolic tangent activation function for the hidden layer and an identity function for the output layer (Mondal et al., 2022a). The ANN approach allowed for a deeper understanding of the influence of environmental conditions on lead accumulation in different components of the ecosystem. The following equations (Eqs. 9.1 and 9.2) were used to construct the ANN model.

$$Pf \times w = \sum_{n=0}^{i} Pf_n \times w_n \tag{9.1}$$

where; *Pf* = are the physico-chemical factors ($Pf_n = Pf_0 \ldots, Pf_i$), *w* is the synaptic weight assigned for each $Pf\,(w_n = w_0 \ldots, w_i)$.

$$E_{SP} = \Phi(Pf \times w) \tag{9.2}$$

where; E_{SP} = number of species in the higher categorized groups, Φ = Coefficient of activation function.

To assess the health risks associated with lead exposure from gastropod consumption, a survey was conducted among the local population. A total of 300 individuals were sampled from each of two groups—those who regularly consume gastropods from the Banka River and those who do not. The survey aimed to identify health complications such as renal, cardiovascular, and osteological disorders potentially linked to heavy metal exposure.

The human health risk assessment was determined through calculation of average daily intake rate (ADI rt) (Wu et al., 2016). The ADI rt was measured using the US Environmental Protection Agency's formula (USEPA) (Eq. 9.3)

$$ADIrt = \frac{C_c \times I_R \times E_D \times E_F}{B_W \times A_T} \tag{9.3}$$

here, C_c is the *Pb* concentration in the cooked flesh of *Filopaludina bengalensis* (ng kg^{-1}), I_R is the ingestion rate (kg day^{-1}), E_D is the exposure duration (year^{-1}). E_F is the exposure frequency (days year^{-1}), B_W is the body weight (kg), A_T is the time period over which the dose is averaged (day^{-1}).

For each pathway or chemical, the hazard quotient (HQ) was determined by comparing the ADI rt to the reference factor or concentration (R_F) (USEPA) (Eq. 9.4).

$$HQ = \frac{ADIrt}{R_F} \tag{9.4}$$

The ingestion rate critical limit (I_{Rlim}) was determined using Eq. (9.5).

$$I_{Rlim} = \frac{R_F \times B_w}{C_C} \tag{9.5}$$

Principal component analysis (PCA) was conducted using STATISTICA software to identify significant environmental and biological factors influencing lead accumulation. The ANN model and paired *t*-test analyses were performed using the SPSS 20 package.

This study highlights the need for effective pollution control measures, wastewater management strategies, and public health interventions to mitigate the long-term effects of lead contamination in the Banka River ecosystem.

9.3 Result

The physicochemical and heavy metal contamination parameters of the analyzed water body indicate significant environmental concerns. The mean temperature (23.34 °C) remains within an optimal range for aquatic life, with minimal variation (SE = 0.14 °C), ensuring stable thermal conditions. The slightly alkaline pH (7.92 ± 0.08) suggests a well-buffered system, preventing drastic shifts in acidity that could harm aquatic organisms. However, the exceptionally high dissolved oxygen concentration (34.03 ± 0.83 mg/L) may indicate excessive aeration or the presence of photosynthetic organisms, potentially leading to oxygen supersaturation. Total alkalinity (246.26 ± 5.93 mg/L) further reinforces the buffering capacity of the water, maintaining pH stability. Alarmingly, the biological oxygen demand (BOD) is elevated (39.11 ± 2.40 mg/L), signaling substantial organic pollution, likely from anthropogenic sources such as sewage discharge or industrial effluents. Electrical conductivity (1123.09 ± 31.94 μS/cm) and total dissolved solids (TDS) (707.85 ± 25.37 mg/L) indicate high ionic concentration, suggesting potential contamination from dissolved pollutants. Of particular concern is the lead

(Pb) contamination, with concentrations in water (47.24 ± 3.38 μg/L) exceeding WHO's safe limit of 10 μg/L, posing risks to both aquatic life and human health. Furthermore, significant lead accumulation in sediments (52.42 ± 3.98 mg/kg) highlights prolonged contamination, while its presence in biological tissues (7.43 ± 0.67 mg/kg) suggests bioaccumulation and potential biomagnification through the food chain.

The principal component analysis (PCA) reveals that the first two principal components explain a significant portion of the total variance in the dataset. Factor 1 accounts for 59.76% of the variance, while Factor 2 contributes 15.11%, making a cumulative variance of 74.87%. This indicates that most of the variability in the dataset can be effectively summarized by these two principal components (Fig. 9.1).

The variable loadings on Factor 1 show strong negative correlations with pH (−0.915), Dissolved Oxygen (DO) (−0.852), Biological Oxygen Demand (BOD) (−0.927), Electrical Conductivity (EC) (−0.875), and Total Dissolved Solids (TDS) (−0.935), suggesting that these parameters are key drivers of variance in the dataset. Lead concentrations in water (PbW) and sediment (PbS) also load significantly on Factor

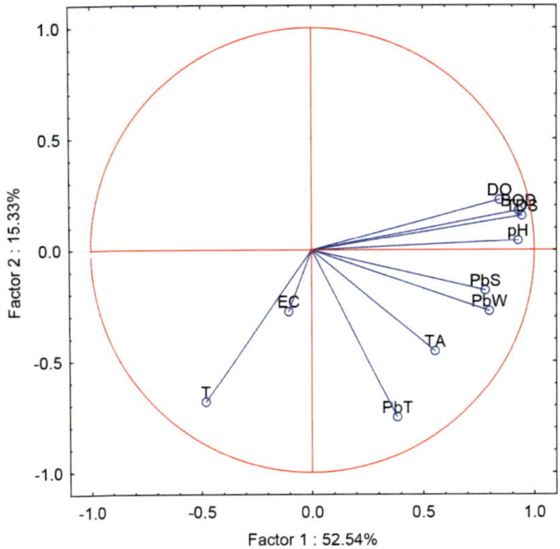

Fig. 9.1 PCA biplot of physicochemical parameters and heavy metal concentration

1 (−0.777 and −0.786, respectively), indicating their strong association with the overall environmental quality parameters.

Factor 2, explaining 15.11% of the variance, has a strong negative correlation with Lead concentration in tissue (PbT: −0.793), distinguishing it as a key parameter along this axis. Temperature (T) shows a moderate positive correlation with Factor 2 (0.294), suggesting a separate trend in its variation compared to other parameters.

The PCA biplot (Fig. 9.1) visually supports these findings, as DO, BOD, EC, TDS, and pH cluster together, indicating interrelated environmental characteristics. PbW, PbS, and TA (Total Alkalinity) show moderate separation, suggesting an influence of lead contamination in water and sediment. PbT and temperature are distinctly separated from the rest, indicating unique variance contributions along Factor 2.

The multilayer perceptron (MLP) artificial neural network (ANN) model (Fig. 9.2) was constructed to predict lead (Pb) concentrations in water (PbW), sediment (PbS), and biotic tissues (PbT) based on environmental parameters. The input layer consisted of seven standardized covariates: temperature (T), pH, dissolved oxygen (DO), total alkalinity (TA), biological oxygen demand (BOD), electrical conductivity (EC), and total dissolved solids (TDS). The model employed a single hidden layer with two units, using a hyperbolic tangent activation function, while the output layer contained three dependent variables with an identity activation function and sum of squares as the error function. The dataset was divided into training (77.8%) and testing (22.2%) subsets.

The model's performance was evaluated using the sum of squares error (SSE) and relative error metrics. In the training phase, the SSE was 15.703, with an average overall relative error of 0.523. Among the scale-dependent variables, PbT exhibited the highest relative error (0.870), while PbW and PbS had comparatively lower errors (0.367 and 0.334, respectively). In the testing phase, the SSE was 3.583, and the overall relative error increased to 0.632, indicating a reduction in prediction accuracy for unseen data. The relative errors for PbW, PbS, and PbT in the testing phase were 0.186, 0.702, and 0.829, respectively, with PbT maintaining the highest prediction error. The parameter estimates of the given master equation have been presented in Table 9.1, where all the values of synaptic weight for activation function have been mentioned.

The parameter estimates revealed that DO had the strongest influence on the hidden neurons, showing a weight of 0.868 for the first hidden unit (H1:1) and −0.392 for the second (H1:2). pH was also a dominant

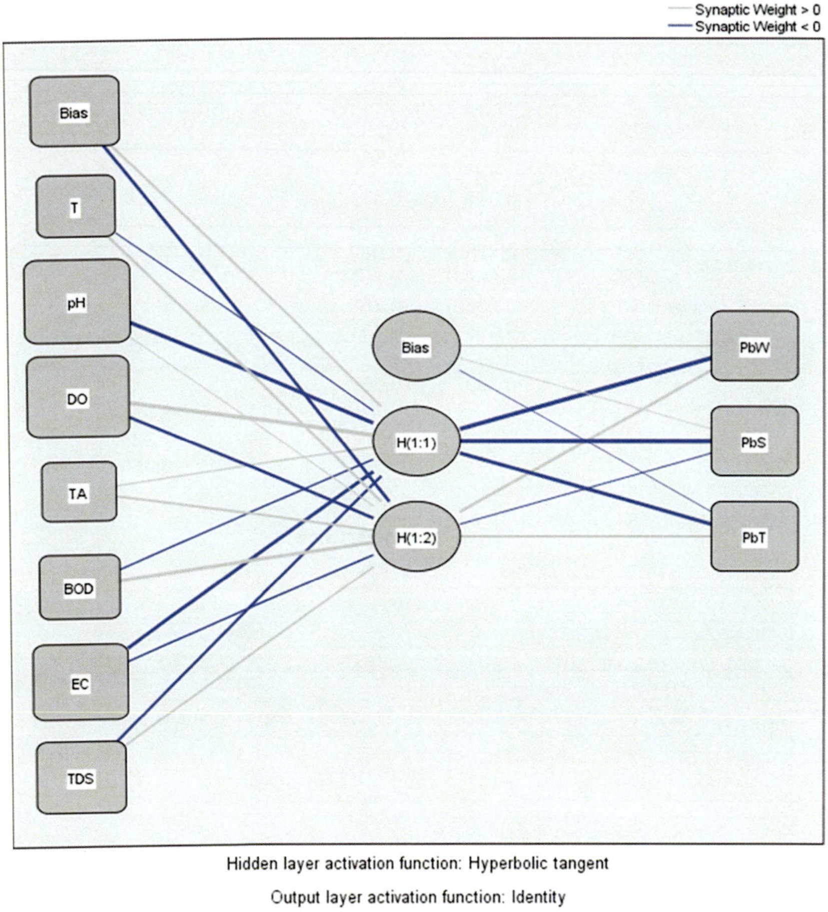

Fig. 9.2 The artificial neural network of physicochemical parameters and heavy metal concentration

predictor, with an importance of 100%, followed by DO (90.2%) and EC (72.2%). Conversely, TA had the lowest contribution (11.7%), suggesting a minimal role in determining Pb concentrations (Fig. 9.3).

The demographic and health-related characteristics of the associated and non-associated populations at the Bank of Banka are presented in

Table 9.1 Parameter estimates of synaptic weight assigned for each node

Predictor		*Predicted*				
		Hidden Layer 1		*Output Layer*		
		H(1:1)	*H(1:2)*	*PbW*	*PbS*	*PbT*
Input layer	(Bias)	0.268	−0.412			
	T	−0.063	0.319			
	pH	−0.786	0.039			
	DO	0.868	−0.392			
	TA	0.083	0.268			
	BOD	−0.165	0.451			
	EC	−0.547	−0.229			
	TDS	−0.323	0.267			
Hidden layer 1	(Bias)			0.122	0.036	−0.029
	H(1:1)			−1.349	−1.325	−0.566
	H(1:2)			0.330	−0.050	0.270

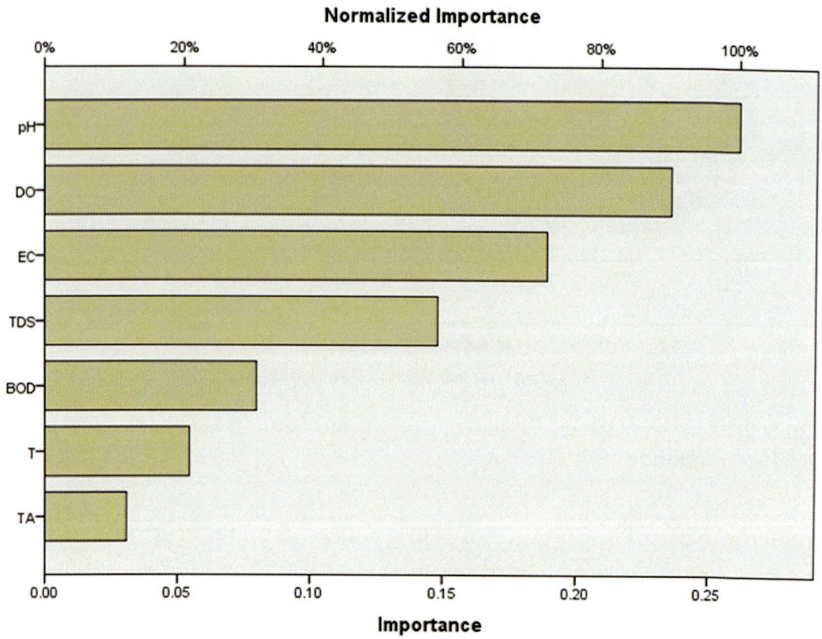

Fig. 9.3 Normalized relative importance analysis of input factors on the model output

Tables 9.2 and 9.3. Table 9.2 provides general anthropometric and educational details, while Table 3 focuses on the average daily intake (ADI rt) of contaminants and the corresponding hazard quotient (HQ). Additionally, the prevalence of various health problems, including neurological, renal, and cardiovascular disorders, has been analyzed across different age groups for both populations. The male-to-female ratio among the associated population was 49:51, whereas in the non-associated population, the ratio was slightly skewed at 55:45. The age distribution showed that individuals aged 11–30, 31–50, and above 50 years constituted 32%, 26%, and 42% of the associated population, while in the non-associated group, the proportions were 28%, 30%, and 42%, respectively. Anthropometric data revealed that the average height of the associated population (164.2 $\pm$ 5.1 cm) was marginally lower than that of the non-associated population (166.1 $\pm$ 6.0 cm). In contrast, the weight of the non-associated group (66.2 $\pm$ 4.2 kg) was significantly higher than that of the associated group (54.8 $\pm$ 3.4 kg). Consequently, the BMI of the associated group (20.3 $\pm$ 1.8) was lower than that of the non-associated population (23.5 $\pm$ 1.9), indicating potential differences in nutritional status or lifestyle between the two groups. Educational qualifications also varied notably, with a higher proportion of the associated population falling into the lower educational categories. Among the associated group, 15% were illiterate, 10% had received education up to 5th grade, 38% had completed 5th–10th grade, 25% had finished 10th–12th grade, and only 12% were graduates or above. In contrast, the non-associated population had a higher proportion of graduates (23%) and individuals with 10th–12th grade education (40%).

Table 9.3 outlines the estimated ADI rt and HQ values across different income groups and age categories. The ADI rt values indicated higher contaminant intake among lower-income groups, with individuals earning less than INR 5000/month exhibiting the highest ADI rt (2.45×10^{-5} mg kg^{-1} day^{-1} for the 11–30-year group), gradually decreasing across higher income groups (>20,000 INR/month: 4.75×10^{-6} mg kg^{-1} day^{-1}). A similar trend was observed for the HQ values, where the highest HQ (0.025) was noted among the lowest-income group aged 11–30 years, while the lowest HQ (0.005) was found in the highest-income group aged above 50 years. This trend suggests a strong inverse relationship between income levels and potential health risks associated with contaminant exposure.

Table 9.2 General information and health problems in relation to the associated and non-associated people at the bank of Banka River

General information		*Associated (N = 300)*	*Not associated (N = 300)*
Male and female ratio		49:51:00	55:45:00
Age group ratio (11–30, 31–50, >50)		32:26:42	28:30:42
Height (cm)		164.2 ± 5.1	166.1 ± 6.0
Weight (kg)		54.8 ± 3.4	66.2 ± 4.2
BMI		20.3 ± 1.8	23.5 ± 1.9
Educational qualification ratio (illiterate, under 5th class, 5th–10th class, 10–12th class, and graduated and above)		15:10:38:25:12	05:12:20:23:40
Neurological problem	40 years above	61.50%	35%
	25–40 year	41%	22.50%
	11–25 year	19%	6%
Renal problem	40 years above	56.60%	32%
	25–40 year	43.20%	18.60%
	11–25 year	17%	5.10%
Cardiovascular problem	40 years above	48.20%	44.50%
	25–40 year	33%	26%
	11–25 year	4%	8.50%

Table 9.3 The average daily intake (ADI rt) and hazard quotient (HQ) of people of different income levels and ages near the bank of Banka River. Here, Ir = 0.075 kg day^{-1} for all income groups and Rf = 0.001 was considered for calculation of ADI rt and HQ, respectively

Income (INR $month^{-1}$)	*ADI rt (mg kg^{-1} day^{-1})*	*HQ*				
	11–30 years	*31–50 years*	*Above 50 years*	*11–30 years*	*31–50 years*	*Above 50 years*
<5000	2.45×10^{-5}	2.21×10^{-5}	2.68×10^{-5}	0.025	0.021	0.026
5000–10,000	1.92×10^{-5}	1.75×10^{-5}	1.98×10^{-5}	0.019	0.018	0.020
10,000–15,000	1.42×10^{-5}	1.29×10^{-5}	1.51×10^{-5}	0.014	0.012	0.015
15,000–20,000	9.62×10^{-4}	8.95×10^{-6}	9.87×10^{-6}	0.010	0.009	0.011
>20,000	4.75×10^{-6}	4.35×10^{-6}	4.95×10^{-6}	0.005	0.004	0.006

The prevalence of health complications was found to be significantly higher in the associated population compared to the non-associated group. Neurological disorders were more common among individuals aged 40 years and above (61.5%) in the associated population compared to 35% in the non-associated group. Similarly, in the 25–40 age group, 41% of the associated individuals reported neurological issues, while only 22.5% were affected in the non-associated group. Among younger individuals (11–25 years), 19% of the associated population reported neurological issues, whereas only 6% of the non-associated group did. Renal problems followed a similar pattern, with a much higher prevalence in the associated group (56.6% for 40+ years) compared to the non-associated group (32%). The prevalence was also higher in younger groups, with 43.2% of the associated individuals aged 25–40 years experiencing renal issues compared to 18.6% in the non-associated group. Cardiovascular problems showed a slightly different trend, where the prevalence was relatively high in both groups among individuals above 40 years (48.2% in the associated group and 44.5% in the non-associated group), though the disparity increased in younger age groups.

9.4 Discussion

The PCA results suggest that water quality parameters (pH, DO, BOD, EC, and TDS) are highly interdependent and significantly influence the primary environmental gradient (Batabyal et al., 2022) (Factor 1). The strong negative loadings of these variables indicate that poor water quality conditions (e.g., high BOD, low DO, and increased conductivity) are major factors driving system variability (Anh et al., 2023; Saha et al., 2022). The clustering of these parameters also suggests that changes in one are likely to affect the others, reflecting typical eutrophication or pollution-driven interactions in aquatic environments (Mandal et al., 2024; Saha et al., 2022).

The presence of lead (Pb) in water and sediment (PbW and PbS) is strongly aligned with Factor 1, indicating that metal contamination is closely linked to water quality degradation. The moderate correlation with Total Alkalinity (TA) suggests potential buffering capacity variations in response to contamination (Ahmad et al., 2024; Stets et al., 2014).

Factor 2 highlights a distinct contribution of PbT (Lead in tissue), which is largely independent of general water quality parameters. This suggests that bioaccumulation of lead in organisms is influenced by a separate process, possibly linked to food web interactions rather than direct water contamination (Collin et al., 2022b; Mousavi et al., 2022). The positive correlation of temperature with Factor 2 indicates that thermal conditions might influence metal uptake or bioavailability in aquatic biota (Lan et al., 2020).

Overall, these findings highlight two key environmental stress gradients: (i) a pollution-driven axis (Factor 1) dominated by water quality degradation and metal contamination, and (ii) a biological uptake axis (Factor 2) influenced by lead bioaccumulation and thermal conditions. The strong negative association of PbT with Factor 2 suggests that lead accumulation in tissues is not solely dependent on its concentration in water or sediment but may be influenced by physiological or ecological factors such as trophic transfer and metabolic activity.

These results underscore the importance of integrated monitoring approaches that account for both abiotic (water and sediment quality) and biotic (bioaccumulation) factors to assess heavy metal pollution and its ecological risks comprehensively.

The ANN analysis demonstrated that the model could reasonably predict Pb concentrations in environmental matrices based on water quality parameters (Zhu et al., 2022). The relatively high SSE in both training and testing phases, particularly for PbT, suggests that biological accumulation processes may be influenced by additional factors not captured in the input variables (Zhu et al., 2022). The high prediction errors for PbT indicate that bioaccumulation mechanisms are more complex and possibly nonlinear, requiring additional predictor variables such as trophic level interactions, metal-binding affinity, or physiological uptake characteristics (Kouadri et al., 2021). pH emerged as the most influential factor, aligning with the established role of pH in controlling metal solubility and bioavailability. Highly acidic or alkaline conditions can significantly alter Pb mobility and bioaccumulation (Sun et al., 2021). The strong contribution of DO suggests that oxygen levels may influence Pb speciation and redox-driven precipitation or dissolution processes. EC and TDS also played significant roles, indicating that ionic interactions and overall salinity influence Pb transport and partitioning (Yang et al., 2024; Zhao et al., 2025).

The model's lower predictive accuracy for PbS and PbT compared to PbW suggests that sediment-bound and bioaccumulated Pb are subject to additional environmental and biological controls, necessitating a more complex model structure (Mondal et al., 2023; Sen et al., 2023). The hyperbolic tangent activation function in the hidden layer likely enhanced the model's ability to capture nonlinearity, yet the relatively small hidden layer (two neurons) may have limited its capacity to fully model intricate relationships. Future model refinements could involve increasing hidden layers, integrating additional predictor variables, or employing alternative optimization algorithms such as gradient descent or Bayesian regularization (Bhattacharya et al., 2024).

Relative importance analysis was used to quantify how strongly each environmental variable influenced the prediction of Pb concentrations in water (PbW), sediment (PbS), and tissue (PbT) (Davis & Jaurand, 2013). Relative importance is a statistical measure that ranks predictor variables based on their contribution to the model's output. In this ANN analysis, pH had the highest relative importance (100%), indicating that changes in pH exerted the strongest influence on Pb concentrations. This result is scientifically valid since pH plays a critical role in metal solubility, adsorption–desorption processes, and bioavailability.

Dissolved oxygen (DO) was the second most influential variable (90.2%), suggesting that oxygen levels affect Pb speciation, redox-driven transformations, and interactions with organic matter. High DO levels may facilitate Pb oxidation and precipitation, while low DO conditions can promote Pb mobilization under reducing conditions.

Electrical conductivity (EC) had a relative importance of 72.2%, indicating a significant role of ionic strength in Pb transport and partitioning. Since EC is a proxy for dissolved ions, higher conductivity levels may reflect greater Pb dissolution or competition with other cations for adsorption sites (Rogacheva et al., 2023).

On the other hand, total alkalinity (TA) had the lowest relative importance (11.7%), suggesting that it played a minimal role in Pb distribution compared to other factors. Alkalinity typically affects buffering capacity and carbonate complexation, but in this study, its influence on Pb mobility appears to be weaker than expected.

The observed variations in demographic, anthropometric, and health parameters between the associated and non-associated populations suggest a substantial impact of environmental exposure on public health. The lower BMI observed in the associated population could be linked to higher physical activity levels or inadequate nutritional intake due to environmental and occupational factors. The higher prevalence of individuals with lower educational qualifications in the associated population may also contribute to reduced health awareness and limited access to healthcare services, further exacerbating health risks.

The analysis of ADIrt and HQ indicates a clear disparity in contaminant exposure levels between different income groups, with lower-income individuals facing higher exposure risks. This is likely due to their increased dependence on local water sources and dietary intake of contaminated food items, as well as occupational exposure. The HQ values suggest that, although none of the estimated values exceed the risk threshold ($HQ > 1$), the relatively higher HQ in lower-income and younger populations suggests a potential for long-term adverse effects, especially with cumulative exposure.

The higher prevalence of neurological, renal, and cardiovascular problems in the associated population strongly indicates a correlation with environmental exposure. The significantly higher rates of neurological disorders among older individuals in the associated group (61.5% vs. 35%) suggest chronic exposure to neurotoxic elements, possibly heavy metals or organic pollutants present in the ecosystem. The renal complications reported in the associated group, particularly among individuals above 40 years (56.6%), align with previous studies indicating nephrotoxicity due to prolonged exposure to contaminants such as heavy metals and industrial pollutants. The higher cardiovascular disease prevalence among both associated and non-associated groups aged above 40 years suggests that age-related risk factors play a crucial role; however, the notably higher prevalence in the younger associated population (33% vs. 26%) points toward environmental stressors as an additional risk factor.

The ADIrt and HQ analysis indicate that lower-income groups are at higher risk due to their increased dependence on contaminated resources (Brouwer et al., 2007). Additionally, age-specific trends in health problems reveal that younger individuals in the associated group are more vulnerable than their non-associated counterparts, indicating an early onset of exposure-related health issues (Brouwer et al., 2007; Nkoh et al., 2022). The relatively higher BMI in the non-associated group suggests differences in lifestyle and dietary patterns, which may also play a role in modulating health outcomes.

The findings suggest that individuals associated with the Banka river ecosystem are at a significantly higher risk of developing neurological, renal, and cardiovascular disorders, likely due to prolonged exposure to environmental contaminants (Quon & Jiang, 2023). The income-related disparities in ADIrt and HQ further highlight the role of socioeconomic status in determining health vulnerability.

9.5 Conclusion

This study provides a comprehensive assessment of lead (Pb) contamination in the freshwater gastropod *Filopaludina bengalensis* and its implications for environmental quality and human health in the Banka River region of Bardhaman, India. The findings indicate a strong interdependence between water quality parameters and Pb accumulation in both sediment and gastropod tissues. PCA results demonstrated that key physicochemical factors, particularly pH and dissolved oxygen (DO), significantly influence Pb mobility and bioavailability in the aquatic system. The ANN model further corroborated the critical role of pH, DO, electrical conductivity (EC), and total dissolved solids (TDS) in governing Pb transport and partitioning. However, the relatively high prediction errors for Pb accumulation in gastropod tissues suggest that bioaccumulation mechanisms are complex and influenced by additional ecological and physiological factors not captured in the current model.

The hazard quotient (HQ) analysis revealed that frequent consumers of *F. bengalensis*, particularly economically disadvantaged individuals and those above the age of forty, are at an elevated risk of lead toxicity. The observed prevalence of neurological, renal, and cardiovascular disorders among the associated population underscores the potential long-term health consequences of chronic Pb exposure. These findings highlight the urgent need for targeted public health interventions, including dietary advisories, routine biomonitoring, and improved healthcare access for vulnerable communities.

From an environmental management perspective, the results emphasize the necessity of stringent pollution control measures to mitigate industrial discharge into the Banka River. Regular monitoring of Pb levels in water, sediment, and aquatic biota is crucial to tracking contamination trends and assessing ecosystem health. Furthermore, future studies should incorporate trophic transfer dynamics and metabolic interactions to refine bioaccumulation models and better understand Pb assimilation in freshwater food webs.

Overall, this study underscores the importance of integrating environmental and human health risk assessments to develop sustainable strategies for mitigating heavy metal pollution. A multidisciplinary approach, combining environmental science, toxicology, and public health policies, is essential for safeguarding both ecological integrity and community well-being in metal-contaminated regions.

References

Ahmad, M., Ahmad, A., Omar, T. F. T., & Mohammad, R. (2024). Current trends of analytical techniques for total alkalinity measurement in water samples: A review. *Critical Reviews in Analytical Chemistry, 54*(8), 2734–2744.

Anh, N. T., Nhan, N. T., Schmalz, B., & Le Luu, T. (2023). Influences of key factors on river water quality in urban and rural areas: A review. *Case Studies in Chemical and Environmental Engineering, 8*, Article 100424.

Batabyal, S., Das, N., Mondal, A., Banerjee, R., Gangopadhyay, S., & Mandal, S. (2022). Land cover changes in green patches and its impact on carbon sequestration in an Urban System, India. *Applied geomorphology and contemporary issues* (pp. 397–409). Springer.

Bhattacharya, S., Ali, T., Chakravortti, S., Pal, T., Majee, B. K., Mondal, A., Pande, C. B., Bilal, M., Rahman, M. T., & Chakrabortty, R. (2024). Application of machine learning and deep learning algorithms for landslide susceptibility assessment in landslide prone Himalayan Region. *Earth Systems and Environment*, 1–19.

Brouwer, R., Akter, S., Brander, L., & Haque, E. (2007). Socioeconomic vulnerability and adaptation to environmental risk: A case study of climate change and flooding in Bangladesh. *Risk Analysis: An International Journal, 27*(2), 313–326.

Collin, M. S., Venkatraman, S. K., Vijayakumar, N., Kanimozhi, V., Arbaaz, S. M., Stacey, R. S., Anusha, J., Choudhary, R., Lvov, V., & Tovar, G. I. (2022a). Bioaccumulation of lead (Pb) and its effects on human: A review. *Journal of Hazardous Materials Advances, 7*, Article 100094.

Collin, S., Baskar, A., Geevarghese, D. M., Ali, M. N. V. S., Bahubali, P., Choudhary, R., Lvov, V., Tovar, G. I., Senatov, F., & Koppala, S. (2022b). Bioaccumulation of lead (Pb) and its effects in plants: A review. *Journal of Hazardous Materials Letters, 3*, Article 100064.

Davis, J. M., and Jaurand, M.-C. (2013). *Cellular and molecular effects of mineral and synthetic dusts and fibres*. Springer Science & Business Media.

Dey, S., Sen, K., Saha, S., Saha, N. C., Snow, D. D., & Hossain, M. Y. (2025). *Safeguarding aquatic ecosystems: innovations in water resource management, sustainable synergy: Harnessing ecosystems for climate resilience* (pp. 117–130). Springer.

Dhara, K., Chukwuka, A. V., Saha, S., Saha, N. C., & Faggio, C. (2022a). Effects of short-term selenium exposure on respiratory activity and proximate body composition of early-life stages of Catla catla, Labeo rohita and Cirrhinus mrigala. *Environmental Toxicology and Pharmacology, 90*, Article 103805.

Dhara, K., Saha, S., Chukwuka, A. V., Pal, P., Saha, N. C., & Faggio, C. (2022b). Fluoride sensitivity in freshwater snail, Bellamya bengalensis (Lamarck, 1882): An integrative biomarker response assessment of behavioral indices, oxygen consumption, haemocyte and tissue protein levels under environmentally relevant exposure concentrations. *Environmental Toxicology and Pharmacology, 89*, Article 103789.

Dhara, K., Saha, S., Pal, P., Chukwuka, A. V., Panigrahi, A. K., Saha, N. C., & Faggio, C. (2022c). Biochemical, physiological (haematological, oxygen-consumption rate) and behavioural effects of mercury exposures on the freshwater snail, Bellamya bengalensis. *Comparative Biochemistry and Physiology Part c: Toxicology & Pharmacology, 251*, Article 109195.

Fulke, A. B., Ratanpal, S., & Sonker, S. (2024). Understanding heavy metal toxicity: Implications on human health, marine ecosystems and bioremediation strategies. *Marine Pollution Bulletin, 206*, Article 116707.

Gawad, S. S. A. (2018). Concentrations of heavy metals in water, sediment and mollusk gastropod, Lanistes carinatus from Lake Manzala. *Egypt. Egyptian Journal of Aquatic Research, 44*(2), 77–82.

Ghosh, D., Banerjee, M., Pal, S., & Mandal, M. (2022). Spatio-temporal variation of channel migration and vulnerability assessment: A case study of Bhagirathi River within Barddhaman District, West Bengal, India. *Drainage Basin Dynamics: An Introduction to Morphology, Landscape and Modelling*, 305–330.

Ghosh, S., Dey, S., Mandal, A. H., Sadhu, A., Saha, N. C., Barceló, D., Pastorino, P., & Saha, S. (2025a). Exploring the ecotoxicological impacts of microplastics on freshwater fish: A critical review. *Journal of Contaminant Hydrology* 104514.

Ghosh, S., & Maji, T. (2011). An environmental assessment of urban drainage, sewage and solid waste management in Barddhaman municipality, West Bengal. *International Journal of Environmental Sciences, 2*(1), 92–104.

Ghosh, S., Mondal, A., Gangopadhyay, S., & Mandal, S. (2020). Cadmium bioaccumulation in Lamellidens marginalis and human health risk assessment: A case study in India. *Human and Ecological Risk Assessment: An International Journal, 26*(3), 713–725.

Ghosh, S., Sadhu, A., Mandal, A. H., Biswas, J. K., Sarkar, D., & Saha, S. (2024). Copper oxide nanoparticles as an emergent threat to aquatic invertebrates and photosynthetic organisms: A synthesis of the known and exploration of the unknown. *Current Pollution Reports, 11*(1), 6. https://doi.org/10.1007/s40726-024-00334-6

Ghosh, S., Sadhu, A., Mandal, A. H., Dey, S., Saha, N. C., & Saha, S. (2025b). Metabolomics profiling and hazard prediction of spinosyn-based agricultural products: A computational approach. *Science of the Total Environment, 959*, Article 178279.

Keeley, N., Laroche, O., Birch, M., & Pochon, X. (2021). A substrate-independent benthic sampler (SIBS) for hard and mixed-bottom marine habitats: A proof-of-concept study. *Frontiers in Marine Science, 8*, Article 627687.

Kouadri, S., Elbeltagi, A., Islam, A. R. M. T., & Kateb, S. (2021). Performance of machine learning methods in predicting water quality index based on irregular data set: Application on Illizi region (Algerian southeast). *Applied Water Science, 11*(12), 190.

Lan, W.-R., Huang, X.-G., Lin, L.-X., Li, S.-X., & Liu, F.-J. (2020). Thermal discharge influences the bioaccumulation and bioavailability of metals in oysters: Implications of ocean warming. *Environmental Pollution, 259*, Article 113821.

Levin, R., Vieira, C. L. Z., Rosenbaum, M. H., Bischoff, K., Mordarski, D. C., & Brown, M. J. (2021). The urban lead (Pb) burden in humans, animals and the natural environment. *Environmental Research, 193*, Article 110377.

Mandal, A. H., Sadhu, A., Ghosh, S., Saha, N. C., Mossotto, C., Pastorino, P., Saha, S., & Faggio, C. (2024). Evaluating the impact of neonicotinoids on aquatic non-target species: A comprehensive review. *Environmental Toxicology and Pharmacology* 104606. https://doi.org/10.1016/j.etap.2024.104606

Mondal, A., Chanda, D., Patra, S., & Barman, M. (2020). Phrynus whitei (Gervais, 1842) from the type locality Burdwan is actually Charinus bengalensis (Gravely, 1911). *ZOO'S PRINT, 35*(7), 1–3.

Mondal, A., Das, N., Banerjee, R., Batabyal, S., Gangopadhyay, S., Ray, H., Biswasd, N., & Mandal, S. (2023). Investigating the health of a rice field ecosystem using thermodynamic extremal principles. *Ecology, Economy and Society-the INSEE Journal, 6*(1), 81–108.

Mondal, A., Das, N., & Mandal, S. (2022a). Domestication of rice reshaped associated biodiversity of rice field ecosytem. *Journal of Animal Diversity, 4*(1), 74–90.

Mondal, A., Mondal, S., & Mandal, S. (2022b). Empirical dynamic model deciphers more information on the nutrient (N)—phytoplankton (P)–zooplankton (Z) dynamics of Hooghly-Matla estuary, Sundarban, India. *Estuarine, Coastal and Shelf Science, 265*, Article 107711.

Mousavi, S. M., Brodie, G., Payghamzadeh, K., Raiesi, T., & Strivastava, A. (2022). Lead bioavailability in the environment: Its exposure and effects. *Journal of Advances in Environmental Health Research, 10*(1), 1–14.

Niloy, H. K., Chowdhury, A. I., Islam, M. S., Pervez, A., Asseri, A. H., Alsohibany, K. S., Molla, M. H. R., Rahman, M. A., & Uddin, M. S.

(2024). Bioaccumulation of heavy metals in water and mollusks in the Karnafully estuary: Potential human health risk and environmental contamination. *Regional Studies in Marine Science, 78*, Article 103752.

Nkoh, J. N., Ajibade, F. O., Atakpa, E. O., Abdulaha-Al Baquy, M., Mia, S., Odii, E. C., & Xu, R. (2022). Reduction of heavy metal uptake from polluted soils and associated health risks through biochar amendment: A critical synthesis. *Journal of Hazardous Materials Advances, 6*, Article 100086.

Oladimeji, T., Oyedemi, M., Emetere, M., Agboola, O., Adeoye, J., & Odunlami, O. (2024). Review on the impact of heavy metals from industrial wastewater effluent and removal technologies. *Heliyon 10*(23).

Quon, H., & Jiang, S. (2023). Decision making for implementing non-traditional water sources: a review of challenges and potential solutions. *npj Clean Water 6*(1), 56.

Raj, K., & Das, A. P. (2023). Lead pollution: Impact on environment and human health and approach for a sustainable solution. *Environmental Chemistry and Ecotoxicology, 5*, 79–85.

Rogacheva, E., Nikolaenko, G., & Nashchekina, O. (2023). Transport and thermoelectric properties of the Pb1-xSnxTe topological crystalline insulator in the vicinity of the band inversion. *Journal of Physics and Chemistry of Solids, 183*, Article 111635.

Saha, S., Chukwuka, A. V., Mukherjee, D., Saha, N. C., & Adeogun, A. O. (2022). Hydrological connectivity, surface water quality and distribution of fish species within sub-locations of an urban oxbow lake, East India. *Watershed Ecology and the Environment, 4*, 44–58.

Sen, P., Das, N., Saha, N. C., Ghosh, S., Mondal, A., & Mandal, S. (2023). Modeling of cadmium bioaccumulation dynamics in Channa punctata (Bloch, 1793) of Dankuni wetland ecosystem and assessment of risk to human health. *Applied Biochemistry and Biotechnology, 195*(6), 3681–3698.

Sharma, S., Dar, O. I., Singh, K., Kaur, A., & Faggio, C. (2021). Triclosan elicited biochemical and transcriptomic alterations in Labeo rohita larvae. *Environmental Toxicology and Pharmacology* 103748. https://doi.org/10.1016/j.etap.2021.103748

Sheng, L., & Wang, L. (2021). The microbial safety of fish and fish products: Recent advances in understanding its significance, contamination sources, and control strategies. *Comprehensive Reviews in Food Science and Food Safety, 20*(1), 738–786.

Singh, B. J., Chakraborty, A., & Sehgal, R. (2023). A systematic review of industrial wastewater management: Evaluating challenges and enablers. *Journal of Environmental Management, 348*, Article 119230.

Stets, E. G., Kelly, V. J., & Crawford, C. G. (2014). Long-term trends in alkalinity in large rivers of the conterminous US in relation to acidification, agriculture, and hydrologic modification. *Science of the Total Environment, 488*, 280–289.

Sun, X., Du, Y., Deng, Y., Fan, H., Tao, Y., & Ma, T. (2021). Contribution of groundwater discharge and associated contaminants input to Dongting Lake, Central China, using multiple tracers (222 Rn, 18 O, Cl−). *Environmental Geochemistry and Health, 43*, 1239–1255.

USEPA, I. R. I. S. (2011). *Integrated risk information system. Environmental protection agency region I*, Washington DC 20460.

Wu, D., Hu, Y., Cheng, H., & Ye, X. (2023). Detection techniques for lead ions in water: A review. *Molecules, 28*(8), 3601.

Yang, X., Qu, S., Wang, J., Zhao, Y., Ren, X., Dong, S., & Yu, R. (2024). Trace element distribution, sources, toxicity risk characteristics associated with lacustrine groundwater discharge in boreal lakes: Implications for the eco-environmental security in the Yellow River Basin, China. *Ecological Indicators, 167*, Article 112600.

Zhang, L., Zhu, Y., Gu, H., Lam, S. S., Chen, X., Sonne, C., & Peng, W. (2024). *A review of phytoremediation of environmental lead (pb) contamination.* Chemosphere: 142691.

Zhao, Y., Ren, X., Qu, S., Liao, F., Zhang, K., Li, M., Wang, J., & Yu, R. (2025). Contrastive mechanisms of lacustrine groundwater discharge and associated pollutant fluxes into two typical inland lakes in Inner Mongolia, Northwest China. *Journal of Environmental Sciences, 159*, 661–669. PMID: 41005924. https://doi.org/10.1016/j.jes.2025.03.018

Zhu, M., Wang, J., Yang, X., Zhang, Y., Zhang, L., Ren, H., Wu, B., & Ye, L. (2022). A review of the application of machine learning in water quality evaluation. *Eco-Environment and Health, 1*(2), 107–116.

CHAPTER 10

cxsPlastic Footprints in Sacred Waters: Assessing Microplastic Presence in the Himalayan Ganga River Ecosystem

Kusum Pandey and Jitendra Rajput

10.1 Introduction

The Ganga River, often seen as the cradle of Indian civilization, holds a deeply sacred and vital role in the cultural, ecological, and economic fabric of South Asia (Kumar, 2017; Chauhan et al. 2021 Han et al., 2023). Rising from the Gangotri Glacier in the Himalayas, it winds its way over 2,500 km across northern India before emptying into the Bay of Bengal. Along its course, it sustains the lives of over 400 million people and nourishes one of the most fertile and densely populated river basins

K. Pandey (✉)
G. B. Pant National Institute of Himalayan Environment, Garhwal Regional Centre, Srinagar, Uttarakhand, India
e-mail: khushipandey166@gmail.com

J. Rajput
ICAR-Indian Agricultural Research Institute, New Delhi, India

S. Ghosh (ed.), *Reimagining Indian Rivers for Sustainability*, Palgrave Studies in Environmental Sustainability,
https://doi.org/10.1007/978-3-032-04754-0_10

on Earth (Kumar et al., 2021; Singh et al., 2025). Despite being venerated in Hinduism as Ganga Ma, the divine mother who purifies and gives life to the river is under growing threat from human activity. Among the most alarming pollutants are microplastics, a modern menace hiding in plain sight (Yang et al., 2023; Han et al., 2023).

Microplastics defined as plastic particles smaller than 5 millimetres have emerged as a critical environmental hazard. They stem from two main sources: primary microplastics, which are intentionally manufactured for use in products like cosmetics and industrial cleaners, and secondary microplastics, which are the result of larger plastic debris breaking down under environmental stress (Auta et al., 2017; Yang et al., 2023). Because of their minute size, durability, and ability to absorb and transport toxic chemicals, microplastics have become a stealthy but serious danger to freshwater systems, aquatic life, and human health (Yang et al. 2023; Matavos-Aramyan, 2024).

While much of the attention around microplastics has focused on the oceans, recent research is uncovering their alarming presence in major freshwater systems, including the Ganga. Even the Himalayan stretch of the river long thought to be pristine is increasingly polluted. Factors such as booming tourism, urban sprawl, religious pilgrimages, and poor waste management are contributing to the crisis (Li et al., 2018; Thakur et al. 2021; Chaudhary et al., 2025). Sacred towns like Rishikesh and Haridwar see millions of visitors each year, many of whom leave behind mountains of plastic waste (Free et al., 2014; Raha et al., 2021). With limited infrastructure to manage this influx, much of the plastic ends up in the river, either directly or through stormwater runoff (Gupta & Kumar, 2025; Kumar et al., 2021).

What makes this situation especially troubling is the spiritual reverence Indians hold for the Ganga. It's worshipped as a holy cleanser of sins, yet it is being overwhelmed by the very waste produced by modern lifestyles. This contradiction highlights the urgent need to rethink our relationship with the river. Microplastics have been found not just in the river's surface waters and sediments, but also in its wildlife even in the high Himalayan zones disproving the belief that remote ecosystems are immune to pollution (Talukdar et al. 2023; Witczak et al. 2024).

This chapter delves into the microplastic problem within the Himalayan Ganga region. It investigates where these particles come from, how they move through the ecosystem, and what effects they have on both the environment and public health. The chapter also evaluates current policy efforts and emphasizes the need for more holistic, culturally aware strategies. By framing plastic pollution as not just an environmental challenge but a socio-cultural and ethical issue, this work calls for solutions that combine science, tradition, and grassroots action.

10.2 What Are Microplastics?

Microplastics are tiny plastic particles that are less than 5 millimetres in size (about the size of a sesame seed or smaller). They are a form of plastic pollution and are found in oceans, rivers, soil, and even the air we breathe.

There are two main types of microplastics:

Primary microplastics: These are manufactured at a small size for specific uses, such as microbeads in cosmetics and personal care products, industrial abrasives, or plastic pellets used in manufacturing (Singh et al., 2025).
Secondary microplastics: These are formed from the breakdown of larger plastic items like bottles, bags, and fishing nets due to exposure to sunlight, wind, water, and other environmental factors (Sarkar et al. 2022).

10.2.1 Microplastic size classification

Microplastic size classification (Fig. 10.1) is essential for understanding their behaviour, transport, and ecological impacts in different environments (Table 10.1). The classification is generally based on particle dimensions and can be categorized as follows.

10.2.2 Physical and Chemical Characteristics of Microplastics

Microplastics vary widely in their size, shape, colour, density, and chemical composition (Table 10.2), which influence their behaviour in the environment and interactions with organisms.

10.2.3 Sources of Microplastics in the Ganga River

Understanding the sources of microplastics in the Ganga River is crucial because it helps identify the main contributors to pollution and informs

Fig. 10.1 Microplastic size classification

Table 10.1 Microplastic size classification

Type	*Size range*	*Description*
Macroplastics	> 25 mm	Large plastic debris (e.g., bottles, bags)
Mesoplastics	5–25 mm	Intermediate fragments (e.g., plastic pellets)
Microplastics	< 5 mm	Small particles from wear, fragmentation, or microbeads

Table 10.2 Physical and chemical characteristics of microplastics

Characteristics	*Description*
Physical	
Size	< 5 mm; ranges from visible fragments to microscopic particles
Shape	Fragments, fibres, beads, films, foams
Colour	White, transparent, black, blue, red, and other bright colours
Density	Varies by polymer; PE and PP float, PET and PVC sink
Chemical	
Polymer type	Polyethylene (PE), polypropylene (PP), polystyrene (PS), polyethylene terephthalate (PET), polyvinyl chloride (PVC)
Additives	Plasticizers, flame retardants, UV stabilizers
Adsorbed pollutants	Heavy metals, PAHs, PCBs, pesticides from the surrounding environment
Surface texture	Can be smooth, rough, cracked, or biofilm-coated, affecting pollutant adhesion

Fig. 10.2 Source of microplastics in Ganga River

effective control strategies (Fig. 10.2). Table 10.3 listed the most important sources of microplastics in the Ganga River. It is increasingly contaminated by microplastics from various human activities, urban sewage, industrial effluents, tourism and religious practices, agricultural runoff etc.

10.2.4 Microplastics: An Emerging Environmental Threat

Microplastics, tiny plastic fragments less than 5 mm in size have emerged as one of the most insidious pollutants of the twenty-first century (Borriello et al., 2023; Hale et al., 2020). Found across terrestrial, freshwater, and marine ecosystems, these particles originate from the fragmentation of larger plastic waste or are deliberately produced for use in cosmetics, textiles, and industrial applications (Kosior & Crescenzi, 2020). Their small size allows them to bypass conventional filtration systems, making their way into waterways, soils, the atmosphere, and even drinking water. Once released, they are persistent, non-biodegradable, and highly mobile, making their containment and removal extremely challenging (Montes et al., 2022; Wheatley, 2024).

The ecological consequences of microplastic pollution are far-reaching. In aquatic ecosystems, fish, birds, and invertebrates often mistake microplastics for food, leading to ingestion, internal injuries, digestive blockages, and even death. These particles also serve as carriers for

Table 10.3 Major sources of microplastics in the Ganga River

Source	*Description*
Urban waste and sewage discharge	Household sewage carries synthetic fibres from laundry, microbeads from cosmetics, and degraded plastic waste due to inadequate wastewater treatment
Industrial effluents	Effluents from textile, packaging, and plastic manufacturing industries may contain plastic pellets, resins, and microbeads
Tourism and religious activities	Pilgrims and tourists leave behind plastic packaging, bottles, and ritual waste like plastic wrappers, which degrade into microplastics
Agricultural runoff	Plastic mulch films and agrochemical packaging break down in fields and are washed into the river via rainwater and runoff
Solid waste mismanagement	Open dumping and poor segregation of plastic waste on riverbanks lead to fragmentation and entry of plastics into the river system
Atmospheric deposition	Microplastics present in air can settle into the river directly or onto surrounding land and be washed into the river later

toxic chemicals like heavy metals, pesticides, and persistent organic pollutants (POPs), which adsorb onto their surfaces (Bodke et al., 2024). As these contaminated plastics travel through food webs, they bioaccumulate and biomagnify, threatening not just wildlife but also human health through the consumption of seafood, water, and agricultural produce (Olmo-Gilabert et al., 2024; Pironti et al., 2021).

What makes microplastics particularly dangerous is their invisibility to the naked eye and their deceptive pervasiveness. From the peaks of the Himalayas to the depths of the oceans, no environment remains untouched. The Ganga River, a sacred and lifeline waterway for millions, is now burdened by this invisible menace, highlighting the urgent need for regulatory frameworks, sustainable waste management practices, public awareness, and scientific innovation (Singh et al., 2025). Combating microplastic pollution requires a multi-sectoral approach, involving communities, industries, and policymakers to ensure the protection of both ecosystems and human well-being (Munhoz et al., 2022).

10.3 Findings from the Key Studies

In the lower and estuarine stretches of the Ganga River, Sarkar et al. (2019) reported a high abundance of microplastics (MPs) in sediment samples. These particles were mainly secondary microplastics, indicated by their irregular fragment shapes, and were primarily composed of polyethylene (PE), polypropylene (PP), and polystyrene (PS), as confirmed through ATR-FTIR analysis. Moving upstream to the central and lower stretches of the Ganga, Singh et al. (2025) found MPs in 87% of surface water samples and 92% of sediment samples. The most common forms were fibres and fragments, with PE, PP, and polyethylene terephthalate (PET) being the dominant polymers.

A broader spatial study by Napper et al. (2021), covering urban and semi-urban locations including Sahibganj, Patna, Varanasi, Kannauj, Rishikesh, and Harsil, detected MPs at all sampling sites. Concentrations were highest in urban areas, and the MPs were predominantly fibres, mostly originating from synthetic textiles and packaging materials. Similarly, Badola et al. (2023) studied the Ganga River from Rishikesh to Farakka and confirmed the ubiquitous presence of MPs across all sampled locations. The density was highest near urban and industrial areas, with polyethylene and polystyrene being the most common types.

Beyond the Ganga basin, a study by Farooq et al. (2023) in the Jhelum River in Kashmir revealed that 94% of surface water samples contained MPs. These were mainly fragments and films, especially in urban stretches of the river, with PE, PP, and polyamide (nylon) as dominant constituents. Finally, in the Alaknanda River in Garhwal, Uttarakhand, Chauhan et al. (2021) observed MPs in both surface water and sediment, with a higher concentration in sediments. Fragments and fibres were prevalent forms, and SEM–EDS analysis indicated signs of weathering along with the presence of heavy metals such as iron (Fe), zinc (Zn), and aluminium (Al) adhered to MP surfaces. The studies related to microplastic pollution in the Ganga River is given in Table 10.4.

10.4 Ecological Impacts of Microplastics in the Ganga River

Microplastics in the Ganga River pose serious ecological threats by contaminating aquatic food chains and harming fish, plankton, and invertebrates through ingestion and bioaccumulation (Yu & Singh, 2023).

Table 10.4 Studies related to microplastic pollution in the Ganga River

Serial Number	*Location*	*Sample type*	*Methodology*	*Findings*	*References*
1	Lower and estuarine stretches of Ganga River	Sediment	ATR-FTIR	High abundance of microplastics (MPs) observed, with dominance of polyethylene (PE), polypropylene (PP), and polystyrene (PS); particles were irregular fragments indicating secondary MPs	Sarkar et al. 2019
2	Central and lower stretch of Ganga River	Surface water, sedi- ment	Stereomicroscope and ATR-FTIR	MPs found in 87% of water samples and 92% of sediment samples; fibres and fragments were most common; FTIR identified PE, PP, and PET as dominant polymers	Singh et al. 2021
3	Ganga River (Sahibganj, Patna, Varanasi, Kannauj, Rishikesh and Harsil)	Water sample	Light microscope and FTIR	MPs detected in all sites, with urban sites showing highest concentrations; fibres were predominant; MPs composed mainly of synthetic textiles and packaging materials	Napper et al., 2021

(continued)

Table 10.4 (continued)

Serial Number	*Location*	*Sample type*	*Methodology*	*Findings*	*References*
4	Ganga River, Rishikesh to Farakka	Surface water, sediment	Microscope, ATR- FTIR	Ubiquitous presence of MPs across all locations; highest MP density recorded near urban and industrial zones; polyethylene and polystyrene were prevalent polymers	Badola et al., 2023
5	Jhelum River, Kashmir	Surface water	Microscope, ATR- FTIR	MPs observed in 94% of samples; fragments and films dominant in urban zones; MPs primarily composed of PE, PP, and polyamide (nylon)	Farooq et al., 2023
6	Alaknanda River, Garhwal, Uttarakhand	Surface water, sediment	Stereomicroscope, SEM and EDS	MPs identified in both media, with higher concentration in sediment; fragments and fibres common; SEM–EDS showed weathering and presence of heavy metals (Fe, Zn, Al) on MP surfaces	Chauhan et al., 2021

They degrade riverbed habitats, disrupt nutrient cycling, and host harmful microbes, impacting biodiversity and ecosystem functions (Rajan et al., 2023). Microplastics also enter adjacent wetlands, affecting amphibians and riparian vegetation (Helcoski et al., 2020; Qian et al., 2021). Their long-term persistence makes the damage chronic and difficult to reverse. Here's a concise overview:

10.4.1 Ingestion and Bioaccumulation in Aquatic Fauna

One of the most immediate and visible ecological impacts of microplastics is their ingestion by aquatic organisms.

- **Fish and invertebrates**: Studies have shown that fish in the Ganga often ingest microplastics, mistaking them for food (Badola et al., 2023; Kumari et al., 2023). This leads to intestinal obstruction, reduced feeding efficiency, and even starvation. Microplastics have been found in the stomachs of fish such as *Labeo rohita* and *Catla catla*, which are commonly consumed by humans (Rashid et al., 2024).
- **Zooplankton and filter feeders**: Lower trophic organisms like zooplankton and mussels are particularly vulnerable (Setala et al., 2014). As they ingest microplastics, it affects their feeding behaviour and reproduction, disrupting the entire food web.
- **Trophic transfer**: Microplastics, along with the toxic chemicals they adsorb (like PCBs, dioxins, and heavy metals), bioaccumulate in organisms and bio magnify through the food chain, ultimately posing health risks to top predators including humans (Pandey, 2024; Roy, 2024).

10.4.2 Alteration of Habitat and Ecosystem Functions

Microplastics not only affect individual species but also disrupt entire ecosystems and natural processes.

- **Habitat disruption**: Accumulation of plastic debris alters the physical structure of habitats, especially in riverbeds and delta regions. It affects breeding grounds, nesting sites, and shelters for various species like turtles, amphibians, and benthic invertebrates (Isangedighi et al., 2020).
- **Sediment quality**: Microplastics alter the texture and composition of river sediments, affecting soil porosity, water infiltration, and microbial activities (Yang et al., 2023).
- **Interference with nutrient cycling**: Microplastics provide surfaces for microbial colonization including pathogenic or invasive microbes. These biofilms, known as the "plastisphere," interfere

with normal microbial community functions such as denitrification, organic matter decomposition, and oxygen cycling (Debroy et al., 2022).

10.4.3 Ecotoxicological Effects and Physiological Stress

Microplastics induce a range of sub-lethal physiological effects in aquatic life:

- **Oxidative stress**: Laboratory studies show that microplastics induce oxidative damage in fish liver and gills (Barboza et al., 2018).
- **Endocrine disruption**: Additives in plastic (like BPA and phthalates) act as endocrine disruptors, altering hormone regulation and affecting growth, metabolism, and reproduction (Godswill & Godspel, 2019).
- **Immune suppression**: Chronic exposure to microplastics weakens the immune response in aquatic organisms, making them more susceptible to diseases (Godswill & Godspel, 2019).

10.4.4 Impact on Wetland and Riparian Ecosystems

The Ganga River is lined with numerous floodplain wetlands that serve as biodiversity reservoirs.

- **Microplastic accumulation in wetlands**: During floods, microplastics are transported into adjacent wetlands, where they accumulate in the sediment and water column, affecting amphibians, birds, and aquatic plants (Qian et al., 2021).
- **Impact on riparian vegetation**: Microplastics in soil affect seed germination, root growth, and water absorption in riparian flora, which in turn impacts soil stability and erosion control mechanisms (Li et al., 2023).

10.4.5 Spread of Invasive Species and Pathogens

Microplastics are now known to serve as vectors for microbial pathogens and invasive species:

- **The plastisphere**: Microplastic surfaces in the Ganga have been found to host diverse microbial communities, including antibiotic-resistant bacteria and toxigenic cyanobacteria. This has serious implications for aquatic and human health (Chaturvedi et al., 2024).
- **Biological invasions**: These artificial substrates can transport alien species over long distances, disturbing native species balance and competition (Tyrrell & Byers, 2007).

10.4.6 Long-Term and Irreversible Ecological Consequences

The persistence of microplastics in the Ganga ecosystem has cumulative impacts:

- **Chronic pollution load**: Unlike organic pollutants, microplastics do not degrade easily. They persist for decades, continually accumulating in sediments, aquatic organisms, and floodplains.
- **Irreversible ecosystem shifts**: Over time, the continual stress caused by microplastics may lead to regime shifts permanent changes in ecosystem structure and function, such as reduced species diversity, dominance of tolerant species, and collapse of sensitive ecological communities (Horton & Barnes, 2020).

10.4.7 Human Health Impacts of Microplastics in the Ganga River

The implications of microplastic contamination extend beyond ecology to human health, especially for communities relying directly on the Ganga for water, agriculture, and fish consumption (Bodke et al., 2024).

i **Contaminated Drinking Water**

Surface and groundwater sourced from the Ganga are widely used for human consumption, both directly and indirectly. Microplastics have been detected in tap water and bottled water samples in India, indicating inadequate filtration during treatment. Long-term exposure, even in low concentrations, may pose risks due to the potential leaching of toxic additives such as bisphenol A (BPA), phthalates, and flame retardants (Vo et al., 2024).

ii **Food Chain Transfer**

Humans consuming fish and shellfish from the Ganga may be exposed to microplastics and associated contaminants. Although gastrointestinal tracts are often removed before consumption, smaller particles may translocate into edible tissues. Additionally, rice and salt from areas irrigated with microplastic-contaminated water have shown traces of microplastic content (Kumari et al., 2023).

iii **Health Risks from Associated Chemicals and Pathogens**

Microplastics can absorb and concentrate harmful chemicals from the environment. Once ingested, these chemicals can leach into the body, potentially causing endocrine disruption, inflammation, genotoxicity, and carcinogenesis. Moreover, biofilms on microplastics can harbour pathogens such as E. coli and Vibrio spp., raising the risk of gastrointestinal diseases among river-dependent populations (Debroy et al., 2022).

iv **Occupational Exposure and Gender Implications**

Women engaged in washing clothes along riverbanks, and children playing in or near the water, are particularly vulnerable to skin contact and incidental ingestion of contaminated water. Occupational groups such as fishermen and sanitation workers may also face disproportionate exposure, highlighting a need for gender-sensitive and occupation-specific health assessments (Adams et al., 2022).

10.4.8 Persistent and Future Environmental Crisis in the Ganga Basin

The Ganga Basin is at a tipping point where persistent pollution and future climate-driven changes could converge into an irreversible environmental crisis. Coordinated, science-based interventions such as stricter pollution control, sustainable land and water management, and climate resilience planning are crucial to protect the ecological and cultural sanctity of this lifeline of northern India.

i. **Persistent environmental crises**

- **Water pollution:** Untreated domestic sewage, industrial effluents, agricultural runoff, and religious offerings have turned large stretches of the Ganga into highly polluted zones. Major cities like

Kanpur, Varanasi, and Patna discharge vast amounts of wastewater directly into the river (Barsay, 2022). Faecal coliform levels regularly exceed safe limits, affecting both aquatic life and human health.

- **Declining water flow:** Excessive water extraction for irrigation and dams has reduced the natural flow of the river, particularly during dry seasons. This disrupts sediment transport and degrades the river's self-cleaning ability.
- **Plastic and microplastic pollution:** Plastic waste in the Ganga is a growing concern, with much of it originating from urban runoff, tourism, and religious practices (Kumar, 2017). Over time, plastics break down into microplastics, which are ingested by aquatic organisms and enter the human food chain (Pironti et al., 2021). These pollutants disrupt ecosystems and pose risks to biodiversity and public health. Without improved waste management, the plastic load will keep rising.
- **Loss of biodiversity:** The Ganga supports a rich variety of aquatic and wetland species, but many are endangered due to pollution, habitat loss, and flow alterations. Species like the Ganges River dolphin, otters, and native fish are under serious threat (Singh & Singh, 2020). Encroachment and changing water regimes reduce breeding and feeding grounds. The decline in biodiversity also affects the livelihoods of river-dependent communities, such as fishers and farmers (Warren & Steenbergen, 2021).

ii. **Future environmental threats**

- **Climate change impacts:** Climate change is expected to severely alter the hydrological cycle of the Ganga Basin (Jain & Singh, 2020). Melting glaciers in the Himalayas, the river's primary source, are shrinking, threatening long-term water availability (Miller et al., 2012). Unpredictable monsoons may bring erratic rainfall, causing more intense floods or prolonged droughts. This instability can impact drinking water supply, agriculture, and hydropower. Rising temperatures also affect crop yields and water demand.
- **Groundwater depletion:** The Ganga Basin is heavily dependent on groundwater for irrigation and domestic use. Excessive withdrawal has led to rapidly falling water tables, particularly in the Indo-Gangetic plains (Miller et al., 2012). This depletion affects

base flow to the river and risks drying up wells in rural areas. It also increases the danger of arsenic and fluoride contamination in aquifers (Chakraborti et al., 2018). Without sustainable water management, a crisis of water insecurity is likely.

- **Urbanization and encroachment:** Rapid urban growth along the Ganga's banks is encroaching on floodplains and wetlands. Cities are expanding without adequate infrastructure for sewage and waste treatment. This leads to large-scale discharge of pollutants into the river (Bassi et al., 2014). Industrial zones, especially around Kanpur and Varanasi, release toxic chemicals, worsening water quality. Unregulated development is increasing the basin's ecological stress and flood vulnerability.
- **Agricultural intensification:** To feed a growing population, agriculture in the basin has become more intensive, relying heavily on chemical inputs. Excessive use of fertilizers and pesticides is contaminating soil and water bodies. Over-irrigation, often with groundwater, adds pressure on already-stressed water sources. These practices reduce soil fertility and contribute to long-term productivity loss. Climate-sensitive farming is urgently needed for sustainable food security.

10.5 Policy, Research, and Public Action

Effectively mitigating microplastic pollution in the Himalayan Ganga River ecosystem calls for an integrated approach that aligns governance frameworks, scientific inquiry, and community-level action.

i. **Policy reforms and institutional action**

While the *Namami Gange Mission* has made significant strides in sewage treatment, solid waste management, and riverfront development, microplastics remain an unaddressed pollutant in regulatory frameworks (Balkrishna et al., 2024). To bridge this gap, the following actions are essential:

- Incorporate microplastic monitoring into routine river water and sediment quality assessments, especially under the Central Pollution Control Board's (CPCB) National Water Quality Monitoring Programme.

- Ban or phase out single-use plastics in eco-sensitive and pilgrimage areas like Rishikesh, Haridwar, and Gangotri through state-level amendments and municipal by-laws.
- Enforce Extended Producer Responsibility (EPR) for plastic manufacturers, ensuring that plastic packaging waste is collected and processed properly.
- Establish plastic waste interception units at major drainage outfalls and river confluences, with technologies adapted to mountainous hydrology.

ii **Research and innovation needs**

There is a significant data deficit on microplastic presence and behaviour in high-altitude river systems. Strengthening research will require:

- Long-term monitoring programs to study spatial, seasonal, and altitudinal variations in microplastic concentrations from glacier-fed headwaters to the plains.
- Development of low-cost, portable FTIR or Raman spectroscopy tools to enable field-based microplastic detection in remote areas.
- Toxicological studies to assess the impacts of ingested MPs on fish, macroinvertebrates, and even humans through water and food chain contamination.
- Interdisciplinary collaborations between hydrologists, ecologists, material scientists, and social scientists to understand both ecological and socio-cultural implications.

iii **Public engagement and behavioural shifts**

Cultural reverence for the Ganga can serve as a powerful motivator for environmental action. Public outreach should focus on:

- Plastic literacy programs tailored for different stakeholders' pilgrims, tourists, school children, and local vendors using storytelling, visual aids, and religious narratives.
- Promotion of sustainable pilgrimage by encouraging the use of cloth bags, metal containers, and biodegradable offerings, supported by local governance and temple trusts.
- Launching citizen science platforms for monitoring plastic waste (e.g., mobile apps, clean-up reporting tools) to democratize data collection and build awareness.

- Supporting NGO-led river clean-up campaigns, youth-driven Swachh Ganga movements, and spiritual leaders as ambassadors for zero-plastic behaviour.

10.6 Conclusion: Restoring the Sacred Flow

The Ganga River continues to flow as a lifeline for millions and a sacred emblem in Indian culture, its waters are now tainted with the invisible menace of microplastic products of human negligence, consumer excess, and systemic disconnect from nature. The Himalayan stretch of the Ganga, once pristine and revered as divine origin, now bears the early imprints of anthropogenic pressures, from mass tourism and pilgrimage to urban expansion and plastic proliferation. Microplastic contamination, although microscopic in size, reflects a much larger crisis of sustainability, stewardship, and cultural accountability. Scientific assessments have underscored the ecological, human health, and socio-cultural implications of microplastic pollution. From aquatic biodiversity degradation to bioaccumulation in food chains, from community health risks to the erosion of cultural sanctity every dimension of the Ganga's being is affected. It requires robust monitoring of microplastics, strengthened waste governance, and innovative policy frameworks. But above all, it calls for community-centred engagement, where behavioural shifts are nurtured through environmental literacy, inclusive participation, and empowerment particularly of youth, women, and spiritual leaders. Faith-based environmentalism must evolve from ceremonial rhetoric to climate action. Religious institutions, local governments, NGOs, scientists, and citizens must form a collective consciousness, a new form of pilgrimage that seeks not just the blessings of the river, but actively contributes to its healing. Through science, policy, and community action, it is possible to restore the sanctity of these sacred waters for generations to come.

References

Adams, E. A., Byrns, S., Kumwenda, S., Quilliam, R., Mkandawire, T., & Price, H. (2022). Water journeys: Household water insecurity, health risks, and embodiment in slums and informal settlements. *Social Science & Medicine, 313*, Article 115394.

Ajay, K., Behera, D., Bhattacharya, S., Mishra, P. K., Ankit, Y., & Anoop, A. (2021). Distribution and characteristics of microplastics and phthalate esters from a freshwater lake system in Lesser Himalayas. *Chemosphere, 283*, Article 131132.

Auta, H. S., Emenike, C. U., & Fauziah, S. H. (2017). Distribution and importance of microplastics in the marine environment: A review of the sources, fate, effects, and potential solutions. *Environment International, 102*, 165–176.

Badola, N., Sobhan, F., & Chauhan, J. S. (2023). Microplastics in the River Ganga and its fishes: study of a Himalayan River. *Science and Total Environment, 901*, 1.

Balkrishna, A., Singh, S. K., Ghosh, S., Banerjee, S., Verma, S., & Arya, V. (2024). An analytical review on the integrated management of river resources through Namami Gange. *Water Policy, 26*(5), 462–479.

Barboza, L. G. A., Vieira, L. R., Branco, V., Carvalho, C., & Guilhermino, L. (2018). Microplastics increase mercury bioconcentration in gills and bioaccumulation in the liver, and cause oxidative stress and damage in Dicentrarchus labrax juveniles. *Scientific Reports, 8*(1), 15655.

Barsay, V. P. (2022). Responsibility and Liability of Upstream Cities for the Public Nuisance to the Downstream Cities along the River Ganga. *NUJS Journal of Regulatory Studies, 7*, 66.

Bassi, N., Kumar, M. D., Sharma, A., & Pardha-Saradhi, P. (2014). Status of wetlands in India: A review of extent, ecosystem benefits, threats and management strategies. *Journal of Hydrology: Regional Studies, 2*, 1–19.

Bhardwaj, L. K., Rath, P., Yadav, P., & Gupta, U. (2024). Microplastic contamination, an emerging threat to the freshwater environment: A systematic review. *Environmental Systems Research, 13*(1), 8.

Bodke, V., Tekade, B., & Mali, A. (2024). Health impacts and detection challenges of human exposure to microplastics. *Cancer plus, 6*(4), 4954.

Borriello, L., Scivicco, M., Cacciola, N. A., Esposito, F., Severino, L., & Cirillo, T. (2023). Microplastics, a global issue: Human exposure through environmental and dietary sources. *Foods, 12*(18), 3396.

Chakraborti, D., Singh, S. K., Rahman, M. M., Dutta, R. N., Mukherjee, S. C., Pati, S., & Kar, P. B. (2018). Groundwater arsenic contamination in the Ganga River Basin: A future health danger. *International Journal of Environmental Research and Public Health, 15*(2), 180.

Chaturvedi, S., Chakraborty, B., Min, L., Kumar, A., Pathak, B., Kumar, R., & Yu, Z. G. (2024). Exploring the dynamic microbial tapestry of South Asian rivers: Insights from the Ganges and Yamuna ecosystems. *Ecohydrology, 17*(6), Article e2662.

Chaudhary, M., Rawat, S., & Suthar, S. (2025). Microplastic in upper Himalayan Ganga river: Occurrence, seasonal dynamics and ecological risk. *Science of the Total Environment, 967*, Article 178824.

Chauhan, J. S., Semwal, D., Nainwal, M., Badola, N., & Thapliyal, P. (2021). Investigation of microplastic pollution in river Alaknanda stretch of Uttarakhand. *Environment, Development and Sustainability, 23*, 16819–16833.

Debroy, A., George, N., & Mukherjee, G. (2022). Role of biofilms in the degradation of microplastics in aquatic environments. *Journal of Chemical Technology & Biotechnology, 97*(12), 3271–3282.

Farooq, M., Nisa, F. U., Manzoor, Z., Tripathi, S., Thulasiraman, A. V., Khan, M. I., Khan, M. Y. A., & Gani, K. M. (2023). Abundance and characteristics of microplastics in a freshwater river in northwestern Himalayas, India - Scenario of riverbank solid waste disposal sites. *Science and Total Environment, 886*, 1.

Free, C. M., Jensen, O. P., Mason, S. A., Eriksen, M., Williamson, N. J., & Boldgiv, B. (2014). High-levels of microplastic pollution in a large, remote, mountain lake. *Marine Pollution Bulletin, 85*, 156–163.

Godswill, A. C., & Godspel, A. C. (2019). Physiological effects of plastic wastes on the endocrine system (Bisphenol A, Phthalates, Bisphenol S, PBDEs, TBBPA). *International Journal of Bioinformatics and Computational Biology, 4*(2), 11–29.

Gupta, P., & Kumar, M. (2025). Is the River Ganga a gigantic contributor to microplastic contamination in the groundwater ecosystem under the aegis of climatic extremes? *Journal of Contaminant Hydrology, 1*, 104623.

Hale, R. C., Seeley, M. E., La Guardia, M. J., Mai, L., & Zeng, E. Y. (2020). A global perspective on microplastics. *Journal of Geophysical Research Oceans, 125*(1), e2018JC014719.

Han, B., Yacoub, M., Li, A., Nicholson, K., Gruver, J., Neumann, K., & Sharma, S. (2023). Human activities increased microplastics contamination in the himalaya mountains. *Hydrology, 11*(1), 4.

Helcoski, R., Yonkos, L. T., Sanchez, A., & Baldwin, A. H. (2020). Wetland soil microplastics are negatively related to vegetation cover and stem density. *Environmental Pollution, 256*, Article 113391.

Horton, A. A., & Barnes, D. K. (2020). Microplastic pollution in a rapidly changing world: Implications for remote and vulnerable marine ecosystems. *Science of the Total Environment, 738*, Article 140349.

Isangedighi, I. A., David, G. S., & Obot, O. I. (2020). Plastic waste in the aquatic environment: impacts and management. In *Analysis of nanoplastics and microplastics in food* (pp. 15–43). CRC Press.

Jain, C. K., & Singh, S. (2020). Impact of climate change on the hydrological dynamics of River Ganga, India. *Journal of Water and Climate Change, 11*(1), 274–290.

Jolaosho, T. L., Rasaq, M. F., Omotoye, E. V., Araomo, O. V., Adekoya, O. S., Abolaji, O. Y., & Hungbo, J. J. (2025). Microplastics in freshwater and marine ecosystems: Occurrence, characterization, sources, distribution dynamics, fate, transport processes, potential mitigation strategies, and policy interventions. *Ecotoxicology and Environmental Safety, 294*, Article 118036.

Kosior, E., & Crescenzi, I. (2020). Solutions to the plastic waste problem on land and in the oceans. In *Plastic waste and recycling* (pp. 415–446). Academic Press.

Kumar, D. (2017). River Ganges–historical, cultural and socioeconomic attributes. *Aquatic Ecosystem Health & Management, 20*(1–2), 8–20.

Kumar, R., Verma, A., Shome, A., Sinha, R., Sinha, S., Jha, P. K., & Vara Prasad, P. V. (2021). Impacts of plastic pollution on ecosystem services, sustainable development goals, and need to focus on circular economy and policy interventions. *Sustainability, 13*(17), 9963.

Kumari, N., Yadav, D. K., Khan, P. K., & Kumar, R. (2023). Occurrence of plastics and their characterization in wild caught fish species (Labeo rohita, Wallago attu and Mystus tengara) of River Ganga (India) compared to a commercially cultured species (L. rohita). *Environmental Pollution, 334*, 122141.

Lamichhane, G., Acharya, A., Marahatha, R., Modi, B., Paudel, R., Adhikari, A., & Parajuli, N. (2023). Microplastics in environment: Global concern, challenges, and controlling measures. *International Journal of Environmental Science and Technology, 20*(4), 4673–4694.

Li, K. X., Jin, M., & Shi, W. (2018). Tourism as an important impetus to promoting economic growth: A critical review. *Tourism Management Perspectives, 26*, 135–142.

Li, X., Wang, R., Dai, W., Luan, Y., & Li, J. (2023). Impacts of micro (nano) plastics on terrestrial plants: Germination, growth, and litter. *Plants, 12*(20), 3554.

Matavos-Aramyan, S. (2024). Addressing the microplastic crisis: A multifaceted approach to removal and regulation. *Environmental Advances, 1*, 100579.

Miller, J. D., Immerzeel, W. W., & Rees, G. (2012). Climate change impacts on glacier hydrology and river discharge in the Hindu Kush-Himalayas. *Mountain Research and Development, 32*(4), 461–467.

Montes, A. M., Navarro-Pedreño, J., Almendro-Candel, M. B., Lucas, I. G., & Zorpas, A. A. (2022). Microplastics in soils as a source of pollution and environmental risk. *Emerging pollutants in sewage sludge and soils* (pp. 37–59). Springer International Publishing.

Munhoz, D. R., Harkes, P., Beriot, N., Larreta, J., & Basurko, O. C. (2022). Microplastics: A review of policies and responses. *Microplastics, 2*(1), 1–26.

Napper, I. E., Baroth, A., Barrett, A. C., Bhola, S., Chowdhury, G. W., Davies, B. F. R., Duncan, E. M., Kumar, S., Nelms, S. E., Hasan Niloy, M. N., Nishat, B., Maddalene, T., Thompson, R. C., & Koldewey, H. (2021). The abundance and characteristics of microplastics in surface water in the transboundary Ganges River. *Environmental Pollution, 274*, Article 116348.

Olmo-Gilabert, R., Fagiano, V., Alomar, C., Rios-Fuster, B., Compa, M., & Deudero, S. (2024). Plastic webs, the new food: Dynamics of microplastics in a Mediterranean food web, key species as pollution sources and receptors. *Science of the Total Environment, 918*, Article 170719.

Pandey, K. (2024). Invisible threats in Himalayan region: Unmasking the impact of microplastic pollution on human health today and tomorrow. In *Microplastics in African and Asian Environments: The Influencers, Challenges, and Solutions* (pp. 131–144). Springer Nature Switzerland.

Pironti, C., Ricciardi, M., Motta, O., Miele, Y., Proto, A., & Montano, L. (2021). Microplastics in the environment: Intake through the food web, human exposure and toxicological effects. *Toxics, 9*(9), 224.

Qian, J., Tang, S., Wang, P., Lu, B., Li, K., Jin, W., & He, X. (2021). From source to sink: Review and prospects of microplastics in wetland ecosystems. *Science of the Total Environment, 758*, Article 143633.

Raha, U. K., Kumar, B. R., & Sarkar, S. K. (2021). Policy framework for mitigating land-based marine plastic pollution in the Gangetic Delta Region of Bay of Bengal- a review. *Journal of Cleaner Production, 278*, Article 123409.

Rajan, K., Khudsar, F. A., & Kumar, R. (2023). Urbanization and population resources affect microplastic concentration in surface water of the River Ganga. *Journal of Hazardous Materials Advances, 11*, Article 100342.

Rashid, E., Hussain, S. M., Ali, S., Sarker, P. K., & Al-Ghanim, K. A. (2024). Microplastics accumulation in gut and revealing their impacts on nutritional quality and health of freshwater carp, Catla Catla. *Aquaculture Reports, 38*, Article 102299.

Roy, T. (2024). Aquatic ecosystem toxicity and food chain. In *Toxicity of aquatic system and remediation* (pp. 25–36). CRC Press.

Setälä, O., Fleming-Lehtinen, V., & Lehtiniemi, M. (2014). Ingestion and transfer of microplastics in the planktonic food web. *Environmental Pollution, 185*, 77–83.

Singh, P. K., Singh, A., Srivastava, A. K., Chauhan, R., Basniwal, R. K., & Chauhan, A. (2025). Microplastic pollution in the Ganga River: A state-of-the-art review of pathways, mechanisms, and mitigation. *Water Supply*, ws2025009.

Singh, R., & Singh, G. S. (2020). Integrated management of the Ganga River: An ecohydrological approach. *Ecohydrology & Hydrobiology, 20*(2), 153–174.

Tyrrell, M. C., & Byers, J. E. (2007). Do artificial substrates favor nonindigenous fouling species over native species? *Journal of Experimental Marine Biology and Ecology, 342*(1), 54–60.

Vasse, G. F., & Melgert, B. N. (2024). Microplastic and plastic pollution: impact on respiratory disease and health. *European Respiratory Review, 33*(172), 1.

Vo, D. H. T., Nguyen, V. T., Nguyen, T. B., Le, L. T., Mukhtar, H., Nguyen, H. V., & Bui, X. T. (2024). Microplastics contamination in water supply system and treatment processes. *Science of the Total Environment, 1*, 171793.

Warren, C., & Steenbergen, D. J. (2021). Fisheries decline, local livelihoods and conflicted governance: An Indonesian case. *Ocean & Coastal Management, 202*, Article 105498.

Wheatley, M. C. (2024). Microplastics: Environmental Presence and Ecological Impact. *Science, 1*, Article 100010.

Yang, X., Zhang, Z., & Guo, X. (2023). Impact of soil structure and texture on occurrence of microplastics in agricultural soils of karst areas. *Science of the Total Environment, 902*, Article 166189.

Yu, R. S., & Singh, S. (2023). Microplastic pollution: Threats and impacts on global marine ecosystems. *Sustainability, 15*(17), 13252.

CHAPTER 11

Exploring the Ecological Risks of Urban River Water Quality Through Risk Quotient Assessment: A Case Study of the Banka River

Surajit Ghosh, Shubhajit Saha, Ahamadul Hoque Mandal, Nimai Chandra Saha, and Ayan Mondal

11.1 Introduction

The growing global population requires expanding reliable water networks, putting pressure on current systems (Mishra et al., 2021). Water quality, especially surface water, is crucial for human survival, providing drinking water and irrigation for homes and farms (Khatri & Tyagi, 2015). Rapid industrialization and urban growth threaten the fragile balance of river ecosystems. The encroachment of development

S. Ghosh · S. Saha · A. H. Mandal
Ecotoxicology Research Laboratory, Department of Zoology, The University of Burdwan, Bardhaman, West Bengal, India

N. C. Saha
School of Life Science, Swami Vivekananda University, North 24 Parganas, West Bengal, India

S. Ghosh (ed.), *Reimagining Indian Rivers for Sustainability*, Palgrave Studies in Environmental Sustainability,
https://doi.org/10.1007/978-3-032-04754-0_11

into riparian zones and watersheds enables the uncontrolled release of pollutants, radically changing the chemical and biological integrity of these vital waterways (Dhara et al., 2021; Ghosh et al., 2024, 2025a; Mandal et al., 2024a). River water quality degrades through two main pathways: point source pollution from identifiable sites like industrial pipes and sewage plants, and nonpoint pollution from diffuse sources like farm runoff, urban stormwater, and airborne deposition (Akhtar et al., 2021). Both contribute a complex mix of contaminants with distinct impacts on aquatic environments. Point sources introduce concentrated loads of specific pollutants, while nonpoint sources spread a wider range of substances over larger areas, complicating control efforts (Singh & Gupta, 2016). Pollution threatens aquatic ecosystems. Beyond just looking bad, it hurts biodiversity and ecological processes, and risks human health (Ghosh et al., 2025b; Mandal et al., 2024b). Heavy metals, pollutants, and germs can make water unfit for drinking, farming, and recreation, undermining society's well-being (Sadhu et al., 2024). Many nations struggle to meet water quality goals despite water management plans and regulations. Challenges include poor enforcement, lack of funding, and absence of integrated watershed approaches (Wang et al., 2016). The dynamic nature of pollution and complex hydrology demand flexible and responsive management. Protecting and restoring rivers requires a major shift in water resource strategies. This includes sustainable practices to reduce pollution, investing in advanced wastewater treatment, and raising public awareness of responsible water stewardship (Wang et al., 2016). Collaborative efforts among government, industry, and communities are vital for a holistic approach to river conservation. Comprehensive monitoring programs are essential to detect water quality issues early, enabling timely interventions and adaptive management. Prioritizing river health ensures long-term sustainability of these critical resources for present and future generations.

The Banka River basin is situated in the Bardhaman district of West Bengal. Originating from farmland near Ramgopalpur village in the Galsi Police Station area, the river meanders eastward, running parallel to the

A. Mondal (✉)
Department of Zoology, Government General Degree College, Mohanpur, Paschim Medinipur, West Bengal, India
e-mail: mondalayan.zoo@gmail.com

Damodar River through the town of Burdwan. The river basin stretches up to Kanchannagar, covering an expansive area of approximately 181 square kilometers (Chakraborty et al., 2011). It flows 125 kilometers before meeting the Bhagirathi River in Kalna. Burdwan's growing population greatly worsens Banka River pollution. More waste overwhelms sewage and disposal systems, causing untreated or partly treated waste to flow into the river. This pollutes the water, lowers oxygen, harms wildlife, and risks public health, degrading the river's environment. Hazardous heavy metal pollution in aquatic habitats is a pressing issue due to its toxic and enduring nature (Dhara et al., 2022; Saha et al., 2023). These contaminants amass in living beings and amplify up the food web, leading to grave consequences (Saha et al., 2023). The likelihood of intensified river water pollution in Burdwan town is high, potentially inflicting substantial harm on the riverine ecosystem and nearby residents through direct consumption and skin exposure. Therefore, extensive investigation is essential to assess the water quality and quantify the concentrations of trace and heavy metals in the Banka River, which may pose ecological and human health risks.

The water quality index is a straightforward technique used to evaluate the fitness of surface waters for human use (Naubi et al., 2016; Syeed et al., 2023). It condenses various water quality indicators, such as dissolved oxygen, pH, nitrate, and electrical conductivity, into a single numerical metric (Akkoyunlu & Akiner, 2012; Kannel et al., 2007; Sánchez et al., 2007). This method proves advantageous for assessing river water conditions and comprehending the overall status across different sampling points (Fathi et al., 2018; Kumar & Dua, 2009; Naubi et al., 2016). This study employs risk quotient analysis to assess water quality and evaluate the potential environmental and public health implications (Mohtar et al., 2019; Varol & Tokatlı, 2023). The research focuses on the impact of urbanization on the Banka River, which serves as a critical habitat for aquatic flora and fauna that are a vital source of nutrition and protein for the local population. The study's primary objective is to determine the river's water quality using the WQI method, thereby assessing its suitability. The present investigation is centered around conducting a comprehensive risk assessment for six key water quality parameters: pH, electrical conductivity (EC), suspended solids (TDS), biochemical oxygen demand (BOD), chemical oxygen demand (COD), and ammoniacal nitrogen (NH_3-N), as well as two heavy metal concentrations in the water, namely lead (Pb) and cadmium (Cd).

11.2 Methodology

11.2.1 Study Area and Sampling Locations

The Banka River basin, situated in the Bardhaman district of West Bengal, originates from agricultural lands near Ramgopalpur village in the Galsi Police Station area. Flowing eastward, parallel to the Damodar River, the river passes through the urban center of Burdwan. Its catchment up to Kanchannagar spans approximately 181.30 km^2. Extending for around 125 kilometers, the Banka River ultimately converges with the Bhagirathi River in Kalna. Sampling locations, ranging from upstream pre-urban to urban and downstream sections within Bardhaman District, were strategically chosen based on assumed water quality, population status, and prior knowledge of contamination levels in the investigation area. The riverine landscape is predominantly characterized by agriculture, built-up areas, and residential and commercial developments. Figure 11.1 and Table 11.1 shows the landmark coordinates of the selected study area. The study area had three stations: S-1, S-2, and S-3. S-1 was upstream in Kanchan Nagar, near the Rathtala bridge, and the main pollution sources were human settlements. S-2 was in a densely urbanized area, with municipal discharge as the main pollution source. S-3 was downstream, where the town ended and rural areas began (Fig. 11.1).

11.2.2 Sampling Procedure

To examine river water quality, 500 mL water samples were collected using plastic bottles. These bottles underwent a rigorous cleaning process: first, they were washed with a detergent solution, followed by a 24-hour soak in 5% nitric acid (HNO_3). The sample bottles were thoroughly rinsed with purified water and dried prior to use. During the sampling procedure, the bottles were pre-rinsed at least three times with the river water itself. The collected samples were submerged approximately 10 cm deep within the water body before being retrieved (Kabir et al., 2020). Without delay, the samples were transported to the laboratory for comprehensive chemical analysis. For the heavy metal assessment, Whatman filter paper was employed to filter the samples. Subsequent to filtration, 150 mL aliquots were extracted, and a small quantity of concentrated nitric acid was added to preserve the samples. These preserved samples were then stored in a freezer. Throughout the laboratory procedures, utmost care was taken to maintain the samples in a clean, cool,

Fig. 11.1 Locations of study area and sampling stations in the Banka River around the urbanized area in Burdwan town of West Bengal, India

Table 11.1 Coordinates of the selected locations

Landmark	*Division*	*Location*	*Coordinates*
S-1	Upstream point	Kanchan Nagar, West Bengal	N 23° 14.682' and E 87° 49.560'
S-2	Urban point	Perbirhata, West Bengal	N 23° 14.151', E 87° 51.287'
S-3	Downstream point	Bamchandaipur, West Bengal	N 06° 06.839', E 100° 21.887'

and dry setting, thereby preventing any potential contamination (APHA, 1998).

11.2.3 Sample Analysis

The acidity (pH), electrical charge concentration (EC), and dissolved particle quantities (TDS) were determined at the sample collection location. A digital pH gauge was leveraged to approximate the acidity, while sophisticated EC and TDS instruments were utilized to assess the electrical charge concentration and dissolved particle quantities, respectively. All the digital instruments were calibrated using deionized water and buffer solutions prior to the investigation (APHA, 1998). Biochemical oxygen demand was assessed through a two-step procedure: the initial DO1 was recorded promptly after sample gathering, and the ensuing DO5 was approximated following a 5-day incubation of the samples in the dark at 20 °C. The Huq and Alam (2005) approach was then employed to compute the BOD. As for the heavy metal evaluation, the prepared samples were placed in vials and scrutinized for lead and cadmium using an Atomic Absorption Spectrometer (APHA, 1998).

11.2.4 Quality Control

Highly-purified reagents and chemicals were utilized to maximize accuracy and precision. Water specimens were gathered and evaluated using calibrated equipment. Quality control measures included duplicate sampling, triplicate data collection, and repeated testing. Procedures like blank measurement, reagent standard monitoring, and reference material tests were also performed. Each analytical method had a quality assurance program. Detailed documentation of each step was carried out and maintained for data management, troubleshooting, and review purposes.

*11.2.5 Calculation of Water Quality Index (*WQI*)*

To determine the water quality index, six critical factors were evaluated. The data used for each variable reflected the mean across the three sampling locations studied. The *WQI* was calculated using the conventional method for evaluating drinking water quality. The weighted arithmetic index technique was employed according to Brown et al. (1972) to assess the water body's water quality index, following these steps.

*11.2.5.1 Calculation of Sub Index of Quality Rating (*Qn*)*
To determine the quality rating (Qn) for each water quality parameter, the following formula is used:

$$Qn = 100[(Vn - Vio)/(Sn - Vio)]$$

where, Qn represents the quality rating for the nth water quality parameter; Vn is the measured value of the nth parameter at the sampling location; Sn is the maximum acceptable value for the nth parameter, based on established standards; and Vio is the ideal value of the nth parameter in pure water. For most drinking water parameters, the ideal value (Vio) is assumed to be zero. However, exceptions exist for pH, where Vio is 7.0, and dissolved oxygen (DO), where Vio is 14.6 mg/L (Pathak et al., 2015).

*11.2.5.2 Calculation of Unit Weight (*Wn*)*
The individual weight assigned to each water quality measurement is determined by its inverse relationship to the established regulatory standard. This relationship can be expressed as:

$$Wn = K/Sn$$

where Wn represents the weight of the nth parameter, Sn is the standard value for that same parameter, and K is a constant of proportionality.

The overall water quality index (WQI) is then calculated by summing the products of each parameter's quality rating (RQn) and its corresponding weight (Wn), and dividing that sum by the total sum of all weights. This can be expressed as:

$$WQI = \sum RQnWn / \sum Wn$$

11.3 Evaluation of Potential Human Health Impacts

To determine the water quality index, six critical factors were evaluated. The values used for each variable represented the mean across the three sampling locations under study. The water quality index was established using the conventional approach for assessing the potability of water. The weighted arithmetic index technique was employed to calculate the water quality index of the water source, adhering to the following steps (Proshad et al., 2021):

$$CDI(Oral) = \frac{C * IR * EF * ED}{BW * AT}$$

$$CDI(Dermal) = \frac{C * SA * Kp * ET * EF * ED}{BW * AT}$$

Specifically, '*C*' represents the heavy metal's level in water. '*IR*' indicates how much water is consumed. '*EF*' denotes how often exposure occurs. '*ED*' is the length of time an individual is exposed. '*AT*' is the average time period over which exposure is considered. '*BW*' is the individual's body weight. '*SA*' is the body's surface area in contact with the water. '*Kp*' is a measure of how easily the metal passes through the skin. Finally, '*ET*' is the duration of skin contact.

The risk quotient (Marquis et al.) is employed to determine the likelihood of non-carcinogenic health effects. This assessment is based on the chronic daily intake (*CDI*) and is used to characterize potential health hazards, following the methodology outlined in Akter et al. (2020).

$$RQ(Oral) = \frac{CDI(Oral)}{RfD(Oral)}$$

$$RQ(Dermal) = \frac{CDI(Dermal)}{RfD(Dermal)}$$

11.4 Results and Discussion

11.4.1 Physicochemical Characteristics of Riverine Environments

Table 11.2 shows the chemical and physical characteristics of the Banka River water, including heavy metal levels. Electrical conductivity (EC), which indicates the amount of dissolved salts, varied between 375 and 995 mS/cm, averaging 747.67 mS/cm. The lowest value was at location S-1 and the highest at S-3. While this is below the inland surface water standard of 1200 mS/cm, it exceeds the recommended EC for recreational use (500 mS/cm) and some irrigation practices (750 mS/cm), but falls within the range for aquaculture (800–1000 mS/cm) (Dugan, 2024). Because EC is directly related to total dissolved solids (TDS), these high EC values suggest elevated salt concentrations (Thirumalini & Joseph, 2009). The concentration of total dissolved solids is crucial for supporting aquatic ecosystems, with recommended thresholds of 500 ppm for drinking, 1500 ppm for industrial, and 2000 ppm for irrigation (Firmansyah et al., 2020). Measurements along the Banka River revealed TDS values ranging from 76.42 to 171.65 parts per million, with an average of 130.31 ppm. The lowest and highest concentrations were observed at sampling sites S-1 and S-3, respectively. These findings align with observations reported by Awomeso et al. (2010) and Sharma et al. (2020). Similarly, pH, a key factor in water quality affecting chemical solubility and toxicity, ranged from 6.63 to 8.12, with a mean of 7.40, also showing extremes at S-1 and S-3. All measured pH values fell outside the World Health Organization's (WHO) acceptable range of 6.5–7.5 for drinking water and aquatic life. Notably, samples from urban and downstream areas exhibited pH levels above 7, likely due to municipal runoff introducing carbonate and bicarbonate compounds. Biological Oxygen Demand (BOD), which measures the oxygen used by microbes to break down organic pollutants, inversely correlates with dissolved Oxygen (DO) (Maddah, 2022). In this study, BOD levels ranged from 137 to 356 ppm, averaging 235 ppm, with the lowest value at sampling site S-1 and the highest at S-3, mirroring the DO pattern. These BOD levels align with findings from research done by Islam et al. (2015b) and Jha et al. (2007). Table 11.2 outlines the heavy metal concentrations, reported in parts per million, across the three sampling locations along the Banka River. These toxic substances enter the water through diverse mechanisms, such as sediment disturbance, chemical reactions, and biological processes (Singh et al., 2022). Notably, the lead and cadmium levels in the sampled waters surpassed the permissible guidelines for drinking water. Islam et al. (2015a) have reported that high concentrations of most metals are associated with wastewater discharges from surrounding municipal sources. Although the analysis of variance conducted in our study at a significance level of 0.05 did not reveal significant differences in the concentrations of Pb and Cd across the three sites, the ANOVA for other parameters showed statistically significant differences between the sites at the 0.05 level. The descriptive statistics and the P-value close to 0.05 suggest that there may be potential differences worth further investigation. Site 3 exhibited the highest values across the measured parameters, followed by Site 2 and then Site 1. The study revealed that the examined river exhibited substantial levels of pollution, with notably elevated levels of contaminants detected across multiple measurements and sampling locations. These findings align with findings from Zhao et al. (2013), Zhou et al. (2022), and Uddin and Jeong (2021).The findings indicate a significant degree of environmental degradation in this waterway, likely due to a combination of industrial and municipal waste discharges into the river system.

11.4.2 *Water Quality Index (*WQI*) Analysis*

The study assessed water quality at three locations along the Banka River, revealing average Water Quality Index (WQI) values of 1092.37, 2056.4, and 2650.1, respectively, indicating generally poor water quality unsuitable for both domestic and agricultural use based on established standards by Brown et al. (1972). The downstream site, situated proximal to municipal waste management facilities, exhibited the most extreme levels of EC, TDS, pH, BOD, COD, NH_{3}-N, Pb, and Cd, resulting in its substantially compromised water quality (Khalil et al., 2023; Suthar et al., 2010). Conversely, the upstream site (S-1), in a less populated area, showed comparatively better water quality. The increased WQI at sites with higher heavy metal concentrations, like lead and cadmium, highlights the significant impact of these pollutants, even at low levels. Domestic sewage from settlements was identified as a major pollution source, particularly at the downstream sites, exceeding the impact of agricultural runoff observed upstream. Overall, the average WQI of 1932.96 confirms the river's compromised condition, necessitating treatment for safe utilization.

11.4.3 *Potential Causes of Banka River Water Contamination*

Our studies regarding the Banka River's water quality painted a vivid and compelling portrait. The correlation matrix, our initial window into the data, unveiled a remarkable interconnectedness: all measured water quality metrics, including the worrying heavy metal concentrations, demonstrated a substantial positive correlation. This observation immediately suggested that a common underlying factor was influencing the overall water quality, prompting us to delve deeper into the data structure. To unravel these complex interrelationships and pinpoint potential pollution sources, we employed Principal Component Analysis (PCA). The scree plot, a visual representation of the eigenvalues, provided a critical insight. It demonstrated that the first principal component (PC1) dominated the variance, explaining a substantial 97.7% of the total variation with an eigenvalue of 7.82. This dominance strongly indicated the presence of a powerful, overarching factor affecting water quality across the sampling sites. Notably, PC1 exhibited moderate positive loadings for all measured parameters – pH, EC, TDS, BOD, COD, NH_3-N, Pb, and Cd. This suggests that sites with higher PC1 values are characterized by elevated levels of all these parameters, effectively establishing PC1 as a robust general pollution or contamination index. However, the second principal component (PC2), while accounting for a much smaller 2.3% of the variance, offered a nuanced perspective. These results also aligned with correlation analysis confirmed the impact of both inorganic and organic pollution (Ustaoğlu et al., 2020). It highlighted a distinct contrast between organic and chemical pollution. Specifically, BOD, a marker of organic pollution, showed a positive loading, while EC, Pb, and COD, indicative of chemical pollution, displayed negative loadings. This divergence suggests that PC2 differentiates sites based on the relative influence of these two pollution types. This differentiation could reflect the influence of distinct pollution sources. For instance, sites characterized by high BOD values might be predominantly affected by organic waste from sewage or agricultural runoff. Conversely, sites with elevated EC, Pb, and COD levels could be more susceptible to industrial discharges or leaching from contaminated soils (Yuan et al., 2017). The separation of these pollution types by PC2 is crucial for targeted remediation and pollution control strategies. By pinpointing the primary drivers of pollution at each site, we can implement more effective and site-specific interventions. This study's findings highlight the pivotal role of PCA in unraveling the intricate connections between water quality factors and pinpointing pollution sources, which greatly enhances our understanding of the Banka River's environmental well-being.

11.4.4 Risk Quotient Assessment

To assess the potential health hazards stemming from extended exposure to heavy metal-polluted water through ingestion and dermal absorption, the Chronic Daily Intake and Risk Quotient were quantified for both adults and children depending on the Banka River's surface water (Mohtar et al., 2019). RQ > 1 indicates a non-carcinogenic health risk from exposure to a hazardous element, while RQ < 1 indicates acceptable risk (Kapoor et al., 2024). The RQ, indicating non-carcinogenic risk, shows that adults face minimal risk as their RQ values for both lead (Pb) and cadmium (Cd) are below the safe threshold of 1 across all sampling sites. However, children exhibit a significant risk from Pb exposure at all sites, with RQ values exceeding 1. Additionally, children are at risk from Cd exposure at site-2 and site-3, where RQ values are also above 1, suggesting varying levels of vulnerability based on age and location (Mgbenu & Egbueri, 2019).

11.4.5 Impact of Water Pollution on Human Health and Environmental Functionally in the City

The Banka River's severe pollution, driven by untreated municipal sewage and heavy metal contamination (lead and cadmium), poses significant risks to human health (Ghosh et al., 2020) and disrupts environmental and fluvial functionality in Burdwan (Mondal et al., 2020, 2022). Elevated levels of lead and cadmium, exceeding WHO drinking water standards, bioaccumulate in aquatic organisms like fish, a dietary staple for local communities, leading to chronic health issues such as neurological impairment, kidney dysfunction, and developmental disorders, particularly in children, as evidenced by risk quotient values exceeding 1 for lead exposure across all sampling sites and cadmium at urban sites (Ghosh et al., 2020; Sen et al., 2023). High biochemical oxygen demand (BOD: 137–356 ppm) and chemical oxygen demand (COD: 53.42–307.56 ppm) indicate organic pollution, depleting dissolved oxygen and creating hypoxic conditions that impair fish survival and biodiversity, critical for fluvial ecosystems (Majhi et al., 2018). This degradation disrupts nutrient cycling and sediment transport, reducing the river's capacity to support aquaculture and irrigation (Mondal et al., 2023), vital for Burdwan's economy. Urban runoff and industrial discharges further exacerbate water quality decline, with electrical conductivity (375–995 μS/cm) and total dissolved solids (76.42–171.65 ppm) signaling high salinity, unsuitable for domestic use. The compromised fluvial functionality limits the river's role as a natural drainage system, increasing flood risks in urban areas (Batabyal et al., 2022; Ghosh & Maji, 2011). Comprehensive sewage treatment and continuous monitoring are essential to restore ecological balance and safeguard public health.

11.4.6 Remedial Measures to Prevent Environmental Crisis

To avert the environmental crisis in the Banka River basin, characterized by high organic pollution (BOD: 137–356 ppm, COD: 53.42–307.56 ppm) and heavy metal contamination (lead: 0.1–0.44 ppm, cadmium: 0.03–0.09 ppm), a multifaceted remediation strategy is imperative (Ghosh & Maji, 2011). Implementing centralized sewage treatment plants (STPs) with advanced technologies, such as activated sludge processes and membrane bioreactors, can effectively reduce untreated municipal sewage discharge, a primary pollution source (Chakraborty et al., 2011). Constructed wetlands, leveraging phytoremediation with plants like *Phragmites australis*, or making net-bund with *Eichhornia crassipes* can naturally filter organic pollutants and heavy metals, enhancing water quality. Regular dredging and sediment management are essential to remove contaminated sediments (Ghosh & Maji, 2011), preventing metal bioaccumulation in aquatic organisms. Enforcing strict regulations on industrial discharges, coupled with real-time monitoring using automated sensors for pH, EC, and heavy metals, ensures compliance and early detection of pollution spikes. Community-based initiatives, including public awareness campaigns on waste disposal and sustainable agriculture (Gangopadhyay et al., 2023), can minimize nonpoint source pollution from urban and farm runoff (Banerjee et al., 2020). Riparian buffer zones with native vegetation should be established to reduce contaminant inflow and stabilize riverbanks. Integrating these measures with continuous water quality index (WQI) assessments and principal component analysis (PCA) will guide adaptive management, prioritizing upstream rural segments (e.g., S-1, WQI: 1092.37) for preservation and urban sites (S-2, S-3, WQI: 2056.4–2650.1) for intensive restoration, ensuring ecological recovery and public health safety.

11.5 Conclusion

The investigation into the Banka River's water quality revealed a distressing scenario, characterized by substantial pollution primarily stemming from the burgeoning population pressures of the surrounding regions. Specifically, the discharge of untreated municipal sewage emerged as a dominant contributor, a fact starkly illustrated by the data collected. The analysis of various physicochemical parameters unveiled that a significant number of these metrics exceeded the recommended limits for both domestic and agricultural utilization, underscoring the severity of the contamination. Notably, the water quality deteriorated significantly at urban and downstream sampling points, a phenomenon directly attributable to the accumulation of waste generated from domestic activities. The presence of excessive concentrations of toxic heavy metals, including lead and cadmium, further exacerbated the situation, leading to a substantial decline in the Water Quality Index (WQI). This deterioration rendered the river water unsuitable for critical applications such as fisheries, industrial processes, and irrigation, severely limiting its utility. Moreover, the elevated levels of organic pollutants, such as COD, BOD, and NH_{3}-N pose a direct threat to the aquatic ecosystem. These pollutants have the potential to significantly reduce dissolved oxygen levels within the river, creating hypoxic conditions that are detrimental to fish and other aquatic organisms, which rely on oxygen for survival. Furthermore, the contamination of the river water leads to the bioaccumulation of heavy metals within the tissues of aquatic organisms, particularly fish. These organisms absorb and retain metals like lead and cadmium, effectively concentrating these toxins within their bodies. When humans consume fish contaminated in this manner, the accumulated metals transfer up the food chain, resulting in increased exposure levels. Prolonged exposure to these heavy metals can cause a variety of severe health problems, such as neurological impairment, kidney malfunction, and developmental issues, particularly in vulnerable groups like children and pregnant women. The calculated risk quotient further substantiated this concern, suggesting a high likelihood of detrimental health risks associated with the use of the river water. Consequently, the study determined that the Banka River's water is unsuitable for utilization in its current condition and presents significant adverse consequences if utilized. To address this critical situation, a comprehensive and well-structured restoration plan is imperative. This plan should focus on reducing pollution loadings from various

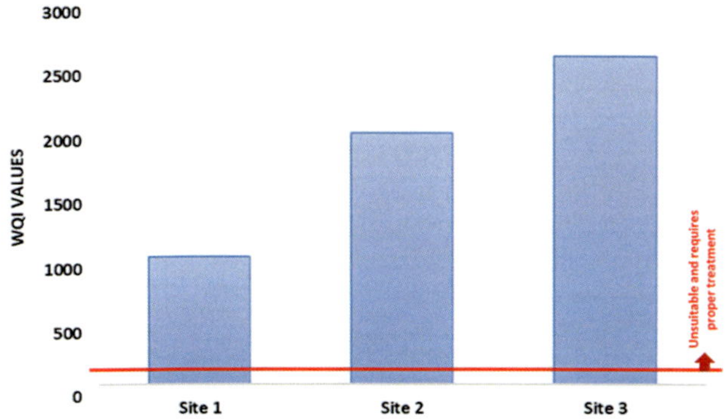

Fig. 11.2 Suitability status of Banka River water among sampling stations for utilizing in various purposes

sources, thereby improving the river's management system. The implementation of centralized sewage treatment plants is crucial, as they offer a more effective solution compared to the current decentralized practices in residential areas. Moreover, the application of appropriate treatment technologies for all types of wastewaters before discharge into the river is essential. However, the success of these interventions hinges on the establishment of a robust and continuous monitoring program. Consistent surveillance and protective measures are essential to maintain the ecological well-being of the Banka River. Simultaneously, it is crucial to curtail harmful human activities that exacerbate the river's degradation. By enacting these actions, we can strive to revive the Banka River to a flourishing and sustainable condition, safeguarding both the ecosystem and the communities reliant upon it (Figs. 11.2, 11.3, 11.4, 11.5, 11.6, and Tables 11.2, 11.3).

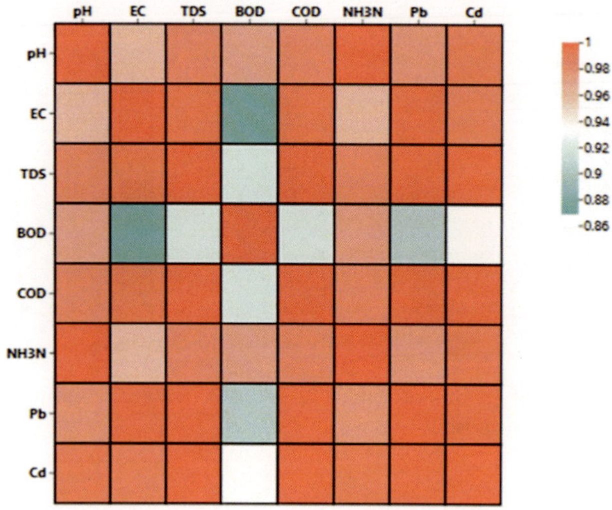

Fig. 11.3 Correlation matrix between physicochemical parameters and heavy metals of water

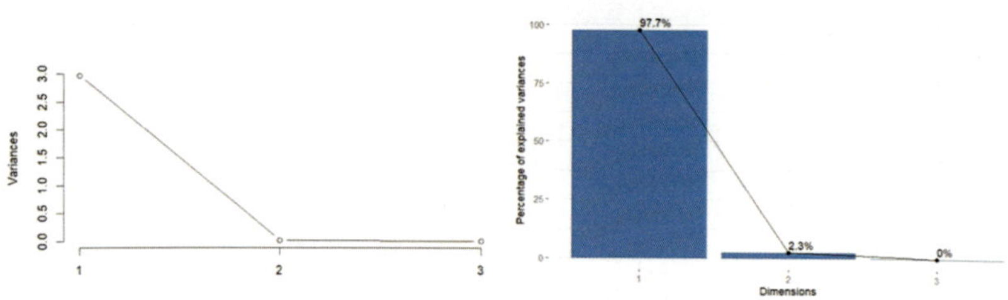

Fig. 11.4 Scree plot describing variance for each PC for determining the number of principal components (PCs) to retain

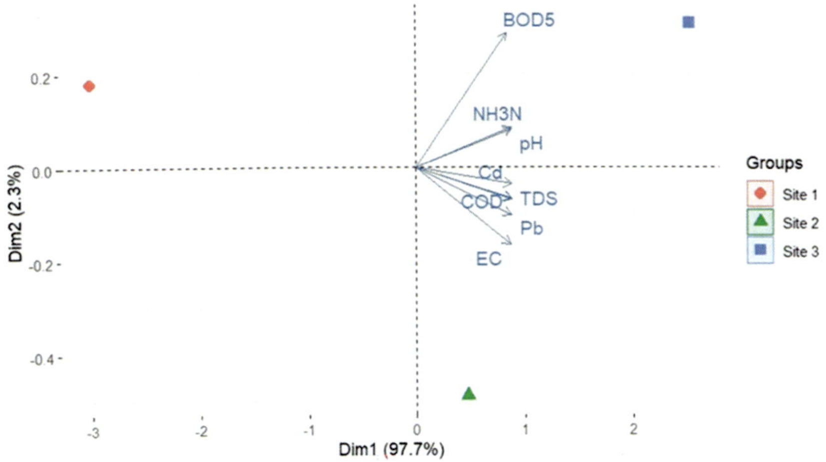

Fig. 11.5 PCA plot highlighting valuable insights into the underlying factors influencing water quality variations across the three sampling sites

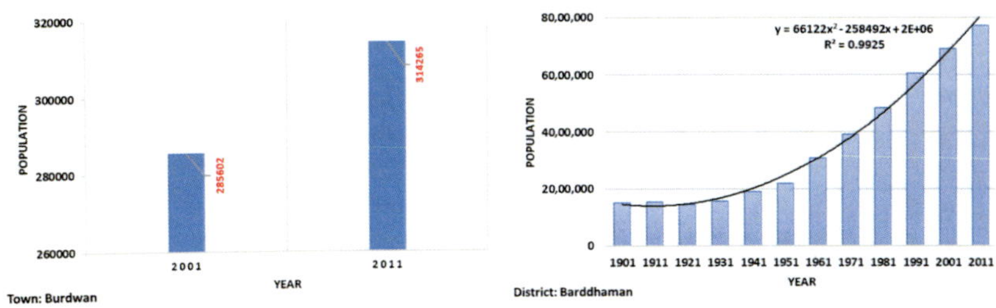

Fig. 11.6 Diagram showing population status of the Burdwan town and Bardhaman district (*Source of data* Census of India, 2011)

Table 11.2 Observed physicochemical properties of water of the Banka River at different sampling points with the corresponding standard

Parameters	*S-1*	*S-2*	*S-3*	*Mean ± SD*	*Standard*
pH	6.63	7.45	8.12	7.40 ± 0.75	6.5–8.5 (ECR 1997)
EC (μS/cm)	375	873	995	747.67 ± 328.45	300 (WHO 2011)
TDS (ppm)	76.42	142.87	171.65	130.31 ± 48.84	500 (WHO 2011)
BOD (ppm)	137	212	356	235.0 ± 111.30	5 (WHO 2011)
COD (ppm)	53.42	231.16	307.56	197.38 ± 130.39	10 (WHO 2011)
NH_3-N	0.35	.83	1.23	0.80 ± 0.44	0.5 (WHO 2011)
Pb (ppm)	0.1	0.35	0.44	0.30 ± 0.18	0.01 (WHO 2011)
Cd (ppm)	0.03	0.07	0.09	0.06 ± 0.03	0.003 (WHO 2011)

Table 11.3 Observed CDI and RQ for both oral and dermal exposure, for adults and children, at each site

		Adults				*Children*			
Heavy Metal	*Site*	*CDI (Martinez-Tavera et al.) (mg/kg/day)*	*CDI (Dermal) (mg/kg/day)*	*RQ (Martinez-Tavera et al.)*	*RQ (Dermal)*	*CDI (Martinez-Tavera et al.) (mg/kg/day)*	*CDI (Dermal) (mg/kg/day)*	*RQ (Martinez-Tavera et al.)*	*RQ (Dermal)*
Pb	1	0.000219	0.0000458	0.0626	0.0916	0.000795	0.000154	1.59	1.54
Pb	2	0.000766	0.0001604	0.2189	0.3208	0.002782	0.000539	5.56	5.39
Pb	3	0.000963	0.0002012	0.2751	0.4024	0.003498	0.000677	6.99	6.77
Cd	1	0.0000657	0.00000682	0.0657	0.0682	0.000239	0.0000231	0.478	0.462
Cd	2	0.0001533	0.00001591	0.1533	0.1591	0.000557	0.0000539	1.114	1.078
Cd	3	0.0001969	0.00002046	0.1969	0.2046	0.000718	0.0000693	1.436	1.386

References

Akhtar, N., Syakir Ishak, M. I., Bhawani, S. A., & Umar, K. (2021). Various natural and anthropogenic factors responsible for water quality degradation: A review. *Water, 13*, 2660.

Akkoyunlu, A., & Akiner, M. E. (2012). Pollution evaluation in streams using water quality indices: A case study from Turkey's Sapanca Lake Basin. *Ecological Indicators, 18*, 501–511.

Akter, M., Zakir, H. M., Sharmin, S., Quadir, Q. F., & Mehrin, S. (2020). Heavy metal bioaccumulation pattern in edible tissues of different farmed fishes of Mymensingh Area, Bangladesh and health risk assessment. *Advanced Research, 21*(4), pp. 44–55.

APHA. (1998). *Standard methods for the examination of water and wastewater*. American Public Health Association.

Awomeso, J., Taiwo, A., Gbadebo, A., & Adenowo, J. (2010). Studies on the pollution of water body by textile industry effluents in Lagos, Nigeria. *Journal of Applied Sciences in Environmental Sanitation, 5*, 353–359.

Banerjee, R., Batabyal, S., Guha, S., Mondal, A., Gangopadhyay, S., Das, N., Ghosh, P., & Mandal, S. (2020). Temporal stratification of weeds under System of Rice Intensification and conventional rice cultivation. *International Journal of Chemical and Environmental Sciences, 2*(1), 41–52.

Batabyal, S., Das, N., Mondal, A., Banerjee, R., Gangopadhyay, S., & Mandal, S. (2022). Land cover changes in green patches and its impact on carbon sequestration in an urban system, India. In *Applied geomorphology and contemporary issues* (pp. 397–409). Springer International Publishing.

Brown, R. M., Mcclelland, N. I., Deininger, R. A., & O'connor, M. F. (1972, December 26–31, 1971). A water quality index—Crashing the psychological barrier. In *Indicators of Environmental Quality: Proceedings of a symposium held during the AAAS meeting in Philadelphia, Pennsylvania* (pp. 173–182). Springer.

Chakraborty, K., Mukherjee, I., & Ghosh, S. (2011). Historical and geo-environmentalappraisal of changing growth centre, urbanization and behavioural dynamics of the rivers of Barddhaman Town West Bengal. *Internaltional Journal Current Research, 3*, 216–224.

Dhara, K., Saha, S., Pal, P., Chukwuka, A. V., Panigrahi, A. K., Saha, N. C., & Faggio, C. (2022). Biochemical, physiological (haematological, oxygen-consumption rate) and behavioural effects of mercury exposures on the freshwater snail, Bellamya bengalensis. *Comparative Biochemistry and Physiology Part C: Toxicology & Pharmacology, 251*, Article 109195.

Dhara, K., Saha, S., & Saha, N. C. (2021). Sensitivity of Common Carp, Cyprinus carpio (Linnaeus, 1758) to the Grey List Metal, Zinc under Laboratory Condition. *Asian Journal of Biological Life Sciences, 10*, 133.

Dugan, H. A. (2024). Salinity and ionic composition of inland waters. *Wetzel's Limnology*. Elsevier.

Fathi, E., Zamani-Ahmadmahmoodi, R., & Zare-Bidaki, R. (2018). Water quality evaluation using water quality index and multivariate methods, Beheshtabad River, Iran. *Applied Water Science, 8*, 210.

Firmansyah, D. A., Ibadurrohman, K. K., Aji, B. R., & Suprijanto, S. (2020). Measuring instrument for refilled drinking water using a tds sensor. *Spektra: Jurnal Fisika dan Aplikasinya, 5*, 153–162.

Gangopadhyay, S., Saha, A., Pal, S. C., Banerjee, R., Batabyal, S., Das, N., Ghosh, P. B., Mondal, A., & Mandal, S. (2023). A new methodological approach to the establishment of sustainable agricultural ecology in drought vulnerable areas of eastern India. *Ecological Informatics, 75*, Article 102013.

Ghosh, S., Dey, S., Mandal, A. H., Sadhu, A., Saha, N. C., Barceló, D., Pastorino, P. & Saha, S. (2025a). Exploring the ecotoxicological impacts of microplastics on freshwater fish: A critical review. *Journal of Contaminant Hydrology*, 104514.

Ghosh, S., & Maji, T. (2011). An environmental assessment of urban drainage, sewage and solid waste management in Barddhaman municipality, West Bengal. *International Journal of Environmental Sciences, 2*(1), 92–104.

Ghosh, S., Mondal, A., Gangopadhyay, S., & Mandal, S. (2020). Cadmium bioaccumulation in Lamellidens marginalis and human health risk assessment: A case study in India. *Human and Ecological Risk Assessment: An International Journal, 26*(3), 713–725.

Ghosh, S., Sadhu, A., Mandal, A. H., Biswas, J. K., Sarkar, D., & Saha, S. (2024). Copper oxide nanoparticles as an emergent threat to aquatic invertebrates and photosynthetic organisms: A synthesis of the known and exploration of the unknown. *Current Pollution Reports, 11*, 6.

Ghosh, S., Sadhu, A., Mandal, A. H., Dey, S., Saha, N. C., & Saha, S. (2025b). Metabolomics profiling and hazard prediction of spinosyn-based agricultural products: A computational approach. *Science of the Total Environment, 959*, Article 178279.

Huq, S. I., & Alam, M. (2005). *A handbook on analyses of soil, plant and water* (p. 246). BACER-DU, University of Dhaka.

Islam, M. S., Ahmed, M. K., Raknuzzaman, M., Habibullah-Al-Mamun, M., & Islam, M. K. (2015a). Heavy metal pollution in surface water and sediment: A preliminary assessment of an urban river in a developing country. *Ecological Indicators, 48*, 282–291.

Islam, M. S., Uddin, M. K., Tareq, S. M., Shammi, M., Kamal, A. K. I., Sugano, T., Kurasaki, M., Saito, T., Tanaka, S., & Kuramitz, H. (2015b). Alteration of water pollution level with the seasonal changes in mean daily discharge in three main rivers around Dhaka City, Bangladesh. *Environments, 2*, 280–294.

Jha, R., Ojha, C., & Bhatia, K. (2007). Development of refined BOD and DO models for highly polluted Kali River in India. *Journal of Environmental Engineering, 133*, 839–852.

Kabir, M. H., Islam, M. S., Tusher, T. R., Hoq, M. E., Muliadi, M., & Al Mamun, S. (2020). Changes of heavy metal concentrations in Shitalakhya river water of Bangladesh with seasons. *Indonesian Journal of Science and Technology, 5*, 395–409.

Kannel, P. R., Lee, S., Lee, Y.-S., Kanel, S. R., & Khan, S. P. (2007). Application of water quality indices and dissolved oxygen as indicators for river water classification and urban impact assessment. *Environmental Monitoring and Assessment, 132*, 93–110.

Kapoor, K., Kumar, S., Vishwakarma, D. K., Obaidullah, A. J., & Yadav, K. K. (2024). Health risk assessment of heavy metals in groundwater sources: Carcinogenic and non-carcinogenic evaluation. *Journal of Water and Health, 22*, 1972–1987.

Khalil, M. M., Aboueldahab, S. M., Abdel-Raheem, K. H., Ahmed, M., Ahmed, M. S., & Abdelhady, A. A. (2023). Mixed agricultural, industrial, and domestic drainage water discharge poses a massive strain on freshwater ecosystems: A case from the Nile River in Upper Egypt. *Environmental Science and Pollution Research, 30*, 122642–122662.

Khatri, N., & Tyagi, S. (2015). Influences of natural and anthropogenic factors on surface and groundwater quality in rural and urban areas. *Frontiers in Life Science, 8*, 23–39.

Kumar, A., & Dua, A. (2009). Water quality index for assessment of water quality of river Ravi at Madhopur (India). *Global Journal of Environmental Sciences, 8*(1).

Maddah, H. A. (2022). Predicting optimum dilution factors for BOD sampling and desired dissolved oxygen for controlling organic contamination in various wastewaters. *International Journal of Chemical Engineering, 2022*, 8637064.

Majhi, B. M., Nath, A. M., Dey, C., Mondal, A., & Saha, N. M. (2018). Ecological assessment of hooghly-bhagirathi river system through the study of diversity of bivalves and gastropods in relation to physico-chemical parameters. *International Journal of Current Microbiology and Applied Sciences, 7*(7), 2700–2715.

Mandal, A. H., Ghosh, S., Adhurjya, D., Chatterjee, P., Samajdar, I., Mukherjee, D., Dhara, K., Saha, N. C., Piccione, G., Multisanti, C. R., Saha, S., & Faggio, C. (2024a). Exploring the impact of zinc oxide nanoparticles on fish and fish-food organisms: A review. *Aquaculture Reports, 36*, Article 102038.

Mandal, A. H., Sadhu, A., Ghosh, S., Saha, N. C., Mossotto, C., Pastorino, P., Saha, S., & Faggio, C. (2024b). Evaluating the impact of neonicotinoids on aquatic non-target species: A comprehensive review. *Environmental Toxicology and Pharmacology*, 104606.

Mondal, A., Chanda, D., Patra, S., & Barman, M. (2020). Phrynus whitei (Gervais, 1842) from the type locality Burdwan is actually Charinus bengalensis (Gravely, 1911). *Zoo's Print, 35*(7), 1–3.

Mondal, A., Das, N., & Mandal, S. (2022). Domestication of rice reshaped associated biodiversity of rice field ecosytem. *Journal of Animal Diversity, 4*(1), 74–90.

Mondal, A., Das, N., Banerjee, R., Batabyal, S., Gangopadhyay, S., Ray, H., Biswas, N., & Mandal, S. (2023). Investigating the health of a rice field ecosystem using thermodynamic extremal principles. *Ecology, Economy and Society-the INSEE Journal, 6*(1), 81–108.

Marquis, B. J., Love, S. A., Braun, K. L., & Haynes, C. L. (2009). Analytical methods to assess nanoparticle toxicity. *The Analyst, 134*, 425–439.

Martinez-Tavera, E., Rodriguez-Espinosa, P., Shruti, V., Sujitha, S., Morales-Garcia, S., & Muñoz-Sevilla, N. (2017). Monitoring the seasonal dynamics of physicochemical parameters from Atoyac River basin (Puebla), Central Mexico: Multivariate approach. *Environmental Earth Sciences, 76*, 95.

Mgbenu, C. N., & Egbueri, J. C. (2019). The hydrogeochemical signatures, quality indices and health risk assessment of water resources in Umunya district, southeast Nigeria. *Applied Water Science, 9*, 22.

Mishra, B. K., Kumar, P., Saraswat, C., Chakraborty, S., & Gautam, A. (2021). Water security in a changing environment: Concept, challenges and solutions. *Water, 13*, 490.

Mohtar, W. H. M. W., Maulud, K. N. A., Muhammad, N. S., Sharil, S., & Yaseen, Z. M. (2019). Spatial and temporal risk quotient based river assessment for water resources management. *Environmental Pollution, 248*, 133–144.

Naubi, I., Zardari, N. H., Shirazi, S. M., Ibrahim, N. F. B., & Baloo, L. (2016). Effectiveness of water quality index for monitoring malaysian river water quality. *Polish Journal of Environmental Studies, 25*(1), 231–239.

Pathak, S., Prasad, S., & Pathak, T. (2015). Determination of water quality index river Bhagirathi in Uttarkashi, Uttarakhand, India. *International Journal of Research Granthaalayah, 3*, 1–7.

Proshad, R., Islam, S., Tusher, T. R., Zhang, D., Khadka, S., Gao, J., & Kundu, S. (2021). Appraisal of heavy metal toxicity in surface water with human health risk by a novel approach: A study on an urban river in vicinity to industrial areas of Bangladesh. *Toxin Reviews, 40*, 803–819.

Sadhu, A., Mandal, A. H., Ghosh, S., Lakdawala, P., & Saha, S. (2024). Dimethoate-induced toxicity: A comprehensive review of its aquatic environmental hazards. *Toxicology and Environmental Health Sciences, 16*, 243–258.

Saha, S., Dhara, K., Chukwuka, A. V., Pal, P., Saha, N. C., & Faggio, C. (2023). Sub-lethal acute effects of environmental concentrations of inorganic mercury on hematological and biochemical parameters in walking catfish, Clarias batrachus. *Comparative Biochemistry and Physiology Part C: Toxicology & Pharmacology, 264*, Article 109511.

Sánchez, E., Colmenarejo, M. F., Vicente, J., Rubio, A., García, M. G., Travieso, L., & Borja, R. (2007). Use of the water quality index and dissolved oxygen deficit as simple indicators of watersheds pollution. *Ecological Indicators, 7*, 315–328.

Sen, P., Das, N., Saha, N. C., Ghosh, S., Mondal, A., & Mandal, S. (2023). Modeling of cadmium bioaccumulation dynamics in Channa punctata (Bloch, 1793) of Dankuni wetland ecosystem and assessment of risk to human health. *Applied Biochemistry and Biotechnology, 195*(6), 3681–3698.

Sharma, R., Kumar, R., Satapathy, S. C., Al-Ansari, N., Singh, K. K., Mahapatra, R. P., Agarwal, A. K., Le, H. V., & Pham, B. T. (2020). Analysis of water pollution using different physicochemical parameters: A study of Yamuna River. *Frontiers in Environmental Science, 8*, Article 581591.

Singh, A., Sharma, A., Verma, R. K., Chopade, R. L., Pandit, P. P., Nagar, V., Aseri, V., Choudhary, S. K., Awasthi, G., & Awasthi, K. K. 2022. Heavy metal contamination of water and their toxic effect on living organisms. In *The toxicity of environmental pollutants*. IntechOpen.

Singh, M. R., & Gupta, A. (2016). *Water pollution-sources, effects and control* (pp. 1–16). Centre for Biodiversity, Department of Botany, Nagaland University.

Suthar, S., Sharma, J., Chabukdhara, M., & Nema, A. K. (2010). Water quality assessment of river Hindon at Ghaziabad, India: Impact of industrial and urban wastewater. *Environmental Monitoring and Assessment, 165*, 103–112.

Syeed, M. M., Hossain, M. S., Karim, M. R., Uddin, M. F., Hasan, M., & Khan, R. H. (2023). Surface water quality profiling using the water quality index, pollution index and statistical methods: A critical review. *Environmental and Sustainability Indicators, 18*, Article 100247.

Thirumalini, S., & Joseph, K. (2009). Correlation between electrical conductivity and total dissolved solids in natural waters. *Malaysian Journal of Science, 28*, 55–61.

Uddin, M. J., & Jeong, Y.-K. (2021). Urban river pollution in Bangladesh during last 40 years: Potential public health and ecological risk, present policy, and future prospects toward smart water management. *Heliyon, 7*.

Ustaoğlu, F., Tepe, Y., & Taş, B. (2020). Assessment of stream quality and health risk in a subtropical Turkey river system: A combined approach using statistical analysis and water quality index. *Ecological Indicators, 113*, Article 105815.

Varol, M., & Tokatlı, C. (2023). Evaluation of the water quality of a highly polluted stream with water quality indices and health risk assessment methods. *Chemosphere, 311*, Article 137096.

Wang, G., Mang, S., Cai, H., Liu, S., Zhang, Z., Wang, L., & Innes, J. L. (2016). Integrated watershed management: Evolution, development and emerging trends. *Journal of Forestry Research, 27*, 967–994.

Yuan, Y., Xiang, M., Liu, C., & Theng, B. K. (2017). Geochemical characteristics of heavy metal contamination induced by a sudden wastewater discharge from a smelter. *Journal of Geochemical Exploration, 176*, 33–41.

Zhao, H., Duan, X., Stewart, B., You, B., & Jiang, X. (2013). Spatial correlations between urbanization and river water pollution in the heavily polluted area of Taihu Lake Basin, China. *Journal of Geographical Sciences, 23*, 735–752.

Zhou, W., Zhang, Y., Yin, J., Zhou, J., & Wu, Z. (2022). Evaluation of polluted urban river water quality: A case study of the Xunsi River watershed, China. *Environmental Science and Pollution Research, 29*, 68035–68050.

CHAPTER 12

The Interplay of Changing Environment and Water Management on the Degradation of Indian River Systems: Navigating Aspects of Fragility, Resiliency, and Sustainability

Subhankar Bera

12.1 Introduction

Rivers in the tropical region are the most powerful agent of geomorphic process and can be define as the veins of the landscape. This could be a powerful metaphor that signifies the crucial role rivers play in supporting and maintaining aquatic ecosystem and shaping the landforms. Like veins in the human body that transport blood and nutrients, rivers carry water, nutrients, and sediments, connecting different parts of the planet and sustaining ecosystems, including human populations. Rivers from the ancient time, have played a foundational role in the development of human civilizations and provided essential resources such as water,

S. Bera (✉)
Burdwan Model School, Purba Bardhaman, West Bengal, India
e-mail: subhankar.geo@gmail.com

S. Ghosh (ed.), *Reimagining Indian Rivers for Sustainability*, Palgrave Studies in Environmental Sustainability,
https://doi.org/10.1007/978-3-032-04754-0_12

Table 12.1 Major Civilization of the World and Rivers

Civilization	*River(s)*	*Present Region*	*Time Period*
Mesopotamia	Tigris & Euphrates	Iraq and Syria	~3500–500 BCE
Ancient Egypt	Nile	Egypt	~3100–30 BCE
Indus Valley	Indus	Pakistan and NW India	~2600–1900 BCE
Ancient China	Huang He	Northern China	~2000 BCE onward
Mesoamerican	Tigris and Euphrates	Central America	~1500 BCE onward

fertile soil, and transportation routes, making them natural centers for settlement and growth (Table 12.1).

Tropical rivers like Amazon (**Annual discharge**: **~6600 km^3/year** and sediment load: ~**1.2 billion tons/year)** and Ganga-Brahmaputra (Annual discharge: ~**1200–1300 km^3/year** and sediment load: ~**1.5–2.0 billion tons/year)** carry **enormous volumes of water and sediment** annually due to high rainfall, intense weathering, and large drainage areas (Milliman & Farnsworth, 2011; Syvitski & Milliman, 2007) and developed worlds some largest and most diversified deltas. Tropical rivers are typically surrounded by dense vegetation such as rainforests or Sunderbans, which not only influence the ecology of the river but also contribute organic matter and nutrients to the water. Despite their significance, rivers are increasingly threatened by anthropogenic activities such as pollution, dam construction, and over-extraction (Table 12.2). But in the twenty-first century rivers are losing their natural flow, ecological balance, or economic and cultural importance due to human or environmental factors (Table 12.3). This paper explores the importance, causes of decaying, resiliency, and sustainability, using data-driven insights to highlight their indispensable contributions to global ecosystems and human society. This study has been focused on the major rivers of India (Fig. 12.1) and some basic information about these rivers is given in Table 12.2.

12.2 The Importance of Rivers

Rivers are essential components of the Earth's hydrological cycle, influencing the distribution and availability of freshwater resources. Representing a small fraction of the planet's total water resources (approximately 0.01% of Earth's water), rivers support over 80% of the human

Table 12.2 Major Indian rivers, their importance, and threats

River Name	*Length (in km)*	*Area (in sq Km)*	*Climatic Condition*	*Importance*	*Threat*
Ganga	2525	861,404	All basin experiences a monsoon climate	Source of water, a cultural and religious symbol, and a driver of economic activity	Pollution and over-extraction of water
Mahanadi	900	132,100		Agriculture, industries, and ecosystems	Silt deposition, flood management challenges, pollution
Godavari	1465	312,812		Agriculture, industries, and cultural practices	Threats from pollution and urbanization
Krishna	1400	258,948		Region's agriculture, water supply, and biodiversity	Faces significant threats from pollution, unsustainable practices, and climate change
Cauvery	800	81,155		Vital source of water for irrigation, drinking, and electricity generation	Pollution, over-extraction, and inter-state disputes over water sharing
Narmada	1312	97,410		Essential water resources for agriculture, drinking, and industrial use	Pollution from industrial and domestic waste, agricultural runoff, and over-extraction of water
Tapti	724	65,145		Water for agriculture, industry, and Religious and Cultural Significance	Pollution, and dam operations

Table 12.3 Some decaying rivers of the world

Rivers	*Cause of decaying*
Yamuna River (India)	Polluted with industrial and domestic waste, especially around Delhi
Ganga River (India)	Overused and polluted; efforts like Namami Gange launched for its revival
Colorado River (USA)	No longer reaches the sea due to overuse
Murray-Darling Basin (Australia)	Faces ecological stress from over-extraction and drought
River Thames (Historical)	Was biologically dead in the 1950s; since revived through cleanup efforts

population's freshwater needs (Gleick 2014). Historically, rivers have been foundational to the development of human civilization, providing water, transportation, and fertile land for agriculture. In this part, we investigate the ecological, economic, and cultural importance of rivers.

12.2.1 Ecological Significance of Rivers

Rivers are crucial for maintaining ecological balance in freshwater and riparian ecosystems. Freshwater habitats, despite covering less than 1% of Earth's surface, host about 10% of all known animal species (World Wildlife Fund). They are key habitats for a diverse array of species, particularly freshwater fish, which represent more than 40% of the world's total vertebrate species (Dudgeon et al., 2006a, 2006b). The Amazon River, for instance, hosts around 2200 species of fish, while the Mekong River in Southeast Asia is the second most biodiverse river system globally, with more than 2500 species of fish (Dudgeon et al., 2006a, 2006b). The hydrological dynamics of rivers—such as flow regimes, sediment transport, and water quality—have a profound impact on aquatic biodiversity and ecosystem function. River systems also serve as migratory corridors for species like salmon (Oncorhynchus spp.) and provide critical breeding and feeding grounds for various amphibians and waterfowl. Rivers also play a vital role in nutrient cycling. The transport of organic matter, sediment, and dissolved nutrients from upstream to downstream regions supports downstream ecosystems, including wetlands and floodplains, which are among the most productive ecosystems on Earth. Floodplains, for example, are replenished by nutrient-rich sediment deposits,

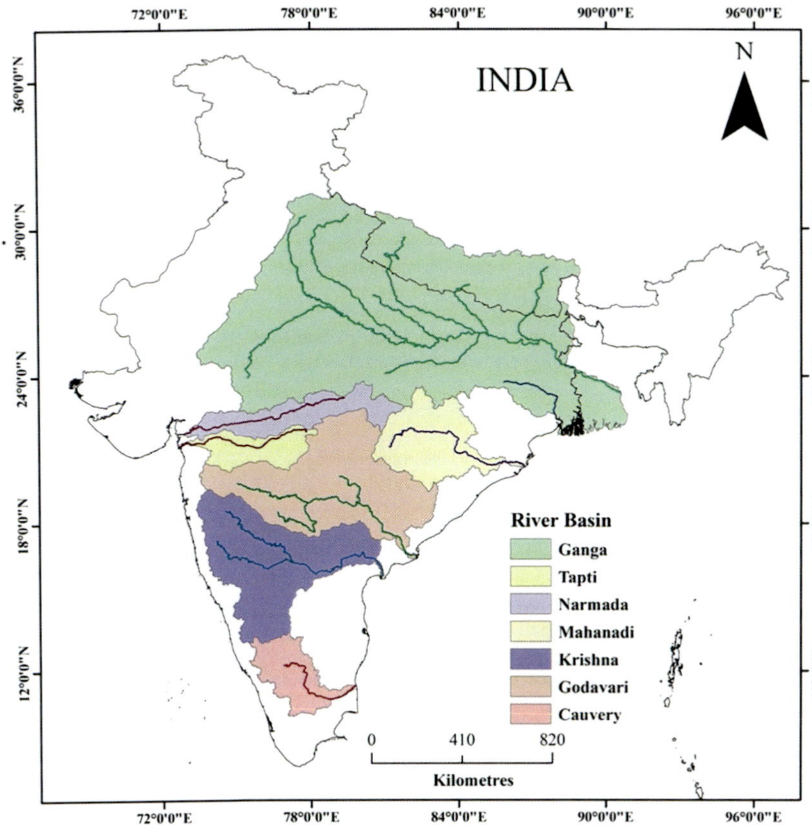

Fig. 12.1 Rivers of India under study

making them highly fertile areas for agriculture (Tockner & Stanford, 2002). Rivers and their riparian zones (vegetated areas along the banks) provide critical ecosystem services, including water purification, erosion control, and flood regulation (Table 12.4). For example, wetlands in river systems, such as those found in the Mississippi and Nile basins, act as natural filters, improving water quality and mitigating pollutants (UN Water). In addition, these ecosystems play a role in carbon sequestration. Wetlands associated with rivers can store up to 30% of the world's carbon, significantly contributing to climate change mitigation (IPCC).

Table 12.4 Ecological importance of rivers

Ecological role	*Scientific explanation*	*Reference*
Biodiversity Hotspots	Rivers and their floodplains host high biodiversity, supporting ~10% of global fish species in freshwater ecosystems. Rivers like the Amazon and Congo are global biodiversity hotspots	Dudgeon et al. (2006a, 2006b)
Nutrient Cycling	Rivers transport key nutrients (e.g., nitrogen, phosphorus) from land to sea, driving primary productivity in both freshwater and coastal ecosystems	Vanni (2002)
Groundwater Recharge	Percolation from riverbeds through alluvial aquifers contributes to maintaining groundwater levels, especially in arid and semi-arid regions	Winter et al. (1998)
Floodplain Fertility	Periodic floods deposit alluvium, enhancing floodplain soil fertility and supporting agriculture. This natural fertilization is crucial in deltas like the Ganges and Nile	Junk et al. (1989)
Wetland Formation & Stability	Rivers help form and sustain wetlands, which act as carbon sinks and breeding grounds for many species. Wetlands also filter pollutants	Mitsch and Gosselink (2000)
Climate Regulation	Rivers influence local humidity and temperature and contribute to the global water and carbon cycles through evapotranspiration and carbon storage in sediments	Raymond et al. (2013)
Water Purification	Through microbial activity, sedimentation, and uptake by aquatic plants, rivers naturally break down pollutants such as nitrates, phosphates, and heavy metals	Palmer et al. (2004)

(continued)

Table 12.4 (continued)

Ecological role	*Scientific explanation*	*Reference*
Sediment Transport	Rivers shape landscapes by transporting sediments downstream, forming deltas, beaches, and wetlands, which are critical for buffering sea-level rise	Syvitski et al. (2005)
Species Migration Corridors	Rivers serve as longitudinal corridors facilitating the seasonal migration of fish (e.g., salmon, eel) and the dispersal of aquatic and semi-aquatic organisms	Pringle (2003)
Riparian Vegetation Support	Riverbanks support riparian forests that stabilize banks, filter runoff, and provide habitat. These zones act as ecotones—ransition zones of high ecological productivity	Naiman and Décamps (1997)

12.2.2 Hydrological and Agricultural Roles

Rivers have a significant impact on global water management and agricultural production. Approximately 20% of the world's cultivated land relies on irrigation from river systems, and these areas contribute to around 40% of global food production (FAO, 2017). Major rivers such as the Nile, Ganges, and Yangtze provide irrigation for millions of hectares of cropland, supporting both subsistence and commercial agriculture. Globally, approximately 70% of all freshwater withdrawals are used for agriculture, with a significant portion drawn from rivers (FAO, 2017). Rivers also transport and deposit sediments that enrich soils, making them fertile for agriculture. The Ganges River, for example, is known for depositing large quantities of fertile silt onto its floodplains, which has made it one of the most agriculturally productive regions in the world (Sinha et al., 2018). Rivers also contribute to the generation of hydroelectric power. The construction of dams on rivers like the Three Gorges Dam (China) and the Itaipu Dam (Brazil/Paraguay) has facilitated the generation of large amounts of electricity, serving as an important source of renewable energy. For example, the Three Gorges Dam, with an installed capacity of 22,500 MW, is one of the largest hydroelectric power stations in the world

Table 12.5 Hydrological and agricultural significance of major indian rivers

Rivers name	*Hydrological importance*	*References*
Ganga	– Forms one of the world's most fertile alluvial plains, crucial for agriculture and aquifer recharge – Maintains deltaic sedimentation and sustains monsoon-driven surface–groundwater interaction	Tare et al. (2017), *Science of The Total Environment*
Mahanadi	– High monsoon variability impacts floodplain recharge and sedimentation – Supports major irrigation and hydropower schemes like Hirakud	Patra et al. (2020), *Water Science & Technology Library*
Godavari	– Largest monsoon-fed river in peninsular India, enabling extensive sediment transport and delta nourishment – Important for ecological flows and irrigation	Jain et al. (2007), *Current Science* 93(1):90–96
Krishna	– Critical for water supply in semi-arid zones – Subject to climate-driven stream flow reductions and altered seasonal hydrology	Ghosh and Dutta (2012), *Water Resources Management*
Cauvery	– Perennial due to regulated flows; supports inter-state irrigation disputes – Strong interaction between hydrology, land use, and policy	Sridhar et al. (2009), *Environmental Management*
Narmada	– West-flowing river with deep gorges aiding aquifer recharge – Supports large hydropower projects and interlinking potentials	Singh et al. (2012), *Journal of Earth System Science*
Tapti	– Supports surface water availability in dry areas – One of the key westward draining rivers contributing sediment to Arabian Sea shelf	Kale and Hire (2007), *Journal of the Geological Society of India*

(Shao, 2019). Hydrological and agricultural importance are summarized in Table 12.5.

12.2.3 Economic and Transportation Functions

Rivers have historically been vital to transportation and trade. Inland waterway systems such as the Mississippi River (USA), the Rhine (Europe), and the Amazon River (South America) facilitate the movement of goods and people, often providing more efficient and environmentally sustainable alternatives to road and rail transport. For instance, river transport is responsible for moving approximately 20% of the total freight tonnage in Europe (European Commission, 2020). This mode of transportation reduces greenhouse gas emissions and congestion associated with road traffic. Rivers also contribute to economic development through tourism, fishing, and industry. Urban settlements along rivers benefit from proximity to these water sources for industrial cooling, sanitation, and drinking water. The economic significance of rivers is, however, dependent on the sustainable management of these resources, as over-extraction and pollution can disrupt their ability to serve these functions.

12.2.4 Sociocultural and Religious Significance

Beyond their economic and ecological value, rivers hold profound sociocultural and religious importance. Numerous cultures have centered their religious and cultural practices around rivers, with the Ganges River in India, the Nile River in Egypt, and the Jordan River in the Middle East being particularly revered in major religious traditions. For example, the Ganges is considered a goddess in Hinduism, and millions of people participate in pilgrimages and rituals centered on its waters. Rivers also shape settlement patterns, influencing the development of ancient civilizations that thrived in proximity to river systems due to their abundant resources. Archaeological evidence from river valleys such as the Mesopotamian, Indus, and Egyptian civilizations demonstrates the central role of rivers in the establishment of urban centers, trade routes, and agricultural productivity (Childe, 1950).

12.3 Causes of River Degradation and Sustainability of Rivers

River degradation denotes the progressive decline in the structural and functional integrity of fluvial systems, driven by both natural variability and, more critically, anthropogenic pressures. Scientifically, it encompasses alterations to hydrological regimes, geomorphic stability, biogeochemical cycling, and ecological resilience, often manifesting as reduced base-flows, disrupted sediment transport, eutrophication, and biodiversity loss. Key drivers include unsustainable water abstraction, land-use change, industrial and domestic effluent discharge, and flow regulation via dams and barrages, all of which impair a river's capacity to sustain ecosystem services such as nutrient cycling, habitat provision, and water purification. In India, diverse river basins—from Himalayan glacial-fed systems like the Ganga and Yamuna to peninsular rivers like the Cauvery and Mithi—exhibit acute symptoms of degradation, including hypoxia, species invasions, and fluvial habitat fragmentation. These disruptions are further exacerbated by climate-induced hydrometeorological variability, necessitating integrative, basin-scale restoration frameworks grounded in hydrological, ecological, and geomorphological sciences.

12.3.1 Water Pollution

Pollution remains a critical factor in the degradation of India's rivers, with industrial, domestic, and agricultural waste significantly impacting water quality. The Ganges (Ganga) River, revered by millions, continues to face severe pollution challenges. Despite substantial investments under initiatives like the National Mission for Clean Ganga, the river's health remains compromised, particularly after large gatherings such as the 2025 Maha Kumbh Mela in Prayagraj, which attracted approximately 660 million pilgrims. Such events, while culturally significant, contribute to increased waste and strain on the river's ecosystem.

The Yamuna River, especially the stretch flowing through Delhi, is experiencing alarming levels of pollution. Recent reports indicate that fecal coliform concentrations have soared to 7.9 million units per 100 ml at Asgarpur, far exceeding the permissible limit of 2500 units per 100 ml. This surge is primarily attributed to untreated sewage and industrial discharges. Additionally, ammonia levels have risen above 2.0 ppm, surpassed the treatment capacity of existing water treatment plants and led to water shortages in several parts of Delhi. The presence of toxic foam, resulting from high phosphate content due to detergents and industrial effluents, has been observed during festivals like Chhath Puja, posing health risks to devotees. Despite efforts, the Delhi government missed its 2023 deadline to treat all sewage, making the 2025 goal of cleaning the Yamuna to bathing standards increasingly challenging. Water quality index and some important parameters which are making the rivers unhealthy and decaying are given in Table 12.6. The analysis rivers reveal significant spatiotemporal variability in water quality, with the Ganga exhibiting the highest pollution levels (WQI: 25 to >100), primarily due to anthropogenic inputs including industrial effluents, agricultural runoff, sewage, and solid waste. Key pollutants include elevated levels of biochemical oxygen demand (BOD: 3.0–6.0 mg/L), turbidity

Table 12.6 Water quality level and Some important water pollution parameters of the studied rivers. All the rivers in majority case crossed the quality threshold

River Name	*Pollution level (WQI)*	*Type of pollutants*	*Sources of pollution*	*Water pollution parameters*
Ganga	25 to >100	Biochemical Oxygen, Nitrates (NO_3), Phosphates, Heavy metals (like, Pb, As, Hg), Total Suspended Solids, DDT, Polyethylene, Packaging, Plastics, Pathogens etc	Industrial effluents, Agricultural runoff, Sewage, Domestic waste, Dumping into river bed etc	Turbidity = 40 – 600 NTU pH = 7.2 to 9.5, BOD = 3.0 to 6.0 mg/l Heavy metals = µg/L Pb = 20 – 150, Cd = 1 – 5, Cr = 20 – 100
Mahanadi	30 to 45			Turbidity = 30 – 300 NTU pH = 6.97 to 7.95, BOD = 1.8 to 39 mg/l Heavy metals = µg/L Pb = 10 – 40, Cd = 1 – 2, Cr = 10 – 60
Godavari	40 to 70			Turbidity = 40 – 250 NTU pH = 7.51 to 7.9, BOD = upto 8.0 mg/l Heavy metals = µg/L Pb = 15 – 50, Cd = 1 – 3, Cr = 15 – 70
Krishna	35 to 65			Turbidity = 35 – 220 NTU pH = 6.5 to 8.9, BOD = 2.0 to 4.4 mg/l Heavy metals = µg/L Pb = 10 – 45, Cd = 1 – 2, Cr = 10 – 65

(continued)

Table 12.6 (continued)

River Name	*Pollution level (WQI)*	*Type of pollutants*	*Sources of pollution*	*Water pollution parameters*
Cauvery	30 to 50			Turbidity = 20 – 120 NTU pH = 6.7 to 8.68, BOD = upto 6.0 mg/l Heavy metals = μg/L Pb = 8 – 30, Cd = 1 – 2, Cr = 8 – 60
Narmada	25 to 45			Turbidity = 15 – 100 NTU pH = 7.1 to 8.0, BOD = 0.4 to 9.8 mg/l Heavy metals = μg/L Pb = 5 – 25, Cd = 1 – 1.5, Cr = 5 – 50
Tapti	40 to 70			Turbidity = 25 – 180 NTU pH = 7.01 to 8.05, BOD = 4.6 to 21.5 mg/l Heavy metals = Pb = 6 – 28, Cd = 1 – 2, Cr = 6 – 55

Data based on the Central Pollution Control Board (CPCB), National Water Quality Monitoring Programme (NWMP) and State Pollution Control Board (SPCB) (water quality <25 = Excellent, 26 – 50 = Good, 51 – 75 = Moderate, 76 – 100 = Poor, >100 = Very Poor)

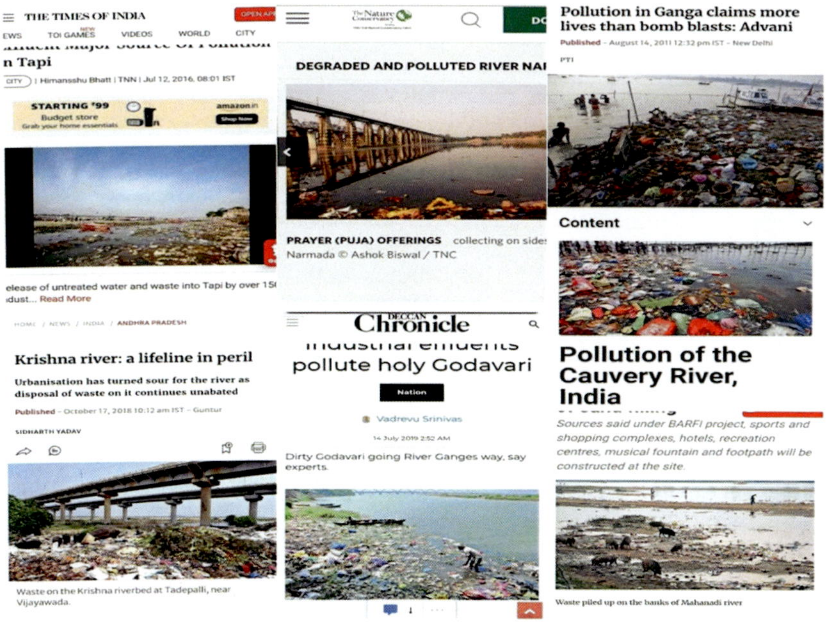

Fig. 12.2 River Pollution is always a burning environmental issue and these are some articles on news papers

(40–600 NTU), and heavy metals such as Pb (20–150 μg/L), Cd (1–5 μg/L), and Cr (20–100 μg/L). Other rivers such as the Mahanadi, Godavari, Krishna, Cauvery, Narmada, and Tapti show moderate WQI values (25–70), with varying physicochemical parameters including pH (6.5–9.5), turbidity (15–300 NTU), and BOD (0.4–39 mg/L), along with detectable concentrations of heavy metals. These findings indicate widespread contamination from both point and non-point sources, posing ecological and public health risks (Fig. 12.2).

12.3.2 Over-Extraction, Damming, and Alteration of River Flow

Over-extraction of water for agricultural, industrial, and urban use is another major cause of river degradation. Many Indian rivers, such as the Godavari, Narmada, and Krishna, are subjected to excessive water extraction through dams, canals, and borewells (Table 12.7). This leads

Table 12.7 Damming and flow regulation in Studied River Basins

River Name	*Numbers of construction (Dams, Barrage, Weirs, & Lift)*	*Water regulate, million cubic meters (MCM)*	*Most purpose to regulate*
Ganga	Around 784	Approx 60,000	Irrigation, water supply to Urbans, hydroelectric power
Mahanadi	Around 254	Approx 13,000	
Godavari	Around 921	Approx 19,500	
Krishna	Around 100	Approx 35,600	
Cauvery	Around 98	Approx 8000	
Narmada	Around 277	Approx 26,000	
Tapti	Around 70	Approx 6000	

to reduced river flow, especially in the dry season, and can result in the drying up of rivers or a significant reduction in water levels. Over-extraction also disrupts aquatic ecosystems, endangering species that depend on riverine habitats. Damming and the construction of barrages alter the natural flow of rivers, disrupting the delicate balance of river ecosystems. While dams are crucial for water storage and power generation, they often have adverse effects on rivers. The construction of large dams on rivers like the Narmada (Sardar Sarovar Dam) and the Teesta (Teesta Barrage) in India has led to the displacement of communities, loss of biodiversity, and reduced sediment flow downstream. This alters the river's natural sediment deposition patterns, impacting agriculture, fisheries, and overall ecosystem health. Encroachment along riverbanks and urbanization contribute to river degradation and changing the flow pattern of the river. As cities expand, rivers are encroached upon for housing, infrastructure, and industrial development. In cities like Bangalore, where the Arkavathy and Vrishabhavathi rivers are threatened by urban sprawl, construction along the riverbanks increases the risk of flooding, changing course, and water pollution. Deforestation in river catchments also leads to increased soil erosion, which results in sedimentation of riverbeds, further reducing the river's ability to carry water.

12.3.3 *Natural Disasters and Solid Waste Disposal*

Climate change has had a profound impact on river systems, exacerbating the severity of floods, droughts, and other natural disasters. Changing rainfall patterns affect the seasonal flow of rivers. For instance, rivers like the Brahmaputra and the Ganga experience erratic floods due to unpredictable monsoon rains and glacial melt. Additionally, the retreat of glaciers in the Himalayas has led to the reduction of freshwater inflow into major rivers like the Ganga, which threatens their long-term sustainability. The degradation of river ecosystems often leads to a loss of biodiversity. Rivers are home to various aquatic species, many of which are endangered due to pollution, habitat destruction, and over-exploitation of resources. The Cauvery River in southern India, which is home to the endangered Gangetic dolphin, faces a significant decline in dolphin populations due to habitat loss and water pollution. Similarly, the Ganga River, which was once teeming with diverse species, now has fewer fish and aquatic life due to human intervention. Not only this, improper disposal of solid waste, including plastics, organic waste, and religious offerings, adds to river degradation. For example, the Ganga River is often found to be clogged with a variety of materials such as plastics, foam, and floral waste. These contribute to the blocking of river channels, disrupt the flow of water, and degrade the esthetic and ecological value of the river. Many cities along the banks of the Ganga and Yamuna Rivers face similar challenges due to inadequate waste management infrastructure. Some Examples of River Degradation in India.

- **Ganges (Ganga)**: The Ganga is one of the most significant rivers in India, yet it faces severe pollution. Efforts like the Namami Gange project aim to clean the river, but challenges like the influx of untreated sewage, religious offerings, and industrial effluents continue to threaten its health.
- **Yamuna**: The Yamuna, especially in Delhi, is considered one of the most polluted rivers. Despite being a crucial water source for millions, it suffers from high levels of industrial waste, sewage, and solid waste, leading to a dramatic decline in water quality.
- **Cauvery**: The Cauvery River, which serves as a lifeline for southern India, is under threat from both water scarcity and pollution. Industrial waste, agricultural runoff, and encroachment have significantly impacted the river's water quality and aquatic biodiversity.

12.4 Conclusion

River degradation in India is a multifaceted issue involving the decline of the physical, chemical, and ecological integrity of river systems due to both natural and anthropogenic factors. Key drivers include pollution, over-extraction of water, damming, urban encroachment, climate change, and poor waste management. Pollution from industrial, agricultural, and domestic sources severely impacts water quality in rivers like the Ganga and Yamuna, despite government interventions such as the National Mission for Clean Ganga. Events like the 2025 Maha Kumbh Mela intensify waste loads, while the Yamuna continues to experience high levels of fecal contamination and ammonia, largely due to untreated sewage and industrial discharge. Over-extraction for agriculture and urban use significantly reduces river flows in rivers such as the Godavari and Krishna, while large dams like the Sardar Sarovar on the Narmada and the Teesta Barrage alter sediment transport and aquatic habitats, causing biodiversity loss and displacing communities. Rapid urban expansion, as seen in Bangalore, leads to encroachment on riverbanks, increasing flood risk and pollution. Climate-induced changes, including erratic rainfall and glacial retreat, further disrupt river hydrology, notably in the Ganga and Brahmaputra basins. Moreover, improper disposal of solid waste—plastics, organic matter, and religious offerings—clogs rivers and degrades their ecological value. Iconic species like the Gangetic dolphin are now endangered due to habitat degradation. Addressing these challenges requires integrated, basin-level restoration strategies that incorporate hydrological, ecological, and geomorphological insights, along with stronger governance, improved waste management, and community involvement. Only such holistic approaches can ensure the long-term sustainability of India's rivers and the essential ecosystem services they provide.

References

Carpenter, S. R., Caraco, N. F., Correll, D. L., Howarth, R. W., Sharpley, A. N., & Smith, V. H. (1998). Nonpoint pollution of surface waters with phosphorus and nitrogen. *Ecological Applications, 8*(3), 559–568.

Childe, V. G. (1950). The Urban Revolution. *Town Planning Review, 21*(3), 3–17.

Dudgeon, D., Arthington, A. H., Gessner, M. O., Kawabata, Z. I., Knowler, D. J., & Leveque, C. (2006a). Freshwater biodiversity: Importance, threats, status and conservation challenges. *Biological Reviews, 81*(2), 163–182.

Dudgeon, D., et al. (2006b). Freshwater biodiversity: Importance, threats, status and conservation challenges. *Biological Reviews, 81*(2), 163–182.

European Commission. (2020). *Inland waterways: a sustainable mode of transport*. https://ec.europa.eu/transport/modes/inland_en

FAO. (2017). *Water for Sustainable Food and Agriculture*. Food and Agriculture Organization of the United Nations.

Ghosh, S., & Dutta, S. (2012). Impact of climate change on streamflow in Krishna river basin. *Water Resources Management, 26*(9), 2539–2555.

Gleick, P. H. (2014). *The world's water Volume 8: The biennial report on freshwater resources*. Island Press.

Gregory, K. J. (2006). *The basics of geomorphology: Key concepts*. SAGE Publications.

Grill, G., Lehner, B., Lumsdon, A., Macdonald, T., & Zarfl, C. (2019). Mapping the world's free-flowing rivers. *Nature, 569*(7755), 215–221. https://doi.org/10.1038/s41586-019-1111-9

Jain, S. K., Agarwal, P. K., & Singh, V. P. (2007). Hydrology and water resources of India. *Current Science, 93*(1), 90–96. [Referenced using INCOH data].

Junk, W. J., et al. (1989). The flood pulse concept in river-floodplain systems. *Canadian Special Publication of Fisheries and Aquatic Sciences, 106*, 110–127.

Kale, V. S., & Hire, P. S. (2007). Temporal variations in suspended sediment transport in the Tapi River, India. *Journal of the Geological Society of India, 70*(5), 873–880.

Milliman, J. D., & Farnsworth, K. L. (2011). *River discharge to the coastal ocean: A global synthesis*. Cambridge University Press.

Mitsch, W. J., & Gosselink, J. G. (2000). *Wetlands*. Wiley.

Naiman, R. J., & Décamps, H. (1997). The ecology of interfaces: Riparian zones. *Annual Review of Ecology and Systematics, 28*(1), 621–658.

Palmer, M. A., et al. (2004). Ecological science and sustainability for the 21st century. *Frontiers in Ecology and the Environment, 2*(6), 298–306.
Patra, K. C., Das, M., & Sahu, M. (2020). Hydrological assessment and water management in Mahanadi River Basin, India. In B. Sahoo & M. Vairavamoorthy (Eds.), *Urban water security* (pp. 117–141). Springer.
Poff, N. L., Olden, J. D., Merritt, D. M., & Pepin, D. M. (2007). Homogenization of regional river dynamics by dams and global biodiversity implications. *Proceedings of the National Academy of Sciences, 104*(14), 5732–5737.
Pringle, C. M. (2003). What is hydrologic connectivity and why is it ecologically important? *Hydrological Processes, 17*(13), 2685–2689.
Raymond, P. A., et al. (2013). Global carbon dioxide emissions from inland waters. *Nature, 503*(7476), 355–359.
Shao, Y. (2019). China's Three Gorges Dam and its role in sustainable energy generation. *Renewable and Sustainable Energy Reviews, 98*, 121–134.
Singh, A., Panda, S. N., & Mohapatra, P. K. (2012). Optimal land and water resources allocation policies for sustainable development of Narmada Basin in India. *Journal of Earth System Science, 121*(1), 167–181.
Sinha, R., et al. (2018). *Sediment deposition and agriculture in the Ganges Basin.* Agricultural Systems.
Sridhar, V., Naik, V., & Ghosh, S. (2009). Land-use change and water resources management in the Cauvery Basin, India. *Environmental Management, 44*(5), 853–865.
Syvitski, J. P. M., et al. (2005). Impact of humans on the flux of terrestrial sediment to the global coastal ocean. *Science, 308*(5720), 376–380.
Syvitski, J. P. M., & Milliman, J. D. (2007). Geology, geography, and humans battle for dominance over the delivery of fluvial sediment to the coastal ocean. *Journal of Geology, 115*(1), 1–19.
Tare, V., Roy, P. K., Bose, P., Singh, L., & Ghosh, A. (2017). A planning and decision-making framework for rejuvenation of the Ganga River. *Science of the Total Environment, 582*, 388–403.
Tockner, K., & Stanford, J. A. (2002). Riverine flood plains: Present state and future trends. *Environmental Conservation, 29*(3), 308–330.
Vanni, M. J. (2002). Nutrient cycling by animals in freshwater ecosystems. *Annual Review of Ecology and Systematics, 33*, 341–370.

Winter, T. C., et al. (1998). *Ground water and surface water: A single resource*. US Geological Survey.

GRDC major river basin of the world. https://mrb.grdc.bafg.de/#:~:text=They%20refer%20to%20a%20hydrological,Hydrology%20(BfG)%20in%20Germany

Open Government data platform, Govt of India. https://www.data.gov.in/catalog/boundaries-water-resources-projects

DIVA-GIS. https://diva-gis.org/data.html

PART II

Fluvial Dynamics and Risk in Anthropocene

CHAPTER 13

Capturing Dimensions of Dam-Induced Flow Alteration in East-Flowing Rivers of Peninsular India Through Temporal Dynamics of Channel Vegetation Cover

Joyeta Sarkar and Joy Sanyal

13.1 Introduction

Dams and reservoirs serve as integral systems for effective water resource management and conservation, aiding in domestic and industrial water supply and flood risk mitigation, supporting agricultural irrigation, enabling hydroelectric power generation (Rosenberg et al., 2000; Tortajada, 2015) and facilitating recreational activities (Mácová & Kozáková, 2023), therefore contributing significantly in a nation's economy (Shi et al., 2019). Due to the diverse contributions of dams to societies and economies, there have been a rapid increase in dam development and operation in the last century worldwide, especially in the developing

J. Sarkar · J. Sanyal (✉)
Department of Geography, Presidency University, Kolkata, India
e-mail: joy.geog@presiuniv.ac.in

S. Ghosh (ed.), *Reimagining Indian Rivers for Sustainability*, Palgrave Studies in Environmental Sustainability,
https://doi.org/10.1007/978-3-032-04754-0_13

nations (Mallik & Richardson, 2009) and almost 60% of the large river systems of the world have already been impacted by large dams (Nilsson et al., 2005). However, research works from all over the world show that these dams can have profound impact on the hydro-morphology and ecology of the river systems, including extensive impacts on the flow regime in the form of increased low-flow and decreased high flows (Annys & Frankl, 2025), alterations in mean monthly flows and extreme flow events (Achini Ishankha et al., 2024), reduction of maximum flood peak and high-flow durations (Deng et al., 2025) while increasing the low-flow duration (Binh et al., 2020), as well as reduced flood peaks, loss of longitudinal connectivity (Gopal, 2013) and reduction of magnitude and frequency of flood pulses (Ekka et al., 2022). Studies also show that in many rivers, the effects of these large dams even surpass those induced by climate change (El Aoula et al., 2021; Luo et al., 2023).

These alterations in the hydrological regime of rivers downstream of dams also lead to significant changes in their morphological characteristics. Researches from all over the world have reported marked channel incision (Choi et al., 2005; Gordon & Meentemeyer, 2006) due to an imbalance between sediment supply and transport capacity after dam construction (Słowik et al., 2018). Bed coarsening and armouring are also reported as a result of deficit in the amounts of fine sediments, as well as due to the prevalence of constant discharge rates over longer time-periods (Wolf et al., 2022). Other studies show a decline in the relative relief of the channel bed and cross-sectional area downstream of dams (Sanyal, 2017). Upstream flow regulation can also result into marked changes in river planforms and delta morphology, including alterations in channel width and thalweg (Liaghat et al., 2017), channel narrowing and replacement of meandering and sinuous channel forms by anabranching and multi-thread patterns (Słowik et al., 2018), as well as recession of deltas downstream of these large dams (Anthony et al., 2015).

The rivers of peninsular India, particularly the east-flowing ones have been subjected to extensive dam building in the post-colonial period during the later half of the twentieth century. Consequently, these rivers faced widespread impacts from dam construction. Various local and catchment-scale studies document these effects across different river systems. In the Damodar River, dams and reservoirs significantly alter the flow regime (Ghosh et al., 2022), affecting flow duration and flood frequency. These changes often result shift in peak flows from July to August–September (Bhattacharyya & Wiley, 2014; Ghosh & Guchhait,

2014) and disconnection of the river from its distributaries (Ghosh, 2011). The riparian ecosystem and geomorphology also face considerable alterations (Ghosh et al., 2022; Purnanjali & Kasiviswanathan, 2022). The Kansabati River shows similar patterns, with dams reducing downstream flow variability and absorbing high flows (Mittal et al., 2014, 2016). Studies also report extensive sediment trapping (Ghorai & Ghosh, 2024) and geomorphic changes in downstream areas (Mura & Gope, 2020).

Case studies on the Godavari and Krishna Rivers show that upstream flow regulation has led to significant hydrological alterations in the Godavari River (Singh et al., 2021), especially in terms of mean monthly and mean annual streamflow (Singh & Jain, 2021), as well as significant decline in suspended sediment load (Singh et al., 2025). For instance, the Godavari River—the largest in peninsular India—has experienced a decline in sediment flux from 142×10^6 tonnes before 1990 to approximately 59×10^6 tones post-1990 (Das et al., 2021). This sediment deficit from the hinterland is accelerating the erosion of fertile deltas due to marine action in this region. Other studies have also reported an increase in the streamflow drought periods in the regulated parts of the central and eastern Godavari basin, compared to the non-regulated parts of the basin (Ghosal et al., 2024). Similarly, a significant declining trend in mean annual flow and in the mean monthly flows have been reported for Krishna (Uday Kumar & Jayakumar, 2018), especially for the monsoon months, with the most impacts caused in the downstream of the Nagarjuna Sagar dam (Uday Kumar & Jayakumar, 2020, 2022).

These hydrological alterations also have had a profound impact on the morphology of these rivers. Studies in the Godavari basin show a significant lowering of the channel beds of the Godavari River and its tributaries in the post-dam period (Sanyal et al., 2020), while another study on a tributary of the Damodar River shows an increase in the channel gradient downstream of dams (Biswas & Pani, 2021). Other researches on these regulated streams across peninsular India have reported bed coarsening and armouring (Pal, 2016), as well as alterations in river planforms, including an increase in the braiding index and the creation of multiple cut-offs and mid-channel bars (Pradhan et al., 2019). Studies have also reported the recession of the delta shorelines in Godavari and Krishna basins due to the retention of sediments by large dams across these streams (Rao et al., 2010).

All these hydro-morphic alterations have led to distinct environmental concerns in the peninsular river basins, including a rapid decline in the population of various wild species, fish and other aquatic creatures due to the loss of natural habitats, spawning grounds, alterations in water quality and temperatures, as well as the loss of longitudinal connectivity creating obstacles in the dispersion of these species (Bhaumik et al., 2017; Verma et al., 2022).

However, the outcomes of all the above-mentioned hydrological researches are solely dependent on the analysis of hydrologic data derived from river gauging stations, or based on hydrological models again validated by these datasets (as in case of Huang et al., 2024; Liu et al., 2022a, 2022b). The major problem regarding this matter is that in spite of extensive dam construction and operation worldwide, these flow gauging stations have remained limited in number, especially in the developing countries like India, and sometimes they are available at an interval of hundreds of kilometres along a particular river channel. Apart from that, they are able to provide insights regarding the flow regime of a river at specific cross-sections only, not for the entire longitudinal length of the channel. Moreover, there is a scarcity of data particularly for the pre-dam conditions (Sanyal, 2023), that creates a major problem while comparing between the pre-dam and post-dam flow conditions, therefore hindering the analysis of dam-induced flow alterations in these rivers. Keeping all this in mind, we can say that there is a pressing need for an indicator that would be useful to examine the effect of upstream dams in the alteration of the flow conditions of a river along its entire longitudinal length.

In-channel vegetation response to the alterations in a river regime in a very specific manner (Corenblit et al., 2007). Frequent high-flow conditions and repeated inundation can inhibit the growth of channel vegetation cover. According to various studies, hydrological alterations in a river can have significant impacts on riparian habitats (Belmar et al., 2013) and even a slight modification in the historic flow conditions can lead to marked alterations in the structure of riparian plants (Tonkin et al., 2017). While increased duration and frequency of floods can lead to reduced shrub cover in the riparian zone (Pettit et al., 2001), increased flood intensity can also cause significant alterations in the vegetation composition, leading to a shift towards species that have less ability to stabilize sand, therefore hindering the overall growth of riparian plants (Stromberg et al., 2010). On the other hand, a decline in the frequency, duration and intensity of flood events can lead to a helpful condition for settling of sand and formation of a quasi-soil layer suitable for growth of grasses and other plants, while parts of the channel bed can also become suitable for agriculture, orchards and similar land uses. Studies have also reported increased grass cover in the riparian zone below dams (Elderd, 2003), that again confirms their influence on a river's hydrological regime in the downstream parts.

A significant development of riparian vegetation cover and an increasing trend in NDVI values have been reported after dam construction for large parts of Tibetan Plateau (Li et al., 2018; Liu et al., 2022a, 2022b), as well as in rivers at the northern part of India (Beale et al., 2022). Other authors have reported altered spatial–temporal dynamics in vegetation covering the riparian zone largely driven by anthropogenic factors in a Dam-Reservoir-River (DRR) system (Zhang et al., 2020). The extent of growth of riparian vegetation cover in these zones are reported to be largely determined by factors such as dam height, valley slope and distance from the upstream dams (Yi et al., 2019), with the impacts of cascading dams being notably higher than single dam along a river channel (Li et al., 2012). Thus, the temporal dynamics of the channel vegetation cover can be taken as an indicator of the flow conditions of a river and therefore can be used to depict the flow alterations induced by upstream dams along the entire longitudinal length of a river.

From the above discussion, it can be seen that all the mentioned studies on dam-induced flow alterations have been conducted in a local or catchment level scale, and there is a lack of such studies in a vast regional scale. Again, due to the lack of gauging sites in large parts of the world, we cannot rely upon only hydrological datasets to create a regional or global picture. On the other hand, it can be seen that there is a general conclusion drawn by all of the above-mentioned literatures that indicates that the river flow conditions have been affected significantly by dam construction and operation, and that riparian vegetation cover is largely affected by upstream flow regulation. Additionally, channel vegetation cover is more or less homogeneously distributed along the longitudinal dimension of a river, and data regarding their presence or absence in a particular region can again be obtained from Remote Sensing datasets, that are available from around 1980s, covering parts of the pre-dam period in case of most rivers. Also, remote sensing and riparian vegetation data are universally available.

Therefore, this study used remote sensing—supplemented by field observations—to assess in-channel vegetation cover as a means to evaluate the extent of dam-induced flow alterations in the east-flowing rivers across peninsular India. This approach also helps address the challenge of limited gauging data, particularly for the pre-dam period, by exploring the potential of in-channel vegetation cover as a proxy for understanding historical flow regimes. The main objectives of our study are: (a) to provide a regional perspective on the effect of dam construction on the major east-flowing rivers of peninsular India, (b) investigating whether similar levels of flow alteration have taken place across all these rivers and (c) examining whether the flow alteration has been homogeneous along the entire longitudinal length of a particular river channel.

13.2 Materials and Methods

13.2.1 Study Area

The peninsular plateau of India is a vast and prominent geological formation covering a significant portion of southern India. Since post-colonial times, this region has become a hotspot for dam construction due to three primary reasons: (a) Unlike northern India's plains, the absence of deep soil layers and the prevalence of impervious crystalline rocks in the substrate make groundwater-based irrigation impractical; (b) the availability of natural slopes and valleys enclosed by low hills provides ideal sites for large dams and reservoirs, particularly for hydropower generation—features absent in the plains south of the Himalayas; (c) the presence of several mighty rivers—such as the Godavari, Krishna, Cauvery, Mahanadi, Narmada and Tapi—and their numerous fast-flowing tributaries.

With the exception of the Narmada and Tapi, all these rivers flow in an east-southeast direction, following the general slope of this cratonic block. The east-flowing rivers carry significant sediment loads and form the great deltas of the Mahanadi, Godavari, Krishna and Cauvery Rivers. These predominantly sandy-bedded, east-flowing rivers are often colonized by channel vegetation, especially during low-flow seasons. Consequently, the east-flowing rivers of peninsular India provide an ideal experimental setting to investigate the impact of dams on downstream flow regimes, using the growth and decline of channel vegetation as an indicator of such impact.

To examine dam-induced flow alterations in downstream reaches, we selected eight river segments (one from each of the Damodar, Kansabati, Subarnarekha, Baitarani, Brahmani, Mahanadi, Godavari and Krishna

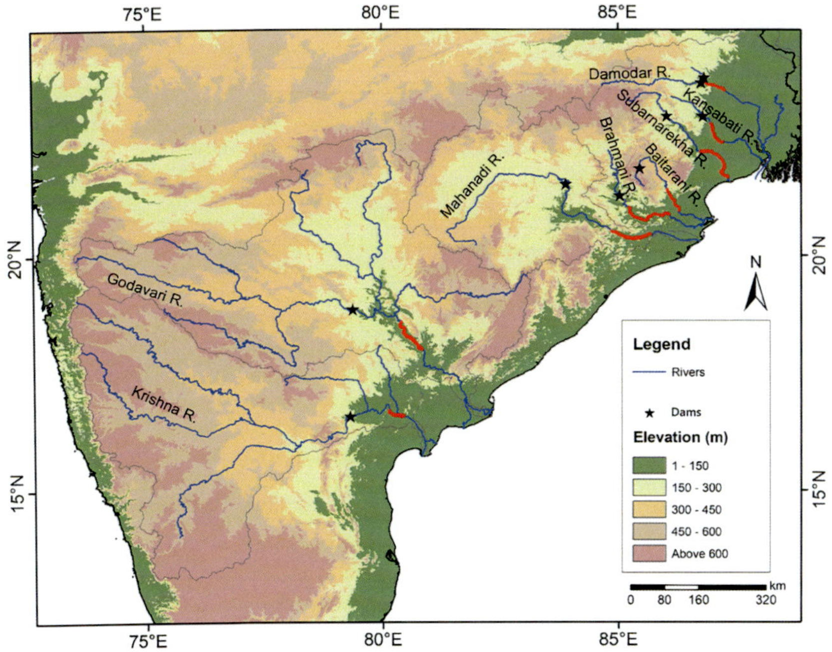

Fig. 13.1 Map of the study area

rivers) located below nine different dams (Fig. 13.1). Detailed information about these dams and their respective downstream reaches is provided in Table 13.1.

13.2.2 Methodological Framework and Its Ground Validation

We hypothesize that frequent high flows and repeated inundation hinder the growth of channel vegetation cover, while low-flow conditions facilitate their growth within the riparian zone, by helping in the settling of sand and creating a stable quasi-soil layer suitable for the growth of plants. This natural process is the primary reason for using channel vegetation as an indicator for changes in flow regime, especially downstream of dams. Our study empirically validated this hypothesis over a reach in the Brahmani River. We analysed 65 cloud-penetrating Sentinel-1 Synthetic Aperture Radar (SAR) images from the European Space Agency (ESA)

Table 13.1 Details of the dams and the corresponding downstream section of the river under consideration

River	*Name of upstream dam*	*Year of construction*	*Storage capacity (in million m^3)*	*Length of the reach under consideration (km)*	*Distance of the reach from upstream dam (km)*
Damodar	Maithon, Panchet	1957, 1959	1062, 1200	12.18	43.90
Kansabati	Mukutmanipur	1959	1200	30.3	70.10
Subarnarekha	Chandil	1982	250	143.2	129.12
Baitarani	Kanjhari	1975	164	65.9	67.80
Brahmani	Rengali	1987	3600	36.5	125.25
Mahanadi	Hirakud	1957	5600	209.15	95.10
Godavari	Sri Ram Sagar	1963	1600	195.3	112.12
Krishna	Nagarjuna Sagar	1967	11,553	123.56	30.10

to determine the extent of water within the bankfull width of the river during monsoon seasons (June to mid-October) from 2016 to 2023. Calm water surfaces act as specular reflectors of radar waves, returning minimal backscatter and appearing distinctly dark in images compared to dry areas. This characteristic of SAR imagery has been widely used for water identification (Sanyal & Chowdhury, 2023; Sanyal & Lu, 2009). We used Google Earth Engine (GEE) to classify the Sentinel-1 SAR scenes into water and dry areas. Wet pixels were assigned a value of 1 while the dry pixels were identified as 0. For this analysis, we utilized the Sentinel-1 SAR GRD: C-band Synthetic Aperture Radar collection available in GEE. Figure 13.2 represents such a Sentinel-1 SAR scene for the monsoon period (June) of the year 2023 showing the inundation extent in parts of Brahmani Channel and associated landcover characteristics of the same region for the post-monsoon season (November) of the same year.

A database of inundated pixels over the study reach was created for eight years (2016–23). A landcover data of the same region was also derived by supervised classification of Landsat-9 image of 2023 post-monsoon (November) season. We attempted to examine whether there is any existent relationship between the channel vegetation cover and the frequency and extent of inundation in the preceding monsoon periods.

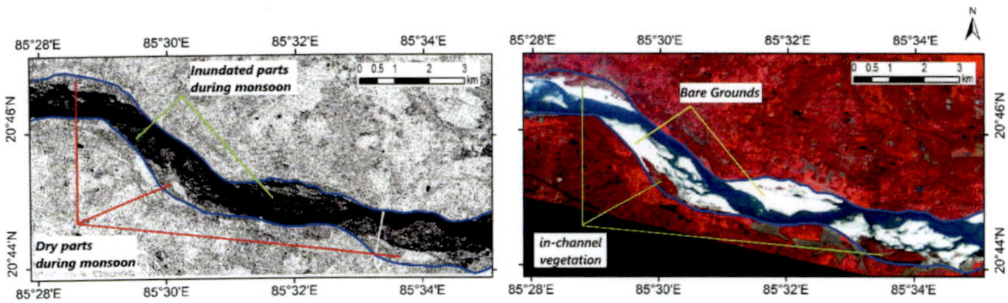

Fig. 13.2 SAR image of parts of Brahmani Channel (monsoon 2023) and Landsat-9 image of the same parts (post-monsoon 2023)

To achieve our goal a layer containing the total frequency of inundation for each of the pixels during this time period was developed by summing up the values in all the classified inundation layers. As the inundated pixels in each year were assigned a value of 1, summing all the classified SAR images pixel-wise provided the frequency of inundation at pixel level within the channel. Another layer was developed by extracting the vegetation pixels from the landcover image of the region for the year 2023. The vegetation layer and the frequency of inundation map were superimposed to analyse the relationship between the two.

The different pixel sizes for Sentinel-1 (10 m × 10 m) and Landsat-9 (30 m × 30 m) were an issue, as each pixel of the vegetation layer covered 9 pixels of the inundation frequency layer. It was resolved by calculating the total number of water pixels within one vegetation pixel. If the number of water pixels within a vegetation pixel was 4 or less out of 9, the vegetation pixel was assigned an inundation value of 0, whereas if this number was 5 or more out of 9, we would assign the vegetation pixel with an inundation value of 1. This process was repeated for each of the 65 classified inundation layers, and the total frequency of inundation between 2016 and 2023 was calculated by summing up these inundation values derived from the 65 classified images. Apart from calculating the inundation frequency for the vegetation pixels, we also calculated the same for the bare-ground pixels of the landcover layer of 2023 using the same method mentioned earlier. This database enabled us to examine how the inundation frequency for the past eight years (2016–23) has shaped the in-channel LULC characteristics.

Accounts of the in-channel LULC types, particularly vegetation versus bare sand, revealed the following features: (1) a vast majority of the pixels

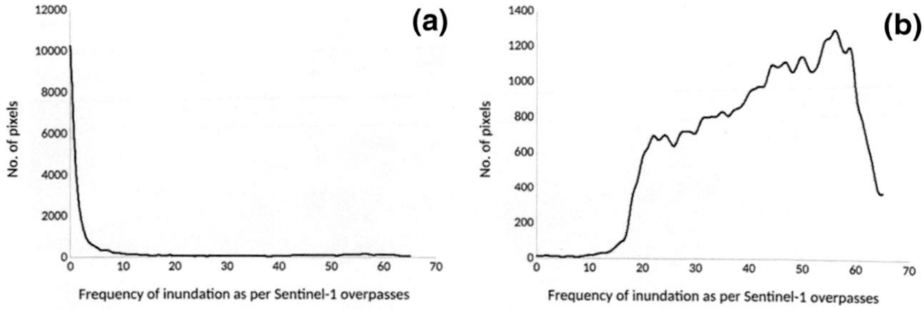

Fig. 13.3 Number of pixels experiencing different frequency of inundation (as per Sentinel-1 overpasses) for (a) vegetation and (b) bare-ground land cover categories (as per Landcover Classification of Brahmani region for 2023)

identified as in-channel vegetation in the post-monsoon season of 2023 never experienced inundation in the preceding eight years (Fig. 13.3a) as depicted by very large number of pixels with inundation frequency value of less than 2, (2) moderate to high number of bare sand pixels in 2023 experiencing 20 to 60 instances of inundation out of 65 satellite overpasses (Fig. 13.3b). This result validates the core assumption for this study, portraying that frequent high-flow conditions and repeated inundation can hinder the growth of channel vegetation cover, thus keeping the frequently inundated parts of the channel bare. The subsequent empirical portion of this study relies on this assumption.

13.3 Data Preparation and Analysis

13.3.1 *Detecting In-Channel Land Cover Changes in the Major East-Flowing Rivers of Peninsular India*

The temporal trends in in-channel vegetation cover across eight selected river reaches were quantified using supervised landcover classification of Landsat images from 1987–88 to 2022–23. For this analysis, we selected Landsat 5, 8 and 9 Top-of-Atmosphere (TOA) images from the Google Earth Engine (GEE) data catalogue. We specifically used Landsat images from the post-monsoon period (December-January, one image per year) due to the abundant availability of cloud-free imagery during this timeframe. We opted for supervised landcover classification to identify vegetation cover and other landcover classes rather than vegetation indices such as NDVI, as these indices can be significantly influenced by leaf cover, vegetation density and vegetation type or health within individual pixels (Beale et al., 2022). The Random Forest supervised landcover classification algorithm was implemented through the GEE platform to classify these images into three categories: water, vegetation and bare earth. Figure 13.4 presents an example of the land cover change data generated for sections of the Brahmani River. We ultimately created a comprehensive database documenting the areal proportions occupied by each of these three landcover types for each year from 1987–88 to 2022–23. The accuracy of these classified images was tested through various accuracy assessment indices and through field surveys done in different river reaches during the post-monsoon period of 2023–24. The field validation process is described in Fig. 13.5(a) and field photos from different river channels have been displayed in Fig. 13.5(b).

Depending on the reference location points derived during field survey and other locations derived from Google Earth, the accuracy of a few randomly selected classified images was calculated by using a range of accuracy assessment indices—namely, overall accuracy, user's and producer's accuracy and Cohen's Kappa. Among them, the User's and Producer's accuracy values range between 0 and 100, with 100 being the perfect score, while the overall accuracy and Cohen's Kappa values range between 0 and 1, with unity being the perfect score. The formulae used in this process, as well as the accuracy values of selected classified images are given in Tables 13.2 and 13.3.

13.3.2 Assessing the Homogeneity of Dam's Impact in a Selected Downstream Reach

To assess the spatial uniformity of a dam's impact throughout its downstream reach, we examined a section of the Brahmani River immediately downstream of Rengali Reservoir. We divided this reach into 25 sub-reaches, each 5 km in length (Fig. 13.6). Using the landcover dataset generated from supervised classifications of Landsat imagery spanning from 1989–90 to 2022–23, we calculated the vegetation fraction for each sub-reach at two points in time: 1989 and 2022. Subsequently, we determined the percentage growth in channel vegetation fraction within each sub-reach during this time period. The equation used for this calculation

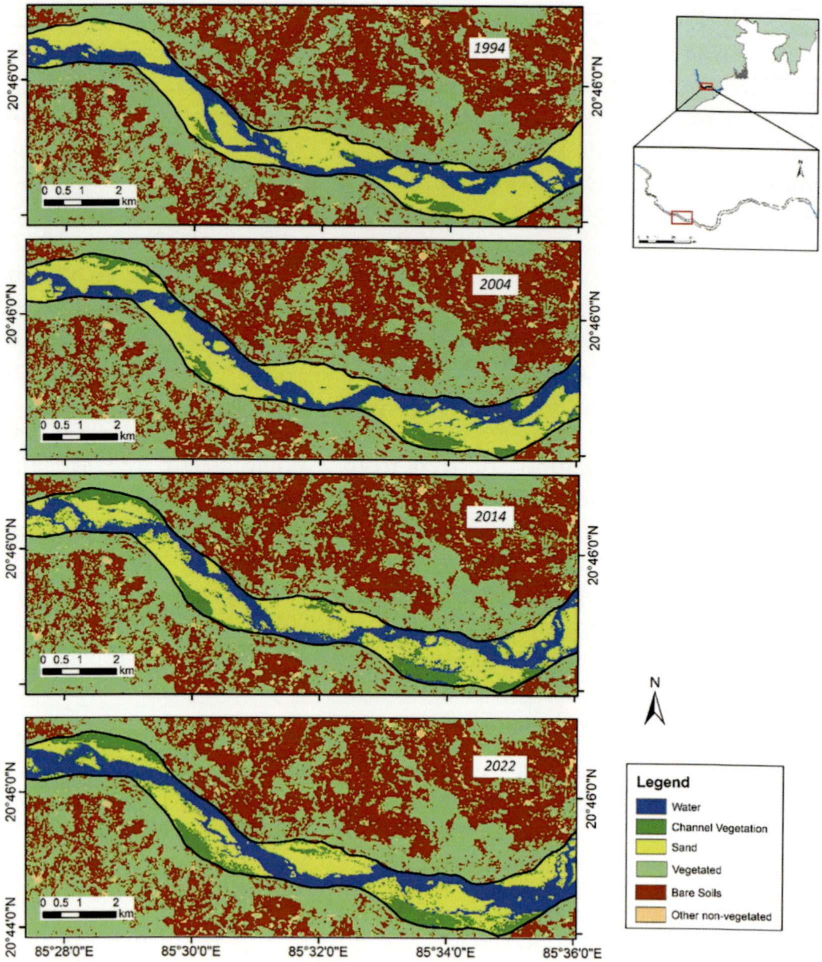

Fig. 13.4 Classified landcover images for parts of the Brahmani River channel showing the spatial pattern of channel vegetation cover for the years 1994, 2004, 2014 and 2022

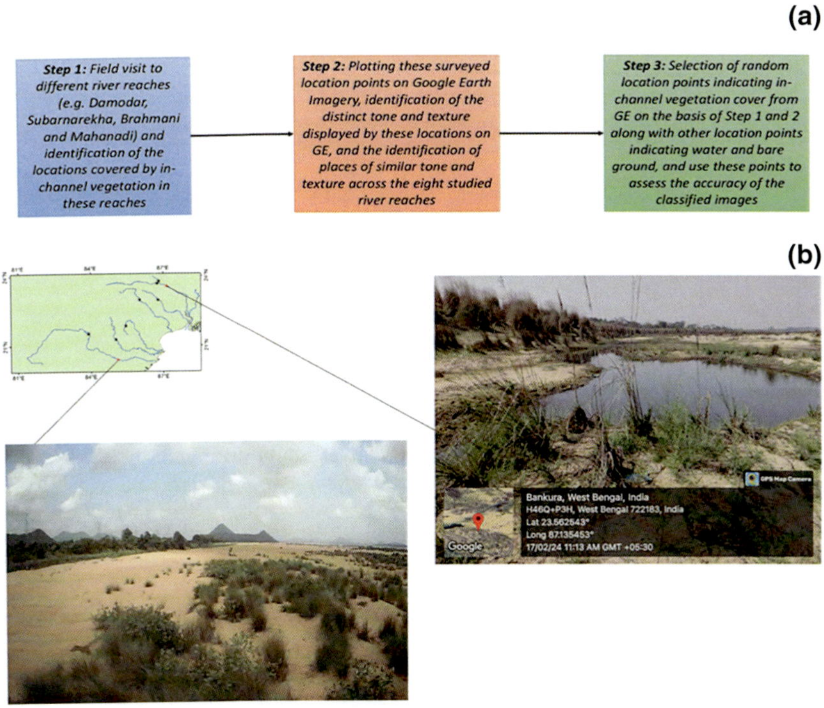

Fig. 13.5 (a) Steps of for acquiring training samples and performing accuracy assessment of the land cover classification through a combination of field visit and the use of high-resolution images available in Google Earth; (b) field photos from Damodar and Mahanadi River Channels

is presented below:

$$\% \text{ changes vegetation fraction}(1989-2022) = \frac{(\text{Vegetation fraction in } 2022) - (\text{Vegetation fraction in in}1989)}{\text{Vegetation fraction in } 1989} \times 100 \quad (13.1)$$

The percentage change in vegetation cover fraction for each sub-reach quantified the extent of the changes in the channel vegetation cover at different parts of the river channel along its longitudinal length.

Table 13.2 Different accuracy assessment indices used and the formulae used

Accuracy Assessment Indices	*Formula*
Overall Accuracy	$\frac{\text{Total no. of correctly classified points}}{\text{Total no. of reference points}} \times 100$
User's Accuracy	$\frac{\text{Total no. of correctly classified points in a class}}{\text{Total no. of reference points in that class (Row total)}} \times 100$
Producer's Accuracy	$\frac{\text{Total no. of correctly classified points in a class}}{\text{Total no. of reference points in that class (Column total)}} \times 100$
Cohen's Kappa	$\frac{\text{Observed accuracy} - \text{Expected Accuracy}}{1 - \text{Expected Accuracy}}$ Where Expected accuracy = $\frac{\text{Sum of diagonal of the error matrix}}{\text{Sum of all elements in the error matrix}}$

Table 13.3 Accuracy assessment indices values for randomly selected classified images

Classified Image		*Overall Accuracy*	*User's Accuracy*			*Producer's Accuracy*			*Cohen's Kappa*
River	*Year*		*Water*	*Veg*	*Others*	*Water*	*Veg*	*Others*	
Damodar	2006	0.86	78.26	86.96	88.89	75.00	83.33	92.31	0.77
Kansabati	2008	0.82	82.76	80.43	84.00	82.76	86.05	75.00	0.72
Subarnarekha	2008	0.83	89.29	80.00	81.25	75.76	84.21	89.66	0.74
Baitarani	2009	0.77	67.65	81.48	82.05	79.31	78.57	74.42	0.65
Brahmani	2016	0.81	68.97	89.29	83.72	80.00	75.76	85.71	0.71
Mahanadi	2014	0.85	80.95	89.47	82.93	73.91	89.47	87.18	0.77
Godavari	2022	0.85	69.57	68.75	95.08	80.00	91.67	85.29	0.71
Krishna	2022	0.86	86.21	88.57	83.33	83.33	88.57	85.71	0.79

13.4 Results

13.4.1 Land Cover Changes in the Studied River Reaches

We analysed the temporal trends (1987–2022) of in-channel land cover types (vegetation, water, and bare sand) for selected segments of eight east-flowing rivers in peninsular India. Annual cover fractions for each land cover type were plotted as scatter plots (Fig. 13.7), and we quantified the temporal trends through linear regression equations (Table 13.4). The reliability of these regression analyses was evaluated using the coefficient of determination (R^2) for each river reach.

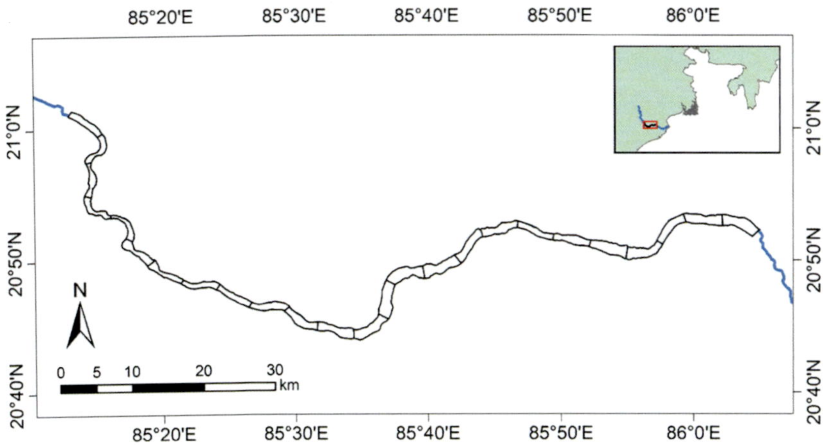

Fig. 13.6 Longitudinal divisions of the Brahmani River reach

Figure 13.7 demonstrates that vegetation cover increased in all rivers, though at varying rates, typically at the expense of bare sand cover—a pattern particularly evident in the Kansabati River. The extent of water surface showed no consistent temporal pattern across these east-flowing rivers. Notably, some rivers such as the Subarnarekha and Mahanadi exhibited significantly low R^2 values (Table 13.4), reducing confidence in our assessment of vegetation cover change rates.

Since our central hypothesis focuses on temporal trends in in-channel vegetation, we required more accurate trend determination. To address the relatively weak linear temporal trends in vegetation cover fraction, we applied the Mann-Kendall non-parametric test along with Sen's Slope to more robustly quantify the rate of vegetation cover change within these river channels.

The Mann-Kendall test is a statistical test that determines if a set of data is increasing or decreasing over time. It's a non-parametric test, meaning it doesn't assume anything about the data's distribution. This test can be used to determine the monotonic trend in a series of values even where there is no linear trend present. While the Mann-Kendall test is used to determine the direction of trend in a series of data, Sen's Slope estimator is used to quantify the trend. The Sen's slope estimator formula is calculated by taking the median of all possible slopes between pairs of data points in a time series, where the slope for each pair is calculated as (yj −

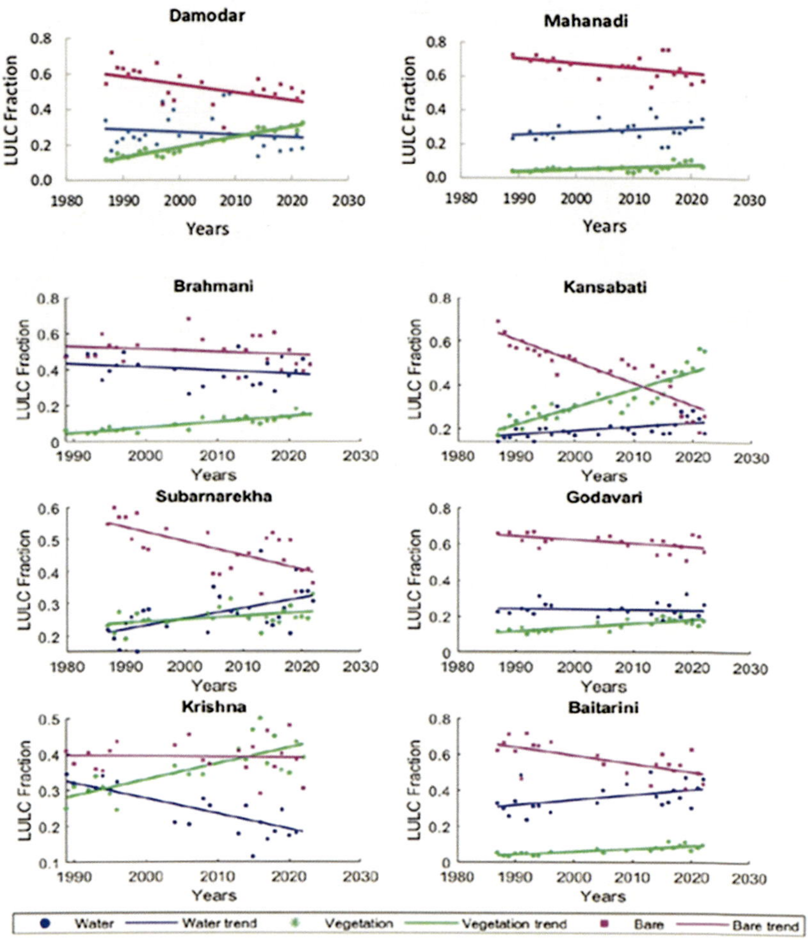

Fig. 13.7 Temporal trends of landcover changes in the eight studied river reaches during the time period 1987–2022

yi) / (xj − xi), with (xi, yi) and (xj, yj) being two data points where j > i; essentially, it is the median of all the "slopes" between every possible pair of data points in a dataset. It is also a non-parametric test, meaning it does not require assumptions about the underlying data distribution. The Sen's slope values for the vegetation cover fraction time series data (Fig. 13.8)

Table 13.4 Slope of trendlines and their respective R^2 values for the three landcover types in the eight studied river reaches

River	*Slope of Trendlines*			R^2 *values*		
	Water	*Vegetation*	*Bare grounds*	*Water*	*Vegetation*	*Bare grounds*
Damodar	−0.0014	0.0059	−0.0044	0.029	0.9344	0.2416
Kansabati	0.0018	0.0081	−0.01	0.2642	0.7987	0.7905
Subarnarekha	0.0033	0.0011	−0.0044	0.2959	0.1883	0.4698
Baitarani	0.0029	0.0017	−0.0046	0.2446	0.7052	0.4225
Brahmani	−0.0017	0.0032	−0.0014	0.0701	0.7855	0.0416
Mahanadi	0.0015	0.0013	−0.0028	0.0818	0.3274	0.2267
Godavari	−0.0003	0.0022	−0.0019	0.0072	0.6257	0.269
Krishna	−0.0042	0.0044	−0.0002	0.5375	0.5846	0.0026

provide a basis for the comparison of the rates of vegetation cover growth (and thus, the extent of dam-induced flow regime alterations) between the eight studied reaches. The positive values in Fig. 13.8 indicate that the channel vegetation cover has grown in each of these reaches during 1987 to 2022. These findings highlight the effectiveness of using in-channel vegetation as a highly sensitive indicator of altered flow regimes.

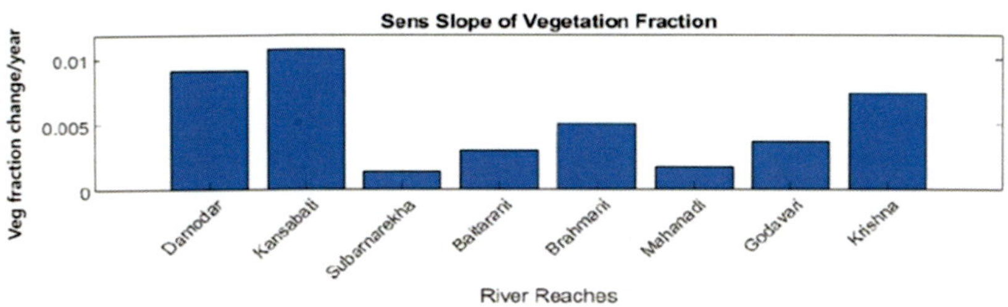

Fig. 13.8 Sen's Slope values quantifying the extent of channel vegetation cover growth in the eight studied reaches

13.4.2 *Factors Affecting the Extent of Dam-Induced Flow Alterations*

While in-channel vegetation showed an increasing trend during the post-dam period across east-flowing rivers of peninsular India, both the vegetation growth rate (Fig. 13.8) and the extent of dam-induced flow alteration varied significantly between sites. After investigating various geophysical parameters, our study identified that the distance from the dam to the examined reach is the most influential factor. Figure 13.9 clearly illustrates the positive correlation between the distance of the studied reach from its upstream dam and the rate of in-channel vegetation increase (as measured by Sen''s slope).

River reaches situated closer to their respective upstream dams (such as Damodar, Kansabati and Brahmani) have undergone the most substantial changes in channel vegetation cover compared to other investigated reaches. Conversely, reaches located farther downstream from their respective dams (such as Mahanadi and Subarnarekha) demonstrate less significant changes in channel vegetation cover over approximately the last three decades, suggesting that dam-induced flow alterations have been less pronounced in these more distant river segments.

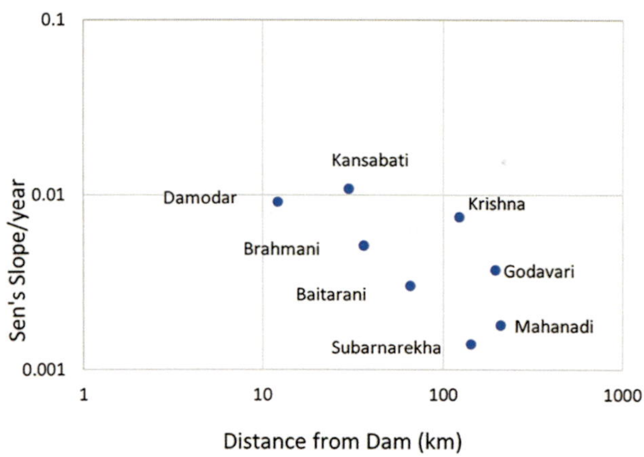

Fig. 13.9 Scatter plot showing the relationship between the distance of the studied river reaches from their respective upstream dams and the growth rates of channel vegetation cover in these regions during 1987–2022, here indicated by Sen's Slope values

13.4.3 *Investigating the Homogeneity in Streamflow Response to Upstream Dam Building*

The introduction section notes that numerous studies have attempted to quantify changes in fluvial hydrology following dam construction by analysing historical river gauge records downstream of dams. However, these studies couldn't illuminate the spatial variability of these impacts along the river's length because gauge data only represent conditions at specific cross-sections of the stream. Channel vegetation cover, on the other hand, is highly responsive to river flow conditions. In previous parts of our study, we have demonstrated how it can serve as a proxy for examining alterations in the river's flow regime. Additionally, vegetation is relatively homogeneously distributed along the longitudinal extent of a river channel, making it particularly valuable for assessing local variations in flow conditions throughout the river's length. In this part of our study, we attempted to examine whether a river channel's streamflow response to upstream flow regulations remains consistent along its entire longitudinal extent. We also investigated whether channel vegetation cover could serve as a proxy to detect variations in the magnitude of dam-induced flow alterations along the river's length.

A reach of the Brahmani River downstream of Rengali dam was selected as our case study. We divided this study reach into 25 longitudinal sections to conduct a comparative analysis of the Rengali dam's impact across different segments of the downstream river portion. We calculated the percentage change in channel vegetation cover between 1987–2022 for each section and plotted these values against their respective distances from the upstream dam (Fig. 13.10). The impact of dam-induced flow alteration was not found to be homogeneous (as indicated by vegetation colonisation) for entire river reach. Instead, we observed a general decrease in channel vegetation growth percentage across the 25 sub-sections as distance from the dam increased.

13.5 Discussion

The overall outcome of this study (Figs. 13.7 and 13.8) aligns with our core hypothesis that reduced flows and infrequent inundations promote plant growth and stability within the channel. This further demonstrates that channel vegetation cover is highly responsive to dam-induced flow alterations and the proliferation of channel vegetation in east-flowing

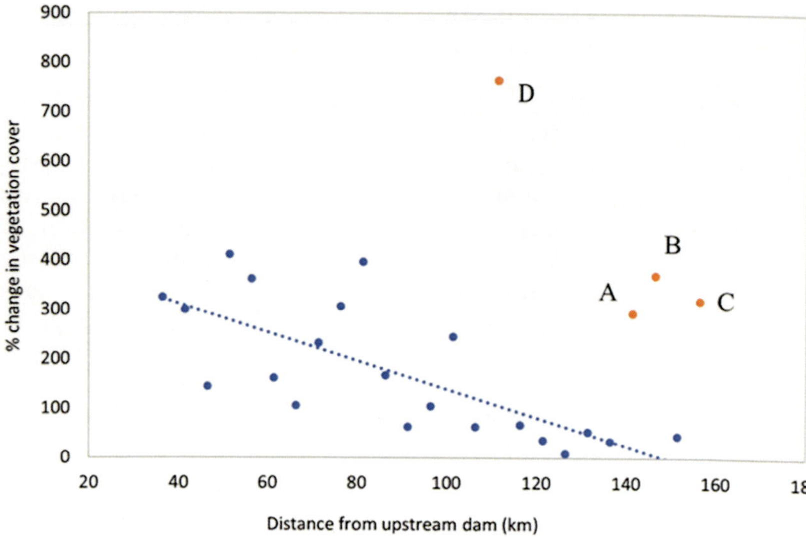

Fig. 13.10 Relationship between Downstream length and percentage change in channel vegetation cover in the Brahmani River channel; each point in the scatter representing one of the 25 sub-sections of the reach

river channels of India's peninsular plateau region, indicating flow regime modifications caused by upstream dams. Notably, this pattern is consistent across all studied river reaches throughout the peninsular plateau region, despite local variations in lithology, topography, climate, soil and natural vegetation types. Therefore, this study establishes that channel vegetation cover can serve as a reliable indirect indicator of dam-induced flow alterations in sand-bed river channels downstream of large dams across diverse environmental conditions.

Although we were unable to perform statistical analysis on the relationship between distance from upstream dam and channel vegetation growth rate (Fig. 13.8) due to our limited sample of eight east-flowing rivers, visual interpretation of the diagram suggests that distance from the upstream dam plays an important role in determining the extent of flow alterations in downstream channel sections.

The degree of impact of an upstream dam on downstream flow alteration across various sections of the longitudinal profile roughly follow the overall pattern where the distance from the dam plays the most crucial role (Fig. 13.10). However, some sub-sections, identified as Point A, B, C and D in Fig. 13.10 reveals notable deviation from the overall inverse relation between the increasing rate of vegetation colonization of the channels and the distance from the upstream dam. Upon further investigation, we derived the slope of each longitudinal segment along the Brahmani River channel by extracting the elevations at the upstream and downstream ends of each subdivision from the Shuttle Radar Topographic Mission (SRTM) DEM. We observed that the segments exhibiting notable deviations from the general trend—where the rate of percentage vegetation cover growth decreases with increasing distance from the dam—had significantly lower slope values (points A, B and D: 0.2 m/km; point C: 0.4 m/km) compared to the average slope across all segments (0.63 m/km). This finding suggests that local slope plays an important role in influencing river stage at specific cross-sections. The slope angle can influence channel vegetation cover by altering flow velocity and the stream's capacity to erode bed and bank materials. Steeper slopes, characterized by faster water flow, typically support less in-channel vegetation, as the increased velocity can uproot plants and hinder their establishment. In contrast, gentler slopes promote more diverse and abundant vegetation, as slower flows facilitate the settling of sediment and create stable substrata, providing favourable conditions for vegetation growth within the channel. This factor probably amplified the effect of upstream dam-induced flow alterations at segments A, B, C and D (Fig. 13.10) relative to the distance of these segments from the upstream dam. It explains the deviation from the overall trend of decreasing in-channel vegetation growth in the post-dam period with increasing distance from the dam.

13.6 Conclusion

This study adopted a novel approach by using channel vegetation cover, rather than hydrograph records from gauging stations, to assess the impacts of upstream dam construction on downstream flow conditions across major east-flowing rivers of peninsular India. It examined how proximity to upstream dams influences the degree of flow alteration and explored the spatial heterogeneity of these effects along river channels. The assumption that increased channel vegetation indicates reduced frequency of high flows, while bare surfaces suggest frequent inundation, was validated through empirical evidence, supporting the effectiveness of channel vegetation as a proxy for flow conditions. Overall, the findings align with existing literature, which reports a dampening of flood peaks and an increase in low-flow conditions during the post-dam period across the region.

At the reach scale, results from the Brahmani River further emphasized the critical role of distance from upstream dams in shaping downstream flow regimes. They also highlighted the limitations of relying solely on site-specific hydrograph data, which often fails to capture the spatially heterogeneous response of river systems to flow regulation. Together, these insights underscore the importance of considering longitudinal variability when assessing the ecological and hydrological impacts of dam building.

References

Achini Ishankha, W. C., Shrestha, S., Van Binh, D., & Kantoush, S. A. (2024). Comprehensive assessment of cascading dams-induced hydrological alterations in the lancang-mekong river using machine learning technique. *Journal of Environmental Management, 371*, Article 123082. https://doi.org/10.1016/j.jenvman.2024.123082

Annys, S., & Frankl, A. (2025). A systematic review of the hydrogeomorphological impacts of large dams in Africa. *Earth-Science Reviews, 262*, Article 105048. https://doi.org/10.1016/j.earscirev.2025.105048

Anthony, E. J., Brunier, G., Besset, M., Goichot, M., Dussouillez, P., & Nguyen, V. L. (2015). Linking rapid erosion of the Mekong River delta to human activities. *Scientific Reports, 5*(1), 14745. https://doi.org/10.1038/srep14745

Beale, J., Grabowski, R. C., Long'or Lokidor, P., Vercruysse, K., & Simms, D. M. (2022). Vegetation cover dynamics along two Himalayan rivers: Drivers and implications of change. *Science of The Total Environment, 849*, 157826. https://doi.org/10.1016/j.scitotenv.2022.157826

Belmar, O., Bruno, D., Martínez-Capel, F., Barquín, J., & Velasco, J. (2013). Effects of flow regime alteration on fluvial habitats and riparian quality in a semiarid Mediterranean basin. *Ecological Indicators, 30*, 52–64. https://doi.org/10.1016/j.ecolind.2013.01.042

Bhattacharyya, K., & Wiley, M. J. (2014). Dams, riparian settlement and the threat of climate change in a dynamic fluvial environment: A case study of the Damodar River, India. In M. Nüsser (Ed.), *Large dams in Asia* (pp. 75–100). Springer Netherlands. https://doi.org/10.1007/978-94-007-2798-4_5

Bhaumik, U., Mukhopadhyay, M. K., Shrivastava, N. P., Sharma, A. P., & Singh, S. N. (2017). A case study of the Narmada River system in India with particular reference to the impact of dams on its ecology and fisheries. *Aquatic*

Ecosystem Health & Management, 20(1–2), 151–159. https://doi.org/10.1080/14634988.2017.1288529

Binh, D. V., Kantoush, S. A., Saber, M., Mai, N. P., Maskey, S., Phong, D. T., & Sumi, T. (2020). Long-term alterations of flow regimes of the Mekong River and adaptation strategies for the Vietnamese Mekong Delta. *Journal of Hydrology: Regional Studies, 32*, Article 100742. https://doi.org/10.1016/j.ejrh.2020.100742

Biswas, S. S., & Pani, P. (2021). Changes in the hydrological regime and channel morphology as the effects of dams and bridges in the Barakar River, India. *Environmental Earth Sciences, 80*(5), 209. https://doi.org/10.1007/s12665-021-09490-0

Choi, S., Yoon, B., & Woo, H. (2005). Effects of dam-induced flow regime change on downstream river morphology and vegetation cover in the Hwang River, Korea. *River Research and Applications, 21*(2–3), 315–325. https://doi.org/10.1002/rra.849

Corenblit, D., Tabacchi, E., Steiger, J., & Gurnell, A. M. (2007). Reciprocal interactions and adjustments between fluvial landforms and vegetation dynamics in river corridors: A review of complementary approaches. *Earth-Science Reviews, 84*(1–2), 56–86. https://doi.org/10.1016/j.earscirev.2007.05.004

Das, S., Sangode, S. J., & Kandekar, A. M. (2021). Recent decline in streamflow and sediment discharge in the Godavari basin, India (1965–2015). *CATENA, 206*, Article 105537. https://doi.org/10.1016/j.catena.2021.105537

Deng, C., Zhang, H., Hamilton, D. P., & Hollingsworth, A. (2025). Impacts of changes in climate and water demand on flow in a subtropical river catchment below a major dam. *Journal of Environmental Management, 375*, Article 124137. https://doi.org/10.1016/j.jenvman.2025.124137

Ekka, A., Keshav, S., Pande, S., Van Der Zaag, P., & Jiang, Y. (2022). Dam-induced hydrological alterations in the upper Cauvery river basin, India. *Journal of Hydrology: Regional Studies, 44*, Article 101231. https://doi.org/10.1016/j.ejrh.2022.101231

El Aoula, R., Mhammdi, N., Dezileau, L., Mahe, G., & Kolker, A. S. (2021). Fluvial sediment transport degradation after dam construction in North Africa. *Journal of African Earth Sciences, 182*, Article 104255. https://doi.org/10.1016/j.jafrearsci.2021.104255

Elderd, B. D. (2003). The impact of changing flow regimes on riparian vegetation and the riparian species mimulus guttatus. *Ecological Applications, 13*(6), 1610–1625. https://doi.org/10.1890/02-5371

Ghorai, A., & Ghosh, K. (2024). Enhanced river regulation following dam reoperation: Impacts on flow regime, sediment transport, and planform adjustment in a plateau river basin in the lower Ganga plains. *CATENA, 237*, Article 107834. https://doi.org/10.1016/j.catena.2024.107834

Ghosal, M., Swarnkar, S., Poonia, V., Kundu, S., & Pandey, S. (2024). Hydrological dichotomy: Streamflow drought in dammed vs. undammed regions of the largest Indian Peninsular basin. *Journal of Arid Environments, 225*, 105273. https://doi.org/10.1016/j.jaridenv.2024.105273

Ghosh, S. (2011). Hydrological changes and their impact on fluvial environment of the lower damodar basin over a period of fifty years of damming The Mighty Damodar River in Eastern India. *Procedia - Social and Behavioral Sciences, 19*, 511–519. https://doi.org/10.1016/j.sbspro.2011.05.163

Ghosh, S., & Guchhait, S. K. (2014). Hydrogeomorphic variability due to dam constructions and emerging problems: A case study of Damodar River, West Bengal, India. *Environment, Development and Sustainability, 16*(3), 769–796. https://doi.org/10.1007/s10668-013-9494-5

Ghosh, S., Hoque, Md. M., Saha, U. D., & Islam, A. (2022). Assessment of dam-induced changes in ecogeomorphological behaviour and fluvial functionality in the Damodar River, West Bengal, India. *Journal of Water Supply: Research and Technology-Aqua, 71*(6), 722–750. https://doi.org/10.2166/aqua.2022.003

Gopal, B. (2013). *Environmental flows: An introduction for water resource managers* (First). National Institute of Ecology.

Gordon, E., & Meentemeyer, R. K. (2006). Effects of dam operation and land use on stream channel morphology and riparian vegetation. *Geomorphology, 82*(3–4), 412–429. https://doi.org/10.1016/j.geomorph.2006.06.001

Huang, Y., Huang, B., Zhang, K., Qin, T., & Yan, D. (2024). Novel framework for quantifying dam construction, climate change, and land-use change impacts on flow regime: A case study in Ganjiang River Basin, China. *Journal of Hydrology: Regional Studies, 55*, Article 101918. https://doi.org/10.1016/j.ejrh.2024.101918

Liaghat, A., Adib, A., & Gafouri, H. R. (2017). Evaluating the effects of dam construction on the morphological changes of downstream meandering rivers (Case Study: Karkheh River). *Engineering, Technology & Applied Science Research, 7*(2), 1515–1522. https://doi.org/10.48084/etasr.969

Li, J., Dong, S., Yang, Z., Peng, M., Liu, S., & Li, X. (2012). Effects of cascade hydropower dams on the structure and distribution of riparian and upland vegetation along the middle-lower Lancang-Mekong River. *Forest Ecology and Management, 284*, 251–259. https://doi.org/10.1016/j.foreco.2012.07.050

Li, L., Zhang, Y., Liu, L., Wu, J., Wang, Z., Li, S., Zhang, H., Zu, J., Ding, M., & Paudel, B. (2018). Spatiotemporal patterns of vegetation greenness change and associated climatic and anthropogenic drivers on the Tibetan Plateau during 2000–2015. *Remote Sensing, 10*(10), 1525. https://doi.org/10.3390/rs10101525

Liu, S., Pérez-Sánchez, J., Jimeno-Sáez, P., Alcalá, F. J., & Senent-Aparicio, J. (2022). A novel approach to assessing the impacts of dam construction on hydrologic and ecosystem alterations. Case study: Castril river basin, Spain. *Ecohydrology & Hydrobiology, 22*(4), 598–608. https://doi.org/10.1016/j.ecohyd.2022.08.004

Liu, Y., Lu, H., Tian, P., & Qiu, L. (2022b). Evaluating the effects of dams and meteorological variables on riparian vegetation NDVI in the Tibetan Plateau. *Science of the Total Environment, 831*, Article 154933. https://doi.org/10.1016/j.scitotenv.2022.154933

Luo, Z., Zhang, S., Liu, H., Wang, L., Wang, S., & Wang, L. (2023). Assessment of multiple dam- and sluice-induced alterations in hydrologic regime and ecological flow. *Journal of Hydrology, 617*, Article 128960. https://doi.org/10.1016/j.jhydrol.2022.128960

Mácová, K., & Kozáková, Z. (2023). How important for society is recreation provided by multi-purpose water reservoirs? Welfare analysis of the Vltava River Reservoir System. *Water, 15*(10), 1966. https://doi.org/10.3390/w15101966

Mallik, A. U., & Richardson, J. S. (2009). Riparian vegetation change in upstream and downstream reaches of three temperate rivers dammed for hydroelectric generation in British Columbia, Canada. *Ecological Engineering, 35*(5), 810–819. https://doi.org/10.1016/j.ecoleng.2008.12.005

Mittal, N., Bhave, A. G., Mishra, A., & Singh, R. (2016). Impact of human intervention and climate change on natural flow regime. *Water Resources Management, 30*(2), 685–699. https://doi.org/10.1007/s11269-015-1185-6

Mittal, N., Mishra, A., Singh, R., Bhave, A. G., & Van Der Valk, M. (2014). Flow regime alteration due to anthropogenic and climatic changes in the Kangsabati River, India. *Ecohydrology & Hydrobiology, 14*(3), 182–191. https://doi.org/10.1016/j.ecohyd.2014.06.002

Mura, S. N. S., & Gope, A. (2020). Anthropogenic impact on forms and processes of the Kangsabati River Basin. In *Anthropogeomorphology of Bhagirathi-Hooghly River System in India* (1st ed.). CRC Press.

Nilsson, C., Reidy, C. A., Dynesius, M., & Revenga, C. (2005). Fragmentation and flow regulation of the world's large river systems. *Science, 308*(5720), 405–408. https://doi.org/10.1126/science.1107887

Pal, S. (2016). Impact of Massanjore Dam on hydro-geomorphological modification of Mayurakshi River, Eastern India. *Environment, Development and Sustainability, 18*(3), 921–944. https://doi.org/10.1007/s10668-015-9679-1

Pettit, N. E., Froend, R. H., & Davies, P. M. (2001). Identifying the natural flow regime and the relationship with riparian vegetation for two contrasting western Australian rivers. *Regulated Rivers: Research & Management, 17*(3), 201–215. https://doi.org/10.1002/rrr.624

Pradhan, C., Chembolu, V., & Dutta, S. (2019). Impact of river interventions on alluvial channel morphology. *ISH Journal of Hydraulic Engineering, 25*(1), 87–93. https://doi.org/10.1080/09715010.2018.1453878

Purnanjali, C., & Kasiviswanathan, K. S. (2022). Assessment of impacts of altered environmental flow on fishing in lower Damodar river basin. *Acta Geophysica, 70*(2), 833–846. https://doi.org/10.1007/s11600-022-00752-4

Rao, K. N., Subraelu, P., Kumar, K. Ch. V. N., Demudu, G., Malini, B. H., Rajawat, A. S., & Ajai. (2010). Impacts of sediment retention by dams on delta shoreline recession: Evidences from the Krishna and Godavari deltas, India. *Earth Surface Processes and Landforms, 35*(7), 817–827. https://doi.org/10.1002/esp.1977

Rosenberg, D. M., Mccully, P., & Pringle, C. M. (2000). Global-scale environmental effects of hydrological alterations: Introduction. *BioScience, 50*(9), 746. https://doi.org/10.1641/0006-3568(2000)050[0746:GSEEOH]2.0.CO;2

Sanyal, J. (2017). Predicting possible effects of dams on downstream river bed changes of a Himalayan river with morphodynamic modelling. *Quaternary International, 453*, 48–62. https://doi.org/10.1016/j.quaint.2017.03.063

Sanyal, J. (2023). Assessing the influence of upstream basin and climatic characteristics on post-dam downstream streamflow changes: Empirical insights from Peninsular India. *Water Resources, 50*(6), 842–856. https://doi.org/10.1134/S0097807823600729

Sanyal, J., & Chowdhury, A. (2023). Reassessment of reservoir sedimentation rates under a monsoon climate with combined optical and microwave remote sensing: A case study of three reservoirs in the Upper Godavari Basin, India. *Journal of Hydrologic Engineering, 28*(11), 05023022. https://doi.org/10.1061/JHYEFF.HEENG-5995

Sanyal, J., Dasgupta, P., & Kanae, S. (2020). Predicting dam-related downstream geomorphic response with widely available stream gauge data: A case study of the Godavari River Basin, India. *Singapore Journal of Tropical Geography, 41*(2), 284–298. https://doi.org/10.1111/sjtg.12323

Sanyal, J., & Lu, X. X. (2009). Ideal location for flood shelter: A geographic information system approach. *Journal of Flood Risk Management, 2*(4), 262–271. https://doi.org/10.1111/j.1753-318X.2009.01043.x

Shi, H., Chen, J., Liu, S., & Sivakumar, B. (2019). The role of large dams in promoting economic development under the pressure of population growth. *Sustainability, 11*(10), 2965. https://doi.org/10.3390/su11102965

Singh, A., Swarnkar, S., Kundu, S., & Prakash, S. (2025). Investigating human-induced hydrological changes in the largest Indian peninsular basin. *Hydrological Sciences Journal, 02626667*(2025), 2486160. https://doi.org/10.1080/02626667.2025.2486160

Singh, A., Tripathi, V. K., Singh, R. M., & Kumar, P. (2021). Assessment of hydrologic alteration caused by human interventions in flow regimes of the Godavari River. *Environment and Ecology, 39*. ISSN: 0970-0420.

Singh, R. K., & Jain, M. K. (2021). Complexity analyses of Godavari and Krishna river streamflow using the concept of entropy. *Acta Geophysica, 69*(6), 2325–2338. https://doi.org/10.1007/s11600-021-00660-z

Słowik, M., Dezső, J., Marciniak, A., Tóth, G., & Kovács, J. (2018). Evolution of river planforms downstream of dams: Effect of dam construction or earlier human-induced changes? *Earth Surface Processes and Landforms, 43*(10), 2045–2063. https://doi.org/10.1002/esp.4371

Stromberg, J. C., Lite, S. J., & Dixon, M. D. (2010). Effects of stream flow patterns on riparian vegetation of a semiarid river: Implications for a changing climate. *River Research and Applications, 26*(6), 712–729. https://doi.org/10.1002/rra.1272

Tonkin, J. D., Merritt, David. M., Olden, J. D., Reynolds, L. V., & Lytle, D. A. (2017). Flow regime alteration degrades ecological networks in riparian ecosystems. *Nature Ecology & Evolution, 2*(1), 86–93. https://doi.org/10.1038/s41559-017-0379-0

Tortajada, C. (2015). Dams: An essential component of development. *Journal of Hydrologic Engineering, 20*(1), A4014005. https://doi.org/10.1061/(ASCE)HE.1943-5584.0000919

Uday Kumar, A., & Jayakumar, K. V. (2018). Assessment of hydrological alteration and environmental flow requirements for Srisailam dam on Krishna River, India. *Water Policy, 20*(6), 1176–1190. https://doi.org/10.2166/wp.2018.203

Uday Kumar, A., & Jayakumar, K. V. (2020). Hydrological alterations due to anthropogenic activities in Krishna River Basin, India. *Ecological Indicators, 108*, Article 105663. https://doi.org/10.1016/j.ecolind.2019.105663

Uday Kumar, A., & Jayakumar, K. V. (2022). Assessment of flow regime alteration in the Krishna River basin. *ISH Journal of Hydraulic Engineering, 28*(2), 212–218. https://doi.org/10.1080/09715010.2020.1852979

Verma, S. K., Sahu, R. T., Singh, H., Prasad, A. D., & Verma, M. K. (2022). A study of environmental and ecological impacts due to construction and operation of Tehri-Polavaram Dam. *IOP Conference Series: Earth and Environmental Science, 1032*(1), Article 012020. https://doi.org/10.1088/1755-1315/1032/1/012020

Wolf, S., Esser, V., Lehmkuhl, F., & Schüttrumpf, H. (2022). Long-time impact of a large dam on its downstream river's morphology: Determined by sediment characteristics, pollutants as a marker, and numerical modelling. *Journal of Sedimentary Environments, 7*(3), 403–424. https://doi.org/10.1007/s43217-022-00103-9

Yi, Y., Zhou, Y., Song, J., Zhang, S., Cai, Y., Yang, W., & Yang, Z. (2019). The effects of cascade dam construction and operation on riparian vegetation. *Advances in Water Resources, 131*, Article 103206. https://doi.org/10.1016/j.advwatres.2018.09.015

Zhang, P., Cai, Y., Yang, W., Yi, Y., Yang, Z., & Fu, Q. (2020). Contributions of climatic and anthropogenic drivers to vegetation dynamics indicated by NDVI in a large dam-reservoir-river system. *Journal of Cleaner Production, 256*, Article 120477. https://doi.org/10.1016/j.jclepro.2020.120477

CHAPTER 14

Tectono-Geomorphological and LULC Changes in the Transboundary Section of the New Brahmaputra River Channel Using Chinese Gaofen-1/6 WFV (16 m) Datasets and Geospatial Techniques

Biswajit Nath, Bo Zhong, Junjun Wu, Aixia Yang, Shanlong Wu, Shukla Acharjee, Nandini Choudhury, and Atin Kumar Mitra

14.1 Introduction

A number of natural processes are involved as river systems migrate across riverine floodplains on the earth's surface, including channel shifting, meandering, braided activity, sandbar development within the channel,

B. Nath (✉) · B. Zhong (✉) · J. Wu · A. Yang · S. Wu
Aerospace Information Research Institute (AIR), National Engineering Laboratory for Satellite Remote Sensing Applications, Chinese Academy of Sciences (CAS), Beijing, P. R. China
e-mail: nath.gis79@cu.ac.bd

S. Ghosh (ed.), *Reimagining Indian Rivers for Sustainability*, Palgrave Studies in Environmental Sustainability,
https://doi.org/10.1007/978-3-032-04754-0_14

erosion of the river valley (Leopold et al., 1964; Pati et al., 2008; Randle, 2006; Yang, 1971), and river movement (Chakraborty & Mukhopadhyay, 2015). Moreover, a river system can be affected by combined anthropogenic and natural activities in the course of time and space (Kawser et al., 2021; Nath et al., 2022a, 2022b; Singh et al., 2020).

Proper assessment of tectonic activity and its progress over a geographical region saves lives and livelihoods. Understanding the processes involved in channel migration in river systems is crucial for assessing the risk factors for people living within the area (Singh et al., 2020). River systems are always sensitized and respond to changes in sediment load, hydrology and active tectonics (Bali et al., 2012; Delcaillau et al., 2006; Goswami et al., 1999; Han et al., 2003; Kanhaiya et al., 2019; Kawser et al., 2021; Keller, 1986; Keller & Pinter, 2002; Pal et al., 2016a; Pati et al., 2008; Schumm et al., 2000; Singh, 2015; Singh et al., 2016, 2017; Valdiya et al., 1992). In the past, active tectonics was also evaluated by a thorough analysis of alluvial river dynamic activities (Jackson et al., 1998; Nandy, 2001; Ouchi, 1985). Among the aspects considered by significant studies across the world are morphotectonic study using satellite datasets (Biswas & Grasemann, 2005; Kale et al., 2014; Lone, 2017; Luirei et al., 2014; Mohaimen et al., 2024; Mohanty & Singh, 2017; Mohanty et al., 2004; Pandey & Sharma, 2021; Philip, 1996; Radaideh et al., 2016) geomorphic and sedimentary response to active tectonics deformations (Bhatt et al., 2007; Holbrook & Schumm, 1999; Jain, 2017; Kumar & Duarah, 2020; Malik & Nakata,

B. Zhong
e-mail: zhongbo@aircas.ac.cn

B. Nath
Lab of Geoinformatics and Earth Observation Research (LGEOR), Department of Geography and Environmental Studies, Faculty of Biological Sciences, University of Chittagong, Chittagong, Bangladesh

S. Acharjee
Centre for Studies in Geography, Dibrugarh University, Dibrugarh, Assam, India

N. Choudhury · A. K. Mitra
Department of Earth Sciences, Indian Institute of Engineering Science and Technology, Shibpur, Howrah, West Bengal, India

2003; Ntokos et al., 2016; Peters & van Balen, 2007; Raj & Bhandari, 2003; Ramírez-Herrera, 1998; Singh et al., 2007; ul Haq et al., 2019), active tectonics and alluvial river responses (Kumar et al., 2003; Luirei et al., 2015; Malik et al., 2008; Morino et al., 2014; Ramasamy et al., 2011; Schumm et al., 2000; Spitz & Schumm, 1997), neotectonic movements (Bali et al., 2012; Elias, 2015; Sarma & Sharma, 2018; Srivastava & Kulshrestha, 2005), geomorphological characterization using river sinuosity index (Ghosh & Mistri, 2012; Horacio, 2014; Kawser et al., 2021; Nath et al., 2022a, 2022b), automatic extraction and geospatial analysis of lineaments and their tectonic significance (Choudhury et al., 2025; Nath et al., 2017, 2022a, 2022b; Thannoun, 2013), structural interpretation of lineaments by satellite image processing (Ahmadi & Pekkan, 2021; Bhakuni et al., 2016; Choudhury et al., 2025; Farahbakhsh et al., 2020; Gannouni & Gabtni, 2015; Neawsuparp & Charusiri, 2004), seismic focal mechanisms on stress drop (Kamra et al., 2023), and land use and land cover (LULC) changes and landscape risk in earthquake-prone areas (Nath et al., 2020).

Many of these studies used digital elevation models (DEMs), aerial photographs, historical topographic maps, and satellite imageries, supported by remote sensing (RS) and geographic information systems (GIS) (Kawser et al., 2021; Nath et al., 2022a, 2022b; Pavano et al., 2016, 2019; Radaideh & Mosar, 2019). Integration of RS and GIS techniques has created a new field of geomorphic indices and neotectonics study (Ali et al., 2021; Elias et al., 2019; Nath et al., 2022a, 2022b; Singh & Kanhaiya, 2015; Wells et al., 1988) focusing in particular on landscape evolution, active tectonics, uplift, and valley morphology subsidence (Ahmad et al., 2014; Ezati et al., 2021; Kawser et al., 2021; Nath et al., 2022a, 2022b; Sarma et al., 2015; Rastogi et al., 1973). Lineament extraction and its changes have recently been studied globally by earth scientists analyzing surface and subsurface structural change and monitoring stress change in typical areas (Gannouni & Gabtni, 2015; Choudhury et al., 2025; Nath et al., 2017, 2020, 2022a, 2022b; Pal et al., 2016b). Geology, sedimentology, and geomorphology of fault lines and fracture systems have been studied using satellite data at spatio-temporal scale (Horák & Orliková, 2019). Lineaments extraction has been carried out in loess-covered areas in the northern part of Jinglinghe basin, China using Chinese GF-1 images (Han et al., 2018), and fractal and multi-fractal characteristics of lineaments in the Qianhe Graben

(China) and its tectonic significance (Liu et al., 2021) have also been studied.

Quantitative analysis techniques, especially when different geomorphologic indices are applied, have been found to be effective for evaluating tectonic activity (Kawser et al., 2021; Nath et al., 2022a, 2022b; Pavano et al., 2016, 2019; Pérez-Peña et al., 2010). Images from the Chinese GF-1/6 satellites have been used effectively in numerous applications since their inception. The GF-1 and GF-6 satellite missions with a wide field of view (WFV) sensor were launched by the Chinese government in 2013 and 2018 respectively for the China High-resolution Earth Observation System (CHEOS) (available at: https://chinaspaceeport.wordress.com/spacecraft/gaofen/). The spatial resolution is 16 meters and the revisit period is four days (two days in between these GF-1/6). The four bands in the GF-1 and the eight bands in the GF-6 cameras have a range of 830 km. Wide-view zenith angle observations have been performed using sensor-based time series calibration and analysis-ready data (ARD) from the GF-1/6 WFV satellite (Fan et al., 2023). However, there has been a lack of practical application in south Asia, especially in Bangladesh which has recently faced geo-environmental challenges in its transboundary areas, and to date no studies have targeted the tectono-geomorphic and environmental issues. The present study therefore uses multi-temporal Chinese GF-1/6 WFV (16 m) satellite images and geospatial techniques to analyze changes in the Brahmaputra River Basin (BRB) in the light of tectono-geomorphological changes and the associated LULC impact.

The BRB, a vital and dynamic river basin and a lifeline for millions of people, is located in the eastern Himalayas and borders China, India and Bangladesh. It is significantly affected by natural processes and tectonic forces, impacting regional socio-economic activities and environmental sustainability. The BRB is of irregular shape; its maximum east-west length is 1540 km and the maximum north-south width is 682 km. Morphotectonic studies using geoinformatics techniques (Sarma et al., 2015) have recently been carried out in the Indian section of the BRB, and geomorphic signatures have been analyzed to support the active tectonics of the Subansiri river (Kumar & Duarah, 2020).

In this chapter, the recent increase in seismic activity in a transboundary section of the new Brahmaputra River channel between India and Bangladesh (hereafter NBRB-IB) and the nearby zone of influence along multiple fault segments is studied. Collecting ground-based data over the dynamic fluvial channel section is time-consuming and complex. Satellite remote sensing has the advantage of rapidly collecting data which can be quickly analyzed using RS and GIS techniques. This chapter reports the first analysis of tectono-geomorphological changes and induced LULC impact between 2014 and 2025 that has been carried out using multi-temporal Chinese Gaofen-1/6 WFV analysis-ready datasets with 16 m resolution. The study has four main objectives: (1) to utilize tectonic information and its intensity to understand tectonic activity in the basin; (2) to identify geomorphological changes on the left river bank through multiple sinuosity indices calculations based on multi-temporal 2014–2025 data; (3) to construct a longitudinal river profile identifying knickpoint locations, and vertical transects for elevation changes using Google Earth Pro (GEP); and (4) to perform qualitative and quantitative analysis, using multiple software applications together with geospatial techniques, to understand structural disturbance and dynamic lineament changes with orientation analysis.

14.2 Materials and Methods

14.2.1 Geographic and Geological Characteristics of the Study Area

The BRB is an irregular basin located in the seismically most active zone V (ISR 2025). According to Immerzeel (2008), the BRB has three different physiographic zones: (a) the Tibetan Plateau (TP), (b) Himalayan Belt (HB), and (c) the floodplain (FP). These three zones respond differently to tectonic activity. The area of this study lies between 25°45′N and 26°10′N latitude and 89°45′E to 90°05′E longitude (Fig. 14.1), with an elevation varying from 76 m to below 0 m in the river and adjoining waterbodies, and 0–150 m a.s.l. over the land in the Indian and Bangladesh territory.

The subtropical hot-humid climate in this area is characterized by monsoon rainfall and cold winters. The physiography is typically braided and dynamic, sensitized to channel migration, avulsion, and seismic activity (Bandyopadhyay et al., 2021).

Multiple geological types that play an important role in hydro-morphodynamic activity are found in this channel segment: sedimentary deposits of afy = young gravelly sand, asd = alluvial sand, asl = alluvial silt; H_2O = water bodies (rivers and multiple channels), lake; and U = undefined geological categories (Fig. 14.2). The BRB carried extensive

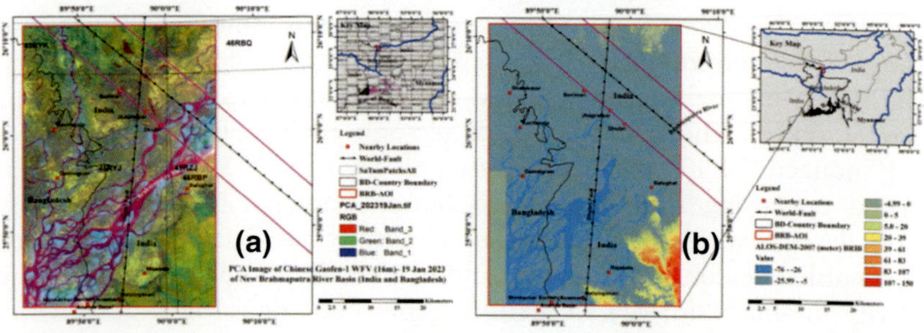

Fig. 14.1 Location map of the study area. (**a**) New Brahmaputra river channel section and surroundings (India and Bangladesh) with background, PCA image generated from Chinese Gaofen-1 WFV (referred here as an example image), 19 January 2023; (**b**) Background image, ALOS PALSAR digital elevation model (12.5 m resolution) representing the NBRB-IB elevation

sediments throughout the quaternary period, and the section between India and Bangladesh has been extensively modified by fluvial actions and nearby intra-basinal seismic activity.

The section of the complex and winding braided river under study is divided into six reaches. From the Indian part to the entry point of the Bangladesh border, the NBRB channel flows NE to SW, and the valley is bounded by the Dhubri-Chungthang fault zone (DCFZ) in the NW to SE direction, and the vertical Dhurbi fault at approx. 90° aligned crossing over this DCFZ (Fig. 14.2). From that transboundary border point towards the Bangladesh section, it turns NE–SW with some deflection from the entry level, which is possibly due to the presence of the two nearby faults The Dapsi fault is also actively present in the NW to SE direction.

14.2.2 Datasets Used

The study used multi-temporal Chinese Gaofen-1/6 wide field of view (WFV) analysis-ready data for the years 2014, 2018, 2023, and 2025 with 16 m resolution (Fig. 14.3a–d), obtained from the China High-resolution Earth Observation System (CHEOS) archive (available at: https://www.cnsa.gov.cn/english/n6465715/n6465716/c6840357/content.html).

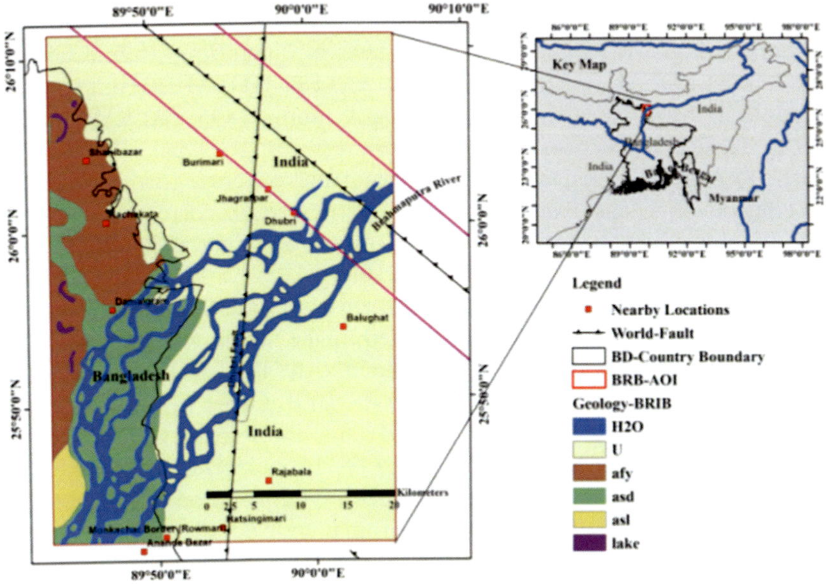

Fig. 14.2 Geology of the study area with fault distribution. U = undefined, afy = young gravelly sand, asd = alluvial sand, asl = alluvial silt, H_2O = water bodies (rivers and multiple channels); red rectangle represents the study area on the insert map

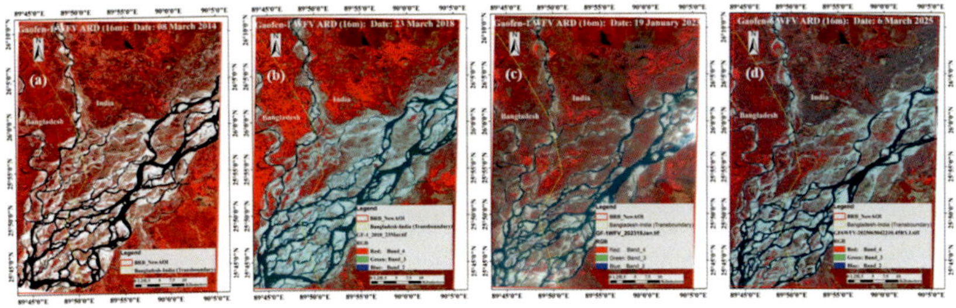

Fig. 14.3 Multi-temporal Chinese GF-1/6 WFV satellite datasets (16 m resolution) of BRIB-IB. (**a**) GF-1 WFV image, 8 March 2014, (**b**) GF-1 WFV image, 23 March 2018, (**c**) GF-1 WFV image, 19 January 2023, and (**d**) GF-6 WFV image, 06 March 2025. *Note* All the base images are displayed in standard false color composite (FCC) with band combinations R, G, B = 4, 3, 2 (NIR, red, and green bands) at 1: 2,20,000 scale

Details of the multi-sensor processed Chinese Gaofen1/6 satellite datasets along with other ancillary datasets, including digital elevation model (DEM) data from USGS ASTER GDEM 2010 (30 m), available at https://earthexplorer.usgs.gov, and ALOS PALSAR radiometric terrain correction (12.5 m) datasets for the year 2007 from Alaska Satellite Facility-Distributed Active Archive Center (https://asf.alaska.edu/datasets/daac/alos-palsar-radiometric-terrain-correction/) were found highly effective, particularly in identification of elevation differences in the study area between 2007 and 2010.

Datasets of earthquake epicenter information (1930–2025) from the USGS earthquake catalog (https://earthquake.usgs.gov/earthquakes/) in kml format supported by Google Earth Pro (GEP) was later converted to shapefile format with ArcGIS 10.8. Other GIS-ready layers in shapefile format included geological fault lines from Geotech Data Library Center's geology layer (http://techcenter.jefferson.kctcs.edu/data), important location points with attribute information retrieved from GEP, and digital geological data of India and Bangladesh (geo6ag), downloaded from the US Geological Survey Data Catalog (available at: http://certmapper.cr.usgs). Note that all these data layers are geometrically corrected and projected into a World Geodetic System (WGS) 1984, Universal Transverse Mercator (UTM), Zone 45N coordinate system (Table 14.1).

14.2.3 Methodology

There are no applications of GF-1/6 WFV 16 m-resolution satellite data over Bangladesh in the published literature. This is the first study to consider GF-1/6 satellite coverage of multiple time domains from 2014 to 2025, performed using multi-integrated remote sensing, GIS, and geological software with the six-step processing technique. The methodological steps are described in the following sub-sections.

14.2.3.1 Seismic Source Data Download and Processing

Earthquake data from 1 January 1930 to 25 April 2025 have been used to establish the seismicity of the study area and its surrounding zone, downloaded as kml GEP native format from the Earthquake Search Catalog option of the USGS Earthquake Hazards Program Server (source: https://earthquake.usgs.gov/earthquakes/search/). Earthquakes from 2.5–9.0 M_W were searched with the user-defined AOI bounding box of

Table 14.1 Details of multi-sensor processed satellite datasets used in the present study

Sl. No.	*Satellite Sensor*	*Date of Image Acquisition*	*Image Grid UTM Mark No*	*Resolution (m)*	*Band Considered*	*SCC (%)*	*Source*
1	Gaofen-1 WFV	6 March 2014	45RYJ, 45RZJ	16	1,2,3,4	0.00	CEOS (China)
2	Gaofen-1 WFV	19 January 2018	45RYJ, 45RZJ	16	1,2,3,4	0.00	CEOS (China)
3	Gaofen-1 WFV	19 January 2023	45RYJ, 45RZJ	16	1,2,3,4	0.00	CEOS (China)
4	Gaofen-6 WFV	6 March 2025	45RYJ, 45RZJ	16	1,2,3,4,5,6,7,8	0.00	CEOS (China)
5	ASTER GDEM-2010	2010	–	30 (16 bit)	1		USGS (USA)
6	ALOS PALSAR	31 March 2007 Published in October 2014	AP_05450_FBS_F0500_RTC1, and AP_05450_FBS_F0510_RTC1	12.5 (16 bit)	1 (GeoTIFF)		ASF (USA)
7	Geologic Map of South Asia	April 1998 (Published on 11 June 2021	geo8ag		GIS Shapefile		Certmapper (USGS, USA)

Note WFV—wide field of view, SSC—scene cloud cover, CEOS—China Earth Observation Satellite; ASTER GDEM—global digital elevation model, ALOS-PALSAR—Alaska Satellite Facility - phased array L-band synthetic aperture radar, FBS—fine beam scenes, RTC—radiometric terrain corrected, USGS—United States Geological Survey the study area (geographic co-ordinate extension 25°29′N–26°31′N latitude and 88°45′E–91°05′E longitude). This kml file, with all its attribute information, was later converted into a shapefile for further use in the GIS system. A 2D depth map with earthquake focal depths was prepared, using the ArcGIS 10.8 inverse distance-weighted (IDW) interpolation method, to visualize the different magnitudes of earthquakes occurring at different depths. These are classified into five classes from blue (low-intensity earthquakes) to red (high-intensity earthquakes) in accordance with the natural breaks classification method.

14.2.3.2 PCA Image Generation, Image Classification and GIS Analysis

GF-1/6 geotiff images were initially used in ENVI 5.3 software to generate principal component analysis (PCA) images with the option of transform/PCA rotation/forward PCA rotation new statistics and rotate option. The PCA images were then converted into a best-band PC1 file from among all available bands of GF-1/6 images following Eigen vectors and Eigen values PCA profile. The original PCA images were used for land use and land cover (LULC) classification. Raster to polygon conversion was used to mask out layers of water bodies from 2014 to 2025 to monitor dynamic river activity, and to calculate channel length, channel polygon area, migration rate, and river sinuosity (SSI, TSI, and HSI).

14.2.3.3 River Dynamic Movements Analysis

Layers of water bodies for the different time periods are masked out from the original PCA images after raster to polygon conversion is performed in ArcGIS10.8 software. After river dynamic activity monitoring is carried out, channel length (km) (L) (left part) and channel polygon area (A) (km^2) are measured on each image. 2014 is counted as the base image year The channel migration rate (km yr^{-1}) (left bank) in the subsequent years to 2025 was calculated using a combination of ArcGIS10.8 and Microsoft Excel 2013 software according to the following formula (Giardino & Lee, 2011):

$$\text{Migration Rate (Rm)} = (A/L)/y \tag{14.1}$$

where L = length of the left channel, A = channel polygon area, and y = time interval.

The polygon areas used for quantitative assessment of channel areal change and channel shifting or migration change rate (MCR) were also determined using the above formula, based on the GF-1/6 PCA river polygon layers. Different time intervals were considered, including 2014–2018 (4 years), 2018–2023 (5 years), 2023–2025 (2 years), and 2014–2025 overall (11 years).

14.2.3.4 Tectono-Geomorphology Indices Analysis

Five important detailed tectono-geomorphologic quantitative analyses were performed using the GF-1/6 WFV PCA images: hypsometric integral (HI), drainage basin asymmetry factor (Af), mountain front sinuosity (Smf), vertical transects and river longitudinal elevation profile, and river sinuosity indices, including standard sinuosity index (SSI), tectonic sinuosity index (TSI), and hydraulic sinuosity index (HSI).

14.2.3.5 Hypsometric Integral (HI) Analysis

The HI curve and its percentage value have been widely accepted for understanding river basin development stages and tectonic activity (Chen et al., 2003; Mohaimen et al., 2024; Pavano et al., 2016, 2019; Pérez-Peña et al., 2010) HI values ranging from 0 to 1 are calculated by the following formula adopted from Strahler (1952):

$$HI = \frac{(\text{Elevmean} - Elevmin)}{(Elevmax - Elevmin)}, \text{ and } HI = h/H, \text{ and } a/A \tag{14.1}$$

where HI = hypsometric integral, and in the relative height (h/H), h indicates the maximum elevation between each pair of contours above the base, and H refers to total basin height.

This calculation was performed in Microsoft Excel after extracting quantitative information from two different DEM images using ArcGIS 10.8 zonal statistics software, and finally, the HI data were plotted in the Y-axis of the graph. ‘a’ represents the area covered by a pair of contours, and ‘A’ indicates the total basin area which is represented on the X-axis. In this graph, concave-upward, convex-upward and concave-convex HI curves suggests extremely low relief indicating mature to late mature landform stage, rugged terrain and deep incision, and an equilibrium condition between fluvial and slope processes, respectively (Bajracharya & Jain, 2021; Chen et al., 2003; Mohaimen et al., 2024; Pavano et al., 2016, 2019; Pérez-Peña et al., 2010; Schumm, 1956; Strahler, 1952).

Where HI values lie between 0.60 and 1.00, a youthful stage susceptible to high erosion activity is indicated; values between 0.35 and 0.60 indicate maturely dissected landform with erosion activity, and less than 0.35 indicates basin equilibrium or old stage of dissection with stabilized landform activity (Keller & Pinter, 2002; Strahler, 1952).

14.2.3.6 Drainage Basin Asymmetry Factor (AF) Analysis

The drainage basin asymmetry factor (AF) is a geomorphic index that measures the asymmetry condition of a particular drainage basin. The Af is calculated by the following formula, which is adopted from Hare and Gardner (1985), Cox (1994), Burbank and Anderson (2001), and Keller and Pinter (2002):

$$AF = (Ar/At) \times 100 \tag{14.2}$$

where AF = asymmetry factor, Ar indicate the area of the basin on the right side of the channel (facing downstream), and At represent the total area of the basin.

Typically, an AF value around 50 indicates the drainage basin is stable, value >50 indicates the main channel is shifted to the left, and AF value <50 means that the channel has shifted to the right. In our study, the NBRB-IB has been arbitrarily divided into three sub-basins: upstream, middle stream, and downstream. Multi-temporal GF1/6 PCA images for the years 2014, 2018, 2023, and 2025 have been used to calculate AF. Temporal analysis of AF represents the basin tilting status in respect of tectonic activity (El Hamdouni et al., 2008).

14.2.3.7 Mountain Front Sinuosity (Smf) Analysis

It is essential to know how the landforms of the NBRB-IB sections are affected by Smf-induced tectonic activity. Smf represents the balance between erosional processes in an uneven mountain front and active tectonics. It is calculated according to the Bull and McFadden (1977) formula:

$$\text{Smf} = (\text{Lmf}/\text{Ls}) \tag{14.3}$$

where Smf = mountain front sinuosity, Lmf = the length of the mountain front along the foot of the mountain, and Ls = the straight-line length of the mountain front. Bull and McFadden (1977) classified Smf < 1.4 as tectonically active, Smf = 1.43.0 as slightly active, and Smf > 3 as inactive. Dehbozorgi et al. (2010) classified Smf as follows: Class1 (Smf < 1.1), Class 2 ($1.1 \leq$ Smf < 1.5), and Class 3 (Smf ≥ 1.5). The lower the Smf value, the more the landform is still influenced by tectonic activity, and vice versa. In this study, multi-temporal GF-1/6 PCA images from 2014 to 2025 were used to calculate the multi-temporal Smf values.

14.2.3.8 Vertical Transects and River Longitudinal Elevation Profiles

First, three vertical transects arbitral lines (red, green and blue) were constructed in the GEP platform over the river channels and surrounding areas to find the elevation or topographic variations of the study area. This elevation profile was constructed automatically by the GEP based on the user-defined distance covered over the surface topography. Next, a prominent mid-channel section-based river longitudinal elevation profile was constructed. Left and right bank channels were selected in the GEP environment and onscreen digitization was performed up to 112 kms from the upper to the lower part of the target AOI, and corresponding elevation profiles were constructed to identify knickpoints of the changing channel system elevation, in order to observe tectonic effects over the NBRB-IB section.

14.2.3.9 River Sinuosity Indices Analysis

In the fifth and final stage of the geomorphic indices analysis, three different river sinuosity indices are examined: the standard sinuosity index (SSI), topographic sinuosity index (TSI), and hydraulic sinuosity index (HSI). These are useful for checking the basin development stage, and are calculated based on the following three formulas adopted by Mueller (1968):

$$\text{SSI} = \text{CI}/\text{VI}, \quad \text{CI} = \text{CL}/\text{Air}, \text{VI} = \text{VL}/\text{Air} \tag{14.4}$$

$$\text{TSI} = \% \text{ of equivalent of } (\text{VI} - 1)/(\text{CI} - 1) \tag{14.5}$$

$$\text{HSI} = \% \text{ of equivalent of } (\text{CI} - \text{VI})/(\text{CI} - 1) \tag{14.6}$$

where SSI = standard sinuosity index, TSI = topographic sinuosity index, and HSI = hydraulic sinuosity index, CI = channel index, VI = valley index, CL = the length of the channel (thalweg), and VL = the valley length along a stream.

In this study, the left channel of the NBRB-IB section was divided into six segments based on meander characteristics. First the PCA images were visually interpreted, then vector polylines were constructed in ArcGIS 10.8 software, and digitization performed on each sub-segment of the left channel, along with the corresponding straight-line distance between the start and end points. Finally, the length was calculated in km, using the geometry calculator option in ArcGIS 10.8 after assigning WGS 1984 45 N projection parameter. These values enabled us to know the temporal characteristics of SSI, TSI, and HSI. The calculations, tabulations, and corresponding line graphs, based on the four selected years, were executed in MS Excel 2013.

14.2.3.10 Lineament Data Extraction, LD, and Orientation Change Analysis

It is important to know the details of geo-tectonic stress developing at the micro level along multiple faults. PCI's Geomatica Banff Catalyst Earth software, specifically the widely used modified line algorithm technique (Nath et al., 2017, 2021, 2022a, 2022b) was used to extract lineament features and hence observe the structural changes of the target area along the complex NBRB-IB section.

The six parameters with their corresponding values are: RADI, the radius of the filter in pixels (10); GTHR, the threshold for edge gradient (15); LTHR, the threshold for curve length in pixels (20); the FTHR threshold for line fitting error in pixels (3); ATHR, the threshold for angular differences in degrees (15); and DTHR, the threshold for linking distance in pixels (10). Next, lineaments were edited to take account of artifacts with the road and rail GIS layers crossing over the lineaments, and the line was split into vertices. Projection parameters were assigned, and lineament overlay, lineament density (LD) and rose diagrams were prepared using ArcGIS 10.8 Model Builder, and RockWorks 17.0 software, respectively.

14.2.3.11 LULC Classification and Change Analysis

In the final part, all four GF-1/6 PCA images (i.e., from the years 2014, 2018, 2023, and 2025) were examined in the ArcGIS 10.8 image classification window to generate unsupervised iso-cluster classification maps of the study area. This was followed by GEP- and ESRI-supported LULC validation techniques. Each unsupervised LULC map included five categories: water bodies, agricultural land, built-up land, barren land, and vegetated island. Corresponding LULC areas (in km^2) were calculated using the geometric calculator option. To understand how the river has been affected by the tectonic activity of this basin, including LULC modification processes, change analysis was performed using the minus operator tool in ArcGIS 10.8 software for 2014–2018, 2018–2023, 2023–2025, and 2014–2025 overall.

14.3 Results and Discussion

14.3.1 Seismicity Analysis of the Study Area (Past and Present)

Most of the current and previous earthquake events of different magnitudes (2.5–7.0M_w) have occurred in the India and Bangladesh trans-boundary zones (Fig. 14.4). A total of 83 earthquakes were reported in the study area and its surrounding zones between 1 January 1930 and 25 April 2025 (source: USGS earthquake catalog, 2025: available at: https://earthquake.usgs.gov/earthquakes/search/).

The earthquake depth in the study area and surrounding area varies significantly between 5.00 and 63.58 km (see Fig. 14.4). The depth map shows that our study area (red bounding box and nearby zones) has experienced shallower earthquake activity of 4.0–4.5 $M_{w,}$ with depths ranging between 5.00 and 26.59 km. The depth map suggests a significant increase in seismicity in the surrounding zones, with moderate to high-intensity earthquake activity occurring at a depth range of 26.89–63.58 km.

However, when we look closely at the multiple fault lines crossing the study region, we observe that in the NW segment, earthquakes occurred at a depth of 5.00–16.26 km, the shallow category. Multiple clusters of earthquakes are also observed in the vicinity of the study area, both in the past and recently, especially in the S, SE, and E directions along the fault zones of DCFZ, and the Oldham, Dapsi, and Dhurbu faults (Fig. 14.4).

14.3.2 River Dynamic Movements Analysis, 2014–2025

This section highlights dynamic river activity over the NBRB-IB section with overlay analysis in each time period (2014–2018, 2018–2023, 2023–2025, and 2014–2025 overall), along with quantitative information on dynamic river movements from 2014 to 2025 (Fig. 14.5a–d and Table 14.2).

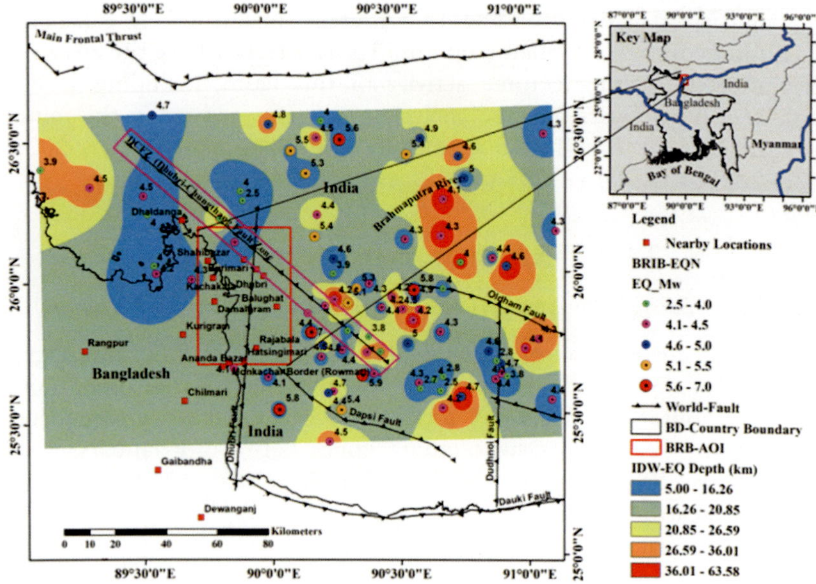

Fig. 14.4 Seismic activity of the study area and surroundings between 1 January 1930 and 25 April 2025, with IDW-interpolated earthquake depth map (earthquake data source: USGS earthquake catalog)

The quantitative results of this dynamic change indicate that the NBRB-IB section of the river changed dramatically at different time periods: during 2014–2018 (4 years) it changed by 12.61 $km^2\ yr^{-1}$, whereas during 2018–2023 (5 years), and 2023–2025 (2 years) it changed at the rate of 5.07 $km^2\ yr^{-1}$, and 91.68 $km^2\ yr^{-1}$. Over the total period of dynamic activity from 2014 to 2025 (11 years), it changed at the rate of 18.95 $km^2\ yr^{-1}$.

Conversely, when we consider the left channel linked with the fault systems, the maximum migration change rate of the NBRB-IB section is calculated as 2.98 km yr^{-1} in the year 2014. However, between 2014 and 2018 (4-year interval) the rate of change was 0.90 km yr^{-1}, between 2018 and 2023 (5-year interval) it was 0.58 km yr^{-1}, between 2023 and 2025 (2-year interval) it was 2.92 km yr^{-1}. Over the whole period of 11 years, the MCR was 0.26 km yr^{-1} (Table 14.2). The results are corroborated by the dynamic process of the river left channel, which

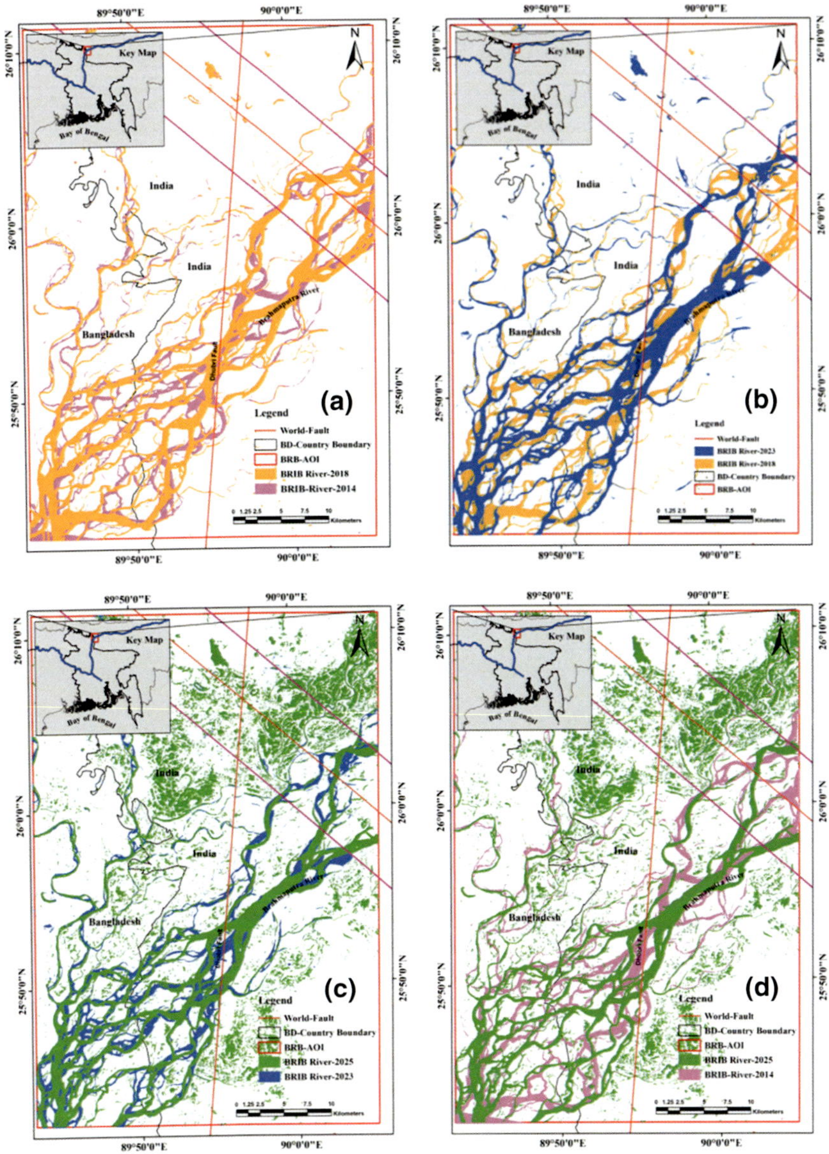

Fig. 14.5 NBRB-IB section and river dynamic activity in the study and surrounding areas, 2014–2025: (**a**) river lines overlay between 2014 and 2018, (**b**) river lines overlay between 2018 and 2023, (**c**) river lines overlay between 2023 and 2025, (**d**) river lines overlay between 2014 and 2025

Table 14.2 The migration rate of the NBRB-IB section, calculated using the Giardino and Lee (2011) method, with 2014 taken as the base year

Year	*Channel Polygon Area (A) (km²)*	*Channel Length Left segment (km) (L)*	*A/L*	*Migration Change Rate (km yr^{-1})*
2014	196.50	66.02	2.98	2.98
2018	246.95	68.45	3.61	4 years: 0.90
2023	221.59	76.67	2.89	5 years: 0.58
2025	404.96	69.25	5.85	2 years: 2.92
				Overall Change Rate (11 years): 0.26

Source Calculations by the authors based on the adopted formula

is aligned with multiple complex active fault systems. Due to the presence of fault systems and ongoing tectonic activity near and within the channel systems, this left channel has forced the NBRB-IB section to flow from NE to SW with multiple branching systems. Our MCR results are supported by the study conducted by Singh et al., (2020), who also observed different MCRs (0.04 to 0.12) in three reaches of the Ganga River at different time intervals.

14.3.3 Tectono-Geomorphological Indices Analysis, 2014–2025

14.3.3.1 Hypsometric Integral (HI) Analysis

It is widely accepted in tectono-geomorphological studies that HI is necessary for the stages of basin development. In our case, two DEM datasets, ALOS PALSAR 12.5 m and GDEM 30 m, HI curves were calculated representing the scenarios of 2007 and 2010, respectively (see Fig. 14.6a, b).

Our data indicate topographic variations ranging from the late youthful stage to the equilibrium or mature phase of basin development. In the NBRB-IB section, the HI value from ALOS PALSAR data (Fig. 14.6a) is 0.50 (50%), representing equilibrium or the mature stage, whereas in the GDEM data, the HI value was 0.76 (76%), representing the youthful stage susceptible to high erosion activity during basin development (Fig. 14.6b). Our results are almost consistent with those of Das and Gupta (2019) whose study of the Sali River Basin, Bankura district, West Bengal (India) found HI values of 48.21%.

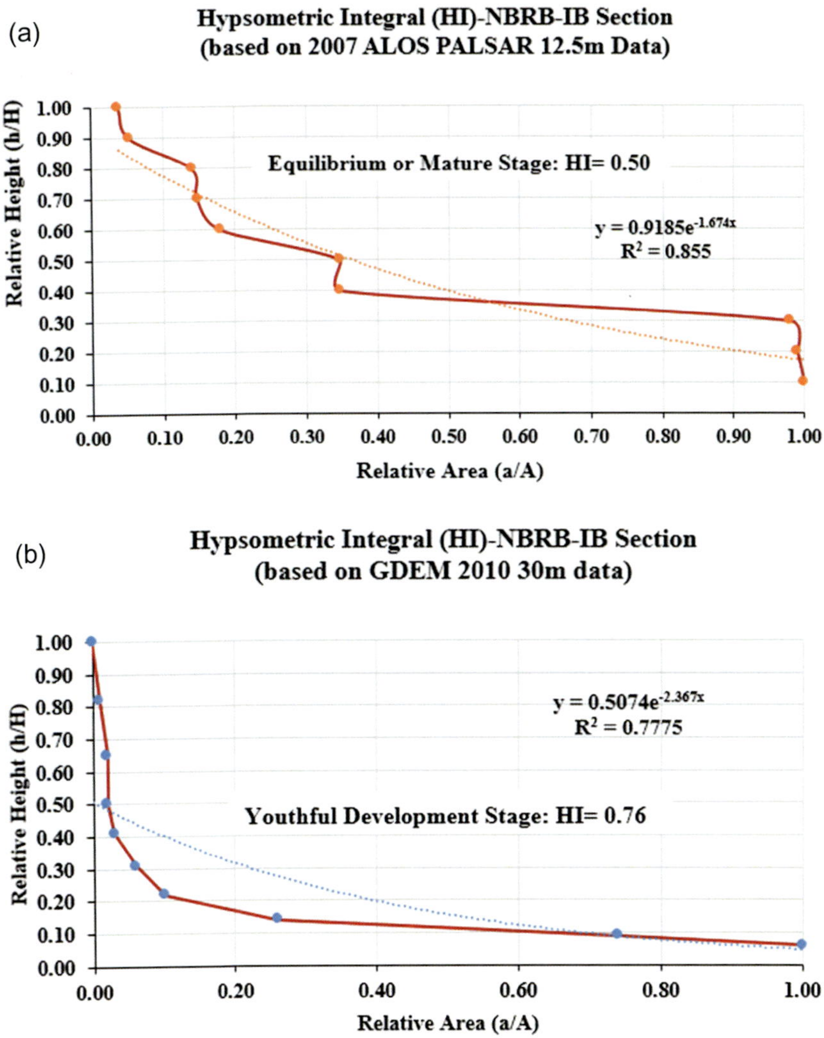

Fig. 14.6 ALOS PALSAR- and GDEM-based HI measurements of the study area: (**a**) HI of NBRB-IB for ALOS PALSAR 12.5 m, and (**b**) HI of NBRB-IB for GDEM 30 m

14.3.3.2 Drainage Basin Asymmetry Factor (AF) Analysis

The AF results indicate whether a river basin is tilted in a specific direction. Our quantified AF results for 2014, 2018, 2023, and 2025 for the three segments and the overall basin are presented in Table 14.3.

Segment 1, the northern segment of the NBRB-IB section, is left tilted with AF values of 89.50, 89.42, 88.42, and 88.72 in the years 2014, 2018, 2023, and 2025, respectively. At the same time points in Segment 2 (middle section), we found AF values of 55.92, 42.33, 44.05, and 48.96, respectively, indicating the basin is left, right, right, and right tilted, respectively. At the same time points in Segment 3, the lower segment, the basin is right tilted with lower AF values of 5.70, 3.76, 3.93, and 3.62, respectively.

Table 14.3 Details of multi-temporal GF-1/6 WFV image-based AF calculations

Basin Segments	*Date of Image Acquisition*	*At*	*Ar*	*AF*	*Comments*
1	6 Mar 2014	850.18	760.91	89.50	Basin is left tilted
2	6 Mar 2014	575.81	322.01	55.92	Basin is left tilted
3	6 Mar 2014	547.08	31.17	5.70	Basin is right tilted
1	19 Jan 2018	850.18	760.23	89.42	Basin is left tilted
2	19 Jan 2018	575.81	243.72	42.33	Basin is right tilted
3	19 Jan 2018	547.08	20.60	3.76	Basin is right tilted
1	19 Jan 2023	850.18	751.70	88.42	Basin is left tilted
2	19 Jan 2023	575.81	253.64	44.05	Basin is right tilted
3	19 Jan 2023	547.08	21.52	3.93	Basin is right tilted
1	6 Mar 2025	850.18	754.26	88.72	Basin is left tilted
2	6 Mar 2025	575.81	281.94	48.96	Basin is right tilted
3	6 Mar 2025	547.08	19.81	3.62	Basin is right tilted

Source Authors' calculations based on the AF formula

Our AF results in the NBRB-IB section are similar to those of a previous survey conducted in the BRB by Sarma et al. (2015), who observed an AF value of 33.00, indicating a tilted basin and right-tilted channel. These results are also consistent with the study by Reday et al. (2023), which focused on alluvial fan along the Dauki fault system, Bengal Basin (Bangladesh) and observed AF values between 35.19 and 67.04, meaning the basin is right to left tilted. Overall, our results suggest that both the second and third segments are right tilted after 2018–2025, having been left tilted in 2014. On the other hand, the northern Segment 1 has always sought to remain left tilted. These results confirm that the NBRB-IB section has a dynamic river movement controlled by tectonic activity, as observed in multiple segments of this basin.

14.3.3.3 Mountain Front Sinuosity (Smf) Index Analysis

The Smf index is widely considered to identify areas prone to tectonic activity and to indicate the degree of activity and the erosional modifications of the tectonic structure (Bull & McFadden, 1977). Based on Dehbozorgi et al. (2010), this study observed different Smf values in the range of 1.74 to 2.06 from 2014 to 2025 (Fig. 14.7).

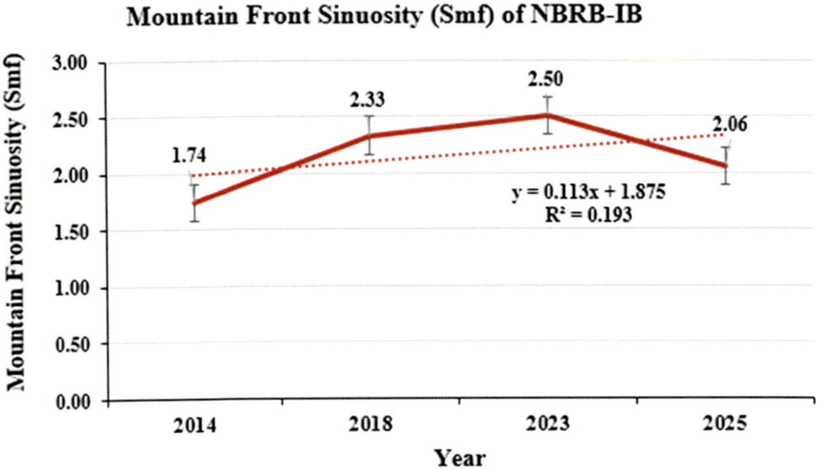

Fig. 14.7 Smf measurements of the study area based on multi-temporal Gaofen 1/6 WFV satellite datasets (16 m)

According to Bull and McFadden (1977), the Smf values in the NBRB-IB section were gradually increasing from 2014 until 2023, while in the current year they have reduced slightly to 2.06 (Fig. 14.7), although this level is higher than in the year 2014, indicating the basin is tectonically active. Our results are consistent with the resultsose of Sarma et al. (2015), and Reday et al. (2023) who calculated Smf values over the entire BRB by dividing it into seven segments, and alluvial fan along the Dauki fault system, Bangladesh, with Smf values of 1.11–1.68, and 1.35, respectively, both of these basins therefore being classed as tectonically active. Our Smf results indicate that the NBRB-IB section is sensitive to tectonic activity, which is affected to a greater or lesser extent by the surrounding seismicity and multiple fault lines.

14.3.3.4 Vertical Transects and Longitudinal Profiles-based Elevation Change Analysis

In our study, three vertical transect lines (red, blue, and green) up to 112 km distance from the north to south vertical direction over the NBRB-IB section were constructed (Fig. 14.8a), and automatic transects elevation profiles are generated in the GEP system (Fig. 14.8b–d). Each of the vertical elevation profiles suggest diversity of topography or elevation change, which represents either erosion or tectonic activities operated by various geomorphological agents in recent time. Similar results were found in the Dawki fault adjoining areas over the NE frontier of Sylhet Basin (Bangladesh) by Mohaimen et al. (2024).

The longitudinal river profile helps us to understand the geological, structural and tectonic deformation of the study basin (Pavano et al., 2016; Roy & Sahu, 2015; Singh & Awasthi, 2011). In general terms, it indicates the channel gradient of the river from its source to the mouth, and in our case, two longitudinal profiles have been constructed, marked in red and blue and representing left and right channels (Fig. 14.8e). These two profiles were constructed over GEP following the method described in the relevant methodology sub-section, and several abrupt elevations with multiple knickpoints were observed in both the left (19 in total) (Fig. 14.8f), and right bank channels (22 in total) (Fig. 14.8g) of the NBRB-IB section which measures 112 km. These two longitudinal profiles over two channel banks confirm active tectonic activity with upliftment processes present on both banks, and these results are consistent with studies by Roy & Sahu (2015), Das & Gupta (2019), and Mohaimen et al. (2024).

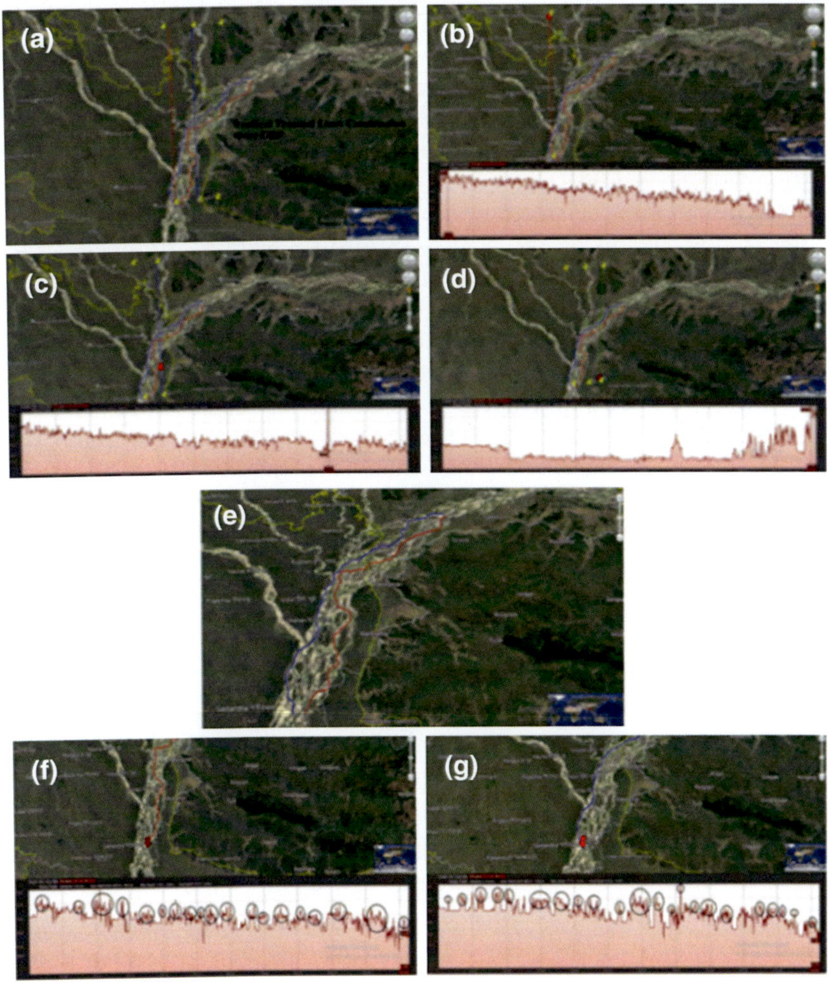

Fig. 14.8 Vertical transects and longitudinal-based elevation change profiles of the NBRB-IB section: (**a**) three vertical transect lines (red, blue and green) constructed over the study area, (**b**) vertical transect lines with elevation profile in outside main NBRB right-hand channel, (**c**) vertical transect lines with elevation profile in the middle section crossing over the main channel, (**d**) vertical transect lines with elevation profile in the lefthand section crossing over the main channel, (**e**) longitudinal elevation profiles construction over left and right channels of NBRB-IB section, (**f**) longitudinal elevation profiles construction over lefthand channel (red), and (**g**) construction of longitudinal elevation profile over the righthand channel (blue)

14.3.3.5 River Sinuosity Indices (Multiple Index-based) Change Analyses

The Chinese GF-1/6 WFV surface reflectance products measured different types of river sinuosity indices—SSI, TSI, and HIS—from 2014 to 2025 (Table 14.4). The NBRB-IB section flows over multiple cross-cutting fault lines, impacting on the flow and making this transboundary section more prone to tectonic activity. As explained in our methodology section, we divide the NBRB-IB into six sections, numbered 1–6 (Fig. 14.9). The sinuosity index measures the upward to downward deviation of the actual river path from a straight line between two specific points (Pan, 2013).

The SSI value shows variations among the six segments. It is high in seg-6 during 2018, whereas it is at its lowest in 2025 at 1.01 (Fig. 14.10a). In 2014 and 2023, the SSI value is 1.31 in seg-2, compared to 1.15 (sinous) in seg-2, and 1.26 (sinous) in seg-1 in 2023, respectively. TSI showed greater variation in the six sections, with values ranging from 69.21% to 83.96% (Fig. 14.10b). The TSI value was significantly higher in

Table 14.4 Details of multi-temporal GF-1/6 WFV image-based river sinuosity calculations

Standard Sinuosity Index (SSI)	*Seg-1*	*Seg-2*	*Seg-3*	*Seg-4*	*Seg-5*	*Seg-6*
2014	1.31	1.31	1.37	1.34	1.16	1.15
2018	1.23	1.08	1.26	1.07	1.12	2.29
2023	1.26	1.15	1.13	1.20	1.18	1.04
2025	1.41	1.42	1.18	1.17	1.14	1.01
Comments/Status	Simuos	Simuos	Simuos	Simuos	Simuos	Simuos
Topographic Sinuosity Index (TSI) (%)	**Seg-1**	**Seg-2**	**Seg-3**	**Seg-4**	**Seg-5**	**Seg-6**
2014	75.10	75.21	71.21	71.48	84.38	81.85
2018	79.94	91.62	78.30	92.65	88.13	82.40
2023	79.54	87.13	88.65	83.43	85.01	95.85
2025	69.21	69.13	83.96	83.39	83.27	89.96
Hydraulic Sinuosity Index (HSI) (%)	**Seg-1**	**Seg-2**	**Seg-3**	**Seg-4**	**Seg-5**	**Seg-6**
2014	24.90	24.79	28.79	28.52	15.62	18.15
2018	20.06	8.34	21.70	7.94	11.87	17.60
2023	20.46	12.87	11.35	16.57	14.99	4.15
2025	30.79	30.87	16.04	16.61	16.73	10.04

Source Authors' calculations based on the formula of three river sinuosity indices: SSI, TSI, and HSI

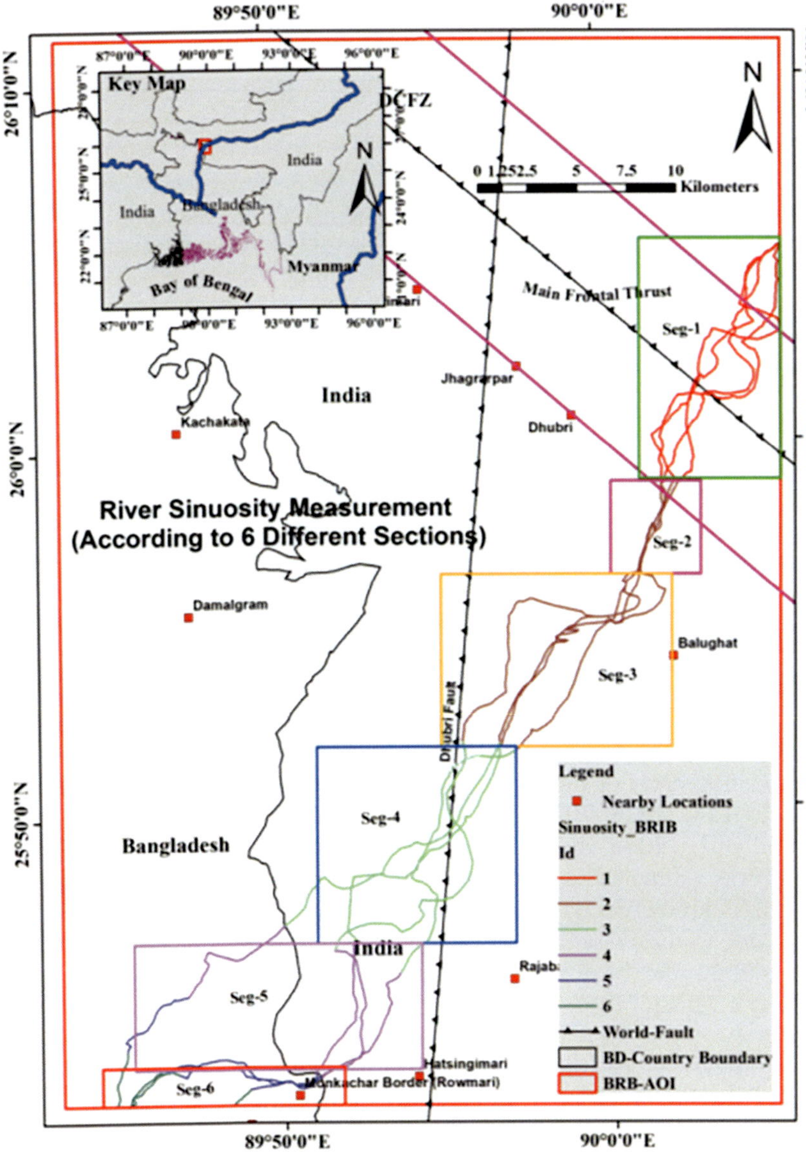

Fig. 14.9 Sinuosity measurement of six segments of the NBRB-IB section, based on temporal river line digitization GF-1 WFV PCA images from 2014 to 2025

2018, at 92.65% in seg-4 and 91.62% in seg-2 than the 69.13% observed in seg-2 during 2025 (Fig. 14.10c).

On the other hand, in 2025, the maximum HSI was 30.87% in seg-2, with the lowest value in seg-6 (10.04%) (Fig. 14.10c). The HSI values also varied sharply at other times: in 2014, it ranged from 15.62% (seg-5) to 28.79% (seg-3), and in 2018 and 2023, from 7.94% (seg-4) to 21.70% (seg-3), and 4.15% (seg-6) to 20.46% (seg-1), respectively (Fig. 14.10c). The results overall indicate that the TSI values in all six segments are higher than the HSI and SSI values, confirming the major dominance of tectonic control over the NBRB-IB section.

14.3.4 Dynamic Lineaments, LD and Orientation Change Analyses

Based on the four multi-temporal Chinese GF-1/6 WFV ARD images, significant lineament changes, LD-based stress changes, and directional changes were observed along the NBRB-IB section and its neighboring zones from 2014 to 2025. Figures 14.11(a–h) and 14.12(a–d), and Table 14.5 show the number of extracted lineaments with the detailed derivative statistical information.

On 6 March 2025, the total lineaments recorded were 8972, an increase on 2014 (8836), 2018 (8571), and 2023 (6420) (Table 14.3 and Fig. 14.11a–d). The dynamic changes in the number of lineaments between different years is clearly depicted through the overlay technique. The ongoing structural disturbance in the NBRB-IB section and its adjoining areas in the different time periods can be seen: 2014–2018 (Fig. 14.11a), 2018–2023 (Fig. 14.11b), 2023–2025 (Fig. 14.11c), and overall, 2014–2025 (Fig. 14.11d). The maximum length of the lineaments was estimated at 3.68 km in 2014, with a sudden reduction to 2.40–2.21 km between 2018 and 2023, a clear indication of crustal shortening due to the presence of dynamic stresses in this area. However, on 6 March 2025, the maximum lineament length was greater than in all other years at 4.21 km (Table 14.3).

However, the LD analysis reveals a progressive decline in overall LD from 2014 to 2025 (Fig. 14.11e–h). The 2014 LD image indicates a dominant NE–SW orientation of lineaments, characterized by a relatively high number of short, discrete segments. Over time these lineaments exhibit increasing spatial dispersion and irregularity, although the prevailing NE–SW structural grain remains consistent. By 2023, the number of lineaments appears reduced, yet their average length has

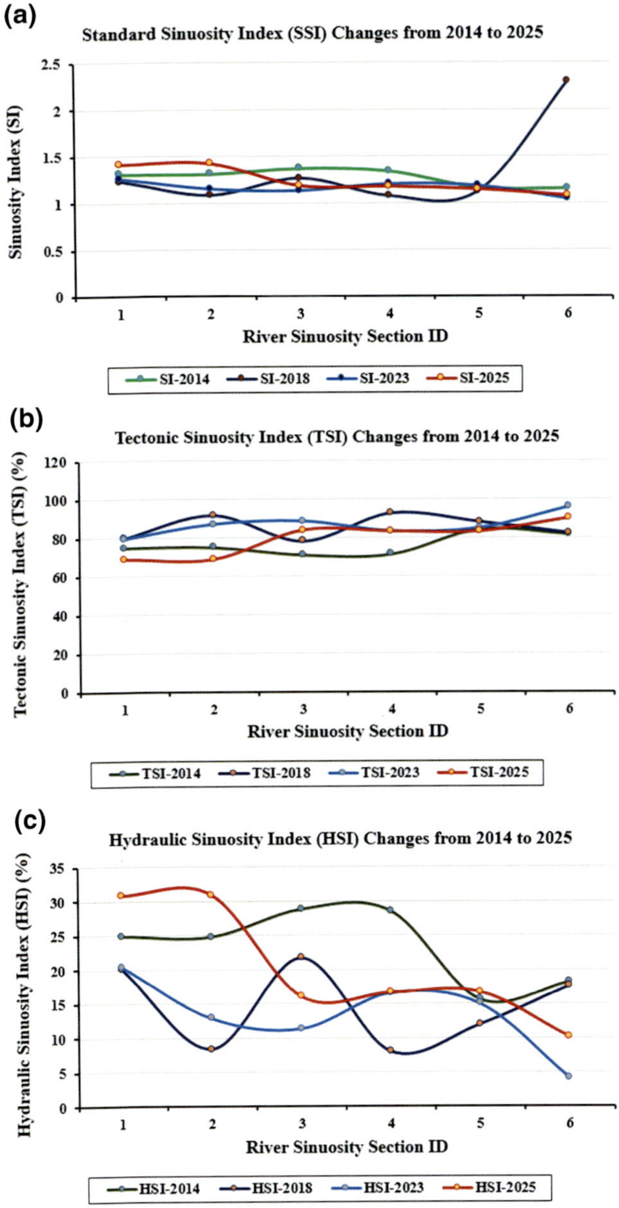

Fig. 14.10 Temporal river sinuosity trends based on GF-1/6 WFV reflectance images between 2014 and 2025: (**a**) SSI changes, 2014–2025, (**b**) TSI changes, 2014–2025, (**c**) HSI changes, 2014–2025

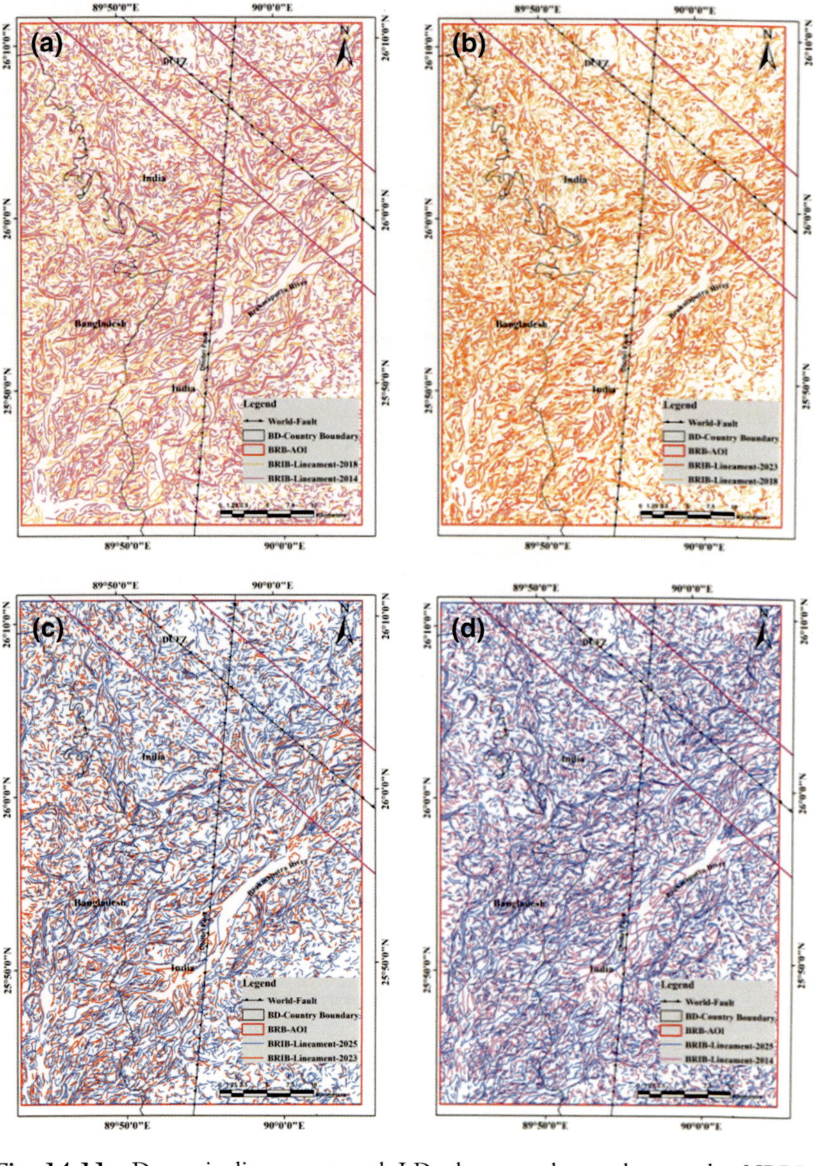

Fig. 14.11 Dynamic lineament and LD changes observed over the NBRB-IB section and its adjoining areas based on GF-1/6 WFV reflectance images between 2014 and 2025: (**a**) lineament overlay changes, 2014–2018, (**b**) lineament overlay changes, 2018–2023, (**c**) lineament overlay changes, 2023–2025, (**d**) overall lineament changes, 2014–2025, (**e**) LD in 2014, (**f**) LD in 2018, (**g**) LD in 2023, and (**h**) LD in 2025

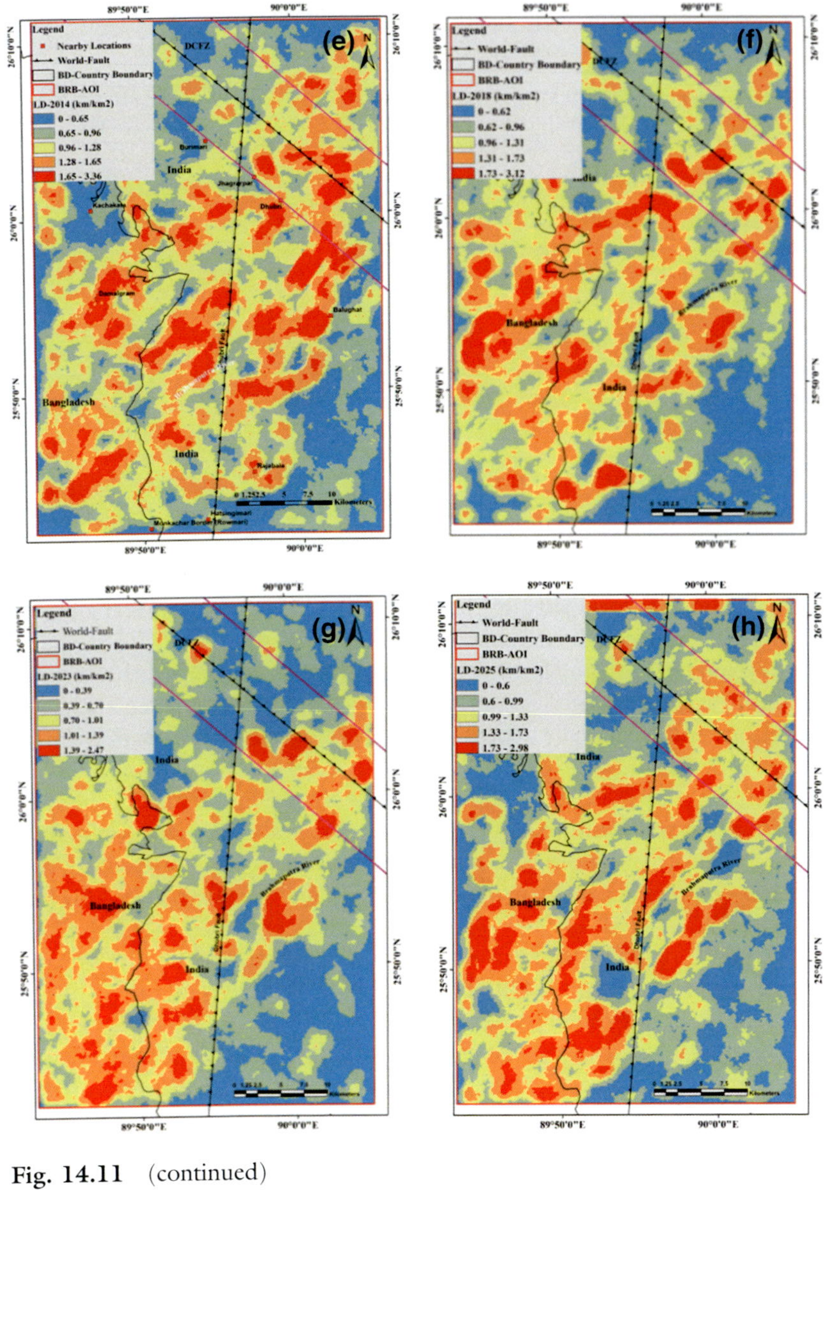

Fig. 14.11 (continued)

Table 14.5 Details of multi-temporal GF-1/6 WFV image-based lineaments extraction with statistical analysis

Basic lineaments statistics	*GF-1 WFV (6 Mar 2014)*	*GF-1 WFV (19 Jan 2018)*	*GF-1 WFV (19 Jan 2023)*	*GF-1 WFV (6 Mar 2025)*
No. of lineaments Observed	8836	8571	6420	8972
Maximum length (km)	3.68	2.40	2.21	4.21
Mean vector total length (μ)	3785.57	3698.96	2781.84	3872.13
SD	0.23	0.23	0.20	0.24
Mean vector (μ)	47.9°–227.86°	48.4°–228.36°	40.8°–220.85°	46.4°–226.35°

Source Authors' calculation using ArcGIS10.8 software

increased—likely due to the coalescence of spatially aligned segments—resulting in more continuous, densely packed features oriented predominantly in the NE–SW direction. This temporal evolution is corroborated by rose diagram analyses of each dataset.

An additional noteworthy observation is the temporal change in NNW–SSE trending lineaments. In 2014, a distinct NNW–SSE lineament set emerges from the northwestern portion of the study area with maximum LD values of 3.36 km/km^2 (Fig. 14.11e). However, this trend diminishes and is virtually absent in the 2018 and 2023 datasets with LD values of 3.12 km/km^2 (Fig. 14.11f), and 2.47 km/km^2 (Fig. 14.11g). Intriguingly, the 2025 LD map (Fig. 14.11h) with the maximum value of 2.98 km/km^2 indicates a re-emergence of lineament development in this orientation, suggesting renewed structural activity along that trend. The LD changes are at their lowest in the year 2023 at 0–0.39 km/km^2 (Fig. 14.11g).

Furthermore, rose diagrams reflect different directional patterns based on the 16 m-resolution GF-1/6 data from four different times—2014, 2018, 2023, and 2025 (Fig. 14.12a–d).

Figure 14.12(a–d) show changes in the orientation of lineaments over time. Figure 14.12 (a–c), where the lineaments are mostly oriented in NE–SW and E–W directions, are generated by GF-1 satellite sensors. In Fig. 14.12 (d) the majority of the lineaments are NE–SW, E–W, and a few N–S based on the GF-6 satellite sensor, showing the dynamic surface and sub-surface activity present in the NBRB-IB.

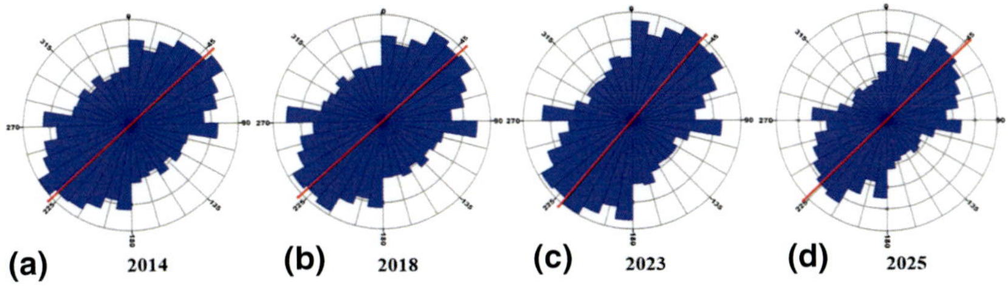

Fig. 14.12 Rose diagram for directional change pattern observations over the NBRB-IB section and its adjoining areas using GF-1/6-derived lineaments datasets

It is inferred from the above analysis that the dominant NE–SW orientation closely aligns with the regional fluvial morphology, as shown by the river channel map (Fig. 14.6a–d), suggesting tectonic control of the river's evolution. When the seismicity data from January 1930 to April 2025 are superimposed on the geological framework (Sarma et al., 2015), two principal fault orientations become evident: (a) E–W trending structures such as the main frontal thrust (Srivastava et al., 2018), and Dauki Fault, and (b) NE–SW trending faults including the Barapani–Tyrsad Fault (Dhuandhar et al., 2020). Additionally, the Dhubri and Dudhnoi faults exhibit a NNW–SSE orientation (Kamra et al., 2023), while several minor faults like the Kopili (Sharma & Baruah, 2017), Dapsi (Baruah, 2011) and Samin faults have an NW–SE direction.

The resurgence of NNW–SSE-oriented lineaments in 2025 may signify reactivation of these pre-existing fault systems under ongoing tectonic stress. Seismicity patterns further support this inference, as epicentral clusters display two dominant alignments, one of which is NNW–SSE (Fig. 14.4). The spatial correlation between this seismic trend and the newly forming lineaments strengthens the hypothesis of active tectonism along the Dhurbi–Dudhnoi fault zone (Baruah, 2011). Furthermore, the lateral migration of river channels observed over the study period (Fig. 14.5a–d) provides additional geomorphic evidence of sustained tectonic deformation.

Conversely, earthquake data from the study region and surrounding areas reflects a clear concentration of seismicity spreading in different directions: NS, NE, N, and NW. From S–N and E–W, the focal depth of the different earthquake magnitudes vary considerably, increasing from near surface to a depth of 30 km (Fig. 14.6). The histogram of the hypocentral depth shows a significant concentration at a depth of 10–15 km. Both lateral and vertical variations are observed in the NBRB-IB section, as in the study by Uthama et al. (2024).

14.3.5 Induced LULC Impact Analysis over the NBRB-IB Section

The LULC maps derived from multi-temporal GF-1/6 images, with area statistics and % of distribution over the 11-year period 2014–2025 are shown in Fig. 14.13(a–c), Table 14.6, and Fig. 14.14. The five major categories of land use and land cover in this study are water bodies (WB), built-up land (BUL), agricultural land (AL), barren land (BL), and vegetated island (VI).

The results demonstrate the dynamic LULC changes in the NBRB-IB section over the short period of 11 years, showing that the LULC change was strongly affected by the tectonic and geomorphic disturbances of the study area. Of all the categories, WB has changed its courses in multiple directions – by approximately 23% in 2023, and the same for 2025. However, the LB changes were low in 2014 (9.98%) (Table 14.6 and Fig. 14.14), with gradual change continuing in the current year, 2025. The share of BUL is growing rapidly in 2025, at 34.89% compared to the lowest observation of 16.83% in 2018 (Table 14.6). On the other hand, AL has reduced significantly recently compared to its 2014 (32.06%) and 2018 (34.55%) levels, changing at the rate of 1.10% yr^{-1}. The remaining

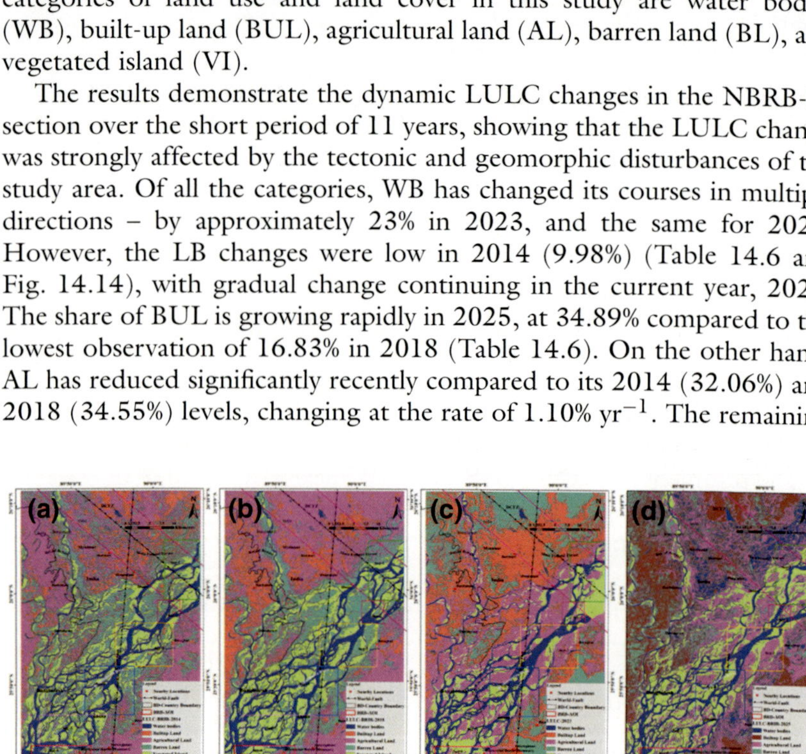

Fig. 14.13 LULC change observations over the NBRB-IB section and its adjoining areas using GF-1/6 PCA-based images (K-means iso-clusters unsupervised classification) from 2014 to 2025: (**a**) LULC 2014, (**b**) LULC 2018, (**c**) LULC 2023, (**d**) LULC 2025

Table 14.6 Details of LULC change statistics in the NBRB-IB section extracted from iso-cluster unsupervised classification of multi-temporal GF-1/6 WFV images

SL.No	*LULC Category*	*2014 Area (km^2)*	*% of Share*	*2018 Area (km^2)*	*% of Share*	*2023 Area (km^2)*	*% of Share*	*2025 Area (km^2)*	*% of Share*
1	WB	196.50	9.98	243.49	12.36	455.29	23.10	458.30	23.26
2	BUL	364.32	18.49	331.49	16.83	686.51	34.84	687.50	34.89
3	AL	631.71	32.06	680.72	34.55	386.78	19.63	393.25	19.96
4	BL	482.15	24.47	394.12	20.00	221.59	11.25	220.60	11.20
5	VI	295.52	15.00	320.38	16.26	196.50	11.18	210.55	10.69
Total		1970.20	100.00	1970.20	100.00	1970.20	100.00	1970.20	100.00

Note WB: Water bodies, BUL: Built-up land, AL: Agricultural land, BL: Barren land, VI: Vegetated Island
Source Authors' calculation based on ArcGIS10.8 software

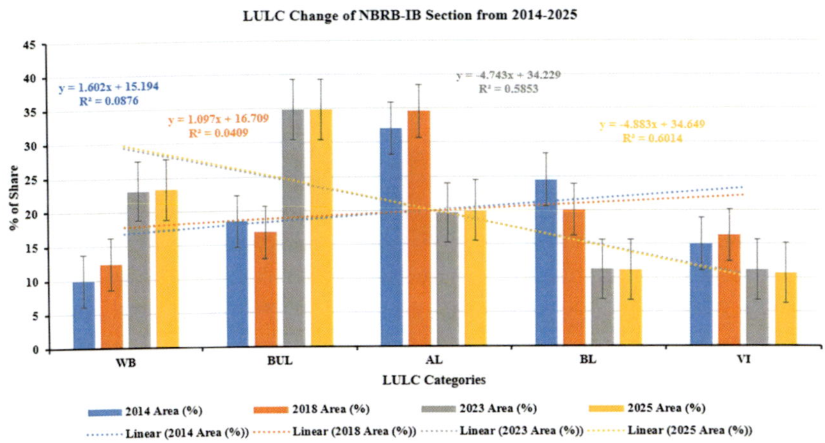

Fig. 14.14 LULC change observation over the NBRB-IB section and its adjoining areas from 2014–2025 (% share of each category)

two categories, BL and VI, have both reduced sharply in recent time, when we consider the 11-year timespan (2014–2025), changing at the rate of 1.49% yr^{-1}, and 0.39% yr^{-1}, respectively.

14.4 Conclusion and Future Directions of the Study

In this chapter, we studied multiple tectono-geomorphic signatures derived from Chinese Gaofen-1/6 WFV 16 m-resolution satellite images from 2014 to 2025. In the river channel especially, digitization was best performed at the 1:10,000 scale. Multiple sinuosity measurements showing the dynamic activity of the NBRB-IB section were made from the channel lines extracted for multiple years. Pronounced changes in lineaments and LD changes were observed that reflect stress transfer in this basin, confirmed by the directional movements of the lineaments within the river channel systems as well as its surrounding areas.

The multiple tectonic events near these zones, together with the complex fault systems, contributed to forcibly modify the river channels. These changes are confirmed by rapid LULC changes in the study area, especially in WB and BUL which have changed more rapidly than BL and VI.

Finally, this study has shown the successful integration of multiple remote-sensing and geospatial software techniques, with tectono-geomorphic applications of GF-1/6 WFV analysis-ready data used for qualitative and quantitative assessment. The study also recommends the use of Chinese GF-1/6 WFV images. Together with field validation results these can be effectively used to assess tectono-geomorphic changes in complex hydro-geomorphology basins. These assessments are important in avoiding any further tectonic-induced risk, promoting both the safety of millions of people and the sustainability of these areas.

Acknowledgements The first author is grateful to the Aerospace Information Research Institute (AIR), Chinese Academy of Sciences (CAS), Beijing, China for providing Chinese GF-1/6 datasets to test their effectiveness for tectono-geomorphological applications, including 2024 CAS-PIFI Visiting Scientist (Associate Category) Program, and the International Collaboration Project supported by the Ministry of Science and Technology (MOST), Beijing, China. The authors also offer sincere thanks to the editor of this book for his invitation to write this chapter.

Competing Interests This study was funded by the Ministry of Science and Technology (MOST), China and CAS-PIFI Visting Scientist Program, the National Key Research and Development Program (Grant No. 2017YFA0603004, Grant No. 2021YFE0194700, and Grant No.

2021YFE0117400), and grant number 2024VTB0004. Biswajit Nath has a received research grant from AIRCAS, China with Ministry of Science and Technology, China and CAS-PIFI Visting Scientist Program. Bo Zhong has received an International Collaboration Project of AIRCAS-GESCU from Ministry of Science and Technology, China. Junjun Wu and Aixia Yang are an Associate and Assistant Research Professor at AIRCAS, China. Shanlong Wu is a system engineer at the National Engineering Laboratory for Satellite Remote Sensing Applications AIRCAS, China; Shukla Acharjee is an Assistant Professor at Dibrugarh University, Assam, India; Nandini Choudhury is a PhD candidate at IIEST, Shibpur, West Bengal, India; and Atin Kumar Mitra is an Assiatant Professor at IIEST, Shibpur, West Bengal, India. The authors have no conflicts of interest to declare that are relevant to the content of this chapter.

Ethics Approval The authors of this chapter have no involvement in human perspective surveys at the field level related to this study; therefore, no ethics approval was required.

References

Ahmadi, H., & Pekkan, E. (2021). Fault-based geological lineaments extraction using remote sensing and GIS – A review. *Geosciences, 11*(5), 183. https://doi.org/10.3390/geosciences11050183

Ahmad, S., Bhat, M. I., Madden, C., & Bali, B. S. (2014). Geomorphic analysis reveals active tectonic deformation on the eastern flank of the Pir Panjal Range, Kashmir Valley, India. *Arabian Journal of Geosciences, 7*, 2225–2235. https://doi.org/10.1007/s12517-013-0900-y

Ali, U., Ali, S. A., Yousuf, M., Rasool, Q. A., Ahmad, M., & Ahmad, I. (2021). Consideration of geomorphic indices in assessment of relative active tectonics in a part of seismogenic compressional Kashmir Basin. *Arabian Journal of Geosciences, 14*, 1266. https://doi.org/10.1007/s12517-021-07523-3

Bali, R., Agarwal, K. K., Nawaz Ali, S., Rastogi, S. K., & Krishna, K. (2012). Drainage morphometry of Himalayan Glacio-fluvial basin, India: Hydrologic and neotectonic implications. *Environmental Earth Sciences*, *66*, 1163–1174. https://doi.org/10.1007/s12665-011-1324-1

Bajracharya, P., & Jain, S. (2021). Hydrologic similarity based on width function and hypsometry: An unsupervised learning approach. *Computers & Geosciences*, *163*, 105097. https://doi.org/10.1016/j.cageo.2022.105097

Baruah, S. (2011). Seismotectonics in Chedrang valley and its vicinity. *Science and Culture, 77*(11–12). https://doi.org/10.1007/s00024-023-03420-w

Bhakuni, S. S., Philip, G., & Suresh, N. (2016). Structural analysis of Nalagarh lobe, NW Himalaya: Implication to thrusting across tectonic edge of

NW limb of Nahan salient, Himachal Pradesh, India. *International Journal of Earth Sciences (Geologische Rundschau)*. https://doi.org/10.1007/s00531-016-1385-1

Bhatt, C. M., Chopra, R., & Sharma, P. K. (2007). Morphotectonic analysis in Anandpur Sahib area, Punjab (India) using remote sensing and GIS approach. *Journal of the Indian Society of Remote Sensing, 35*, 129–139. https://doi.org/10.1007/BF02990777

Biswas, S., & Grasemann, B. (2005). Quantitative morphotectonics of the southern Shillong Plateau (Bangladesh/India). *Australian Journal of Earth Sciences, 97*, 82–93.

Bull, W. B., & McFadden, L. D. (1977). Tectonic geomorphology north and south of the Garlock Fault, California, in arid regions. In *Proceedings of the Eighth Annual Geomorphology Symposium* (pp. 115–138). State University of New York, Binghamton.

Burbank, D. W., & Anderson, R. S. (2001). Tectonic geomorphology. Blackwell Scientific Publications, Oxford, p 274

Chakraborty, S., & Mukhopadhyay, S. (2015). Riverbank Erosion and Channel Width Adjustments across a Meandering Channel of North Bengal, India. *Earth Science India. 8*. 974–8350. https://doi.org/10.31870/ESI.08.3.2015.5

Chen, Y-C., Sung, Q., Cheng, K-Y. (2003). Along-strike variations of morphotectonic features in the Western Foothills of Taiwan: Tectonic implications based on stream-gradient and hypsometric analysis. *Geomorphology, 56*(1–2), 109–137. https://doi.org/10.1016/S0169-555X(03)00059-X

Choudhury, N., Mitra, A. K., Nath, B., Lindsay, M. D. (2025). Lineament extraction and structural mapping using Landsat-9 OLI and Sentinel-1 SAR data in the Proterozoic North Singhbhum Mobile Belt, Eastern India. *Geosystems and Geoenvironment, 4*(3), 100392. https://doi.org/10.1016/j.geogeo.2025.100392

Cox, R. T. (1994). Analysis of drainage-basins symmetry as rapid technique to identify areas of possible quaternary tilt-block tectonics: An example from Mississippi Embayment. *Geological Society of America Bulletin, 106*, 571–581.

Das, S., & Gupta, K. (2019). Morphotectonic analysis of the Sali river basin, Bankura district, West Bengal. *Arabian Journal of Geosciences, 12*, 244. https://doi.org/10.1007/s12517-019-4406-0

Dehbozorgi, M., Pourkermani, M., Arian, M., Matkan, A. A., Motamedi, H., & Hosseiniasl, A. (2010). Quantitative analysis of relative tectonic activity in the Sarvestan area, central Zagros, Iran, *Geomorphology, 121*, 329–341. https://doi.org/10.1016/j.geomorph.2010.05.002

Delcaillau, B., Carozza, J. M., & Laville, E. (2006). Recent fold growth and drainage development: The Janauri and Chandigarh anticlines in the Siwalik foothills, northwest India. *Geomorphology, 76*, 241–256. https://doi.org/10.1016/j.geomorph.2005.11.005

Dhuandhar, A. P., Pradhan, A. K., Gupta, C. S., Mohan, R., Bhattacharjee, P., & Shullai, W. (2020). Natural environmental radiation mapping in parts of Meghalaya, India. *Environment and Earth Science, 79*, 1–17. https://doi.org/10.1007/s12665-020-09082-4

El Hamdouni, R., Irigaray, C., Fernández, T., Chacón, J., & Keller, E. A. (2008). Assessment of relative active tectonics, southwest border of the Sierra Nevada (southern Spain). *Geomorphology, 96*(1–2), 150–173. https://doi.org/10.1016/j.geomorph.2007.08.004

Elias, Z. (2015). The neotectonic activity along the Lower Khazir River by using SRTM image and geomorphic indices. *Journal of Earth Science, 4*, 50–58. https://doi.org/10.11648/j.earth.20150401.15

Elias, Z., Sissakian, V. K., & Al-Ansari, N. (2019). Assessment of the tectonic activity in northwestern part of the Zagros Mountains, northeastern Iraq by using geomorphic indices. *Geotechnical and Geological Engineering, 37*, 3995–4007. https://doi.org/10.1007/s10706-019-00888-z

Ezati, M., Gholami, E., & Mousavi, S. M. (2021). Tectonic activity level evaluation using geomorphic indices in the Shekarab Mountains, Eastern Iran. *Arabian Journal of Geosciences, 14*, 385. https://doi.org/10.1007/s12517-021-06724-0

Fan, X., Chang, H., Huo, L., & Hu, C. (2023). GF-1/6 Satellite Pixel-by-Pixel Quality Tagging Algorithm. *Remote Sensing*, *15*(7), 1955. https://doi.org/10.3390/rs15071955

Farahbakhsh, E., Chandra, R., Olierook, H. K. H., Scalzo, R., Clark, C., Reddy, S, M, & Muller, R. D. (2020). Computer vision-based framework for extracting tectonic lineaments from optical remote sensing data. *International Journal of Remote Sensing, 41*, 1760–1787.

Gannouni, S., & Gabtni, H. (2015). Structural interpretation of lineaments by satellite image processing (Landsat TM) in the region of Zahret Medien (northern Tunisia). *Journal of Geographic Information System, 7*, 119–127.

Ghosh, S., & Mistri, B. (2012). Hydrogeomorphic significance of sinuosity index in relation to river instability: A case study of Damodar River, West Bengal India. *International Journal of Advanced Earth Science and Engineering, 1*(2), 49–57.

Giardino, J. R., & Lee, A. A. (2011, October). *Rates of Channel Migration on the Brazos River: Final Report, report*. College Station, Texas. https://texashistory.unt.edu/ark:/67531/metapth415066/m1/4/ (accessed June 4, 2025), University of North Texas Libraries, The Portal to Texas History, https://texashistory.unt.edu; crediting UNT Libraries Government Documents Department.

Goswami, U., Sarma, J. N., & Patgiri, A. D. (1999). River channel changes of the Subansiri in Assam, India. *Geomorphology, 30*(3), 227–244. https://doi.org/10.1016/S0169-555X(99)00032-X

Han, L., Liu, Z., Zhao, Z., & Ning, Y. (2018). Analysis and application of lineaments extraction using GF-1 satellite images in Loess covered. *The International Archives of the Photogrammetry, Remote Sensing and Spatial Information Sciences, XLII-3*, 467–473. https://doi.org/10.5194/isprs-archives-XLII-3-467-2018

Han, Z., Wu, L., Ran, Y., & Ye, Y. (2003). The concealed active tectonics and their characteristics as revealed by drainage density in the North China plain (NCP). *Journal of Asian Earth Sciences, 21*(9), 989–998.

Hare, P. W., & Gardner, T. W. (1985). Geomorphic indicators of vertical neotectonism along converging plate margins, Nicoya Peninsula, Costa Rica. In M. Morisawa & J. T. Hack (Eds.), *Tectonic Geomorphology, Proceeding of 15th Annual Binghamton Geomorphology Symposium*, September 1984. Allen & Unwin.

Holbrook, J., & Schumm, S. A. (1999). Geomorphic and sedimentary response of rivers to tectonic deformation: A brief review and critique of a tool for recognizing subtle epeirogenic deformation in modern and ancient settings. *Tectonophysics, 305*(1–3), 287–306.

Horacio, J. (2014). River sinuosity index: Geomorphological characterisation, Technical note 2, CIREF and Wetlands International, 6p.

Horak, J., & Orlıkova, L. (2019). Circular Statistics for Directional and Temporal Data: Case Studies of Lineaments and Noise Violations Offence, 2019 International Conference on Military Technologies (ICMT), 6p. https://doi.org/10.1109/MILTECHS.2019.8870075

Jackson, J., Van Dissen, R., & Berryman, K. (1998). Tilting of active folds and faults in the Manawatu region, New Zealand: Evidence from surface drainage patterns. *New Zealand Journal of Geology and Geophysics, 41*, 377–385.

Jain, V. S. (2017). Geomorphic effectiveness of a long profile shape and the role of inherent geological controls in the Himalayan hinterland area of the Ganga River basin, India. *Geomorphology, 304*, 15–29. https://doi.org/10.1016/j.geomorph.2017.12.022

Kale, V. S., Sengupta, S., Achyuthan, H., & Jaiswal, M. K. (2014). Tectonic controls upon Kaveri River drainage, cratonic Peninsular India: Inferences from longitudinal profiles, morphotectonic indices, hanging valleys and fluvial records. *Geomorphology, 227*, 153–165. https://doi.org/10.1016/j.geomorph.2013.07.027

Kamra, C., Chopra, S., & Yadav, R. B. S. (2023). Tectonic stress of north-eastern Indian region derived from sesimic focul mechanism on stress drop: A comparative analysis with Kachchh interplate region of India. *Geophysical*

Journal International, 234(1), 453–478. https://doi.org/10.1093/gji/ggad083

Kanhaiya, S., Singh, B. P., Singh, S., Mittal, P., & Srivastava, V. K. (2019). Morphometric analysis, bedload sediments, and weathering intensity in the Khurar River Basin, central India. *Geological Journal*, *54*(1), 466–481. https://doi.org/10.1002/gj.3194

Kawser, U., Hoque, A., & Nath, B. (2021). Observing the impacts of 1950s great Assam earthquake in the tectono-geomorphological deformations at the Young Meghna estuarine floodplain of Bangladesh: Evidence from Noakhali Coastal region. *Arabian Journal of Geosciences, 14*, 306. https://doi.org/10.1007/s12517-020-06427-y

Keller, E. A. (1986). Investigation of active tectonics: Use of surficial earth processes. In *Panel on active tectonics* (pp. 136–147). National Academy Press.

Keller, E. A., & Pinter, N. (2002). *Active tectonics: Earthquakes, uplift, and landscape* (2nd ed., pp. 1–362). Prentice Hall.

Kumar, D., & Duarah, B. P. (2020). Geomorphic signatures of active tectonics in Subansiri River Basin, eastern Himalayas. *Journal of Mountain Science, 17*(6), 1523–1540. https://doi.org/10.1007/s11629-019-5492-x

Kumar, R., Ghosh, S. K., Mazari, R. K., & Sangode, S. J. (2003). Tectonic impact on the fluvial deposits of Plio-Pleistocene Himalayan foreland basin, India. *Sedimentary Geology, 158*(3–4), 209–234.

Leopold, L. B., Wolman, M. G. and Miller, J. P. (1964). *Fluvial Processes in Geomorphology*. San Francisco, California: W. H Freeman and Company, 522 p.

Liu, Z., Han, L., Du, C., Cao, H., Guo, J., & Wang, H. (2021). Fractal and Multifractal Characteristics of Lineaments in the Qianhe Graben and Its Tectonic Significance Using Remote Sensing Images. *Remote Sensing*, *13*(4), 587. https://doi.org/10.3390/rs13040587

Lone, A. (2017). Morphometric and morphotectonic analysis of Ferozpur drainage basin, left bank tributary of River Jhelum of Kashmir Valley, NW Himalayas, India. *Journal of Geography and Natural Disasters, 7*. https://doi.org/10.4172/2167-0587.1000208

Luirei, K., Bhakuni, S. S., & Kothyari, G. C. (2015). Drainage response to active tectonics and evolution of tectonic geomorphology across the Himalayan Frontal Thrust, Kumaun Himalaya. *Geomorphology, 239*, 58–72.

Luirei, K., Bhakuni, S. S., Suresh, N., et al. (2014). Tectonic geomorphology and morphometry of the frontal part of Kumaun Sub-Himalaya: Appraisal of tectonic activity. *Zeitschrift für Geomorphologie, 58*, 435–458. https://doi.org/10.1127/0372-8854/2014/0134

Malik, J. N., & Nakata, T. (2003). Active faults and related Late Quaternary deformation along the Northwestern Himalayan Frontal Zone, India. *Annales Geophysicae, 46*, 917–936.

Malik, J. N., Nakata, T., Philip, G., et al. (2008). Active fault and paleoseismic investigation: Evidence of a historic earthquake along Chandigarh fault in the frontal Himalayan zone, NW India. *Himalayan Geology, 29*, 109–117.

Mohaimen, A., Nath, B., & Hasan, M. R. (2024). Geospatial-based tectono-morphometric analyses of the drainage system in the Chengi and Myinee River basins in the Chittagong Hill Tracts, Bangladesh. *GeoGeo, 3*(1), Article 100224. https://doi.org/10.1016/j.geogeo.2023.100224

Mohanty, C., Baral, D. J., & Malik, J. N. (2004). Use of satellite data for tectonic interpretation, NW Himalaya. *Journal of the Indian Society of Remote Sensing, 32*, 241–247. https://doi.org/10.1007/BF03030884

Mohanty, K. P. T., & Singh, J. K. P. S. (2017). Morphotectonics of the Jamini River basin, Bundelkhand Craton, Central India: Using remote sensing and GIS technique. *Applied Water Science, 7*, 3767–3782. https://doi.org/10.1007/s13201-016-0524-y

Morino, M., Kamal, A. S. M. M., Akhter, S. H., Rahman, M. Z., Ali, R. M. E., Talukder, A., Khan, M. M. H., Mastsuo, J., & Kaneko, F. (2014). A paleo-seismological study of the Dauki Fault at Jaflong, Sylhet, Bangladesh: Historical seismic events and an attempted rupture segmentation model. *Journal of Asian Earth Sciences, 91*, 218–226.

Mueller, J. E. (1968). An introduction to the hydraulic and topographic sinuosity indexes. *American Association of Geographers Annals, 58*(2), 371–385.

Nandy, D. R. (2001). *Geodynamics of northeastern India and the adjoining region* (p. 209p). ACB Publication.

Nath, B., Choudhury, N., & Mitra, A. K. (2022a). Observing tectonic-geomorphological changes along the Dawki Fault and adjoining areas of Sylhet, Bangladesh from 1980 to 2020 using remote sensing and GIS techniques. *Journal of Earth System Science, 131*, 160. https://doi.org/10.1007/s12040-022-01900-6

Nath, B., Niu, Z., & Acharjee, S. (2017). Pre-earthquake anomaly detection and assessment through lineament changes observation using multi-temporal Landsat 8-OLI imageries: Case of Gorkha and Imphal. In R. B. Rustamov, S. Hasanova, & M. H. Zeynalova (Eds.), *Multi-purposeful application of geospatial data* (pp. 149–171). IntechOpen, UK. https://doi.org/10.5772/intechopen.72735

Nath, B., Singh, R. P., Gahalaut, V. K., & Singh, A. P. (2022b). Dynamic relationship study between the observed seismicity and spatiotemporal pattern of lineament changes in Palghar, North Maharashtra (India). *Remote Sensing, 14*(1), 135. https://doi.org/10.3390/rs14010135

Nath, B., Wang, Z., Ge, Y., Islam, K., Singh, R. P., & Niu, Z. (2020). Land use and land cover change modeling and future potential landscape risk assessment using Markov–CA model and analytical hierarchy process. *ISPRS International Journal of Geo-Information, 9*(2), 134. https://doi.org/10.3390/ijgi9020134

Neawsuparp, K., & Charusiri, P. (2004). Lineaments analysis determined from Landsat data: Implication for tectonic features and mineral occurrences in northern Loei Area, NE Thailand. *Science Asia, 30*, 269–278.

Ntokos, D., Lykoudi, E., & Rondoyanni, T. (2016). Geomorphic analysis in areas of low-rate neotectonic deformation: South Epirus (Greece) as a case study. *Geomorphology, 263*, 156–169.

Ouchi, S. (1985). Response of alluvial rivers to slow active tectonic movement. *Geological Society of America Bulletin, 96*(4), 504. https://doi.org/10.1130/0016-7606

Pal, R., Biswas, S. S., Pramanik, M. K., & Mondal, B. (2016a). Bank vulnerability and avulsion modelling of the Bhagirathi-Hugli river between Ajay and Jalangi confluences in lower Ganga Plain. India. *Modeling Earth Systems and Environment, 2*, 65. https://doi.org/10.1007/s40808-016-0125-7

Pal, S. K., Narayan, S., Majumdar, T. J, & Kumar, U. (2016b). Structural mapping over the 85E Ridge and surrounding using EIGEN6C4 high-resolution global combined gravity field model: An integrated approach. *Marine Geophysical Research, 37*(3), 159–184. https://doi.org/10.1007/s11001-016-9274-3

Pan, S. (2013). Application of remote sensing and GIS in studying changing river course in Bankura District, West Bengal. *International Journal of Geomatics and Geosciences, 4*(1), 149–163.

Pandey, P., & Sharma, L. N. (2021). Morphotectonic studies of Ghaggar Basin in the northwestern frontal part of the Himalaya based on remote sensing and GIS techniques. *Journal of the Geological Society of India, 97*, 70–78. https://doi.org/10.1007/s12594-021-1627-0

Pati, J. K., Lal, J., Prakash, K., & Bhusan, R. (2008). Spatio-temporal shift of western bank of the Ganga river, Allahabad city and its implications. *Journal of Indian Society of Remote Sensing 36*, 289–297. https://doi.org/10.1007/s12524-008-0030-2

Pavano, F., Pazzaglia, F. J., & Catalano, S. (2016). Knickpoints as geomorphic markers of active tectonics: A case study from northeastern Sicily (southern Italy). *Lithosphere, 8*(6), 633–648. https://doi.org/10.1130/L577.1

Pavano, F., Romagnoli, G., Tortorici, G., & Catalano, S. (2019). Morphometric evidences of recent tectonic deformation along the southeastern margin of the Hyblean Plateau (SE-Sicily, Italy). *Geomorphology, 342*, 1–19. https://doi.org/10.1016/j.geomorph.2019.06.006

Pérez-Peña, J. V., Azor, A., Azañón, J. M., & Keller, E. A. (2010). Active tectonics in the Sierra Nevada (Betic Cordillera, SE Spain): Insights from geomorphic indexes and drainage pattern analysis. *Geomorphology, 119*(1–2), 74–87. https://doi.org/10.1016/j.geomorph.2010.02.020

Peters, G., & van Balen, R. T. (2007). Tectonic geomorphology of the northern Upper Rhine graben, Germany. *Global Planet Change, 58*(1–4), 310–334.

Philip, G. (1996). Landsat Thematic Mapper data analysis for Quaternary tectonics in parts of the Doon valley, NW Himalaya, India. *International Journal of Remote Sensing, 17*, 143–153. https://doi.org/10.1080/01431169608948991

Radaideh, O. M. A., & Mosar, J. (2019). Tectonic controls on fluvial landscapes and drainage development in the westernmost part of Switzerland: Insights from DEM-derived geomorphic indices. *Tectonophysics, 768*, 1–29.

Radaideh, O. M., Grasemann, B., Melichar, R., & Mosar, J. (2016). Detection and analysis of morphotectonic features utilizing satellite remote sensing and GIS: An example in SW Jordan. *Geomorphology, 275*, 58–79.

Raj, R., & Bhandari, S. (2003). Geomorphic indicators of active tectonics in the Karjan River Basin, Lower Narmada Valley, western India. *Journal of the Geological Society of India, 62*, 739–752.

Ramasamy, S. M., Kumanan, C. J., Selvakumar, R., & Saravanavel, J. (2011). Remote sensing revealed drainage anomalies and related tectonics of South India. *Tectonophysics, 501*, 41–51. https://doi.org/10.1016/j.tecto.2011.01.011

Ramírez-Herrera, M. T. (1998). Geomorphic assessment of active tectonics in the Acambay graben, Mexican Volcanic Belt. *Earth Surface Processes and Landforms, 23*, 317–332. https://doi.org/10.1002/(SICI)1096-9837(199804)

Randel, T. J. (2006). Channel Migration Model for Meandering Rivers. PROCEEDINGS of the Eighth Federal Interagency Sedimentation Conference (8thFISC), April2-6, 2006, Reno, NV, USA. PP. 241–248.

Rastogi, B. K., Singh, J., & Verma, R. K. (1973). Earthquake mechanisms and tectonics in the Assam-Burma region. *Tectonophysics*, *18*(3–4), 355–366. https://doi.org/10.1016/0040-1951(73)90052-8

Reday, Md. I., Khan, Md. S. H., Hossain, Md. S., Pati, P., Rahman, M. J. J., & Bari, Z. (2023). Tectono-geomorphic evolution of an alluvial fan along the Dauki fault system, Bengal Basin, Bangladesh: Remote sensing and pedogenic approach. *Geology, Ecology, and Landscapes, 9*(1), 142–160. https://doi.org/10.1080/24749508.2023.2178131

Roy, S., & Sahu, A. S. (2015). Quaternary tectonic control on channel morphology over sedimentary low land: A case study in the Ajay-Damodar interfluve of Eastern India. *Geoscience Frontiers*, *6*(6), 927–946. https://doi.org/10.1016/j.gsf.2015.04.001

Sarma, J. N., Acharjee, S., & Murugante, B. (2015). Morphotectonic study of the Brahmaputra basin using geoinformatics. *Journal of the Geological Society of India, 86*(3), 324–330.

Sarma, J. N., & Sharma, S. (2018). Neotectonic activity of the Bomdila Fault in northeastern India from geomorphological evidences using remote sensing and GIS. *Journal of Earth System Science, 127*, 113. https://doi.org/10.1007/s12040-018-1008-2

Schumm, S. A. (1956). Evolution of Drainage Systems and Slopes in Badlands at Perth Amboy, New Jersey. *Geological Society of America Bulletin. 67*, 597–646. https://doi.org/10.1130/0016-7606

Schumm, S. A., Dumont, J. F., & Holbrook, J. M. (2000). *Active tectonics and alluvial rivers*. Cambridge University Press.

Sharma, S., & Baruah, S. (2017). Modelling of the Kopili Fault based on the slip rate, moment rate and seismic activity in Mikir Hills Plateau of Northeastern India. *Geomatics, Natural Hazards and Risk, 8*(2), 1157–1172. https://doi.org/10.1080/19475705.2017.1300609

Singh, A. K., Jaiswal, M. K., Pattanaik, J. K. & Dev, M. (2016). Luminescence chronology of alluvial fan in North Bengal, India: Implications to tectonics and climate. *Geochronometria*, *43*(1), 102–112. https://doi.org/10.1515/geochr-2015-0037

Singh, A. K., Parkash, B., & Choudhury, P. R. (2007). Integrated use of SRM, Landsat ETM+ data and 3D perspective views to identify the tectonic geomorphology of Dehradun valley, India. *International Journal of Remote Sensing, 28*, 2403–2414. https://doi.org/10.1080/01431160600993397

Singh, A., Thomsen, K. J., Sinha, R. et al., (2017). Counter-intuitive influence of Himalayan river morphodynamics on Indus Civilisation urban settlements. *Nature Communications*, *8*, 1617. https://doi.org/10.1038/s41467-017-01643-9

Singh, D. S., & Awasthi, A. (2011). Implication of Drainage Basin Parameters of Chhoti Gandak River, Ganga Plain, India. *Journal of the Geological Society of India*, *78*(4), 370–378. https://doi.org/10.1007/s12594-011-0102-8

Singh, S., & Kanhaiya, S. (2015). Morphometry of the Barakar River Basin, India: A remote sensing and GIS approach. *International Journal of Current Research, 7*(7), 17948–17955.

Singh, S., Prakash, K. & Shukla, U. K. (2020). Spatiotemporal Migration of the River Ganga in Middle Ganga Plane: Application of Remote Sensing and GIS Technique. *Journal of the Indian Society of Remote Sensing*, *48*, 1495–1507. https://doi.org/10.1007/s12524-020-01170-z

Spitz, W. J., & Schumm, S. A. (1997). Tectonic geomorphology of the Mississippi Valley between Osceola, Arkansas and Friars Point, Mississippi. *Engineering Geology, 46*(3–4), 259–280.

Srivastava, G. S., & Kulshrestha, A. K. (2005). Neotectonic movements of Markanda and Bata blocks, Himachal Pradesh: Inferences from morphometric analysis. *Special Publication of Geological Survey of India, 85*, 217–224.

Srivastava, V., Mukul, M., Barnes, J. B., & Mukul, M. (2018). Geometry and kinematics of Main Frontal thrust-related fault propagation folding in the Mohand Range, northwest Himalaya. *Journal of Structural Geology, 115*, 1–18. https://doi.org/10.1016/j.jsg.2018.06.022

Strahler, A. N. (1952). Hypsometric (area-altitude) analysis of erosional topography. *Geological Society of America Bulletin, 63*, 1117.

Thannoun, R. G. (2013). Automatic extraction and geospatial analysis of lineaments and their tectonic significance in some areas of northern Iraq using remote sensing techniques and GIS. *International Journal of Enhanced Research In Science Technology & Engineering, 2*, 1–11.

ul Haq, A., Pandita, S. K., Singh, Y., et al. (2019). Evidence of active tectonic deformation in Kishtwar area, Jammu and Kashmir, Northwest Himalaya, India. *Journal of Geological Society of India, 93*, 331–342. https://doi.org/10.1007/s12594-019-1181-1

United States Geological Survey. (2001). *Surface Geology of Bangladesh (geo8bg): U.S. Geological Survey data release*. https://doi.org/10.5066/P9HZQA15

Uthaman, M., Singh, C., Singh, A., Hetényi, G., Dutta, A., Kumar, G., & Dubey, A. K. (2024). Complex multi-fault dynamics in Sikkim Himalaya: New insights from local earthquake analysis. *Geochemistry, Geophysics, Geosystems, 25*, e2023GC011363. https://doi.org/10.1029/2023GC011363

Valdiya, K. S., Rana, R. S., Sharma, P. K., & Dey, P. (1992). Active Himalayan frontal fault, Main Boundary Thrust and Ramgarh Thrust in southern Kumaun. *Journal of the Geological Society of India, 40*(6), 509–528.

Wells, S. G., Bullard, T. F., Menges, C. M., Drake, P. G., Karas, P. A., & Kelson, K. I. (1988). Regional variations in the tectonic geomorphology along a segmented convergent plate boundary, Pacific Coast of Costa Rica. *Geomorphology, 1*, 239–265.

Yang, C. T. (1971). Formation of Riffles and Pools. *Water Resources Research. 7*(6), 1567–1574. https://doi.org/10.1029/WR007i006p01567

CHAPTER 15

Unveiling the Nature of Flood Susceptibility Using Bivariate and Multivariate Statistical Approach in Lower Gangetic Floodplain Region, India

Arnab Ghosh and Pankaj Chakraborty

15.1 Introduction

Over the past few decades, natural disasters have become a major obstacle to developed and developing countries. Currently, natural disasters are responsible for almost 60% of socio-economic losses worldwide. Floods are one of the major impediments to human development, one of the floodplain regions (Balica et al., 2013). Due to the river's carrying capacity decrease in the floodplain area, the river water overflows in the monsoon season by elevating water level. It floods the relatively low and dry surrounding area. Although floods are a natural calamity, they cause financial, social, socio-economic damage to the people living along the

A. Ghosh (✉) · P. Chakraborty
21St Century GIS Academy, Kolkata, India
e-mail: drarnab.swre@gmail.com

S. Ghosh (ed.), *Reimagining Indian Rivers for Sustainability*, Palgrave Studies in Environmental Sustainability,
https://doi.org/10.1007/978-3-032-04754-0_15

river and damage their property by destroying crops and killing livestock (Kazakis et al., 2015). The magnitude and severity of flooding have increased with global climate change accelerating irregular rainfall. As the river's navigability decreases, excess water from dams and reservoirs floods the banks of the river, causing floods. From the beginning of civilization, people have chosen the river bank for their living and other facilities. In developing countries, the population increases due to increasing settlements along the river banks, an increase in the amount of arable land and construction of roads to improve the transportation system, and stakeholders are easily affected by the floods (Chakraborty & Mukhopadhyay, 2019; Ghosh & Kar, 2018). Based on the monsoon, the flood rate in developing countries of South-East Asia is more than 80%. Mainly in India, Bangladesh, China, Indonesia, Myanmar, etc., about 13 million people are affected by floods annually. Climate change, increasing population in the riparian areas, uncontrolled and unscientific construction, depletion of river water holding capacity, lack of dredging, and scientific river management are good reasons for flood disasters. Due to the factors, the countries mentioned earlier in Asia face more than 85% of disasters and catastrophes every year, and about 90% of the people face a severe loss of life (Rahman et al., 2019; Vignesh et al., 2021). Since the floodplain region is of great economic importance to the country's prosperity, this increasing flood situation over time is a considerable concern for policymakers. As a result, policymakers need to map flood susceptibility, as these types of maps make it easier to locate and change the location of future flood situations. Flood susceptibility maps allow researchers and policymakers to take the necessary steps to prevent flooding (Baky et al., 2020; Dandapat & Panda, 2017). Moreover, flood-affected areas worldwide, including Southeast Asian countries, are now following the low-cost, sustainable river management system based on risk assessment methods rather than the traditional flood protection method with various structures (Das, 2018; Khosravi et al., 2016).

Geographical Information System (GIS) and Remote Sensing (RS) methods make it easy to determine flood susceptibility areas (Vojtek et al., 2021). Flood susceptibility is determined based on the relative probability of flood, depending on the extent and duration of the previous flood. As the amount of flood organization is much higher in the plains and floodplain areas, methods and statistics on flood vulnerability, risk analysis, and mapping are essential to analyse the flood susceptible character of those areas. These methods include analytical hierarchy process

(AHP) (Goumrasa et al., 2021) as an expert knowledge-based model, frequency ratio (FR) (Sarkar & Mondal, 2020), information value (IV) (Ul Moazzam et al., 2020), certainly factor (CF) (Cao et al., 2020), logistic regression (LR) (Fustos et al., 2017), weights of-evidence (WOE) (Tehrany et al., 2017), fuzzy logic (Perera & Lahat, 2015), neuro-fuzzy logic (Kambalimath & Deka, 2020) as expert-based models; Artificial neural network (ANN) (Pham et al., 2020), Adaptive neuro-fuzz inference system (ANFIS) (Samantaray et al., 2021), decision tree (DT) (Khosravi et al., 2021), support vector machine (SVM) (Costache et al., 2021), random forest (RF) (Chen et al., 2019) as a machine learning model. Hydraulic Engineering Centre-River Analysis System (HEC-RAS) (Namara et al., 2021) and soil water assessment tool (SWAT) (Nasir et al., 2020) models are also used as hydrological models for determining flood vulnerability and risk. Among the various methods above-mentioned, the AHP model has been the most used for flood vulnerability and risk assessment. The reason is that the AHP model can be used very quickly, and this model helps to present the actual picture of floods in any region (Dhekra et al., 2020; Hammami et al., 2019). Flood risk in a region is based on socially, economically, and geographically evolving correlated spatiality, which is easy to express through the AHP model. However, when the AHP model is used as an integrated part with GIS, the efficiency and effectiveness of the AHP model are significantly increased. In many cases, it has been found that using the AHP model integrating with GIS can make excellent predictions and quickly identify future flood-prone areas. While the AHP model is a direct method for determining flood susceptibility, the statistical models act indirectly (Wubalem et al., 2021). These statistical models, widely used worldwide, are based on mathematical expressions to evaluate the relationship between the causes of floods and the occurrence of floods. Determining flood susceptibility through models based on statistical methods is done with perfect, accurate and easy to understand without any specific demand (Bednarik et al., 2012; Regmi et al., 2014). The most common statistical models can be divided into multivariate and bivariate statistics based on reliable results. The bivariate statistical approach assesses the correlation between flood control parameters and past flood organizations, like frequency ratio (FR), information value (IV), certainly factor (CF), Shannon entropy (SE), weights of-evidence (WOE), etc. (Wubalem & Meten, 2020). The role of each input parameter in the flood organization is examined through the bivariate statistical method, with accurate, reliable information provided about the flood and past flood organization statistics. However, even if each parameter is examined separately, their correlation is not examined, making this model much different from the multivariate statistical model. The multivariate statistical approach assesses the correlation between independent and dependent variables important to flood organization (Das & Lepcha, 2019). Logistic regression and discriminant analysis, used for flood susceptibility modelling, are notable examples of multivariate statistical approaches that help to reveal susceptibility through reliable results (Kouhpeima et al., 2017).

India is a country with a monsoon climate, and it not only affects the social, economic, and geographical areas here and plays a significant role in flood management. West Bengal, Bihar, Assam, Uttar Pradesh, and Orissa are among the flood-prone regions of the Gangetic plain to India's north and northeast plain portion (Sahoo & Sreeja, 2015). In addition, in Maharashtra, Gujarat in western India and some flood-prone areas in Tamil Nadu and Karnataka in the south, about 3.3 million hectares of India are inundated by floods (Patel et al., 2017). The Bhagirathi-Hooghly River basin in the floodplain region of South Bengal of eastern India and the Gangetic plain are the flood-prone areas in India. The lives and livelihoods of about 3.5–4 lakh people are lost every year due to these floods. Bhagirathi-Hooghly River flows through West Bengal, India and is recognized as a major tributary of the Ganges. It is separated from the Ganges near the Mithipur block of Murshidabad district and flows into the Bay of Bengal, encompassing Kolkata (Ghosh et al., 2020). This river system is recognized as the lifeline of West Bengal and provides a wide range of its role in developing irrigation, industrialization, and human habitation. However, over time, the water flow from the main river has decreased, and tributaries like Ajay and Damodar from the Chhotanagpur Plateau have accumulated a lot of sediment and deposited it in the Bhagirathi-Hooghly River. Also, the lack of dredging scientifically with the help of continuous urban construction encroachment has dramatically reduced the river's navigability. However, for the agro-economic, industrialization, and communication system, the Nabadwip-Kalyani sub-catchment basin of the Bhagirathi-Hooghly river's riparian zone is socially and economically developed overpopulated. Part of this region is known as the rice bowl of West Bengal for its abundant paddy production. Every year during the monsoon season, severe floods cause catastrophe to the daily livelihoods of the riparian population and ruin crops, which weakens people economically in the aftermath of the floods (Ghosh & Kar, 2018;

Jha & Bairagya, 2013; Mujiburrehman, 2015; Sanyal & Lu, 2005, 2006). For mitigating flood-related problems, flood hazards and the vulnerable regions must identify to ensure the location of flood risk areas.

Although several qualitative and quantitative methods have previously been used to determine flood susceptibility, there is no clear or practical basis for determining the best of them. The model's acceptability and relevance depend on the quality and reliability of the parameters and their uses through technological advancement. However, each model has its limitations. For example, in the case of the LR model, although the flood susceptibility is determined based on the correlation between independent and dependent variables, this model does not shed any light on the flood control parameters in the flood organization. The FR model again overcomes this limitation through the bivariate statistical method, where the role of each parameter in the flood organization is examined. Still, no relationship exists between the flood control parameter and the flood organization. Observing all of the above conditions and reviewing the history of flood organization in the study area, this article analyses the development of a flood susceptibility model generated by three different methods (expert, bivariate, and multivariate) in the current study area. Also, it assists in adopting flood management policy by measuring flood severity, catastrophe, and risk level based on spatiality. The models' effectiveness and performance are evaluated by the receiver operating characteristic curve (ROC) and the area under the curve (AUC). Also, the article compares the performance of the flood susceptibility model with the measurement of the actual inundation scenario of the region, which is expected to increase the acceptability of the models greatly.

15.2 Materials and Methods

15.2.1 Study Area

The selected sub-catchment basin of the middle course of Bhagirathi-Hooghly River is up to 161 km long from Nabadwip (upstream) to Kalyani (downstream), comprising a portion of Burdwan, Nadia and Hooghly district of West Bengal, India (23°24′45″ N/ 88°22′45″ E - 22°58′01″ N/ 88°24′27″ E). This area is geologically located in the Rarh region, the lower part of the moribund deltaic part of the Bengal Basin. It is composed of recent deposits of the Pleistocene period (Ghosh et al., 2020). Initially, the area was covered by sandy clay and sand along the river's course. Fine silt, sandy loam, and loamy soil have been found in the flat portion of the plain (Fig. 15.1). Since British rule, the region has been experiencing severe floods, and subsequent floods have significantly increased property loss (Fig. 15.2). Heavy to weighty rainfall, monsoon tropical cyclones, urban area encroachment on river channels, excessive sedimentation, unscientific levee and building construction along the river bank are some of the causes of riverbank erosion. Although the Bhagirathi Hooghly River is the main river in this basin, the river's course is joined by tributaries such as the Khari from the right and the Jalangi and Churni rivers from the left. The total catchment area, including the main river and three branches, is 1690 sq. km.

The hydrodynamic condition here is very different from other regions as the whole study area belongs to floodplain morphology. Lateral bank line movement can be seen through the meandering flow of the river and vice versa, which gives rise to the riverbank erosion problem. In addition to meandering, floodplain features include wetlands, ox-bow lakes and shoal formations (Jha & Bairagya, 2013; Mujiburrehman, 2015; Sanyal & Lu, 2005, 2006). Due to the shallow slope of the area (maximum 20–40), rainwater falls into the river very quickly as runoff after rainfall. Due to the tropical monsoon climate, the region receives 110–124 rainy days per year, and the average annual rainfall is between 1400 and 1800 mm. An analysis of the trend of average annual precipitation from 1985 to 2020 shows that the recent rain has increased significantly. The reason is that over the past decade, tropical cyclones and depressions have markedly increased due to the whimsy weather. Such downpours can cause 150 to 250 mm of rainfall per day in the rainy season, resulting in a rapid increase in river water and reduced navigability.

The Nabadwip-Kalyani sub-basin consists of several blocks and about 662 mouzas in East Burdwan, Hooghly and Nadia districts (Table 15.1). Some of these mouzas are prone to river erosion, and some areas are prone to floods. Due to embankment along the banks in some areas, the height of places is higher than the natural plain. But in the low-lying areas next to the embankment opposite the river, the water accumulates during the monsoon. Most of the mouzas on the river banks are prone to floods, but only if submergence is some distance away. Since the outlet of the Nabadwip-Kalyani sub-basin is compressed like a funnel enclosing the urban area at the bottom, the runoff causes problems in the discharge of sediments. Sedimentation is seen through the formation of sandbars. As a result, the river's navigability is reduced, and heavy rains in

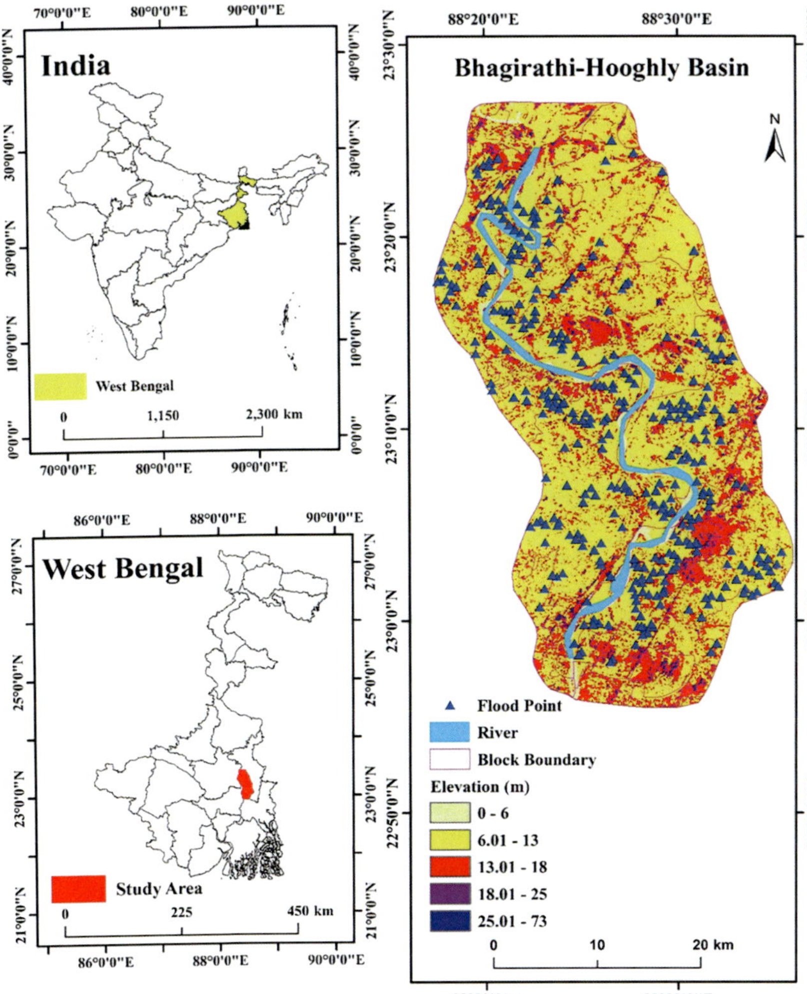

Fig. 15.1 Location of the study area with flood inventory

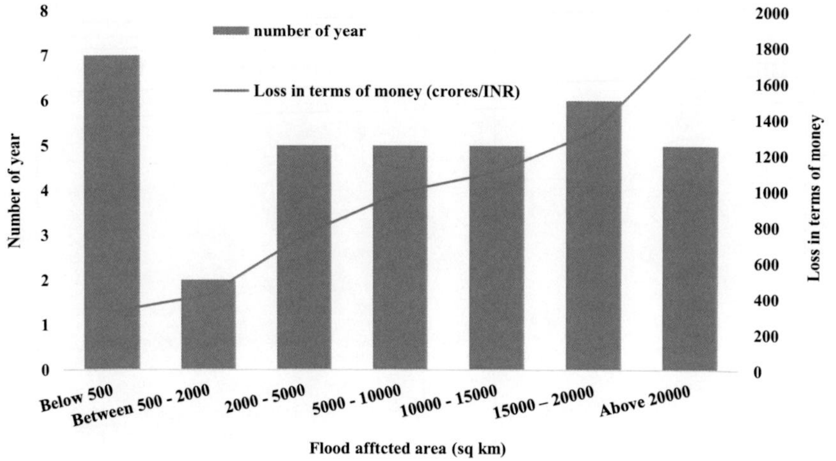

Fig. 15.2 Flood against time periods and loss in study area

the upstream region increase the water and cause floods. The trending nature of peak discharge with the sedimentation of the river (1985–2020) shows the accelerated concentration of sediment from the upstream region towards the downstream segment. The Bhagirathi-Hooghly River is slowly moving towards becoming a rain-fed river in future. At the same time, if we compare the amount of rainfall with the duration of floods in the region through discharge, we can see that the year there is more rainfall, the amount of flooding and the destructive power of discharge increase. Although the Bhagirathi-Hooghly River originates as a tributary of the Ganges, its water flow depends on the Farakka Barrage and the Indo-Bangladesh Water Treaty, reducing its flow of water day by day. Also, unscientific and uncontrolled construction of houses along the riverbank with constant urban encroachment has resulted in changes in land use. The unscientific construction of jetties and rampant sand mining from the riverbed increases the possibility of flood vulnerability in this region.

15.2.2 Data Used

Since floods reveal a multidimensional view of the hydrodynamic state of a region, the flood susceptibility model requires a large amount

Table 15.1 Flood-affected blocks of study area

District Name	*Block name*	*Number of mouzas*	*% of flood affected*
Nadia	Nabadwip	94	61.30%
	Krishnanagar 1	108	65.20%
	Santipur	142	71.30%
	Chakdah	136	67.80%
East Burdwan	Purbasthali 1	47	42.70%
	Kalna 1	84	55.60%
Hooghly	Balagarh	51	39.80%

of information to interpolate in one place to express flood hazards. Land use land cover (LULC) map of pre-monsoon period (February 18, 2020) obtained from geo-rectified USGS Landsat 8 OLI (Operational Land Imager) map (Path and Row—138/44, cell size—30 m) has been used for identification of different geographical and communicational parameter (http://earthexplorer.usgs.gov). In addition, ASTER (Advanced Spaceborne Thermal Emission and Reflection Radiometer) DEM (Digital Elevation Model) (1 arc second horizontal resolution) has been used to determine the altitude, land slope, river flow direction and various other hydrological parameters of the entire region (https://search.earthdata.nasa.gov/search). The ArcGIS Hydrology Toolset was later used to mosaic the whole DEM and prepare it for flood hazard, vulnerability and risk zonation. These images and base maps are projected through Universal Transverse Mercator projection with zone 45 north (UTM 45N) and World Geodetic Survey 1984 (WGS 1984) in Arc-GIS v 10.4.1 software. The images have been retrieved from different sources and not georectified for accuracy. A geometric correction was introduced with a root mean square error (RMSE) of ± 0.5 pixels to correct the image to the image registration technique. Suitable soils of the present study area and their spread have been extracted from the soil map obtained from FAO (https://www.fao.org/soils). Based on data from IMD, Pune, the average annual rainfall in the region has been determined (https://pypi.org/project/imdlib/). The road network of the study area has been obtained from https://www.openstreetmap.org/#map=9/25.8110/89.3513. Also, collected data, including the number of floods organized from 1985 to 2020, the amount of damage and the number of affected areas, and the amount of flooded area, were collected from the District

Disaster Management Plan of the Government of West Bengal (https://censusindia.gov.in/pca/pcadata/pca.html).

15.2.3 Flood Inventory Map

The Bhagirathi-Hooghly River basin has experienced some terrible floods in the past from the British period. In 2000, 2003, 2004, 2005, 2006, 2007, 2013, 2017, the region was severely damaged by floods. Past flood events are a major indicator that if the watershed's climatic, geomorphological, hydrological conditions remain the same in the future, then the possibility of an equal consequence is highlighted. The respective authority records flood information at regular intervals, estimating the probability of flood risk using various spatial modelling techniques. Besides, flood inventory maps can map flood Hazard susceptibility mapping of watersheds. At the same time, non-flood points commonly seen in highland areas are used for accuracy testing of flood hazard susceptibility mapping. In this study, a flood inventory map has obtained from https://code.earthengine.google.com/716b1914850650212c5595e2525d8b37. The flood and non-flood points data into raster format with a spatial resolution of 30 × 30 pixels in Arc GIS interface. A sufficient number of flood and non-flood points has been selected from the raster layer, and the training and testing dataset has formed in the ratio of 70:30 following the space robustness method (Fig. 15.1). The training dataset builds up the model, and the test dataset validates the model by confirming its accuracy. Since flood susceptibility mapping is considered a binary classification in the machine learning method, there is a requirement for positive and negative cases. The flood pixels were assigned a value of 1, and non-flood pixels were regarded as 0. The value of flood-determining factors was extracted from these pixels and organized a text based on that. It is then exported to R Statistical Software (3.2.3) and analysed.

This study used the Google Earth Engine to indicate 650 flood points and 350 non-floods. Of these, 750 randomly selected points are considered training, and the rest are considered testing points for flood hazard susceptibility mapping. Flood inventory maps reveal that the concentration of flood points is higher than another part of the study area, thus easily defining the area of interest. The machine learning method can produce non-linear interpolation, so if the number of training points is insufficient, the accuracy of the prediction model has become unreliable.

15.2.4 Flood Driving Indicators

Hazard and its derivative indicators for disaster are based on geographical and environmental factors. Also, these indicators have been used to determine the weightage, which helps create the flood susceptibility model (Table 15.2 and Fig. 15.3a-l). These indicators are—(a) Elevation (based on data from ASTER-DEM 1 arc second void filled and mosaic of ArcGIS raster tools. However, being a floodplain area, the height here is much lower), (b) Slope (measured using ArcGIS's Surface Analysis Tool based on DEM), (c) Drainage density (measured using DEM's ArcGIS density analysis tool), (d) Distance from the river (measured by the ring of radial buffer at 500 m from the river up to 3945 m), (e) TWI or Topographical wetness index (created in Arc GIS with the help of a raster calculator based on DEM and the water flow equation), (f) MNDWI or modification of normalized difference water index, and (g) NDVI or normalized difference vegetation index (LANDSAT OLI band 4 and 5) has been used to calculate NDVI, whereas bands 3 and 7 have been used to calculate MNDWI. If the value of NDVI decreases and at the same time the value of MNDWI increases, there is the possibility of flood), (h) Roughness (measured by DEM. Since this is a floodplain region, the roughness is much less), (i) Average rainfall (based on weather information from IMD, Pune. This region belongs to the tropical monsoon; therefore, the rainfall is much higher here), (j) MFI or Modified Fournier Index (mainly used to measure rainfall intensity. During the rainy season, this region receives heavy rainfall in a brief time.), (k) Soil (has been extracted from the FAO soil map), (l) STI or Sediment transport index (based on unit stream-power theory and length-slope factor in the revised universal soil loss equation (RUSLE) (where the slope of the land is less than 100 m and the slope is less than 140).

15.2.5 Methodology

15.2.5.1 Analytical Hierarchy Process (AHP)

AHP is used as a qualitative method for determining the probability of flooding in a region, formed by the interrelationships between the various components that help in flood management and the occurrence of floods occurring promptly. The AHP model has previously been used to solve various problems in mathematics and psychology as a specific structural and decision-making component (Saaty, 1977; Zhang et al.,

Table 15.2 Indicator's weightage of flood susceptibility or hazard

Parameter	*Class*	*Range*	*AHP Weight*	*FR*	*PR*	W_j *(%)*	*LR Coefficient*
Elevation	1	0–6	0.2666	0.2384	9.1836	0.0995	−23.508
	2	6.01–13		1.1424			
	3	13.01–18		0.2944			
	4	18.01–25		0.2410			
	5	25.01–73		0,3478			
Slope	1	0–3	0.1920	1.3486	4.6967	0.0329	−9.959
	2	3.01–6		0.6816			
	3	6.01–9		1.3450			
	4	9.01–12		1.7697			
	5	12.01–49.11		0.5972			
Drainage Density	1	0–0.15	0.1234	0.7521	2.3630	0.0073	−5.892
	2	0.16–0.43		0.9268			
	3	0.44–0.69		1.1249			
	4	0.7–0.95		0.8542			
	5	0.96–1.67		1.2570			
Distance from river	1	0–500	0.1222	1.2940	2.6352	0.0119	−11.097
	2	500–1000		1.3351			
	3	1000–1500		1.1909			
	4	1500–2000		0.7191			
	5	2000–3945		0.8373			
TWI	1	3.45–7.94	0.0867	1.0757	2.3397	0.0085	−3.761
	2	7.95–10.39		0.9516			
	3	10.4–12.51		1.5406			
	4	12.52–15.45		1.6220			
	5	15.46–24.26		1.4011			
MNDWI	1	−0.49–−0.11	0.0627	0.7261	3.2515	0.0138	37.006
	2	−0.1–−0.05		0.8274			
	3	−0.04–0.02		0.8383			
	4	0.03–0.13		1.1505			
	5	0.14–0.31		1.4289			
Roughness	1	0.11–0.31	0.0444	1.1057	1.0000	0.0014	3.319
	2	0.32–0.43		0.9246			
	3	0.44–0.51		0.9874			
	4	0.52–0.61		0.9736			
	5	0.62–0.89		1.1480			
Average Rainfall	1	996.57–1,035.44	0.0284	3.3031	5.3648	0.0369	−3.258
	2	1,035.45–1,069.57		4.1609			
	3	1,069.58–1,106.54		2.4666			

(continued)

Table 15.2 (continued)

Parameter	*Class*	*Range*	*AHP Weight*	*FR*	*PR*	W_j *(%)*	*LR Coefficient*
	4	1,106.55–1,155.84		1.8208			
	5	1,155.85–1,238.32		1.1513			
MFI	1	202.04–215.01	0.0284	0.5707	4.8110	0.0399	45.592
	2	215.02–230.47		1.1845			
	3	230.48–245.1		2.3061			
	4	245.11–260.25		2.1499			
	5	260.26–277.94		2.0853			
NDVI	1	−0.13–0	0.0151	1.5558	4.7996	0.0369	19.198
	2	0.01–0.11		3.6663			
	3	0.12–0.18		1.5807			
	4	0.19–0.25		3.3002			
	5	0.26–0.48		4.6303			
Soil	1	Meadow soil	0.0151	0.3529	6.7226	0.0608	2.968
	2	Young soil with alluvial deposit		1.0934			
	3	Loamy soil		1.0870			
STI	1	0–969.76	0.0151	0.9654	22.7070	0.6502	−1.20
	2	969.77–3,771.28		74.8466			
	3	3,771.29–7,758.06		0.0000			
	4	7,758.07–15,408.37		0.0000			
	5	15,408.38–27,476.46		0.0000			

2016). Comparative compound methods according to the ranking scale (Saaty, 2008) have been used to determine the balance and weightage of the auxiliary elements in flood organization (Luu et al., 2018; Saaty, 1990). Also, consistency ratio (CR) is used to check the weightage of each flood factor and determine their order of quality. If the value of CR is less than 0.1, then the method of determining the comparative matrix is consistent. However, if the CR value is greater than 0.1, the matrix is completely inconsistent, which prevents the successful AHP model from being created (Luu et al., 2018; Saaty, 2008).

$$CR = CI/RI \tag{15.1}$$

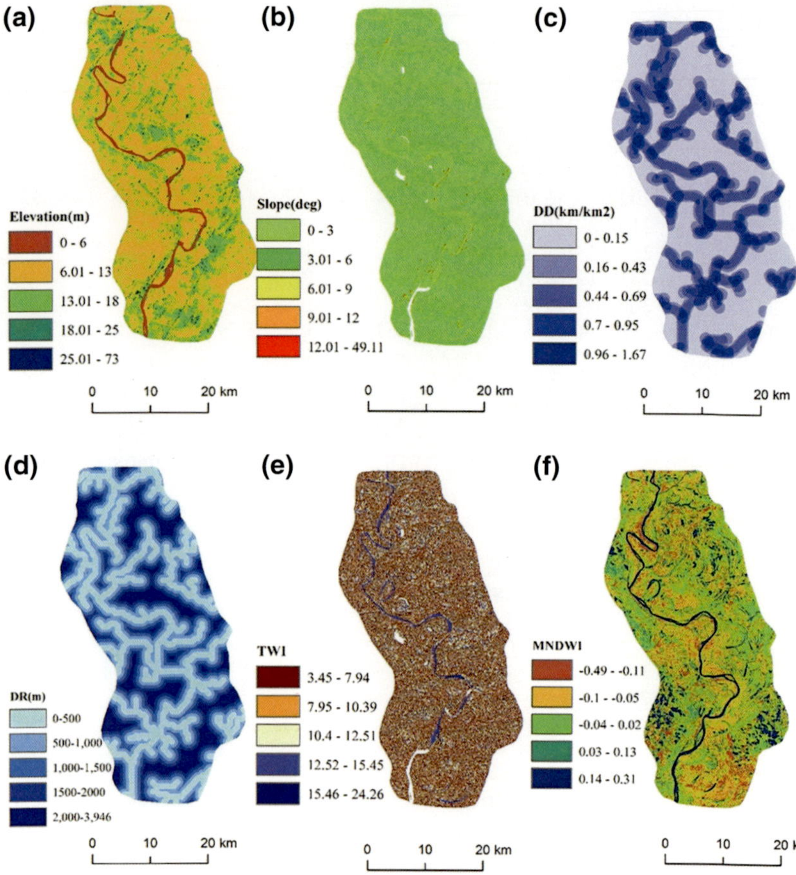

Fig. 15.3 Spatial distribution of flood susceptibility indicator in the study area

where, CI is the consistency index, RI is the random index (basically, it is the number of indicator or parameter). The CI is also defined as,

$$CI = (\lambda\max - n)/(n - 1) \tag{15.2}$$

where, λmax is the largest eigen value of comparison matrix table, n is the number of criteria. After determining the weightage of each component that helps in flood organization, it is possible to examine the flood

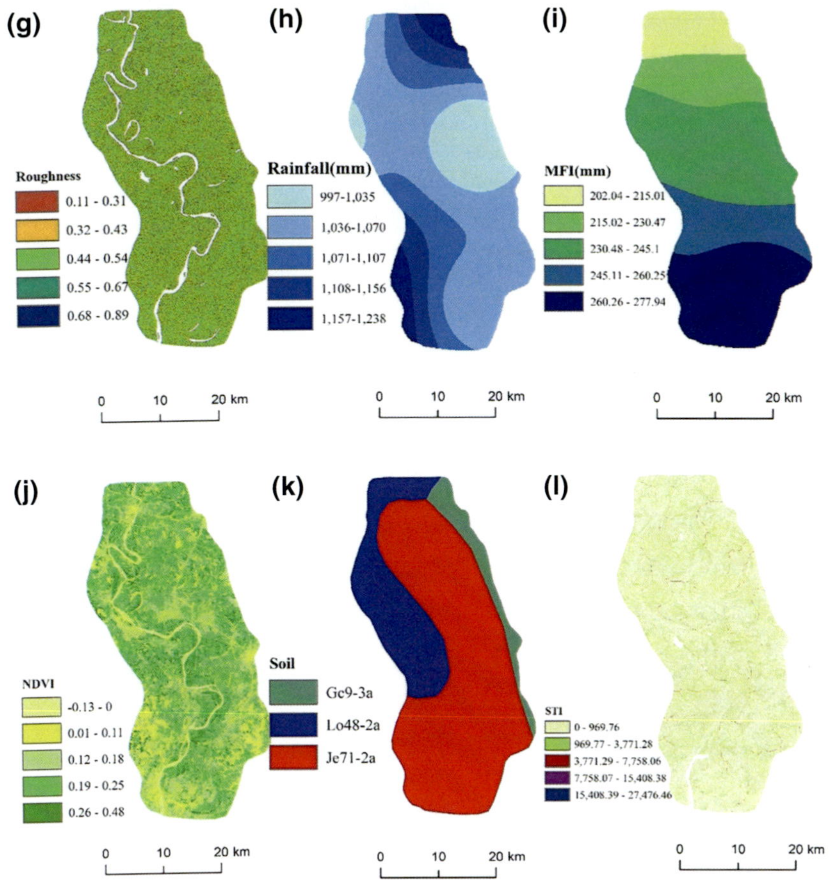

Fig. 15.3 (continued)

susceptibility areas of a particular region (Rahmati et al., 2015).

$$\mathrm{FSI} = \sum_{i=1}^{n} \mathrm{W}_i * \mathrm{X}_n \quad (15.3)$$

Where, FSI is flood susceptibility index, W_i is the weightage of flood factors and X_n is the flood factors.

15.2.5.2 *Frequency Ratio (FR)*

FR is a popular, well-accepted, quantitative methodology for flood hazard inventory mapping. Several studies have used the FR technique to map the flood susceptibility (Samanta et al., 2018; Tehrany et al., 2014). Estimating past flood records is necessary to predict future flooding in a region. FR is a bivariate statistical method that builds a quantitative correlation between the frequency of flooding and the factors influencing the flood. Equation of FR Index is described as follows,

$$\text{FSI} = \sum \text{FR} \tag{15.4}$$

FSI refers to the flood susceptibility index, and FR assigns the frequency rank of each conditioning factor. FR refers to the ratio of flood-prone areas to the whole study area. It refers to the probability of flood and non-flood zone subject to a given attribute (Lee & Talib, 2005).

$$\text{FR} = \frac{\left(\frac{E}{F}\right)}{\left(\frac{M}{L}\right)} \tag{15.5}$$

E is the flood event of each factor, F depicts the total number of the flood event, M refers to the no. of the pixel of each class of any factor, L is the total no of the pixel of the study area. The value of FR refers to the correlation between the occurrence of flood and the conditioning factors influencing it. If the value of FR is less than 1, it means weak correlation, and if it is more than 1, it assigns to strong correlation (Shirzadi et al., 2017).

15.2.5.3 *Shannon Entropy (SE)*

The Shannon Entropy (SE) model is used as an augmented or modified version of the FR model. This model is based on instability, chaos, imbalance, and uncertainty. This model refers to the quality of the most optimal variable of the flood organization (Yufeng & Fengxiang, 2009). Above all, the SE model measures the value of unnecessary variables among the various variables mainly responsible for flood organization. As a result, to study the nature of flood susceptibility, the SE model discusses the contribution aspect according to the weightage of each parameter. It is possible to derive flood data from the mathematical expression of the weightage of each parameter determined by the SE model (Milaghardan et al., 2020).

$$P_{ij} = \frac{FR}{\sum_{j=1}^{sj} FR} \tag{15.6}$$

Where, P_{ij} is the probability density, and FR is the frequency ratio. On basis of this equation, different rank is initiated with respect to different classes of the factor (Milaghardan et al., 2020).

$$H_j = -\sum_{j=1}^{sj} P_{ij} \cdot \log_2 P_{ij} \quad j = 1, \ldots, n \tag{15.7}$$

$$H_{j\max} = \log_2 sj \tag{15.8}$$

$$I_j = \frac{H_{j\max} - H_j}{H_{j\max}} \tag{15.9}$$

$$W_j = I_j P_j \tag{15.10}$$

Where, H_j and $H_{j\max}$ are the entropy values, sj is the number of classes, I_j is the information coefficient, and W_j is the final weightage value of the variables.

15.2.5.4 Logistic Regression (LR)

The LR method has been used to estimate flood susceptibility (Hosmer et al., 2013). It is the equivalent of multilinear regression, and it is considered the best-fitted model for assessing the correlation between independent and dependent variables in flood hazard mapping (Shafapour Tehrany et al., 2017). It depicts the possibility of flood events over the occurrence. This model analyses the probability of flooding by considering independent and dependent variables (Omid Rahmati et al., 2016). This variable can be continuous and discrete. In this case, the presence and absence of flood potential are expressed through binary digits. The analysis of flood susceptibility by LR is valid only if the dependent variable is binary. 0% refers to the probability of non-flood, and 100% means the high chances of flood occurrence. The occurrence of a flood event and its influencing variables can be expressed as (Schicker & Moon, 2012),

$$P = \frac{1}{1 + e^{-z}} \tag{15.11}$$

P refers to the probability of flood occurrence ranging from 0 to 1.

$$Z = B_0 + \sum_{i=1}^{n} B_i X_I \tag{15.12}$$

$X1 ....Xn$ refers to each flood influencing factor, n is the total number of input factor, and Bi depicts the coefficient estimate from sample data.

15.2.6 *Testing of Multicollinearity*

Multicollinearity among all the factors was checked using the tolerance (TOL) and variance inflation factor (VIF). Multicollinearity occurs when independent variables correlate in a regression model. If the correlation of the variables is too high, then there is difficulty in fitting the model and interpreting the result. In case of occurrence, some of the highly correlated independent variables should be removed considering the variance inflation factor (VIF) value (Arabameri et al., 2019). The coefficient value of TOL should be less than 0,1, and the value of VIF must be less than 10.0. If the VIF value of the maximum two variables is more than 10, it can assume that the correlation problem exists between the two indicators and should remove one of them (Pourghasemi et al., 2017).

$$\mathrm{TOL} = 1 - R_j^2 \tag{15.13}$$

$$\mathrm{VIF} = \frac{1}{\mathrm{TOL}} \tag{15.14}$$

However, we should look at the collinearity diagnostics. So, we can say that the variables in the current study are free from linear correlation. By assessing multicollinearity among the thematic layers, 1000 random points have been selected in the study area using the Create Random tool in ARC GIS 10.4 interface. Data for each point are extracted from each thematic layer through the 'Extract Value to points' tool in the GIS atmosphere. Then multicollinearity test of extracted values has been performed in SPSS software.

15.2.7 *Model Performance*

Validity and accuracy assessment are necessary for analysing the predictive results of flood susceptibility mapping through machine learning (Arora et al., 2019). This study used sensitivity, specificity, accuracy, and kappa to evaluate predictive performance (Fig. 15.4). Sensitivity refers to the proportion of correctly classified flood pixels; specification refers to the ratio of non-flood pixels that are correctly classified; accuracy refers to the ratio of pixels that are correctly classified (Onan, 2015). Their value is between 0 and 1; if its value is close to 1, then the model is more accurate and vice versa.

$$\text{Sensitivity (P)} = \frac{\text{TP}}{\text{TP} + \text{FN}} \tag{15.15}$$

$$\text{Specificity (R)} = \frac{\text{TN}}{\text{FP} + \text{TN}} \tag{15.16}$$

$$\text{Accuracy} = \frac{\text{TP} + \text{TN}}{\text{TP} + \text{TN} + \text{FP} + \text{FN}} \tag{15.17}$$

$$\text{F}_{\text{score}} = \frac{2 \times \text{P} \times \text{R}}{(\text{P} + \text{R})} \tag{15.18}$$

TP refers to true positive, defined as the number of correctly classified flood pixel. FP is false positive, depicts as the number of the incorrectly classified pixel of the flood event. TN determines true negative, which is number of the correctly classified pixel of non-flood event. FN is false negative refers to the number of the wrongly classified pixel of a non-flood occurrence. Kappa is a widely used statistical measure defined as the proportion of agreement after the chance agreement is removed. The following formula expresses it,

$$\text{K} = \frac{\text{Observed Agreement } (\text{P}_o) - \text{Expected Agreement } (\text{P}_e)}{1 - \text{Observed Agreement } (\text{P}_o)} \tag{15.19}$$

P_o and P_e are calculated as follows,

$$\text{P}_o = \frac{\text{TP} + \text{TN}}{n} \tag{15.20}$$

$$\text{P}_e = \frac{(\text{TP} + \text{FN}) + (\text{TP} + \text{FP}) + (\text{FP} + \text{TN}) + (\text{FN} + \text{TN})}{\sqrt{\text{N}}} \tag{15.21}$$

N refers to the total number of training pixel, and n is the proportion of incorrectly classified flood and non-flood occurrence pixels. The

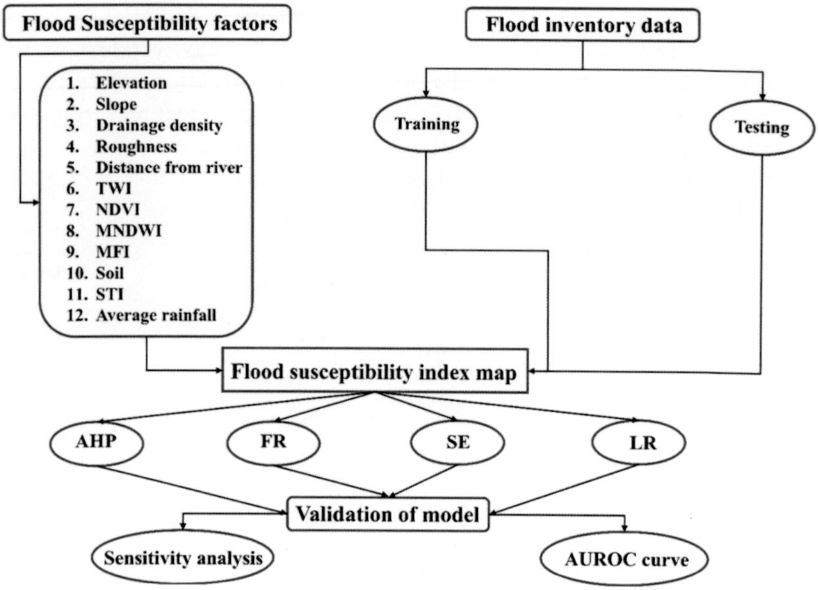

Fig. 15.4 Methodological framework of the study

value of K and its respective performance ranges between 0 (poor) and 1 (perfect). The ROC curve is widely used to evaluate the overall performance (Hong et al., 2016; Pourghasemi & Kerle, 2016). It is plotted as 2D, showing specificity on the X-axis and sensitivity on the Y-axis. Its value ranges from 0.5 to 1, which indicates low performance to the high level of performance, respectively.

$$S_{AUC} = (X_{k+1} + X_k)\left(S_k + 1 - S_{k+1} - \frac{S_k}{2}\right) \tag{15.22}$$

S_{AUC} is area under curve, X_k refers to the 1 − Specificity and S_k is 1 − Sensitivity of ROC. In addition to the AUROC curve, the accuracy of these four models is validated by the Taylor diagram (Taylor, 2001).

15.3 Result and Discussion

15.3.1 Multicollinearity Test

This study tests the multicollinearity of 12 factors affecting flooding based on the quality of VIF and TOL. It is seen that the value of VIF ranges from 1.028 to 2.681, highest and lowest VIF associated with distance from the river and average rainfall, respectively. In the case of TOL, the range is between 0.373 and 0.973. The highest and lowest values of TOL are seen in the average rainfall and distance from the river (Table 15.3). It is seen that there is no multicollinearity problem among the 12 above-mentioned factors in demarcating the flood susceptibility zone. Therefore, 12 factors have been used for flood modelling in this study area. This study assesses the linear correlation of each pair of conditioning factors using Pearson's correlation. The results show that the value does not exceed 0.7, which confirms that all the factors are free from multi-collinearity, and there is no influencing factor among the 12 factors. If there is a high correlation value among any two factors, the analysis has been repeated by deleting one of the factors from the dataset (Fig. 15.5).

Table 15.3 Multicollinearity on susceptibility indicator

Susceptibility Parameter	*Collinearity Test*	
	TOL	*VIF*
Elevation	0.854	1.301
Slope	0.786	1.272
Drainage density	0.383	2.612
Distance from river	0.373	2.681
TWI	0.717	1.396
MNDWI	0.679	1.473
Roughness	0.964	1.038
Average rainfall	0.973	1.028
MFI	0.754	1.32
NDVI	0.667	1.500
Soil	0.814	1.741
STI	0.810	1.234

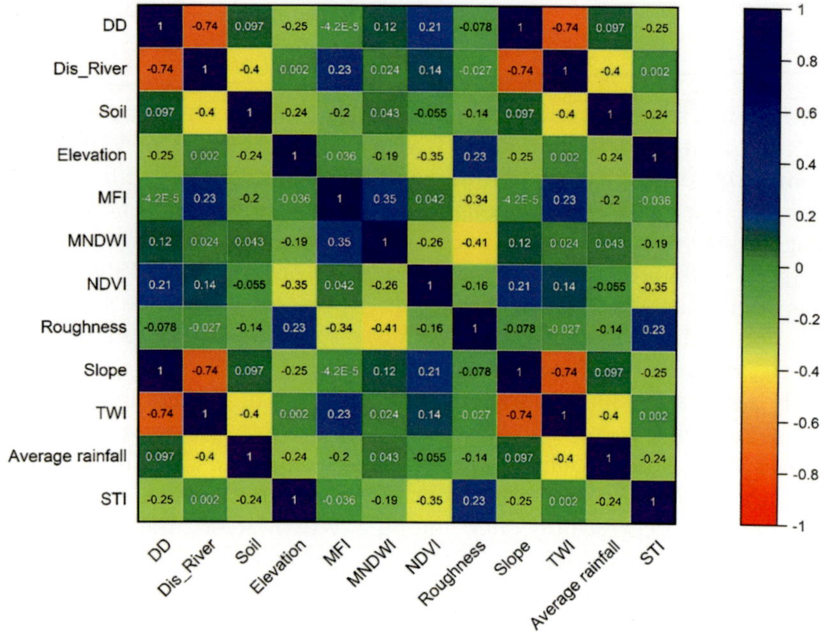

Fig. 15.5 Pearson correlation among the flood conditioning factors

15.3.2 Flood Susceptible Map

15.3.2.1 Application of AHP Comparison Matrix on Flood Susceptibility

AHP is a multidimensional, multi-purpose multi-criteria decision analytics (MCDA) criterion that allows one to arrive at the proper conclusion by comparing different dimensional decisions and choices. A hierarchical structure is created by combining the different paths obtained through group discussions and is fulfilled through the AHP model to achieve a particular purpose. The primary activity regulators are modelled according to their efficiency with equal importance. The AHP model develops an alternative causation approach by reviewing the purpose and subject matter of each specific activity and variable. The weightage of susceptibility indicators in the current study area is determined using the AHP model (Table 15.2). The indicators are compared pairwise, and the matrix generates a specific score based on their importance. Each factor is measured according to its relative importance. A priority vector table is then created according to the normalized Eigenvector of the matrix. The weight of the objective sequence is calculated based on the average obtained from it to measure flood susceptibility (Tables 15.4 and 15.5). In the created comparison matrix table, consistency ratio (CR) is used to see if all the parameters are compatible with the AHP model. According to this index, if CR => 0.10, the performance of that model is consistent. With 12 susceptible indices criteria in this study, the CR and CI values are calculated as 0.077 and 0.115 (Eqs. 15.1–15.2). But surprisingly, despite the difference in the number of criteria, the RI value is the same (1.48) in sensitive cases. The susceptibility index is created after ranking according to the weightage of the susceptibility indicators from the comparison matrix table. The flood susceptibility index (FSI) is measured as (Rahmati et al., 2015),

$$\begin{aligned}\mathrm{FSI} = {} & (w_E \times E) + (w_S \times S) + (w_{DD} \times DD) + (w_{DR} \times DR) \\ & + (w_{TWI} \times TWI) + (w_{MNDWI} \times MNDWI) + \left(w_{Roughness} \times Roughness\right) \\ & + (w_{Rainfall} \times Rainfall) + (w_{MFI} \times MFI) + (w_{NDVI} \times NDVI) \\ & + (w_{Soil} \times Soil) + (w_{STI} \times STI) \qquad (15.23)\end{aligned}$$

Where, E, S, DD, DR refer to the elevation, slope, drainage density, and distance from river, respectively. w_E, w_S, w_{DD}, w_{DR}, w_{TWI}, w_{MNDWI},

Table 15.4 Comparison matrix table for flood susceptibility

Matrix	*E*	*S*	*DD*	*DR*	*TWI*	*MNDWI*	*Rough*	*Rain*	*MFI*	*NDVI*	*Soil*	*STI*
E	**1.00**	3.00	4.00	4.00	5.00	6.00	7.00	8.00	8.00	9.00	9.00	9.00
S	0.33	**1.00**	3.00	4.00	4.00	5.00	6.00	7.00	7.00	8.00	8.00	8.00
DD	0.25	0.33	**1.00**	1.00	3.00	4.00	5.00	6.00	6.00	7.00	7.00	7.00
DR	0.25	0.25	1.00	**1.00**	3.00	4.00	5.00	6.00	6.00	7.00	7.00	7.00
TWI	0.20	0.25	0.33	0.33	**1.00**	3.00	4.00	5.00	5.00	6.00	6.00	6.00
MNDWI	0.17	0.20	0.25	0.25	0.33	**1.00**	3.00	4.00	4.00	5.00	5.00	5.00
Rough	0.14	0.17	0.20	0.20	0.25	0.33	**1.00**	3.00	3.00	4.00	4.00	4.00
Rain	0.13	0.14	0.17	0.17	0.20	0.25	0.33	**1.00**	1.00	3.00	3.00	3.00
MFI	0.13	0.14	0.17	0.17	0.20	0.25	0.33	1.00	**1.00**	3.00	3.00	3.00
NDVI	0.11	0.13	0.14	0.14	0.17	0.20	0.25	0.33	0.33	**1.00**	1.00	1.00
Soil	0.11	0.13	0.14	0.14	0.17	0.20	0.25	0.33	0.33	1.00	**1.00**	1.00
STI	0.11	0.13	0.14	0.14	0.17	0.20	0.25	0.33	0.33	1.00	1.00	**1.00**
SUM	2.93	5.86	10.55	11.55	17.48	24.43	32.42	42.00	42.00	55.00	55.00	55.00

Table 15.5 Normalized matrix table for flood susceptibility

Matrix	*E*	*S*	*DD*	*DR*	*TWI*	*MNDWI*	*Rough*	*Rain*	*MFI*	*NDVI*	*Soil*	*STI*	*Sum*	*Avg*
E	0.34	0.51	0.38	0.35	0.29	0.25	0.22	0.19	0.19	0.16	0.16	0.16	3.20	0.27
S	0.11	0.17	0.28	0.35	0.23	0.20	0.19	0.17	0.17	0.15	0.15	0.15	2.30	0.19
DD	0.09	0.06	0.09	0.09	0.17	0.16	0.15	0.14	0.14	0.13	0.13	0.13	1.48	0.12
DR	0.09	0.04	0.09	0.09	0.17	0.16	0.15	0.14	0.14	0.13	0.13	0.13	1.47	0.12
TWI	0.07	0.04	0.03	0.03	0.06	0.12	0.12	0.12	0.12	0.11	0.11	0.11	1.04	0.09
MNDWI	0.06	0.03	0.02	0.02	0.02	0.04	0.09	0.10	0.10	0.09	0.09	0.09	0.75	0.06
Rough	0.05	0.03	0.02	0.02	0.01	0.01	0.03	0.07	0.07	0.07	0.07	0.07	0.53	0.04
Rain	0.04	0.02	0.02	0.01	0.01	0.01	0.01	0.02	0.02	0.05	0.05	0.05	0.34	0.03
MFI	0.04	0.02	0.02	0.01	0.01	0.01	0.01	0.02	0.02	0.05	0.05	0.05	0.34	0.03
NDVI	0.04	0.02	0.01	0.01	0.01	0.01	0.01	0.01	0.01	0.02	0.02	0.02	0.18	0.02
Soil	0.04	0.02	0.01	0.01	0.01	0.01	0.01	0.01	0.01	0.02	0.02	0.02	0.18	0.02
STI	0.04	0.02	0.01	0.01	0.01	0.01	0.01	0.01	0.01	0.02	0.02	0.02	0.18	0.02

$w_{Roughness}$, $w_{Rainfall}$, w_{MFI}, w_{NDVI}, w_{Soil}, and w_{STI} are the corresponding weightage of the susceptibility indicators. The flood susceptibility map through AHP is available by arranging the indicators responsible for the flood hazard according to their compliance according to the comparison matrix table with spatiality in the ArcGIS environment (Fig. 15.7a). Classes are arranged from very low to very high by preparing all the maps according to the symbology tool. The maps show that most flood hazard areas are confined to the study area's east, north, and southeast, mainly beside the riparian zone. The elevation of the land is much lower than the area covered by the floodplain. Due to constant channel change in the meandering flow of the river, most of the old abandoned channels have remained on both sides of the river, with some oxbow lakes. Floodwaters easily inundate low-lying areas, marshy land and wetlands, causing waterlogging and submergence problems. As a result, the flood situation here quickly became catastrophic. From the point of view of regional hazard distribution, most of the region has been part of the moderate to very high flood susceptibility index (Table 15.6 and Fig. 15.6). From the administrative point of view, some blocks like Shantipur, Chakdah, Nabadwip, Krishnanagar 1 (Nadia), Purbasthali 1 (East Burdwan) and a large part of Kalna had been considered as flood-prone areas according to the flood susceptibility index of AHP. Since the hazard index is mainly dependent on physical parameters, low elevation and slope, high rainfall, excessive sedimentation is primarily responsible for flood hazard and disaster.

Table 15.6 Flood susceptibility level in different classification methods

FSI	*AHP*		*FR*		*SE*		*LR*	
	Area (^{2} km)	*Area (%)*	*Area (^{2} km)*	*Area (%)*	*Area (^{2} km)*	*Area (%)*	*Area (^{2} km)*	*Area (%)*
Very Low	137.15	12.07	57.90	4.76	0.27	0.02	7.12	0.70
Low	238.14	20.23	100.43	6.98	0.00	0.00	58.40	5.27
Moderate	367.25	29.02	213.41	17.77	12.45	1.02	188.80	15.67
High	290.77	24.01	454.89	40.33	490.31	39.93	450.56	40.52
Very High	165.55	14.68	372.10	31.86	695.83	59.03	393.41	37.84

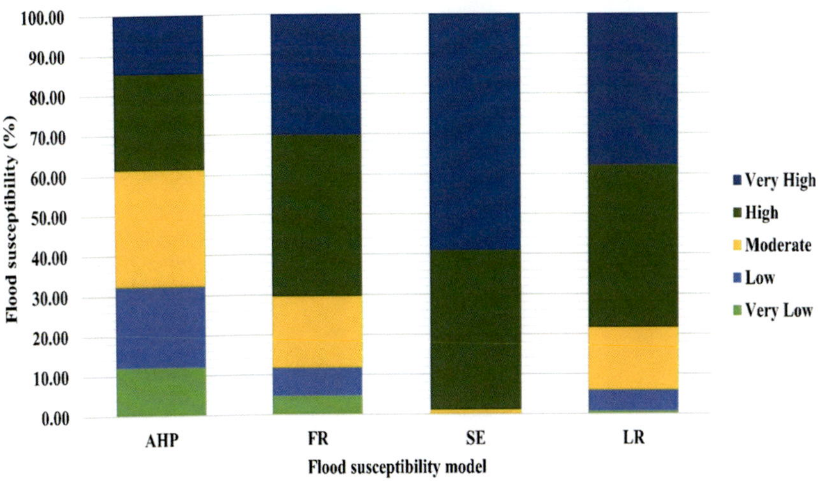

Fig. 15.6 Model-wise flood susceptibility percentage

15.3.2.2 Application of FR on Flood Susceptibility

BSA (Bivariate Statistical Analysis) has to be used to develop the frequency ratio (FR) model of flood arrival through different classified and independent components. The BSA factors examine the proportional probability of each component's presence and absence for the development of flooding conditions. The dual potential of the component-wise independent variable is measured based on the duration of the flood organization with the variable. It is possible to establish a relationship between the flood organization and the flood-supporting elements or variables

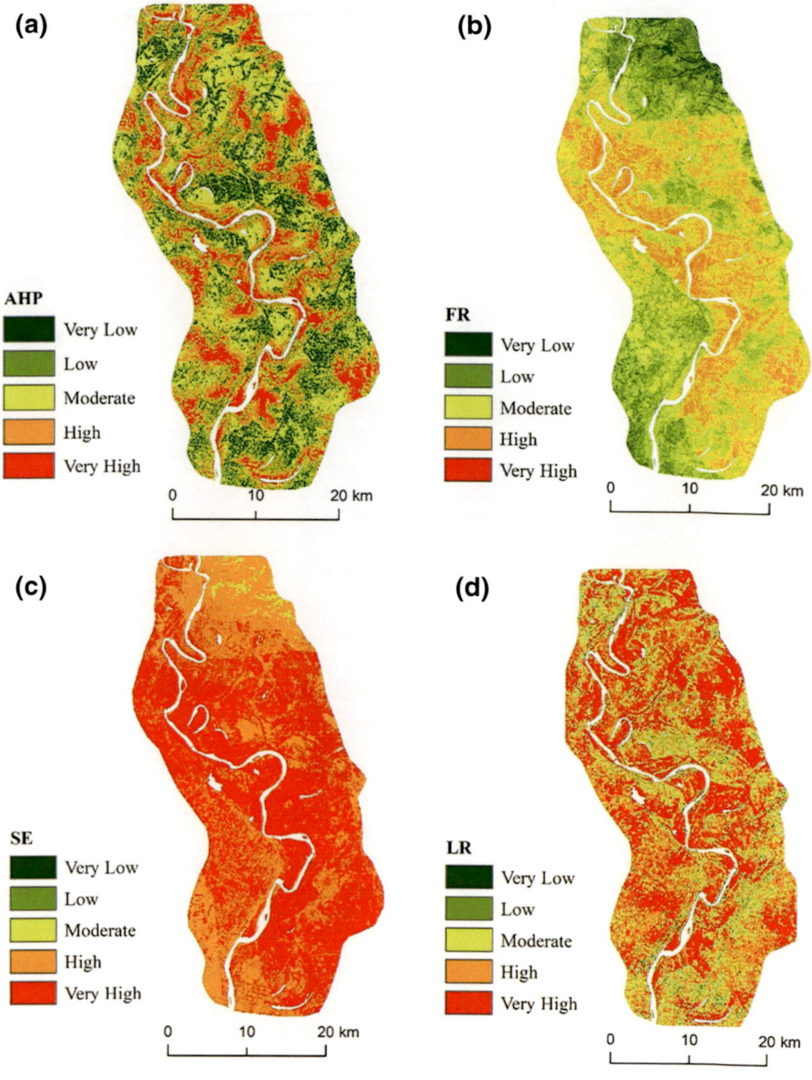

Fig. 15.7 Flood susceptibility map of the study area

through various statistical calculations. The FR method examines the role of subdivisions of each flood-forming element, comparing the proportion of the amount of flood organization with the specific area of a region. The FR model is developed by examining the contribution to flood organization according to the subdivision of each flood organizing component. The FR model measures the predictive ratio (PR) based on the relative proportions of the flood organization (Eqs. 15.4–15.5). In the present study area, the value of PR is much higher for elevation, soil, and STI. On the other hand, in the case of the slope, NDVI, MNDWI, roughness, drainage density, distance from the river, TWI, MFI, and average rainfall etc., the value of PR is much lower, and in case of all these elements, the value of PR is between 1 and 5.36. These relatively low-amount PRs are important in determining flood susceptibility, and the highest of these values indicates the highest flood susceptibility.

$$\begin{aligned} \text{FSI} = {} & (9.18 \times \text{E raster}) + (4.70 \times \text{S raster}) + (2.36 \times \text{DD raster}) \\ & + (2.64 \times \text{DR raster}) + (2.34 \times \text{TWI raster}) + (3.25 \times \text{MNDWI raster}) \\ & + \left(1 \times \text{Roughness raster}\right) + (5.36 \times \text{Rainfall raster}) + (4.81 \times \text{MFI raster}) \\ & + (4.79 \times \text{NDVI raster}) + (6.72 \times \text{Soil raster}) + (22.71 \times \text{STI raster}) \end{aligned} \tag{15.24}$$

The FR value (Table 15.2) of each flood conditioning factor shows that the low elevated range (6.01–13 m) has the highest FR value (1.142). Areas higher than 25 meters do not contribute to flooding in the study area. Low land slope areas are prone to flooding, so low slope range areas (0°–3°) have the highest FR values (1.769). The higher the drainage density, the greater the flood propensity. FR values are highest (1.257) in high-value DD areas (0.96–1.67). TWI with the highest range (12.52–15.45) is the most susceptible to flooding (Fig. 4), so this class's FR value is high (1.622). MNDWI also shows that the highest FR value (1.43) is at the maximum class range (0.14–0.31). FR analysis shows higher susceptibility to flooding in lower roughness ranges (0.11–0.31) (1.148). The table shows that rainfall with >1200 mm substantially impacts flood occurrence; similarly, the highest value of MFI and NDVI is also most susceptible to flood events. The region near the river with alluvial soil deposits has a greater correlation with flood occurrence than far land. After calculating the FR value of all classes of each variable, Flood susceptibility is mapped in the GIS interface using the formula mentioned earlier (Fig. 15.7b). Finally obtained map is categorized into five classes ranging from Very low (4.76%), low (6.98%), moderate (17.77%), high (40.33%)

and very high (31.86%), respectively. However, according to the FR model, the relatively low elevation plains along the river are considered more flood-prone (Table 15.6 and Fig. 15.6). The lower elevated study area on the left and right banks of the entire study area is naturally inundated very easily. The wetlands on both sides of the river along the middle of the entire region and the relatively low swamps east and southeast are rapidly inundated (Ranaghat 1 and 2 blocks of Nadia district and Kalyani sub-division). From the administrative point of view, it is realized that riparian portions of the blocks like Shantipur, Chakdah, Krishnanagar 1 (Nadia), Purbasthali 1 (East Burdwan) and a large part of Kalna, Balagarh (Hooghly) had been considered as flood-prone areas through FR model.

15.3.2.3 Application of SE on Flood Susceptibility

The SE model creates a realistic flood susceptibility model by eliminating the disparity between the auxiliary elements in flood organization. This model identifies the relationship between the flood control group and the flood management period and identifies the most significant component of the flood organization. The SE model is based on probability density, which can measure from the FR model, and for that reason, the SE model is referred to as an extension of the FR model. The SE model is determined based on their weightage by dividing the various components that help flood organizations into entropy values according to their values (Eq. 15.6–15.10). According to the SE model, almost every component except STI is more important than the components that help flood organization in the present study area. According to the results obtained from the SE model, factors such as elevation, slope, soil, average rainfall, MFI, NDVI, etc., have played an important role in determining flooding throughout the region (Table 15.2). As a result, these elements are expected to play a significant role in flood management throughout the region by the SE model (Constantine et al., 2011).

$$\begin{aligned}\text{FSI} = {} & (0.09 \times \text{E raster}) + (0.03 \times \text{S raster}) + (0.007 \times \text{DD raster}) \\ & + (0.012 \times \text{DR raster}) + (0.008 \times \text{TWI raster}) \\ & + (0.014 \times \text{MNDWI raster}) + \left(0.0014 \times \text{Roughness raster}\right) \\ & + (0.037 \times \text{Rainfall raster}) + (0.04 \times \text{MFI raster}) \\ & + (0.037 \times \text{NDVI raster}) + (0.061 \times \text{Soil raster}) \\ & + (0.65 \times \text{STI raster}) \end{aligned} \tag{15.25}$$

After calculating the weightage value of each variable through the SE method, Flood susceptibility is mapped in the GIS interface using the SE model's formula (Fig. 15.7c). Finally obtained map is categorized into five classes ranging from Very low (0.02%), low (0%), moderate (1.02%), high (39.93%), and very high (59.03%), respectively. Due to the important role of elevation and slope in the weightage of components of the SE model, the relatively flat areas of the entire floodplain topography have been identified as flood-prone (Table 15.6 and Fig. 15.6). According to the SE model, the region is likely to be flooded, except for relatively high north-east and south-west sections. All the areas except Krishnanagar 1 and 2 blocks and some parts of the Balagarh block belong to the flood susceptibility zone as per the SE model.

15.3.2.4 *Application of LR on Flood Susceptibility*

This study uses R Studio software to perform logistic regression analysis. LR results show that drainage density, distance from the river, elevation, slope, and TWI are important factors in flood susceptibility modelling as their significant value is less than 0.5. Other factors are statistically insignificant in the case of model building as their signifiable value is more than 0.5 (Shafapour Tehrany et al., 2017). The positive value of a decision variable indicates the greater contribution of this indicator in flood events and vice versa. Based on the logistical coefficient, the probability of flooding is calculated using the formula to develop flood susceptibility mapping in the ArcGIS interface (Table 15.2). In this regression, z value was obtained using the following formula,

$$\begin{aligned} z = & -21.453 + (-23.508 \times \text{E}) + (-9.959 \times \text{S}) + (-5.892 \times \text{DD}) \\ & + (-11.097 \times \text{DR}) + (-3.761 \times \text{TWI}) + (37.006 \times \text{MNDWI}) \\ & + \left(3.319 \times \text{Roughness}\right) + (-3.258 \times \text{Rainfall}) \\ & + (45.592 \times \text{MFI}) + (19.198 \times \text{NDVI}) + (2.968 \times \text{Soil}) \\ & + (-1.20 \times \text{STI}) \end{aligned} \tag{15.26}$$

After measuring the coefficient of logistic regression, flood susceptibility has been determined according to the formula of the LR model at the interface of Arc GIS (Fig. 15.7d). Finally obtained map is categorized into five classes ranging from Very low (0.70%), low (5.27%), moderate (15.67%), high (40.52%), and very high (37.84%), respectively (Table 15.6 and Fig. 15.6). Distance from river, elevation, slope and TWI etc., are important factors determining flood like other models. According to the LR model, floodwaters inundated the low-lying areas of the entire study area. The vast upper parts of Krishnanagar 1 and 2 and Ranaghat 1 and 2 blocks are also likely to be inundated by floods, analysed through the LR model.

15.3.3 *Performance Evaluation of Flood Susceptibility Models*

In this study, a detailed statistical analysis among four classifiers has been done, showing (Table 15.6) that all the models are fitted with high goodness fit. AHP has the highest sensitivity (0.945), which indicates that 94.5% pixel was correctly classified of the flood as a flood event, followed by LR (0.941), FR (0.934), and SE (0.913) (Table 15.7). However, LR has the highest specificity (0.978), followed by AHP (0.966), FR (0.934), and SE (0.928). The result shows that AHP yields the highest accuracy (0.951). It depicts that 95.1% pixel of flood and non-flood was correctly classified as flood and non-flood events, followed by LR (0.944), FR (0.920), and SE (0.907). Kappa is almost the same in both AHP (0.864) and LR (0.856), indicating a good proportion of model and training data agreement. For FR and SE, the k value is lower (0.841 and 0.812). Based on the result, AHP is the best method for training data, followed by LR, FR, and SE.

The predictive capability of the four classifiers is assessed by the testing data shown in Table 15.7. Highest sensitivity has shown in LR (0.984),

Table 15.7 Evaluation of model performance based on selected parameters

Model	*Stage*	*Parameters*						
		Sensitivity	*Specificity*	*PPV*	*NPV*	*Accuracy*	*F-Score*	*Kappa*
AHP	Training	0.945	0.966	0.998	0.754	0.951	0.943	0.864
	Testing	0.977	0.971	0.980	0.690	0.928	0.947	0.854
FR	Training	0.934	0.934	0.900	0.710	0.920	0.927	0.841
	Testing	0.951	0.941	0.963	0.789	0.912	0.968	0.839
SE	Training	0.913	0.928	0.921	0.732	0.907	0.927	0.812
	Testing	0.924	0.931	0.900	0.765	0.897	0.946	0.832
LR	Training	0.941	0.978	0.987	0.767	0.944	0.941	0.856
	Testing	0.984	0.981	0.991	0.712	0.931	0.975	0.846

followed by AHP (0.977), FR (0.951) and SE (0.924). In terms of specificity, LR (0.981) and AHP (0.971) acquire the highest value, followed by FR (0.941) and SE (0.931). LR obtained the highest level of accuracy (0.931), followed by AHP (0.928) and FR (0.912). SE yields the lowest accuracy among the four classifiers (0.897). In the case of kappa, the higher to lower value was associated with AHP (0.854), LR (0.846), FR (0.839), and SE (0.832), respectively. Since the model's statistical values are very close, it is difficult to determine which method performs best. Therefore, the local government should adopt the map obtained from these classifiers to reduce possible damage due to flood events.

The accuracy, quality, and predictability of four individual models (AHP, FR, SE, LR) were calibrated and implemented for the flood susceptibility map. So, first of all, it is necessary to verify the validity and prediction of these four indexes based on actual and false-positive rates. All four models are validated based on their predictive ability (Fig. 15.8). Success rates are observed using training datasets (650 points), which depend on flood locations in 65% of the region. The prediction accuracy depends on the rest of the data set that was not used in training (350 points) and guided the flood situation in 35% of the region. The AUROC was used to evaluate the performance of each model. All the AUROC obtained from the individual model was overloaded into a single frame for visual interpretation. Furthermore, the fact that the AUROC value close to 1 shows that the model is reliable. Result reveals that the best performance among all the individual models belongs to AHP (0.897), followed by LR (0.889), FR (0.647) and SE (0.638), respectively. In addition to the AUROC, the accuracy of the models is shown in the Taylor diagram (Fig. 15.9). All the statistical analyses in a single frame are shown with the help of the Taylor diagram. It is seen that the performance, accessibility, and accuracy of AHP are better here, like AUROC, comparing the accuracy of the models with the correlation coefficient and standard deviation.

15.3.4 Validation of Flood Inundation Through Real Data

Although flood susceptibility is determined and validated according to bivariate and multivariate statistical methods, the actual validation of the models is possible with the actual flood picture of the entire region. The Irrigation and Waterways Department of the Government of West Bengal measures the rising water level (from May to October) of rivers in this

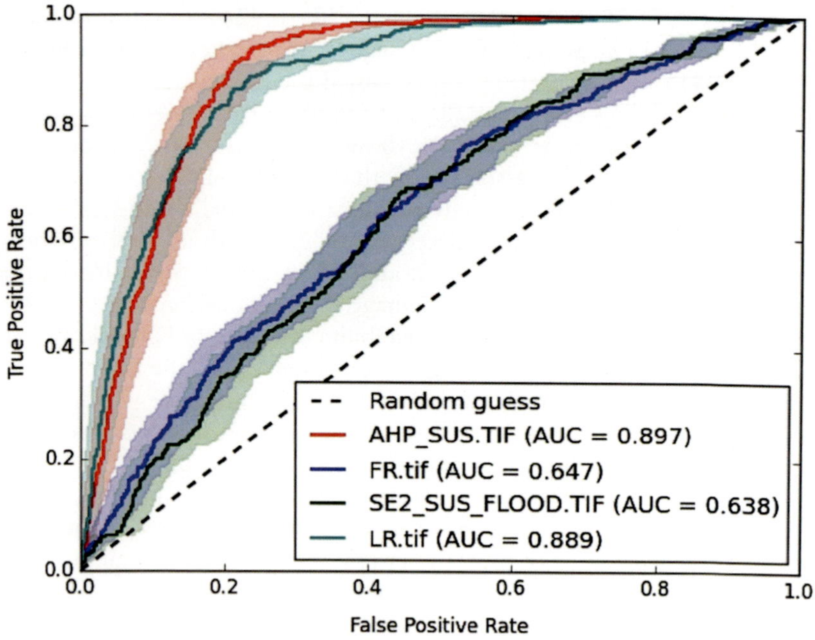

Fig. 15.8 Performance analysis of four models in study area

region during monsoons through their gauge stations. Based on the information received from their department, the elevated water level has been compared with the elevation of the land upstream of the entire region, and numerical measurements of the flooded area have been made. Data obtained from the field survey, Kolkata Port Trust and Irrigation and Waterways Department of the Government of West Bengal, Swarupganj near Nabadwip, have been taken as RL 10.442 m (Fig. 15.10). Based on that RL, the flooded area is measured based on the change in water level per meter compared to the normal water level. 7 mouzas in Krishnanagar 1 block will be completely flooded, and 3 mouzas in Shantipur, Ranaghat 1 and Nabadwip will be partially flooded if the water level rises 2 m above normal. With the rise in water level of 4 m, vast areas of Krishnanagar 1 (11 mouzas), Nabadwip (6 mouzas), Shantipur and Ranaghat 1 (3 mouzas) and Chakdah (2 mouzas) blocks were completely inundated. 16 mouzas of Krishnanagar 1 block, 9 mouzas of Nabadwip block,

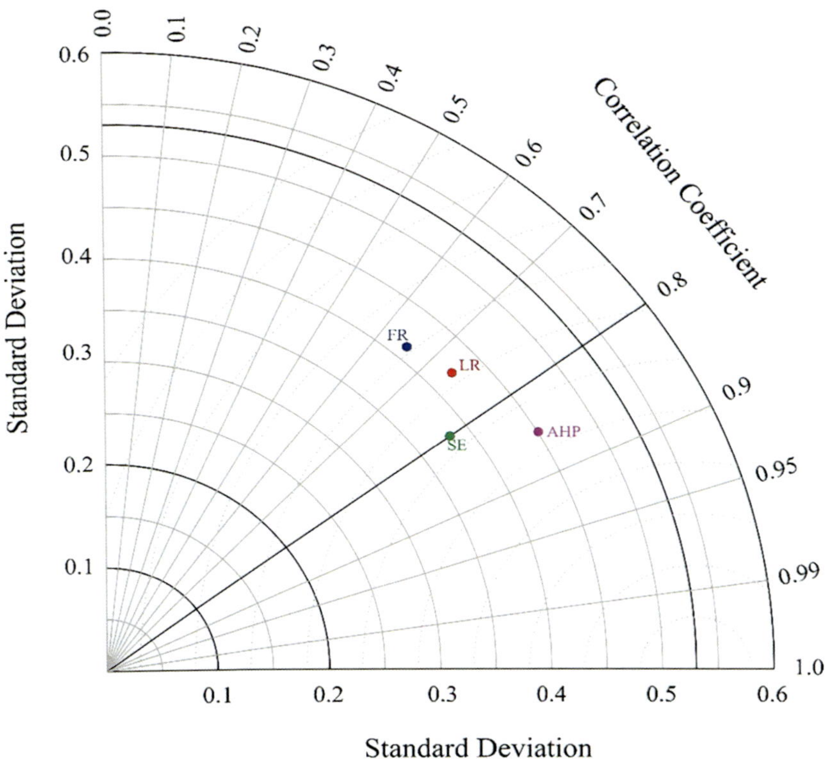

Fig. 15.9 Performance analysis of four models through Taylor diagram

4 mouzas of Purbasthali 1 block, 4 mouzas of Shantipur block, 5 mouzas of Ranaghat 1 block, 4 mouzas of Chakdaha block and 4 mouzas of Balagarh block is completely flooded. About 15,000 people in all the blocks were severely affected. As the water level rose to 8 meters, the floodwaters became more destructive. This type of flooding is not commonly seen. However, suppose the water level increases by 8 meters in future. In that case, 22 mouzas of Krishnanagar 1 block, 14 mouzas of Nabadwip block, 9 mouzas of Purbasthali 1 block, 12 mouzas of Shantipur block, 11 mouzas of Ranaghat 1 block, 9 mouzas of Chakdaha block and 7 mouzas of Balagarh block will be flooded. About 30–40 thousand people will be affected.

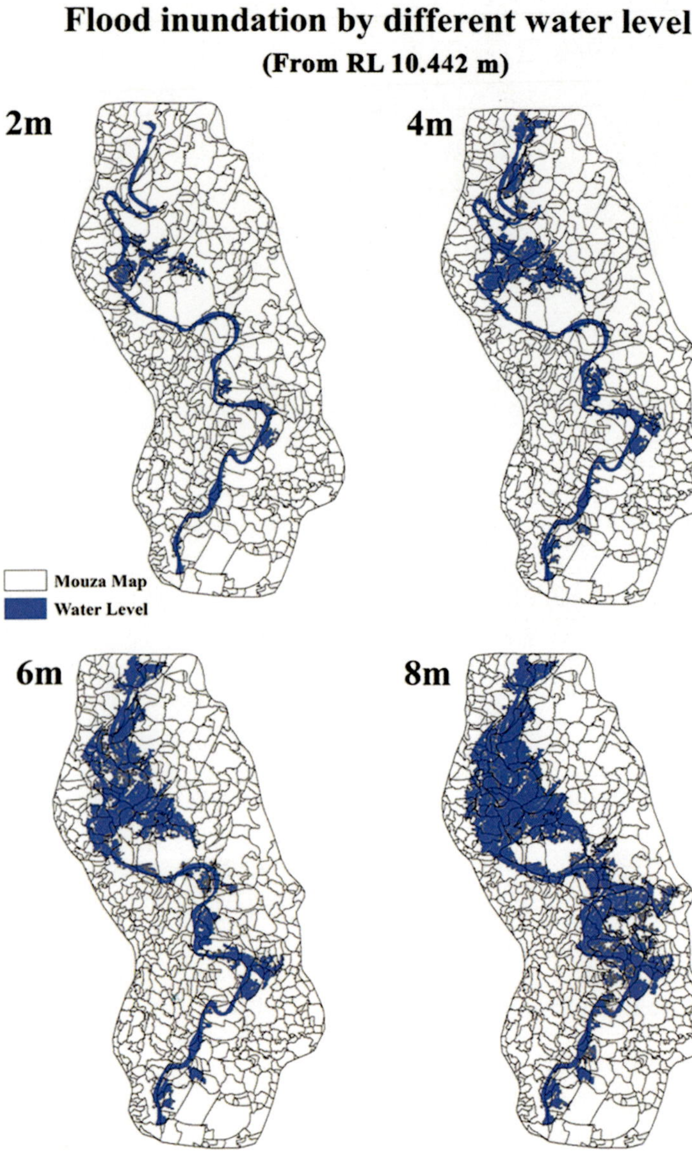

Fig. 15.10 Flood inundation area by different water levels

Analysing the flood models shows that the AHP and LR values are much better than other models in terms of sensitivity and accuracy. These models have been discussed in several sections according to flood susceptibility. These sections highlight flood and disaster areas along the entire region. A closer look at the flood inundation maps obtained due to rising water levels reveals that the extent to which 2 and 4 meters of rising water levels will inundate areas is equivalent to the AHP flood susceptibility model. Again, an increase of 6 m of water level has resulted in the total amount of flooded area being equal to the size of the FR model and the amount of flooded area under 8 m under the LR model. The mouzas flooded by these models are the same mouzas that are inundated as the water level rises. Also, the AHP and FR models introduce local stakeholders to the current flood situation, and in that case, the flood situation can be seen in the 2-to-6-meter water level rise. The LR model highlights the overall spatial change in the future flood situation in the Bhagirathi-Hooghly floodplain region. The 8-meter water level rise is not seen at present, but if the flood situation worsens in the future, the spatial pattern of the huge area inundated by the flood is very similar to the LR model. It turns out that there is a positive correlation of mousse-based inundation with the increase in the water table with different flood susceptibility methods modelled.

15.4 Conclusion and Recommendation

From the earliest times of civilization, the floodplain area of the river was easily crowded with food, clothing and shelter. By declining navigability and carrying capacity of the river in this region, the excess water overtopping the riverbank in monsoon season results in flood. It has resulted in flooding of the plains along the river, resulting in loss of life, home and property, leading to catastrophe. Therefore, using the flood susceptibility model to know about the temporary and spatial changes of floods in a floodplain area becomes useful for the people's livelihood in that area. The present article highlights the flood situation in the Navadwip-Kalyani sub-catchment basin beside the Bhagirathi-Hooghly River in West Bengal and the need for a flood susceptibility model to determine the nature and extent of floods, which is expected to play an important role in government policymaking. In this study, 4 bivariate and multi-variate flood susceptibility methods, i.e., AHP, FR, SE, and LR, have been used to determine the flood susceptibility zonation of the Nabadwip-Kalyani sub-catchment basin. Based on 12 flood control elements, these flood susceptibility models divide the entire region into 5 flood zone categories. According to this category, the overall flood situation in the region has been best highlighted through the AHP model. While the rest of the models also accurately cover flood situations, the SE model is not considered suitable for determining flood susceptibility in this region. The performance of each model is assessed by using several statistical techniques integrated with the R and GIS interface. Although the AHP model is sensitive enough to determine the true picture of flooding in the above sub-catchment basin, according to a survey based on training and testing data, there is an opportunity to explore the future flood potential of this region through the LR model. AUROC curve evaluated through the performance of four individual models. The results show that AHP provides the best performance in AUC and training datasets followed by LR. Also, for the presence of gauge stations, the water level information available from the Irrigation and Waterways Department, Government of West Bengal, has been used to create an inundation map and compare it with the FSI models. Accurate measurements of the amount of flooding caused by an increase in the water level by 2–4 meters during floods can be made with the help of flood susceptibility models such as AHP and FR. In future catastrophic floods, the water level will rise to around 8 meters, and the spatial arrangement and measurement of the flooded area at that time is analysed with the help of the LR model. The output of each model shows that the low-lying part of the Nabadwip-Kalyani sub-catchment basin are most susceptible to flood. For all these reasons, during the monsoon season, the river swells quickly and floods its banks. Based on regionalism, more than 60% of Nabadwip, Krishnanagar 1, Shantipur, Chakdaha block of Nadia district and Purbasthali 1 block of East Burdwan district are considered as flood-prone areas.

- The maps on flood susceptibility along the Nabadwip-Kalyani sub-basin beside the Bhagirathi-Hooghly River are expected to play a significant role in future flood management and control. As flood susceptibility methods are discussed based on 4 different models, it is expected that government policy on flood control will determine according to the AHP model based on success, which will benefit the local stakeholders.

- There is an urgent need to construct an artificial embankment in the mouzas along the river to control the flood situation. According to the RRZ, settlements 500 meters away from the river are less likely to be damaged by floods. However, there are doubts about how widespread this policy will be among the riparian people in this densely populated region.
- As the navigability of the river decreases and floods occur, there is an excellent need for dredging in the river. In addition, illegal and unscientific sand mining from the river bed should be restricted. There is a need to create awareness about floods, prevent them and build more healthy flood shelters.
- This study combines geomorphological and hydrological attributes to assume flood hazard intensity. It integrates socio-cultural, demographic, structural factors to determine the magnitude of flood susceptibility. Thus, priority areas are recognized for efficient management and proper measures. The local government needs to have flood risk information to implement flood management projects and inundation management easily. It is useful to take appropriate measures to mitigate the negative effects of food events.

References

Arabameri, A., Pradhan, B., Rezaei, K., Sohrabi, M., & Kalantari, Z. (2019). GIS based landslide susceptibility mapping using numerical risk factor bivariate model and its ensemble with linear multivariate regression and boosted regression tree algorithms. *Journal of Mountain Science, 16*, 595–618. https://doi.org/10.1007/s11629-018-5168-y

Arora, A., Pandey, M., Siddiqui, M. A., Hong, H., & Mishra, V. N. (2019). Spatial flood susceptibility prediction in Middle Ganga Plain: Comparison of frequency ratio and Shannon's entropy models. *Geocarto International, 36*(18), 2085–2116. https://doi.org/10.1080/10106049.2019.1687594

Baky, M. A. A., Islam, M., & Paul, S. (2020). Flood hazard, vulnerability and risk assessment for different land use classes using a flow model. *Earth Systems and Environment, 4*, 225–244. https://doi.org/10.1007/s41748-019-00141-w

Balica, S. F., Popescu, I., Beevers, L., & Wright, N. G. (2013). Parametric and physically based modelling techniques for flood risk and vulnerability assessment: A comparison. *Environmental Modelling and Software, 4*, 84–92. https://doi.org/10.1016/j.envsoft.2012.11.002

Bednarik, M., Yilmaz, I., & Marschalko, M. (2012). Landslide hazard and risk assessment: A case study from the Hlohovec–Sered' landslide area in south-west Slovakia. *Natural Hazards, 64*, 547–575. https://doi.org/10.1007/s11069-012-0257-7

Cao, Y., Jia, H., Xiong, J., Cheng, W., Li, K., Pang, Q., & Yong, Z. (2020). Flash flood susceptibility assessment based on geodetector, certainty factor, and logistic regression analyses in Fujian Province, China. *ISPRS International Journal of Geo-Information, 9*(12), 748. https://doi.org/10.3390/ijgi9120748

Chakraborty, S., & Mukhopadhyay, S. (2019). Assessing flood risk using analytical hierarchy process (AHP) and geographical information system (GIS): Application in Coochbehar district of West Bengal, India. *Natural Hazards, 99*, 247–274. https://doi.org/10.1007/s11069-019-03737-7

Chen, W., Hong, H., Li, S., et al. (2019). Flood susceptibility modelling using novel hybrid approach of reduced-error pruning trees with random subspace and random subspace ensembles. *Journal of Hydrology, 575*, 864–873. https://doi.org/10.1016/j.jhydrol.2019.05.089

Constantin, M., Bednarik, M., Jurchescu, M. C., & Vlaicu, M. (2011). Landslide susceptibility assessment using the bivariate statistical analysis and the index of entropy in the Sibiciu Basin (Romania). *Environment and Earth Science, 63*, 397–406. https://doi.org/10.1007/s12665-010-0724-y

Costache, R., Arabameri, A., Elkhrachy, I., Ghorbanzadeh, O., & Pham, Q. B. (2021). Detection of areas prone to flood risk using state-of-the-art machine learning models. *Geomatics, Natural Hazards and Risk, 12*, 1488–1507. https://doi.org/10.1080/19475705.2021.1920480

Dandapat, K., & Panda, G. K. (2017). Flood vulnerability analysis and risk assessment using analytical hierarchy process. *Modeling Earth Systems and Environment, 3*, 1627–1646. https://doi.org/10.1007/s40808-017-0388-7

Das, G., & Lepcha, K. (2019). Application of logistic regression (LR) and frequency ratio (FR) models for landslide susceptibility mapping in Relli Khola river basin of Darjeeling Himalaya, India. *SN Applied Sciences, 1*, 1453. https://doi.org/10.1007/s42452-019-1499-8

Das, S. (2018). Geographic information system and AHP-based flood hazard zonation of Vaitarna basin, Maharashtra, India. *Arabian Journal of Geosciences, 11*, 576. https://doi.org/10.1007/s12517-018-3933-4

Dhekra, S., Lahcen, Z., Salma, H., et al. (2020). GIS-based MCDM – AHP modeling for flood susceptibility mapping of arid areas, south eastern Tunisia. *Geocarto International, 35*(9), 991–1017. https://doi.org/10.1080/10106049.2019.1566405

Fustos, I., Abarca-del-Rio, R., Ávila, A., & Orrego, R. (2017). A simple logistic model to understand the occurrence of flood events into the Biobío River

Basin in central Chile. *Journal of Flood Risk Management, 10*(1), 17–29. https://doi.org/10.1111/jfr3.12131

Ghosh, A., & Kar, S. K. (2018). Application of analytical hierarchy process (AHP) for flood risk assessment: A case study in Malda district of West Bengal, India. *Natural Hazards, 94*, 349–368. https://doi.org/10.1007/s11069-018-3392-y

Ghosh, A., Roy, M. B., & Roy, P. K. (2020). Estimation and prediction of the oscillation pattern of meandering geometry in a sub-catchment basin of Bhagirathi-Hooghly river, West Bengal, India. *SN Applied Sciences, 2*, 1497. https://doi.org/10.1007/s42452-020-03275-z

Goumrasa, A., Guendouz, M., Guettouche, M. S., & Belaroui, A. (2021). Flood hazard susceptibility assessment in Chiffa wadi watershed and along the first section of Algeria North-South highway using GIS and AHP method. *Applied Geomatics, 13*, 565–585. https://doi.org/10.1007/s12518-021-00381-4

Hammami, S., Zouhri, L., Souissi, D., et al. (2019). Application of the GIS based multi-criteria decision analysis and analytical hierarchy process (AHP) in the flood susceptibility mapping (Tunisia). *Arabian Journal of Geosciences, 12*, 653. https://doi.org/10.1007/s12517-019-4754-9

Hong, H., Pourghasemi, H. R., & Pourtaghi, Z. S. (2016). Landslide susceptibility assessment in Lianhua County (China): A comparison between a random forest data mining technique and bivariate and multivariate statistical models. *Geomorphology, 259*, 105–118. https://doi.org/10.1016/j.geomorph.2016.02.012

Hosmer, D. W., Lemeshow, S., & Sturdivant, R. X. (2013). *Applied logistic regression*. Wiley.

Jha, V. C., & Bairagya, H. (2013). Flood and flood plains of West Bengal, India: A comparative analysis. *Revista Eletro^nica Geoaraguaia. Barra do Garc¸as-MT. Edic¸a˜o Especial*, 1–10.

Kambalimath, S., & Deka, P. C. (2020). A basic review of fuzzy logic applications in hydrology and water resources. *Applied Water Science, 10*, 191. https://doi.org/10.1007/s13201-020-01276-2

Kazakis, N., Kougias, I., & Patsialis, T. (2015). Assessment of flood hazard areas at a regional scale using an index-based approach and Analytical Hierarchy Process: Application in Rhodope–Evrosregion, Greece. *Science of the Total Environment, 538*, 555–563. https://doi.org/10.1016/j.scitotenv.2015.08.055

Khosravi, K., Nohani, E., Maroufinia, E., et al. (2016). A GIS-based flood susceptibility assessment and its mapping in Iran: A comparison between frequency ratio and weights-of-evidence bivariate statistical models with multi-criteria decision-making technique. *Natural Hazards, 83*, 947–987. https://doi.org/10.1007/s11069-016-2357-2

Khosravi, K., Pham, B. T., Chapi, K., Shirzadi, A., et al. (2021). A comparative assessment of decision trees algorithms for flash flood susceptibility modeling at Haraz Watershed, Northern Iran. *Science of the Total Environment, 627*, 744–755. https://doi.org/10.1016/j.scitotenv.2018.01.266

Kouhpeima, A., Feyznia, S., Ahmadi, H., & Moghadamnia, A. R. (2017). Landslide susceptibility mapping using logistic regression analysis in Latyan catchment. *Desert, 22*(1), 85–95. https://doi.org/10.22059/jdesert.2017.62181

Lee, S., & Talib, J. A. (2005). Probabilistic landslide susceptibility and factor effect analysis. *Journal of Environmental Geology, 47*, 982–990. https://doi.org/10.1007/s00254-005-1228-z

Luu, C., Von Meding, J., & Kanjanabootra, S. (2018). Assessing food hazard using food marks and analytic hierarchy process approach: A case study for the 2013 food event in Quang Nam Vietnam. *Natural Hazards, 90*, 1031–1050. https://doi.org/10.1007/s11069-017-3083-0

Milaghardan, A. H., Abbaspour, R. A., & Khalesian, M. (2020). Evaluation of the effects of uncertainty on the predictions of landslide occurrences using the Shannon entropy theory and Dempster-Shafer theory. *Natural Hazards, 100*, 49–67. https://doi.org/10.1007/s11069-019-03798-8

Mujiburrehman, K. (2015). Preparation of flood inundation map in Ganga River at Farakka Bridge, Malda, West Bengal India. *International Journal of Research in Geography, 1*(1), 1–17.

Namara, W. G., Damisse, T. A., & Tufa, F.G. (2021). Application of HEC-RAS and HEC-GeoRAS model for Flood Inundation Mapping, the case of Awash Bello Flood Plain, Upper Awash River Basin, Oromiya Regional State, Ethiopia. *Modeling Earth Systems Environment, 8*(2). https://doi.org/10.1007/s40808-021-01166-9

Nasir, M. J., Iqbal, J., & Ahmad, W. (2020). Flash flood risk modeling of swat river sub-watershed: A comparative analysis of morphometric ranking approach and El-Shamy approach. *Arabian Journal of Geosciences, 13*, 1082. https://doi.org/10.1007/s12517-020-06064-5

Omid Rahmati, O., Pourghasemi, H. R., & Zeinivand, H. (2016). Flood susceptibility mapping using frequency ratio and weights-of-evidence models in the Golastan Province, Iran. *Geocarto International, 31*(1), 42–70. https://doi.org/10.1080/10106049.2015.1041559

Onan, A. (2015). A fuzzy-rough nearest neighbour classifier combined with consistency-based subset evaluation and instance selection for automated diagnosis of breast cancer. *Expert Systems with Applications, 42*(20), 6844–6852. https://doi.org/10.1016/j.eswa.2015.05.006

Patel, D. P., Ramirez, J. A., Srivastava, P. K., Bray, M., & Han, D. (2017). Assessment of flood inundation mapping of Surat city by coupled 1D/2D hydrodynamic modeling: A case application of the new HEC-RAS 5. *Natural Hazards, 89*, 93–130. https://doi.org/10.1007/s11069-017-2956-6

Perera, E. D. P., & Lahat, L. (2015). Fuzzy logic-based flood forecasting model for the Kelantan River basin, Malaysia. *Journal of Hydro-environment Research, 9*(4), 542–553. https://doi.org/10.1016/j.jher.2014.12.001

Pham, B. T., Phong, T. V., Nguyen, H. D., Qi, C., et al. (2020). A comparative study of Kernel Logistic Regression, Radial Basis Function Classifier, Multinomial Naïve Bayes, and Logistic Model Tree for Flash Flood Susceptibility Mapping. *Water, 12*(1), 239. https://doi.org/10.3390/w12010239

Pourghasemi, H. R., & Kerle, N. (2016). Random forests and evidential belief function-based landslide susceptibility assessment in Western Mazandaran Province, Iran. *Environmental Earth Sciences, 75*, 185. https://doi.org/10.1007/s12665-015-4950-1

Pourghasemi, H. R., Yousef, S., Kornejady, A., & Cerdà, A. (2017). Performance assessment of individual and ensemble data-mining techniques for gully erosion modeling. *Science of the Total Environment, 609*, 764–775. https://doi.org/10.1016/j.scitotenv.2017.07.198

Rahman, M., Ningsheng, C., Islam, M. M., et al. (2019). Flood susceptibility assessment in bangladesh using machine learning and multi-criteria decision analysis. *Earth Systems and Environment, 3*, 585–601. https://doi.org/10.1007/s41748-019-00123-y

Rahmati, O., Zeinivand, H., & Besharat, M. (2015). Flood hazard zoning in Yasooj region, Iran, using GIS and multi-criteria decision analysis. *Geomatics, Natural Hazards and Risk.* https://doi.org/10.1080/19475705.2015.1045043

Regmi, A. D., Yoshida, K., Pourghasemi, H. R., et al. (2014). Landslide susceptibility mapping along Bhalubang-Shiwapur area of mid-western Nepal using frequency ratio and conditional probability models. *Journal of Mountain Science, 11*, 1266–1285. https://doi.org/10.1007/s11629-013-2847-6

Saaty, T. L. (1977). A scaling method for priorities in hierarchical structures. *Journal of Mathematical Psychology, 15*(3), 234–281. https://doi.org/10.1016/0022-2496(77)90033-5

Saaty, T. L. (1990). How to make a decision: The analytic hierarchy process. *European Journal of Operational Research, 48*(1), 9–26. https://doi.org/10.1016/0377-2217(90)90057-I

Saaty, T. L. (2008). Decision making with the analytic hierarchy process. *International Journal of Services Sciences, 1*, 83–98. https://doi.org/10.1504/IJSSCI.2008.017590

Sahoo, S. N., & Sreeja, P. (2015). Development of flood inundation maps and quantification of flood risk in an Urban catchment of Brahmaputra River. *ASCE-ASME Journal of Risk and Uncertainty in Engineering Systems, Part*

A: Civil Engineering, 3, A4015001. https://doi.org/10.1061/AJRUA6.0000822

Samantaray, S., Sahoo, A., & Agnihotri, A. (2021). Assessment of Flood Frequency using Statistical and Hybrid Neural Network Method: Mahanadi River Basin, India. *Journal of the Geological Society of India, 97*, 867–880. https://doi.org/10.1007/s12594-021-1785-0

Samanta, S., Pal, D. K., & Palsamanta, B. (2018). Flood susceptibility analysis through remote sensing, GIS and frequency ratio model. *Applied Water Science, 8*, 66–89. https://doi.org/10.1007/s13201-018-0710-1

Sanyal, J., & Lu, X. X. (2005). Remote sensing and GIS-based flood vulnerability assessment of human settlements: A case study of Gangetic West Bengal, India. *Hydrological Processes, 19*, 3699–3716. https://doi.org/10.1002/hyp.5852

Sanyal, J., & Lu, X. X. (2006). GIS-based flood hazard mapping at different administrative scales: A case study in Gangetic West Bengal, India. *Singapore Journal of Tropical Geography, 27*, 207–220. https://doi.org/10.1111/j.1467-9493.2006.00254.x

Sarkar, D., & Mondal, P. (2020). Flood vulnerability mapping using frequency ratio (FR) model: A case study on Kulik river basin, Indo-Bangladesh Barind Region. *Applied Water Science, 10*, 17. https://doi.org/10.1007/s13201-019-1102-x

Schicker, R., & Moon, V. (2012). Comparison of bivariate and multivariate statistical approaches in landslide susceptibility mapping at a regional scale. *Geomorphology, 161–162*, 40–57. https://doi.org/10.1016/j.geomorph.2012.03.036

Shafapour Tehrany, M., Shabani, F., Jebur, M. N., et al. (2017). GIS-based spatial prediction of flood prone areas using standalone frequency ratio, logistic regression, weight of evidence and their ensemble techniques. *Geomatics, Natural Hazards and Risk, 8*, 1538–1561. https://doi.org/10.1080/19475705.2017.1362038

Shirzadi, A., Chapi, K., & Shahabi, H. (2017). Rock fall susceptibility assessment along a mountainous road: An evaluation of bivariate statistic, analytical hierarchy process and frequency ratio. *Environment and Earth Science, 76*, 152. https://doi.org/10.1007/s12665-017-6471-6

Taylor, K. E. (2001). Summarizing multiple aspects of model performance in a single diagram. *Journal of Geophysical Research, 106*, 7183–7192. https://doi.org/10.1029/2000JD900719

Tehrany, M. S., Lee, M.-J., Pradhan, B., Jebur, M. N., & Lee, S. (2014). Flood susceptibility mapping using integrated bivariate and multivariate statistical models. *Environment and Earth Science, 72*, 4001–4015. https://doi.org/10.1007/s12665-014-3289-3

Tehrany, M. S., Shabani, F., Jebur, M. N., Hong, H., Chen, W., & Xie, X. (2017). GIS-based spatial prediction of flood prone areas using standalone frequency ratio, logistic regression, weight of evidence and their ensemble techniques. *Geomatics, Natural Hazards and Risk, 8*(2), 1538–1561. https://doi.org/10.1080/19475705.2017.1362038

Ul Moazzam, M. F., Lee, B. G., Ur Rahman, A., Farid, N., & Rahman, G. (2020). Spatio-statistical analysis of flood susceptibility assessment using bivariate model in the floodplain of River Swat, District Charsadda, Pakistan. *Journal of Geoscience and Environment Protection, 8*, 159–175. https://doi.org/10.4236/gep.2020.85010

Vignesh, K. S., Anandakumar, I., Ranjan, R., et al. (2021). Flood vulnerability assessment using an integrated approach of multi-criteria decision-making model and geospatial techniques. *Modeling Earth Systems and Environment, 7*, 767–781. https://doi.org/10.1007/s40808-020-00997-2

Vojtek, M., Vojteková, J., Costache, R., Pham, Q. B., et al. (2021). Comparison of multi-criteria-analytical hierarchy process and machine learning- boosted tree models for regional flood susceptibility mapping: A case study from Slovakia. *Geomatics, Natural Hazardsand Risk, 12*(1), 1153–1180. https://doi.org/10.1080/19475705.2021.1912835

Wubalem, A., & Meten, M. (2020). Landslide susceptibility mapping using information value and logistic regression models in Goncha Siso Eneses area, northwestern Ethiopia. *SN Applied Sciences, 2*, 807. https://doi.org/10.1007/s42452-020-2563-0

Wubalem, A., Tesfaw, G., Dawit, Z., Getahun, B., Mekuria, T., & Jothimani, M. (2021). Comparison of statistical and analytical hierarchy process methods on flood susceptibility mapping: In a case study of the Lake Tana sub-basin in northwestern Ethiopia. *Open Geosciences, 13*(1), 1668–1688. https://doi.org/10.1515/geo-2020-0329

Yufeng, S., & Feng Xiang, J. (2009). Landslide stability analysis based on generalized information entropy. In *Proceedings of the 2009 International Conference on Environmental Science and Information Application Technology, Wuhan, China, 4–5 July 2009* (Vol. 2, pp. 83–85). Institute of Electrical and Electronics Engineers (IEEE). https://doi.org/10.1109/ESIAT.2009.258

Zhang, W., Lu, J., & Zhang, Y. (2016). Comprehensive evaluation index system of low carbon road transport based on fuzzy evaluation method. *Procedia Engineering, 137*, 659–668. https://doi.org/10.1016/j.proeng.2016.01.303

CHAPTER 16

Satellite-Based Land Cover Mapping in Narmada River Basin: Understanding the Transition from Natural Ecosystems to Anthropogenic Landscapes

Ankita Singh, Nishant Singh, and Pratiksha Singh

16.1 Introduction

Land use and land cover (LULC) are two related but distinct concepts that are commonly used in the fields of geography, environmental science, urban planning, and remote sensing to describe and classify the way land is utilized and its physical characteristics (Sharma et al., 2024). Land use refers to the human activities and purposes to which a specific area

A. Singh (✉)
Department of Environmental Science and Limnology, Barkatullah University, Bhopal, Madhya Pradesh, India
e-mail: singh29ankita@gmail.com

N. Singh
Department of Civil Engineering, NRI Institute of Information Science and Technology (NIIST), Bhopal, Madhya Pradesh, India

S. Ghosh (ed.), *Reimagining Indian Rivers for Sustainability*, Palgrave Studies in Environmental Sustainability,
https://doi.org/10.1007/978-3-032-04754-0_16

of land is put. It describes how humans use the land for various activities, such as residential, commercial, industrial, agricultural, recreational, transportation, and institutional purposes. Land use classification helps to understand the spatial distribution of human activities and the functional purposes of different areas within a region or landscape (Chowdhury et al., 2020). For example, a land use map might show areas designated for housing, shopping centers, farmland, factories, parks, roads, etc. Land cover, on the other hand, describes the physical and biological cover of the Earth's surface, including natural elements like vegetation, water bodies, bare soil, rock, and artificial features like buildings and roads. Land cover classification focuses on the physical characteristics of the land and the types of materials or vegetation that cover it (Bayramoglu et al., 2020; Nagler et al., 2001). For example, a land cover map might show forests, grasslands, wetlands, lakes, urban areas, deserts, etc. In summary, land use focuses on human activities and the functional purposes of land, while land cover concentrates on the physical characteristics and natural/built features covering the land. These two concepts are essential for understanding and managing landscapes, environmental changes, urban development, and the impact of human activities on the Earth's surface. Human activities including urbanization, deforestation, agriculture practices, mining and extractive industries, infrastructure development, and industrial pollution cause climate change, wildfires, natural disasters, population growth, and land demand (Shukla et al., 2014). Addressing land use and land cover disturbance requires a combination of sustainable land management practices, conservation efforts, and policies aimed at mitigating the negative impacts of human activities on the environment.

Remote sensing technologies, such as satellite imagery and aerial photography, are commonly used to create land use and land cover maps, aiding in land management and environmental planning efforts (Kaul & Sopan, 2012). It is a powerful tool used in the analysis of LULC on the Earth's surface. It involves the acquisition of information about the Earth's surface using sensors mounted on aircraft or satellites, allowing us to observe and study the Earth from a distance. These sensors can capture

P. Singh
Department of Zoology and Applied Aquaculture, Barkatullah University, Bhopal, Madhya Pradesh, India

data in various parts of the electromagnetic spectrum, such as visible light, infrared, and microwave, enabling us to differentiate and classify different land features (Gondwe et al., 2021; Nath et al., 2018; Roy & Roy, 2010). During the analysis of land use and land cover, remote sensing is used for the Data Acquisition—which provides valuable information about the land surface and its changes over time; Image Processing—involves filtering, geometric correction, and radiometric calibration to enhance the quality of data and extract relevant information (Saadat et al., 2011); Change Detection—identify and monitor changes in land use and land cover over time and by comparing multiple images taken at different times, one can detect alterations such as deforestation, urban expansion, or agricultural changes; Vegetation Analysis—used to assess vegetation health, density, and distribution. Indices such as the Normalized Difference Vegetation Index (NDVI) are commonly employed to quantify the amount of green vegetation in an area, Urban Growth Monitoring—urban areas are dynamic and continuously changing (Abebe et al., 2022; Ay et al., 2014; Fezzi et al., 2015; Gaur & Singh, 2023). Remote sensing helps track urban growth, monitor infrastructure development, and analyze the expansion of built-up areas (Gaur & Singh, 2023).

Image classification is the primary task in land use and land cover analysis. This process involves categorizing the pixels in the image into different land cover classes (e.g., forests, agricultural land, urban areas, and water bodies) based on their spectral properties (Seyam et al., 2023; Vivekananda et al., 2021). There are two main approaches to image classification: supervised and unsupervised classification. Supervised Classification is the method that requires the analyst to select representative training samples of each land cover class (Kafi et al., 2014; Keuchel et al., 2003). The algorithm then uses these samples to classify the entire image based on their spectral signatures. Unsupervised Classification in the algorithm groups pixels with similar spectral characteristics into clusters without prior knowledge of the land cover classes (Bayramoglu et al., 2020; Chowdhury et al., 2020). The analyst then interprets and assigns labels to these clusters based on their spectral properties.

Numerous academics believe it to be a significant phenomenon that has local, regional, and worldwide effects on both natural environments and socioeconomic status (Hassan et al., 2016; Paul et al., 2018). The objective of this paper is to evaluate the land use pattern in the Harda District and the effect of human inferences or activities on natural resources, which play a crucial function in maintaining the socioeconomic values of the district. Harda is one of the most developed districts of Madhya Pradesh in agricultural practices because of the fertile soils along the river Narmada and its tributaries. The district extends over an area of 3,334 sq. km, and the forest covers 1,130.21 sq. km. out of the total area of the district (Saigal et al., 2016). Its hilly area is covered with forests, which are predominantly seen in the district, and along the river Ganjal, which has a major waterbody cover area, but due to dominating human activities and practices, they started to decline and lose their environmental and socioeconomic values.

16.2 Landuse/Landcover

Land use and land cover mapping and monitoring were crucially needed in the study area to preserve the existing natural resources and comprehend the causes and effects of overusing soil, forest, and water resources (Hasan et al., 2021; Patel et al., 2019). Landsat data is a valuable resource for the analysis of land use and land cover due to its ability to provide detailed and consistent information about the Earth's surface over time. Landsat is a series of Earth-observing satellites operated by the United States Geological Survey (USGS) and NASA. Landsat satellites are equipped with multiple spectral bands, including visible, near-infrared, and thermal infrared bands. These bands allow for the measurement of various surface properties, including vegetation health, water content, and temperature (Yang & Lo, 2002). One of the significant advantages of Landsat data is its high temporal resolution. New images are collected roughly every 16 days, providing a rich dataset for tracking changes in land use and land cover over time. Landsat sensors offer different spatial resolutions, with Landsat 7 and 8 having a 30-meter spatial resolution for most bands (Toure et al., 2018). Landsat 8 also includes a 15-meter panchromatic band and a 100-meter thermal band. Landsat data is often used for multispectral analysis, which involves combining different spectral bands to highlight specific land cover features (Sekertekin et al., 2017; Singh & Vyas, 2022). For instance, the Normalized Difference Vegetation Index (NDVI) is commonly derived from Landsat data to assess vegetation health (Paul et al., 2018). Landsat data can be used to classify land cover into different categories, such as forests, agriculture, urban areas, water bodies, and more. This classification is often performed using remote sensing techniques and machine learning algorithms. Landsat data is particularly useful for change detection analysis. By comparing Landsat images from different periods, can identify and quantify changes in land use and land cover, such as urban expansion or deforestation (Reis, 2008; Toure et al., 2018).

To perform land use and land cover analysis using Landsat data, it would typically need remote sensing and GIS software, along with knowledge of image processing techniques (Onur et al., 2009; Szuster et al., 2011). Additionally, combining Landsat data with other data sources like aerial imagery, field surveys, and demographic data can enhance the accuracy and relevance of the analysis (Cheruto et al., 2016; Regasa et al., 2021). Overall, remote sensing plays a crucial role in analyzing land use and land cover changes over large areas and can provide valuable insights for land management, urban planning, environmental conservation, and disaster response (Loukika et al., 2021).

16.3 Narmada River Basin

The Narmada River is India's third-holiest and fifth-longest west-flowing river. With a total watershed size of 98,797 square kilometers, it is also the biggest and one of the 13 most well-known rivers in the nation. It is the third-largest river that flows inside the boundaries of India, behind the Ganges and the Godavari. The little pond known as Narmada Kund, the source of the Narmada River, is situated on the 1,057-meter-high Amarkantak hill in the Shahdol District in eastern Madhya Pradesh. The Narmada River basin, with a mean elevation of 760 m, is situated in the geographic center of India between 72° 20" and 81° 45" E longitude and 21° 20" to 23° 45" N latitude (Gupta et al., 2020). With its extraordinarily rich natural legacy sustaining ancient cultures, the Narmada basin plays a significant role in our nation.

The Narmada River is revered by Indians as the picture of holiness and freedom (Mandal et al., 2022). The Narmada region is distinguished geographically by its rich and varied landscape, which includes lush woods, fertile plains, steep terrains, and the vast Narmada Valley. It has a broad variety of habitats and biodiversity due to its varied climatic conditions, which vary from tropical to subtropical. Natural resources in the area, such as its woods, rivers, and fertile soils, have been essential in sustaining human settlements, agriculture, and industry for many years. The Narmada region's diverse indigenous cultures, numerous historical and archaeological monuments, and melting pot of cultural traditions make it a culturally rich area. It is well-known for its spiritual significance, as seen by the numerous temples and places of worship that dot the Narmada River's scenery.

16.4 Climatic Conditions of the Narmada River Basin

The Narmada region's climate, which is located in the center of India, is as varied as its landscapes and encompasses a variety of climatic conditions. This area, which includes the states of Gujarat, Maharashtra, and Madhya Pradesh, has a mix of tropical, subtropical, and even semi-arid climate zones. The Narmada River, which runs through the center of the area, has a big impact on the local climate. The Narmada area has sweltering summertime temperatures, particularly in the western plains where the mercury may get well beyond 40 degrees Celsius (Singh & Vyas, 2024). This dry and semi-arid environment may be found on Madhya Pradesh's western borders and in some areas of Gujarat. The local economy and agriculture are strained throughout the summer by the sweltering, dry winds and decreasing water sources. The Tropic of Cancer, which passes across the upper plains section of the Narmada basin, is situated just below a sizeable chunk of it. The basin's climate generally tropical and humid, however, there are several locations where heat and cold extremes are frequent (Table 16.1). In the basin, there are four different seasons throughout the year. The south-west monsoon, hot weather, post-monsoon, and cold weather are stated in that order.

16.5 Methodology

16.5.1 Study Area

In Narmada basin, there are 41 main tributaries that feed the Narmada. 22 of these tributaries join from the left bank and 19 from the right, with a catchment area of more than 500 km for the majority of these tributaries (Fig. 16.1). Except for the Orsang and Karjan, every one of these tributaries originates in M.P. The length of the 8773 kilometers of Narmada's tributaries. There are 50 minor rivers that join the Narmada in all. From a fisheries perspective, the deep pools of the Narmada are crucial for local fish spawning and development. The significant tributary, Hiran,

Table 16.1 Description of climate conditions in the Narmada River basin

Serial number	*Climate parameters*	*Description*
1	Climatic zones	Tropical monsoon; semi-arid in west, and humid in east
2	Average annual rainfall	1,000–1,500 mm
3	Rainfall pattern	Seasonal—about 90% during southwest monsoon, i.e., June to September
4	Rainfall distribution	Eastern basin receives more than western parts
5	Temperature range	Summer: 30–45 °C; Winter: 10–25 °C
6	Summer season	March to June—hot and dry
7	Monsoon season	June to September—heavy rainfall
8	Winter season	November to February—mild and dry
9	Humidity levels	High during monsoon; low during winter and summer
10	Evapotranspiration rate	High due to elevated temperatures
11	Flood risk period	Monsoon season, in the lower basin regions
12	Drought-prone areas	Western parts of the basin (especially in Khandwa and Barwani districts)

joins it after Jabalpur. Before Narmadapuram, Barna and Sher, two tributaries from the north, join the Narmada. From the south: Shakkar, Sunar, Dudhi, Baru Reva, and Tawa. These rivers have a sizable catchment area. Tawa and its tributary, Denwa, both deliver a lot of water and degraded debris. With the exception of Hiran, the Vindhyan hills provide less water to Barna, Tendoni, Kolar, and Temar, all of which meet the Narmada from the north. Important Narmada tributaries Sunar, Sher, and Shakkar flow through steep gorges in the Narsinghpur area. The main river Ganjal and its tributary Morand have a combined catchment area of 1930 km^2 in Harda after Narmadapuram. Hathra, Palakmati, Indra, Agal, and Machak, Uri, Hathni, and Kunthi and Goi are other rivers that join the main stream on the alluvial plains. The river finally drains into the Gulf of Khambhat (Swain et al., 2021).

The study area, i.e., Harda District, is part of the Narmada basin, located in Madhya Pradesh state of central India (Fig. 16.2). It has a longitude of 77°6′8.96″E and a latitude of 22°11′54.2″N and is bounded by the Sehore and Khandwa districts. It is situated in the Narmada River basin, which is known as a lifeline for the population of Madhya Pradesh.

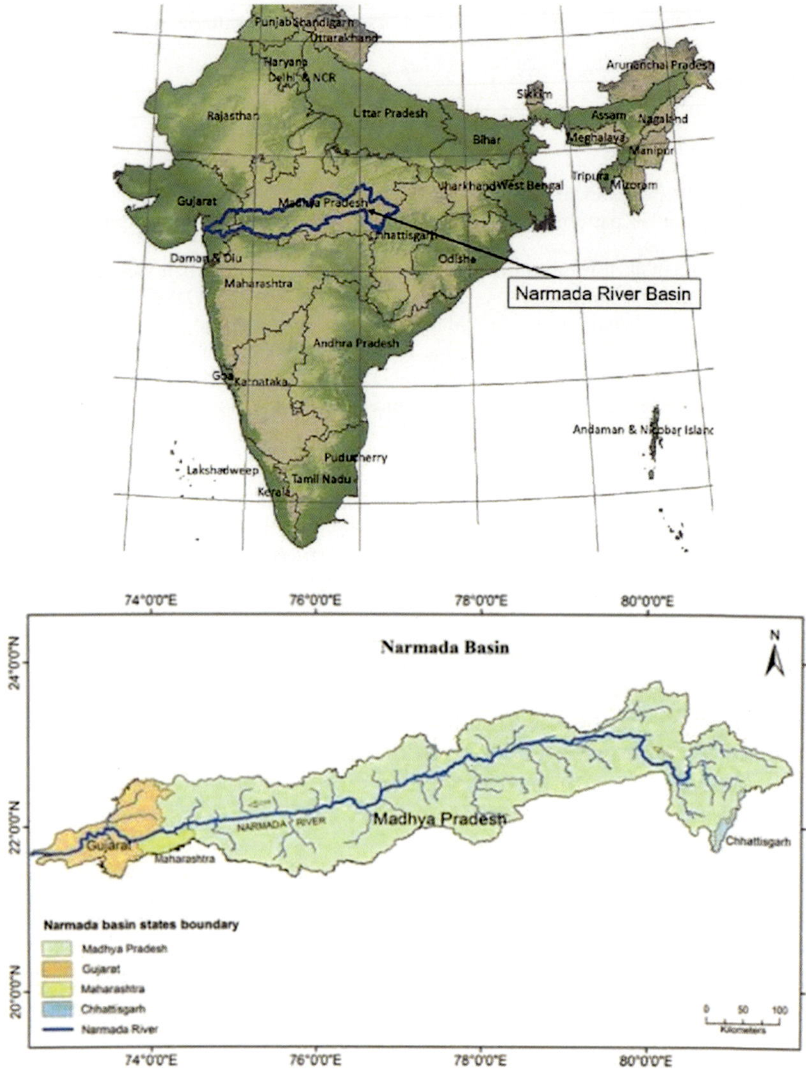

Fig. 16.1 Geographical map of the Narmada Basin (*Source* https://indianapublications.com/articles/IJAL_4%287%29_611_64c4d8d8d9c4c8.06126048.pdf)

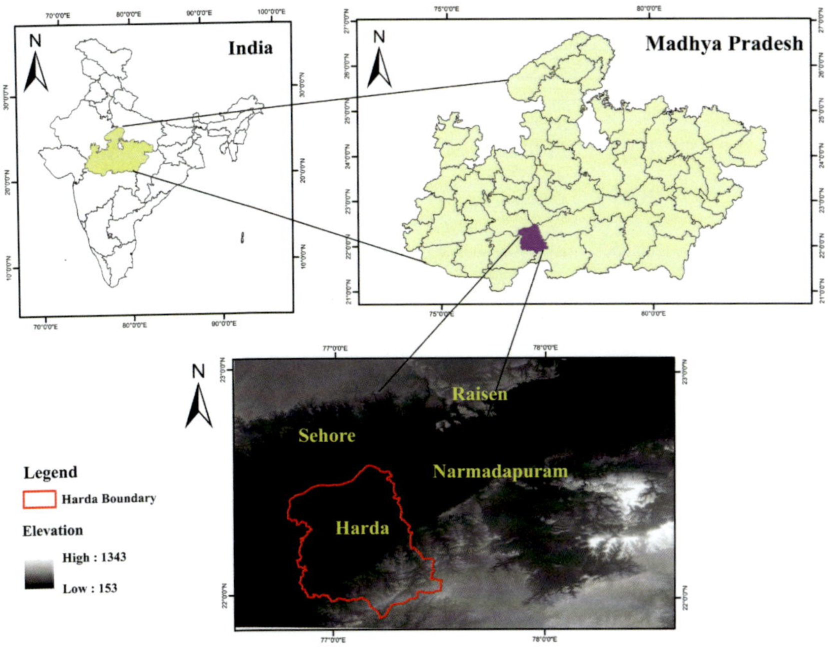

Fig. 16.2 Map showing the study area of Harda District

Harda has more than 570,465 population who mostly depend on agricultural practices for their livelihood. It has a Tropical wet and dry or savanna, i.e., summer dry season climate with a temperature of 30.04 °C yearly. It receives a maximum of 5 mm of rainfall during the southwest monsoon period. The forest type is a Tropical Deciduous Forest of Sagwan with scattered Bamboo and Satkata trees. The teak forest is the dominant species, which covers 50% area of the forest division.

16.5.2 Remote Sensing Data

For this work, the LU/LC distribution maps and their associated analysis were prepared using the time series satellite datasets for the years 1995, 2005, 2015, and 2021. For image categorization and analysis, the software programs ERDAS Imagine and ArcGIS were employed (Abd El-Kawy et al., 2011; Rejaur Rahman & Saha, 2008). It was anticipated that

Table 16.2 Satellite images and their characteristics

Serial number	*Year*	*Image type*	*Sensor*	*Spatial resolution*	*Spectral resolution*
1	1995	Landsat	Landsat 5 TM	30 m	Multispectral (8 bands)
2	2005	Landsat	Landsat 7 ETM +	30 m	Multispectral (8 bands)
3	2015	Landsat	Landsat 5 ETM + LISS III	30 m	Multispectral (8 bands)
4	2021	Landsat	Landsat 8 (OLI/TIRS)	30 m	Multispectral (11 bands)

the outcome of the LULC mapping would include data on the spatial distribution of land use categories as well as identification and estimation of land use changes over the previous 25–30 years. This study required the following datasets: time series satellite data, i.e., Landsat 5 TM for 1980, Landsat 7 ETM + for 2005, Landsat 5 ETM + LISS III for 2015, and Landsat 8 OLI/TIRS for 2021 with a spatial resolution of 30 m, respectively (Table 16.2).

16.5.3 Land Use and Land Cover

The National Remote Sensing Centre's standard LU/LC classification system was followed for the classification. Level 1 classes of the standard classification system were used for the image classification. Each satellite image was classified into five land use land cover classes using the Maximum Likelihood Classifier (MLC) of the supervised classification approach (Onur et al., 2009; Patil et al., 2012). Five major LU/LC classes identified in the study area were built-up lands, forests, agricultural lands, waterbodies, and wastelands. The classification was followed by the field verification and accuracy assessment (Fig. 16.3). Post classification change detection method was used to study the changes in LU/LC of Harda District.

16.5.4 Accuracy Assessment

Accuracy assessment is the comparison of classification with ground-truth data to evaluate how well the classification represents the real world.

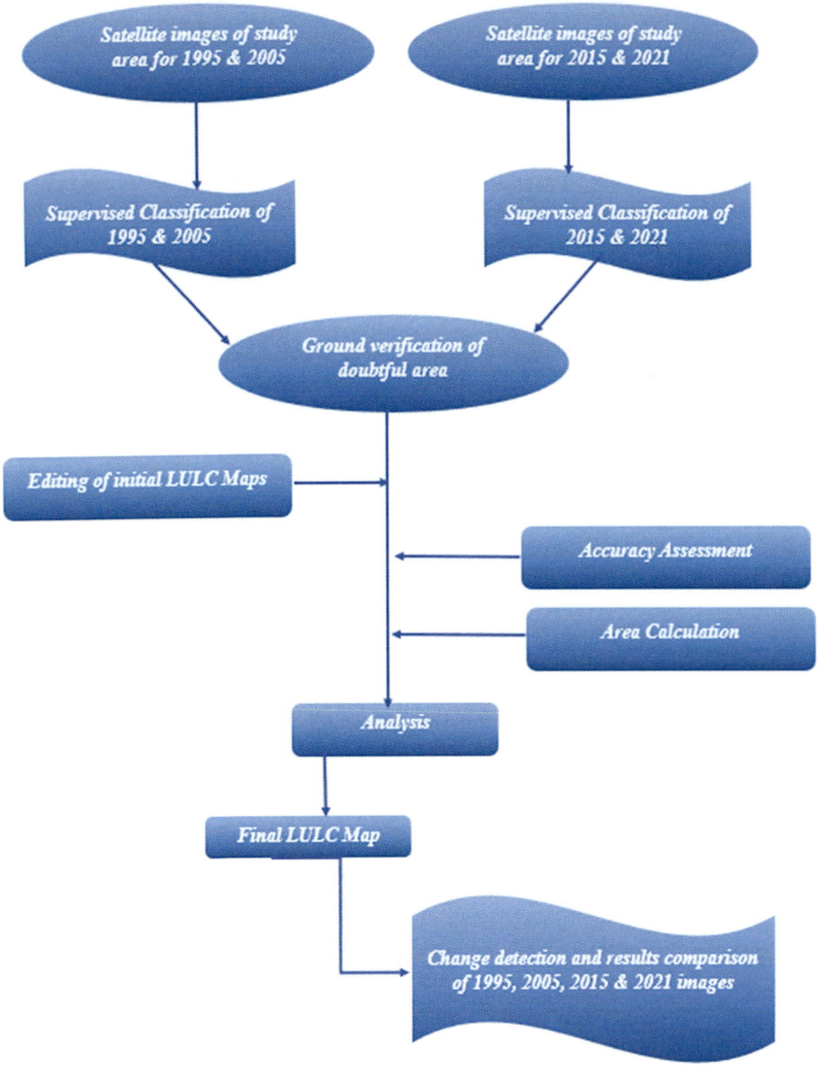

Fig. 16.3 Map showing the methodology for LULC assessment

In this study, accuracy assessment of the resultant classified images was carried out to determine the quality of information derived from the data using a random sampling method (Rwanga & Ndambuki, 2017) using 150 points. This assessment was carried out using an error matrix. In addition, Kappa statistics along with the total accuracy of the classified images were also performed to measure the extent of classification accuracy from the report section of ERDAS Imagine and through field validation (Marangoz et al., 2017; Tehrany et al., 2014).

16.6 Result and Discussion

16.6.1 Landuse and Landcover Mapping and Change Detection

The Harda District is showing slight change in total area from the year 1995 to 2021, i.e., in 1995—3345.34 km^2, 2005—3345.33 km^2, 2015—3334.07 km^2, and in 2021—3333.60 km^2, respectively. To determine the Landscape pattern and temporal changes in LULC in the study area, seven different classes have been selected for classification, i.e., Waterbody, Trees (Forest), Shrub/scrub, Grassland, Agriculture, Settlement, and Bareland or bare soil. After performing the supervised classification, it was found that the forest area and the settlement area show greater variation between 1995 and 2021, as shown in Fig. 16.4 (a–d), and the summary of retrieved data is presented in Tables 16.3 and 16.4.

In 1995, the Harda District was mostly covered by an agricultural area followed by forest cover but with the change of time, the landcover pattern also changed. This region consists of a plain surface with rich black soil, which provides good fertility and supports cultivation due to which agricultural practices are dominant in this area. According to Table 16.3, the calculated area for agriculture cover was 2113.83 km^2, i.e., 63.19% in 1995, 2125.13 km^2, i.e., 63.53% in 2005, 2194.15 km^2, i.e., 63.53% in 2015, 2233.82 km^2, i.e., 67.01% in 2021.

From 1995 to 2005 after agricultural land, forest was the second crucial landcover in which major change was found during the years 2005–2015, i.e., 960.67 km^2–459.75 km^2. The forest area transformed into grasslands and shrub/scrub land with time due to which the forest floras and faunas became extinct or declined in this area. Some areas of forest are also shifted to settlement development and agricultural land, which are the basic needs of human beings at present time.

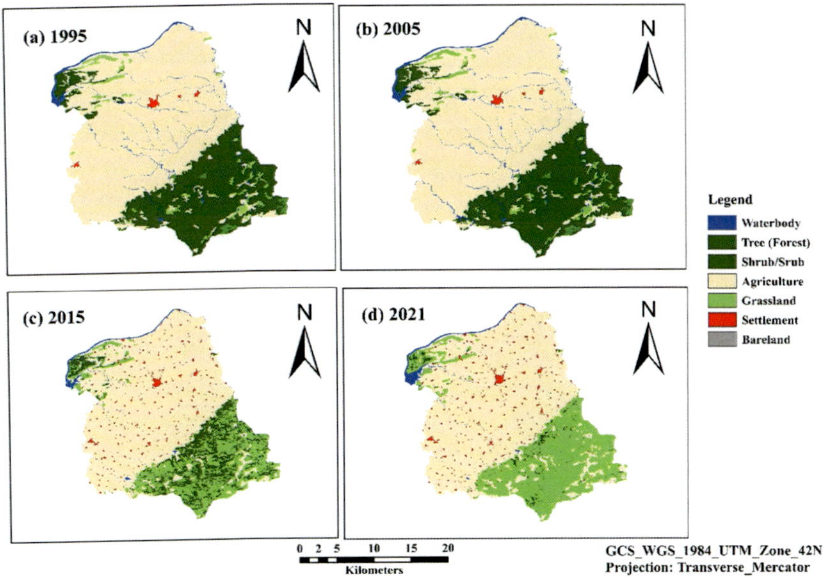

Fig. 16.4 Showing the LULC change from 1995 to 2021

Table 16.3 Landuse/landcover classes in Harda District

Perimeter	*1995*		*2005*		*2015*		*2021*	
	Area (km^2)	*% Cover*	*Area (km^2)*	*% Cover*	*Area (km^2)*	*% Cover*	*Area (km^2)*	*% Cover*
Waterbody	77.88	2.33	84.79	2.53	36.42	1.09	48.05	1.44
Tree (forest)	988.41	29.54	960.67	28.72	459.75	13.79	63.4	1.9
Shrub/scrub	79.27	2.37	80.42	2.4	576.85	17.3	910.53	27.31
Grassland	73.2	2.19	56.69	1.69	0.27	0	0.0087	0
Agriculture	2113.83	63.19	2125.13	63.53	2194.15	63.53	2233.82	67.01
Settlement	12.04	0.36	12.14	0.63	66.86	2.29	77.68	2.33
Bareland	0.71	0.02	1.73	0.05	0.026	0	0.086	0

Table 16.4 Change detection for classification in the Harda District

% Change				
Perimeter	*1995–2021*	*1995–2005*	*2005–2015*	*2015–2021*
Waterbody	−0.89	−0.2	1.44	−0.35
Tree (forest)	−27.64	0.82	14.93	11.89
Shrub/scrub	−24.94	−0.03	−14.9	−10.01
Grassland	2.19	0.5	1.69	0
Agriculture	−3.82	−0.34	−2.28	−1.2
Settlement	−1.97	−0.27	−1.66	−0.04
Bareland	0.02	−0.03	0.05	0

During 1995, the agriculture practices raised their areas (Fig. 16.5), which was followed by settlement area cover by 2021. At the end of the year 2021, the area of settlement increased compared to previous years, which shows the cause of population growth. According to the census, the population was 380,762 in 1991, 4,74,916 in 2001, 5,70,565 in 2011, and 6,68,600 in 2021 (https://en.wikipedia.org/wiki/Harda_district), which utilized a major portion of land cover for their livelihood.

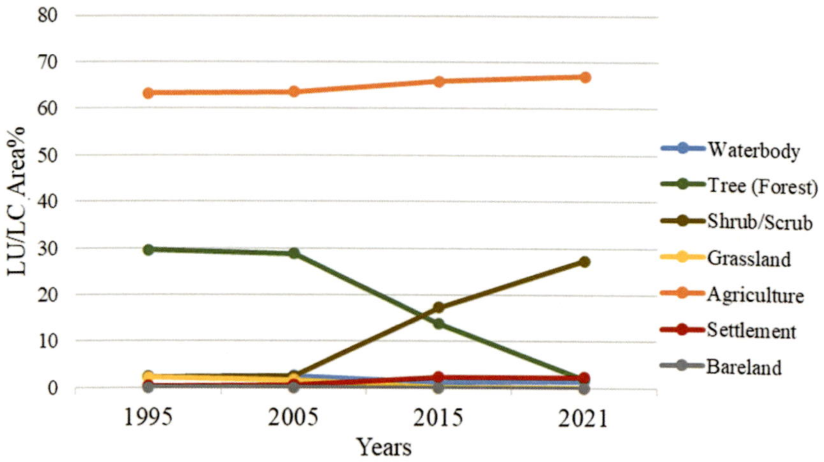

Fig. 16.5 Temporal pattern of land use/land cover area in Harda District

Four rivers, i.e., Narmada, Ganjal, Ajnal, and Machak, pass through Harda District, which supports agriculture, applicable for irrigation purposes, and livelihood. Between 2005 and 2015, an area of water-bodies showed a major decline, i.e., 84.79 km^2 (2.53%) to 36.42 km^2 (1.09%) that was occupied by agricultural land and settlement. Some areas of water bodies also changed into Bareland due to climate change or less rainfall but after 2015, the area of waterbody cover started increasing, and at the end of 2021, the total area of waterbody cover was 48.05 km^2. The rainfall data for the year 1993 were 290.2 mm, in 2005, 134.7 mm, and in 2015, it was 391.0 mm (http://mpwrd.gov.in/wp-content/uploads/2018/01/Harda.pdf). The maximum rainfall in Central India was 350.82–427.7 mm recorded in 2019 (https://hydro.imd.gov.in/hydrometweb/(S(vo3wtanhjge1rk5500ogip55))/PRODUCTS/Publications/Rainfall%20Statistics%20of%20India%20-%202021/Rainfall%20Statistics%20of%20India%202021.pdf) in comparison to the year 2015, which played an important role in expanding waterbody area cover in this District. However, an overall change in waterbodies reveals that this land cover fell off from the year 1995 to 2021 (Table 16.4).

Due to the decline in waterbodies in the district, the growth of forest trees became limited, which supported the growth of shrubs or bushes instead of large forest trees, and it shifted into grasslands or shrublands. After 2005, Shrublands or scrublands became a major land cover that occupied the major portion of forest cover and converted into shrublands, due to which at the end of the year 2021, the area of shrubland increased in comparison to previous years. The calculated area for shrubland was 79.27 km^2, i.e., 2.37% in 1995, 80.42 km^2, i.e., 2.4% in 2005, 576.85 km^2, i.e., 17.3% in 2015, 910.53 km^2, i.e., 27.31% in 2021.

16.6.2 Landuse/ Landcover Changes and Accuracy Assessment in Harda District of Narmada Basin

For accuracy assessment, ERDAS Imagine randomly searched for 150 selected points for a random point table, and ground verification was also done for a few pieces of data. These points were produced to assess errors in the final classification. A summary of the accuracy assessment of Harda District for the years 1995, 2005, 2015, and 2021 is shown in Table 16.5, respectively. The overall accuracy for supervised classification images of 1995, 2005, 2015, and 2021 was found to be 89.71%, 89.00%, 85.50%, and 84.34%, and the Kappa coefficient (KC) value for those years

Table 16.5 Accuracy assessment for LULC 1995–2021

S. No	Classes	1995		2005		2015		2021	
		Producer Accuracy	User Accuracy	Producer Accuracy	User Accuracy	Producer Accuracy	User Accuracy	Producer Accuracy	User Accuracy
1	Waterbody	93.87	99.32	50.00	100.00	99.55	94.87	99.21	83.00
2	Tree (forest)	96.01	93.03	85.71	85.71	89.17	63.53	91.07	90.38
3	Shrub/ scrub	82.4	77.70	85.00	80.95	92.31	88.89	48.51	83.11
4	Grassland	72.34	79.70	71.06	87.30	71.06	87.30	47.19	38.47
5	Agriculture	78.62	83.82	100.00	82.35	86.58	93.67	89.93	90.65
6	Settlement	90.13	94.59	89.66	83.87	98.01	92.63	83.7	95.79
7	Bareland	98.17	73.88	95.65	95.65	86.40	88.16	87.40	61.36
Overall accuracy		89.71		89.00		85.50		84.34	
Kappa coefficient		0.82		0.86		0.78		0.58	

Table 16.6 Kappa coefficient rating category

Serial number	*Kappa coefficient*	*Rate*
1	Less than 0.4	Poor
2	0.4 < KC < 0.5	Fair
3	0.55 < KC < 0.7	Good
4	0.7 < KC < 0.85	Very good
5	KC > 0.85	Excellent

was 0.82, 0.86, 0.78, and 0.58, respectively. According to Table 16.6, the values of KC between 0.55 and 0.85 are a good to very good class, between <0.4 and 0.55 is a poor to fair class, and if it is >0.85 shows an excellent class of assessment. The assessment in this study exists between >0.55 and >0.85, i.e., a fair to excellent class, which indicates that these Kappa values depict that land use classification accuracy is acceptable.

16.6.3 Consumption of Water Resources

India has always been an agrarian economy since its inception. This can be seen in water consumption data of M.P., according to which Agriculture is the largest consumer of water (via canal irrigation systems, lift irrigation) as crops grown—wheat, rice, cotton are generally water intensive. Another field where water is used in a large percentage is Domestic use in urban and rural areas, where drinking water supplied is from nearby reservoirs (e.g., Indira Sagar). The third largest consumer of water is Industries such as thermal power plants and Heavy industries in cities like Jabalpur, Narmadapuram, Harda, etc., which exclusively rely on river water. The fourth largest consumer of water is Hydropower generation in large dams like Sardar Sarovar, Indira Sagar, and Omkareshwar, which generate hydroelectricity. However, the water is reusable but the volume of water required to generate usable electricity is very high.

16.6.4 Human Activities Responsible for LULC Change

In the present days, rapid increase in population growth leads to dynamic change in livelihood needs, which directly or indirectly depend on our natural resources. Agricultural expansion, which is driven by population growth and food demand, results in the conversion of forests and grasslands into farmland. Dams and Irrigation projects like Sardar Sarovar, Indira Sagar, and Omkareshwar altered the landscape pattern in the basin. Urbanization and infrastructure show the rapid growth of towns like Jabalpur, Narmadapuram, Harda, and Barwani, and road and railway construction. Tree felling for fuel, timber, and land clearance results in deforestation is mainly responsible for LULC change in the Narmada basin.

16.6.5 Floodplain Modification (Road Networks and Urban Expansion)

M.P. has only two major rivers, i.e., Narmada and Tapi, but they in turn have numerous tributaries whose floodplains are evolving due to numerous artificial factors. One such factor is Infrastructure Development like roads, highways, and rail lines, which often interferes with floodplains and restricts the natural flow of rivers. Another factor is urbanization as encroachment on floodplains (towns like Harda and Narmadapuram) increases the risk of flash floods in such areas. Irregular Planning of Location for Embankments and Dams also plays a significant role in floodplain modification as such hindrances alter the sediment flow of a river and force it to meander and promote the use of artificial reservoirs. Bridges and Causeways at critical junctures of a city contribute to narrowing or channelization of the rivers, which consequently impact aquatic corridors. All these human factors have resulted in loss of wetlands, creation of oxbow lakes, and side channels filled with sediments.

16.6.6 Effects of LULC Changes on Riverine Landscapes

Riverine landscapes are the areas that are most affected by land use and land cover modification, as such changes have various detrimental effects. Deforestation in the upper catchment area is one of the most significant among them, leading to increased runoff and soil erosion. Agriculture expansion is another factor that results in altered surface runoff, reduction in groundwater recharge, and increased nutrient and sediment loading. Another factor having adverse effects on such landscapes is urbanization, which increases impervious surfaces that in turn causes a reduction in infiltration and an increase in surface runoff. Under urbanization, industrial development specifically leads to thermal and chemical pollution, which affects the local microclimate, and Sand & Bauxite mining disturb riverbeds, causing bank erosion, and habitat degradation.

16.6.7 Environmental Crisis in the Narmada Basin

The Narmada River Basin has undergone significant land-use and land-cover (LULC) changes over recent decades, driven largely by human activity and associated processes. These changes have led to considerable environmental crises, especially affecting the ecological balance, water quality, and socio-economic fabric of the region (Table 16.7).

Table 16.7 Details of the environmental crisis due to LULC change and its impact on the Narmada basin

Serial number	*Environmental crisis*	*Description*	*Impact*
1	Deforestation and biodiversity loss	Forest loss due to agriculture, dams, and settlements has fragmented ecosystems	Endangered flora and fauna species decline (e.g., in Satpura ranges)
2	River flow disruption	Large dams have altered the natural flow regime	Reduced sediment transport; changed fish migration; impact on downstream delta ecology
3	Water pollution	Discharge of untreated sewage, agrochemicals, and industrial waste into river and tributaries	Degraded water quality; threat to aquatic biodiversity and human health
4	Displacement and social unrest	Over 2 million people displaced due to dam projects like Sardar Sarovar	Livelihood loss; inadequate rehabilitation; tribal marginalization
5	Flooding and drainage issues	Encroachment on floodplains and wetland areas increases flood vulnerability	Increased frequency of flash floods in low-lying areas; damage to crops and settlements
6	Groundwater depletion	Excessive irrigation and poor recharge practices	Lowered water table in many areas; dry wells during summer
7	Soil degradation and erosion	Intensive agriculture, deforestation, and mining	Reduced fertility; sedimentation of rivers and reservoirs
8	Wetland loss	Wetlands are often drained or encroached for farming or construction	Decline in natural flood control and water purification functions

16.7 Conclusion

This study reveals that the Harda District mostly suffers from anthropogenic activities, which cause a major change in LULC. Over the last 25–30 years, the population growth rate increased drastically, which became a major factor in influencing the natural processes in our environment, which draws people's awareness for the protection of the environment and natural resources. For the assessment of the natural condition of landscape patterns in the study, different satellite data were used, which confirm that the different products of Landsat satellites that are Landsat 5, Landsat 7, and Landsat 8 are suitable data and the quality of images and their spatial resolution play a vital role in the accuracy assessment of the classification (Marangoz et al., 2017; Sekertekin et al., 2017). By using GIS tools, it becomes easier to analyze the LULC data of previous years in the present time, and after interpretation, it is found that the Harda District is classified into seven classes, in which agricultural land dominates land cover. Overall, major loss is observed in Forest cover, and the major change is found in shrubland or scrubland, increasing year by year. Due to an increase in irrigation needs and disturbing the path of river drainage patterns, the watershed area (Atay Kaya & Kut Görgün, 2020; Dewan & Yamaguchi, 2009) also declined, which directly or indirectly affects the health of the major river, i.e., the river Narmada. It is also found that the major portion of the forest area has been converted to grasslands, agricultural land, and settlement development, which is a matter of great concern.

Based on the output of the study and its conclusion, few recommendations and management are necessary for the conservation of natural resources like forest, water, soil quality, and biodiversity, which play an important role in maintaining the ecological cycles and services.

- Controlled and planned urbanization, implementing best management practices during agriculture, and conservation of the forest are highly recommended for the sustainable development of the river basins.
- For biodiversity conservation, educational or awareness efforts are necessary for the community members to learn the importance of the critical biodiversity area and their sustainable management.
- The results of this research apply to the decision-makers in the Municipal Corporation sector, the State Pollution Control Board (SPCB), and the Ministry of Environment, Forest, and Climate Change (MoEFCC).
- Prepare an action plan for the restoration of riparian zones. More plantations are needed on river sites and in forest areas, which can reduce soil erosion, hold moisture for maintaining the other herbs and shrubs' growth, and improve the water quality and water level,

which is directly or indirectly applicable to the villagers and our society.
- The data retrieved from this study also played an additional role in supporting the National Land Utilisation Policy.

Acknowledgements We would like to thank the anonymous reviewers of this manuscript for their feedback, which helped us to improve the paper in multiple ways.

Author's Contribution: All authors contributed to the study's conception and design. Material preparation, data collection, and analysis were performed by Ankita Singh, Nishant Singh, and Pratiksha Singh. All authors read and approved the final manuscript.

Funding: No funding was received for this work.

Consent to publish: This declaration does not apply to this study.

Availability of data and materials: The datasets and figures available in this manuscript (Tables 16.1–16.4 and Fig. 16.1–16.4) can be accessed.

Ethical approval: This is an observational study, so this declaration is not applicable.

Conflicts of Interest: The authors declare no conflict of interest.

References

Abd El-Kawy, O. R., Rød, J. K., Ismail, H. A., & Suliman, A. S. (2011). Land use and land cover change detection in the western Nile delta of Egypt using remote sensing data. *Applied Geography, 31*(2), 483–494.

Abebe, G., Getachew, D., & Ewunetu, A. (2022). Analysing land use/land cover changes and its dynamics using remote sensing and GIS in Gubalafito district, Northeastern Ethiopia. *SN Applied Sciences, 4*(1), 30.

Atay Kaya, İ, & Kut Görgün, E. (2020). Land use and land cover change monitoring in Bandırma (Turkey) using remote sensing and geographic information systems. *Environmental Monitoring and Assessment, 192*(7), 430.

Ay, J. S., Chakir, R., Doyen, L., Jiguet, F., & Leadley, P. (2014). Integrated models, scenarios and dynamics of climate, land use and common birds. *Climatic Change, 126*(1), 13–30.

Bayramoglu, B., Chakir, R., & Lungarska, A. (2020). Impacts of land use and climate change on freshwater ecosystems in France. *Environmental Modeling & Assessment, 25*(2), 147–172.

Cheruto, M. C., Kauti, M. K., Kisangau, D. P., & Kariuki, P. C. (2016). *Assessment of land use and land cover change using GIS and remote sensing techniques: A case study of Makueni County*, Kenya.

Chowdhury, M., Hasan, M. E., & Abdullah-Al-Mamun, M. M. (2020). Land use/land cover change assessment of Halda watershed using remote sensing and GIS. *The Egyptian Journal of Remote Sensing and Space Science, 23*(1), 63–75.

Dewan, A. M., & Yamaguchi, Y. (2009). Land use and land cover change in Greater Dhaka, Bangladesh: Using remote sensing to promote sustainable urbanization. *Applied Geography, 29*(3), 390–401.

Fezzi, C., Harwood, A. R., Lovett, A. A., & Bateman, I. J. (2015). The environmental impact of climate change adaptation. *Nature Climate Change*, 1–18.

Gaur, S., & Singh, R. (2023). A comprehensive review on land use/land cover (LULC) change modeling for urban development: Current status and future prospects. *Sustainability, 15*(2), 903.

Gondwe, J. F., Lin, S., & Munthali, R. M. (2021). Analysis of land use and land cover changes in urban areas using remote sensing: Case of Blantyre City. *Discrete Dynamics in Nature and Society, 2021*(1), 8011565.

Gupta, D., Shukla, R., Barya, M. P., Singh, G., & Mishra, V. K. (2020). Water quality assessment of Narmada River along the different topographical regions of the central India. *Water Science, 34*(1), 202–212.

Hasan, M. M., Islam, R., Rahman, M. S., Ibrahim, M., Shamsuzzoha, M., Khanam, R., & Zaman, A. K. M. M. (2021). Analysis of land use and land cover changing patterns of Bangladesh using remote sensing technology. *American Journal of Environmental Sciences, 17*, 64–74.

Hassan, Z., Shabbir, R., Ahmad, S. S., Malik, A. H., Aziz, N., Butt, A., & Erum, S. (2016). Dynamics of land use and land cover change (LULCC) using geospatial techniques: A case study of Islamabad Pakistan. *Springerplus, 5*(1), 812.

Kafi, K. M., Shafri, H. Z. M., & Shariff, A. B. M. (2014, June). An analysis of LULC change detection using remotely sensed data; A Case study of Bauchi City. In *IOP conference series: Earth and environmental science* (Vol. 20, No. 1, p. 012056). IOP Publishing.

Kaul, H. A., & Sopan, I. (2012). Land use land cover classification and change detection using high resolution temporal satellite data. *Journal of Environment, 1*(4), 146–152.

Keuchel, J., Naumann, S., Heiler, M., & Siegmund, A. (2003). Automatic land cover analysis for Tenerife by supervised classification using remotely sensed data. *Remote Sensing of Environment, 86*(4), 530–541.

Loukika, K. N., Keesara, V. R., & Sridhar, V. (2021). Analysis of land use and land cover using machine learning algorithms on google earth engine for Munneru River Basin, India. *Sustainability, 13*(24), 13758.

Mandal, B., Dixit, M. M., Kumar, S., Karuppannan, P., Laxminaryana, K., Catchings, R. D., Behera, L., & Kumar, P. (2022). Detecting sub-basalt Mesozoic sediments and active tectonics in the Narmada-Tapti Rift Zone, central India. *Geosystems and Geoenvironment, 1*(4), Article 100047.

Marangoz, A. M., Sekertekin, A., & Akçin, H. (2017). Analysis of land use land cover classification results derived from sentinel-2 image. *Proceedings of the 17th International Multidisciplinary Scientific GeoConference surveying geology and mining ecology management, SGEM* (pp. 25–32).

Nagler, P. L., Glenn, E. P., & Huete, A. R. (2001). Assessment of spectral vegetation indices for riparian vegetation in the Colorado River delta, Mexico. *Journal of Arid Environments, 49*(1), 91–110.

Nath, B., Niu, Z., & Singh, R. P. (2018). Land Use and Land Cover changes, and environment and risk evaluation of Dujiangyan city (SW China) using remote sensing and GIS techniques. *Sustainability, 10*(12), 4631.

Onur, I., Maktav, D., Sari, M., & Kemal Sönmez, N. (2009). Change detection of land cover and land use using remote sensing and GIS: A case study in Kemer, Turkey. *International Journal of Remote Sensing, 30*(7), 1749–1757.

Patel, R., Vadher, B. M., Waikhom, S., & Yadav, V. G. (2019). Change detection of Land Use/Land Cover (LULC) using remote sensing and GIS in Surat City. *Global Research Development Journal of Engineering, 24*, 573.

Patil, M. B., Desai, C. G., & Umrikar, B. N. (2012). Image classification tool for land use/land cover analysis: A comparative study of maximum likelihood and minimum distance method. *International Journal of Geology, Earth and Environmental Sciences, 2*(3), 189–196.

Paul, S. S., Li, J., Wheate, R., & Li, Y. (2018). Application of object oriented image classification and Markov chain modeling for land use and land cover change analysis. *Journal of Environmental Informatics, 31*(1), 30–40.

Regasa, M. S., Nones, M., & Adeba, D. (2021). A review on land use and land cover change in Ethiopian basins. *Land, 10*(6), 585.

Reis, S. (2008). Analyzing land use/land cover changes using remote sensing and GIS in Rize, North-East Turkey. *Sensors, 8*(10), 6188–6202.

Rejaur Rahman, M., & Saha, S. K. (2008). Multi-resolution segmentation for object-based classification and accuracy assessment of land use/land cover classification using remotely sensed data. *Journal of the Indian Society of Remote Sensing, 36*, 189–201.

Roy, P. S., & Roy, A. (2010). Land use and land cover change in India: Aremote sensing & GIS prespective. *Journal of the Indian Institute of Science, 90*(4), 489–502.

Rwanga, S. S., & Ndambuki, J. M. (2017). Accuracy assessment of land use/land cover classification using remote sensing and GIS. *International Journal of Geosciences, 8*(04), 611.

Saadat, H., Adamowski, J., Bonnell, R., Sharifi, F., Namdar, M., & Ale-Ebrahim, S. (2011). Land use and land cover classification over a large area in Iran based on single date analysis of satellite imagery. *ISPRS Journal of Photogrammetry and Remote Sensing, 66*(5), 608–619.

Saigal, S., Kumar, C., & Chaturvedi, R. (2016). Nadi bachao samriddhi lao–a forest landscape restoration initiative in Harda district, Madhya Pradesh, India. *World Development Perspectives, 4*, 1–4.

Sekertekin, A., Marangoz, A. M., & Akcin, H. (2017). Pixel-based classification analysis of land use land cover using Sentinel-2 and Landsat-8 data. *The International Archives of the Photogrammetry, Remote Sensing and Spatial Information Sciences, 42*, 91–93.

Seyam, M. M. H., Haque, M. R., & Rahman, M. M. (2023). Identifying the land use land cover (LULC) changes using remote sensing and GIS approach: A case study at Bhaluka in Mymensingh, Bangladesh. *Case Studies in Chemical and Environmental Engineering, 7*, Article 100293.

Sharma, S. K., Sinha, R. K., Eldho, T. I., & Patel, H. M. (2024). Individual and combined impacts of land use/cover and climate change on water balance components of a tropical river basin. *Environmental Modeling & Assessment, 29*(1), 67–90.

Shukla, S., Khire, M. V., & Gedam, S. S. (2014). Monitoring land use/land cover changes in a river basin due to urbanization using remote sensing and GIS approach. *The International Archives of the Photogrammetry, Remote Sensing and Spatial Information Sciences, 40*, 949–953.

Singh, A., & Vyas, V. (2022). A Review on remote sensing application in river ecosystem evaluation. *Spatial Information Research, 30*(6), 759–772.

Singh, A., & Vyas, V. (2024). Geomorphological and GIS-based analysis of catchment areas in River Narmada, Central India. *Rivers of India: Past, Present and Future* (pp. 1–21). Springer International Publishing.

Swain, S., Mishra, S. K., & Pandey, A. (2021). A detailed assessment of meteorological drought characteristics using simplified rainfall index over Narmada River Basin, India. *Environmental Earth Sciences, 80*, 1–15.

Szuster, B. W., Chen, Q., & Borger, M. (2011). A comparison of classification techniques to support land cover and land use analysis in tropical coastal zones. *Applied Geography, 31*(2), 525–532.

Tehrany, M. S., Pradhan, B., & Jebuv, M. N. (2014). A comparative assessment between object and pixel-based classification approaches for land use/land cover mapping using SPOT 5 imagery. *Geocarto International, 29*(4), 351–369.

Toure, S. I., Stow, D. A., Shih, H. C., Weeks, J., & Lopez-Carr, D. (2018). Land cover and land use change analysis using multi-spatial resolution data and object-based image analysis. *Remote Sensing of Environment, 210*, 259–268.

Vivekananda, G. N., Swathi, R., & Sujith, A. V. L. N. (2021). Retracted article: Multi-temporal image analysis for LULC classification and change detection. *European Journal of Remote Sensing, 54*(sup2), 189–199.

Yang, X., & Lo, C. P. (2002). Using a time series of satellite imagery to detect land use and land cover changes in the Atlanta, Georgia metropolitan area. *International Journal of Remote Sensing, 23*(9), 1775–1798.

CHAPTER 17

The Desert River's Dilemma: A Hydro-Geomorphic Exploration of Flood Dynamics in the Luni River Across the Arid Landscapes of Rajasthan, India

Arup Bishram and Padmini Pani

17.1 Introduction

Flooding is one of the most damaging natural disasters in our world and it creates a large impact on ecosystems and human life worldwide (Guddo, 2023; Jonkman et al., 2024; Zhang et al., 2024). The desert River has a unique hydrological feature that traverses the semi-arid landscape, often characterised by the ephemeral channel flow. Unlike the world's perennial rivers, which continue to maintain a flow of water,

A. Bishram (✉) · P. Pani
Centre for the Study of Regional Development, Jawaharlal Nehru University, New Delhi, India
e-mail: arupbishram@gmail.com

P. Pani
e-mail: padmini@jnu.ac.in

S. Ghosh (ed.), *Reimagining Indian Rivers for Sustainability*, Palgrave Studies in Environmental Sustainability,
https://doi.org/10.1007/978-3-032-04754-0_17

desert rivers typically experience seasonal or periodic discharge, primarily driven by infrequent but intense rainfall events. Those rivers are very vital to arid ecology a significant part in the formation of the geomorphology (Grodek et al., 2024). Drought is the most significant event in arid environments, but some specific semi-arid and arid regions experience several flash floods because of rainfall, and desert soil is often compacted as well as impermeable, which means precipitation leads to runoff rapidly and tends to extreme flooding. Such flood phenomena reshape the physical landscape as well as the river planform (Abd-Elhamid et al., 2023). A striking example of the particular phenomenon of floods in the Luni River basin in July 1979, when the occurrence of an instance of a monsoonal depression resulted in heavy rainfall over western Rajasthan. Flooding is a seasonal process that occurs every year in the monsoonal phase in India (Verma & Rai, 2023). The Luni River basin is located in the north-western part of India, covering the Rajasthan and Gujarat. Radically, it is a semi-arid desert region, but the catastrophic rainfall creates some instances, and the basin's sensitivity accelerates damage. It is smaller than the other significant rivers, and it receives water from a complex river system. In addition, a difference exists between the tributaries of the Luni River, which have different topographies and morphometries, and its effect on how to react to runoff and precipitation over the study region of desert India. River sensitivity plays a vital role in frequent experiences of flooding (Dhir et al., 1982; Kushwaha et al., 2024; Sharma et al., 1982). The Luni River's tributaries are found on both the left and right sides; however, there are unique hydrology as well as the geomorphology that play an important role in determining the dynamic flood nature. The present research focuses on the lack of systematic comparative studies of the hydro-geomorphological behaviour of flooding between the two banks of the Luni River. The river basin is also anthropogenically modified, creating seasonal water blockage and flood conditions. This study examines the significant morphometric and topographic indices of the geomorphology and geology, including structural conditions (lineaments) and the anthropogenic site of the river, to find the pattern of floods caused in the Luni River basin. Numerous morphological metrics, including the length of the stream, order of the stream, density of the river, and overland flow, have a strong correlation with the geology and geomorphology of the river and are considered. The important topographic factor, are like TPI, TWI, and slope are also taken into account in the study, for determining the regional potentiality of flooding, which regulates how the water moves through a river basin and how the topography reacts to the water and its effect on the flood. The water course of the Luni River has a complex interaction between the climate, hydrology, and geomorphology in a desert environment. Managing the water resource and reducing the flood risk requires for an understanding of the river behaviour, as the climate is very uncertain. The uniqueness of the research lies in characterising the basin characteristics and comparative analysis of both bank tributaries of the Luni River to demonstrate the varying effects of these parameters on flood causality. The objective of my research is (1) Mapping of flood occurrences to understand the flood sensitivity of the river systems (2) Understanding the nature and magnitude of possible causes of floods in the Luni River.

17.2 Study Area

The Luni River basin covers the states of Rajasthan and Gujarat and is considered to be a thirsty region of India. The basin is approximately 70,672 km^2, and the river originates in the Aravalli Hills near Ajmer city. After that, the river drains into the wetland of Gujarat (Chatterji et al., 1978; Deopa & Resmi, 2024). The basin is very rugged terrain in the east with the Aravalli Mountain ranges with rocky undulating surfaces and the narrowing the alluvial plan in the north and western part of the river characterised the westward slope of the river that joins to the Luni River has a differential drainage system and the control tectonically and lithological respect over the right and left side of the river basin. The length of the river is approximately 450 km, and the tributaries range from 76 to 150 km. The major tributaries are Sukri, Jojori, Guhiya, Bandi, Khari, Jawa, and the unorganised overland flow occurs over the desert sand sheet region (right bank) of the Luni River system. The region has much variability in rainfall conditions and 90% of rainfall occurs in 3 months of the monsoonal seasons (Sharma et al., 1984), and the seasonal fractionation is very high in that region, the monsoonal current are intensified by depression and low-pressure system that occurs over the Luni river basin creating flash flood with different magnitudes, the other magnitude of the flooding is the outcome of topographic and geological properties of the region creating an integrated drainage network, and interaction of monsoonal wind to the Aravalli ranges contribute the differential rainfall over the region. The rivers flow throughout various geomorphic regions, including mountains, terrain, valleys, and undulating plains some of the commonplace river or interdune areas having different sensitivity to flood, and creating the region so diversified, and the basin's hydrological response is shaped by the underlying geology and geomorphology, presence of impervious calcium carbonate pan and the shallow soil, about 50% of the area is rock exposed that accelerated the flash flood (Dhir et al., 1982). The river system and the catchment area are vital for understanding the hydrological dynamics, particularly during extreme flood events; the basin's historical flood record and the sedimentary evidence indicate the recordable of extreme floods (Kale et al., 2000). Even though floods have a significant influence on this region, the scientific literature has not thoroughly examined the causality of these floods, particularly across many tributaries of the Luni River. The foundation of this study is the necessity to comprehend the geographically distinct causes that contribute to floods, specifically between the Luni River's left and right bank tributaries (Fig. 17.1).

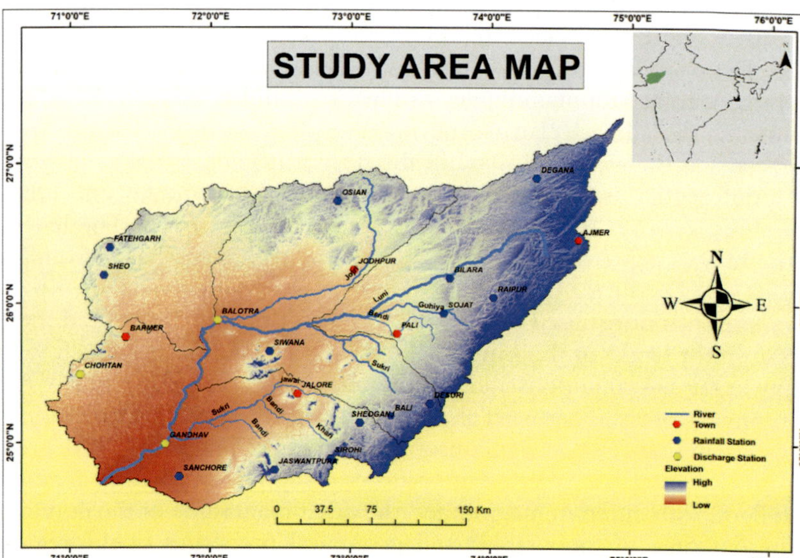

Fig. 17.1 Map of the study area

The climatic region of Luni is an arid region of India, having high temperatures, low precipitation (Kar, 1994). Mean maximum temperature during the winter season from 24.6 °C in the eastern parts to 26.5 °C in the western parts of the basin, while in summer, it ranges from 36.1 °C in the east to 38.3 °C in the west. The mean annual rainfall varies from 300 mm in the west to 550 mm in the east (Moharana & Kar, 2010). Precipitation is highly dependent on the cyclonic depression generally comes in in the monsoonal period, 80 to 90% of rainfall occurs in this annual period, and there is large inter-annual variability, leading to climatic extremes of drought and floods. On average, the Thar Desert in India, which includes parts of the Luni basin, experiences flood-causing rainfall (~300 mm or more in two to three consecutive days) in two to three years in a decade (Kale et al., 2000) with a maximum of five such years recorded during the 1950s, 1970s, and 1990s (Dhir et al., 1982). Heavy rains triggered flood damage in the Luni basin in July and August 1990 and August 2006 (Deopa & Resmi, 2024). The region experiences high moisture deficiency, creating conditions more favourable for aeolian processes than fluvial processes, although many areas have thick fluvial deposits from periods with different climatic and hydrologic conditions.

17.3 Methodology

17.3.1 Data Source

The data incorporates collecting various sources and analysing them to understand the hydro-morphological behaviour of the Luni River basin. The dataset involves conducting research that incorporates multiple national and international agencies, including the Geological Survey of India (GSI), and United States Geological Survey (USGS), HydroSheds data, and NASA Earth data. Each spatial and attribute data plays a critical role in analysing the river basin structure, drainage pattern, and associated geology and geomorphology. The foundational dataset started with a basic description of the area of study, which consisted of basin and sub-basin outlines that are available as a SHP file from HydroSheds. Primarily, it's essential for outlining the study area and understanding its extent and shape. Additionally, river network and stream order data are used for general understanding. The final study area was prepared using the Arc SWAT basin delineation tool, utilising the digital elevation model, and those datasets were processed and analysed using remote sensing and GIS software such as ArcGIS 10.4.1 and Global Mapper, which allows spatial analysis for visual interpretation.

Topographical maps provide detailed information on the study region and its physical features. Collecting open series maps from the GSI and using topographical maps and Google Earth Pro (GEP) for more detailed information enhances terrain analysis, geomorphology, and landform structure.

The digital elevation model (DEM) spatial data structure was crucial for the hydrological and topographical study, using ASTER DEM 30 m spatial resolution collected from NASA Earth data. Those data aid in the preparation of morphometric analysis, terrain modelling, as well as the computation of topographical indices. GIS software processes and analyses those DEMs for the identification of elevational changes, slope pattern changes, and watershed delineation.

The research utilises Landsat 8–9 OLI and Landsat ETM+, both with 30 m resolution. For such analysis, using software such as ERDAS Imagine 9.1 is used to allow satellite image classification into major classification groups. The Bhukosh portal provides additional thematic data on geology and geomorphology. The geological map helps to study rock formation and their impact on flood scenarios, geomorphological maps classify the landforms based on their process of development, which is essential for understanding the morphological and hydrological evolution of rivers. The lineaments map is used for the structural control of the river course, it reveals the faults and fractures, and how their effect on the flow path and direction is analysed using ArcGIS 10.4.1, ArcGIS Pro, and Global Mapper.

Google Earth imageries are helpful as a supporting role by providing high-resolution images. This data aids in understanding, validation, and configuration of geospatial datasets (Table 17.1).

17.3.2 Drainage Basin Delineation and Generation of Drainage Network

For generating the drainage network using arc Gis 10.4.1 platform in this study the acquisition drainage network using ASTER Dem (30 m) but the dem has inaccurate pixel values depicted the elevational information for removing that noise fill the dem by using the arc GIS software interface, for the evaluation of the final drainage network using Google Earth Pro for checking the drainage network.

Table 17.1 Data sources

Data type	*Details*	*Uses*	*Source*	*Links*
Features data	*File format*			
1. Basin and Sub-basin Outline	Vector File (.Shp)	For a basic outline of the river basin and pre-study and general understanding of the study region,	HydroSheds	https://www.hydrosheds.org/products/hydrorivers
2. River Network, Stream Order	Vector File (.Shp)	Understanding the branching pattern of the river and the order of the river	HydroSheds	https://www.hydrosheds.org/products/hydrorivers
Topographical map	*RF*			
Open Series Map	1:50,000	The topographical sheet reveals all the necessary information about these landforms. As well as their exact locations, proximal information	GSI	http://164.100.58.63/pages/display/189-download-open-series-maps(osm)
Toposheet From GEP	1:25,000		GEP	https://earth.google.com/web/@25.19914709,72.16486402

(continued)

Table 17.1 (continued)

Data type	*Details*	*Uses*	*Source*	*Links*
Dem data	*Resolution*			
1. ASTER Dem 2. SRTM Dem	30 m 30 m	Basin Preparation, Morphometric Analysis, Topographic Indices Analysis	NASA earth data USGS	https://asterweb.jpl.nasa.gov/index.asp https://Earthexplorer.Usgs.Gov/
Satellite remote sensing data	*Resolution*			
1. Landsat 8–9 Oli (2013) 2. Landsat Etm+	30 m 30 m	Digital Image Classification, Land Use Landcover Classification,	USGS	https://Earthexplorer.Usgs.Gov/
Google Earth Pro				
Images of GEP	–	General configuration of the region and verify the used Landsat Images, Flood Point Analysis,	GEP	https://earth.google.com/web/@25.19914709,72.16486402
Survey of India Data	*Resolution*			
1. Geology 2.Geomorphology 3. Lineament	50 k 250 k –	Geological Map (rock formation of ancient age to present and studied its importance over flood scenario, Geomorphological Map (studied the geomorphological landforms in the site of the river and the classification of landforms based of their process of development, Lineament Map (studied because of how the rivers are controlled over the lineaments)	GSI /Bhukosh GSI /Bhukosh GSI /Bhukosh	https://bhukosh.gsi.gov.in/Bhukosh/MapViewer.aspx

For the delineation of basin boundary and sub-basin delimitation using a SWAT project setup, which was added as an add-on with ArcGIS 10.4.1, after the pore point analysis, the basin was delimited. Also, verify and recheck the delineated basin by Google Earth Pro.

17.3.3 Preparation of the Geodatabase for the Thematic Layer

For the preparation of the thematic map, after using data from the GSI, thematic layers and elevation layers (geology, geomorphology, land use and land cover, and lineaments) were prepared using different remote sensing and GIS software, including ArcGIS 10.4.1, ArcGIS Pro, Global Mapper, and ERDAS Imagine 9.1 for the preparation of the map by using the coordinates of WGS_1984_UTM_Zone_43N (Fig. 17.2).

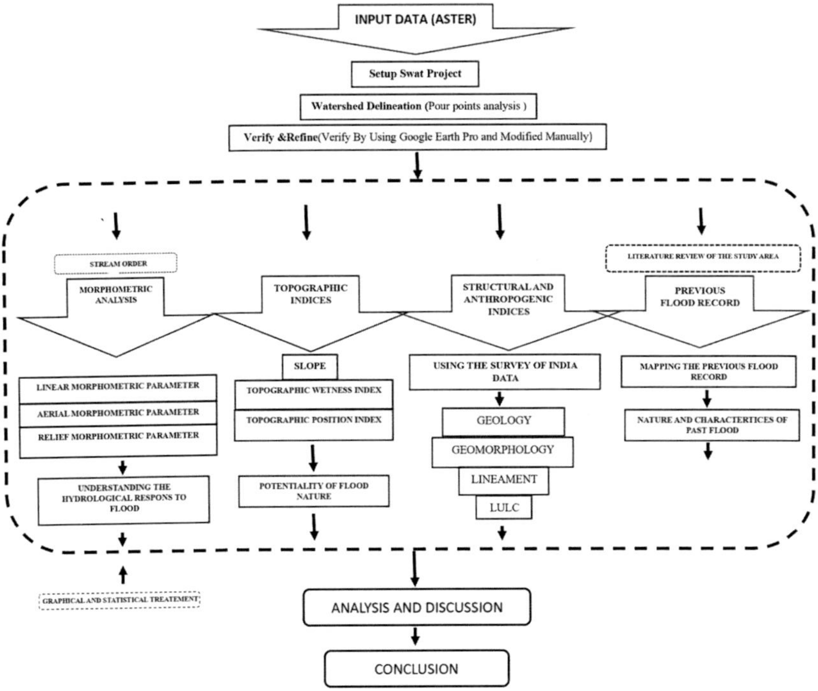

Fig. 17.2 Methodological flow chart

17.3.4 Computation of Morphometric Parameters

The study particularly uses the parameter that affects the flood's dependability, as the morphometric parameter and topographic indices are both significant. The morphometrics.

The morphometrics were performed in three aspects: linear aspect, aerial aspect, and relief aspect. For each of the six sub-basins,

17.3.5 Computation of Topographic Indices

Slope is one of the important topographic indices. Topographic indices are the matrices derived from elevational data, and they help in understanding the relief topographic features (Table 17.2).

And analysing the geomorphic and environmental processes that occur over the region (Gupta & Dixit, 2024). Landforms in the region is highly influencing the variability of floods and play a dynamic role in directing flow of water and affecting the acceleration of discharge. The use of topographic indices to collect physical topographic information of a particular region and for categorisation of that region, as well as understanding the regional potentiality of flooding, also defines the nature of the flood through topographic indices (Merz et al., 2020).

The topographic wetness index is also extracted as a hydrogeomorphic feature of the basin (Beven & Kirkby, 1979). Beven and Kirkby introduced the model and evaluated how surface topography controls the movement of water, following the formula for TWI taken from (Günther et al., 2014).

$$\mathrm{TWI} = \ln(\frac{\alpha}{tan\beta})$$

In addition, α is contributing to the area of the upslope per unit contour length in that particular basin area, and the β represents the gradient of local slope conditions to know the capacity of local drainage (Fig. 17.3).

The topographic position index signifies the height of each cell relative to the average height of the surrounding cells according to the radius (Costache & Bui, 2020).

$$\mathrm{TIP} = \frac{E_m - \sum_{n-1} E_n}{n}$$

Table 17.2 Conceptualisation of morphometric and topographic indices

Parameter/indices	*Formula/symbols*	*References*
Morphometric parameter		
No stream order	(Nμ)	
Total number of the steams within the basin	(ΣNμ)	Calculated from DEM
Stream order	Hierarchical rank	Strahler (1964)
Bifurcation ratio	Rb = Nμ/Nμ + 1 No of stream of the order /no of the stream of the next given order	Schumm (1956)
Total stream length of order μ	Lμ	Horton (1945)
Mean stream length of the order	$\bar{l}\mu$ = Lμ/Nμ total length of the stream of order/no of the stream of order	Strahler (1964)
Stream length ratio	RL = $\bar{l}\mu/\bar{l}\mu - 1$ The total length of the stream of an order /total stream length of the next higher order	Horton (1945)
Area of the order μ	Aμ	Schumm (1956)
Length of the basin of order μ	Lbμ	Schumm (1956)
Width of basin of order μ	Brμ	Schumm (1956)
Basin perimeter of order μ	Pμ	Schumm (1956)
Relative perimeter (Pr)	Pr = A/P Area /basin perimeter	Schumm (1956)
Drainage density (Dd)	Dμ = (ΣLμ/ΣAμ) Length of the stream of order/ area of the stream of an order	Horton (1945)
Stream frequency (Fs)	F μ = (ΣNμ/Σ Aμ) No of stream of an order/area of the stream of an order	Horton (1945)
Infiltration Number (If)	If = Fs * Dd Stream frequency/drainage density	Strahler (1964)
Length of overland flow (Lg)	Lg = 1/Dd * 2	Horton (1945)
Constant channel maintenance (sq. Kms/km)	C = 1/Dμ Half of the drainage density	Schumm (1956)
Height of the highest point on the watershed	Z	Calculated from DEM
Height on basin mouth	z	Calculated from DEM
Total basin relief	H = Z-z	Strahler (1952)
Relative Relief	Rr = H1–H2	Melton (1957)

(continued)

Table 17.2 (continued)

Parameter/indices	*Formula/symbols*	*References*
Relief ratio	Rh = H/Lb Total relief of the basin/basin length	Schumm (1956)
Dissection Index	DI = H1–H2 / H1	Horton (1945) and Schumm (1956)

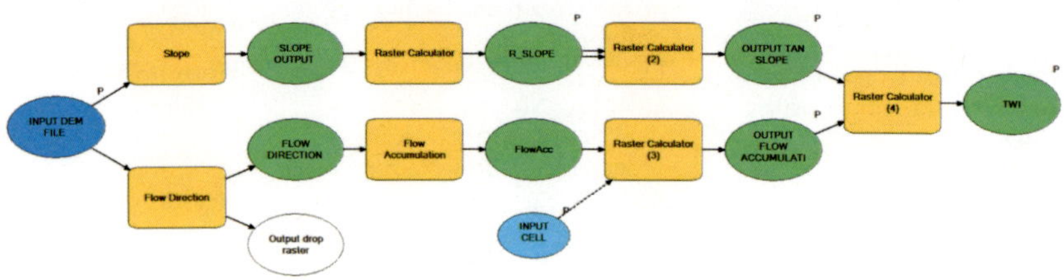

Fig. 17.3 Conceptual model builder of the TWI index

where the elevation of the centre is, and the average elevation of the surrounding radius is E_n The positive values indicate the elevation is higher than the surrounding elevation, and lower TWI signifies the elevation than those lower surroundings (De Reu et al., 2012).

17.4 Result and Discussion

17.4.1 Nature and Characteristics of Past Floods in the Luni River

The Luni River has experienced an increase in floods in the past few years due to a shift in precipitation across Rajasthan. In July 1975, an intensifying flood caused lateral spreading and migration of channels in the Mathri, Jojri, Guhiya River, and Bandi Rivers, which were the upstream tributaries of the Luni River, mainly due to upstream rainfall in the Pali District and south of Jodhpur, which were severely affected by the floods (Moharana & Kar, 2013). People in the area recall the 1979 flood in March, which was as disastrous as the one that occurred 177 years ago. The level of the water was 25 to 30 cm, and it rose 500 cm in one day of rainfall. The Bhavi region is situated in the upper course of the Luni River basin water level rose from 75 to 400 cm in just 5 hours of rainfall and spread to both sides by about 1 km. Bendi River at Jetpur rose from 75 to 400 cm in just 5 hours. At the gauging station of Bisalpur, the river is flowing at 70 cm, whereas it gradually rose to 5.66 m in 4 to 5 hours and declined in 3 hours, that is, at a faster rate of Lateral spreading. The upstream River (Mathir, Jojri) spread to both sides about 0.5 km and the other side up to Piper city, with an average lateral spreading of 1.75 km in the flood year of 1979 (Dhir et al., 1982). India's second satellite program was launched in 1979 using a satellite-prepared flood map, showing the Jalor and Pali regions affected by the flood (Pandey, 1980). According to (Kale et al., 2000), the average rainfall of the basin is 300 to 400 mm, but the region received high precipitation above 500 mm in 5 days, creating a catastrophic flood. This unprecedented flood devastated 93,000 houses, causing losses of about 370 people. More than one lakh animals were washed out, and more than 50,000 hectares of land were rendered unsuitable because of flood deposition on both sides of the river (Dhir et al. 1982; Sharma et al., 1984). The level of the flood was 179.52 m above the MSL, but in the year 1979, on 15 July, it rose to nearly 2 m above the previous height of the flood (Dhir et al., 1982). In 1979, floods also occurred in the regions of Bamer Jailor and western Pali, mainly the Lower Luni River basin. In 1990, floods also happened two times in the first week of July and August because of the monsoonal depression. In, flood in July, the left side of the river was more affected because flood water coming from the Sukri River affected the Sirohi, Pali Jalor. The Sanchor region of Bali recorded high rainfall that is above 700 mm in five days, the second time of a flood that affected the Jodhpur and Beamer regions of the right tributary of the Luni River in June (Singh et al., 1992).

In 1996, the Luni River flooded the regions of Jalore, Jodhpur, Jaisalmer, and Balotra-Barmer. During flash floods, high amounts of sediment are deposited on both sides of the river, creating wide and shallow river channels (Kar, 1994). Additionally, in 2001, rainfall over the basin created moderate flood conditions (Deopa & Resmi, 2024).

In 2006, for four consecutive days, rainfall poured into the regions of Bermer and Jaisalmer due to the flood water not having proper outlets over the shallow soil conditions and flat topography. However, the 2006 flood in Bermer was unique because the area lacked well-connected drainage outlets for flood discharge, resulting in Kawas village being under 10 to 20 feet of water.

In July 2015, July experienced 60% more rainfall, causing floods that affected the Jaisalmer, Barmer, and Jalor regions (Kotecha et al., 2023).

According to the IMD, the 2019 flood affected the Jodhpur, Nagaur, and Pali regions due to intensifying rainfall. The recent flood that occurred on 10 August 2024, according to reports, was a flash flood that occurred in the middle of the basin, mainly affecting Bermer, Jodhpur, and Balotra (Fig. 17.4).

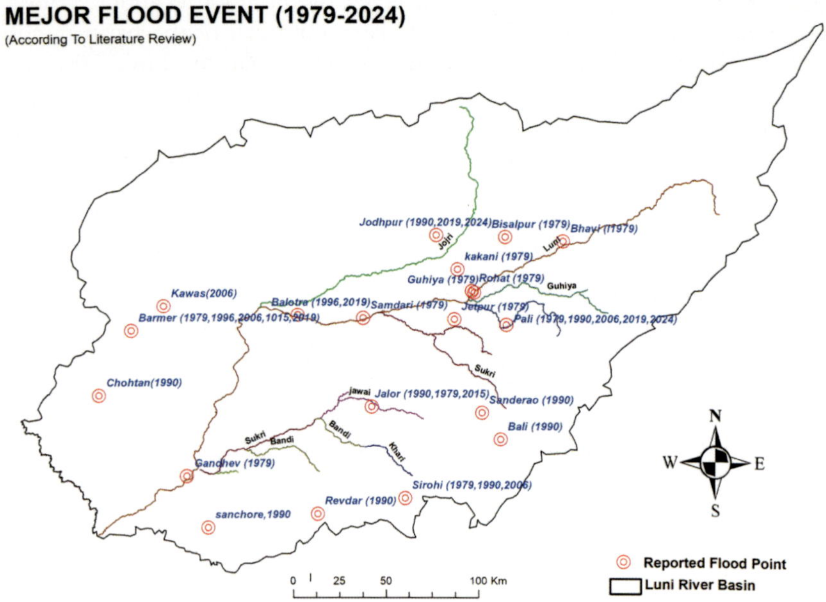

Fig. 17.4 Demonstrates the year of flood occurrence according to the previous year's flood report and literature review of the basin, comparing the left side of the basin, which experienced more frequent floods and multiple reoccurrences, especially in the Pali, Jalor, and Sirohi areas, and the right side of the basin, which has a long record of flood occurrences in only Barmer and Jodhpur

17.4.2 Geomorphological, Structural, and Anthropocentric Influences on Floods

From a global perspective, the magnitude, frequency, duration, and flood volume depend on the source of external forces as well as internal mechanisms such as local geology and physiography (O'Connor et al., 2011). Geomorphology and geology play a significant role in influencing floods and shaping hydrological behaviour (Thompson & Clayton, 2002). They determine the suitability for flooding. Geologically, approximately 80% of the Luni River area is a young geological formation, primarily a Quaternary formation (Aravalli and patches of older formations), while an impervious underlayer covers approximately 50% of the basin. The terrain is composed of rugged small mountains featuring igneous rock and crystalline metamorphic rock. The entire basin is situated on hard and massive rock formations with crystalline and metamorphic processes, as the depth of the alluvium is less than 1 m to 4 m (Kar, 1994). Since the Luni River flows over unconsolidated and impervious bedrock, hardpan, and a large exposure of rock that hinders the creation of a thick river profile and groundwater reservoir, the flood process is accelerated with a very high magnitude, often resulting in flash floods or catastrophic events (Kale et al., 2000). The aridification process involves the development of massive impermeable formations of calcium carbonate pan with varying depths (Sharma et al., 1984). From a geomorphological perspective, factors such as drainage pattern and basin morphometry are heavily controlled by the geological structure (Jain & Sinha, 2003; Lastra et al., 2008). The occurrence of lime carbonate pan below a shallow soil layer limits soil infiltration on the river course, with sediment depth varying from 13 to 112 cm, and outside of the main river channel course. Geology complements the geomorphological factor of subsurface runoff; the porosity and permeability of rock, soil, and rock formation, as well as cohesiveness and strength, determine the drainage pattern. Owing to geological advantages, structural drainage is developed on the left side of the river basin. The right-side catchments and left-side catchments of the basin are very different geologically because the older formation is on the left side of the river, and the newer formation on the right side of the river produced different types of terrain conditions. These conditions influence the stability and resistance (left of the Luni basin having the extended part of Aravalli mountain and the fragmented short mounts create the gradient for the runoff potentiality of the river) and also impact the infiltration and runoff of the geomorphological landform to the left of the Luni River, the domination of fluvial process resulted in there is a high percentage of alluvial plain and well-connected river higher relative relief produces high magnitude and low duration flash flood, but the right side is mostly the aeolian process. The landforms are flat, and a low gradient produced the landforms like the interdune plan, aeolian sand sheet, and aeolian dune.

The lineament has the role of shaping the flood over the geological structure, it was assumed that such movement led to the formation of horst graben faulting and folding. The lineament is suspected to have extended from the Rann of Kutch to Sambar Lake (Dhir et al., 1982). Several lineaments cross the Luni River mid-plain and Jawai plain, and many of them are suspected of controlling the present-day channel processes and floods (Dhir et al., 1982; Kar, 1994). Jawai is a notable tributary of the Luni River basin (Kar, 1994). The Jawai plain, as seen from satellite imagery from different directions, crosses several lineaments that control the flood as well as the flooding behaviour. For this reason, the left basin has a specific region (Pali, Jalor, Sirohi, Bali) where a flood will reoccur repetitively with high magnitude because it has a limited area for any type of change in the river course (Fig. 17.5).

Human activity plays an important role in not only the effects of the river channel but also the floodplains at a global scale, which strongly depends on the hydrological and geomorphological characteristics of the whole basin (Friedrich et al., 2019; Fryirs, 2016; Wolman & Gerson, 1979). In some instances, human disturbance creates a disbalance in the highly modified river system or landscape, leading to irreverent alteration of catchment-scale discharge and sediment fluxing, and insolvent alteration of the river creates system sensitivity, which disbalances the lower system of a river. An increase in human activity leads to an increase in built-up areas, leading to more runoff. There is very limited space for human activity, but dam construction and the development of roads modify the river discharge.

17.4.3 Spatial Variation in Morphometry Influence Over the Flood Pattern

Morphometry is a quantitative investigation of the drainage basin (Strahler, 1964). The basic drainage parameters are basin area, basin perimeter, basin length, max elevation, and min elevation. According to

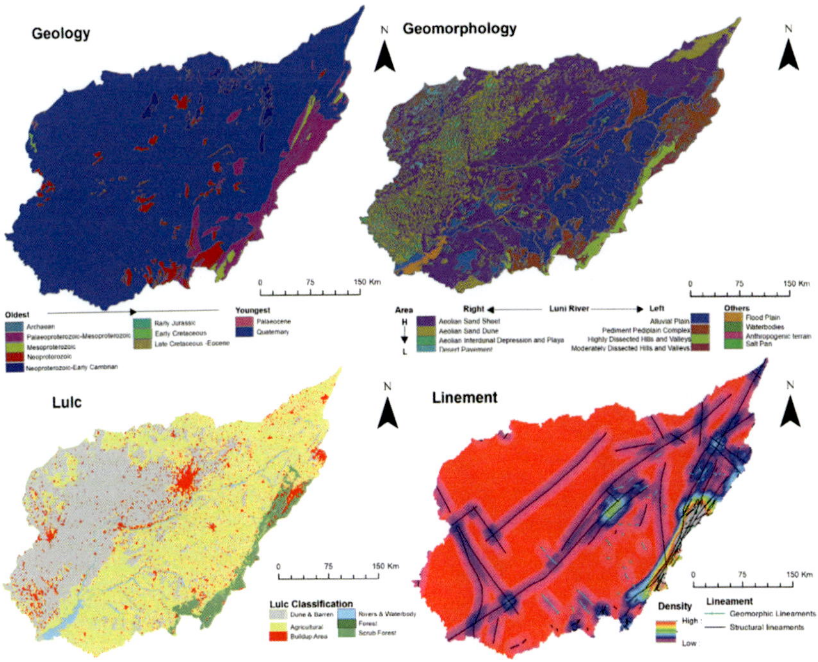

Fig. 17.5 Geology: Map is prepared according to the oldest to newer formations, whereas the majority of the basin is a composite of the youngest geological formations having fragmented older geology the erosional mounts and hills and left respectively older formation giving the landscape more resistance having the formation of metamorphic and ingenious rock, Geomorphology: the right and left geomorphology is different right are more prone to aeolian geomorphology and left are more concentrated towards fluvial geomorphology; LU LC: land use and landcover also been visible the difference of classification respect to the left and right of the basin; Lineament: left of the basin are high density of lineament respect to left and do influence over the flow pattern

Strahler, the number of the stream order is directly interlinked with the size and dimensions of the drainage basin (Strahler, 1964). Then, no of the stream and length will be measured according to their order.

17.4.3.1 Sub-basin Delineation

Six sub-basins as delineated by using ArcGIS 10.4.1, three basins right side of the Luni River and three basins on its left side of the Luni River (Figs. 17.6 and 17.7).

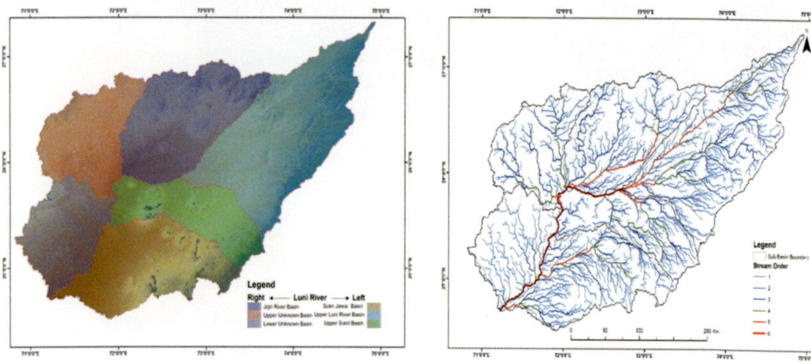

Fig. 17.6 Sub-Basin of Luni River

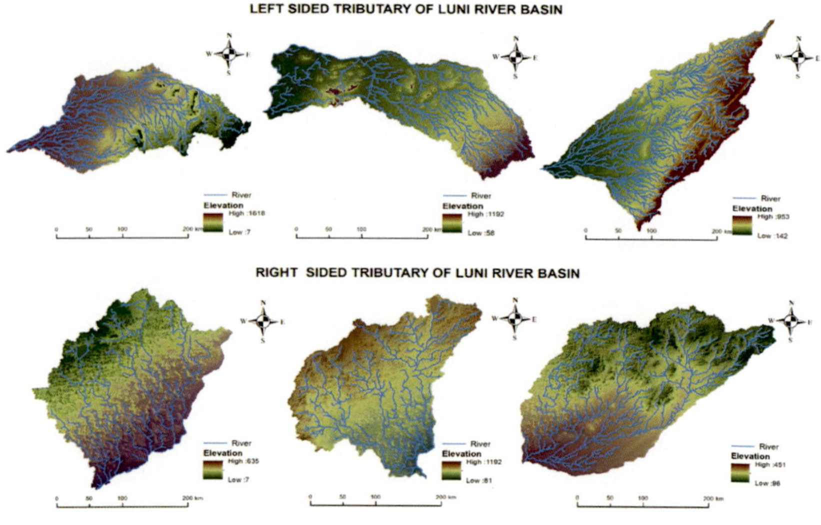

Fig. 17.7 Sub-basin elevational difference

17.4.3.2 Stream Order, Bifurcation Ratio

According to Strahler, the number of stream orders is directly linked to the size and dimensions of the drainage basin (Strahler, 1964). The bifurcation ratio helps to figure the basin geometry along with the numbers and order of the stream, according to the law of stream numbers. The stream in each order is inversely related to the geometric sequence with order numbers (Strahler, 1964). When the geological structure is not distorted in the basin, the range of the bifurcation ratio will be 3.0 to 5.0; if there is any geological control, it will have higher ratio values, and the lowest possible value is 2.0 when it approaches a basin under very natural conditions (Strahler, 1964). In the case of flooding, the combined stream order and bifurcation ratio play important roles in reflecting how much the geology and lithology control the river system (Horton, 1945). The ratio is termed the bifurcation ratio ($Nu/N\underline{u} + 1$), where Nu is the number of streams needed for one stream in the higher order (Strahler, 1964) (Table 17.3).

Across the Luni River basin, both sides of the catchment have different drainage networks. The number of rivers is five times greater on the

Table 17.3 Stream order

Catchment		*I Order*	*II Order*	*III Order*	*IV Order*	*V Order*	*VI Order*	*Total*	
Left	*Upper Luni Bain*	905	109	15	5	2	1	1037	2090
	Upper Sukri Basin	304	86	19	2	1		412	
	Sukri Jawai Basin	550	70	16	4	1		641	
Right	*Jojri River Basin*	239	49	12	3	1		304	488
	Jojri River Basin 1	89	24	7	3	1		123	
	Jojri River Basin 2	46	11	3	1			61	

left bank site compared to the right bank site. The left bank streams have relatively higher capacity, and the drainage area accumulates more water. Left-bank rivers impact the lower reaches, causing catastrophic flash floods. Also, the slope is high, and the response time is subsequently very low for those tributaries. The well-organised drainage of the left bank has led to the development of a wide flood plain, as the right bank of the river has few streams to drain the water. The nature of the catchment is more prone to localised flash floods (Table 17.4).

Over the Luni River basin, the left-sided catchment's higher bifurcation ratio suggested that steep and dissected terrain, tectonically controlled. The drainage network resulted in higher flood risk in the upstream and lower stream area compared to the right-sided catchment of the Luni River basin, which had a low bifurcation ratio and many flatter streams, indicating stable water accumulation and fewer outlets for water discharge.

In the upper Sukri basin, approximately ten streams of third order streams are joined to maintain one fourth order stream, and the river also

Table 17.4 Bifurcation ratio

Catchment		*I Order*	*II Order*	*III Order*	*IV Order*	*V Order*	*VI Order*	*TotalAverage*	
Left	*Upper Luni Bain*		8.3		7.263	2.5	2	23.064.612	4.85
	Upper Sukri Basin		3.53		4.529.5	2		19.554.8875	
	Sukri Jawai Basin		7.85		4.374	4		20.225.055	
Right	*Jojri River Basin*		4.87		4.084	3		15.953.9875	3.47
	Jojri River Basin 1		3.7		3.422.33	3		12.453.1125	
	Jojri River Basin 2		4.18		3.663			6.663.33	

has geological control, but the large volume of upstream water results in catastrophic floods in the lower region; for this reason, Balotra flooded catastrophically in 1996 and 2019, and Samdari in 1979.

17.4.3.3 Stream Length, Mean Stream Length, and Channel Storage Capacity

Stream length is an important morphometric parameter; it decreases according to the geometric sequence (Horton, 1945). The low stream length also reflects the channel storage capacity of the river, which is proportional to the total length of the stream. The stream length is calculated by using the measurements on the ArcGIS platform. The stream length in kilometres is a very notable hydrological parameter for understanding the basin runoff characteristics; longer stream lengths and shorter stream lengths reflect the steepness, gradient, and texture of the basin (Strahler, 1964). Mean stream length is a dimensionless property of a river basin, characteristic of size and a component of the drainage contributing to the basin surface (Strahler, 1964). The mean stream length of the channel of a given order is greater than that of the next order (Horton, 1945) (Table 17.5).

The stream length ratio of the sub-basin is calculated by dividing the mean length of the stream of order and the next lower order of the stream for each sub-basin; higher values depict the young geological formation, and lower values are vice versa. (Kaliraj et al., 2014) As mentioned above about total stream length (sum of all streams in order) a longer total stream indicates the most extensive drainage network its facilitation better water dispersion, therefore, a longer total stream reduces flood risk but leads to the water in the lower sub-basin the total stream length and mean stream length determine the capacity and efficiency of a basin to Hendel the basin runoff and mean stream length reflects the development and efficiency of the drainage branching, which is also important for flood prevention (Table 17.6).

Mean stream length is a critical parameter for understanding the geomorphology and hydrological behaviour of the river in the left-side catchment. The values suggest that a steeper upstream area and a flatter downstream tectonic factor indicate a river with a high potential for **flash flooding** in comparison to the right-sided catchment. The right-side tributary values reflect that the simpler drainage network with limited connectivity and limited terrain causes low flood levels when monsoonal depressions occur (Fig. 17.8).

Table 17.5 Mean stream length

	Catchment	*I Order*	*II Order*	*III Order*	*IV Order*	*V Order*	*VI Order*	*Total*	*Average*	
Left	*Upper Luni Bain*	3.9	6.21	32.62	80.24	111.19	13.14	247.3	41.2	51.2
	Upper Sukri Basin	3.78	6.65	13.04	113.38	161.52		298.37	59.6	
	Sukri Jawai Basin	3.52	14.8	28.17	36.86	180.6		263.95	52.7	
Right	*Jojri River Basin*	4.95	15.86	23.65	56.42	97.39		198.27	39.6	20.9
	Lower Jojri River Basin 1	2.06	3.66	9.4	16.33	18.00		31.45	7.8	
	Lower Jojri River Basin 2	2.95	12.61	20.24	25.96			61.76	15.4	

Table 17.6 Stream length ratio

Catchment		*I Order*	*II Order*	*III Order*	*IV Order*	*V Order*	*VI Order*	*Average*	
Left	*Upper Luni Bain*		1.59	5.25	2.46	1.39	0.12	1.35	11.81
	Upper Sukri Basin		1.76	1.96	8.7	1.42		2.62	
	Sukri Jawai Basin		4.2	1.9	1.31	4.9		1.47	
Right	*Jojri River Basin*		3.2	1.49	2.39	1.73		0.59	0.74
	Lower Jojri River Basin 1		1.77	2.57	1.73	1.1		0.38	
	Lower Jojri River Basin 2		4.27	1.61	1.28			1.25	

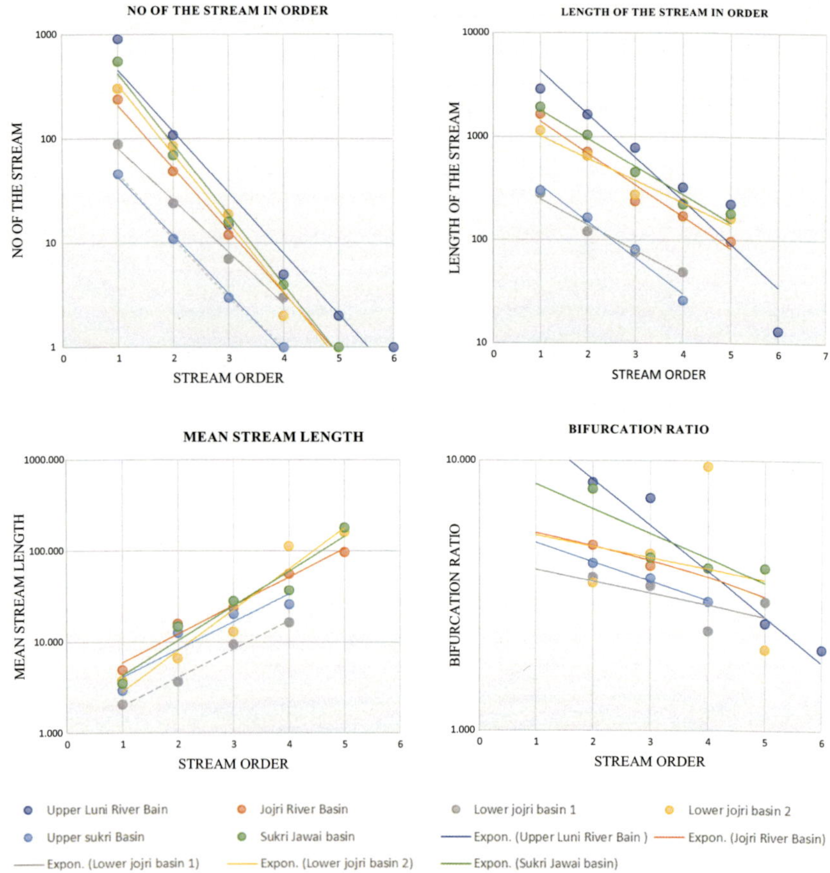

Fig. 17.8 Graphical representation of linear morphometric parameters

17.4.3.4 Drainage Density

Drainage density explains how the rivers are spaced and close together in a basin, and defines the lag time concentration of the basin (Horton, 1945). Slope and gradient are the main morphometric controls on drainage density, which is measured by the total stream length of the basin divided by the area of the basin (Strahler, 1964). The drainage density reflects the landscape dissection, climatic conditions, infiltration capacity, and runoff potential of a basin (Alam et al., 2020). Drainage basins with small drainage density have the potential for higher flash floods, but higher drainage density has a low flood potential, and water flows towards the lower sub-basin with high discharge (Abdel-Lattif & Sherief, 2010; Smith, 1950). High values of drainage density reflect the high relief and impermeable rock surfaces that contribute more runoff in the lower stream of the basin.

Over the Luni River basin, the average drainage density is higher in the left-sided river basin, with a high interconnection of stream water flows that results in water flowing very quickly throughout the river basin and higher drainage density. This condition often results in a higher likelihood of flash flooding in the lower stream regions like Sanchore, Ravdev, Sirohi, Bali, Sanderao, Pali, Bhavi, and Jalor. Specifically, Pali and Jalor experience floods every time a monsoonal depression occurs.

Because the system quickly conveys rainwater downstream from the left side. The lower drainage density in the right-sided river basin takes more time to infiltrate, but ground conditions are different in that water does not much infiltrate; for that reason, it **increases the duration** and **lowers the magnitude of flooding**.

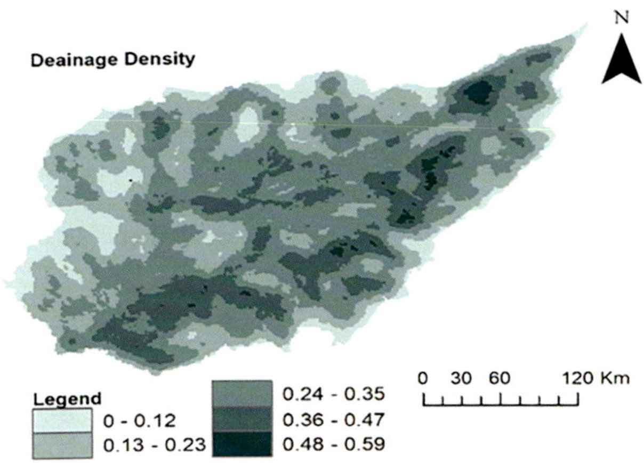

17.4.3.5 *Stream Frequency*

Stream frequency or channel frequency is the number of streams per unit area (Horton, 1945). Stream frequency results in a high concentration of runoff (Alam et al., 2020; Kaliraj et al., 2014). Generally, high stream frequency is related to unconsolidated rock surface material, high relief, and low infiltration conditions (Ozdemir & Bird, 2008).

The Luni River basin's highest stream frequency in the left catchments has a significantly higher number of streams per unit of area, characterised by a quick river response, causing flash floods with immediate flood peaks. A lower stream frequency in the right catchment indicates a lower number of streams, limited outlets, blockage of water, and slower onset of the flood, with a fragmented flood area.

17.4.3.6 Texture Ratio

The texture ratio measures the number of streams and the perimeter of the basin over the Luni River. The left basin has higher texture rates, indicating a more channel per unit of perimeter. The basin rapidly conveys the water into the drainage network, which leads to a higher risk of flash flooding, also associated with a steeper slope, and more impervious surface, while the right catchment of the Luni River basin has fewer streams per unit of perimeter, reflecting a coarser drainage structure, which indicates a lower likelihood of flooding with longer water retention times.

17.4.3.7 Infiltration Number, Length of Overland Flow, and Constant Channel Maintenance

The length of overland flow represents the average distance that water travels over the ground surface before it concentrates into channels (Horton, 1945). This length affects runoff generation and is a key factor in watershed and streamflow modeling related to the sub-basin's slope and gradient (Rudraiah et al., 2008). The maintenance of the channel is defined as the reciprocal of drainage density (Schumm, 1956a, 1956b). Constant channel maintenance is expressed as the lower values with our higher potential for flood.

Constant channel maintenance is expressed as lower values, indicating a higher potential for flood, low infiltration, and impervious soil conditions over the basin, the left basin has lower infiltration, shotter overland flow (allows the water for quick runoff) lesser area for constant channel maintenance of the channels and the other side has a lower flood potentiality with a longer duration, high infiltration, longer overland flow, longer time for water to infiltrate, and longer channel instances (Table 17.7).

Table 17.7 Catchment characteristics of Luni River (Left and Right Bank Tributary)

Basin Characteristics	*Left*				*Right*			
	Upper Luni Bain	Upper Sukri Basin	Sukri Jawai Basin	*Average*	Jojri River Basin	Lowe Jojri River Basin 1	Lower Jojri River Basin 2	*Average*
Drainage Density	0.227	0.293	0.291	**0.270**	0.166	0.198	0.166	**0.177**
Stream Frequency	0.050	0.049	0.049	**0.049**	0.015	0.044	0.006	**0.022**
Texture Ratio	1.501	0.858	1.050	**1.136**	0.620	0.299	0.160	**0.360**
Infiltration Number (If)	5.650	5.995	5.970	**5.872**	9.442	4.384	9.372	**7.733**
Length Of Overland Flow (Lg)	1.777	1.703	1.717	**1.733**	3.515	2.611	8.697	**4.941**
Constant Channel Maintenance	3.555	3.406	3.435	**3.465**	7.030	5.223	17.394	**9.882**
Total Basin Relief (H)	811	1134	1611	**1185**	355	368	628	**450**
Relief Ratio (Rh)	2.885	7.247	7.265	**5.799**	1.775	4.054	4.751	**3.527**
Ruggedness Number (Rn)	0.228	0.333	0.469	**0.343**	0.051	0.070	0.036	**0.052**
Gradient Ratio (Rg)	2.885	7.247	7.265	**5.799**	1.775	4.054	4.751	**3.527**

17.4.3.8 Total Basin Relief, Relief Ratio, and Ruggedness Number

An essential indicator of a watershed's runoff, landform change, and denudation is basin Relief. The relief ratio is the height relief to horizontal distance along the basin's longest dimension parallel to the main drainage line. It evaluates a drainage basin's overall steepness and acts as a gauge of the intensity of erosion processes on the basin's slopes. Ruggedness is defined as the degree of surface irregularity and the smoothness and roughness of the basin topography. Steep slopes are suggested by rough terrain with a high roughness score, which could lead to flash floods and erosion. A low roughness value denotes a less steep slope, which traps flooding water at the surface. The gradient is the change of vertical elevation by a change of unit of horizontal distance. The gradient ratio is the slope of the drainage basin by the subduction of the highest elevation and the lowest elevation of the basin, for which runoff evaluation is estimated (Sreedevi et al., 2005). A higher gradient may indicate a higher acceleration of the runoff-induced flash floods, along with the drainage density and drainage frequency of the basin (Abdel-Latif & Sherief, 2010; Kaliraj et al., 2014).

Over the left basin of the Luni River, more significant elevational differences and the steeper terrain and denser channel network lead to high flash floods due to rapid runoff, and the right basin of the Luni River has less elevational difference and less rugged terrain, which indicates a lower magnitude of flood.

17.4.3.9 Gradient Ratios

It is defined as the change in vertical elevation by a change in the unit of horizontal distance. A low roughness value denotes a less steep slope, which traps flooding water at the surface. The higher the gradient is, the greater the acceleration of the runoff-induced flash floods, along with the drainage density and drainage frequency of the basin (Abdel-Latif & Sherief 2010; Kaliraj et al., 2014). On the left bank of the Luni River, a higher gradient ratio has a steeper terrain that makes a trend that tends towards flash flooding due to rapid water movement, and the right bank has a gentle slope, which reduces the speed of the runoff, resulting in slower but more prolonged flooding (Fig. 17.9).

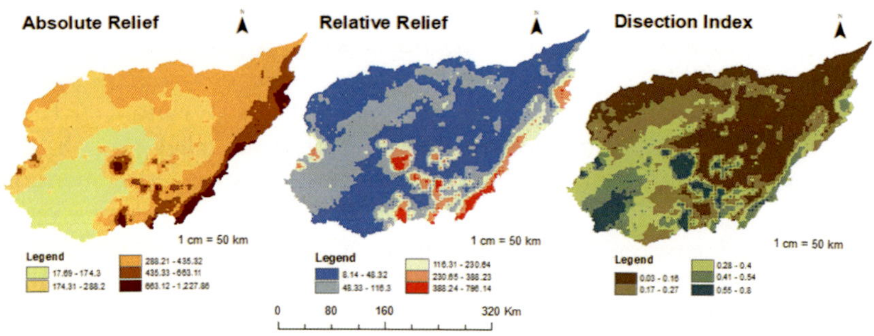

Fig. 17.9 Relief parameter of Luni River basin

17.5 Topographic Indices and Its Influence Over the Flood

17.5.1 Slope

Slope value provides important information on the topographic steepness, which directly influences water runoff, erosion, and flood behaviour. A high slope value indicates a steeper area leading to fast runoff and less water retention, while a slope value close to zero is associated with flatter terrain where water accumulates, increasing the flood potential.

For the right basin, the value ranges from 0 to 60 degrees, but most of the region has a flatter terrain, resulting in slower runoff and fewer instances of flooding the left basin of the Luni River result in significantly higher slopes, indicating steep and rugged topography These conditions result in the faster runoff, greater potential for flash floods, and increased erosion risk. However, the presence of flat zones (slope = 0) makes these areas vulnerable to localised water pooling.

In particular, for the Luni Jawai River basin, the relief difference is very high significantly higher acceleration of discharge. A flash flood occurred in 1990 in Sanchore and Ravdev. Upstream and downstream of Gandhav were affected by flash floods in 1979.

17.5.2 Topographic Wetness Index

The Topographic Wetness Index (TWI) describes the location and size of the area that may be subjected to retention of water caused by overland flow generation. TWI was developed by Beven and Kirkby as a runoff model. The effects of local topographic conditions on the generation of runoff and short-term and long-term moisture in the landscape are shown. Water accumulation is described as the trend of water at a point where the water accumulates over the local slope and gravity (Fitra et al., 2024; Pourali et al., 2014).

Along with the topographic configuration, the TWI provides insight into soil moisture distribution and areas prone to water accumulation. High TWI values indicate zones likely to retain water (valleys, low-gradient areas), while low TWI values indicate areas with higher runoff potential (ridges, steep slopes). Here are the TWI values for the catchments on the left and right banks of the Luni River basin. The difference is of much significance, the effect on the nature of flooding, e.g. left of the basin suffering from flash flood, and the other is a prolonged flood. The left bank of the Luni River has a significant accumulation zone, especially in the low gradient areas, and those areas are prone to water retention. The lower part of the hill area has potential for flash floods.

17.5.3 Topographic Position Index (TPI)

TPI is a geomorphic-based automated landform classification methodology that provides detailed and comprehensive classification and categorisation of landforms within the area of study (Gupta & Dixit, 2024; Thannoun & Ismaeel, 2024). Therefore, for the TPI analysis the high neighbouring size value it's for major landform classification and lower neighbouring cell is major landform disappeared and it is focused on the minor landforms (Al-Sababhah, 2023) The positive of TPI I values the representing the positive relief feature ridge and hilltop, and the negative TPI values and reflects the negative relief feature that is plan area, shallow slope and mid-slope (Thannoun & Ismaeel, 2024).

There is a significant difference in TPI; the right bank has comparatively low TPI values; that is total range of their variation is from − 12.66 to 28.22, which generally indicates the relief variation is not very high, with gentle slope conditions and moderate runoff. The region prominent lower ridge and very shallow valley and the gentle terrain might lead to slower rate of runoff, reducing erosive potential and increasing the flood extent. And the context to the left bank of the river basin the TPI values the ranging between − 50.66 to 52.55 extremely steep terrain increases the potential for rapid water flow, making the area is highly susceptible for flood and the high terrain leads to strong erosion.

The right basins of the Luni River have a gentle TPI, where floods may cover broader areas but have reduced flow velocity. Flood risk may be more related to waterlogging or inundation, and for the left bank, Larger TPI variations reflect steeper, more rugged terrain. Floods here are likely faster, more erosive, and potentially destructive, especially in valleys with deep depressions (Fig. 17.10 and Table 17.8).

In summary, floods in the Luni River basin are influenced mainly by the differences in slope, Topographic Wetness Index (TWI), and Topographic Position Index (TPI) across different basins. High slope values, such as 70.37 in the Sukri Jawai Basin, lead to faster runoff, reduced water retention, and limited flood duration, whereas the lower slopes, especially those in right bank basins, create flat terrain, and water accumulates more

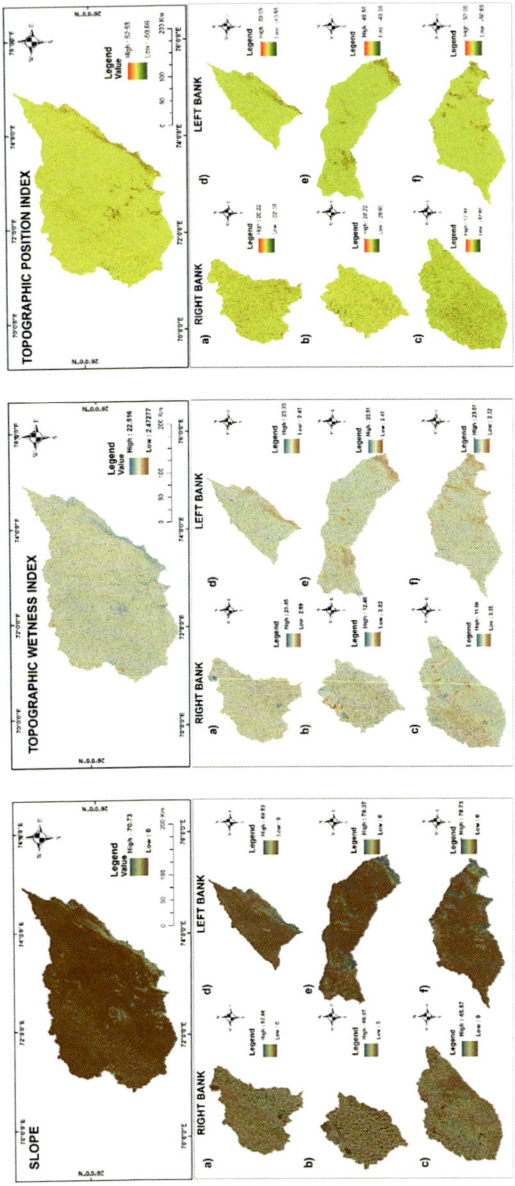

Fig. 17.10 **a** Lower Jojri River Basin **1b** Lower Jojri River Basin 2, **c** Jojri River Basin… **d** Upper Luni River Basin **e** Upper Sukri Basin…. **f** Sukri Jawai Basin

Table 17.8 Basin wise topographic indices

Basin characteristics	*Left*						*Right*					
	Upper Luni Bain		Upper Sukri Basin		Sukri Jawai Basin		Jojri River Basin		Lower Jojri River Basin 1		Lower Jojri River Basin 2	
Values	*High*	*Low*	*High*	*Low*	*High*	*Low*	*High*	*Low*	*High*	*Low*	*High*	*Low*
SLOPE	69.52	0	70.37	0	70.73	0	45.57	0	57.66	0	60.07	0
TWI	23.33	2.43	22.51	2.45	23.51	2.32	11.98	3.35	23.45	2.99	12.46	2.82
TPI	39.55	–40.55	49.55	–43.55	52.55	−50.66	17.44	-12.66	20.22	−22.33	28.22	−28.66

if there is sufficient rainfall. TWI values further infuse those patterns, with the high values in the left basin indicating greater moisture retention and making a key flood-generating zone, and the lower values. In the right basin, the opposite conditions occur. Over the TPI values, the left basin primarily experiences flash floods due to the steep slope and higher runoff. Whereas the right bank basin has the potential for prolonged and regional flooding due to poor drainage and topographic depression.

17.6 Flood as a Resource for Water Resource Management

In arid and semi-arid regions, flash floods are considered to be the most devastating natural hazard, particularly in terms of human life and infrastructure. However, the large volume of water generated by the intense and short duration of the rainfall events is a potential water resource for future use and sustainable water uses. Floods provide renewable freshwater, primarily by recharging the groundwater. Proper utilisation and strong management of those renewable surface water sources in the arid regions is an optimal choice to overcome the water deficiency problem.

Luni River basin in the Thar Desert of India is an example of such an arid region, and the Luni River is described as a lifeless river, where any flow of water is notable news. Stream flow is almost zero in all of them due to high evaporation. The river is typically dry, with measurable flow occurring only in response to high magnitudes of rainfall during the monsoonal season; those events are also short-lived (sometimes lasting not more than two days), can cause the channel to widen and overflow, resulting in a flood. Historically, the Luni River experienced several floods due to unusual rainfall patterns.

For an arid basin like the Luni, capturing and utilising those infrequent flash flood events is crucial for material resource management, especially given to challenges of water scarcity. Harvesting such events can significantly improve ground water resource and provide water for crop production over longer periods, and dependency over the dry farming is reduced and which can help compensate for frequency occurring drought of that region, water harvesting is significant approach for the water resource management and can provide the water for human and livestock uses and agricultural development, this approach decreases peak flows, reducing flooding and damages and transforms the water into a uses full resource for agriculture.

17.7 Key Findings

1. The left bank of the Luni River Basin is more suitable for flash floods due to its steeper gradient, rugged terrain, and higher drainage density and the right bank has slower runoff localised pooling and less intensive or magnitude of the flood because of gentle slope and lower drainage density (Through Morphometric and Topographic Analysis).
2. The flood of the left river basin is a lineament control, but the right bank is a non-lineament control because the left basin has a specific region where a flood will reoccur repetitively with high magnitude, e.g. Pali, Sirohi, Bali, etc. The right bank nature of the flood is fragmented and limited reoccurrence with very low magnitude. (Through lineament and morphometric analysis)
3. The left bank of the Luni River basin has older formations and high relief, leading to catastrophic flash floods for the regions of Pali, Jailor, and Balotra (Through geology and geomorphology)
4. The nature of the flood on the left bank is a highly catastrophic high peak and low duration as in the region of Bhavi Jetpur Bilaspur (for 2 to 4 days of rainfall, the water rose 40 cm to 4 m, and after 2 days water level again fails) because of impervious soil and geological control.
5. Floods that occur on the right of the Luni River are prominent in the interdune and highly volatile and flood-sensitive channel, leading to more sheet flow and waterlogging floods. (Through morphometry and topographic analysis)
6. The Flood of the right bank nature has a low flood peak and long duration because of its flatter basin relief structure. (Through morphometric analysis)
7. Overland flow is greater on the right bank, and with the more non-channelised flow on the left bank of the Luni River, the river has good drainage branching (Through morphometric analysis).
8. Anthropogenic dams and embankments create barriers to flood discharge. Mainly, the cities of Pali and Jalor collect the water for their city water storage. Additionally, after Lulc analysis, the area of buildup on the left of the drainage basin leads to flow acceleration for flash floods (Through land use and land cover analysis).

9. The river has a lower drainage density, and higher overland flow. Also, the topography of the lower right side of the Luni River basin is a highly dynamic and volatile channel that flows only in response to floods (Through previous flood occurrences, Morphometric and Topographic Analysis).

17.8 Conclusion

The study of the Luni River basin using the morphometric and topographic indices creates valuable insights into the causes and characteristics of the flood occurrence, as a regional comparison of the left and right bank tributaries revealed notable differences in their flood dynamics influenced by geomorphology, geology, and human intervention. The left bank tributary, characterised by a steeper slope, higher drainage density, and greater interconnection of the drainage network, is prone to flash floods due to rapid runoff. The context of the right bank tributaries and their flatter terrain is only a few streams, and basic water accumulation capacity resulting in fewer but more prolonged floods through the interdune channel.

The conclusion is that the flood that occurs on the left of the Luni River basin (Jawai, Sukri, Khari River, Bandi, Guhiya) is geological and lineament control, whereas the geological structure and geomorphology plays an important role. The right bank of the Luni River has a more arid climatic control where the interdune channel, more non-channel overland flow, and regional water blockage occur because fewer outlets for flood water create prolonged floods.

Flash flood comes through a destructive way, offering a vital opportunity for water resource development in the region like the Luni River basin, after measuring and capturing the phenomena and their magnitude, it's possible to recharge the groundwater aquifers of lower Luni Rivers, support agriculture in semi-arid region and reduce or compensate the compatibility of drought, effective flood water harvesting transform a natural hazard into a sustainable resource for long-term regional resilience.

Acknowledgements The first author would like to acknowledge the University Grant Commission (UGC), New Delhi, India, for the financial support under the Junior Research Fellowship (UGC JRF) Scheme.

References

Abd-Elhamid, H. F., Zeleňáková, M., Soľáková, T., Saleh, O. K., & El-Dakak, A. M. (2023). Monitoring flood and drought risks in arid and semi-arid regions using remote sensing data and standardized precipitation index: A case study of Syria. *Journal of Flood Risk Management*, *17*(1). https://doi.org/10.1111/jfr3.12961

Abdel-Lattif, A., & Sherief, Y. (2010). Morphometric analysis and flash floods of Wadi Sudr and Wadi Wardan, Gulf of Suez, Egypt: Using digital elevation model. *Arabian JoUrnal of Geosciences*, *5*(2), 181–195. https://doi.org/10.1007/s12517-010-0156-8

Alam, A., Ahmed, B., & Sammonds, P. (2020). Flash flood susceptibility assessment using the parameters of drainage basin morphometry in SE Bangladesh. *Quaternary International*, *575–576*, 295–307. https://doi.org/10.1016/j.quaint.2020.04.047

Al-Sababhah, N. (2023). Topographic position index to landform classification and spatial planning, using GIS, for Wadi Araba, South West Jordan. *Environment and Ecology Research*, *11*(1), 79–101. https://doi.org/10.13189/eer.2023.110106

Beven, K. J., & Kirkby, M. J. (1979). A physically based, variable contributing area model of basin hydrology / Un modèle à base physique de zone d'appel variable de l'hydrologie du bassin versant. *Hydrological Sciences Bulletin*, *24*(1), 43–69. https://doi.org/10.1080/02626667909491834

Chatterji, P. C., Singh, S., & Qureshi, Z. H. (1978). Hydrogeomorphology of the central Luni basin, Western Rajasthan, (India). *Geoforum*, *9*(3), 211–224. https://doi.org/10.1016/0016-7185(78)90011-8

Costache, R., & Bui, D. T. (2020). Identification of areas prone to flash-flood phenomena using multiple-criteria decision-making, bivariate statistics, machine learning and their ensembles. *The Science of the Total Environment*, *712*, Article 136492. https://doi.org/10.1016/j.scitotenv.2019.136492

De Reu, J., Bourgeois, J., Bats, M., Zwertvaegher, A., Gelorini, V., De Smedt, P., Chu, W., Antrop, M., De Maeyer, P., Finke, P., Van Meirvenne, M., Verniers, J., & Crombé, P. (2012). Application of the topographic position index to heterogeneous landscapes. *Geomorphology*, *186*, 39–49. https://doi.org/10.1016/j.geomorph.2012.12.015

Deopa, H., & Resmi, M. R. (2024). Appraisal of long-term hydrological variability in the Luni River Basin: A quantitative statistical approach. *Theoretical and Applied Climatology*, *155*(7), 6709–6726. https://doi.org/10.1007/s00704-024-05022-5

Dhir, R. P, Kolarkar, Sharma, K. D, Vangani, N. S, Saxena, S. K., Sen, A. K., Ramakrishna, Y. S., & Murthy, K. (1982). *Flash flood in the Luni a case study Central Arid Zone Research Institute*. Central Arid Zone Research Institute.

Fitra, J., Debataraja, S. M. T., & Lismawaty, N. (2024). Identification of flood vulnerability using the topographic wetness index method in Pantai Labu Baru village, Deli Serdang, North Sumatera. *E3S Web of Conferences*, *483*, 01014. https://doi.org/10.1051/e3sconf/202448301014

Friedrich, T., Reinartz, R., & Gessner, J. (2019). Sturgeon re-introduction in the Upper and Middle Danube River Basin. *Journal of Applied Ichthyology, 35*(5), 1059–1068. https://doi.org/10.1111/jai.13966

Fryirs, K. A. (2016). River sensitivity: A lost foundation concept in fluvial geomorphology. *Earth Surface Processes and Landforms, 42*(1), 55–70. https://doi.org/10.1002/esp.3940

Grodek, T., Cloete, G., Vázquez-Tarrío, D., & Benito, G. (2024). The hydromorphological effects of dams in an extreme desert environment—The Fish River Canyon, Namibia. *Geomorphology*, *453*, 109124. https://doi.org/10.1016/j.geomorph.2024.109124

Guddo. (2023). Conceptualization of flooding disaster: A theoretical perspective of management. *Disaster Advances*, *17*(2), 40–45. https://doi.org/10.25303/172da040045

Günther, A., Van Den Eeckhaut, M., Malet, J.-P., Reichenbach, P., & Hervás, J. (2014). Climate-physiographically differentiated Pan-European landslide susceptibility assessment using spatial multi-criteria evaluation and transnational landslide information. *Geomorphology, 224*, 69–85. https://doi.org/10.1016/j.geomorph.2014.07.011

Gupta, L., & Dixit, J. (2024). Investigating the spatial distribution of flood inundation and landforms using topographic position index (TPI) and geomorphon-based automated landform classification methods. *Journal of Earth System Science*, *133*(3). https://doi.org/10.1007/s12040-024-02343-x

Horton, R. E. (1945). Erosional development of streams and their drainage basins; hydrophysical approach to quantitative morphology. *Geological society of America bulletin, 56*(3), 275–370.

Jain, V., & Sinha, R. (2003). Geomorphological manifestations of the flood hazard: A remote sensing based approach. *Geocarto International, 18*(4), 51–60. https://doi.org/10.1080/10106040308542289

Jonkman, S. N., Curran, A., & Bouwer, L. M. (2024). Floods have become less deadly: An analysis of global flood fatalities 1975–2022. *Natural Hazards, 120*(7), 6327–6342. https://doi.org/10.1007/s11069-024-06444-0

Kale, V. S., Singhvi, A. K., Mishra, P. K., & Banerjee, D. (2000). Sedimentary records and luminescence chronology of Late Holocene palaeofloods in the Luni River, Thar Desert, northwest India. *CATENA, 40*(4), 337–358. https://doi.org/10.1016/s0341-8162(00)00091-6

Kaliraj, S., Chandrasekar, N., & Magesh, N. S. (2014). Morphometric analysis of the River Thamirabarani sub-basin in Kanyakumari District, South west coast of Tamil Nadu, India, using remote sensing and GIS. *Environmental Earth Sciences, 73*(11), 7375–7401. https://doi.org/10.1007/s12665-014-3914-1

Kar, A. (1994). Lineament control on channel behaviour during the 1990 flood in the southeastern Thar Desert. *International Journal of Remote Sensing, 15*(13), 2521–2530. https://doi.org/10.1080/01431169408954263

Kotecha, M. J., Tripathi, G., Singh, S. K., Kanga, S., Meraj, G., Sajan, B., & Rai, P. K. (2023). GIS-Based novel ensemble MCDM-AHP modelling for Flash Flood susceptibility mapping of Luni River Basin, Rajasthan. In *River conservation and water resource management* (pp. 267–313). Springer Nature Singapore.

Kushwaha, A. P., Solanki, H., Vegad, U., Mahto, S. S., & Mishra, V. (2024). Land and atmospheric drivers of the 2023 flood in India. *Earth and Space Science, 11*(10). https://doi.org/10.1029/2024ea003750

Lastra, J., Fernández, E., Díez-Herrero, A., & Marquínez, J. (2008). Flood hazard delineation combining geomorphological and hydrological methods: An example in the Northern Iberian Peninsula. *Natural Hazards, 45*(2), 277–293. https://doi.org/10.1007/s11069-007-9164-8

Melton, M. A. (1957). *An analysis of the relations among the elements of climate, surface properties and geomorphology*. Columbia University.

Merz, B., Kuhlicke, C., Kunz, M., Pittore, M., Babeyko, A., Bresch, D. N., Domeisen, D. I. V., Feser, F., Koszalka, I., Kreibich, H., Pantillon, F., Parolai, S., Pinto, J. G., Punge, H. J., Rivalta, E., Schröter, K., Strehlow, K., Weisse, R., & Wurpts, A. (2020). Impact forecasting to support emergency management of natural hazards. *Reviews of Geophysics, 58*(4). https://doi.org/10.1029/2020rg000704

Moharana, P. C., & Kar, A. (2013). Fluvial process measurement of arid ephemeral channels in western Rajasthan. In M. Gaur (Ed.), *Remote sensing applications in dryland natural resource management* (pp. 37–51). New India Publishing Agency.

O'Connor, J. E., Grant, G. E., & Costa, J. E. (2011). The geology and geography of floods. In *Water science and application* (pp. 359–385). https://doi.org/10.1029/ws005p0359

Ozdemir, H., & Bird, D. (2008). Evaluation of morphometric parameters of drainage networks derived from topographic maps and DEM in point of floods. *Environmental Geology, 56*(7), 1405–1415. https://doi.org/10.1007/s00254-008-1235-y

Pandey, P. C. (1980). Rajasthan flood as viewed by Samir Onboard Bhaskara. *Mousam, 31*(4), 561–566.

Pourali, S. H., Arrowsmith, C., Chrisman, N., Matkan, A. A., & Mitchell, D. (2014). Topography wetness index application in flood-risk-based land use planning. *Applied Spatial Analysis and Policy, 9*(1), 39–54. https://doi.org/10.1007/s12061-014-9130-2

Qiao, C., Luo, J., Sheng, Y., Shen, Z., Zhu, Z., & Ming, D. (2011). An Adaptive water extraction method from remote sensing image based on NDWI. *Journal of the Indian Society of Remote Sensing, 40*(3), 421–433. https://doi.org/10.1007/s12524-011-0162-7

Rudraiah, M., Govindaiah, S., & Vittala, S. S. (2008). Morphometry using remote sensing and GIS techniques in the sub-basins of Kagna river basin, Gulburga district, Karnataka, India. *Journal of the Indian Society of Remote Sensing, 36*(4), 351–360. https://doi.org/10.1007/s12524-008-0035-x

Schumm, S. A. (1956). Evolution of drainage systems and slopes in badlands at Perth Amboy, New Jersey. *Bulletin of the Geological Society of America, 67*(5), 214–236, 597–646.

Sharma, K. D., & Vangani, N. S. (1982). Flash flood of July 1979 in the Luni basin—A rare event in the Indian desert. *Hydrological Sciences Journal, 27*(3), 385–397. https://doi.org/10.1080/02626668209491117

Sharma, K. D., Choudhari, J. S., & Vangani, N. S. (1984). Transmission losses and quality changes along a desert stream: The Luni Basin in NW India. *Journal of Arid Environments, 7*(3), 255–262.

Sharma, K. D., Vangani, N. S., Chatterji, P. S., & Singh, G. (1982). A severe: Flood in Luni basin, western Rajasthan during July 1979—A case study. *Mausam, 33*(3), 377–384. https://doi.org/10.54302/mausam.v33i3.2851

Singh, R. S., Ramakrishna, Y. S., Sharma, K. D., Purohit, R. S., Bohra, D. N., & Central Arid Zone Research Institute. (1992). Some hydrometeorological aspects of an unprecedented rainstorm in the Indian desert. In *Annals of Arid Zone* (Vol. 31, Issue 3, pp. 205–210).

Smith, K. G. (1950). Standards for grading texture of erosional topography. *American J* Strahler, A.N., 1964. *"Quantitative Geomorphology of Drainage Basins and Channel Networks"*. in Handbook of Applied hydrology, Ven Te Chow (ed.), McGraw Hill, New York, Section 4-11, pp. 4-39.

Sreedevi, P. D., Subrahmanyam, K., & Ahmed, S. (2005). The significance of morphometric analysis for obtaining groundwater potential zones in a structurally controlled terrain. *Environmental Geology, 47*(3), 412–420.

Strahler, A. N. (1964). Quantitative geomorphology of basins and channel networks. In V. T. Chow (Ed.), *Handbook of applied hydrology*. McGraw Hill Book Company.

Thannoun, R. Gh., Thannoun, & Ismaeel, O. A. (2024). Flood Risk Vulnerability Detection based on the Developing Topographic Wetness Index Tool in Geographic Information System. *IOP Conference Series Earth and Environmental Science, 1300*(1), Article 012012. https://doi.org/10.1088/1755-1315/1300/1/012012

Thompson, A., & Clayton, J. (2002). The role of geomorphology in flood risk assessment. *Proceedings of the Institution of Civil Engineers—Civil Engineering, 150*(5), 25–29. https://doi.org/10.1680/cien.2002.150.5.25

Verma, R., & Rai, S. C. (2023). Evaluation of spatio-temporal variability of a flood using the hydrological process of flood frequency in Ghagra River Basin, India. *Journal of the Geological Society of India, 99*(12), 1671–1682. https://doi.org/10.1007/s12594-023-2523-6

Wolman, M. G., & Gerson, R. (1978). Relative scales of time and effectiveness of climate in watershed geomorphology. *Earth Surface Processes and Landforms, 3*, 189–208.

Zhang, Y., Li, Z., Xu, H., Ge, W., Qian, H., Li, J., Sun, H., Zhang, H., & Jiao, Y. (2024). Impact of floods on the environment: A review of indicators, influencing factors, and evaluation methods. *The Science of the Total Environment, 951*, Article 175683. https://doi.org/10.1016/j.scitotenv.2024.175683

CHAPTER 18

Dynamics of Floodplain Vegetation Under Diverse Inundation Regimes: A Multi-Temporal Geospatial Analysis of the Torsha-Raidak Interfluve, India

Biyas Dey, Olena Dubovyk, and Debajit Datta

18.1 Introduction

Flooding has been one of the leading natural causes of worry for mankind for as long as records have been kept (Dottori et al., 2018). Yet, flood disturbance has a crucial role to play when it comes to shaping river ecosystems. Flood spells can impose anaerobic conditions through prolonged inundation or create bare patches by scouring or burying away

B. Dey · D. Datta (✉)
Landscape Ecology Laboratory, Department of Geography, Jadavpur University, Kolkata, India
e-mail: debajit.geo@gmail.com

O. Dubovyk
Department of Earth System Sciences, Geography Institute, University of Hamburg, Hamburg, Germany

S. Ghosh (ed.), *Reimagining Indian Rivers for Sustainability*, Palgrave Studies in Environmental Sustainability,
https://doi.org/10.1007/978-3-032-04754-0_18

existing vegetation substrates (Hughes, 1997). Such processes ramify floodplain vegetation composition and diversity in between wet and dry phases, through elimination of intolerant species followed by recolonization of pioneer riparian obligates in freshly deposited alluvium (Bendix, 1997). This biological and biogeochemical response of riverine ecosystem has been recognized in the '*flood pulse concept*,' which promulgated the view that flow (and its variability) is the primary driver of floodplain ecological functioning, i.e., lateral exchange of flow pulses between aquatic and terrestrial phases determines the degree of connectivity of matter and organisms across river-floodplain gradients (Junk et al., 1989; Tockner et al., 2000). Thus, floodplain ecosystems are blessed with an inherent capacity to absorb natural transitions (disturbances) and maintain consistent ecosystem structure over long term (Capon et al., 2009). Nevertheless, there are likely to be thresholds to such adaptive resilience which are now confounded by human interventions like land clearing, grazing, water resource development, and most importantly, climate change (Colloff & Baldwin, 2010).

India being 'a land of monsoon' receives rainfall primarily from June to September, when the southwest monsoon winds prevail (Subrahmanyam, 1988). Owing to its physical and geographical settings, the state of West Bengal is also no exception. Thus, the state mean annual precipitation is 1750 mm, with over 75% of it occurring during monsoon. This is heaviest in case of sub-Himalayan regions, where the rainfall ranges from 2500 to 4000 mm (Government of West Bengal, 2020; Hazra et al., 2024). Consequently, rivers like Torsha and Raidak which carries immense load from the Himalayas in monsoon, finally debouch them in the low-lying floodplains (Chakraborty & Mukhopadhyay, 2019). Although it is initially attenuated, a continuous spell of heavy to extremely heavy rainfall often causes rivers to swell and inundate adjacent floodplains. This, in succession, often affects the ecological structure, alters vegetation composition and distribution, and disrupts habitat connectivity across floodplain landscapes. Moreover, recent 'priority zonation' of Indian landscapes on the basis of biodiversity conservation, ecosystem services, and human well-being have highlighted North Bengal districts of West Bengal as one of the high priority areas (Srivathsa et al., 2023).

Nevertheless, almost no study on 'flood ecology' have been conducted previously in sub-Himalayan floodplains, as per the present researchers' knowledge, thereby making the concerned research a pioneering one in this respect. Against this backdrop, we have envisaged a hypothesis that recurrent flood pulses can significantly modulate existing vegetation structure of a floodplain. Accordingly, our specific objectives were to (i) identify the major flood events that have occurred over the past decade; (ii) assess spatio-temporal changes in inundation extent associated with each flood spell; and (iii) analyze the response dynamics between flood pulses and floodplain vegetation.

18.2 Methods

18.2.1 Study Region

Braided river systems of eastern Himalayan foreland, with their prolonged history of flow-mediated fluvial complexity, have appeared as quaternary fluvial deposits across north Bengal floodplains of Alipurduar and Coochbehar districts (Saha et al., 2024). Embedded within such fluvio-geomorphic setting, the Torsha-Raidak interfluve (TRI) covering an area of approximately 1009.76 km^2, is being demarcated by meandering courses of R. Torsha and R. Raidak-II on the west and east respectively, while getting constrained by dissected alluvial fans in the north (Geological Survey of India, 2023) (Fig. 18.1).

However, the concerned stretch has been deliberately confined in its south-eastern extent owing to the locational sensitivity associated with Assam (national)-Bangladesh (international) border. Besides, a host of smaller tributaries and distributaries like Gadadhar, Kaljani, Ghargharia, Ghoramara, etc. interlinking the major channels of Torsha and the anabranches of Raidak, further complement to the hydrological complexity of the interfluve. While these rivers remain predominantly desiccated for most of the year, experience a marked resurgence with the onset of southwest monsoon (Government of West Bengal, 2020). Such elevated flows in the upper catchment compel the river discharge to surpass its channel capacity, thereby triggering flood inundation across in adjacent lowlands (Shil Sharma, 2011). In addition, pronounced meandering of rivers like Ghargharia and Gadadhar having sinuosity indices of 1.98 and 2.13 respectively, often drives these channels to abandon their original courses and breach neighboring concave banks, thus accelerating flood-induced devastation (Chakraborty & Mukhopadhyay, 2019). As a result, flood has become a seasonal norm of TRI, with a recurrence of two to three times sometimes within the same year. However, not all these events are equivalent in terms of their magnitude, intensity and severity.

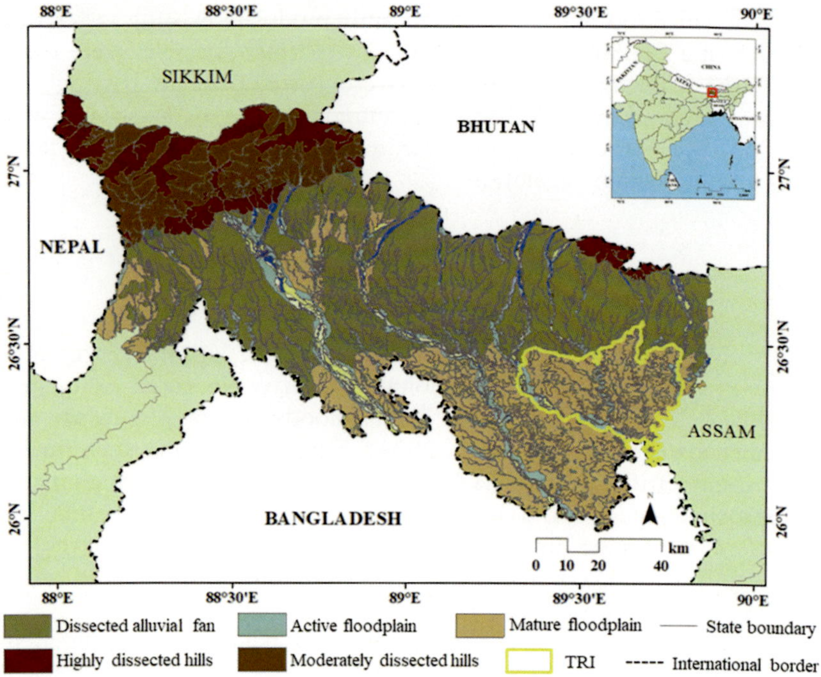

Fig. 18.1 Floodplain stretches of the Torsha-Raidak Interfluve (TRI)

The catastrophic flood of August 12–16, 2017 has been designated as the largest flood of the last two and a half decades, followed by that of 2016 and 2020 over the past decade (Chakraborty & Mukhopadhyay, 2019; Government of West Bengal, 2020). This, in turn, has affected the inherent vegetation structure and composition of the floodplains. Consequently, TRI has emerged as an ideal natural laboratory to study spatio-temporal variations of flood-vegetation response.

18.2.2 Data Sources

Recent developments in space-borne microwave imaging have enabled continuous as well as cost-effective mapping and monitoring of inundation regimes along with their consequent impacts on vegetation dynamics (Datta et al., 2022; Uddin et al., 2019). Such high-resolution images

Table 18.1 Details of the geospatial dataset utilized in this study to analyze flood-vegetation response

Sensor	*Flight direction*	*Purpose*	*Image acquisition date*	
			Before flood	*During/ after flood*
Sentinel-1 C-SAR	Descending	Inundation mapping	April 1–30	July 20–30, 2016
				August 5–20, 2017
				September 5–20, 2020
		RVI	April 19, 2016	August 17, 2016
			April 26, 2017	August 24, 2017
			April 22, 2020	September 19, 2020

now manage to capture earth surface at virtually all-weather, day and night conditions by penetrating cloud cover and detecting real-time landscape features (Psomiadis, 2016). In this context, the present study has employed Level-1 Ground Range Detected (GRD) datasets of Sentinel-1 C-band Synthetic Aperture Radar (C-SAR) for the years 2016, 2017, and 2020 (Table 18.1). Sourced primarily from the Copernicus Open Access Hub, operated by the European Space Agency (https://browser.dataspace.copernicus.eu/), this active microwave sensor functions in a dual-polarization mode, comprising possible combinations of vertical transmit/vertical receive (VV), horizontal transmit/horizontal receive (HH), and cross-polarizations including VV + VH and HH + HV. Consequently, the images have been acquired for descending orbital passes in Interferometric Wide Swath (IW) mode, and processed in Google Earth Engine.

18.2.3 Geospatial Assessment of Flood Inundation Dynamics

As per Irrigation and Waterways annual report, TRI have witnessed major flood events in 2016, 2017 and 2020, over the past decade (Government of West Bengal, 2020). While, the flood pulse of 2017 was far more pernicious as far as the inundation extents are concerned, flood episodes of 2020 was majorly attributed to short yet continuous spell of rainfall throughout the monsoon. This was so much so that the rainfall statistics had consistently shown above average monthly precipitation in sub-Himalayan West Bengal in 2020 for June (27.5%), July (28.1%) and September (96.0%), displaying a total deviation of 28.8% from its normal average. Correspondingly, VH-polarized C-SAR images of pre- and post-flood periods of the three deluged years have been employed, owing to its sensitivity to changes in surface water roughness (Manavalan, 2018) (Table 18.1). These mosaicked images are then despeckled in Google Earth Engine using an Enhanced Lee filter with a 3 × 3 kernel, to retain the details of texture information (Rana & Suryanarayana, 2019). Thereafter, flooded areas are distinguished from permanent waterbodies through temporal changes in backscatter value and the derived images have been exported for further analysis.

18.2.4 Radar-Based Vegetation Mapping

Monitoring broad-scale floodplain ecological response, encompassing growth, vigor, and dynamics of its existing vegetation to subsequent flood pulses can be facilitated through repeated geospatial image analysis. In this context, optical remotely sensed (RS) images, though frequently used, often fail to incorporate the real-time cloud-penetrating ecotonal characteristics of floodplain vegetation (Hu et al., 2024). Hence, the present study has made use of microwave (radar-based) RS to assess the spatio-temporal shift in dynamics of TRI vegetation assemblages with consecutive flood events. Consequently, the dual-polarimetric Radar Vegetation Index (RVI) has been derived using pre- and post-flood SAR imageries for each representative year to reflect changes in volume scattering associated with vegetation canopies (Mandal et al., 2020) (Eq. 18.1).

$$RVI = \frac{4\sigma^{\circ}_{VH}}{\sigma^{\circ}_{VH} + \sigma^{\circ}_{VV}} \tag{18.1}$$

where, cross-polarized VH and co-polarized VV radar signals are converted from decibels (dB) to linear scale.

RVI values generally vary from 0 to +1, with higher RVI values (closer to 1.0) suggesting dense and healthier vegetation, whereas lower values indicate sparse or non-vegetated surfaces like bare earth or built-up areas.

Accordingly, RVI maps can be used to interpret the loss of vegetation cover as well as its regeneration after recurrent flood disturbances.

18.2.5 *Analysis of Vegetation Response to Flood*

Owing to the high pixel volume of RVI across TRI ($N = 11.31$ million), 385 random points are generated for pixel value extraction of each phase, based on a proportion of 0.5 and a 95% confidence level, following the Australian Bureau of Statistics' sample size determination protocol (https://www.abs.gov.au/). Subsequently, mean and standard error of pre- and post-flood RVI values under inundated and non-inundated areas for respective years have been computed (Nandi et al., 2020). Besides, to examine the differences between mean values of RVI under pre and post-flood scenarios, student's t-tests (two-tailed) have been conducted ($p < 0.05$, $p < 0.01$, $p < 0.001$). Thereafter, individual as well as combined effects of flood year, inundation status and vegetation phase (independent or control variables) on RVI dynamics (dependent or response variable) have been assessed by three-way analysis of variance (ANOVA) in SPSS® software version 23.0 (IBM Corporation, Armonk, NY) (Datta et al., 2024).

18.3 Results

18.3.1 *Flood Inundation Extent and Variability*

Hazards in TRI region is synonymous to flood. Moderate-level floods occur almost every year during monsoon, typically lasting from a few days to about a week (Chakraborty & Mukhopadhyay, 2019). Nevertheless, impacts of floods of 2016, 2017 and 2020 were found to be severe in terms of levels recorded at river gauge stations along with the extent and recurrence of inundation. While, in 2016 and 2020 approximately 29,911 ha and 27,949 haof TRI are observed to be submerged; the situation worsened in 2017 when floodwaters engulfed nearly 32,936 ha of the floodplain (Fig. 18.2).

Moreover, from the inundation maps it can be discerned, that the south-eastern area, within the Tufanganj subdivision, was ravaged by flood in all instances. Besides, the torrential rainfall had triggered the rivers of Raidak-I, Kaljani to remain beyond their pre-fixed EDLs (extreme danger levels) for several days eventuating into widespread flood

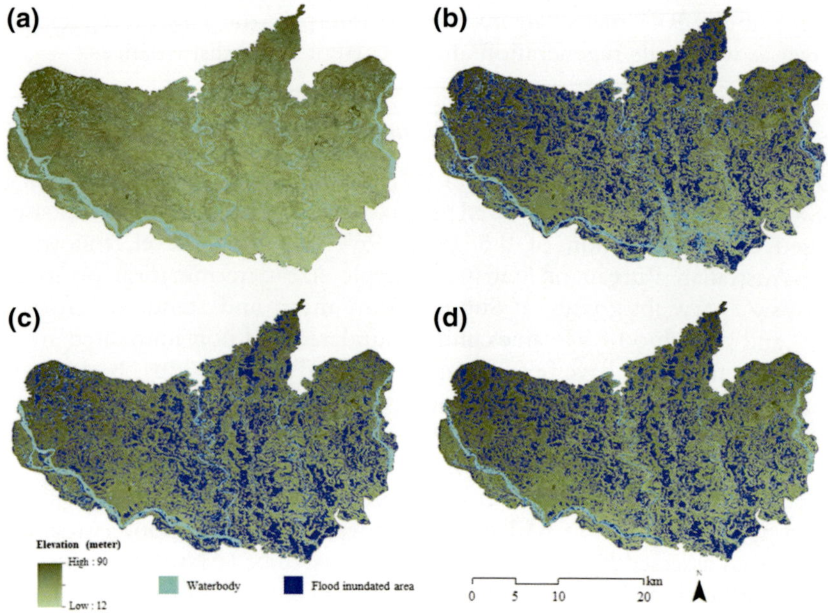

Fig. 18.2 Shifting dynamics of flood inundation regime in the Torsha-Raidak Interfluve (TRI). **a** Non-inundated baseline; inundation extents in **b** 2016, **c** 2017, and **d** 2020

(Table 18.2). Indeed, the flood intensity of 2017 was so overwhelming that river gauge station at Tufanganj had attained its highest flood level till date (0.79 m).

Table 18.2 Peak flood levels attained by the rivers of TRI during flood spells of 2016, 2017, and 2020[a]

Year	*Date*	*River gauge station*	*Peak flood water level*
2016	25-Jul	Alipurduar (R. Kaljani)	Touches EDL
2017	13-Aug	Tufanganj (R. Raidak I)	0.79 m above EDL
	12-Aug	Coochbehar (R. Torsha)	0.22 m above EDL
	12-Aug	Alipurduar (R. Kaljani)	1.15 m above EDL
2020	13-Jul	Tufanganj (R. Raidak I)	0.02 m above EDL

[a] *Data source* IMD-Kolkata

18.3.2 Changing Status of RVI

The present study has considered RVI as the relevant biophysical proxy for assessing transformation in vegetation health in response to flood. As a consequence, it is revealed that RVI values of TRI are uniformly more in every pre-flood scenarios compared to that of post-flood ones (Fig. 18.3). This further confirms that individual flood spell (phase) can have a significant impact on existing structure and composition of floodplain vegetation across the TRI ($p < 0.001$) (Table 18.3).

Moreover, it is observed that RVI has decreased as a whole in 2020 from that of 2016 and 2017, with substantial expanses of TRI exhibiting degraded vegetation health even in pre-flood phase (Fig. 18.3e). Besides, significant variations of flood-vegetation dynamics across inundated and non-inundated zones have been unveiled, which indicated that although mean RVI values consistently declined in inundated areas following flood events, non-inundated zones did not display a similar pattern. Thus, impact of inundation status has also been found to be crucial in modulating floodplain vegetation assemblage ($p < 0.001$).

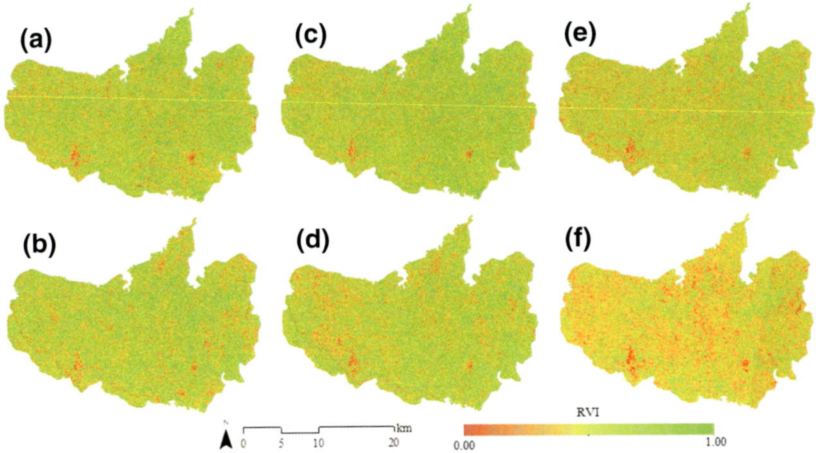

Fig. 18.3 RVI based status of vegetation health in **a** pre-flood 2016, **b** post-flood 2016, **c** pre-flood 2017, **d** post-flood 2017, **e** pre-flood 2020, and **f** post-flood 2020

Table 18.3 Mean Radar Vegetation Index (RVI) (± standard error) values of TRI floodplain and significance values of three-way ANOVA (F) for comparison between vegetation responses under different inundation status over past decade. Degrees of freedom are represented by df. Significant difference of means between phases under different inundation status for each year is shown by student's t-test values. Levels of significance are shown as: $^{*}p < 0.05$, $^{**}p < 0.01$, $^{***}p < 0.001$

Flood year	*Inundation status*	*Phase*	*RVI*
2016	Inundated	Before flood	0.79 ± 0.01
		After flood	0.63 ± 0.01
		t-test (df:768)	8.345***
	Non-inundated	Before flood	0.64 ± 0.01
		After flood	0.66 ± 0.01
		t-test (df:768)	−1.059
2017	Inundated	Before flood	0.81 ± 0.01
		After flood	0.67 ± 0.01
		t-test (df:768)	8.253***
	Non-inundated	Before flood	0.68 ± 0.01
		After flood	0.67 ± 0.01
		t-test (df:768)	.455
2020	Inundated	Before flood	0.73 ± 0.01
		After flood	0.46 ± 0.01
		t-test (df:768)	16.269***
	Non-inundated	Before flood	0.60 ± 0.01
		After flood	0.55 ± 0.01
		t-test (df:768)	2.289
Multivariate analysis	df		2, 1, 1, 2, 2, 1, 2
	Flood year		90.93***
	Inundation status		37.69***
	Phase		167.54***
	Flood year × Inundation status		3.81**
	Flood year × Phase		14.97***
	Inundation status × Phase		133.31***
	Flood year × Inundation status × Phase		3.18**

Overall, the multi-variate analysis affirms that joint effects of distinct flood events along with inundation coverage play a significant role in modifying vegetation distribution across TRI floodplains ($p < 0.01$).

18.4 Discussion

18.4.1 Flood Pulses and Vegetation Modulation

The findings of the present study critically acclaims that changing hydrological conditions, such as inundation duration, intensity, extent, and frequency work in conjunction to adjust floodplain ecosystem response. While the calamitous flood of 2017 is distinctly characterized by its maximum inundation extent and the highest flood level attained till date at some of the river gauge stations; flooding of 2020 was largely driven by persistent, above average monsoonal rainfall resulting in more prolonged submergence rather than extreme river discharge. Besides, the vegetation structure of TRI was already in a degraded condition owing to recurrent flood pulses and escalating anthropogenic interventions. This further reinforces the flood's impact on vulnerability of the exiting floodplain vegetation structure. Thus, ecological responses are not only shaped by the individualistic nature of the short-term high-magnitude flood disturbances, but also by the pre-existing conditions of the landscape.

18.4.2 Climate Change and Flood-Induced Disturbances

Riparian floodplains, known for providing exceptionally high proportions of ecosystem services, are essential biodiversity hotspots that minimize flood's impact by enabling water absorption and nutrient retention (Strassburg et al., 2020). However, in the wake of the present climate crisis, riverine ecosystems are now witnessing biodiversity loss at an alarming pace than that of other terrestrial or marine ecosystems (Jenkins, 2003). Besides, modifications in regional climate can also influence the extent of primary productivity and regeneration of floodplain species (Wassens et al., 2017). This has been amplified further by humanized impacts like channelization, water abstractions, land clearing, and farming developments; which may severely jeopardize the natural pattern of inundation at large regional scales in future. As an effect, the ecological integrity of riparian floodplains are increasingly getting compromised. Such disruptions have not only impaired floodplain key ecosystem functioning but have cumulatively threatened the native vegetation communities. In the present study, we have found strong evidences of how the amplification of flood-pulse frequency and magnitude, either individually or in combination, are modifying the vegetation health characteristics of the region. This is similar to the findings of other studies done in this region that mentioned augmented soil erosion, nutrient loss, and eventual impact of regional biodiversity (Hazra et al., 2024; Mandal et al., 2023).

18.5 Conclusions

This study marks a pioneering effort in illustrating flood as a major agent of disturbance ecology in sub-Himalayan floodplains. The findings tried to explore the inherent spatio-temporal relationship between flood pulses and corresponding vegetation adaptations in the ecologically fragile floodplain ecosystem of the TRI region. Consequently, it has been realized that reiterative flood stresses intensified by profound anthropogenic stresses have augmented the intensity of flood events and thereby, the ecological susceptibility of TRI. Nevertheless, the study can be proceeded further with more in-depth analysis of fragmentation and coalescence patterns of different vegetation zones under consecutive flood events, using different spatial metrics. Moreover, long term monitoring of vegetation pattern changes along selected transects would facilitate a deeper understanding of species-specific flow-response guilds. Hence, periodic monitoring of zone-wise cum site-specific floodplain vegetation dynamics is essential for developing resilient indigenous bio-shields against increasingly exposed floodplain landscape in the face of anticipated climatic disturbances.

References

Bendix, J. (1997). Flood disturbance and the distribution of riparian species diversity. *Geographical Review, 87*(4), 468–483. https://doi.org/10.1111/j.1931-0846.1997.tb00085.x

Capon, S. J., James, C. S., Williams, L., & Quinn, G. P. (2009). Responses to flooding and drying in seedlings of a common Australian desert floodplain shrub: Muehlenbeckia florulenta Meisn.(tangled lignum). *Environmental and Experimental Botany, 66*(2), 178–185. https://doi.org/10.1016/j.envexpbot.2009.02.012

Chakraborty, S., & Mukhopadhyay, S. (2019). Assessing flood risk using analytical hierarchy process (AHP) and geographical information system (GIS): Application in Coochbehar district of West Bengal, India. *Natural Hazards, 99*, 247–274. https://doi.org/10.1007/s11069-019-03737-7

Colloff, M. J., & Baldwin, D. S. (2010). Resilience of floodplain ecosystems in a semi-arid environment. *The Rangeland Journal, 32*(3), 305–314. https://doi.org/10.1071/RJ10015

Datta, D., Dey, M., Ghosh, P. K., & Pal, A. P. (2024). Development of a spatially explicit model of blue carbon storages in tropical mudflat environment through integrated radar-optical approach and ground-based measurements. *Ecological Informatics, 80*, Article 102509. https://doi.org/10.1016/j.ecoinf.2024.102509

Datta, D., Dey, M., Neogy, S., Basu Roy, T., Dutta, D., Kundu, A., & Nandi, G. (2022). Assessing vegetation fragmentation and plantation efficiency in an intertidal mudflat of Eastern India using Radar Forest Degradation Index and spatial metrics. *Geocarto International, 37*(25), 9298–9319. https://doi.org/10.1080/10106049.2021.2017014

Dottori, F., Szewczyk, W., Ciscar, J. C., Zhao, F., Alfieri, L., Hirabayashi, Y., Bianchi, A., Mongelli, I., Frieler, K., Betts, R. A., & Feyen, L. (2018). Increased human and economic losses from river flooding with anthropogenic warming. *Nature Climate Change, 8*(9), 781–786. https://doi.org/10.1038/s41558-018-0257-z

Geological Survey of India. (2023). *Bhukosh*. https://bhukosh.gsi.gov.in/

Government of West Bengal. (2020). Annual flood report of West Bengal, 2020. *Department of Irrigation and Waterways*. Accessed 27 September 2024. https://wbiwd.gov.in/uploads/anual_flood_report/ANNUAL_FLOOD_REPORT_2020.pdf

Hazra, U. N., Mahato, A., Deb, S., Chakraborty, S., Datta, D., Santra, P., Patra, P. S., & Choudhury, A. (2024). Integration of GIS with RUSLE to estimate soil, organic matter and nutrient loss from a watershed of eastern Himalayan Terai. *Environmental Earth Sciences, 83*(24), 668. https://doi.org/10.1007/s12665-024-11971-x

Hu, X., Li, L., Huang, J., Zeng, Y., Zhang, S., Su, Y., Hong, Y., & Hong, Z. (2024). Radar vegetation indices for monitoring surface vegetation: Developments, challenges, and trends. *Science of the Total Environment*, 173974.https://doi.org/10.1016/j.scitotenv.2024.173974

Hughes, F. M. R. (1997). Floodplain biogeomorphology. *Progress in Physical Geography, 21*(4), 501–529. https://doi.org/10.1177/030913339702100402

Jenkins, M. (2003). Prospects of biodiversity. *Science, 302*(5648), 1175–1177. https://doi.org/10.1126/science.1088666

Junk, W. J., Bayley, P. B., & Sparks, R. E. (1989). The flood pulse concept in river-floodplain systems. *Canadian Special Publication of Fisheries and Aquatic Sciences, 106*(1), 110–127.

Manavalan, R. (2018). Review of synthetic aperture radar frequency, polarization, and incidence angle data for mapping the inundated regions. *Journal of Applied Remote Sensing, 12*(2), 021501–021501. https://doi.org/10.1117/1.JRS.12.021501

Mandal, D., Kumar, V., Ratha, D., Dey, S., Bhattacharya, A., Lopez-Sanchez, J. M., McNairn, H., & Rao, Y. S. (2020). Dual polarimetric radar vegetation index for crop growth monitoring using sentinel-1 SAR data. *Remote Sensing of Environment, 247*, Article 111954. https://doi.org/10.1016/j.rse.2020.111954

Mandal, S., Gogoi, M., Deb, S., Datta, D., Sarkar, D., Roy, D., Sinha, A. K., & Choudhury, A. (2023). Surface soil moisture estimation during dry winter in the Terai region of eastern India: Use of Trapezoid models based on Landsat 8 and Sentinel 2 data. *Journal of the Indian Society of Soil Science, 71*(4), 351–361. https://doi.org/10.5958/0974-0228.2024.00001.X

Nandi, G., Neogy, S., Roy, A. K., & Datta, D. (2020). Immediate disturbances induced by tropical cyclone Fani on the coastal forest landscape of eastern India: A geospatial analysis. *Remote Sensing Applications: Society and Environment, 20*, Article 100407. https://doi.org/10.1016/j.rsase.2020.100407

Psomiadis, E. (2016). Flash flood area mapping utilising SENTINEL-1 radar data. In *Earth resources and environmental remote sensing/GIS applications VII* (Vol. 10005, pp. 382–392). SPIE. https://doi.org/10.1117/12.2241055

Rana, V. K., & Suryanarayana, T. M. V. (2019). Evaluation of SAR speckle filter technique for inundation mapping. *Remote Sensing Applications: Society and Environment, 16*, Article 100271. https://doi.org/10.1016/j.rsase.2019.100271

Saha, U. D., Bhattacharya, H. N., & Islam, A. (2024). Neotectonic impact on drainage development in the Eastern Himalayan foreland basin. *Geomorphology, 462*, Article 109328. https://doi.org/10.1016/j.geomorph.2024.109328

Shil Sharma, S. K. (2011). *Causes and effects of flood in Koch Bihar district, West Bengal (Doctoral dissertation)*, University of North Bengal.

Srivathsa, A., Vasudev, D., Nair, T., Chakrabarti, S., Chanchani, P., DeFries, R., Deomurari, A., Dutta, S., Ghose, D., Goswami, V. R., Nayak, R., Neelakantan, A., Thatte, P., Vaidyanathan, S., Verma, M., Krishnaswamy, J., Sankaran, M., & Ramakrishnan, U. (2023). Prioritizing India's landscapes for biodiversity, ecosystem services and human well-being. *Nature Sustainability, 6*(5), 568–577. https://doi.org/10.1038/s41893-023-01063-2

Strassburg, B. B., Iribarrem, A., Beyer, H. L., Cordeiro, C. L., Crouzeilles, R., Jakovac, C. C., Junqueira, A. B., Lacerda, E., Latawiec, A. E., Balmford, A., Brooks, T., Butchart, S. H. M., Chazdon, R. L., Erb, K., Brancalion, P., Buchanan, G., Cooper, D., Díaz, S., Donald, P. F., … Visconti, P. (2020). Global priority areas for ecosystem restoration. *Nature, 586*(7831), 724–729. https://doi.org/10.1038/s41586-020-2784-9

Subrahmanyam, V. P. (1988). Hazards of floods and droughts in India. In: M. I. El-Sabh, & T. S. Murty (Eds.) *Natural and man-made hazards*. Springer. https://doi.org/10.1007/978-94-009-1433-9_24

Tockner, K., Malard, F., & Ward, J. V. (2000). An extension of the flood pulse concept. *Hydrological Processes, 14*, 2861–2883.

Uddin, K., Matin, M. A., & Meyer, F. J. (2019). Operational flood mapping using multi-temporal Sentinel-1 SAR images: A case study from Bangladesh. *Remote Sensing, 11*(13), 1581. https://doi.org/10.3390/rs11131581

Wassens, S., Ning, N., Hardwick, L., Bino, G., & Maguire, J. (2017). Long-term changes in freshwater aquatic plant communities following extreme drought. *Hydrobiologia, 799*, 233–247. https://doi.org/10.1007/s10750-017-3219-y

CHAPTER 19

Sensitivity Analysis of Erosion Factors: Insights from Models Applied in the Dwarkeswar River Basin

Paramita Roy, Alok Pandey, Asish Dhara, Avijit Mahala, Tarig Ali, and Rabin Chakrabortty

19.1 Introduction

Soil erosion due to water actions remains among the most extensive and critical environmental problems throughout the world (Rhodes, 2014). Land degradation occurs due to this major process, leading to adverse effects on agriculture as well as water resources and the ecosystem (Hossain et al., 2020). The erosion process depletes the fertile layer

P. Roy · A. Dhara
Department of Geography, The University of Burdwan, Burdwan, West Bengal, India

A. Pandey
Centre for Remote Sensing and GIS, School of Earth and Environmental Science, Central University of Himachal Pradesh, Shahpur, India

S. Ghosh (ed.), *Reimagining Indian Rivers for Sustainability*, Palgrave Studies in Environmental Sustainability,
https://doi.org/10.1007/978-3-032-04754-0_19

of earth, which results in decreased agricultural yields and creates sediment buildup that contaminates water reservoirs (Rashmi et al., 2022). The combination of soil erosion causes severe damage to natural disasters, including floods and landslides, that disrupt human activities and affect livelihoods. Soil erosion mainly affects areas that experience heavy rainfall and contain sloping landscapes, together with deforested lands (Blanco-Canqui & Lal, 2010). The main research inquiries that this study seeks to address include: The Dwarkeswar River Basin requires assessment regarding the model which delivers superior soil erosion estimates between Universal Soil Loss Equation (USLE), Revised Universal Soil Loss Equation (RUSLE), and Modified Universal Soil Loss Equation (MUSLE). What adjustments would GIS implementation enable to boost predictive accuracy levels of these models when dealing with extensive and data-depleted regions? Soil conservation strategies and land management practices in comparable worldwide regions need clarification concerning the implications of these research findings.

Soil erosion exists as a worldwide problem which generates extensive consequences for environmental systems and human settlements (Owens, 2020). The Food and Agriculture Organization (FAO, 2017) reports that soil erosion depletes 75 billion tons of soil annually, thus creating major challenges for food security and water resources (IFAD 2017). The combination of intense seasonal rainfall with steep slopes and deforestation in subtropical and monsoon regions causes severe soil loss rates (Pal & Chakrabortty, 2022). Wang et al. (2025) established that South Asian agricultural losses from monsoon-driven soil erosion reach 10% annual crop yield reduction because of erosion. The globally used USLE and RUSLE empirical models require extensive field data

---

A. Mahala
Department of Geography, Swamy Shradhanand College, University of Delhi, Delhi, India

T. Ali · R. Chakrabortty (✉)
Department of Civil Engineering, American University of Sharjah, Sharjah, UAE
e-mail: rchakrabortty@aus.edu

T. Ali
e-mail: atarig@aus.edu

collection, yet these models face challenges when implemented in areas with limited infrastructure and resources (Chakrabortty et al., 2025). Soil loss estimation requires affordable solutions in Indian subtropical areas where existing data is either missing or unreliable (Roy et al., 2023). Soil erosion predictions have improved through GIS and remote sensing technology advancements that allow better assessments across extensive areas (Sepuru & Dube, 2018). The research explores the suitability of USLE, RUSLE and MUSLE, integrated with GIS, for estimating soil erosion levels in the Dwarkeswar River Basin, which faces substantial data constraints.

The international field of soil erosion research has progressed substantially through the development of multiple empirical models used for estimating soil loss. The Universal Soil Loss Equation (USLE) developed by Wischmeier and Smith (1978) serves as a widely utilized model which researchers apply across different parts of the Mediterranean (Kouli et al., 2009) and African regions (Angima et al., 2003). Soil erosion estimation through USLE depends on five essential factors that include rainfall erosivity and soil erodibility and slope length and land cover, and support practices (Ali & Hagos, 2016). The USLE shows weaknesses when measuring slope length and steepness variations because these factors remain crucial for regions with complex terrain (Kruk et al., 2020). The Revised Universal Soil Loss Equation (RUSLE) introduced improved topographic factors, including slope length and steepness, to overcome this limitation (Renard et al., 1991). The RUSLE has proven effective in both temperate regions of the United States (Foster et al., 1977) and European Union territories (Van der Knijff et al., 2000). The Modified Universal Soil Loss Equation (MUSLE) was developed to replace rainfall factors with runoff factors for improved sediment yield estimation in storm-driven erosion areas (Williams & Berndt, 1977). The application of these models faces difficulties when used in subtropical areas with minimal available data. The combination of these models with GIS and remote sensing technologies represents a promising method to address data scarcity problems while enhancing model prediction accuracy (Keesstra et al., 2018).

The Dwarkeswar River Basin exists within the subtropical monsoon climate zone of India, where steep terrain and heavy rainfall create significant soil erosion conditions. Soil erosion within this region has become more severe during recent decades due to climatic factors, together with human activities involving deforestation and unregulated agricultural practices (Chakrabortty et al., 2020; Pal et al., 2021). Research on soil erosion in India exists in great numbers, but most studies focus on single regions or employ specific models without performing a complete model comparison (Ghosh & Guchhait, 2012). The implementation of USLE, RUSLE, and MUSLE throughout India remains restricted, along with the limited number of studies that combine these models with GIS for catchment-scale soil erosion assessment. The application of GIS for soil erosion estimation has grown according to Prasannakumar et al. (2011) but the research results fail to compare different models and thus provide no model selection guidance for Dwarkeswar. The research addresses this deficiency by evaluating and comparing USLE, RUSLE and MUSLE models when integrated with GIS within the Dwarkeswar River Basin. The research findings from this study will demonstrate international value for subtropical monsoon areas dealing with steep topography and intense rainfall, like Southeast Asia and parts of Africa and Latin America.

Soil erosion modeling research in subtropical monsoon regions requires this study because it fills an important data availability gap, which prevents accurate soil loss prediction. Multiple empirical models, including USLE, RUSLE and MUSLE, have become standard for many parts of the world, yet few studies have evaluated their suitability in data-limited areas through integrated GIS technology. This research stands out because it evaluates three prominent soil erosion models in the Dwarkeswar River Basin through field observations linked to GIS applications for soil loss prediction. The main goal of this research is to determine the most precise model for predicting soil erosion within the Dwarkeswar River Basin because it will help develop better soil conservation methods and land management approaches for similar areas. This research presents an evaluation of different models to determine which one provides the best solution for extensive soil erosion analysis in subtropical areas. The study's results will enable both policymakers and land planners to develop proper strategies that fight soil erosion while enhancing land use management and sustaining agricultural productivity in vulnerable areas.

19.2 Materials and Methods

19.2.1 Study Area

The Dwarakeswar River starts at Tilabani Hill, which stands at 445 meters elevation (86° 31′ E to 87° 02′ E) near Kashipur in Puruliya District (Fig. 19.1). The river starts in the Bankura District at Chhatna CD Block before moving into Purbamedinipur District through Khandaghosh CD Block. Soil erosion remains a major issue across the entire basin area of the river which extends from 22° 42′ 30″ N to 23° 31′ N in West Medinipur District. The river's current, together with seasonal precipitation, leads to bank and agricultural land erosion mainly in steep gradient areas such as the Puruliya District. Multiple major tributaries, including Gandeshwari, Joypanda, Kanshachora, Indus Nala, Arkosa Birai, and Amador, create localized soil erosion during the monsoon period because of their fast-moving currents and changing water flow patterns. Soil erosion intensifies in the Ghatal area of West Medinipur because the Shilai River merges with the Dwarakeswar to create the Rupnarayan River, which interacts with alluvial plains, resulting in sediment deposits and rising water levels. The area undergoes continuous cycles of erosion and sedimentation that worsen the existing problem. The Rupnarayan River emerges from the Dwarakeswar and Shilai confluence before it reaches the Hooghli River near Gadiara in the Howrah District (22° 57′ N, 88° 05′ E). The complete river system experiences erosion dynamics which threaten both soil quality and agricultural production and riverbank stability, thus requiring immediate research and remediation efforts.

19.2.2 Database

The database containing information for average annual soil loss estimates and soil erosion predictions includes multiple essential datasets which were merged to support modeling analysis. The datasets contribute individually to improving the accuracy of three prediction models: Universal Soil Loss Equation (USLE), Revised Universal Soil Loss Equation (RUSLE) and Modified Universal Soil Loss Equation (MUSLE). The combination of different data types enables researchers to analyze soil erosion patterns extensively throughout the study area, which leads to predictions from models that are trustworthy for local implementation. The research uses three main data types, which consist of remotely sensed data and field-based observations, together with temporal climatic data.

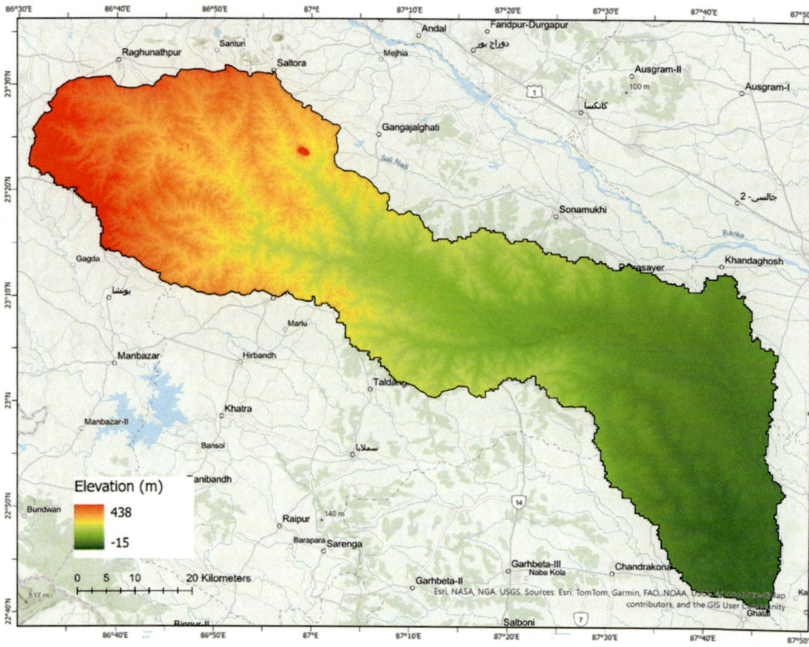

Fig. 19.1 Location of the study region

The SRTM DEM (Shuttle Radar Topography Mission Digital Elevation Model) obtained from USGS serves as a fundamental component for determining topographic characteristics that affect soil erosion. The slope length and slope steepness play essential roles in studying how landscape features control water runoff. Water flow distance is represented through the slope length factor and slope steepness determines how terrain angles affect flow velocity. Water movement speed down slopes depends heavily on these two factors because they control the rate of soil erosion. Soil erosion models heavily depend on slope length and steepness factor calculations which are possible through the SRTM DEM dataset. The Landsat 8 OLI (Operational Land Imager) data serve as a critical dataset for estimating the Cover and Management Factor (C factor) through its high-resolution satellite imagery. Soil erosion effects from vegetation and land management practices can be studied through the essential C factor. Soil resistance to erosion depends on land management practices and vegetation cover because these elements create barriers that control water runoff during rainfall. The Landsat 8 OLI imagery enables the processing of data that leads to C factor derivation for land use analysis and vegetation type identification, and cover density measurement. The collected data proves most beneficial for areas that experience substantial soil erosion effects from vegetation and land management activities, including agricultural fields and forests.

Primary sources of observed data, including local meteorological stations and storm records, provide rainfall data required to determine the Rainfall and Runoff Erosivity Factor (R factor). Rainfall intensity and erosive capacity are measured through the R factor and depend on both the precipitation amount and the way it is distributed throughout time. Storms with high-intensity levels produce greater runoff amounts that lead to increased soil particle dislodgement through erosion. The rainfall data obtained from the study area serves to determine temporal rainfall pattern variations along with their impact on storm-triggered soil erosion potential. Soil erosion intensifies during heavy rainfall events because these storms affect bare or uncovered areas and steep terrain. The Soil Texture dataset, which results from field sample analysis and laboratory testing, enables the calculation of the Soil Erodibility Factor (K factor). The K factor determines soil erosion vulnerability through analysis of its texture that consists of sand, silt and clay proportions. The presence of clay particles in the soil increases its erodibility, while sandy soils with their larger particle size remain less susceptible to rainfall-induced detachment and runoff processes. The K factor estimation for all soil erosion models depends on the results of soil texture analysis that includes organic matter content determination and particle size distribution assessment. Soil organic content, together with permeability and structure elements, influences the ease with which water removes soil particles from the ground. The K factor estimation receives additional precision through the Soil Chemical Properties dataset that results from laboratory testing of soil samples. Soil erodibility depends on three main characteristics: organic carbon content, cation exchange capacity (CEC), and soil permeability. Soil containing high levels of organic matter develops better structural cohesion that protects it from erosion. Soils containing low organic content, together with poor structure, show higher susceptibility to erosion. The dataset improves K factor accuracy by including soil chemical and physical properties which affect cohesion strength and erosion resistance. The Support Practice Factor (P factor) is estimated through field observations and local stakeholder surveys that provide the Support Practice Information. The P factor represents various conservation practices, including contour ploughing and terracing and buffer strip implementation, which decrease water runoff velocity while protecting soil from erosion. Field visits and surveys inside the study area document the extent and types of conservation practices directly. The P factor calculation utilizes this data to modify potential soil erosion predictions because managed areas with proper soil conservation practices tend to have reduced erosion rates. Soil erosion mitigation strategies can be identified using this essential data to protect specific areas from further damage. The Amount of Soil Loss measurements collected from field observations confirms and validates the prediction results from the models. The soil erosion models (USLE, RUSLE, MUSLE) receive validation through empirical data obtained through field measurements and historical soil erosion records in the study area. Model calibration processes and identification of discrepancies between predicted and observed soil loss become possible by comparing model output with field measurements. The addition of this validation process enables the models to achieve higher accuracy and reliability, which leads to predictions that better match real-world erosion dynamics.

19.2.3 Soil Erosion Factors

19.2.3.1 R Factor

The database containing information for average annual soil loss estimates and soil erosion predictions includes multiple essential datasets which were merged to support modeling analysis. The datasets contribute individually to improving the accuracy of three prediction models: Universal Soil Loss Equation (USLE), Revised Universal Soil Loss Equation (RUSLE) and Modified Universal Soil Loss Equation (MUSLE). The combination of different data types enables researchers to analyze soil erosion patterns extensively throughout the study area, which leads to predictions from models that are trustworthy for local implementation. The research uses three main data types, which consist of remotely sensed data and field-based observations, together with temporal climatic data. The SRTM DEM (Shuttle Radar Topography Mission Digital Elevation Model) obtained from USGS serves as a fundamental component for determining topographic characteristics that affect soil erosion. The slope length and slope steepness play essential roles in studying how landscape features control water runoff. Water flow distance is represented through the slope length factor, and slope steepness determines how terrain angles affect flow velocity. Water movement speed down slopes depends heavily on these two factors because they control the rate of soil erosion. Soil erosion models heavily depend on slope length and steepness factor calculations, which are possible through the SRTM DEM dataset. The Landsat 8 OLI (Operational Land Imager) data serve as a critical dataset for estimating the Cover and Management Factor (C factor) through its high-resolution satellite imagery. Soil erosion effects from vegetation and land management practices can be studied through the essential C factor. Soil resistance to erosion depends on land management practices and vegetation cover because these elements create barriers that control water runoff during rainfall. The Landsat 8 OLI imagery enables the processing of data that leads to C factor derivation for land use analysis and vegetation type identification and cover density measurement. The collected data proves most beneficial for areas that experience substantial soil erosion effects from vegetation and land management activities including agricultural fields and forests.

Primary sources of observed data including local meteorological stations and storm records provide rainfall data required to determine the Rainfall and Runoff Erosivity Factor (R factor). Rainfall intensity and erosive capacity are measured through the R factor and depend on both the precipitation amount and the way it distributes throughout time. Storms with high-intensity levels produce greater runoff amounts that lead to increased soil particle dislodgement through erosion. The rainfall data obtained from the study area serves to determine temporal rainfall pattern variations along with their impact on storm-triggered soil erosion potential. Soil erosion intensifies during heavy rainfall events because these storms affect bare or uncovered areas and steep terrain. The Soil Texture dataset which results from field sample analysis and laboratory testing enables the calculation of the Soil Erodibility Factor (K factor). The K factor determines soil erosion vulnerability through analysis of its texture that consists of sand, silt, and clay proportions. The presence of clay particles in the soil increases its erodibility while sandy soils with their larger particle size remain less susceptible to rainfall-induced detachment and runoff processes. The K factor estimation for all soil erosion models depends on the results of soil texture analysis that includes organic matter content determination and particle size distribution assessment. Soil organic content together with permeability and structure elements influence the ease with which water removes soil particles from the ground. The K factor estimation receives additional precision through the Soil Chemical Properties dataset that results from laboratory testing of soil samples. Soil erodibility depends on three main characteristics including organic carbon content and cation exchange capacity (CEC) and soil permeability. Soil containing high levels of organic matter develops better structural cohesion that protects it from erosion. Soils containing low organic content together with poor structure show higher susceptibility to erosion. The dataset improves K factor accuracy through its inclusion of soil chemical and physical properties which affect cohesion strength and erosion resistance. The Support Practice Factor (P factor) is estimated through field observations and local stakeholder surveys that provide the Support Practice Information. The P factor represents various conservation practices including contour ploughing and terracing and buffer strip implementation which decrease water runoff velocity while protecting soil from erosion. Field visits and surveys inside the study area document the extent and types of conservation practices directly. The P factor calculation utilizes this data to modify potential soil erosion predictions because managed areas with proper soil conservation practices tend to have reduced erosion rates. Soil erosion mitigation strategies can be identified using this essential data to protect specific areas from further damage. The Amount of Soil Loss measurements collected from field observations confirm and validate the prediction results from the models. The soil erosion models (USLE, RUSLE, MUSLE) receive validation through empirical data obtained through field measurements and historical soil erosion records in the study area. Model calibration processes and identification of discrepancies between predicted and observed soil loss become possible through the comparison of model output with field measurements. The addition of this validation process enables the models to achieve higher accuracy and reliability which leads to predictions that better match real-world erosion dynamics (Figs. 19.2, 19.3).

19.2.3.2 K Factor

The K Factor stands as an essential variable within USLE and RUSLE, and MUSLE models to describe how well soils resist erosion through their physical makeup and chemical composition. The K Factor originates from soil texture analysis to measure the detachment capability of soil particles by rainfall and runoff. Higher K Factor values indicate increased susceptibility to erosion because water flow during rainfall and

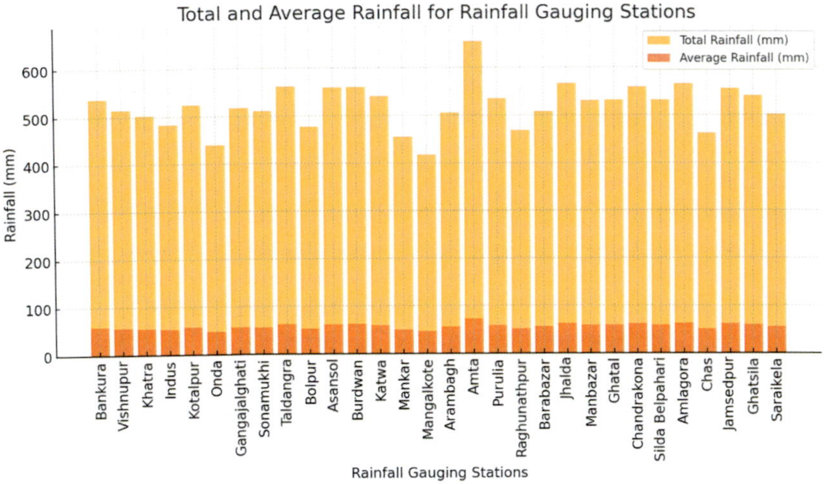

Fig. 19.2 Details information of the rainfall gauging station

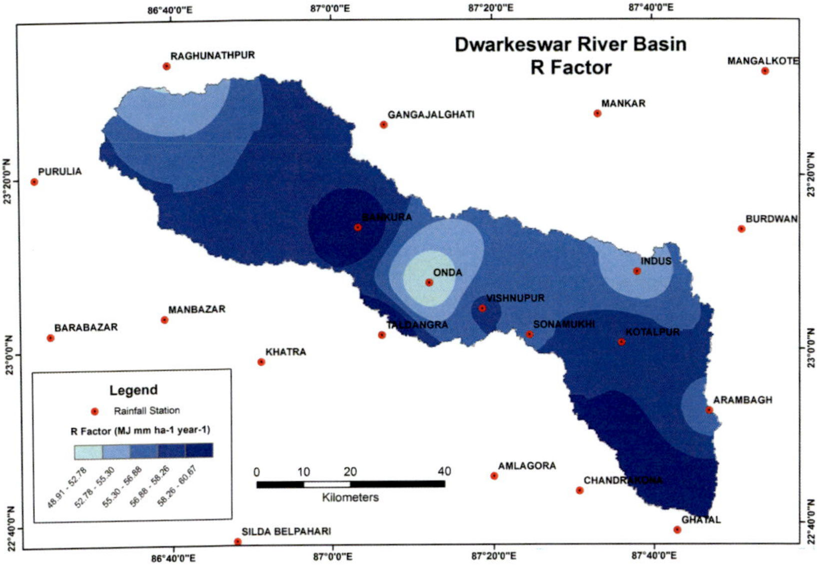

Fig. 19.3 Rainfall and runoff erosivity factor

runoff easily removes the soil, while lower K Factor values demonstrate better erosion resistance. The K Factor depends mainly on soil texture, which represents the proportions of sand, silt and clay in the soil. Soils containing clay and silty clay textures experience higher K Factor values because their smaller particle sizes make them more vulnerable to rainfall-induced detachment. The K Factor values of sand and loamy sand soil remain low because their coarse nature provides better soil cohesion, which reduces the potential for erosion. The K Factor evaluation in this research utilized laboratory-tested soil texture data from sampled soils, which are presented in Fig. 19.4. Soils containing higher amounts of fine clay particles demonstrate elevated K Factor values according to the data collected. The K Factor for clay soils exists between 0.22 to 0.24 when organic matter content remains below 2% and between 0.21 to 0.23 when organic matter content exceeds 2%. The K Factor of sand and similar coarser soils measures at approximately 0.02 to 0.03 because these types of soil resist erosion. Figure 19.5 presents the K Factor map of Dwarkeswar River Basin, which demonstrates how soil erodibility varies across the research area. The K Factor values in the map differentiate between areas of high and low soil erodibility through color coding. The green and yellow areas of the map show lower K Factor values in the soils, while red and purple regions represent higher erodibility soils. Higher K Factor values in specific areas create regions that are more prone to soil erosion according to the spatial distribution analysis. The K Factor map functions as a vital tool for identifying soil erosion hotspots across the basin because it enables focused conservation strategies. Areas with elevated K Factor values need urgent erosion control measures, which include vegetation cover implementation along with contour ploughing and other soil conservation practices.

19.2.3.3 LS Factor

The LS Factor represents an essential element for predicting soil erosion through USLE, RUSLE and MUSLE models. Slope length, together with slope steepness, determines the intensity of soil erosion through this factor. The fundamental concept behind this factor shows that longer slopes with increased steepness produce higher runoff amounts that enhance soil erosion potential. Higher slope length and steepness in a region result in stronger water flow that enhances both soil detachment and transport processes. Steep slopes create more erosion risks than flat regions because of their increased susceptibility to erosion. The LS Factor

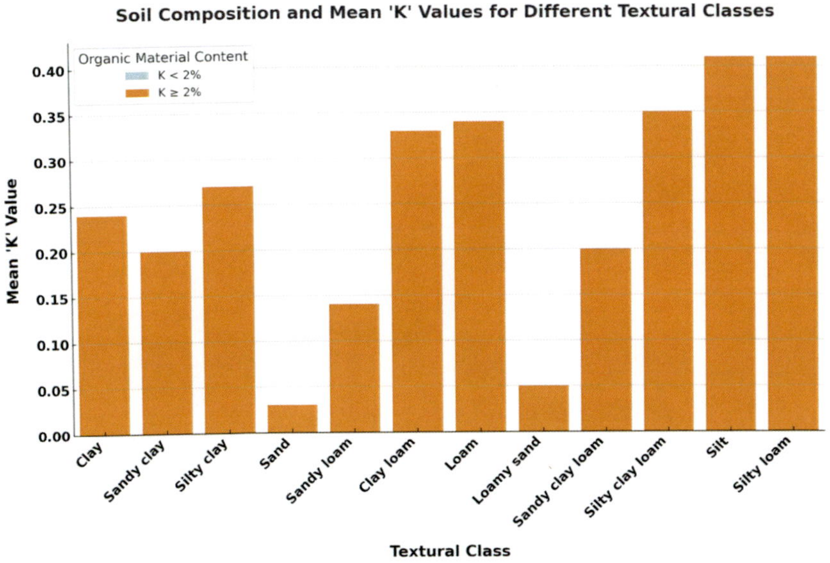

Fig. 19.4 Parameters for estimating the K factor

calculation requires landscape topography data obtained from Digital Elevation Models (DEMs). The models reveal different slope gradients that become the basis for estimating water flow patterns and related erosion threats. A spatial map represents the LS Factor through which areas display different levels of erosion risk based on their elevation. Figure 19.6 shows the LS Factor map of the Dwarkeswar River Basin. The LS Factor distribution on the map utilizes a color spectrum which shows green areas represent regions with gentle slopes, while red areas represent areas with steep slopes. Within the basin area, the LS Factor demonstrates different levels of soil erosion vulnerability through its distribution between regions that have minimal erosion risk and others that face increased risk due to steep slopes and extended flow paths. The basin shows mostly gentle slopes in most of its regions yet steep topography exists in limited areas that create significant erosion risks. The areas with elevated LS Factor values require specific attention for erosion control since they face increased soil loss risks during heavy rainfall events. Soil conservation initiatives should target priority areas first by analyzing the LS Factor distribution of the basin to maximize their effectiveness.

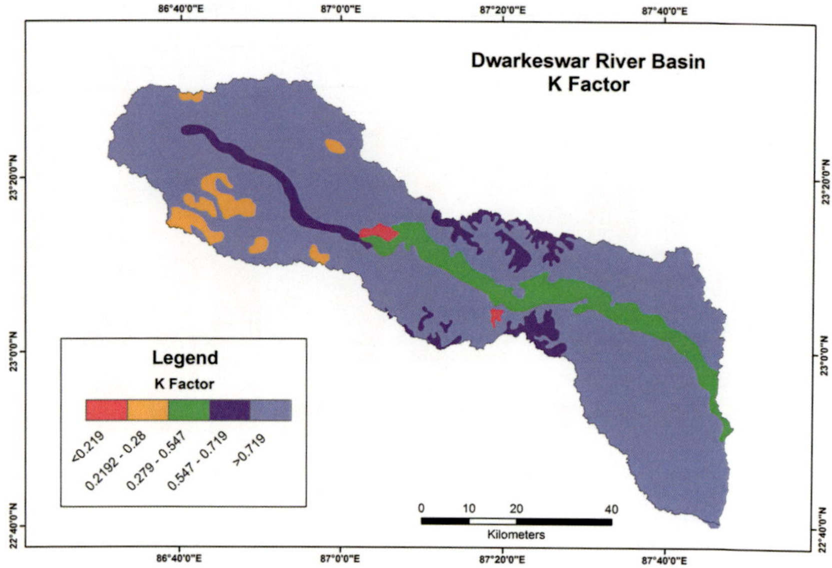

Fig. 19.5 K factor

19.2.3.4 C Factor

Soil erosion prediction models, including USLE, RUSLE and MUSLE, use the C Factor (Cover and Management Factor) as a central variable. The C Factor describes how vegetation covers, together with land management practices, influences soil erosion. The C Factor represents the extent to which land use practices, along with crop cover and vegetation and management methods, decrease rainfall and runoff erosive power. Soil protection from erosion decreases when the C Factor value increases, yet effective land management practices and vegetation reduce the C Factor value. The C Factor depends on vegetation density and land use type, together with land management practices, which include crop rotation and tillage methods. The C Factor in dense vegetation and well-managed agricultural fields remains low because these land cover types reduce water erosion by controlling runoff speed and improving soil infiltration and ensuring soil particle cohesion. Soil exposed without vegetation or poorly maintained through practices like deforestation or overgrazing will experience increased erosion risk because of its elevated C Factor values. The Normalized Difference Vegetation Index (NDVI)

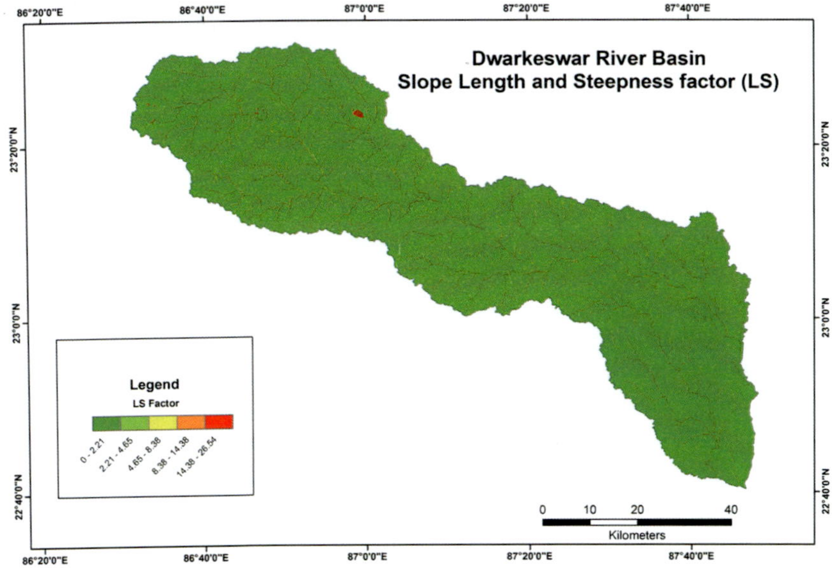

Fig. 19.6 LS factor

serves as a common method for determining the C Factor. The NDVI functions as a remote sensing index which extracts vegetation data from satellite imagery. The red and near-infrared bands show a measurable difference through NDVI to determine vegetation from non-vegetation areas. The NDVI measurement shows dense vegetation through elevated values, yet sparse vegetation produces minimal values. The figure shows the Normalized Difference Vegetation Index (NDVI) of the Dwarkeswar River Basin in Fig. 19.7. The NDVI values extend from -0.13 to 0.44 throughout the figure, where denser vegetation areas show higher values, yet bare land areas and areas with sparse vegetation show lower values. NDVI functions as an accurate indicator for C Factor evaluation because it shows direct correspondence between vegetation density and soil erosion reduction effectiveness. The C Factor estimation for the study area becomes possible through analysis of NDVI data, as illustrated in Fig. 19.8. The map displays how vegetation density affects the C Factor values, where dense vegetation areas show lower C Factor values, which protect the soil, while sparse vegetation areas demonstrate higher C Factor values that make the soil more prone to erosion. The C Factor map shows how different vegetation levels across the basin affect potential soil erosion rates. Soil management quality and conservation practices improve in areas where C Factor values remain low, but erosion risks increase in regions with higher C Factor measurements because of insufficient vegetation cover or poor land management. The direct connection between vegetation cover and erosion potential emerges from the relationship between NDVI and the C Factor as represented in Fig. 19.9. The NDVI values show a direct relationship with the C Factor, where higher vegetation cover corresponds to lower soil erosion potential. The observed connection demonstrates that preserving existing vegetation or implementing new vegetation growth serves as an essential method to control soil erosion in the basin. The C Factor stands as a vital predictor of soil erosion because it demonstrates how vegetation along with proper land management practices protect the soil. Analyzing the NDVI and its connection to the C Factor enables land managers to create effective strategies that fight soil erosion and safeguard the land. The conservation priority should focus on regions exhibiting low C Factor values while implementing protective measures in areas showing high C Factor values to stop further environmental degradation.

19.2.3.5 P Factor

Soil erosion models USLE, RUSLE, and MUSLE heavily depend on the P Factor (Support Practice Factor). The P Factor measures the ability of different soil preservation methods to decrease erosion through practices including contour ploughing and strip cropping and terracing, and buffer strip implementation. Soil protection against erosion becomes possible through these practices, which alter runoff movement and decrease its erosive strength. The P Factor remains vital because it reflects the land management approaches that determine soil erosion levels. Soil conservation practices that are properly implemented decrease the P Factor value, which indicates reduced erosion risks in those areas. Soil conservation measures absence in specific areas results in elevated P Factor values, which indicate substantial erosion threats. The P Factor shows different levels of effectiveness due to varying types and degrees of implemented regional practices. The P Factor receives its assigned values through different slope categories as shown in Fig. 19.10. The implementation of contour ploughing becomes more effective in reducing runoff because gentle slope areas are more suitable for this practice. The implementation of terracing and strip cropping becomes necessary for reducing water

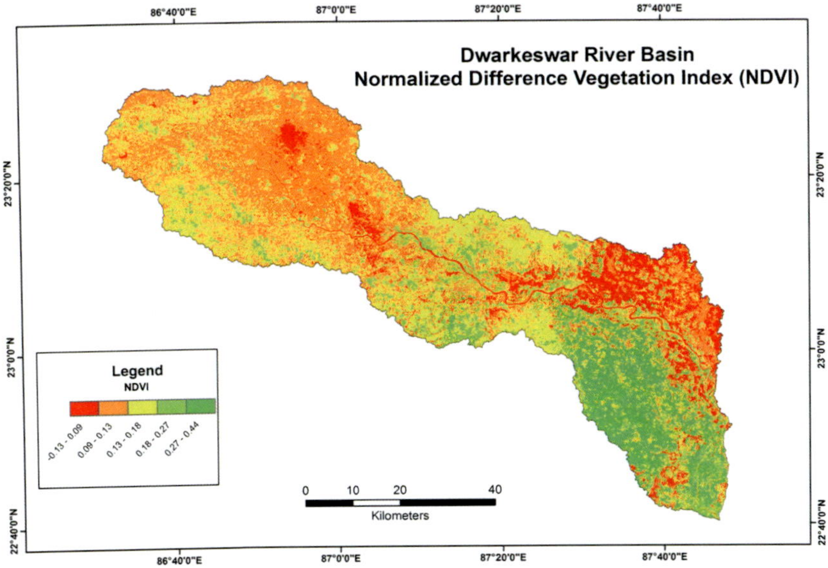

Fig. 19.7 Normalized difference vegetation index

flow effects when slopes increase in steepness. The Dwarkeswar River Basin shows the P Factor distribution pattern in Fig. 19.11. The P Factor values are represented through a color gradient on the map while lighter areas denote regions where conservation practices prevail thus minimizing erosion risks. The darker the shade of color represents areas without widespread soil conservation practices thus indicating higher P Factor values and increased soil erosion risk. The outputs present visual data about erosion potential distribution throughout the study area to identify specific locations needing additional erosion control measures. The implementation of adequate support practices alongside slope steepness and soil erodibility and rainfall intensity factors helps decrease soil loss thereby maintaining sustainable land and water resources in the region.

19.2.3.6 Surface Runoff

Surface runoff stands as an influential factor that determines how topsoil erosion occurs (Hartanto et al., 2003). The erodibility of surface materials depends heavily on both slope direction and slope amount from this perspective. LULC directly affects surface areas that determine runoff

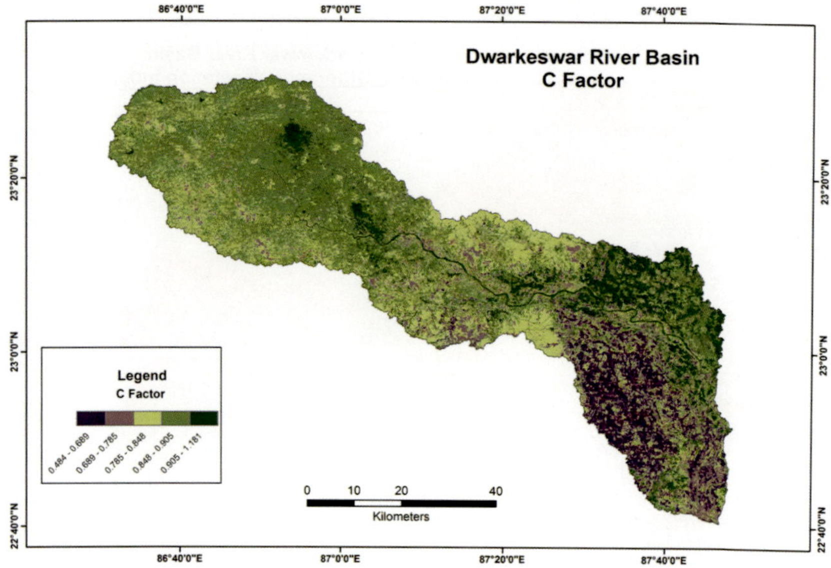

Fig. 19.8 Cover and management factor

quantity and intensity. The erosion intensity of the top surface rises proportionally with increasing amounts of direct surface runoff. The catchment level analysis reveals that runoff directly affects erosion potential. Surface layer runoff and infiltration show different patterns because of the combined effects of vegetation cover and fallow land area, and other land features. Different relationships exist between the quantity of runoff and infiltration rates. The amount of infiltration determines groundwater storage because it leads to positive outcomes for sub-surface water resources accumulation.

The United States SCS formulated the equation to measure the rate of surface runoff:

$$Q = \frac{(P - I_a)^2}{(P - Ia + S)} \tag{19.1}$$

where Q is the runoff in mm, P is the total amount of precipitation, and S is the highest infiltration rate. Ia is the specific parameter which is largely concentrated in soil and LULC, another observation that Ia is the eventual rate of infiltration ($Ia =$ ℷs); ℷ is the abstraction rate in the

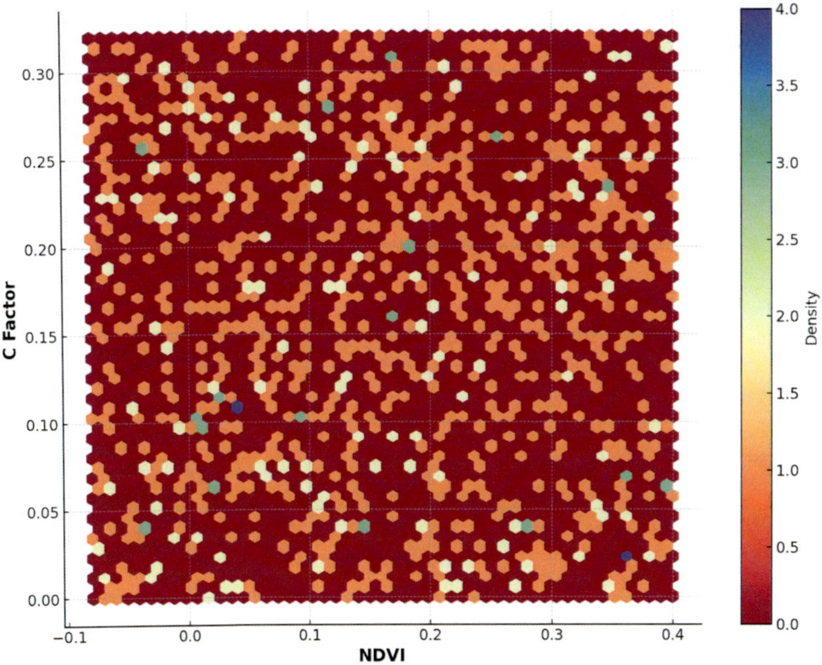

Fig. 19.9 Correlation between NDVI and C factor

initial stage. The ℷ values range between 0 to 0.3.

$$S = \frac{25400}{CN} - 254 \tag{19.2}$$

where S is the Soil retention rate, and CN is the curve number (Pal & Chakrabortty, 2019).

$$Q = \frac{(P - 0.2S)^2}{(P + 0.8S)} \tag{19.3}$$

where 20% (0.2 s) of the potential maximum retention is the initial abstraction before runoff begins, and the other 80% (0.8 s).

But in Indian conditions, the modification of this specific equation has been done for better accuracy and authentic outcomes (Tiwari et al.,

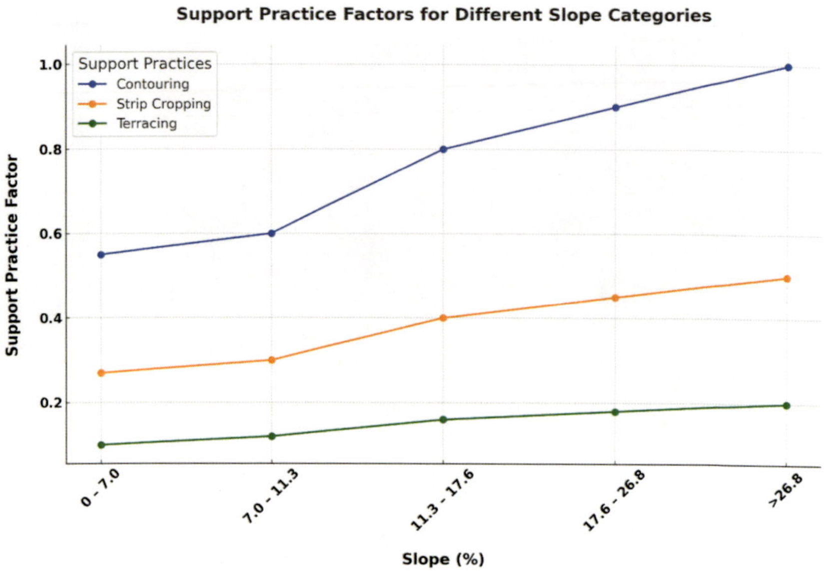

Fig. 19.10 Assign values for support practice

1991):

$$Q = \frac{(P - 0.3S)^2}{(P - 0.7S)} \tag{19.4}$$

where Q is the actual direct runoff from precipitation, P is the total rainfall, and S is the possible maximum retention.

$$\text{Weighted CN} = \frac{\text{CN1} \times \text{A1} \times \text{CN2} \times \text{A2}.\ldots\ldots\ldots\ldots\ldots\ldots.\text{CNn} \times \text{AN}}{\text{A1} + \text{A2} + \ldots\ldots\ldots\ldots\ldots\ldots.\text{An}} \tag{19.5}$$

where CN is the Curve number of the individual land use and A1…..An is the Percentage (%) area under individual land use.

Surface Runoff distribution in the Dwarkeswar River Basin appears through Fig. 19.12 based on Weighted Curve Number (CN) calculations. The map shows surface runoff variations through a color gradient which transitions from light blue for minimum values to dark blue for maximum values. The Surface Runoff value depends on the curve number, which

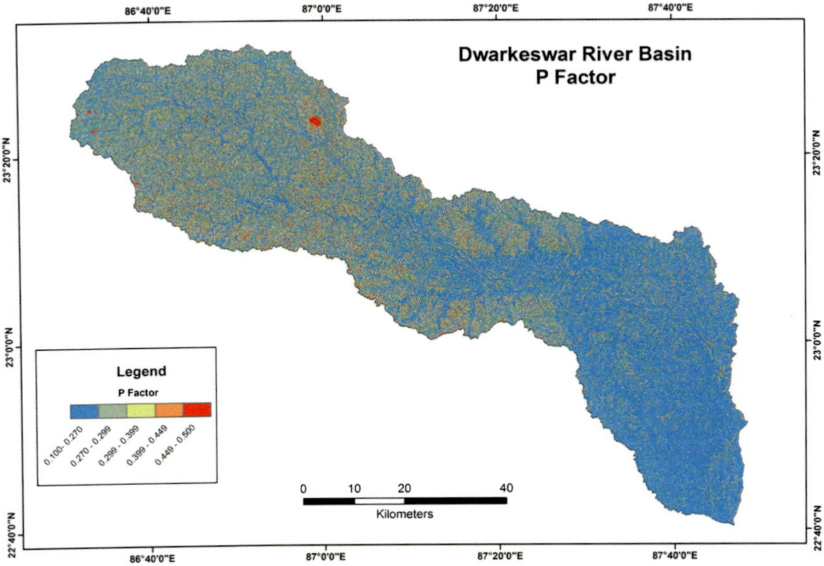

Fig. 19.11 P Factor

gets its value from land use and soil type and hydrological conditions and cover management practices. The infiltration capacity decreases when curve numbers increase, which results in higher runoff amounts. Darker shaded areas in the map demonstrate greater runoff potential because they correspond to urbanized areas with poor soil quality and impervious surfaces. The regions appearing lighter in color show better infiltration capabilities and reduced runoff because they contain agricultural and forested areas with sufficient soil cover. The runoff data serves as a vital input for modeling soil erosion because it enables calculations of surface water movement that directly affects the amount of soil lost to erosion. The soil particles from higher runoff areas tend to erode due to water movement, especially when slopes are steep or have little vegetation. The figure offers essential knowledge about how water flow patterns in the landscape affect soil erosion processes, thus enabling better strategies for protecting soil and watersheds. Surface runoff mapping with the CN method provides an effective tool to evaluate river basin drainage characteristics while enabling better planning of erosion control measures.

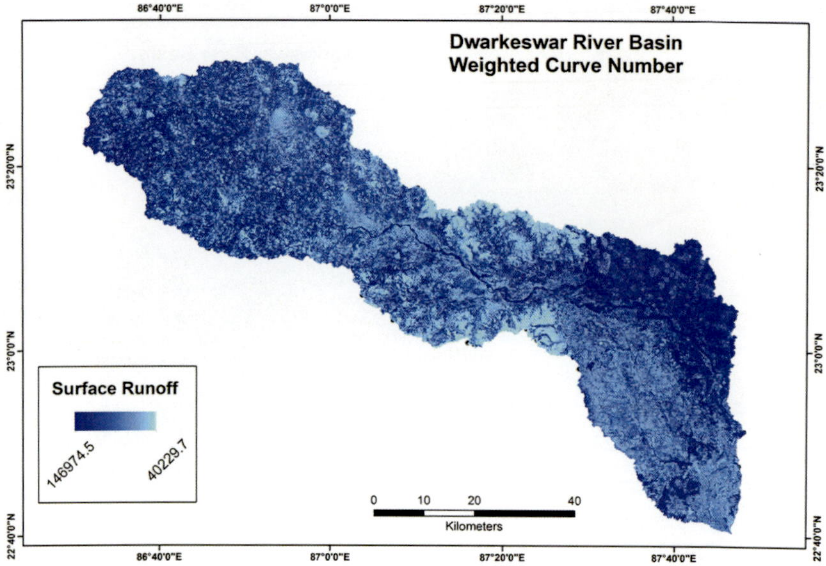

Fig. 19.12 Surface runoff

19.2.4 Application of Different Soil Erosion Models

19.2.4.1 USLE

The Universal Soil Loss Equation (USLE), created by the United States Department of Agriculture (USDA), functions as a tool with worldwide applicability. The statistical model originated from United States-based datasets, which included 10,000 plot-years of natural runoff measurements and 2,000 plot-years of artificial rainfall simulation results (Wischmeier & Smith, 1978). The USLE model provides essential information about basin sediment erosion by analyzing natural basin features. The USLE model now benefits from GIS integration, which has improved its capability to determine appropriate land management conservation areas (Srinivas et al., 2002). The model uses field condition evaluation to determine soil erosion levels, which supports land use planning activities (Nearing et al., 2017). Research shows that land use and land cover dynamics create a direct link to the relative status of soil erosion (Morgan et al., 1998). The USLE model becomes more effective when combined with GIS technology to run the Spatially Distributed Sediment Delivery

Ratio (SDR) model, which provides assessment capabilities for soil erosion and deposition patterns across landscapes. The USLE model demonstrates its effectiveness for detecting island ecosystem soil loss potential and measuring micro-watershed soil loss. The model serves as a tool to identify specific locations which require the most urgent erosion control interventions. Zonal statistics tools enable statistical analysis to study soil loss patterns in different geographical areas. The USLE equation requires five fundamental input factors, including soil erodibility, rainfall erosivity and slope length and steepness and cover management and support practices, which are usually mapped as raster data layers. The model functions as a strong tool for analyzing and managing soil erosion through its extensive methodology, which operates effectively in multiple geographical environments.

$$A = R \times K \times L \times S \times C \times P \tag{19.6}$$

where,

A = estimate gross soil erosion, t/ha/year
R = rainfall erosivity factor, joules/(ha/year), (t-m/ha) (mm/h) per year
K = Soil erodibility factor (t/ha)/erosivity factor (R), t/joules, t/ha-year
L = Slope length factor
S = Slope gradient factor
C = Crop cover or crop management factor
P = Supporting conservation practice factor.

19.2.5 RUSLE

The Revised Universal Soil Loss Equation (RUSLE) functions as a dependable empirical model which experts widely used to determine soil erosion levels. RUSLE uses multiple environmental factors, such as topography and surface canopy cover and conservation practices and rainfall characteristics, and runoff, to determine erosion-caused soil loss (Renard et al., 1991). RUSLE functions as an effective tool to quantify soil erosion through its simplified yet accurate method for evaluating how different variables affect erosion rates. RUSLE demonstrates exceptional capability in predicting annual soil erosion through simple data collection, which makes it suitable for diverse geographical areas. The model's basic input parameters make RUSLE efficient enough to be used across various scenarios, despite other models needing more complex data and advanced calculations. The model stands out because it provides predictions for both short-term storm-based soil loss and long-term annual soil loss. RUSLE demonstrates flexibility through its ability to work across various geographical areas while needing only minor adjustments for calibration. The model proves effective for measuring soil erosion across tropical, subtropical, arid, temperate, and forested regions because it delivers precise erosion assessments in areas with different climate patterns and land management practices. RUSLE shows flexibility through minor adjustments to suit regional characteristics, which makes it ideal for complex topographical areas, including mountains and areas with rough terrains. The RUSLE core equation calculates average annual soil loss through multiplication of five essential factors, which include rainfall erosivity (R), soil erodibility (K), slope length and steepness (LS), cover management (C), and support practice (P). The various factors appear in the following format:

$$A = R \times K \times LS \times C \times P \tag{19.7}$$

Quantitative data availability with good quality remains insufficient in most third world countries thus limiting the application of this model type. Each empirical and semi-empirical model contains its own distinct function that makes a single model insufficient to achieve all objectives. Acquiring and fitting RUSLE variables requires time-consuming and expensive data collection procedures.

19.2.5.1 MUSLE

MUSLE represents an updated version of USLE that specifically measures potential sediment yields from soil loss. MUSLE serves as a widely accepted tool for calculating annual soil loss and sedimentation measurements at average scales. MUSLE incorporates the runoff factor as a replacement for the rainfall factor from USLE because it addresses the unpredictability of rainfall patterns across different periods. The modified model functions better when the rainfall data faces inconsistencies or unreliability. The sediment yield model expressed through MUSLE demonstrates a simpler application because it relies on a direct outlet data structure, which enhances its meaning. The tool provides an efficient method to determine watershed outlets and sediment transport rates, which makes it valuable for watershed management purposes (Sadeghi et al., 2007). MUSLE offers a significant benefit through its ability to predict sediment yield by using runoff characteristics because it establishes clear connections between soil erosion processes and sediment transport mechanics (Nearing et al., 2005).

The MUSLE model uses an erosivity factor to validate its ability to predict runoff characteristics. The model accuracy improves through this factor because it considers runoff intensity and duration, which determine sediment detachment. MUSLE evaluates both runoff volume and peak discharge as essential factors for single-storm soil detachment that produces dynamic sediment yield assessments. The MUSLE calculates storm event sediment yield through this mathematical expression:

$$S = 11.8(Q_{qp})^{0.56} K \times L \times S \times C \times P \tag{19.8}$$

where, S is sediment yield, Q is volume of runoff, qp is peak flow, and K, L, S, C and P are, respectively, the soil erodibility, slope length, slope steepness, crop management and soil erosion control practice factors, as in the USLE model (Williams & Berndt, 1977).

19.3 Results

19.3.1 Average Annual Soil Erosion Using USLE

The Universal Soil Loss Equation (USLE) functions as a common empirical model which calculates average annual erosion rates through analysis of environmental and land management factors. The model includes all aspects of rainfall intensity, together with soil characteristics and topographical features and land usage patterns and operational procedures. The equation determines soil erosion potential through multiplication of R Factor (rainfall erosivity) with K Factor (soil erodibility) and LS Factor (slope length and steepness) and C Factor (cover and management), and P Factor (support practices). The product of these factors creates an estimated total of soil loss expressed in tons per hectare annually. The USLE model creates a vital base to analyze large-scale soil erosion patterns across the Dwarkeswar River Basin. The R Factor serves as the main determinant for identifying erosion-prone areas because it measures rainfall intensity and quantity during heavy precipitation events. The K Factor assesses soil detachment susceptibility under rainfall conditions to identify areas where the soil has natural tendencies to break apart. The LS Factor in the USLE provides a method to evaluate slope and length characteristics that affect water movement and erosion rates. The combination of steep slopes with extended lengths produces more erosion because water flows rapidly while transporting higher quantities of soil material. The C Factor evaluates land cover and management practices because it recognizes vegetation or crop cover as critical elements for protecting soil from rainfall impact. The P Factor includes conservation practices, including contour ploughing or terracing, which modify runoff direction to reduce erosion effects. The USLE model maintains its widespread use because it is simple, yet it fails to consider changing environmental conditions and lacks capabilities for tracking temporal variations like seasonal rainfall patterns and land use modifications. The tool functions effectively to detect areas with high soil loss potential in the basin, thus helping land managers direct their soil conservation efforts. USLE-derived soil erosion predictions direct policy decisions and soil conservation plans to concentrate protective measures in locations with the highest risk of erosion (Fig. 19.13).

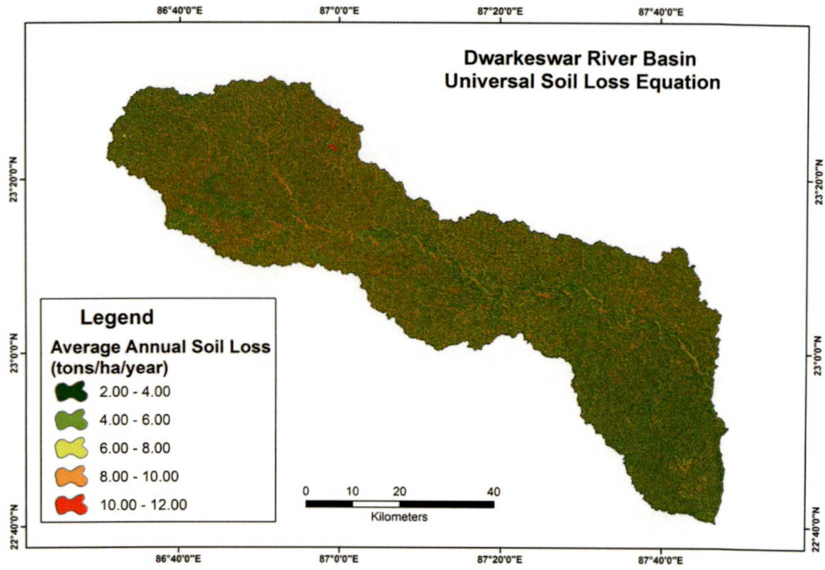

Fig. 19.13 Average Annual Soil Loss using Universal Soil Loss Equation

19.3.2 Average Annual Soil Erosion Using RUSLE

RUSLE advances the principles of USLE by implementing multiple enhancements that produce more precise results for real-world applications. The basic framework of RUSLE remains like USLE, while it introduces enhanced calculation approaches for the C Factor (land cover and management) and the P Factor (support practices). The main benefit of RUSLE stems from its capability to handle seasonal fluctuations and land use changes, which makes it suitable for diverse environmental conditions and time-dependent elements. RUSLE delivers enhanced dynamic capabilities for soil erosion estimation in the Dwarkeswar River Basin beyond what USLE offers. The enhanced C Factor calculation capability enables the model to distinguish between soil erosion effects from different land management practices by evaluating the erosion differences between fields with different cover and land use patterns. RUSLE provides enhanced capabilities for studying how human activities affect soil erosion patterns throughout the basin. RUSLE demonstrates an important strength through its ability to incorporate time-dependent factors when making erosion predictions. The RUSLE model includes a feature to handle changing land use conditions, unlike the USLE model, which operates under static conditions. The dynamic nature of RUSLE enables precise erosion predictions in areas with fast-developing land use patterns, including agricultural regions that experience major seasonal changes in their crop cover or conservation practices. The Dwarkeswar River Basin can benefit from RUSLE implementation to obtain better soil erosion predictions because these estimates support the development of specific conservation approaches suitable for seasonal rainfall patterns and land use changes. Soil managers utilize RUSLE to evaluate extended erosion patterns and immediate changes in erosion rates because this tool enables essential decision-making through adaptive land management strategies (Fig. 19.14).

19.3.2.1 Average Annual Soil Loss and Its Associated Sediment Yield Using MUSLE

The Modified Universal Soil Loss Equation (MUSLE) represents an advanced version of USLE which focuses on areas with intense rainfall patterns and irregular runoff conditions. MUSLE transforms USLE through actual rainfall event characteristics instead of using only long-term average conditions from USLE. MUSLE demonstrates exceptional

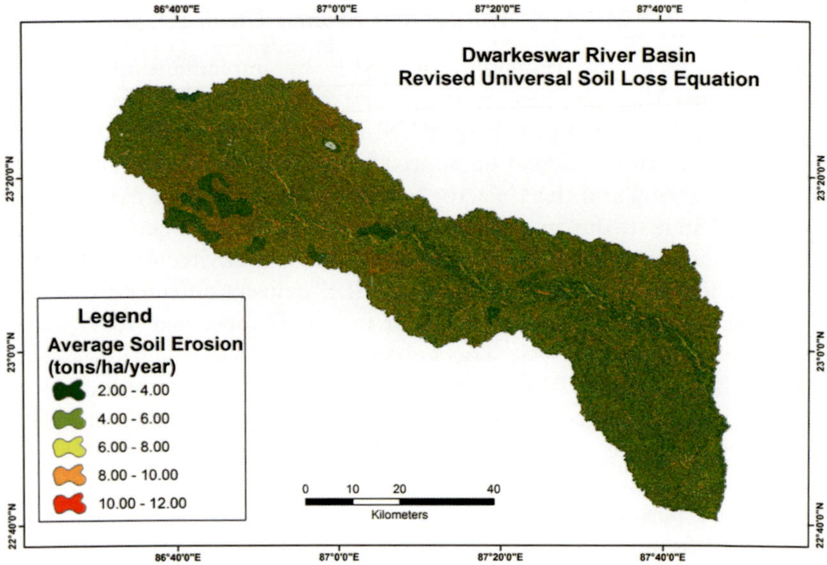

Fig. 19.14 Average Annual Soil Loss using Revised Universal Soil Loss Equation

suitability for dynamic environments like the Dwarkeswar River Basin because it handles the wide range of rainfall intensity and storm events that occur particularly during the monsoon season. MUSLE provides improved calculation methods through its assessment of peak runoff from individual storms rather than using yearly rainfall averages. MUSLE provides better results than USLE because it detects intense localized rainfall that causes unbalanced soil erosion. Regions with minimal ground cover experience significant soil erosion when heavy monsoon rains occur even though the annual rainfall totals remain moderate. MUSLE incorporates rainfall event dynamics through its calculation method which combines event volume with intensity measurements. The MUSLE model delivers improved predictions of soil erosion during heavy storms in the Dwarkeswar River Basin because it outperforms USLE and RUSLE in recognizing these extreme weather events. The implementation of MUSLE depends on real storm data and peak runoff analysis to generate precise erosion predictions for areas with intense rainfall events that drive

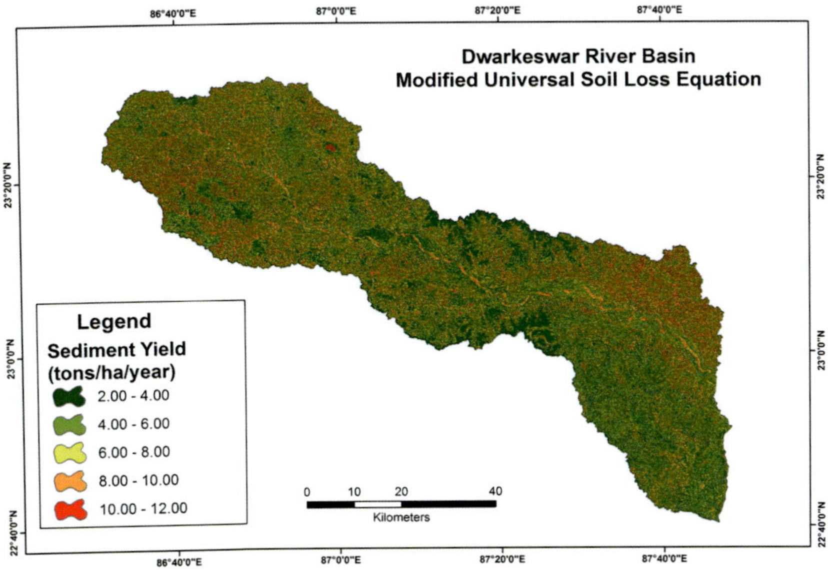

Fig. 19.15 Average Annual Soil Loss Using Modified Universal Soil Loss Equation

most soil loss. The MUSLE model includes sediment yield as an essential factor because it measures the specific amount of soil that water transports across the landscape. The assessment of complete soil erosion effects demands this crucial measurement because it advances beyond basic soil loss predictions to quantify the soil amounts that end up in water bodies leading to river and lake and reservoir sedimentation. MUSLE offers improved high-intensity storm soil erosion assessment capabilities which enables land managers to create better erosion control strategies for flood-prone regions. The development of dynamic erosion control strategies becomes possible through stormwater management and buffer zone implementation and rainwater harvesting systems which reduce soil loss during intense storm events (Fig. 19.15).

19.3.3 Sensitivity Analysis

The evaluation of soil erosion models depends heavily on sensitivity analysis, which helps determine their performance and resistance to change.

The analysis evaluates the impact of input variable fluctuations, including rainfall and soil erodibility and slope and vegetation cover and conservation practices, on the soil erosion models' predicted outcomes. Soil erosion model predictions become more understandable through this analysis because it reveals which input factors create the biggest impact on estimates and shows prediction uncertainty levels. The sensitivity analysis results for Universal Soil Loss Equation (USLE) and Revised Universal Soil Loss Equation (RUSLE) and Modified Universal Soil Loss Equation (MUSLE) appear in Fig. 19.16. The Receiver Operating Characteristic (ROC) curve provided the basis for sensitivity analysis to evaluate model discrimination between different soil erosion classes. The ROC curve demonstrates how model performance relates sensitivity to specificity while offering an effective method to evaluate model accuracy. MUSLE demonstrated the best accuracy level in this analysis by achieving an Area Under the Curve (AUC) value of 0.981, while RUSLE followed with an AUC value of 0.924, and USLE achieved 0.891. A model achieves better accuracy in soil erosion prediction when its AUC value reaches higher levels. MUSLE demonstrated superior performance because it considers rainfall intensity together with runoff event frequency, which are essential factors for regions experiencing unpredictable weather patterns and severe storms. RUSLE demonstrates better performance than USLE, but MUSLE remains superior because it better accommodates dynamic environmental factors, particularly in regions with intense rainfall occurrences. The oldest model, USLE, relies on constant rainfall assumptions, which limits its accuracy for variable environments due to its lack of model flexibility. The sensitivity analysis demonstrates why models which can respond to rainfall and runoff changes should be used. The data collection methods and model parameter refinement need improvement to achieve accurate predictions, specifically in areas that face high erosion risks because of their topography and land use patterns, and climate variability. Soil conservation programs in vulnerable areas will receive better guidance through improved model precision that results from ongoing model refinement.

19.4 Discussion

The study of soil erosion through models such as USLE, RUSLE, and MUSLE provides vital insights into the dynamics of land degradation in the Dwarkeswar River Basin. The results of these models highlight

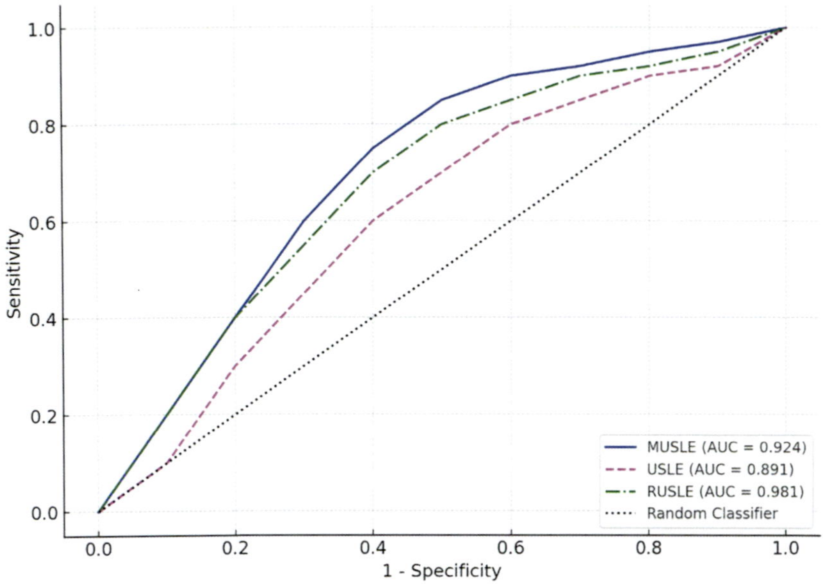

Fig. 19.16 Sensitivity analysis

the varying rates of soil loss across the basin, emphasizing the need for targeted soil conservation strategies in areas most prone to erosion. However, beyond the specific results, this research presents an opportunity to delve into broader discussions regarding the effectiveness and adaptability of these models, comparing their findings with international studies, and reflecting on their implications for future erosion management strategies.

Firstly, the distinction between the three models (USLE, RUSLE, and MUSLE) is important for understanding their comparative utility in predicting soil erosion. USLE, being the oldest and simplest of these models, assumes relatively stable conditions. This assumption can limit its application, especially in regions with variable rainfall, land use, or management practices. The limitations of USLE have been well documented in the literature, with researchers like (Wischmeier & Smith, 1978) noting its inefficiency in regions with dynamic weather conditions. This study aligns with those critiques, showing that USLE may not be as accurate in regions experiencing high variability in climatic and hydrological factors.

In contrast, the Revised Universal Soil Loss Equation (RUSLE) improves upon USLE by incorporating factors such as land cover changes, seasonal variations, and improved estimates for the management practices. RUSLE has been widely adopted in several studies across the globe, particularly in regions with diverse land uses and more advanced data availability (Renard et al. 1997). The RUSLE model's improved adaptability makes it a more reliable tool in areas where conservation practices and vegetation cover fluctuate over time. This study corroborates findings from other regional studies, demonstrating the applicability of RUSLE for better understanding soil erosion processes in tropical and subtropical regions.

The Modified Universal Soil Loss Equation (MUSLE) takes erosion prediction a step further by accounting for high-intensity rainfall events and runoff, providing a more accurate reflection of real-world conditions, especially in places prone to storms. The model's ability to adjust for intense rainfall events has been validated in various studies worldwide (Williams et al., 1996), making it a suitable choice for dynamic environments like the Dwarkeswar River Basin. The higher AUC value for MUSLE in this study is consistent with its reputation for offering superior accuracy in predicting soil erosion under extreme weather conditions. Globally, MUSLE has been recommended as the preferred model in regions where rainfall intensity is a significant variable.

Additionally, the sensitivity analysis conducted in this study, which evaluates the influence of each model's input variables (R, K, LS, C, P), demonstrates the varying significance of these factors in different contexts. This is particularly evident in the Dwarkeswar River Basin, where rainfall intensity (R factor) and slope characteristics (LS factor) play a substantial role in soil erosion dynamics. These findings align with international studies that have emphasized the importance of rainfall as a major driver of erosion, particularly in regions with steep topography and irregular rainfall patterns (Li et al., 2017). The sensitivity of erosion models to changes in the R and LS factors further supports the notion that these variables should be given high priority when designing erosion control measures.

Furthermore, the integration of remote sensing data, such as NDVI for the C factor and the use of high-resolution DEMs for the LS factor, is increasingly being seen as a crucial innovation in soil erosion modeling. These advancements allow for more accurate spatial and temporal mapping of soil erosion risk, a trend that has been observed in other global studies on erosion risk mapping (Lal, 2001). The increasing use of geographic information systems (GIS) to enhance the spatial resolution and accuracy of soil erosion predictions is a significant leap forward in erosion management. As GIS technology continues to evolve, the potential for even more accurate models that integrate real-time data and complex environmental variables increases.

The results of this study also contribute to the ongoing discussion regarding soil conservation practices. While the models demonstrate the area's most vulnerable to soil erosion, the next logical step is to apply the findings to inform land management policies. This requires deeper engagement with local stakeholders, including policymakers, landowners, and conservationists, to design and implement effective soil conservation measures. The study supports the view expressed by various international researchers that soil conservation efforts must be tailored to local conditions, considering factors such as land use, climate, and socio-economic conditions (Prager & Posthumus, 2010).

Finally, the implications of this study are globally in scope. Although the focus was on the Dwarkeswar River Basin, the findings can be generalized to other regions facing similar erosion problems. The global importance of tackling soil erosion cannot be overstated, as it directly impacts food security, water quality, and overall land productivity. With ongoing climate change and increased anthropogenic activity, regions around the world are seeing more erratic weather patterns and intensifying soil erosion. This underscores the need for robust, adaptable models like RUSLE and MUSLE to predict and mitigate erosion impacts, ensuring sustainable land use practices for the future.

In conclusion, this study demonstrates the utility of advanced erosion models such as RUSLE and MUSLE in assessing soil loss and guiding conservation efforts. By comparing the findings with international literature and models, the study contributes valuable knowledge to the global discourse on soil erosion management and offers practical insights for policymakers and land managers in erosion-prone areas.

19.5 Conclusion

The Dwarkeswar River Basin received thorough soil erosion assessments through three established models, including the USLE, RUSLE, and MUSLE. The study combines remote sensing data with field observations to identify specific areas with high soil erosion risk, which enables strategic conservation activities to receive proper attention. Soil erosion estimation begins with USLE, yet RUSLE and MUSLE provide enhanced understanding through their ability to consider seasonal conditions and rainfall patterns. The study results show that accurate erosion prediction requires detailed consideration of environmental factors that change both temporally and spatially, especially rainfall intensity and land use patterns. The key elements that affect soil erosion became clear through sensitivity analysis because rainfall intensity, coupled with soil erodibility and slope characteristics and vegetation cover, proved to be the most influential factors. The research establishes the need for specific soil conservation methods, including terracing and contour ploughing, and reforestation to protect agricultural production from soil degradation. This research provides essential knowledge to policymakers and land managers by supporting current land management and sustainable soil conservation programs. RUSLE and MUSLE advanced models combined with their application in dynamic conditions at the Dwarkeswar River Basin generate precise foundations for designing efficient erosion control strategies. Future studies should enhance these models using actual time data and climate change projections to enhance their ability to predict soil erosion accurately. The research has several limitations because it depends on fixed remote sensing data, together with theoretical model assumptions, which might not represent changing environmental conditions effectively. The future of this research should include real-time monitoring systems and climate change scenario modeling to improve predictive accuracy and adaptation capabilities.

References

Ali, S. A., & Hagos, H. (2016). Estimation of soil erosion using USLE and GIS in Awassa Catchment, Rift valley, Central Ethiopia. *Geoderma Regional, 7*, 159–166.

Angima, S., Stott, D., O'neill M, et al. (2003). Soil erosion prediction using RUSLE for central Kenyan highland conditions. *Agriculture, Ecosystems & Environment, 97*, 295–308.

Blanco-Canqui, H., & Lal, R. (2010). Soil Erosion Under Forests. *Principles of Soil Conservation and Management* (pp. 321–344). Springer.

Chakrabortty, R., Ali, T., Pal, T., et al. (2025). Climate Change and Land Use Dynamics: Modeling Soil Erosion Scenarios to Achieve Sustainable Development Goals. *Earth Syst Environ*. https://doi.org/10.1007/s41748-025-00631-0

Chakrabortty, R., Pal, S. C., Sahana, M., et al. (2020). Soil erosion potential hotspot zone identification using machine learning and statistical approaches in eastern India. *Natural Hazards, 104*, 1259–1294. https://doi.org/10.1007/s11069-020-04213-3

Foster, G. R., Meyer, L., & Onstad, C. (1977). A runoff erosivity factor and variable slope length exponents for soil loss estimates. *Transactions of the ASAE, 20*, 683–0687.

Ghosh, S., & Guchhait, S. K. (2012). Soil loss estimation through USLE and MMF methods in the lateritic tracts of eastern plateau fringe of Rajmahal traps, India. *Ethiopian Journal of Environmental Studies and Management, 5*, 529–541.

Hartanto, H., Prabhu, R., Widayat, A. S., & Asdak, C. (2003). Factors affecting runoff and soil erosion: Plot-level soil loss monitoring for assessing sustainability of forest management. *Forest Ecology and Management, 180*, 361–374.

Hossain, A., Krupnik, T. J., Timsina, J., et al. (2020). Agricultural land degradation: Processes and problems undermining future food security. In S. Fahad, M. Hasanuzzaman, M. Alam, et al. (Eds.), *Environment, climate, plant and vegetation growth* (pp. 17–61). Springer International Publishing.

IFAD U. (2017). *The state of food security and nutrition in the world 2017*. FAO.

Keesstra, S., Mol, G., de Leeuw, J., et al. (2018). Soil-Related sustainable development goals: Four concepts to make land degradation neutrality and restoration work. *Land, 7*, 133. https://doi.org/10.3390/land7040133

Kouli, M., Soupios, P., & Vallianatos, F. (2009). Soil erosion prediction using the Revised Universal Soil Loss Equation (RUSLE) in a GIS framework, Chania, Northwestern Crete, Greece. *Environmental Geology, 57*, 483–497. https://doi.org/10.1007/s00254-008-1318-9

Kruk, E., Klapa, P., Ryczek, M., & Ostrowski, K. (2020). Influence of DEM elaboration methods on the USLE model topographical factor parameter on steep slopes. *Remote Sensing, 12*, 3540.

Lal, R. (2001). Soil degradation by erosion. *Land Degradation and Development, 12*, 519–539. https://doi.org/10.1002/ldr.472

Li, P., Mu, X., Holden, J., et al. (2017). Comparison of soil erosion models used to study the Chinese Loess Plateau. *Earth-Science Reviews, 170*, 17–30.

Nearing, M., Jetten, V., Baffaut, C., et al. (2005). Modeling response of soil erosion and runoff to changes in precipitation and cover. *CATENA, 61*, 131–154.

Owens, P. N. (2020). Soil erosion and sediment dynamics in the Anthropocene: A review of human impacts during a period of rapid global environmental change. *Journal of Soils and Sediments, 20*, 4115–4143. https://doi.org/10.1007/s11368-020-02815-9

Pal, S. C., & Chakrabortty, R. (2022). *Climate change impact on soil erosion in sub-tropical environment: Application of empirical and semi-empirical models*. Springer.

Pal, S. C., & Chakrabortty, R. (2019). Simulating the impact of climate change on soil erosion in sub-tropical monsoon dominated watershed based on RUSLE, SCS runoff and MIROC5 climatic model. *Advances in Space Research, 64*, 352–377.

Pal, S. C., Chakrabortty, R., Roy, P., et al. (2021). Changing climate and land use of 21st century influences soil erosion in India. *Gondwana Research, 94*, 164–185. https://doi.org/10.1016/j.gr.2021.02.021

Prager, K., & Posthumus, H. (2010). Socio-economic factors influencing farmers' adoption of soil conservation practices in Europe. *Human Dimensions of Soil and Water Conservation, 12*, 1–21.

Prasannakumar, V., Shiny, R., Geetha, N., & Vijith, H. (2011). Spatial prediction of soil erosion risk by remote sensing, GIS and RUSLE approach: A case study of Siruvani river watershed in Attapady valley, Kerala, India. *Environmental Earth Sciences, 64*, 965–972.

Rashmi, I., Karthika, K. S., Roy, T., et al. (2022). Soil erosion and sediments: A source of contamination and impact on agriculture productivity. In M. Naeem, J. F. J. Bremont, A. A. Ansari, & S. S. Gill (Eds.), *Agrochemicals in soil and environment* (pp. 313–345). Springer.

Renard, K. G., Foster, G. R., Weesies, G. A., & Porter, J. P. (1991). RUSLE: Revised universal soil loss equation. *Journal of Soil and Water Conservation, 46*, 30–33.

Renard, K. G., USA, USA (Eds.) (1997). Predicting soil erosion by water: a guide to conservation planning with the revised universal soil loss equation (RUSLE).

Rhodes, C. J. (2014). Soil erosion, climate change and global food security: Challenges and strategies. *Science Progress, 97*, 97–153. https://doi.org/10.3184/003685014X13994567941465

Roy, P., Pal, S. C., Chakrabortty, R., et al. (2023). The role of indigenous plant species in controlling the erosion of top soil in sub-tropical environment:

In-situ field observation and validation. *Journal of Hydrology, 625*, Article 129993. https://doi.org/10.1016/j.jhydrol.2023.129993

Sepuru, T. K., & Dube, T. (2018). An appraisal on the progress of remote sensing applications in soil erosion mapping and monitoring. *Remote Sensing Applications: Society and Environment, 9*, 1–9.

Tiwari, K. N., Kumar, P., Sebastian, M., & Pal, D. K. (1991). Hydrologic modelling for runoff determination: Remote sensing techniques. *International Journal of Water Resources Development, 7*, 178–184. https://doi.org/10.1080/07900629108722510

Van der Knijff, J., Jones, R., & Montanarella, L. (2000). *Soil erosion risk assessment in Europe, EUR 19044 EN* (p. 34). Office for official publications of the European communities.

Wang, G., Hu, H., Islam, A. R. M. T., et al. (2025). Spatiotemporal rainfall concentration and erosivity in a tropical monsoon country. *Theoretical and Applied Climatology, 156*, 251. https://doi.org/10.1007/s00704-025-05479-y

Williams, J., & Berndt, H. (1977). Sediment yield prediction based on watershed hydrology. *Transactions of the ASAE, 20*, 1100–1104.

Williams, J., Nearing, M., Nicks, A., et al. (1996). Using soil erosion models for global change studies. *Journal of Soil and Water Conservation, 51*, 381–385.

Wischmeier, W. H., Smith, D. D. (1978). *Predicting rainfall erosion losses: A guide to conservation planning*. Department of Agriculture, Science and Education Administration

CHAPTER 20

Unearthing Fluvial Anomalies: The Impact of Alluvial Composition on Drainage Patterns in the Kunur River Basin (West Bengal, India)

Arpita Saha, Prabir Kumar Kundu, and Arup Kr Saha

20.1 Introduction

A drainage network is an open geomorphic system shaped by the interaction of geological structure, surface processes and time. Structural factors like faults and rock types influence water flow, while erosion, sediment transport and deposition continuously modify the network.

A. Saha (✉)
Department of Geography, Chanchal College, Malda, West Bengal, India
e-mail: arpita.saha.mldt@gmail.com

P. K. Kundu
Department of Geography, Siliguri Mahila Mahavidyalaya, Siliguri, Darjeeling, West Bengal, India

A. K. Saha
Chandrapur College, Purba Bardhhaman, West Bengal, India

S. Ghosh (ed.), *Reimagining Indian Rivers for Sustainability*, Palgrave Studies in Environmental Sustainability,
https://doi.org/10.1007/978-3-032-04754-0_20

Climate changes, tectonic activity and human interventions further impact its evolution. As a dynamic system, drainage networks provide insights into landscape development, hydrology and erosion susceptibility, making them essential for geomorphological studies (Davis & Johnson, 1909). The procedure of drainage development by the methodical organization of streamlets is modified by the terrain changes known by geologic or anthropogenic interferences (Gupta, 2003). A drainage network is the primitive element that carries out the water and sediments generate within the drainage basin (Chorley, 1957). The structure of the drainage system is carefully shaped to regulate the flow of water and sediments, influenced by factors such as stream discharge and its fluctuations, sediment load distribution, channel width and depth, slope, climate and geological conditions. (Gadre, 2006). The composition of the drainage network development over a span of time can impress the procedure generated in the drainage basin (Horton, 1945). Drainage anomalies indicate inconsistencies between the structural characteristics of a basin and the processes affecting it over time (Davis & Johnson, 1909). The morphometric analysis of the hierarchical arrangement of drainage networks facilitates the assessment of the effects of external factors, particularly tectonics, on the evolution of basins (Bahrami, 2013). A quantitative method for calculating stream's hierarchical anomaly number is introduced in the study. Morphometric parameters such as hierarchical anomaly index (Δa), stream's bifurcation ratio (Rb), bifurcation index (R), drainage density (Dd), were used here. The hierarchical anomaly index reflects the organiztion of single streamlets of each order and anomalies in sub-basins which may be improved through tectonically induced diffusive procedure or active fluvial erosion (Anish, 2019). This study aims to analyse the irregularities in the drainage network by examining the bifurcation index (R), drainage density anomalies and hierarchical anomalies. Fluvial morphometry is the quantitative study of river basin characteristics such as drainage patterns, stream order and slope. It plays a vital role in applied research by helping to assess erosion risks, flood potential and watershed behaviour. In river basin management, morphometric analysis supports effective planning for soil conservation, flood control and water resource development. It also aids in identifying priority areas for intervention and designing suitable conservation structures. With the use of GIS and remote sensing, morphometry has become a key tool for sustainable and data-driven river basin management.

20.2 Study Area

The Kunur River, classified as a fifth-order tributary of the Ajay River, holds significance within the riverine system of Rarh Bengal (Fig. 20.1). Its origin can be traced to a spring near Jhanjra village in the Burdwan district. As it progresses, the river flows through several administrative regions, including the areas under the jurisdiction of the Durgapur-Faridpur, Kanksa and Ausgram-I & II police stations. Eventually, it merges with the Ajay River near Kogram village in the Mongalkote block. Geographically, the study area is positioned between latitudes 23° 25′ 50.4″ N and 23° 39′ 21″ N, and longitudes 87° 16′ 26.4″ E to 87° 54′ 12.6″ E, encompassing a total area of approximately 753.90 square kilometres. The main river channel extends over a length of 112 km, and the basin exhibits a predominantly elongated form. The upper portion of the study area consists of a few protected and reserved forested regions. In terms of soil composition, the area features a mixture of laterite and clay interspersed with sand, kankar and alluvium. From a physiographic perspective, the upper segment of the basin represents an extension or fringe region of the Chhotanagpur Plateau. The highest elevation recorded in this region reaches approximately 124 m above mean sea level. The Kunur River is a small tributary of the Ajay River, flowing through parts of the Birbhum and Bardhaman districts in West Bengal, India. The river is primarily rain-fed and exhibits a seasonal flow pattern. It experiences high water levels during the monsoon season from June to September, while its flow significantly decreases or even dries up in some stretches during the dry months from November to May. The river's catchment area is relatively small, and the terrain consists mostly of lateritic and alluvial soils. During the rainy season, soil erosion can cause sediment to be carried into rivers, particularly from agricultural and deforested areas. Localized flooding may occur during heavy monsoon rains, particularly in low-lying farmland areas along riverbanks.

The climate of the Kunur River basin is classified as tropical monsoon. Summers, lasting from March to May, are hot and dry, with temperatures often ranging between 30 °C and 40 °C. The monsoon season brings heavy rainfall and high humidity, with the average annual precipitation ranging from 1,200 to 1,400 millimetres. Winters, from November to February, are cooler and dry, with temperatures ranging from 10 °C to 25 °C. Most of the basin's rainfall occurs during the southwest monsoon,

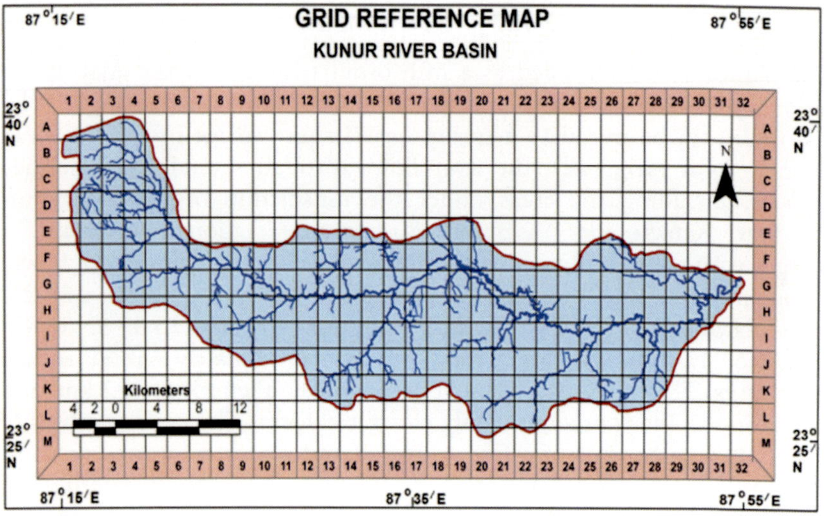

Fig. 20.1 Grid reference map

which plays a crucial role in recharging the river and the local groundwater. In recent years, changing climate patterns have introduced more variability in rainfall, potentially impacting river flow, agriculture and water availability in the region.

20.3 Methodology

The methodology for this study involves the use of various datasets obtained from multiple sources. Topographic maps from the Survey of India (SOI) were utilized, specifically Map No. 73 M/6, 73 M/7, 73 M/10, 73 M/11, 73 M/14 and 73 M/15 with a scale of 1:50,000. Geological maps from the Geological Survey of India, specifically from sheet 73 M, were employed to analyse lithological and structural features of the study area. Additionally, satellite imagery from the United States Geological Survey (USGS) was used, particularly Landsat TM data with a spatial resolution of 30 meters. The spatial analysis and mapping processes were conducted using ArcGIS 9.3 and ERDAS IMAGINE 9.2 software. The base data (stream number for each stream orders and basin area) for computing the hierarchical anomaly number and hierarchical anomaly index are derived from vectorized topo-maps. Kunur River basin is divided into sixteen sub-basins; namely sub-basins 1, 2, 3, 4, 5, 6, 7, 8, 9, 10, 11, 12, 13, 14, 15 and 16 for calculating the hierarchical anomaly index.

The arrangement of the drainage network in a hierarchy is intimately connected to the development of the drainage basin and is influenced by climatic and geological factors. (Avena et al., 1967). The length of a drainage network increases as the basin evolves, unless it is influenced by structural controls or diffusive processes. Sub-basins that have the same total number of streams can display different hierarchical arrangements. The degree of compositional anomalies in the drainage network is assessed through the analysis of the bifurcation index (R) and hierarchical anomaly factors. This analysis provides insight into the factors involved in the development of the drainage basin and its geomorphic evolutionary stage.

20.3.1 Bifurcation Index (R)

Bifurcation index (R) serves as a modification of the drainage basin's branching and dissection throughout its evolution (Horton, 1945). R is calculated by the difference between the bifurcation ratio (R_b) and the direct bifurcation ratio (R_{db}) (Guarnieri & Pirrotta, 2008). Horton (1945) measured Rb as an index of relief and dissection. It is defined as the ratio of the number of the stream segments of given order 'Nu' to the number of streams in the next higher order (Nu + 1). It is a dimensionless index and the values generally range from 3.0 to 5.0 (Strahler, 1957). demonstrated that the bifurcation ratio shows a small wander of variation for other regions or for a dissimilar environment, except where very significant geological control exists. Direct bifurcation ratio (R_{db}) is the proportion between the number of fluvial segments of a given order that connect the next higher order stream and the entire number of streams in the next higher stream order (Avena et al., 1967), Guarnieri & Pirrotta, 2008). It is resolved from the Eq. (20.1).

$$R_b = N_\mu / N_{\mu+1}$$

where N_μ = number of streams of a given order; $N_{\mu+1}$ = number of streams of the next higher order; Mean Bifurcation Ratio (R_b) is calculated as the Arithmetic Mean Bifurcation Ratio.

Drainage density, explained as the total length of stream channels per unit area, may be applied as the most usable index to describe a particular stream network.

As a feature in catchment studies, drainage density can be explained in three main ways. Firstly, it is narrated to water divide or physiographic characteristics such as relief ratio, rock type and basin shape. Secondly, it is narrated to input and output of the drainage basin system. Chorley and Morgan (1962) concluded that greatest runoff, which may consider high intensity rainfall, is a activity of drainage density. Drainage density, since first used by Horton (1945) has been extensively utilized in many hydrological studies. Horton (1945) defined drainage density as a ratio of total length of all stream segments in a given drainage basin to the total area of that basin and thus it can be derived as follows

$$Dd = \frac{L_K}{A_K}$$

where L_K = total lengths of all stream segments of a basin; A_K = total area of the basin.

20.3.1.1 Grid Specific Approach

The grid wise stream length and area of the basin of each grid has been calculated. Drainage density is then calculated. And then calculate the anomaly ($X-\overline{X}$) and RVA ($\overline{X} \pm SD_1$) by using the above formula.

20.3.2 Hierarchical Anomaly Number

Drainage system forms an interrelated network of nodes and links; the stream tips and stream junctions attend as nodes and the stream segment joining the nodes acts as the link (Huggett, 2007). The initial stages of hydrographic network disclosure are characterized by a simple drainage network structure composed of a few numbers of channels and are not so hierarchical in the drainage development arrangement.

As the evolution of drainage basin advances, numerous streams develop and the degree of hierarchism of drainage network increases with thinning of inter-fluvial divides and establishment of mainstream channels (Avena et al., 1967). The number of channels developing in the basin decreases and the degree of hierarchy decreases as the stream enters its old age stage (Avena et al., 1967). Avena et al. intended the concept of the hierarchical system of the drainage network and the consequence of hierarchical anomalies in describing the evolutionary characteristics of the basin. Various researchers interpreted the hierarchical anomalies in the drainage network to describe the influence of tectonics in the evolution of drainage basins (Guarnieri & Pirrotta, 2008; Grecu & Comanescu, 2001).

The drainage system of a basin is divided into fluvial portion of increasing hierarchical order (Horton, 1945). In the Strahler ordering system (Strahler, 1957), all headwater tributaries (nearest the basin divide) are referred to as first-order streams. The two first-order streams generate a second order stream and the two second-order stream supply a third order, and so forward. The numeral stream junctions of basins extend with increasing the stream order at the outlet of basins. For example, a basin with the highest stream order of 4 has 6 stream junctions (1 → 2, 1 → 3, 1 → 4, 2 → 3, 2 → 4 and 3 → 4 junctions) whereas a basin with the highest stream order of 5 has 10 stream junctions (1 → 2, 1 → 3, 1 → 4, 1 → 5, 2 → 3, 2 → 4, 2 → 5, 3 → 4, 3 → 5 and 4 → 5 junctions).

Avena et al. (1967) explained the hierarchical anomaly number (Ha) as the minimum number of first-order segments necessary to render the network completely hierarchical (Guarnieri & Pirrotta, 2008). In this study for the calculation of hierarchical anomaly number Bahrami's (2013) equations have been used. Hierarchical anomaly number for every stream junction ($Ha_{i \to j}$) is calculated based on following equation–

$$Ha_{i-j} = 2^{(j-2)} - 2^{(i-1)} \tag{20.1}$$

The higher difference between i and j, higher the hierarchical anomaly number. After calculation of $Ha_{i \to j}$, the number of streams for each junction ($Ns_{i \to j}$) of basin was obtained. The hierarchical anomaly number of the whole catchment (Hat) is calculated from the sum of $Hai \to j$ multiplied by $Ns_i \to j$:

Where

$$Ha_t = \sum \left(Ha_{i-j} \times Ns_{i-j} \right) \tag{20.2}$$

Ha = anomalous stream number; *j* = Higher order with which lower order stream emerge; *i* = Lower order of which stream meets with higher order; Ns_{i-j} = the number of streams for each junction of basin.

20.3.3 Hierarchical Anomaly Index (Δa)

The Δa index is calculated from Ha_t divided by the actual number of first-order streams (N_1). The value of Ha_t index seems to be high in comparison with the N_1 parameter and hence the Δa values of basin are high.

$$\Delta a = Ha_t / N_1 \tag{20.3}$$

20.4 Results and Analysis

The bifurcation ratio is a dimensionless property and generally ranges from 3.0 to 5.0. The decreased values of R_b are characteristic of the watersheds, which have sustained fewer tectonic disturbances (Strahler, 1964), and the drainage pattern has not been distorted because of fewer structural disturbances (Nag, 1998). In the present study, the higher values of R_b (3.82) indicate relatively strong structural control on the drainage pattern, while the lower value of R_b (4) indicative of watershed that are affected by structural disturbances. It is clear from the above diagram that there is a negative relationship between the order of stream and number of streams. Here, the value of $R^2 = 0.71$, which indicates that for order increase 1 unit number of stream decrease by 0.71 units. It is observed that the more the number of streams in an area, more the soil erosion and poor soil development and vice versa. There is a positive relationship between stream order and bifurcation ratio. Generally, the total stream length of a given order is inversely related to stream order, i.e. total stream length decreases from the lower order to successive higher orders because a maximum number of first-order streams are present in the first stream segment. From the diagram, it is clear that there is a negative relationship between order of stream and length of stream, indicating that development of drainage is higher for the lower order, but in the 5th order stream, the length is increased than 4th order stream, and this makes an anomalous situation (Table 20.1).

Drainage density is a measurement of the structure of the network and shows the balance between the erosive influence of overland flow and the resistance of surface soils and rocks. The factors that affected drainage density include geology and the density of vegetation. Drainage density refers to total stream lengths per unit area. Here divided the basin was divided into 212 grids, and it calculated the stream length (km) of each

Table 20.1 Summary of drainage basin parameters in the study area

Order	*Stream number*	*Bifurcation Ratio*	*Mean Bifurcation Ratio*	*Stream Length (km)*	*Mean Stream Length (km)*	*Area (sq.km)*	*Drainage Density (sq.km)*
1	241	3.8	3.97	232.4955	0.964712	753.90	4.6
2	63	3.3		101.0955	1.604691		
3	19	4.8		89.52448	4.711815		
4	4	4.0		11.54292	2.88573		
5	1			93.75	93.75		

grid and the basin area (sq.km) of each grid for calculate drainage density (Fig. 20.2). The vegetation density influenced drainage density by binding the surface layer and slowing down the rate of overland flow, and stores some of the water for short periods. The drainage density is found to increase from east to west of the basin (Fig. 20.3). A smaller portion of this basin shows a high drainage density (>3.87 km/sq.km), which indicates that the proportion of stream length is higher than the proportion of basin area. Very high drainage density is associated with the rugged and fractured terrains located on the high elevations. The result of lithology on drainage density is noticeable. The overland flow is reduced due to permeable rocks with a high infiltration rate, and consequently, drainage density is also low here. The lower drainage density (<0.58 km/sq.km) is found in the middle portion of the river basin. The high drainage density region consists of hilly or mountainous regions, and here the number of first-order streams is high. The maximum area of the basin is under the moderate drainage density. The basin with such properties lies in the quartzite units and limestone, having some resistant rock.

Here (Fig. 20.4) found that the high positive drainage density anomaly (km/sq.km) has been found in the northwestern portion and some north eastern portion of the basin than the others portion of the basin which indicates drainage density is high than the mean drainage density (km/sq.km). Some portion indicates negative drainage density anomaly (km/sq.km), which indicates drainage density is lower than the average drainage density. The map (Fig. 20.5) shows that a small portion of this basin the value of anomaly is beyond ± SD_1 is not belong to the ideal group of deviation. But a large portion of this basin belongs to the ideal range or group (463.49 km^2 to −51.39 km^2). In the upper catchment

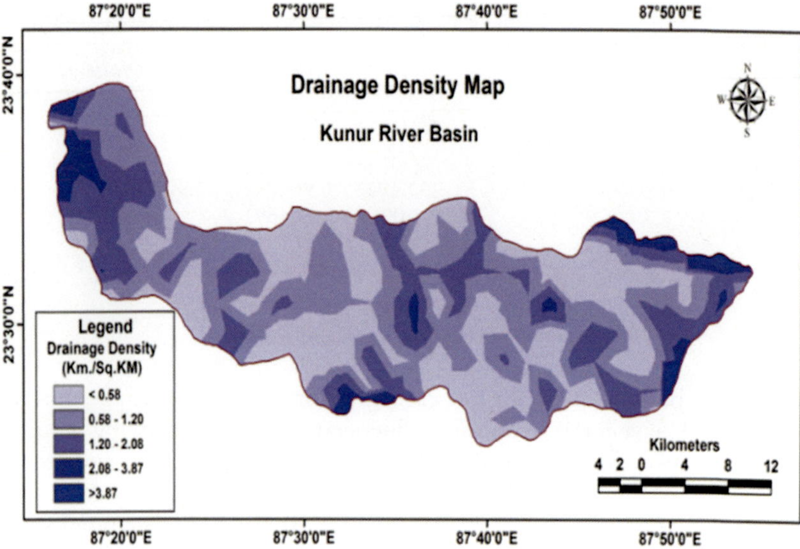

Fig. 20.2 Drainage density of Kunur River basin

and some portion of the lower catchment of the basin drainage density deviation is higher than the drainage density anomaly (Table 20.2).

20.4.1 Hierarchical Anomaly Number and Index

The value of Ha_t seems to be high in comparison with the N_1 parameter if there drainage anomaly, hence the Δa values of the basin will be high. In the Kunur river basin (whole basin), the hierarchical anomaly index is 1.36 (Table 20.3). So, in this basin anomalous condition in the stream junction is prevailing due to spatial variation in geological structures. Kunur River needs 86 more first-order streams to become a perfect hierarchical network.

Kunur River is having a drainage basin area of 753.90 sq. km. To evaluate the hierarchical arrangement of drainage networks it is divided into sixteen sub-basins. The values of Ha_t and Δa indices of different sub-basins are given in Table 20.4. The Ha_t values of basins range from 4 to 186. Table 20.4 shows that, in basins with higher Ha_t values, the number of streams with low order (i.e. 1) joining to higher order (i.e.

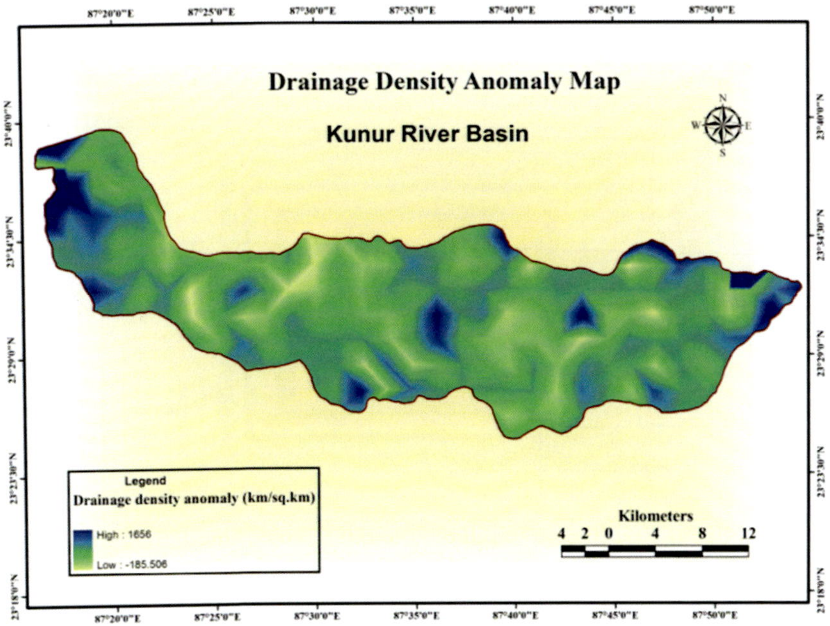

Fig. 20.3 Drainage density anomaly map

5 or 6) is considerably higher. The values of hierarchical anomaly index (Δa) of basins vary from 0.34 to 4.6. So, in different sub-basins perfect hierarchical network is not created (Fig. 20.6).

Hierarchical anomaly index value is high (4.56) in sub-basin 16 because the trunk river receives a lot of first order streams. And in others sub-basins Δa value is low because the trunk river receives low number of 1st order streams and in sub-basin 12 the value is 1.5 which indicates moderate hierarchical network is create here. The presence of hierarchical anomalies in the drainage network reflects the degree of evolution of a basin (Prakash et al., 2016). High values of hierarchical parameters show the poor hierarchical elaboration of drainage basin and commonly, think the control of tectonics or diffusive processes in the drainage basin evolution In different sub-basins hierarchical anomaly number and hierarchical anomaly index are dissimilar because of variation in geological structure, soil texture and slope.

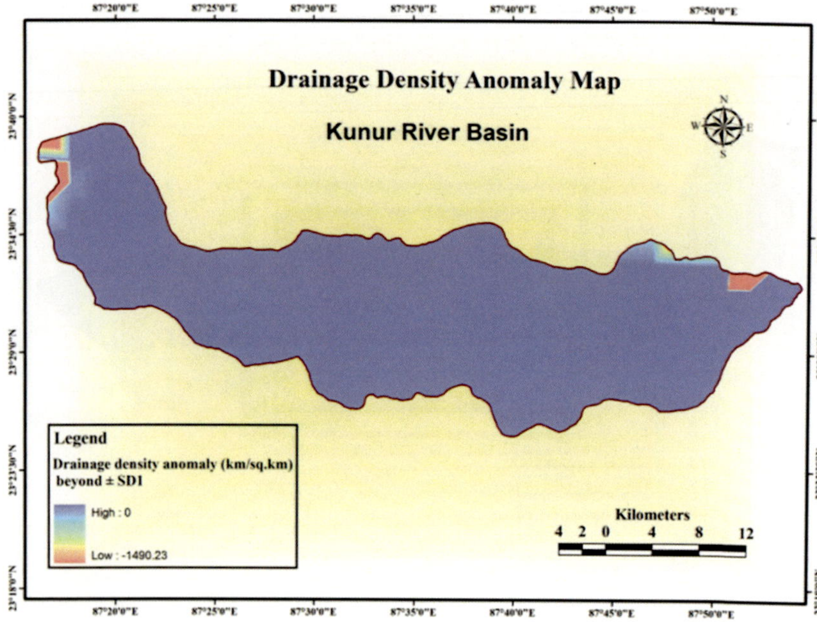

Fig. 20.4 Drainage density anomaly beyond ± SD_1 map

20.4.2 Associationship of Stream Length or Drainage Density Anomaly with Its Determinants

From this relief zone map (Fig. 20.7) it is found that the lower elevation is found in the eastern portion or the lower portion of the basin. In the upper catchment of the basin elevation is high. From this geological map (Fig. 20.8) it is found that a large portion of this basin is covered by older alluvium and also newer alluvium. Older alluvium is found in the middle portion of this basin, and newer alluvium is found in the lower portion of this basin. The upper portion, which belongs to a hilly area, consists of Panchet formations like Sandstone, green and red clay. And the uppermost part of this basin consists of Ranigunj formation, like sandstone, coal and shale. And some little portion of the south western corner consists of super-panchet-ferruginous like sandstone and red siltstone.

From this diagram it is clear that there is a positive relation between drainage density (km/sq.km) and elevation (m). The r^2 value is 0.440

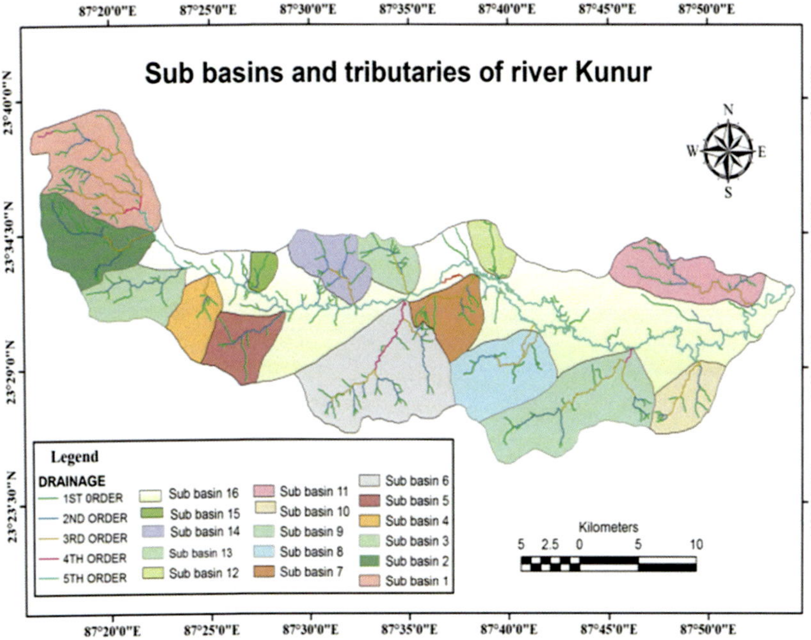

Fig. 20.5 Sub-basins of River Kunur

which indicates that there is a moderate positive relation which means if elevation (m) increases 1 unit, mean drainage density (km/sq.km) increase 0.440 units (Table 20.5). Here found that drainage density is high in high elevation zone or in mountain region and also found an anomaly situation in the lower catchment of this basin due to regional slope (Fig. 20.9).

From this diagram, it is clear that the mean drainage density (km/sq.km) is very high (2.76 km/sq.km) in the panchet formation. And drainage density (km/sq.km) is very low (1.27 km/sq.km) in the Raniganj formation. Drainage density in newer alluvium is high due to the regional slope of the basin, in older alluvium is moderate and in superpanchet formation is low.

Table 20.2 Calculation table for hierarchical anomaly index over the Kunur River

Anomalous stream junctions $(\underset{\rightarrow}{i})i\rightarrow j)$	$Ha_{i\rightarrow j} = 2^{(j-2)} - 2^{(i-1)}$	$N_{s\ i\rightarrow j}$	$(H_{a\ i\rightarrow j} \times Ns_{i\rightarrow j})$
$1 \rightarrow 2$	0	50	0
$1 \rightarrow 3$	1	32	32
$1 \rightarrow 4$	3	2	6
$1 \rightarrow 5$	7	27	189
$2 \rightarrow 3$	0	10	0
$2 \rightarrow 4$	2	0	0
$2 \rightarrow 5$	6	10	60
$3 \rightarrow 4$	0	1	0
$3 \rightarrow 5$	4	10	40
$4 \rightarrow 5$	0	2	0
			$\sum = 327$

Hierarchical anomaly Index $(\Delta a) = Ha_t/N_1$; $Ha_t = 327$; $N_1 = 241$
$= 327/241$
$= 1.36$

20.5 Major Findings

Very high drainage density is associated with the rugged and fractured terrains located on high elevations. The maximum area of the basin is under the moderate drainage density. The basin with such properties lies in the quartzite units and limestone, having some resistant rock.

The high positive drainage density anomaly (km/sq.km) has been found in the northwestern and some northeastern portions of the basin, which indicates that drainage density is higher than the mean drainage density (km/sq.km).

In the Kunur River basin (whole basin), the hierarchical anomaly index is 1.36, which indicates a low stream hierarchical anomaly. The hierarchical anomaly index value is high (4.56) in sub-basin 16 because the trunk river receives a lot of first-order streams. The values of the hierarchical anomaly index (Δa) of the sub-basins vary from 0.34 to 4.6. So, in different sub-basins perfect hierarchical network is not created. Kunur River needs 86 more first-order streams to become a perfect hierarchical network.

In this study, the r^2 value is 0.440, which indicates that there is a moderate positive relation between drainage density (km/sq.km) and

Table 20.3 Calculation table for hierarchical anomaly index over all the sub-basins

Sub-basin	*Anomalous stream junctions* $(\underset{\rightarrow}{i})i \rightarrow j)$	$Ha_{i\rightarrow j} = 2^{(j-2)} - 2^{(i-1)}$	$Ns_{i\rightarrow j}$	$(Ha_{i\rightarrow j} \times Ns_{i\rightarrow j})$	$(\Delta a) = Ha_t/N_1$
1	1 → 3	1	8	8	0.34
	1 → 4	3	1	3	
				$\sum = 11$ N = 32	
2	1 → 3	1	2	2	0.4
	3 → 5	3	1	4	
				$\sum = 6$ N = 15	
3	2 → 5	6	1	6	0.67
				$\sum = 6$ N = 9	
4	1 → 3	1	1	1	0.83
	3 → 5	4	1	4	
				$\sum = 5$ N = 6	
5	2 → 5	6	1	6	1
				$\sum = 6$ N = 6	
6	1 → 3	1	6	6	0.36
	1 → 4	3	2	6	
				$\sum = 12$ N = 33	
7	1 → 3	1	2	2	0.5
	3 → 5	4	1	4	
				$\sum = 6$ N = 12	
8	1 → 3	1	1	2	0.63
	3 → 5	4	1	4	
				$\sum = 5$ N = 8	
9	1 → 3	1	4	4	0.25
				$\sum = 4$ N = 16	
10	1 → 3	1	2	2	0.55
	3 → 5	4	1	4	
				$\sum = 6$ N = 11	

(continued)

Table 20.3 (continued)

Sub-basin	*Anomalous stream junctions* $(\underset{\rightarrow}{i})i \rightarrow j)$	$Ha_{i\rightarrow j} = 2^{(j-2)} - 2^{(i-1)}$	$Ns_{i\rightarrow j}$	$(Ha_{i\rightarrow j} \times Ns_{i\rightarrow j})$	$(\Delta a) = Ha_t/N_1$
11	1 → 3	1	2	2	0.46
	3 → 5	4	1	4	
				$\sum = 6$	
				N = 13	
12	2 → 5	6	1	6	1.5
				$\sum = 6$	
				N = 4	
13	1 → 3	1	1	1	0.38
	3 → 5	4	1	4	
				$\sum = 5$	
				N = 13	
14	1 → 3	1	2	2	0.5
	3 → 5	4	1	4	
				$\sum = 6$	
				N = 12	
15	3 → 5	4	1	4	1
				$\sum = 4$	
				N = 4	
16	1 → 5	7	27	189	4.56
	2 → 5	6	8	48	
				$\sum = 237$	
				N = 52	

Table 20.4 Drainage density (km/sq.km) and Basin relief (m)

Elevation Zone(m)	*Anomaly (km/sq.km)*	*Mean Drainage density km/sq.km)*	*Remarks*	*Direction*
22–37	(0.29–4.91)	2.6	Moderate	Eastern corner
38–57	(0.29–1.64)	0.97	Very low	Middle eastern portion
58–79	(0.89–1.64)	1.27	Low	North western Middle portion
80–94	(2.08–3.87)	2.98	High	North western
95–131	(3.87–4.34)	4.11	Very high	North western corner

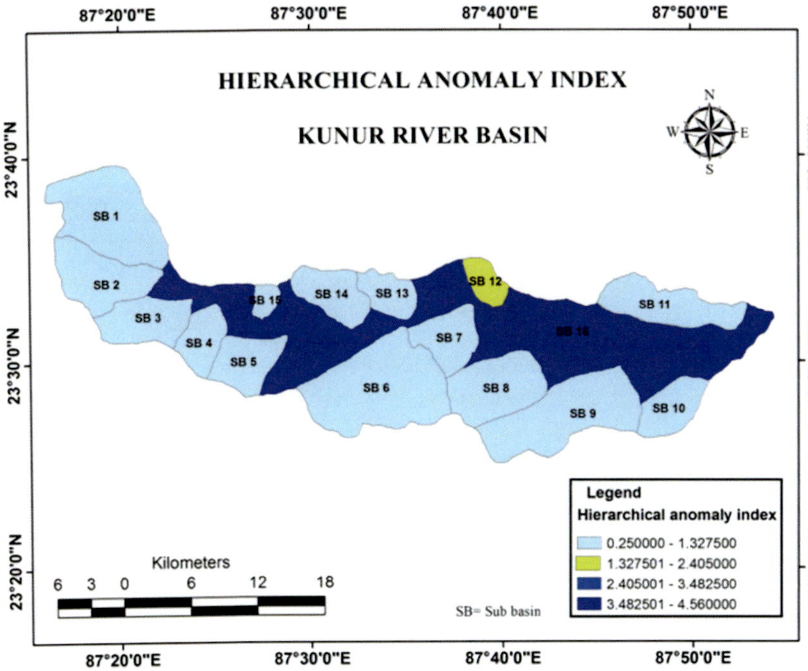

Fig. 20.6 Hierarchical Anomaly Index Map

elevation (m). In the lower catchment drainage density is high, which indicates an anomalous situation.

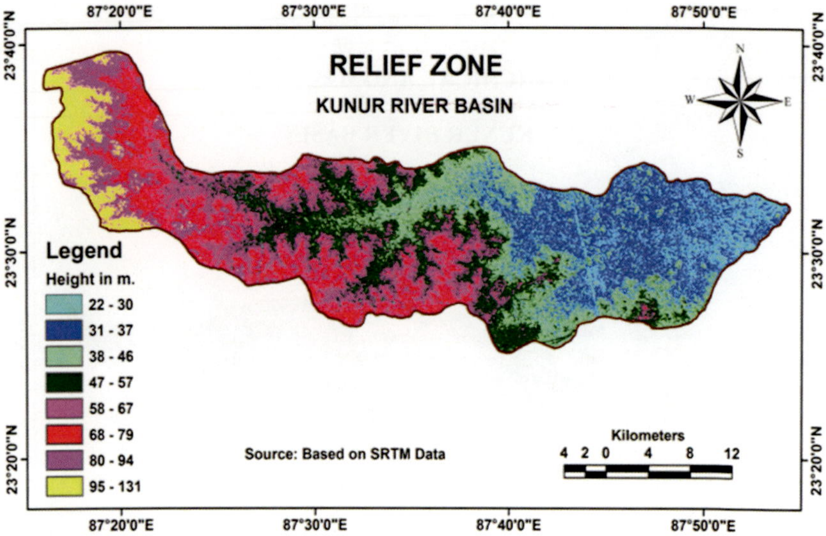

Fig. 20.7 Relief Zone Map

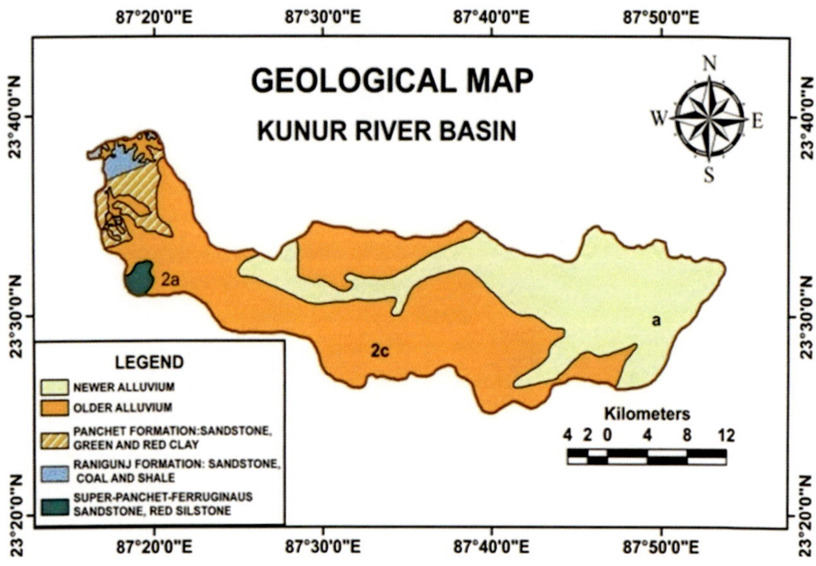

Fig. 20.8 Geological Map

Table 20.5 Geological unit specific drainage density

Geology	*Anomaly (km/sq.km)*	*Mean (Drainage density km/sq.km)*	*Remarks*	*Direction*
Newer alluvium	(0.89–3.89)	2.39	High	Easter part and some middle part
Older alluvium	(0.89–2.98)	1.94	Moderate	Middle part and some western part
Panchet formation (Sandstone,Green and Red clay)	(1.64–3.87)	2.76	Very high	North western part
Raniganj formation(Sandstone, coal and shale)	(0.89–1.64)	1.27	Very low	North western corner
Super-panchet ferruginaus(Sandstone, Red siltstone)	(1.20–2.08)	1.64	Low	South western portion

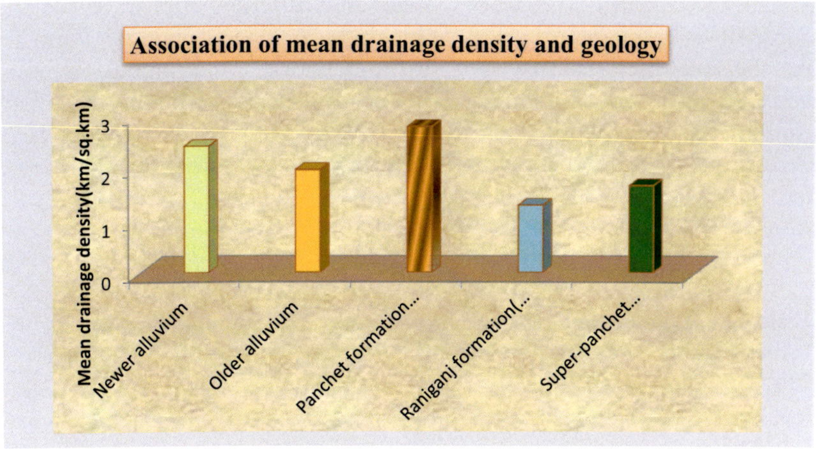

Fig. 20.9 Association of Mean Drainage Density and Geology

20.6 Conclusion

Drainage anomalies are irregularities in river flow patterns caused by geological factors or human activities. They help identify tectonic activity or fault lines, which is important for safe infrastructure planning. Anomalies can affect water flow and increase flood risk, making them useful for flood management. They also influence erosion, sedimentation and groundwater recharge, which are key for water and soil conservation. Understanding drainage anomalies supports better land-use planning and environmental protection within the basin. The morphometric investigation of drainage anomaly and hierarchical association of drainage networks plays an important role in understanding the effects of tectonic activity, geological structure, resistance power of the rocks and slope on basin development. Statistical relations between drainage density with the rugged, geological structure, fractured terrains and elevation have been analysed. Due to the presence of quartzite units and limestone, having some resistant rock in the upper catchment area of the Kunur River basin, the drainage density is relatively high and a positive drainage anomaly is created. The whole basin and different sub-basins of the Kunur River basin are not perfectly hierarchical in the stream network. To take the ideal hierarchical system in the network, the Kunur River basin needs 86 first-order streams. So, such anomalous situations in stream number, drainage density and disordering of the channel network have been induced in this basin due to the presence of variation in the said factors.

References

Abrahams, A. D. (1984). Channel networks: A geomorphological perspective. *Water Resources Research, 20*, 161–188.

Altin, T. B., & Altin, B. N. (2011). Development and morphometry of drainage network in volcanic terrain, Central Anatolia, Turkey. *Geomorphology, 125*, 485–503.

Avena, G. C., Giuliano, G., & Lupia Palmieri, E. (1967). Sulla valutazione quantitativa della gerarchizzazione ed evoluzione dei reticoli fluviali. *Bollettino della Societa Geologica Italiana, 86*, 81–796.

Bahrami, S. (2013). *Analyzing the drainage system anomaly of Zagros basins: Implications for active tectonics* (pp. 914–927). Elsevier B.V, Technophysics.

Burbank, D. W. & Anderson, R. S. (2001). *Tectonic Geomorphology* (p. 274). Blackwell Science.

Deffontaines, B., & Chorowicz, J. (1991). Principles of drainage basin analysis from multisource data: Application to the structural analysis of the Zaire Basin. *Tectonophysics, 194*, 237–263.

Delcaillau, B., Carozza, J. M., & Laville, E. (2006). Recent fold growth and drainage development, the Janauri and Chandigarh Anticlines in the Siwalik Foothills, Northwest India. *Geomorphology, 76*, 241–256.

Delcaillau, B., Deffontaines, B., Floissac, L., Angelier, J., Deramond, J., Souquet, P., Chu, H. T., & Lee, J. F. (1998). Morphotectonic evidence from lateral propagation of active frontal fold; Pakuashan anticline, foothills of Taiwan. *Geomorphology, 24*, 263–290.

Devi, R. K. M., & Bhakuni, BoraPB. (2011). Tectonic implication of drainage set-up in the Sub- Himalaya: A case study of Papumpare district, Arunachal Himalaya, India. *Geomorphology, 127*, 14–31.

Guarnieri, P., & Pirrotta, C. (2008). The response of drainage basins to the late Quaternary tectonics in the Sicilian side of the Messina Strait (NE Sicily). *Geomorphology, 95*, 260–273.

Horton, R. E. (1932). Drainage basin characteristics. *Transactions, American Geophysical Union, 13*, 350–361.

Horton, R. E. (1945). Erosional development of streams and their drainage basin; hydrophysical approach to quantitative morphology. *Geological Society of America Bulletin, 56*(3), 275–370.

Jackson, J., Van Dissen, R., & Berryman, K. (1998). Tilting of active folds and faults in the Manawatu region, New Zealand: Evidence from surface drainage patterns, New Zealand. *Journal of Geology and Geophysics, 41*, 377–385.

Jamieson, S. S. R., Sinclair, H. D., Kirstein, L. A., & Purves, R. S. (2004). Tectonic forcing of longitudinal valleys in the Himalaya, morphological analysis of the Ladakh batholith. *North India, Geomorphology, 58*, 49–65.

Krishnamurthy, J., Srinivas, G., Jayaram, V., Chandrasekhar, M.G. (1996). Influence of rock types and structures in the development of drainage networks in typical hardrock terrain. *ITC Journal, 3–4*, 252–259.

Kumar, R., Kumar, S., Lohani, A. K., Nema, R. K., & Singh, R. D. (2000). Evalution of geomorphological characteristics of a catchment using GIS. *GIS India, 9*, 13–17.

Leopold, L. B. & Miller, J. P. (1956). Ephemeral streams: hydraulic factors and their relation to the drainage network. *U.S. Geological Survey, Professional paper 282-A*.

Ramsey, L. A., Walker, R. T., & Jackson, J. (2008). Fold evolution and drainage development in the Zagros Mountains of Fars province, SE Iran. *Basin Research, 20*, 23–48.

Reddy, G. P. O., Maji, A. K., & Gajbhiye, K. S. (2004). Drainage morphometry and its influence on landform characteristics in a basaltic terrain, Central

India—A remote sensing and GIS approach. *International Journal of Applied Earth Observation and Geoinformation, 6*, 1–16.

Shreve, R. L. (1966). Statistical law of stream numbers. *Journal of Geology, 74*, 17–37.

Simoni, A., Elmi, C., & Picotti, V. (2003). Late Quaternary uplift and valley evolution in the Northern Apennines. *Lamone Catchment, Quaternary International, 101–102*, 253–267.

Singh, S. (1981). Estimation of drainage density. *National Geographer, 16*(2), 81–89.

Strahler, A. N. (1952). Hypsometric (area-altitude) analysis of erosional topography. *Geological Society of America Bulletin, 63*, 1117–1142.

Strahler, A. N. (1957). Quantitative analysis of watershed geomorphology. *Transactions, American Geophysical Union, 38*, 913–920.

Sung, O., & Chen, Y. C. (2004). Geomorphic evidence and kinematic model for quaternary transfer faulting of the Pakuashan anticline, central Taiwan. *Journal of Asian Earth Sciences, 24*, 389–404.

Talling, P. J., & Sowter, M. J. (1999). Drainage density on progressively tilted surfaces with different gradients, Wheeler Ridge, California. *Earth Surface Processes and Landforms, 24*, 809–824.

Zhang, H. P., Liu, S. F., Yang, N., Zhang, Y. Q., & Zhang, G. W. (2006). Geomorphic characteristics of the Minjiang Drainage Basin (Eastern Tibetan Plateau) and its Tectonic implications: New insights from a digital elevation model study. *Island Arc, 15*, 239–250.

CHAPTER 21

Integrating Flood Control Factors for Enhanced Flood Risk Modeling: A Study of the Malaprabha Basin Through Frequency Ratio and Statistical Index Approaches

Uttam Pawar and Mohammad Irfan

21.1 Introduction

A flood is natural hydro-metrological disasters that occur when river water exceeds its natural banks, resulting in inundation of the surrounding areas adjacent to the river channels that are usually dry (Chow, 1956; Luino, 2016). According to statistical records from the United Nations Office for Disaster Risk Reduction (UNISDR), flooding represented 47% of all weather-related disasters from 1995 to 2015, influencing 2.3

U. Pawar (✉)
Department of Geography, Shri Bhausaheb Vartak Arts, Commerce and Science College, Mumbai, India
e-mail: uttampawar12@gmail.com

M. Irfan
Department of Earth Sciences, University of Kashmir, Srinagar, India

S. Ghosh (ed.), *Reimagining Indian Rivers for Sustainability*, Palgrave Studies in Environmental Sustainability,
https://doi.org/10.1007/978-3-032-04754-0_21

billion people, with 95% of those affected living in Asia (Wahlstrom & Guha-Sapir, 2015). The number of fatalities resulting from flooding has increased in various regions globally. In 2007, floods claimed the lives of 3,300 individuals in India and Bangladesh. In 2010, the death toll reached 2,100 in Pakistan and an additional 1,900 in China, while in 2013 floods in India resulted in approximately 6,500 deaths. An estimated 1.81 billion (23%) of the world's population are at risk of flooding with depths greater than 0.15 meters. The largest concentration of this vulnerable population, approximately 1.24 billion, is found in South and East Asia, where China (395 million) and India (390 million) together represent over one-third of the global coverage (Rentschler et al., 2022). Flood occurrence is becoming more frequent due to climate change, rapid and unplanned urbanization along river banks, and encroachment into river channels, which is expected to exacerbate flood risks considerably shortly (Pawar et al., 2022).

An extensive evaluation of flooding for effective hazard management and risk reduction has become a key area of interest for geomorphologists, meteorologists, hydrologists, regional planners, and water resource managers (Pawar et al., 2022). Flood susceptibility maps play a crucial role in flood management strategies as they provide precise information about the spatial extension of floods in the region (Di Baldassarre et al., 2010; Pawar, 2025). An identification of flood-prone areas along with the factors that increase their vulnerability is essential for mitigating the impacts of floods, safeguarding communities, and minimizing economic losses (Towfiqul et al., 2021; Zhao et al., 2019). Flood susceptibility mapping is a comprehensive process that includes recognizing flood-affected zones, determining relevant flood conditioning parameters, and building predictive models (Ebrahimi et al., 2023). The modeling of flood susceptibility encompasses a range of techniques, such as physical and statistical analyses, multi-criteria decision-making (MCDM), and machine learning strategies (Mehravar et al., 2023). The most popular and frequently used flood susceptibility models/ methods are the analytic hierarchy process (AHP) (Khosravi et al., 2016; Nsangou et al., 2022; Parsian et al., 2021; Swain et al., 2020; Vilasan & Kapse, 2022), Shannon's entropy (Arora et al., 2021; Dutta et al., 2023; Haghizadeh et al., 2017; Sharma et al., 2024; Wang et al., 2021), frequency ratio (Arora et al., 2021; Khosravi et al., 2016; Pawar, 2025; Pawar et al., 2022; Shafapour Tehrany et al., 2017; Wang et al., 2021), statistical index (SI)

(Pawar, 2025; Pawar et al., 2022; Suppawimut, 2021), logistic regression (Pradhan, 2010; Shafapour Tehrany et al., 2017; Wu et al., 2024), weights-of evidence (Dutta et al., 2023; Khosravi et al., 2016; Shafapour Tehrany et al., 2017), artificial neural network (ANN) (Khoirunisa et al., 2021; Priscillia et al., 2021; Rudra & Sarkar, 2023), fuzzy logic (Akay, 2021; Parsian et al., 2021) and support vector machine (Tehrany et al., 2015, 2019).

Flood modeling plays a critical role in managing flood risks. Consequently, the objective of the present study is to identify flood-prone areas and the factors contributing to flooding in the Malaprabha Basin, which is essential for effective flood management and risk reduction. This objective is achieved by integrating remote sensing data with geographic information systems (GIS), utilizing frequency ratio (FR) and statistical index (SI) models. Flood risk modeling is a vital aspect of river management and encompasses various flood-controlling factors (FCF), including elevation, slope, aspect, curvature, topographic wetness index (TWI), stream power index, rainfall, distance from rivers, stream density, soil types, and land use/land cover. The Malaprabha River is known for its history of flooding, with significant events recorded in the years 2019 and 2022.

21.2 Study Area

Malaprabha Basin is one of the sub-basins of the Krishna Basin located in the western part of southern India (Fig. 21.1a). The Malaprabha Basin covers part of the Belagavi, Bagalkote, Dharwad, Gadag, Haveri, and Koppal districts in Karnataka (Fig. 21.1b). Malaprabha River is a tributary of the Krishna River rises at an elevation of 792 m near Kanakumbi village in the Belagavi district of Karnataka (Nagraj et al., 2018). It flows for 304 km and joins the Krishna River at Kudalasangama in Bagalkote district at an elevation of 488 meters. The major tributaries of the Malaprabha River are Bennihalla and Tuparihalla (Fig. 21.1b). The catchment area of the Malaprabha basin is 11,549 sq. km. It stretches from 15°05′02″ to 16°20′19″ N latitude and 74°05′43″ to 76°05′33″ E longitude. The geological formations of Malaprabha Basin include rocks belonging to the supergroups of Deccan Trap (Basalt), Dharwad, Kaladgi, and Peninsular Gneiss complex (Fig. 21.1c). The southwest monsoon rainfall controls the climate of the Malaprabha Basin (Nagraj et al., 2018). July is the rainiest month in the basin. The annual average rainfall of the Malaprabha Basin is 737 mm (Fig. 21.1d).

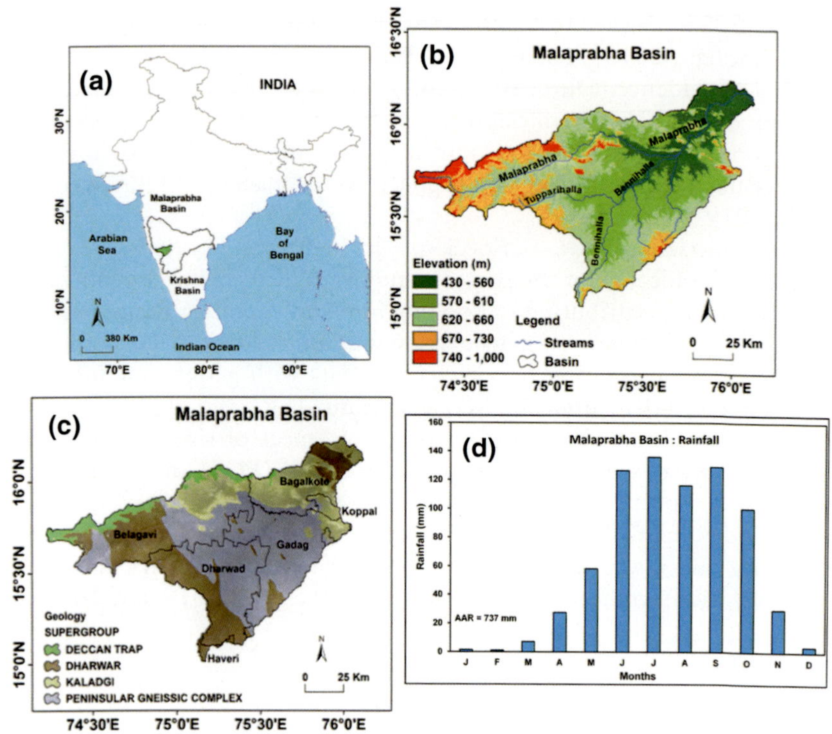

Fig. 21.1 Malaprabha Basin. **a** Location in India, **b** Elevations and drainage network, **c** Geology, **d** Rainfall

21.3 Data and Methodology

Data concerning factors influencing flood control, such as elevation, slope, aspect, curvature, TWI, SPI, precipitation, distance from rivers, stream density, soil types, and land use/land cover (LULC) were collected from various sources as indicated in Table 21.1. Moreover, flood location data were gathered from multiple sources, including reports, newspapers, electronic media (news), and social media, to develop a flood inventory map for the Malaprabha Basin. In the Malaprabha Basin, 296 flood points have been recognized. Among the 296 flood points, 207 (70%) were selected as training samples, whereas 89 (30%) were chosen as testing samples to create and evaluate the reliability of flood models (Fig. 21.2). Figure 21.3 illustrates the flowchart representing the methodology employed in the flood susceptibility analysis (FSA) of the Malaprabha Basin.

Table 21.1 Data sources, data type and map layers

Dataset	*Data types*	*Source*	*Derived map layers*
ASTER DEM (30 m*30 m)	Raster and Vector	https://earthexplorer.usgs.gov	Elevation, Slope, Aspect, Curvature, TWI, SPI, Stream density and distance from rivers
Rainfall	Attribute	https://www.indiawaterportal.org/met_data	Rainfall map
Soil types	Vector	http://www.fao.org	Soil types map
LULC	Vector	https://gee-communitycatalog.org/projects/S2TSLULC/	LULC map

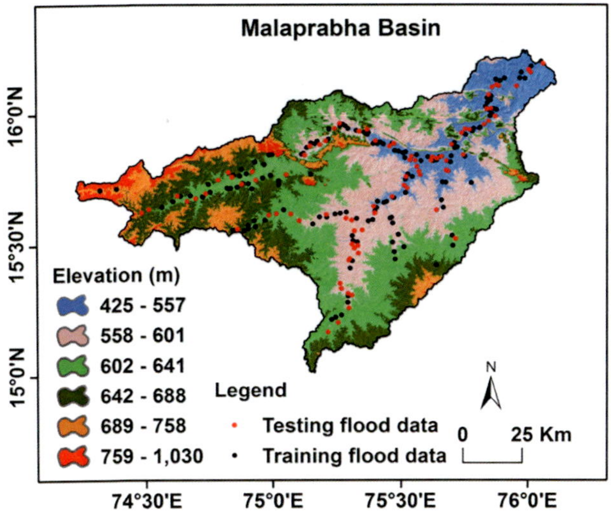

Fig. 21.2 Flood inventory map of the Malaprabha Basin

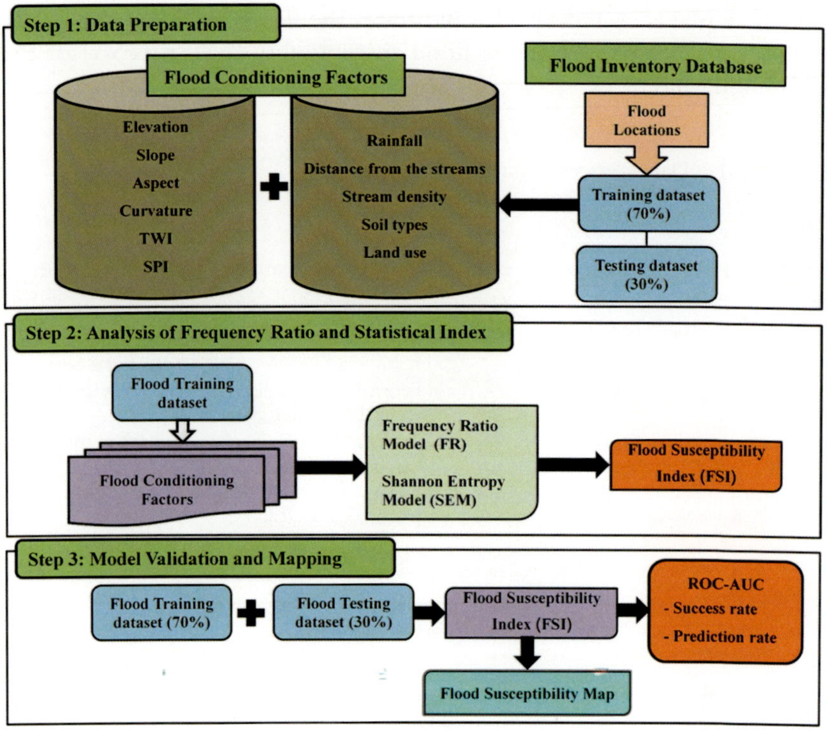

Fig. 21.3 Flowchart of the methodology used in the analysis

21.3.1 Frequency Ratio (FR) Model

The FR model is regarded as one of the most reliable and commonly employed techniques for generating flood vulnerability maps. A higher FR value indicates a stronger correlation, while a lower FR value signifies a weaker correlation (Pawar, 2025; Pawar et al., 2022; Sarkar & Mondal, 2020). The FR value is calculated by the following equation (Eq. 21.1);

$$FR = \frac{Nij}{Pij} / \frac{N}{P} \tag{21.1}$$

where, '*FR*' stands for the frequency ratio; '*Nij*' represents the number of flood points; '*Pij*' is the total count of flood points within the study area; '*N*' refers to the total number of pixels corresponding to the specific class of the factor, while 'P' signifies the total number of pixels in the entire study area. Finally, the flood susceptibility index (FSI) was computed based on the Eq. (21.2).

$$FSI_{FR} = \Sigma FR \tag{21.2}$$

where, 'FSI_{FR}' indicates the flood susceptibility index calculated using the FR method, while '$\sum FR$' denotes the total of the 'FR' values across all categories for each flood-controlling factor.

21.3.2 *Statistical Index (SI) Model*

CJ van Westen introduced the statistical index (SI) Model for landslide hazard assessment in 1997 (van Westen, 1997). SI values are calculated by applying the natural logarithm to the flooding within each class of flood-controlling factors, which is then divided by the total flood quantity for the area being analyzed. A higher positive SI value reflects a strong association between the frequency of floods and the type of conditional factors involved. On the other hand, a negative value implies the potential for minor flooding in a given category. The SI values corresponding to each class of flood control factors were calculated utilizing Eq. (21.3);

$$W_{ij} = ln\left(\frac{D_{ij}}{D}\right) = ln\left[\left(\frac{Nij}{Sij}/\frac{N}{S}\right)\right] \tag{21.3}$$

where, 'Wij' represents the statistical index for the 'i' class of 'j' factors, while 'Dij' denotes the density of floods within the 'i' class of 'j' factors. 'D' refers to the overall flood density in the area being examined. 'Nij' indicates the total count of floods in the 'i' class of 'j' factors within the study area, and 'Sij' signifies the number of pixels corresponding to the 'i' class of 'j' factors. Additionally, 'N' represents the total number of flood pixels in the area under investigation, and 'S' indicates the total number of pixels in that same area. Lastly, The FSI_{SI} was calculated by the Eq. (21.4);

$$FSI_{SI} = \Sigma W_{ij} \tag{21.4}$$

21.3.3 Receiver Operating Characteristic (ROC) Method

One of the most reliable approaches to evaluate the performance of flood susceptibility methods is the Receiver Operating Characteristic (ROC) curve (Bowers & Zhou, 2019; Pawar et al., 2022). The evaluation of the FR and SI models was conducted with 70% of the flood data allocated for training purposes and 30% reserved for testing (Pawar et al., 2022). A higher AUC value signifies superior performance of the models.. The AUC was calculated using the Eq. (21.5);

$$AUC = \sum_{i=0}^{n}(x_i - x_{i-1})y_i - \left[\frac{(x_i - x_{i-1})(y_i - y_{i-1})}{2}\right] \tag{21.5}$$

where, 'x_i' indicates the percentage of the total area, and 'y_i' refers to the area that has been inundated by the flood.

21.4 Results and Discussion

21.4.1 Flood Susceptibility Analysis by Frequency Ratio (FR) Model

The FR model was applied to assess the relationship between locations affected by floods and the factors that regulate flooding, with the results illustrated in Table 21.2. The derived FR values for the eleven FCFs highlight their correlation with flood incidents, as indicated in Table 21.2 and Fig. 21.4. According to Pradhan (2010), the FR value greater than 1 reflects a significant correlation between flood control factors and flooding. In contrast, a value less than 1 suggests a minimal correlation. Examining elevation and flood-prone areas revealed that the flood sites within the Malaprabha Basin, situated at elevations between 425 and 557 meters, exhibited the highest flood risk (FR) value of 2.93 followed by areas at elevations ranging from 558 to 601 meters, which had an FR value of 1.12. Conversely, regions with elevations between 689 and 1030 meters in the basin demonstrated a minimal or nonexistent likelihood of flooding (Table 21.2; Fig. 21.4a). The relationship between the basin's slope and flood sites revealed that the FR value above 1 was associated with slope angles from 2.7° to 13°. The lowest FR value (0.00) was recorded for slope angles ranging from 21° to 56° (Table 21.2; Fig. 21.4b). An analysis of the FR concerning slope aspects and flood sites indicated that the northern, eastern, and western categories had FR values greater than 1. This finding indicates a robust relationship between slope aspects and flood locations in the Malaprabha Basin. Conversely, the northwestern slope aspect displayed the minimum FR value of 0.00 (Table 21.2; Fig. 21.4c). In case of curvature, both convex and concave slopes had a pronounced impact on flooding in the basin, with FR values greater than 1. Convex slopes were responsible for increased runoff, whereas concave slopes were associated with water accumulation. Conversely, flat regions exhibited an FR value of 0.93, indicating a reduced influence on flooding (Table 21.2; Fig. 21.4d).

The highest FR value observed for TWI was 1.33, occurring within the TWI range of 3.2 to 6.7, followed by the TWI class of 6.8 to 8.3. However, the subcategory of TWI value ranges between 12 and 13, showing the lowest FR value (0.35). It reveals that lower TWI had a higher probability of flooding in the Malaprabha Basin (Fig. 21.4e). For the SPI, the FR value greater than 1 was noted for SPI values between −6.2 and −1.1, whereas the minimum FR value recorded for the SPI class varied from 8.2 to 16.0 (Fig. 21.4f). This indicates that a lower SPI correlates with a higher likelihood of flooding in the examined region. The highest FR value (1.19) was observed for the rainfall class 544 to 768 mm, and the lowest value for the high rainfall zone (1671–1890 mm). It indicates that regions with lower rainfall are more susceptible to flooding compared to those with higher rainfall (Fig. 21.4g). The analysis of the distance from river to flood locations revealed that the highest FR value (2.31) was derived for the distance category of 0 to1080 meters followed by the distance category of 1081–2150 meters. It indicates that areas within 2150 meters of streams are more susceptible to flooding compared to those situated farther away (Fig. 21.4h). The evaluation of the stream density factor showed that FR values above 1 were present in areas with stream densities ranging from 0.35 to 0.68 km/sq. km, and low for the stream density class of 0.0–0.11 km/sq. km. It indicates that higher stream densities are associated with an increased risk of flooding (Fig. 21.4i). In the Malaprabha Basin, Vertisols soil types soils exhibited the high FR values of 2.09 for the Pellic Vertisols (Vp) and 1.61 for the Chromic Vertisols (Vc), while Plinthic Acrisols (Ap) soil recorded a value of 1.05 (Fig. 21.4j). Due to their substantial clay content, which fluctuates between 40% and 80%, Vertisols exhibit a remarkable capacity for moisture retention (De Vos & Virgo, 1969; Dudal, 1965). The influence of landuse/landcover (LULC) on flooding is substantial, as evidenced by the built-up areas, which showed the highest FR value of 26.77

Table 21.2 Summary result of the FR and SI models for various flood-controlling factors (FCF)

Factors	*Class*	*Class pixels*	*Area (%)*	*Floods Pixels*	*Flood (%)*	*FR*	*SI*
Elevation	425–557	1,617,671	12.84	111	37.63	2.93	107
	558–601	3,382,282	26.86	89	30.17	1.12	12
	602–641	3,582,595	28.45	50	16.95	0.60	−52
	642–688	2,645,810	21.01	41	13.90	0.66	−41
	689–758	1,051,282	8.35	4	1.36	0.16	−182
	758–1030	314,516	2.50	0	0.00	0.00	0
Slope	0–2.6	3,486,095	27.68	63	21.36	0.77	−26
	2.7–5.2	4,723,375	37.50	127	43.05	1.15	14
	5.3–8.3	2,763,349	21.94	70	23.73	1.08	8
	8.4–13	1,178,189	9.36	32	10.85	1.16	15
	14–20	356,742	2.83	3	1.02	0.36	−102
	21–56	86,406	0.69	0	0.00	0.00	0
Aspect	Flat	2,060,416	16.36	39	13.22	0.81	−21
	North	1,368,430	10.87	42	14.24	1.31	27
	Northeast	2,672,862	21.22	59	20.00	0.94	−6
	East	1,448,249	11.50	35	11.86	1.03	3
	Southeast	1,391,580	11.05	30	10.17	0.92	−8
	South	1,250,824	9.93	22	7.46	0.75	−29
	Southwest	1,239,543	9.84	30	10.17	1.03	3
	West	1,156,511	9.18	38	12.88	1.40	34
	Northwest	5741	0.05	0	0.00	0.00	0
Curvature	Concave	2,311,752	18.36	55	18.64	1.02	2
	Flat	6,774,945	53.79	148	50.17	0.93	−7
	Convex	3,507,459	27.85	92	31.19	1.12	11
Topographic Wetness Index	3.2–6.7	4,693,550	37.27	146	49.49	1.33	28
	6.8–8.3	3,784,950	30.05	91	30.85	1.03	3
	8.4–11.0	1,864,518	14.80	37	12.54	0.85	−17
	12.0–13.0	1,458,429	11.58	12	4.07	0.35	−105
	14.0–17.0	660,004	5.24	7	2.37	0.45	−79
	18.0–27.0	132,705	1.05	2	0.68	0.64	0
Stream Power Index	(−)6.2 – – 1.4	5,327,926	43.47	154	52.20	1.20	18
	(−)1.3 – – 1.1	2,603,484	21.24	63	21.36	1.01	1
	1.2–2.8	2,279,243	18.60	48	16.27	0.87	−13

(continued)

Table 21.2 (continued)

Factors	*Class*	*Class pixels*	*Area (%)*	*Floods Pixels*	*Flood (%)*	*FR*	*SI*
	2.9–4.9	1,290,313	10.53	19	6.44	0.61	−49
	5.0–8.1	610,849	4.98	11	3.73	0.75	−29
	8.2–16.0	144,401	1.18	0	0.00	0.00	0
Rainfall (mm)	544–768	7,190,368	57.09	200	67.80	1.19	17
	769–993	3,775,422	29.98	70	23.73	0.79	−23
	994–1220	446,635	3.55	11	3.73	1.05	5
	1231–1440	364,339	2.89	4	1.36	0.47	−76
	1441–1670	476,165	3.78	6	2.03	0.54	−62
	1671–1890	341,227	2.71	4	1.36	0.50	−69
Distance from rivers (m)	0–1080	3,142,540	24.95	170	57.63	2.31	84
	1081–2150	2,955,318	23.47	94	31.86	1.36	31
	2151–3230	2,516,269	19.98	22	7.46	0.37	−99
	3231–4310	2,066,628	16.41	7	2.37	0.14	0
	4311–5380	1,364,280	10.83	1	0.34	0.03	0
	5381–6460	549,121	4.36	1	0.34	0.08	0
Stream density (km/km^2)	0.0–0.11	1,945,420	15.45	4	1.36	0.09	−243
	0.12–0.23	2,692,841	21.38	13	4.41	0.21	−158
	0.24–0.34	2,779,793	22.07	28	9.49	0.43	−84
	0.35–0.45	2,386,005	18.95	67	22.71	1.20	18
	0.46–0.57	1,920,098	15.25	115	38.98	2.56	94
	0.58–0.68	869,999	6.91	68	23.05	3.34	121
Soils	Vp	531,391	4.22	26	8.81	2.09	74
	Ap	122,145	0.97	3	1.02	1.05	5
	Lc	6,167,860	48.97	99	33.56	0.69	−38
	Bv	85,750	0.68	0	0.00	0.00	0
	Vc	3,128,497	24.84	118	40.00	1.61	48
	Ne	2,558,513	20.32	49	16.61	0.82	−20
Landuse/ Landcover (LULC)	Water	132,578	1.05	1	0.34	0.32	0
	Vegetation	334,673	2.66	0	0.00	0.00	0
	Agriculture	10,537,089	83.67	8	2.71	0.03	−343
	Built area	456,165	3.62	286	96.95	26.77	329
	Rangeland	1,133,651	9.00	0	0.00	0.00	0

(Fig. 21.4k). According to the findings of Pawar et al. (2022), the non-porous quality of surfaces in residential areas contributes to increased surface runoff and decreased water infiltration. Moreover, regions adjacent to rivers face a greater likelihood of flooding (Jung et al., 2011; Pawar et al., 2022).

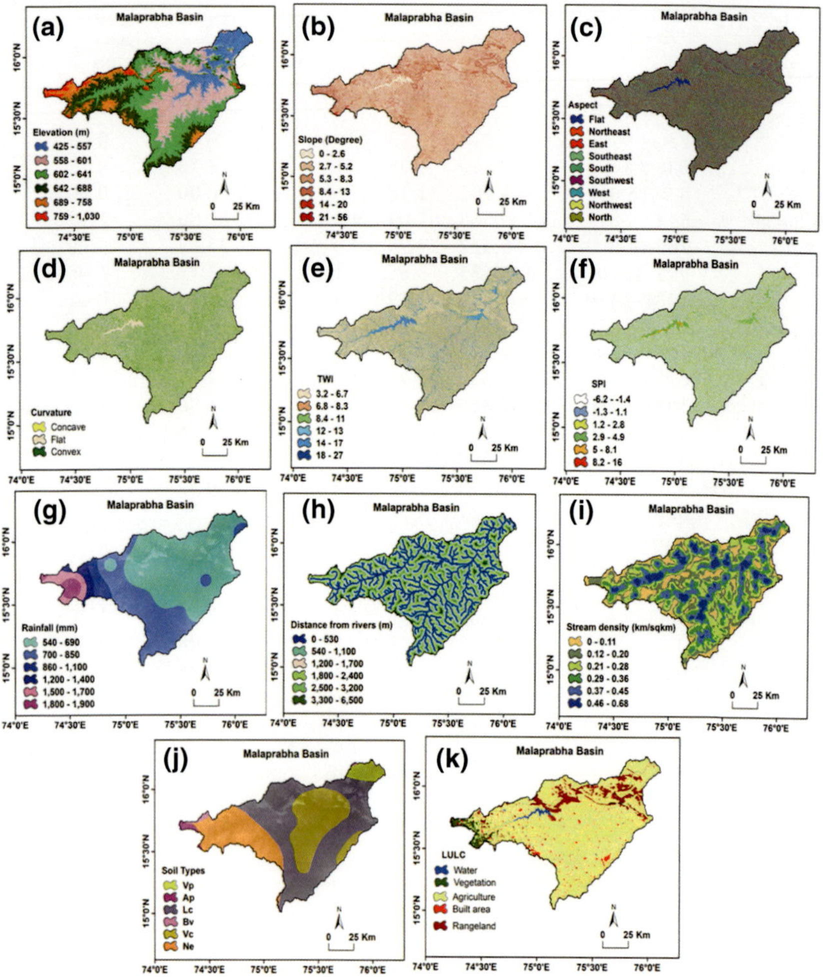

Fig. 21.4 Malaprabha Basin flood-controlling factors (FCF)

A flood susceptibility index (FSI) using the FR model was calculated to generate a flood susceptibility map (FSM) for the Malaprabha Basin in ArcGIS 10.1. The FSM based on the FR model was classified into five distinct categories: very low, low, moderate, high, and very high flood susceptibility (Fig. 21.5a). As illustrated in Fig. 21.5b, only 3% of the basin's area is identified as having high to very high flood risk, while 79% is designated as low to very low flood probability, and 18% is classified as a moderate flood-prone area within the Malaprabha Basin.

21.4.2 Flood Susceptibility Analysis by Statistical Index (SI) Model

The findings of the flood susceptibility analysis were procured based on eleven flood control parameters using the SI model encapsulated in Table 21.2. The SI model revealed positive and negative findings concerning the flood control factors analyzed. Higher positive SI values signified a greater likelihood of flooding, while negative values suggested a reduced risk of flooding in the region (Pawar, 2025). The evaluation of elevation and flood sites showed that locations situated at elevations between 425 and 557 meters are at a greater risk of flooding, as this elevation class recorded the highest SI value (107). As the altitude rises, the SI values decrease to zero or become negative, reflecting a lower likelihood of flooding in those areas (Fig. 21.4a). Slope values range from 2.7° to 13°, reflecting a probability of flooding in regions characterized by positive SI values. However, negative SI values were noted in areas with very low and very high slopes within the basin (Fig. 21.4b). The western slope of the basin exhibited the highest positive SI value of 34, indicating the greatest likelihood of flooding, followed by the northern slope with an SI of 27. In contrast, the negative SI values pointed to a lower probability of flooding about the slope orientation and flood locations (Fig. 21.4c). For the curvature parameter, positive SI values were recorded for both convex and concave curvatures, whereas flat surfaces exhibited negative SI values. It indicates that convex and concave surfaces significantly influence flooding (Fig. 21.4d).

The highest positive SI value (28) was observed for the TWI, which varied between 3.2 and 6.7. In contrast, the negative SI values for TWI classes ranged from 8.4 to 17.0, and zero SI values for the TWI class from 18 to 27(Table 21.2; Fig. 21.4e). For the SPI parameter, the highest positive SI value (18) was found in the SPI class that extends from -6.2 to -1.1. In contrast, negative SI values were observed for other SPI classes, which range from 1.2 to 8.1, while a zero SI value was recorded for SPI varies between 8.2 and 16.0 (Table 21.2; Fig. 21.4f).It implies that reduced stream power contributes to sediment deposition and the blockage of channels, thereby worsening flooding conditions. The highest positive SI value (17) was noted in the lower reaches of the Malaprabha

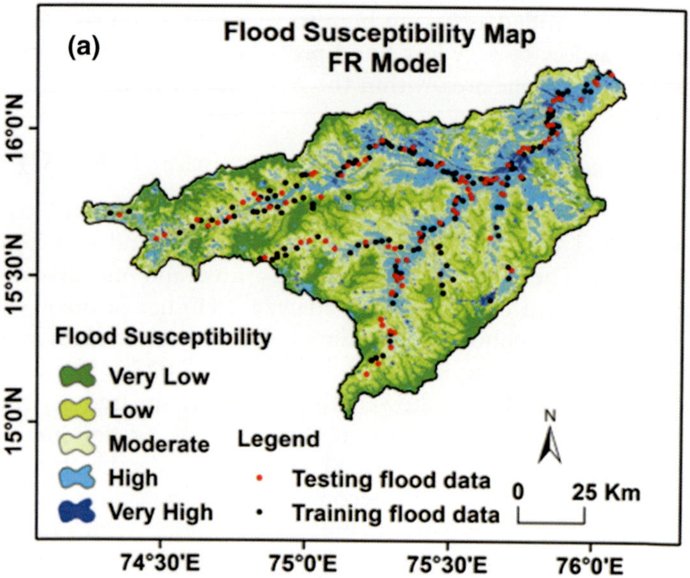

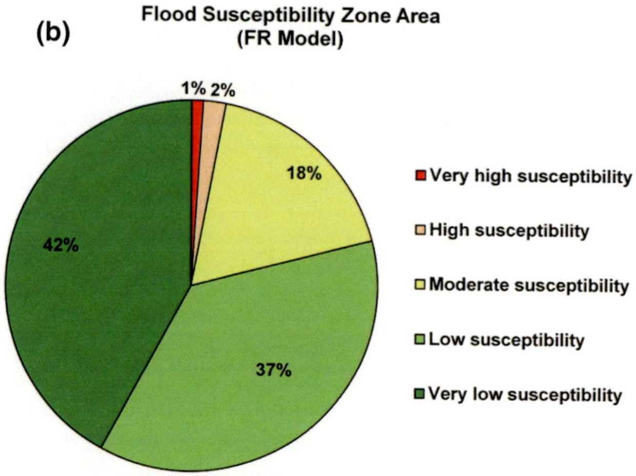

Fig. 21.5 **a** Flood susceptibility map of the Malaprabha Basin using the FR model. **b** Flood susceptible area (%) of the Malaprabha Basin using the FR model

Basin, where rainfall varied from 544 to 768 mm. In contrast, in the western area of the basin, where rainfall ranged from 1231 to 1890 mm, negative SI values were observed (Table 21.2; Fig. 21.4g). The positive SI values were noted for the flood-affected regions situated within 0 to 2150 meters of streams (Table 21.2; Fig. 21.4h). In contrast, locations beyond 2151 meters are not at risk of flooding. It clearly illustrates that the risk of flooding is more pronounced near streams and decreases as the distance from them increases (Khosravi et al., 2016; Pawar et al., 2022).

The highest positive SI value recorded was 121 for the stream density class ranging from 0.58 to 0.68 km/sq. km, followed by the class of 0.46 to 0.57 km/sq. km. In contrast, the most substantial negative SI value of −243 was identified in the area with a stream density class of 0.0 to 0.11 km/sq. km (Fig. 21.4i). This finding indicates that a region characterized by high flow density is more vulnerable to flooding than those with low flow density. The influence of Vertisols on flood occurrence is significant, as evidenced by the high statistical index values for Vp (SI = 74) and Vc (SI = 48) soils. According to Pawar (2025), the high moisture holding capacity in Vertisols due to high clay content causes areas with Vertisols to be more susceptible to flooding (Fig. 21.4j). The flood susceptibility analysis revealed that the landuse and landcover parameter exhibited the highest positive statistical index (SI = 329) for the settlement LULC class. In contrast, the other land use categories displayed either negative or zero SI values, as indicated in (Table 21.2; Fig. 21.4k). At last, the FSI was calculated using the values of the SI model, incorporating all relevant flood control parameters. The flood susceptibility map was classified into five categories: very high, high, moderate, low, and very low (Fig. 21.6a). The analysis revealed that 22% of the basin's area falls within very high and high flood susceptibility categories. Moreover, the moderate flood susceptibility zone encompasses 29% of the basin's area. Overall, 49% of the basin comes within the low and very low flood vulnerability zones (Fig. 21.6b).

21.4.3 *Validation and Comparison of Flood Susceptibility Maps*

The primary objective of hazard mapping is to identify areas susceptible to flooding. To analyze flood susceptibility and develop flood risk zone maps, researchers used various methodologies and models. It is essential to validate the results obtained from the methods or models employed in flood susceptibility analysis and to assess the accuracy of

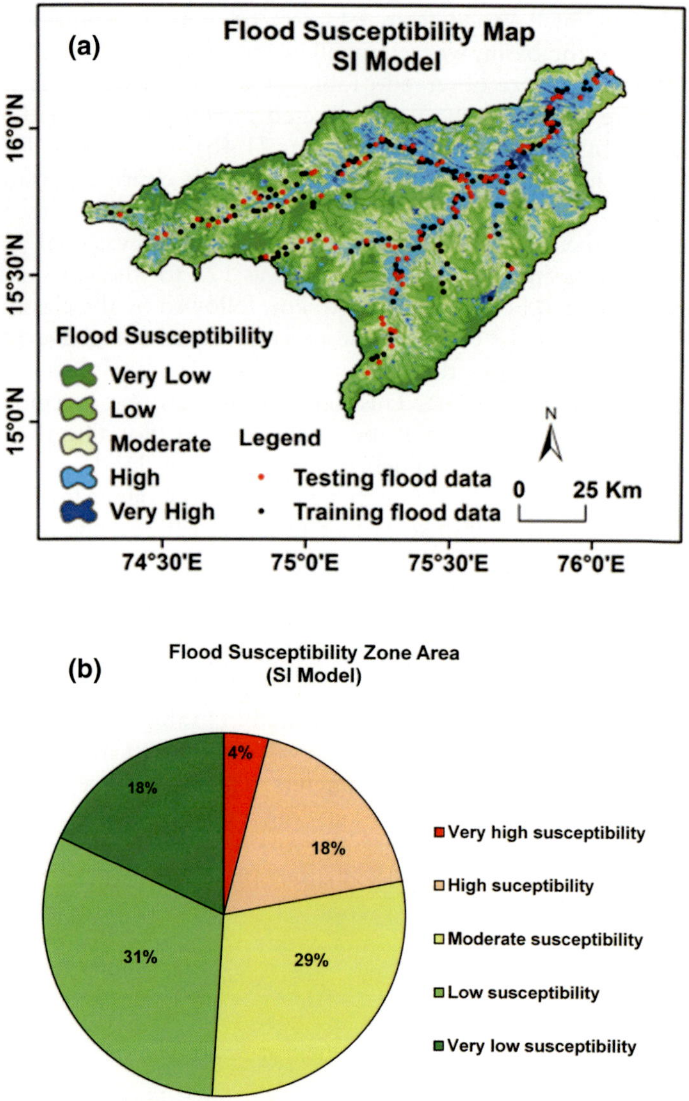

Fig. 21.6 **a** Flood susceptibility map of the Malaprabha Basin using the SI model. **b** Flood susceptible area (%) of the Malaprabha Basin using the SI model

the produced flood maps (Pawar et al., 2022). This study examined the reliability of the flood susceptibility maps produced by the FR and SI models using the ROC-AUC method. Success rates were derived from the training flood dataset, representing 70% of the total data, whereas prediction rates were determined using the testing flood dataset, which constituted 30%. According to Yesilnacar (2005), the accuracy of flood predictions based on the ROC-AUC was divided into five categories: low (50–60%), average (60–70%), good (70–80%), very good (80–90%), and excellent (90–100%). In the analysis presented in Fig. 21.7 and the AUC calculation, the SI model exhibits superior performance compared to the FR model, with a success rate of 82.76% and a prediction rate of 83.16%. The findings indicated that the results for the Malaprabha Basin fall within the excellent category of flood forecasting, as noted by Yesilnacar (2005). Therefore, the flood susceptibility map produced by the SI model serves as an effective tool for planning and mitigating flood hazards in the Malaprabha Basin.

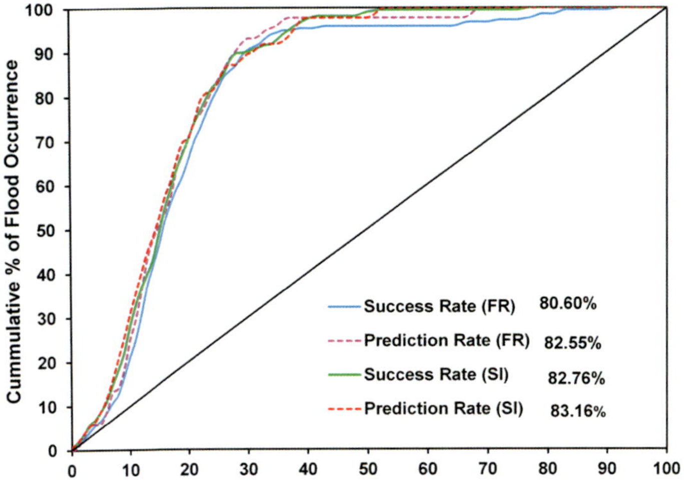

Fig. 21.7 The ROC-AUC outcomes for the success and prediction rates of the FR and SI models

21.5 Conclusions

The main objectives of this study are to analyze flood susceptibility and create a map of areas vulnerable to flooding in the Malaprabha Basin. Geospatial techniques and bivariate statistical models, such as Frequency Ratio (FR) and Statistical Index (SI), were applied to achieve these goals. The analysis was performed on various flood-controlling factors such as elevation, slope, aspect, curvature, the topographic wetness index (TWI), stream power index (SPI), rainfall, proximity to rivers, stream density, soil types, and land use/land cover (LULC), along with information of 296 flood sites in the basin. The random sampling method was applied to designate 70% of the input flood data as training samples, while the remaining 30% was utilized as testing samples for model validation. The analysis revealed that factors such as rainfall, stream density, proximity to rivers, soil types, and land use and landuse/ landcover (LULC) significantly contribute to flooding in the Malaprabha Basin. The outcomes of the model validation derived from the training and testing of flood location points demonstrated that the SI model exhibited a success rate of 82.76% and a prediction accuracy of 83.16%, which is superior to the FR model applied in the current study. The primary challenge in flood vulnerability analysis is the scarcity of reliable long-term extreme event data. Nevertheless, flood susceptibility maps serve as valuable tools for decision-makers, planners, water resource organizations, the Department of Natural Resources, and managers, enabling them to implement effective strategies to manage and improve the flood situation in the Malaprabha Basin.

References

Akay, H. (2021). Flood hazards susceptibility mapping using statistical, fuzzy logic, and MCDM methods. *Soft Computing, 25*(14), 9325–9346.

Arora, A., Pandey, M., Siddiqui, M. A., Hong, H., & Mishra, V. N. (2021). Spatial flood susceptibility prediction in Middle Ganga Plain: Comparison of frequency ratio and Shannon's entropy models. *Geocarto International, 36*(18), 2085–2116.

Bowers, A. J., & Zhou, X. (2019). Receiver operating characteristic (ROC) area under the curve (AUC): A diagnostic measure for evaluating the accuracy of predictors of education outcomes. *Journal of Education Students Placed Risk, 24*(1), 20–46.

Chow, V. T. (1956). Hydrologic studies of floods in the United States. *International Association of Scientific Hydrological Publication, 42*, 134–170.
De Vos, J. H., & Virgo, K. J. (1969). Soil structure in vertisols of the Blue Nile clay plains. *Sudan. European Journal of Soil Science, 20*(1), 189–206.
Di Baldassarre, G., Montanari, A., Lins, H., Koutsoyiannis, D., & Brandimarte, L. (2010). Blöschl G (2010) Flood fatalities in Africa: From diagnosis to mitigation. *Geophysical Research Letters, 37*, L22402. https://doi.org/10.1029/2010GL045467
Dudal, R. (1965). *Dark clay soils of tropical and subtropical regions* (FAO Agricultural Development Paper No. 83). FAO, Rome, Italy, 161pp.
Dutta, M., Saha, S., Saikh, N. I., Sarkar, D., & Mondal, P. (2023). Application of bivariate approaches for flood susceptibility mapping: A district level study in Eastern India. *HydroResearch, 6*, 108–121.
Ebrahimi, E., Araújo, M. B., & Naimi, B. (2023). Flood susceptibility mapping to improve models of species distributions. *Ecological Indicators, 157*, Article 111250.
Haghizadeh, A., Siahkamari, S., Haghiabi, A. H., & Rahmati, O. (2017). Forecasting flood-prone areas using Shannon's entropy model. *Journal of Earth System Science, 126*, 1–11.
Jung, I. W., Chang, H., & Moradkhani, H. (2011). Quantifying uncertainty in urban flooding analysis considering hydro-climatic projection and urban development effects. *Hydrology and Earth System Sciences, 15*(2), 617–633.
Khoirunisa, N., Ku, C. Y., & Liu, C. Y. (2021). A GIS-based artificial neural network model for flood susceptibility assessment. *International Journal of Environmental Research and Public Health, 18*(3), 1072.
Khosravi, K., Nohani, E., Maroufinia, E., & Pourghasemi, H. R. (2016). A GIS-based flood susceptibility assessment and its mapping in Iran: A comparison between frequency ratio and weights-of-evidence bivariate statistical models with multi-criteria decision-making technique. *Natural Hazards, 83*, 947–987.
Luino, F. (2016). Floods. In P. Bobrowsky & B. Marker (Eds.), *Encyclopedia of engineering geology*. Encyclopedia of Earth Sciences Series. Springer. https://doi.org/10.1007/978-3-319-12127-7_126-1
Mehravar, S., Razavi-Termeh, S. V., Moghimi, A., Ranjgar, B., Foroughnia, F., & Amani, M. (2023). Flood susceptibility mapping using multi-temporal SAR imagery and novel integration of nature-inspired algorithms into support vector regression. *Journal of Hydrology, 617*, Article 129100. https://doi.org/10.1016/j.jhydrol.2023.129100
Nagraj, S. P., Nataraja, M., & Omprakash, T. (2018). Evaluation of hydrological components using hydrological model SWAT for Malaprabha sub basin. *Journal of the Geological Society of India, 92*(2), 195–200.

Nsangou, D., Kpoumié, A., Mfonka, Z., Ngouh, A. N., Fossi, D. H., Jourdan, C., Mbele, H. Z., Mouncherou, O. F., Vandervaere, J. P., & Ngoupayou, J. R. N. (2022). Urban flood susceptibility modelling using AHP and GIS approach: Case of the Mfoundi watershed at Yaoundé in the South-Cameroon plateau. *Scientific African, 15*, e01043. https://doi.org/10.1016/j.sciaf.2021.e01043

Parsian, S., Amani, M., Moghimi, A., Ghorbanian, A., & Mahdavi, S. (2021). Flood hazard mapping using fuzzy logic, analytical hierarchy process, and multi-source geospatial datasets. *Remote Sensing, 13*(23), 4761. https://doi.org/10.3390/rs13234761

Pawar, U. (2025). An identification and mapping of flood susceptible areas in the Wardha Basin using frequency ratio and statistical index models, India. *Environmental Science and Pollution Research, 32*, 1565–1580. https://doi.org/10.1007/s11356-024-35871-6

Pawar, U., Suppawimut, W., Muttil, N., & Rathnayake, U. (2022). A GIS-based comparative analysis of frequency ratio and statistical index models for flood susceptibility mapping in the Upper Krishna Basin, India. *Water, 14*(22), 3771. https://doi.org/10.3390/w14223771

Pradhan, B. (2010). Flood susceptible mapping and risk area delineation using logistic regression, GIS and remote sensing. *Journal of Spatial Hydrology, 9*(2).

Priscillia, S., Schillaci, C., & Lipani, A. (2021). Flood susceptibility assessment using artificial neural networks in Indonesia. *Artificial Intelligence in Geosciences, 2*, 215–222.

Rentschler, J., Salhab, M., & Jafino, B. A. (2022). Flood exposure and poverty in 188 countries. *Nature Communications, 13*(1), 3527. https://doi.org/10.1038/s41467-022-30727-4

Rudra, R. R., & Sarkar, S. K. (2023). Artificial neural network for flood susceptibility mapping in Bangladesh. *Heliyon, 9*(6).

Sarkar, D., & Mondal, P. (2020). Flood vulnerability mapping using frequency ratio (FR) model: A case study on Kulik river basin, Indo-Bangladesh Barind region. *Applied Water Science, 10*(1), 1–13.

Shafapour Tehrany, M., Shabani, F., Neamah Jebur, M., Hong, H., Chen, W., & Xie, X. (2017). GIS-based spatial prediction of flood prone areas using standalone frequency ratio, logistic regression, weight of evidence and their ensemble techniques. *Geomatics, Natural Hazards and Risk, 8*(2), 1538–1561.

Sharma, A., Poonia, M., Rai, A., Biniwale, R. B., Tügel, F., Holzbecher, E., & Hinkelmann, R. (2024). Flood susceptibility mapping using GIS-based frequency ratio and Shannon's entropy index bivariate statistical models: A case study of Chandrapur district, India. *ISPRS International Journal of Geo-Information, 13*(8), 297. https://doi.org/10.3390/ijgi13080297

Suppawimut, W. (2021). GIS-based flood susceptibility mapping using statistical index and weighting factor models. *Environment and Natural Resources Journal, 19*(6), 481–493.

Swain, K. C., Singha, C., & Nayak, L. (2020). Flood susceptibility mapping through the GIS-AHP technique using the cloud. *ISPRS International Journal of Geo-Information, 9*(12), 720. https://doi.org/10.3390/ijgi9120720

Tehrany, M. S., Kumar, L., & Shabani, F. (2019). A novel GIS-based ensemble technique for flood susceptibility mapping using evidential belief function and support vector machine: Brisbane. *Australia. Peerj, 7*, Article e7653.

Tehrany, M. S., Pradhan, B., Mansor, S., & Ahmad, N. (2015). Flood susceptibility assessment using GIS-based support vector machine model with different kernel types. *CATENA, 125*, 91–101.

Towfiqul, I. A. R. M., Talukdar, S., Mahato, S., Kundu, S., Eibek, K. U., Pham, Q. B., Kuriqi, A., & Linh, N. T. T. (2021). Flood susceptibility modelling using advanced ensemble machine learning models. *Geoscience Frontiers, 12*, Article 101075. https://doi.org/10.1016/j.gsf.2020.09.006

Vilasan, R. T., & Kapse, V. S. (2022). Evaluation of the prediction capability of AHP and F-AHP methods in flood susceptibility mapping of Ernakulam district (India). *Natural Hazards, 112*(2), 1767–1793.

Wahlstrom, M., & Guha-Sapir, D. (2015). *The human cost of weather-related disasters 1995–2015*. UNISDR.

Wang, Y., Fang, Z., Hong, H., Costache, R., & Tang, X. (2021). Flood susceptibility mapping by integrating frequency ratio and index of entropy with multilayer perceptron and classification and regression tree. *Journal of Environmental Management, 289*, Article 112449.

Westen, C. J. V. (1997). *Statistical landslide hazard analysis. ILWIS 2.1 for Windows application guide* (pp. 73–84). ITC Publication.

Wu, Y., Zhang, Z., Qi, X., Hu, W., & Si, S. (2024). Prediction of flood sensitivity based on logistic regression, eXtreme gradient boosting, and random forest modeling methods. *Water Science & Technology, 89*(10), 2605–2624.

Yesilnacar, E. K. (2005). *The application of computational intelligence to landslide susceptibility mapping in Turkey* (Ph.D Thesis). Department of Geomatics the University of Melbourne 423 pp.

Zhao, G., Pang, B., Xu, Z., Peng, D., & Xu, L. (2019). Assessment of urban flood susceptibility using semi-supervised machine learning model. *Science of the Total Environment, 659*, 940–949. https://doi.org/10.1016/j.scitotenv.2018.12.21

CHAPTER 22

Human Intervention in the Kopai Fluvial System (Birbhum, West Bengal): Overview and Interpretation Through High-Resolution Free Satellite Data and Ground-Truthing Approach

Arindam Sarkar and Shreya Basak

22.1 Introduction

22.1.1 Background

The Kopai River is a seasonal river that plays a critical role in the local ecology and rural livelihoods. However, increasing anthropogenic pressures, including land use changes, riverbank encroachment, sand mining,

A. Sarkar (✉)
Department of Geography, Purash Kanpur Haridas Nandi College, University of Calcutta, Kolkata, India
e-mail: arindam.srkr1@gmail.com

S. Basak
Department of Geography, University of Calcutta, Kolkata, India

S. Ghosh (ed.), *Reimagining Indian Rivers for Sustainability*, Palgrave Studies in Environmental Sustainability,
https://doi.org/10.1007/978-3-032-04754-0_22

and pollution, have disrupted its natural fluvial processes. For the agrarian communities in the eastern Bardhhaman and Birbhum districts, the Kopai River has historically been an essential resource. Its waters have restored aquifers, nourished the land, sustained local livelihoods, and enhanced the area's distinct ecological composition. However, various human activities have seriously altered the river's natural dynamics and flow in recent decades. The natural integrity of the riverbanks has been weakened by increased and frequently uncontrolled activities like sand and laterite mining, and encroachments have limited its borders, limiting the once-thriving ecosystem. The construction of embankments and check dams has altered the river's path, causing erosion in some areas and stagnant water in others. These effects have been exacerbated by intensified agriculture and urbanization. Furthermore, pollutants that threaten aquatic life have been released due to careless waste disposal. Moreover, the indiscriminate waste disposal has released pollutants that endanger aquatic life and deteriorate water quality. Together, these actions have compromised the ecological balance, disrupted sediment transport, and significantly changed the seasonal hydrological patterns that have supported life along the Kopai for generations. Analyzing morphometric parameters quantitatively is highly beneficial for evaluating river basins, prioritizing watersheds for soil and water conservation, and managing natural resources at a micro level (Das, 2016).

22.1.2 *Literature Review*

Fluvial systems are dynamic, influenced by ongoing geomorphic, hydrological, and ecological interactions. However, in recent decades, human interventions have become major factors altering these natural processes, resulting in significant shifts in river morphology, sediment dynamics, and ecological health. Pioneering research by Leopold et al. (1964) and Schumm (1977) established the groundwork for contemporary fluvial geomorphology by highlighting the equilibrium among river energy, sediment load, and channel shape. Human activities such as deforestation, urbanization, sand mining, and agricultural expansion can upset this balance, leading to either channel degradation or aggradation. In India, the deterioration of small and medium rivers due to rampant human activities is well-documented (Ghosh et al., 2018, 2021; Rudra, 2010), with the Kopai River as a notable example. This study employs high-resolution satellite data (e.g., Sentinel-2, Landsat 8) combined with extensive field validation to evaluate changes in channel width, riverbank shape, land use, and sediment deposition patterns. This method follows best practices in river monitoring, as highlighted by Bishop (2013) and Lillesand et al. (2015), Santos et al. (2011), Walia et al. (2018), and Wang et al. (2023), who support the integration of remote sensing and ground-truthing for precise geomorphic evaluations.

Morphometric Diversity on *Kuya* River (combined flow of Kopai and Dwarkeshwar) Basin has been done by Das (2016). A high sinuosity index in the lower catchment suggests a broader river corridor with reduced erosion and increased deposition (Das, 2016). Das and Mukhopadhyay (2022) examined how the Brick Field impacts the riverine morpho-hydrological environment in the Kopai River of the South Bengal Plateau Fringe. The establishment of the brick field results in riverbank erosion, alterations in channel shape, and a displacement of the thalweg near brick factories (Das & Mukhopadhyay, 2022). Regularly transporting heavy vehicles across the riverbed for temporary loading and unloading of raw materials and bricks changes the riverbed's usual characteristics. This practice disturbs the river's natural course (Das & Mukhopadhyay, 2022). Mondal et al., (2022) examined the morphometric features of the Kopai River Basin in West Bengal through geospatial technology. The Kopai River basin features a fifth-order stream, and its drainage network exhibits a dendritic pattern (Mondal et al., 2022). Additionally, the impact of dams and barrages on the flood trends in the lower catchment of the Mayurakshi River basin in eastern India has been studied (Ghosh & Pal, 2015). Analyzing flood history and simulating future flood characteristics using the Gumbel method and Log-Pearson Type III: the case of the Mayurakshi River Basin, India, has been conducted by Islam & Sarkar (2020).

The morphometric analysis of the Kopai River basin focuses on measuring relief, slope, and drainage parameters using geospatial techniques. It shows that integrating remotely sensed datasets with GIS tools helps assess morphometric attributes and their impact on landforms, particularly at the river basin scale (Mondal et al., 2022). Evaluation of Topographic Complexity in a Drainage Basin Utilizing Geographic Information Systems by Nag et al. (2021). Study of Flora in the Rural Areas of Birbhum District, West Bengal, Regarding the Pollination Calendar by Begum and Mandal (2015). Key findings from studies on the Kopai River indicate that channel constriction arises from embankment construction, sediment overload leads to mid-channel and point bars forming, and floodplain connectivity is decreasing. Similar changes are observed in the Rarh Bengal region, where rivers like the Ajay and Brahmani exhibit comparable indicators of anthropogenic stress (Roy & Datta, 2015). The research adds depth to the emerging field of Anthropogeomorphology (Gregory, 2006), which recognizes human influence as a key factor in geomorphic changes. The study fosters a comprehensive understanding of river system evolution amidst anthropogenic pressures by documenting the roles of cultural practices, administrative neglect, and localized development projects.

22.1.3 Rationale

The Kopai River, a seasonal tributary traversing a semi-arid lateritic region, has historically played a vital role in sustaining the area's agriculture, culture, and settlement. However, in recent decades, increased human activities—including encroachments, channel alterations, sand mining, changes in land use, embankment construction, and waste disposal—have disrupted its natural flow processes, adversely affecting its ecological integrity and morphodynamics. Given the magnitude and intensity of these activities, there is an urgent need for a thorough analysis that integrates remote sensing technology with field verification to monitor and understand the nature and scale of these changes. While several studies have examined large Himalayan rivers, smaller rivers like the Kopai are frequently overlooked in academic discussions, despite their ecological and socio-economic significance. This study seeks to fill that void by utilizing freely accessible high-resolution satellite imagery (e.g., Sentinel, Landsat, Google Earth) alongside ground-truthing methods to identify, map, and evaluate human-induced alterations and their impacts on the Kopai River system. This comprehensive strategy can aid in fostering sustainable river management and informing policy at the regional level.

22.1.4 Research Gap

There is a notable scarcity of comprehensive integrated studies that employ satellite-based change detection and ground verification techniques for smaller river systems, such as the Kopai River, particularly in the semi-arid regions of eastern India. This gap highlights the need for more detailed research to better understand the dynamics and environmental changes affecting these vital waterways. Limited focus on specific human interventions presents a challenge due to a lack of comprehensive consequences assessment and regional and temporal variations that must be addressed. Additionally, it is imperative to consider local community and stakeholder perspectives in this context. An analysis of sustainability and long-term impact is essential, alongside exploring technological and policy solutions. Furthermore, the cultural and historical influence of human interventions must be considered.

22.1.5 *Research Question*

How have human activities transformed the morphology, hydrology, and ecological features of the Kopai River, and how effectively can we use high-resolution free satellite data along with ground-truthing methods to evaluate these changes? Understanding how human activities affect the morphodynamics (channel changes, sediment transport, etc.) and ecological functions of the Kopai River is essential for sustainable river management.

22.2 Objectives

The study's primary objectives include identifying and categorizing significant forms of human intervention and spatio-temporal changes in the Kopai fluvial system. It aims to evaluate the types and intensities of anthropogenic activities impacting the river system and analyze the spatial extent and severity of these impacts on river channel structure, floodplain connectivity, and riparian zones. Another objective is to assess the ecological implications, such as biodiversity loss and water quality degradation, resulting from these changes. Furthermore, the study seeks to validate satellite-based findings with detailed field observations and community interviews and confirm remote sensing observations through systematic ground-truthing surveys, local interviews, and photographic documentation. The study will also recommend sustainable management strategies for the river system based on the findings, interpret the geomorphic and ecological consequences of these interventions on river health, hydrological regime, and ecosystem services, and provide recommendations for mitigating adverse human impacts on the Kopai River system, integrating both scientific evidence and local knowledge.

22.3 Study Area

The Kopai River basin (Figs. 22.1 and 22.2) is mainly located in the Birbhum district, West Bengal. Kopai is a significant tributary of the Mayurakshi River basin. Its twin rivers, Kopai and Bakreshwar, form the combined Kuya channel (Nag et al., 2021) in the lower catchment of the Mayurakshi River. The dominant physiographic characteristics (Rarh region) (Fig. 22.3) of the Kopai river basin are upland terrain with lateritic soil and a monsoon-dominated flow regime. The Rarh region in West Bengal is situated between the Western plateau and the highlands that border the Chhotanagpur plateau to the west, and the Ganges Delta to the east. (Ghosal, 2018). Different hydrological parameters, such as the fluctuation of the *Kuya* river (Kopai river in the lower catchment), firmly control the seasonal surface hydrological pattern of the wetland associated with the Kopai river (Das, 2014). The river originates close to the western boundary of Birbhum. The Kopai River is a prominent feature in the region's riverine landscape, serving as a tributary of the Mayurakshi River basin. Also known as the Sal River, it flows eastward, adopting the name Kopai near Bolpur. The river converges with the Bakreshwar River in the *Hasuli Bank* area of Labhpur (Ghosal, 2018). Subsequently, it is referred to as the *Kuya* River as it flows into Murshidabad. Beginning at Khajuri village in Jharkhand, it moves southeast through the Birbhum and Murshidabad districts of West Bengal before joining the Babla River, also known as the Dwarka River, near Sabitrinagar in Murshidabad (Ghosal, 2018). The river spans a total length of 176.4 km, with approximately 24.64 km being non-perennial. The basin area is situated between latitudes 23°26′18″N and 23°56′30″N, and longitudes 87°13′E to 88°09′30″E, covering a total area of 1555.2 sq. km (Das & Mukhopadhyay, 2015a). Agrarian communities of the Kopai Basin area are experiencing increasing urbanization and infrastructure development along the riverbanks. Most of the Kopai river basin is situated in the southern region of Birbhum district in West Bengal. At the same time, the remaining area extends into the southeastern part of the Jamtara district in Jharkhand. The river begins at Khajuri village in the Kundhit C.D. Block of Jharkhand. After flowing just 13 km in Jharkhand, it enters West Bengal, traversing parts of Khoyrasole, Dubrajpur, Illambazar, Bolpur-Sriniketan, Labpur, and Sainthia C.D. Block in Birbhum district. Thus, of its total length of 110 km, approximately 97 km lies within Birbhum district, West Bengal. (Das & Mukhopadhyay,

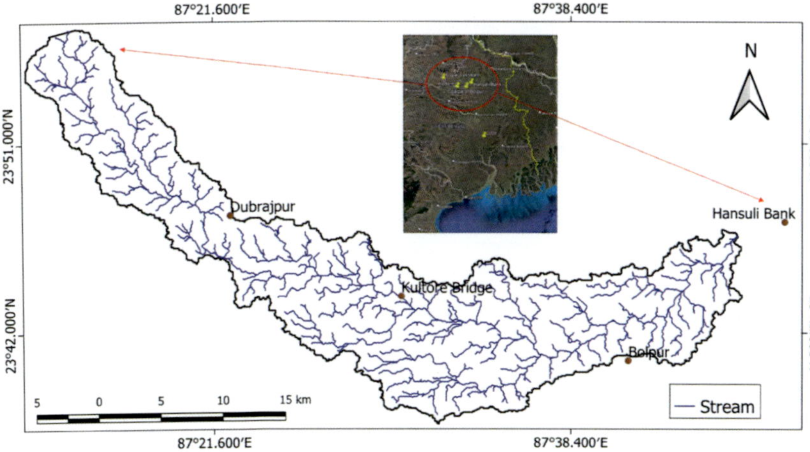

Fig. 22.1 Kopai River basin location. This basin stretches between 23° 37′55″N and 23° 56′45″N latitude, and 87° 16′53″E and 88° 4′42″E longitude. It serves as a significant sub-basin in the Mayurakshi River system. The Kopai basin traverses' areas within the Birbhum, Burdwan, and Murshidabad districts of West Bengal, as well as Jamtara district in Jharkhand, covering 15 Community Development blocks

2022). This region showcases the traces of Late Pleistocene ferruginous formations alongside remnants of tropical deciduous forest (Ghosh & Pal, 2015, 2016). The catchment features a dry, mild, sub-humid, and subtropical monsoonal climate, with rainfall varying by elevation between 829 and 2179 mm from 1901 to 2013 (Ghosh & Pal, 2015, 2016).

22.4 Data and Methodology

22.4.1 Data Type and Source

This study has used, analysed, and interpreted different types of data (Table 22.1).

Fig. 22.2 The Kopai River begins in Khajuri village, Jharkhand, and flows southeast to join the Bakreshwar River in West Bengal, specifically at Hansuli bank near Labpur

22.4.2 Remote Sensing and GIS Analysis

This study utilizes bibliographic history, analyses, and assessments of topographical sheets produced by a survey of India, alongside a chronological evaluation of satellite images (Table 22.1). Outcomes are verified through field visits and measurements using specific instruments. Land Surface temperature has been mapped for the Kopai basin (Fig. 22.4). Water bodies were extracted using the Normalized Difference Water Index (NDWI) (Figs. 22.5 and 22.6) for the Kopai basin area. Additionally, changes of the main channel of the Kopai have been detected and analysed through different temporal images (Fig. 22.7). Normalized Difference Vegetation Index of the Kopai river basin for the years 2011 (Fig. 22.8a) and 2024 (Fig. 22.8b).

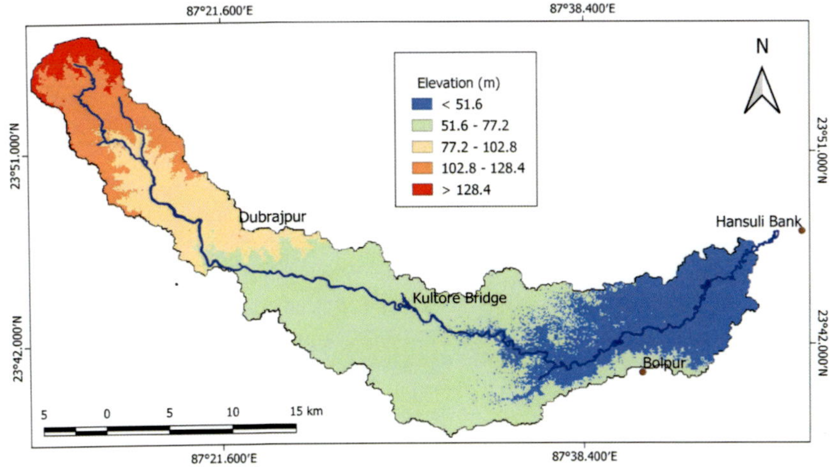

Fig. 22.3 Physiography map of the Kopai River basin. The basin's elevation varies from 129 to 50 meters above sea level. Dubrajpur is situated in the upper catchment area of the Kopai River. At the same time, the Kultore barrage is found in the middle course, and Bolpur is located in the lower catchment region

Table 22.1 Different types of data and their sources

Map and images		*Numbers*	
Topographic map		F45D9, F45D6, F45D5, F45D2, F45D14, F45D13, F45D10, F45D1	
Satellite image	Elevation data	SRTM	30 m
	Geospatial data	Landsat 5 TM 139/44 20/10/2011	Landsat 8 OLI/TIRS 139/44 26/112024
	Google Earth Pro (High-resolution historical imagery)		

22.4.3 *Field-Based Observations*

22.4.3.1 Ground-Truthing

A GPS survey of key river segments and bank profiles has been done during the field survey. In this study, validating interpretations from satellite imagery through ground truth verification is crucial. A ground survey was conducted using a GPS handset to confirm the classification and

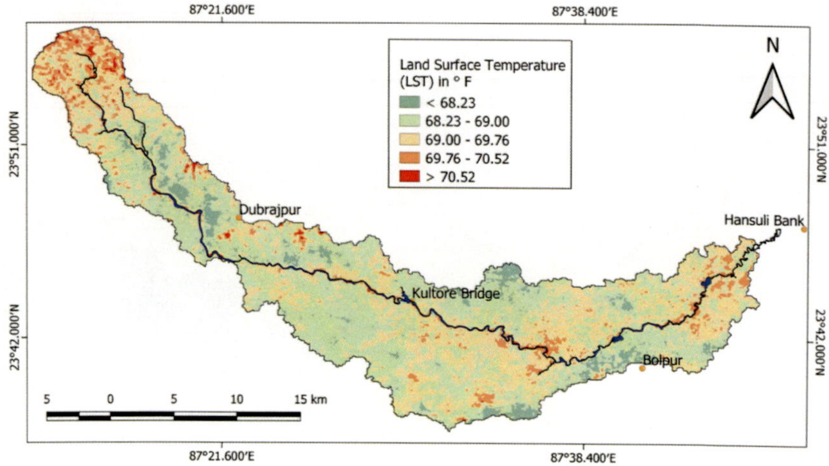

Fig. 22.4 Land Surface Temperature of the Kopai River basin prepared from Landsat images. Maximum temperature remains in the places of the river basin where rapid deforestation occurred near the main channel for agricultural activities and the construction of brick kilns. The lowest land surface temperature is located in forest patches

interpretation of changes in land use/land cover (LULC), channel modifications, embankment structures, and anthropogenic impacts observed through satellite data. The primary objective of the ground survey is to evaluate how human activities, such as sand mining, encroachment, agriculture, and urban expansion, alter the morphology and ecology of the Kopai River. Effective ground truth locations along the main channel of the Kopai River basin were selected, including the stretch from Santiniketan to Kankalitala (for rapid urban expansion), the Labhpur area (for agricultural encroachment), the Bhedia region (for sand mining hotspots), the Kushmore and *Hansulibank* area (for channel shifts and depositional features), and near Illambazar (for traditional water use patterns and pollution sources).

The deviations in the past and present river path of the Kopai main channel have been identified and analysed using GPS mapping and interviews with residents and farmers. Areas with exposed sandy patches, pits, and disturbed textures caused by sand mining along the main course

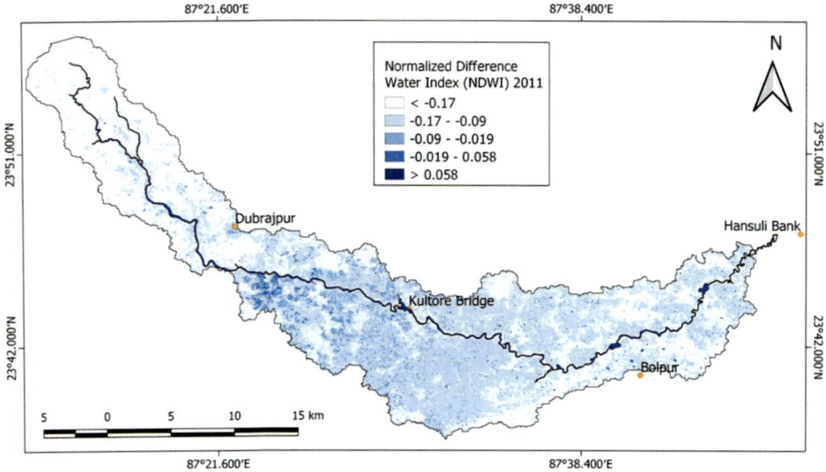

Fig. 22.5 Normalized Difference Water Index of the Kopai river basin for 2011. This map displays the Normalized Difference Water Index (NDWI) for the Kopai River Basin in 2011, highlighting the availability of surface water and wetness conditions throughout the region. The NDWI ranges depicted on the map are as follows: < −0.17 (Dehydrated/vegetated/barren surfaces), −0.17 to −0.09 (Low moisture), −0.09 to −0.019 (Moderate dryness), −0.019 to 0.058 (Moist soils or potential waterlogged areas), and > 0.058 (Surface water body)

of the Kopai have been located and verified through visual confirmation, photographic documentation, and observations by locals. During field surveys, GPS and laser distance meters have measured human encroachment into the Kopai river system, resulting from agriculture and settlement in the floodplain. Urban and infrastructure development linked to the expansion of built-up areas has been identified through ground assessments of new roads and bridges, as well as checks for sewage outlets entering the river via GPS. Agricultural expansion near floodplains or dry channels has been recorded during field surveys, noting crops cultivated, sources of irrigation, and seasonal changes through interviews with farmers on the ground.

22.4.3.2 *Participatory Methods*

According to interviews conducted with 25 local farmers, 36 fisherfolk, and panchayat members during the field survey, human activities have

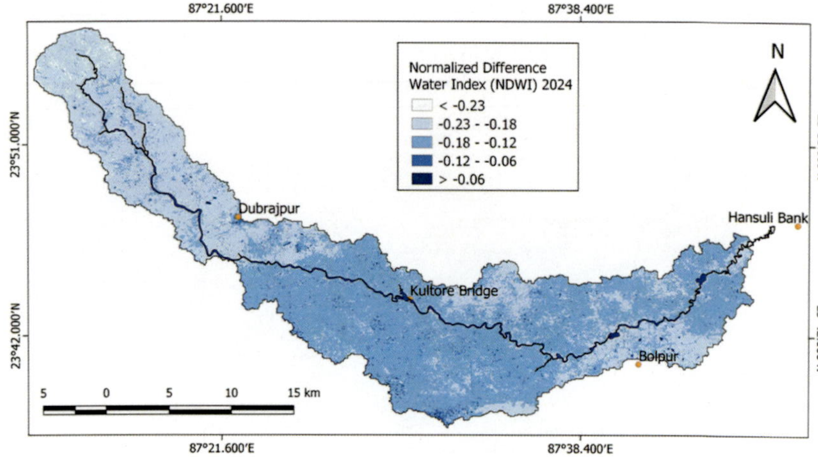

Fig. 22.6 Normalized Difference Water Index of the Kopai river basin for the year 2024. Darker blue regions signify high surface water (such as main river channels, ponds, and wetlands), while lighter blue to white shades indicate minimal or absent surface water. The primary flow of the Kopai River is distinctly marked in dark blue, reflecting positive NDWI values, which indicate the presence of water. Areas surrounding the river that appear light blue to white are linked to seasonal wetlands or floodplains that are moist but not saturated, as well as rain-fed zones that dry rapidly and agricultural areas with varying soil moisture levels

significantly altered the morphodynamics of the Kopai River, particularly in its lower and middle stretches. This has impacted the physical river system and the livelihoods of local stakeholders such as farmers and fisherfolk. The people in this area are closely connected to the Kopai River, but they are overutilizing its water resources.

22.5 Results and Discussion

22.5.1 Nature of Human Intervention in the Kopia Fluvial System

Nature is the intrinsic characteristic of human intervention in the Kopai basin. The nature of human intervention explains the quality, behaviour, and purpose of the activity in the Kopai Basin area. The nature of human intervention also describes how human intervention impacts the Kopai

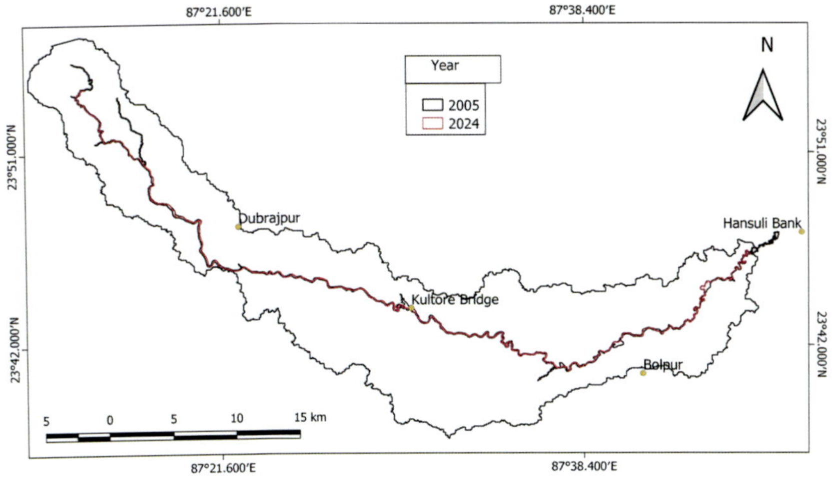

Fig. 22.7 Changes in the main course of the Kopai River from 2005 to 2024. Over the last twenty years, the Kopai River has drastically changed its morphodynamics and hydrology, including lateral shifts, degradation, sediment imbalance, and anthropogenic stress. These factors have combined to create a more seasonal, fragmented, and stressed river system. Furthermore, it is emphasized that sustainable development and the restoration of the river are urgent requirements to preserve its socio-economic and ecological balance. The Kopai River's main course changes from 2005 to 2024 have been analyzed using multi-temporal satellite imagery, field observations, and secondary data. Evidence shows lateral shifting of the river course at Labhpur and Chondrapur, highlighting significant meander migration due to high sediment loads and low bank resistance. The channel is widening in the upper reaches and narrowing downstream, especially in areas affected by embankments and human activities. Intermittent dry periods occur in the basin during the pre-monsoon months, particularly around Nanoor and *Surul*. Aggradation is evident in certain sections of the Kopai River, caused by reduced stream velocity and sediment trapping due to the construction of a dam and barrage for irrigation and water retention. Bed degradation and river incision in the middle and lower reaches of the Kopai are driven by sand and clay mining operations. Additionally, post-monsoon periods see the formation of mid-channel bars and shoals. The floodplain area faces encroachment, significantly reducing riparian vegetation, exacerbating bank erosion and sediment influx into the river. The peak flow magnitude has decreased, and monsoonal surges have become shorter, affecting the river's ability to reshape its channel. Irregular rainfall patterns and human intervention have contributed to the loss of connectivity with oxbow lakes and channel abandonment. The seasonal channel of the Kopai is currently facing a shortage of freshwater

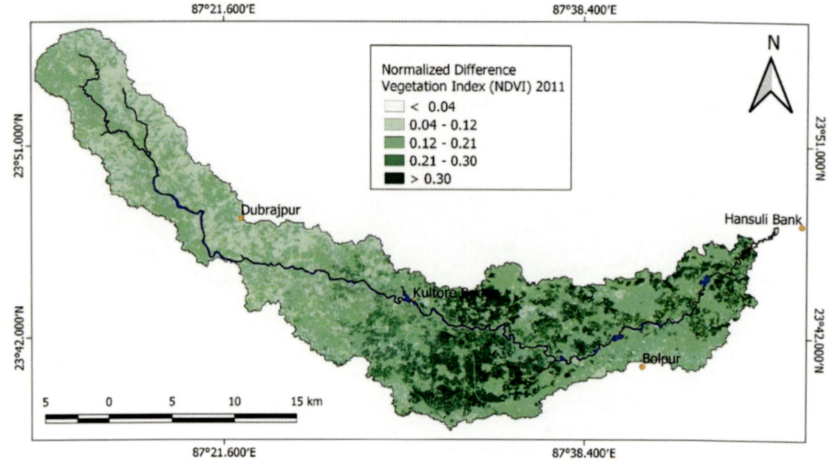

Fig. 22.8a Normalized Difference Vegetation Index of the Kopai river basin for 2011. This map illustrates the Normalized Difference Vegetation Index (NDVI) for the Kopai River Basin in 2011, showcasing the density and vitality of vegetation in the region. An NDVI range on the map of <0.04 denotes barren land, water bodies, or settlements; 0.04–0.12 signifies sparse or stressed vegetation; 0.12–0.21 represents light vegetation; 0.2–0.30 indicates moderate vegetation; and >0.30 reflects dense and healthy vegetation in the basin. Areas of the Kopai basin with high NDWI but low NDVI point to open water or wetlands, while regions exhibiting high NDVI but low NDWI suggest vegetated yet dry lands, such as rainfed fields or uplands

River system. The nature of human intervention, as evidenced in the Kopai River system, is exploratory. Developmental, and Unsustainable or degradation.

22.5.2 Types of Human Intervention and Ways of Human Encroachment in the River System of the Kopai Basin

Human encroachment in the Kopai river basin system has been broadly classified into three main categories: Encroachment of the floodplain, Human impoundment into the fluvial system, and others. People of the agricultural society associated with the Kopai river system encroached on the floodplain through farming activities and the construction of several brick kilns. Human impoundment into the fluvial system is challenging

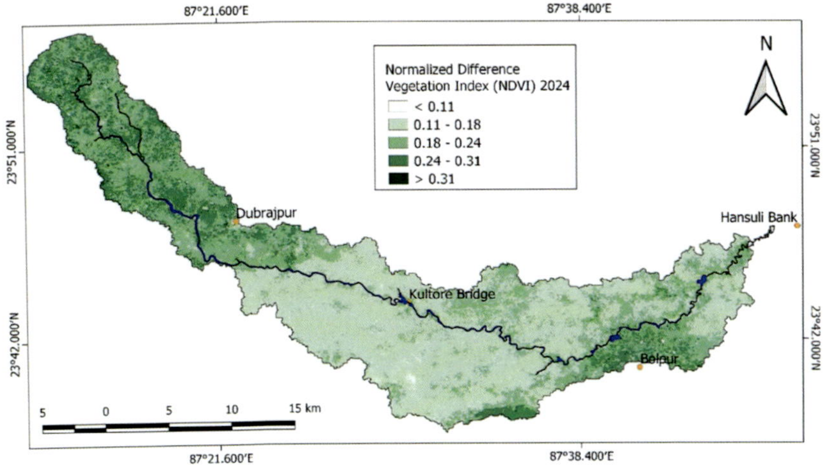

Fig. 22.8b Normalized Difference Vegetation Index of the Kopai river basin for 2024. Dark green regions (NDVI >0.30) indicate likely forests, orchards, or irrigated farmland. In contrast, light green to pale areas (NDVI <0.12) signify farmland, settlements, degraded lands, or places with low moisture and human impact. The Kopai River corridor displays a notably lower NDVI, especially close to the river channel, aligning with common riverine areas where vegetation is sparse or seasonal. Southeast of the river, in the Bolpur and *Hansuli* Bank areas, denser vegetation may result from wetlands, paddy fields, or agroforestry practices. Conversely, the northwest section of the basin near Dubrajpur shows reduced vegetation, likely due to lateritic soil, human settlements, or quarrying activities

for nature, the future, and the next generation. Humans are impounding the Kopai River by constructing dams and bridges, sand mining, and removing bed materials. Other types of human intervention in the Kopai fluvial system are rapid deforestation and the addition of different pollution materials for households and small industries in the basin. Adding large-scale topsoil from agricultural land is also a big concern of the present study.

22.5.3 *Agricultural Activities*

Encroachment of floodplains for cultivation, especially during the dry season, reduces the river's active channel width. In response, the cross-sectional area of the river planform has been changed. Agriculture and its allied activities governed the modification of channel physiology and function. Encroachment of the floodplain for agriculture is a prehistoric event that continues severely in the present time. Conversion of the floodplain and riparian area for agriculture, grazing, and temporary foot tracks for navigation in dry spells. Kopai is a non-perennial river; a sufficient amount of available and usable water is limited. That is why agricultural activities in this river basin are primarily concentrated in the monsoon and post-monsoon seasons. Especially during the post-monsoon season, farmers extract a significant amount of water from the river for Rabi cropping. Extraction of water for irrigation reduces the river discharge. Use of fertilizer and pesticides.

The impact of agricultural activities on the Kopai River channel and its basin involves physical, ecological, hydrological, and socio-economic dimensions. The expansion and intensity of farm activities in the Kopai River basin led to deforestation and alterations of the natural landscape. Irrigation and ploughing during agricultural activity permanently modify the surface and subsurface soil. Post-ploughing periods present a significant risk of surface soil loss as the soil becomes loose after ploughing. Erosion-induced sediment loads increase during the sowing season in this basin area. As a result, sediment loads rise, leading to siltation and aggradation of the Kopai channel. Several locations along the Kopai channel are experiencing flow diversification, the construction of check dams, and local small barriers on the riverbed for irrigation during the post-monsoon dry period. This type of agriculture-based anthropogenic intervention harms the channel. Fragmented flow, altered natural meanders, and reduced water volume in the lower reaches are significant geo-hydrological impacts on the Kopai River basin. High levels of water withdrawal, particularly during the cultivation of Boro rice, significantly diminish streamflow, thereby impacting the perennial characteristics of the river and resulting in the desiccation of channels during lean seasons. Additionally, excessive groundwater extraction for agricultural purposes contributes to a decline in the water table, reducing the base flow contribution to the river.

The primary channel of the Kopai River is impacted by excessive fertilizer runoff, which causes agrochemical pollution within the waterway. Overuse of chemical fertilizers results in high levels of nitrogen and phosphorus that lead to eutrophication in stagnant areas of the river. Persistent organic pollutants infiltrate the food chain, threatening aquatic organisms. Additionally, organic waste from farms, such as cattle manure and decomposed crop residues, raises Biochemical Oxygen Demand (BOD) levels, diminishing dissolved oxygen and negatively influencing aquatic biodiversity. Intensified agricultural practices in the Kopai basin contribute to declining ecosystem services, ecological degradation, and biodiversity loss. This includes reducing native fish species and macroinvertebrates due to pollution and habitat fragmentation. Traditional wetland areas in the basin, which serve as natural filters and flood buffers, are being converted to agricultural land. Consequently, the Kopai River basin is experiencing topsoil loss due to intensive farming conducted without adequate conservation measures. Eroded soil from agrarian fields contributes sediment to the Kopai River's bed, raising its level, reducing cross-sectional area and depth, and decreasing its fluvial sediment carrying capacity.

Agricultural dependence on the Kopai River system has intensified because of seasonal water shortages, resulting in conflicts between upstream and downstream farmers. However, unsustainable practices are causing diminishing returns as soil and water quality deteriorate. Traditional environmentally-friendly techniques, such as mulching and crop rotation, are in decline, having been supplanted by high-input monoculture systems. The stress of monsoon-driven cultivation compels farmers to excessively extract groundwater during the post-monsoon period, especially for the Robi crop. Furthermore, irregular rainfall resulting from climate change exacerbates the stress on river-fed irrigation, increasing reliance on both groundwater and manipulation of river flows.

22.5.4 *Infrastructural Development*

22.5.4.1 *Construction of Barrage, Dam, and Check Dam*

Kopai is an important tributary of the Bakreshwar River system fed by seasonal rain, especially during monsoon. The main characteristic of the rnning water of this river is that it is subject to extreme seasonal flow variability. The Kopai basin faced drying conditions during summer, causing agricultural and meteorological drought conditions throughout the basin. This severe water stress is the prime concern for agricultural activity in the basin area. This irrigation deficit should be addressed while ensuring water availability during the dry season. In this context, several small dams with lock gates and small check dams have been installed in the main course of the Kopai River. Damming in the Kopai River is also subject to the inter-river connection project initiated by the Government of West Bengal, Irrigation and Waterways Department. The dam's construction for irrigation and water storage by altering the natural flow pattern of the river is responsible for the hydrological imbalance of this river. Objectives of construction of barrage, dam, check dam and multi-river linking project in this area are storing monsoonal surface runoff, increasing surface water availability, maintain surface soil moisture, supply water for Kharif crop, recharge of ground water through natural percolation process in adjacent area, enable controlled release water running water system according to need of irrigation in the downstream region of the Kopai river for agriculture.

Kultore barrage (Figs. 22.9, 22.10, 22.11, and 22.12) is located on the river Kopai at Kultore, P.S., Illambazar, District Birbhum. It is a dam and barrage complex part of the larger Mayurakshi Project, a joint venture between Jharkhand and West Bengal to manage water resources in the region. Kultore barrage is locally popular as Kopai barrage with a catchment area of 212.1 sq. km (Linear Water Way-54.86 m). The barrage is designed to discharge 849 cumecs. The Mayurakshi Reservoir Project was initiated in 1951 to provide an optimum water supply for the Kharif and Boro seasons. The Kultore Barrage and dam regulate water flow, particularly for irrigation and flood control in the Lower Kopia catchment area. Two canals have been constructed at Kultore barrage, from the left bank of the Kopai. A canal was built at Kultore barrage, which is linked with the Bakreshwar river in the northeast. Another canal has been constructed from the right bank of the Kopai River at Kultore in the northeast, going through the Bolpur-Sriniketan and meeting with the Bhagirathi-Hooghly River system. The name of this canal is Binuria because it originated at Binuria village, which is 55 km away from Bolpur-Sriniketan.

Moutara Barrage (Figs. 22.13 and 22.14) is another crucial small barrage with two lock gates near Moutara village, in Birbhum district, across the Sal Nadi (Sal River). Sal is a seasonal tributary upstream of the Kopai River basin, known as *Sal* Nadi (river). The Moutara Barrage is on the Sal Nadi, which feeds into the Kopai Nadi system. The Moutara barrage has been constructed to supply water for irrigation during the dry

Fig. 22.9 Temporal images of the Kultore barrage; 2014 (**a**), 2/2016 (**b**), 2018 (**c**) showing impact of construction of Kultore barrage

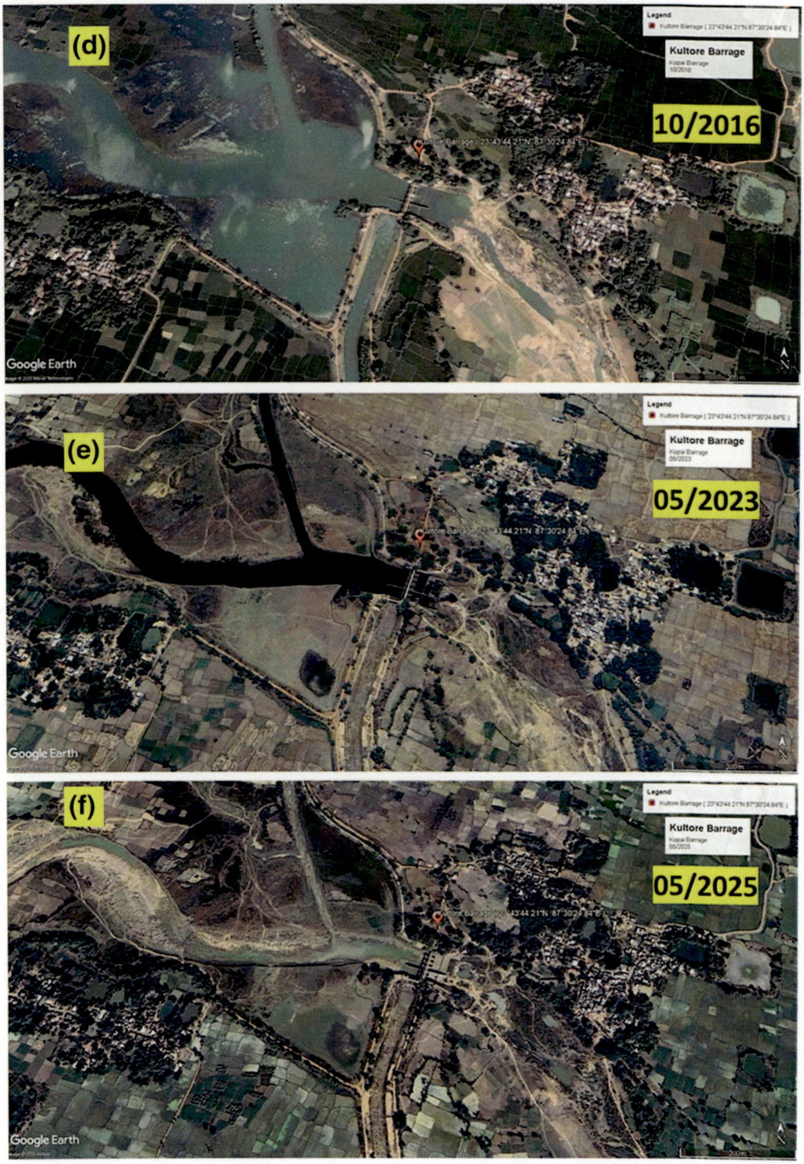

Fig. 22.10 Temporal images of the Kultore barrage; 10/2016 (**d**), 2023 (**e**), 2025 (**f**) showing impact of construction of Kultore barrage

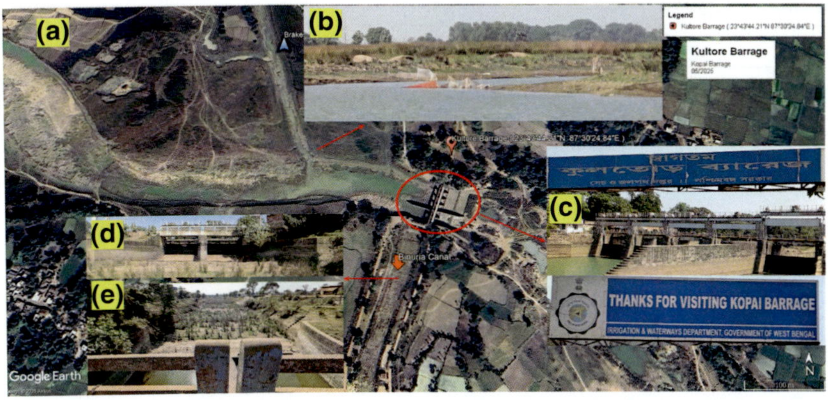

Fig. 22.11 Photo plate of the Kultore barrage and its subsystems. Spatial image of the Kultore barrage (**a**); (**b**) is the connection point of the river linking the canal system to the Bakreshwar River; (**c**) lock gate of the Kultore barrage; (**d**) lock gate of the Binuria canal towards Bolpur, along with a view of the Binuria canal from the top of the lock gate at the Kultore barrage

Fig. 22.12 Lock gate, up and downstream from Kultore barrage

season, helping to regulate seasonal water flow for downstream agricultural lands. The Moutara barrage Acts as a check-dam-like structure that temporarily stores water. It plays a role in local irrigation and groundwater recharge. This barrage is an essential local water management structure for small-scale irrigation and agricultural support.

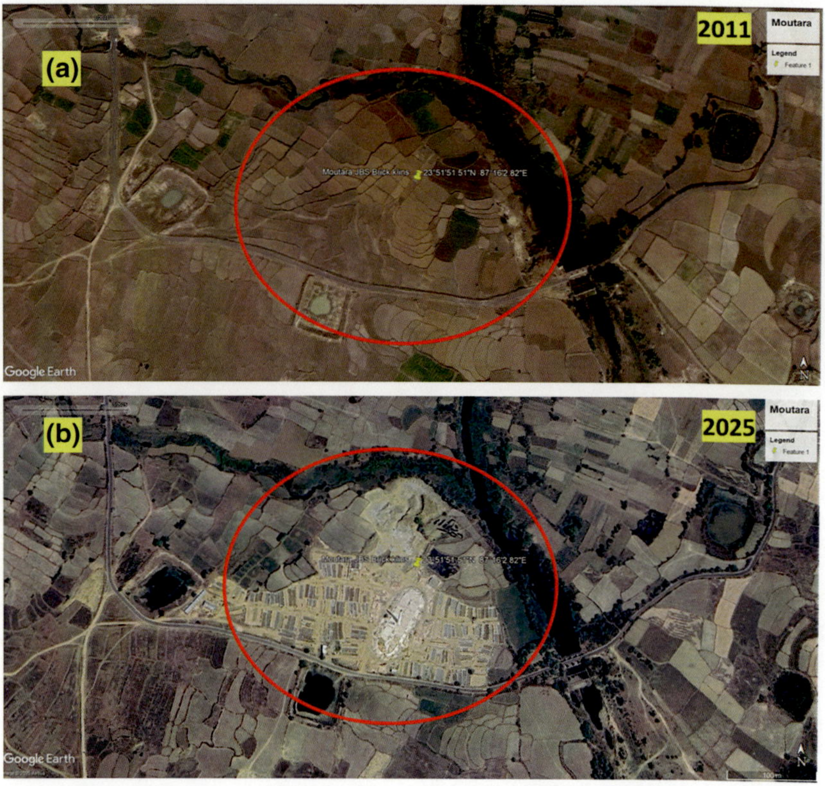

Fig. 22.13 Here is the image of the Moutara bridge and lock gate. The connectivity provided by the nearby roads has transformed agricultural land (**a**) into Brick kilns (**b**)

22.5.4.2 The Construction of Roads and Bridges

The construction of roads and bridges across rivers often disrupts the river's natural course and has significant ecological and geomorphological consequences. Human impoundment into the fluvial system has a greater Impact on the river's natural course. Construction of embankments and pillars of bridges across the river alters the channel by constricting flow, causing changes in velocity, depth of the channel, and depth of the cross-sectional area and sediment transport. This rapid channel aggregation through sediment buildup into the river system results in sediment

Fig. 22.14 Moutara barrage upstream of the Kopai, popularly known as the Sal River. (**a**) spatial orientation of the landscape related to Moutara barrage, (**b**) view of upstream of the Kopai (Sal) river, (**c**) lock gate located at Moutara on Kopai river, (**d**) View of downstream from the top of the lock gate

transport. Several bridges have been constructed across the Kopai River (Fig. 22.15 and 22.16). Bridges with narrow spans and low clearance obstruct flood flows during monsoon, causing flooding and waterlogging in the agricultural land. Construction of roads along the river bed in the Kopai basin is subject to the construction of an embankment. The construction of embankments significantly impacts the dynamic equilibrium of the floodplain. The embankments in the Kopai fluvial system restrict the natural distribution of sediment during floods and overland floods. It helps to change the river dynamics. The construction of artificial levees is responsible for diverting flow from natural meanders, disconnecting floodplains, and altering the hydrological regimes of the river Kopai.

Construction of bridges across the Kopai River has significant negative ecological impacts, such as fragmentation of natural habitat, loss of biodiversity, and declining water quality. The aquatic habitat in the cross-sectional area of the Kopai River, at the point where the bridges are located, has been severely disrupted. The pillars and construction of the bridges badly impact the river planform and riparian vegetation. Bank-side vegetation and natural and artificial water bodies, and their associated micro ecosystem, have also been affected by construction across and along

Fig. 22.15 Shows a Bridge across the Kopai River in the Ballavpur area (Latitude: 23°41′20.29″N, Longitude: 87°38′54.86″E)

the Kopai River. The intensive negative impact of constructing bridges across the Kopai River impedes the movement of fish and macroinvertebrates, affecting their life cycles. Construction induced negative impacts like huge vibration through machinery, noise of construction, numerous debris, and significant human activities in the river cross-sectional area of the Kopai river cause critical disturbance to the natural ecosystem and their habitat. Sensitive species have been displaced and have experienced a significant decline in numbers along the river floodplain of the Kopai River. Increased load of sediment, abandoned oil and chemicals used for construction, mixing of construction waste with river water, and reduced the water quality of the Kopai River in the lower catchment area by altering the oxygen level in the river water. The construction of bridges across the Kopai River has significant negative socio-economic and cultural impacts due to the alteration of traditional water use patterns

Fig. 22.16 Shows a Bridge across the Kopai River in the Goalpara area (Latitude: 23°41′20.29″N, Longitude: 87°38′54.86″E)

like fishing, domestic use, irrigation, etc. It also has a substantial effect on the livelihood of river-dependent communities.

22.5.4.3 Construction of Embankment to Control Flooding

The Kopai River system, a seasonal tributary of the Bakreshwar and Mayurakshi rivers, flows through the lateritic plains of Birbhum and parts of Bolpur-Shantiniketan in West Bengal. The small river Kopai experiences seasonal floods and flash floods during the monsoon (Fig. 22.17). Floods and flash floods in the Kopai River occur due to its limited width, shallow depth, high sediment load, and the impervious lateritic soil of the surrounding catchment area. Inhabitants of the catchment constructed an embankment along the river Kopai in some riverside villages to control floods. An embankment was built on the banks of the river Kopai to prevent the inundation of agricultural land and settlements during high discharge in the monsoon. In this catchment area, embankments help stabilize flood-prone zones, making them usable for agriculture or habitation. In the Kopai catchment area, embankments are constructed in local segments, especially near vulnerable villages and agricultural zones such as Kankalitala, Kirnahar, Labpur, and parts of Bolpur. The height of the embankments in this area ranges between 2 and 3 meters, with a base width of 3 to 4 meters. Some embankments include sluice gates or culverts to allow natural drainage and prevent waterlogging. Most embankments use locally available earth materials (laterite block, soil, sand, etc.). Upstream of the Kopai River are a few embankments, but several have been located downstream due to higher population density and agricultural activity. Seasonal check dams and reinforced bunds have been observed along tributary rivulets or near confluences with high sediment load. In the Kopai River system, the construction of embankments has been successful in protecting against low-frequency floods. The embankments enhance crop security during the monsoon. The embankments on the banks of the Kopai River also protect transportation systems (rail bridges, lines, roads, etc.) and settlements in the basin area, especially those in the floodplain zone of the river.

Fig. 22.17 Left bank of the Kopai River showing sediment deposition sequences and signature of bank erosion and bank failure during flood

Embankments along the banks of the Kopai River (Fig. 22.18) reduce lateral flow and the sediment dispersal mechanism, which is responsible for riverbed aggradation and raising flood levels over time. The lack of management of the embankment and its induced drainage system contributes to long-term waterlogging beside the embankment, especially within the settlements. The embankment disconnects the floodplains from the river channel, causing ecological disruption, loss of wetlands, and reduced groundwater recharge. According to historical flood reports and interviews with local people, poor maintenance of the embankments also leads to breaches during heavy rainfall, resulting in sudden and severe local flooding. A survey conducted among the inhabitants beside the Kopai River reveals that embankments are often built without full consultation, ignoring local communities' traditional flood coping strategies. In villages like Amrai, Kirnahar, and Ghurisha, residents reported mixed experiences—while embankments have reduced direct flood impact, waterlogging and stagnant water in adjacent paddy fields have become chronic.

Embankments are built incrementally without upstream–downstream coordination and suffer from infrequent maintenance. Weak institutional follow-up and a lack of interest among local administrators and policymakers lead to degraded embankments. The incorporation of floodplain restoration, wetland buffers, or afforestation along riverbanks—like nature-based solutions—should be accommodated and implemented alongside the installation of embankments along the bank of the Kopai River in the future. A combination of embankments with catchment treatment processes, such as afforestation programs and installing check dams by involving locals, is essential. Maintenance and long-term stabilization of embankments can be achieved by applying eco-engineering techniques using geotextiles, vetiver grass, and sustainable materials to reinforce embankments. Installing gated culverts and regularly desilting channels is also essential and should be implemented and regulated.

22.5.5 Sand Mining

Excessive and unregulated extraction of sand from the river bed destabilizes the river channel and increases erosion. Kopai is a seasonal tributary that originates from the erosional plateau of Chhotanagpur, carrying substantial sediment from upstream, which fuels small to medium-scale sand mining activities. The bed of the Kopai River becomes exposed

Fig. 22.18 An earthen embankment is constructed along the riverbank, while a small outlet is designated to bypass excess water from the backwater

during the dry months (December to April). The localized and seasonal extraction of sand from the riverbed is primarily illegal. Mining in the Kopai river system (Figs. 22.19 and 22.20) is predominantly manual or semi-mechanized, utilizing spades, carts, tractors, and occasionally small earth movers. Unsustainable sand removal lowers the riverbed, steepens the channel gradients, and destabilizes the banks, particularly during monsoon seasons. This mining disrupts the natural balance between sediment supply and transport, impacting downstream sedimentation and fluvial geomorphology of the Kopai River. The benthic ecosystem, which includes fish spawning grounds and invertebrates, is negatively affected. Channel bars are repeatedly removed or flattened. Sand layers function as natural aquifers, and over-mining may diminish the river's capacity for groundwater recharge, especially in a region like Birbhum that already experiences seasonal water shortages. Sand mining along the banks of the Kopai River provides short-term employment for local laborers and

Fig. 22.19 Sand mining spot at the right bank of the Kopai river at Goalpara area (Latitude: 23°41′20.29″N, Longitude: 87°38′54.86″E). Sand mining is taking place from the point bar of the right bank of the Kopai River

tractor operators. Agricultural land near the Kopai River is frequently utilized as loading points or transit routes, damaging crop fields and causing soil compaction. Encroachments and disputes over sand plots are prevalent, especially in areas with weak institutional oversight.

22.5.6 Enhanced Urbanization

22.5.6.1 Construction of Brick Kilns on the Bank of the Kopai River

Numerous brick kilns have been established along the Kopai River, particularly in Bolpur, Chakdah, Kankalitala, Amjora, and Ilam Bazar. These kilns significantly affect land, ecology, ecosystems, and public health (Fig. 22.13). The mid-catchment area of the Kopai River, characterized by clay-rich alluvial soil in the floodplains, is well-suited for brick production. Ample sand available along the river's bank and floodplain facilitates brick preparation. Locating brick kilns near the river simplifies the transport of finished bricks by tractors, trucks, or bullock carts to nearby towns

Fig. 22.20 This photograph was taken from the Kultore barrage. It signifies sand mining and serves as a marker for sand-carrying vehicles

such as Bolpur and Suri. Seasonal migrant labourers from Jharkhand and Bihar often work under harsh conditions in these kilns, particularly during the dry season.

The topsoil of farmland is removed, converting fertile areas into desolate brickfields. Excavation near riverbanks undermines the floodplain, creating unnatural pools that trap stagnant water during the monsoon. The extraction of clay and sand compromises the stability of the riverbanks. Rainwater runoff from kilns speeds up erosion, changing the channel structure and heightening flood risk. Many kilns operate using coal, rubber waste, or biomass, emitting PM2.5 and PM10 particles along with sulphur dioxide (SO_2), carbon monoxide (CO), and polycyclic aromatic hydrocarbons (PAHs). This dramatically impacts air quality, particularly in winter and dry seasons. Slurry, kiln runoff, and rain-washed debris pollute the river. Temporary settling tanks or pits frequently overflow, discharging ash and chemicals into water channels.

The establishment and functioning of brick kilns along the Kopai River are tied to socio-economic challenges, including the loss of agricultural land, labour disputes, pollution, and health problems. A permanent loss of agricultural space occurs when landowners struggling financially sell or lease their fertile land to kiln operators. These kilns frequently use bonded or informal labour and have poor safety and health regulations. Child labour and the lack of healthcare and education for workers' families are common problems. Skin conditions, eye irritation, and respiratory issues are common in the communities surrounding these kilns. Vulnerable populations, including women and the elderly, are disproportionately impacted by indoor and outdoor air pollution. The unchecked expansion of brick kilns along the Kopai River best illustrates the tension between development and sustainability. These kilns contribute to economic benefits and job creation, but also seriously and permanently damage public health, land, air, and water resources. It is imperative to balance industrial operations and environmental preservation, guided by science, policy, and community involvement.

22.5.6.2 Encroachment for Commercial Purposes

Encroachment along the banks of the Kopai River in Santiniketan, West Bengal, poses a significant environmental and legal challenge, primarily due to unauthorized commercial developments. In 2024, the district administration of Birbhum intervened to stop an illegal construction project initiated by a Kolkata-based real estate agency (Fig. 22.21, 22.22 and 22.23). The agency was accused of constructing a temple and resort along the riverbank without the necessary permissions from local authorities or the state government. Additionally, residents, academics, and social workers protested another illegal construction at Goalpara on the right bank of the Kopai River. Scholars, environmental advocates, and social workers expressed concerns that such developments could alter the river's natural course, as reported by several reputable newspapers. A geospatial analysis identified unauthorized construction along both banks of the Kopai River from 2020 to 2025. With the area becoming a favoured spot for short trips, real estate agencies and local promoters have established hotels and resorts by the riverbanks. The illegal construction of brick kilns along the banks has also significantly contributed to environmental degradation.

Fig. 22.21 The construction designated for recreational purposes, including a resort and restaurant, is situated on the left bank of the Kopai River, near the Kopai River bridge at Goalpara (Latitude: 23°42′31.56″N, Longitude: 87°40′13.06″E)

22.5.6.3 Market-Induced Sediment Source

The Sonajhuri Haat (temporary market of non-sectoral economic activities), also known as Khoai Haat, is a lively weekly market where local artisans showcase and sell handicrafts that reflect their skills and culture. It occurs five days a week in the Sonajhuri (Ballarpur) forest (Fig. 22.24) near Bolpur-Sriniketan (Santiniketan). For many years, the market operated only on Saturdays, but due to its growing popularity on social media, it now runs every day except Wednesdays and Sundays. It has now become a hot spot for market tourism. Situated near the banks of the Kopai River, this open-air bazaar is nestled among the distinctive red laterite soil formations known as "Khoai," shaped by natural wind and water erosion. In post-pandemic times, similar markets are held daily at Goalpara on the right bank of the Kopai River. The Sonajhuri market showcases vibrant cultural and social assimilation, becoming an economic hub for local people, attracting tourists, and supporting local artisans. Marketers demarcate their shops by creating artificial highlands and rills in the protected forest. This process enhances soil erosion and produces

Fig. 22.22 There are claims of illegal construction along the Kopai River (Latitude: 23°42′27.31″N, Longitude: 87°40′13.00″E). The Birbhum district administration has halted this activity in response to the allegations. The claim states that a real estate agency based in Calcutta is building a temple and resort on the riverbank without the necessary approvals from local authorities or the state government. Subsequently, officials from the state government's land and irrigation departments, along with the Birbhum district magistrate, visited the site and confirmed that there is indeed illegal encroachment on the right bank of the Kopai River

sediment contributing to the Kopai River system. Increased foot traffic and human activity in the forest area gradually significantly affect surface soil quality and the fragile forest ecosystem (Fig. 22.25).

22.5.6.4 Rapid Urbanization in the Outskirts of Bolpur

Rapid urbanization on the outskirts of Bolpur, especially in the Bolpur–Sriniketan region and around the Ballavpur forest area (Sonajhuri market), degrades the Kopai River system. The expansion of settlement areas and swift construction, enhanced by tourism development along the right bank of the Kopai River (Figs. 22.21, 22.22, and 22.23), will be a crucial and harmful element to the natural basin hydrological system. Scientists

Fig. 22.23 Illegal construction along the Kopai River (Latitude: 23°42′27.31″N, Longitude: 87°40′13.00″E)

are highly concerned about the construction that is taking place in the low-lying areas of the right bank floodplain between Ballavpur and Goalpara. This encroachment disrupts the river's natural flow, increasing the likelihood of flooding during the monsoon season. These issues are made worse by poor urban planning, which harms the environment in the Kopai River basin.

The rapid urban growth has intensified erosional processes, causing loss of fertile topsoil and the emergence of gullies, further degrading the river's ecosystem. The westward expansion of the Bolpur-Sriniketan urban area has outstripped the growth of sufficient waste management systems. Every day, the town generates about 95 metric tonnes of solid waste, most of which is disposed of incorrectly, polluting the Kopai River. Aquatic life and water quality are adversely affected by this contamination. Ponds and other traditional water bodies in the Kopai River basin, crucial to preserving the area's hydrological balance, have diminished due to urbanization. Their deterioration has increased the region's vulnerability to droughts and is detrimental to groundwater recharge.

Fig. 22.24 Image of Ballavpur forest. This area is popularly referred to as the Sonajhuri forest and the Sonajhuri marketplaces. The boundary line in the center of the photograph was established by the shop owners of the Sonajhuri market to delineate their spaces for temporary daily shops

22.5.7 Drivers of Pollution in the Kopai River Basin

The Kopai River basin is under increasing environmental stress from various human activities. Key pollution sources include untreated sewage and domestic waste directly discharged into the river and industrial outlets that release diverse pollutants. Additionally, plastic, garbage, and other non-biodegradable materials are dumped in the river. Rapid population growth and urbanization have resulted in insufficient sanitation and sewage systems. A large portion of domestic wastewater from homes near the Kopai River is released directly into it without treatment, resulting in elevated levels of biological oxygen demand (BOD) and chemical contamination. Industries such as dyeing, handicrafts, and brick kilns contribute effluents to the Kopai river system. Point sources of pollution include

Fig. 22.25 The low-lying area of the photograph is a road. Beside the road, a slightly elevated spot is the location of temporary shops. These transformations of the Ballavpur forest land have been made by the temporary (non-sectoral) shop owners of Sonajhuri marketplaces. This is a perfect example of a man-made gully. The orientation of the gully is towards the Khoyai channel and Kopai River, following the local slope

heavy metals (like lead and cadmium) from dye and paint workshops, suspended solids, and acidic discharge from kilns and factories. Furthermore, the disposal of plastic and non-biodegradable waste in the river due to tourism and local markets is significant. Accumulations of plastic waste, such as carry bags, packaging, and bottles, are either dumped in the river or nearby areas, leading to blockages in natural drainage and exacerbating microplastic contamination. The Kopai river basin also lacks formal waste collection and disposal, causing residents to use the riverbanks for garbage disposal.

22.5.8 Cultural and Recreational Practices

Cultural and recreational activities surrounding the Kopai River in Birbhum center on ritual practices, social-religious gatherings, and environmental impacts. As it flows through areas like Bolpur-Sriniketan (commonly called Santiniketan), Kankalitala (Fig. 22.26), Ilambazar, and Charka gram, the Kopai River is closely linked to local customs, spiritual traditions, and folk culture. Its banks often become lively cultural hubs, particularly during festivals and seasonal rituals. The river is a popular location for rituals such as idol immersion, with significant Hindu celebrations including Durga Puja, Saraswati Puja, Vishwakarma Puja, and Chhat Puja. Immersion of idols is a crucial aspect of these festivals, involving clay idols sourced from nearby villages and towns (Fig. 22.27) (e.g., Bolpur, Santiniketan), submerged in the Kopai River. Each idol contains heavy metals (lead, mercury, chromium, etc.) and non-biodegradable materials (plastic flowers, thermocol, synthetic fabrics). The immersion process includes painted idols made with chemical dyes, varnish, and artificial decorations, resulting in severe heavy metal pollution of the river. This contributes to rapid eutrophication and a decline in water quality, endangering aquatic life in the ecosystem. During Chhat Puja, rituals are performed at dawn and dusk, including standing in the river, offering food, lighting oil lamps, accumulating offerings, bamboo structures, and incense remnants, further deteriorating the water quality.

Local fairs and melas occur on various banks of the Kopai River, accompanied by different activities and gatherings. Bolpur, situated on the right bank of the Kopai River, is a popular tourist destination for the renowned Nobel laureate Rabindranath Tagore's Santiniketan. Events like Poush Mela and Basanta Utsav create indirect demand through increased tourism. The Sonajhuri hat, a daily market featuring non-sectoral economic activities in the forest area, operates five days a week near the Kopai River. This marketplace is well connected to the Khoai Gully channel. A significant amount of sediment generated from market activities is added to the Kopai River's sediment budget via the Khoai Gully channel. Kankalitala, known for its prominent Kali temple, draws numerous devotees daily, especially during Pujas. Established by the local government, the Kankalitala Arti ghat on the right bank of the Kopai River hosts Sandhya Arti every evening. Small village melas also occur near the river at Chakra gram, utilizing the riverbanks for temporary stalls, open defecation, cooking, and food disposal.

Folk performances and religious gatherings, such as live Baul Fakiri events, sometimes occur along tranquil riverbanks. While culturally vibrant, these events often suffer from inadequate waste management, resulting in litter in food wrappers, alcohol bottles, and plastic containers. Villagers utilize river ghats for morning baths, laundry, and religious

Fig. 22.26 The photograph was taken in the evening from Kankalitola, adjacent to the Kali temple. The local administrator has established a concrete structure on the right bank of the Kopai River to facilitate the performance of Sandhya Arti, a Hindu practice. A significant gathering is presently occurring to observe these cultural practices

purification rituals, all of which contribute to greywater pollution (from detergents and soaps), litter from plastic sachets, shampoo packets, and erosion of riverbanks due to informal access points that suffer from trampling and slope damage. The environmental degradation caused by cultural activities along the primary channel of the Kopai River has become a pressing issue.

Solid waste accumulation in the river's cross-sectional area, including flowers, idols, plastic bags, cloth, and thermocol left after immersion, harms the river's health. This results in declining water quality, with increased BOD and COD levels due to organic matter and paint residues. Informal bathing at the ghats and trampling of vegetation contribute to the loss of the riparian buffer, leading to habitat loss and a declining ecosystem. Many people remain unaware of the ecological impact of their actions. This lack of awareness, infrastructure gaps such as missing waste bins, eco-friendly immersion tanks, and sanitation facilities, and weak

Fig. 22.27 This photograph was captured from the front of Ballavpur Bridge (Latitude: 23°41′20.29″N, Longitude: 87°38′54.86″E). Situated on the right bank of the Kopai River, this location is popular among locals for bathing, as well as for idol immersion and other cultural customs, including Sradhya and pujas

governance by local panchayats and municipal bodies, lacking sufficient human resources and funds, hampers efforts to maintain riverbank cleanliness. The Kopai River is more than just a hydrological feature; it is a cultural artery of Birbhum, essential to its people's lives, beliefs, and daily rituals. However, without ecological considerations, ritualistic and recreational activities drive the river to degradation. Thus, a community-based cultural-ecological approach is vital to protect this sacred and picturesque river while honouring the traditions of those who rely on it.

22.5.9 Human Intervention on River Morphology

Shape of channel (width, depth, sinuosity), river planform like meandering, braided, straight, river profile (Longitudinal profile and cross-section), and especially all the Floodplain features like levees, terraces, oxbow lakes, etc., are considered as river morphology. The form, structure, and Morphology of the Kopai River have been influenced by several factors like regional Geology, Humid-tropical climate, vegetation cover, Sediment type, and load in the upper catchment and entrainment towards the lower catchments. However, human intervention and anthropogenic factors are responsible for disequilibrium in the dynamic river system, alteration of the catchment and hydrological process, and alteration of ecological behaviour in the Kopai river system. Types of human intervention in the Kopai river system are the installation of dams, disruption and diversification of connectivity, construction erosion and breaching of embankments, channel instability through rapid urban development, and the removal of river bed materials rapidly and unscientifically. Downstream flow velocity has been affected by the construction of two major barrages, a canal, a small check dam, and several bridges on the main channel of the Kopai River. These modes of human intervention have greatly affected discharge patterns throughout the Kopai River. In this context, the lower river catchment faced problems like sediment loss and fragmentation of natural habitat in the riparian zone of the Kopai River (Fig. 22.28).

Flow diversion through the canal negatively impacts the downstream flow velocity of the Kopai River. Diversion of flow through an interconnected canal with other rivers decreases the mean flow of the Kopai River, which affects the fluvial ecosystem and connectivity of this dynamic river system. The embankment construction in the Kopai non-perennial river system is subject to erosion and reconstruction. Construction of embankment and its natural consequences over time limit the natural water behaviour in the cross-sectional area, sedimentation and entrainment, and natural depositional character, especially on flood peak during extra discharge during monsoon and torrential rainfall due to nor'easter and western disturbance. Rapid urbanization, especially on the right bank of the Kopai River, adds extra runoff input with a significant amount of soil and sediment, causing channel instability and alteration of flow patterns. Mining riverbed materials along the entire course of the Kopai River is a paramount contemporary concern. Because it affects the normal sediment behaviour of this river, lowering of the bed, instability throughout the cross-sectional area, and collapse of the bank.

The morphology of the Kopai River is the evolving result of morphodynamics processes over time. In differential connectivity, Morphodynamics Processes of the Kopai River Influence its Morphology. River morphology is the form and structure of a river channel and its features, such as meanders, bars, levees, and floodplains. River morphology is

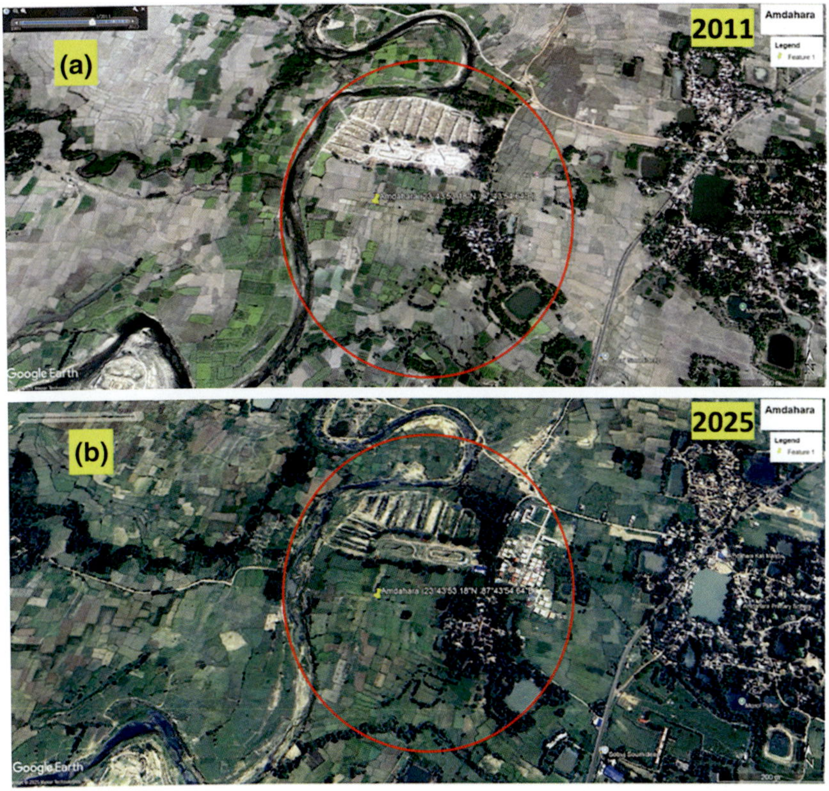

Fig. 22.28 Two temporal Google images show the natural transformation and adjustment due to human modification. The first image (**a**) is from 2011, while the second image (**b**) is from 2025. An active brick line is present in image (a); after the abandonment of the brick line, nature automatically readjusted the area in image (b)

dynamic. It evolves continually in response to morphodynamic processes involving the interactions between flow (hydrodynamics), sediment transport, and boundary materials. Morphodynamics interprets how this transport reshapes morphology. Sand mining from the riverbed enhances the tendency for bank erosion, collapse of the bank, and bed scour in the Kopai River, due to the construction of pillars in the cross-sectional area of the Kopai River in different places. Laying down sediments in slower-flow regions is responsible for forming various channel bars into the Kopai central flow system. Human intervention in the Kopai river system influences the movement and activity of sediment, such as suspended load and bedload. The tendency for the main course of the Kopai River to shift laterally has increased in the last 25 years due to increased human intervention in the catchment area. The position of Cutoffs and the formation of Oxbows indicate unstable conditions in the channel morphology. The lower catchment of the Kopai River experiences these types of signatures. Kopai is a typical meander channel located at the plateau fringe area of the Chhotanagpur. During monsoon, sufficient discharge leads to faster turbulent water flow, causing erosion, flooding, inundation, and sediment deposition in the river migration zone. During pre-dry season, the Kopai River is experiencing a lower flow of running water on the inner bank, which is responsible for maximum deposition into the river bed. Over deposition is enhanced by several human activities in the catchment area, as well as in the river cross-sectional area and the river migration zone. Understanding morphodynamics is crucial for river management, predicting flood risks, controlling erosion, and planning sustainable interventions.

22.5.10 Human Intervention on River Morphodynamics

Morphodynamics of the Kopai River is associated with studying how morphology changes dynamically over time due to interactions between water flow, sediment transport, and human intervention. Human intervention into the Kopai river system is a key concern because it is a major responsible factor for Channel migration, bank erosion, deposition, Bar formation, meander evolution, avulsion, Sediment transport processes (bedload/suspended), and Feedback between flow, sediment, and form (Fig. 22.29, 22.30, and 22.31).

22.5.11 Human Intervention in River Hydrodynamics

River Hydrodynamics of the Kopai River is the fluid flow behaviour in river systems, especially the running water velocity, discharge, turbulence, and energy gradients of in-channel water, which are very much influenced by types of human intervention into the fluvial system. Flow velocity (vertical and horizontal profiles), Discharge rates and annual hydrographs

Fig. 22.29 This comparative satellite image illustrates the changes in the course and surrounding land cover of the Kopai and Bakreshwar Rivers between two dates: the Left Panel shows 07 November 2010, and the Right Panel shows 16 November 2021. The focal point of these images is the confluence area of both rivers in West Bengal. The images are sourced from Google Earth (or similar high-resolution satellite imagery), captured during the post-monsoon season, the ideal time for assessing land use and river morphology. Key observations from 2010 to 2021 indicate an increase in the area and meandering trend of the Kopai River, marked by slight lateral shifts in its course, particularly in the midsection. Notable bank erosion has been detected on the convex side and deposition on the concave side, a classic signature of meander evolution. In addition, the Bakreshwar River appears wider in 2021 compared to 2010, suggesting instances of high flow events, floodplain encroachment, or erosional widening. Some agricultural plots along the riverbanks, seen in 2010, have become more fragmented or indistinct by 2021, potentially due to abandonment, decreased cultivation, increased seasonal flooding, and encroachment of natural vegetation. Agricultural changes in the images show similar trends across the basin area, with the pattern and extent of fields altered by the transition from natural land to human activity. 2010 field boundaries were more distinct than in 2021, indicating land consolidation or irrigation practices. By 2021, both rivers exhibited wider channels and observable sediment islands or bars, particularly near the confluence, which suggests an increase in sediment load, likely from upstream erosion. This reduction in channel confinement has allowed for channel shifting, resulting in measurable changes in the Kopai River's course over the past 11 years, exemplifying active fluvial dynamics and human interactions with the river

Fig. 22.30 Shows the spatial transformation of Ox-bow Lake and the main channel Kopai during 2017–2022

Fig. 22.31 Shows the transformation of the Kopai River island during 2011–2022. Over the past ten years, some patches of vegetation (encircled by red) have been transformed for agriculture and other uses

(flood hydrograph), seasonal Water level variation (depth, stage) and Flow types (laminar, turbulent, subcritical, supercritical) of running water has been influenced by Channel slope and roughness, Cross-sectional area and shape and Upstream catchment hydrology. Humans control channel slope, roughness, cross-sectional area, shape, and upstream catchment hydrology significantly. Hydrodynamics is a key concern for the Kopai river system because it is an essential driver of sediment entrainment.

22.6 Consequences of Human Interventions

22.6.1 *Reduction of Water Flow and Alteration of Natural Courses*

The construction of a dam (Table 22.2) causes a reduction in water flow and alters natural river courses, leading to various mechanisms and consequences. Dams are designed to store water for hydropower, irrigation, and urban supply. During the dry season, the dam decreases flow because it retains water during the monsoon and releases it according to human demands rather than natural cycles. The dam on the Kopai River has resulted in low downstream flow during lean months, a loss of sediment transport, disrupted delta formation, and ecological stress on local flora and fauna that have adapted to natural flow variations. Agricultural expansion encroaches upon the Kopai River, as people reclaim floodplains, straighten meanders, or build embankments near the river (Figs. 22.29 and 22.30). When the Kopai River's natural course changes, its meandering or braided pattern is lost, resulting in channelization and artificial narrowing. This can cause rivers to change their course, which raises the risk of erosion and flooding. Urban growth encroaches on the Kopai River's floodplain, causing wetland areas to disappear and the floodplain's area to shrink. Human activity destroys wetlands in the Kopai basin, which serve as natural filters and hotspots for biodiversity. Human alteration of the drainage basin of the Kopai River diminishes groundwater recharge and water purification capabilities, along with a decline in migratory birds, aquatic life, and local climate regulation. Floodplains are essential buffers that absorb excess water in the river basin, and their disruption adversely affects agriculture and settlements that depend on fertile alluvial soil.

Table 22.2 Mechanism and Consequences of the reduction of water flow and alteration of natural courses

Mechanism	*Consequences*
Construction of a dam	Decrease flow in the dry season
River encroachment by the expansion of agriculture	Alteration of the natural course of the river
River encroachment by urban expansion	Loss of wetland area
	Reduction of floodplain area

Table 22.3 Mechanism and consequences of sedimentation and loss of depth

Mechanism	*Consequences*
Sand mining from the river bed	Reduction of depth
Deforestation in the catchment area	Reduction of carrying capacity
	Increase risk of flooding
	Destruction of the riverine ecosystem
	Sedimentation disturbs aquatic vegetation

22.6.2 Sedimentation and Loss of Depth

The primary causes of sedimentation and the Kopai River's depth loss are described in this section, along with some of their most significant consequences. Sand mining significantly reduces channel depth (Table 22.3) from the riverbed, which alters flow dynamics and results in habitat loss for fish and benthic organisms. A higher sediment load that reduces the channel's carrying capacity results from increased surface runoff and soil erosion brought on by deforestation in the catchment area. Additionally, during periods of high monsoon, there is a greater chance of flooding, which affects photosynthesis and oxygen levels by destroying riverine habitats, upsetting local food webs, and suffocating aquatic vegetation.

22.6.3 Loss of Biodiversity

The main factors contributing to biodiversity loss in the Kopai River basin include habitat destruction (Table 22.4), pollution, and alterations in water flow. Industrial and household waste, alongside nutrient runoff from fertilizers, harm fish populations, leading to their deaths and a decline in invertebrate numbers. Algal blooms result in hypoxic conditions. Additionally, excess silt from upstream erosion and deforestation in the basin suffocates benthic habitats and aquatic plants, diminishing water clarity and negatively affecting visual predators and foragers. Unregulated water extraction for irrigation, along with channel changes caused by sand mining, fragments habitats, creates barriers to fish migration, disrupts seasonal flow patterns, and interferes with breeding cycles. The ecological impact of habitat destruction includes a reduction in native fish spawning grounds and refuge areas and a decrease in nesting sites for riparian birds. This biodiversity loss in the Kopai River basin weakens the fluvial food web and diminishes resilience. Primary consumers, such as aquatic insects and small fish in the river's mid and lower catchment areas, experience declines. In contrast, larger fish and waterbirds at higher trophic levels lose essential food resources. Consequently, the decline of keystone species, like specific fish, hampers the basin's capacity to recover from floods, droughts, or pollution events. Deforestation on the banks of the Kopai River has a greater negative impact on the ecology and biodiversity (Fig. 22.32).

Table 22.4 Mechanism and consequences of loss of biodiversity

Mechanism	*Consequences*
Habitat destruction due to encroachment, pollution, and sedimentation	Decline in populations of native fish, birds, and other aquatic species Disruption of the ecological balance and food web dynamics

Fig. 22.32 This image depicts the right and left banks of the Kopai River at Kankalitola. The image on the left (2011) illustrates the vegetation cover, while the right image (2025) reveals significant deforestation and subsequent transformation. All transformations have occurred following the construction of the check dam near Kankalitola

Table 22.5 Mechanism and consequences of increased flooding and erosion

Mechanism	*Consequences*
Unregulated sand mining reduces the river's ability to manage heavy rainfall	Increased incidence and severity of floods, damaging agricultural fields and local infrastructure
Encroachment on floodplains and natural drainage paths	Severe erosion of riverbanks is threatening nearby settlements and farmlands

22.6.4 Increased Flooding and Erosion

Unregulated sand mining in the Kopai River is causing significant deepening and destabilization of the channel bed. This activity leads to the loss of the natural capacity to attenuate floods, resulting in more frequent and intense flood peaks and threatening the stability of bridges and embankments (Table 22.5). Additionally, due to the conversion of wetlands and paddy fields, the encroachment on floodplains and main channels concentrates flow into narrower corridors, accelerating bank erosion, weakening farm bunds, and damaging nearby homesteads. The impact of unregulated sand mining along the Kopai River is critical, as it intensifies flood peaks and hastens the erosion of banks. Removing natural floodplain buffers directs high-velocity flows against confined banks, leading to soil scouring and tree destabilization.

22.6.5 Pollution and Deterioration of River Health

The Kopai river system is adversely affected by the discharge of untreated sewage and industrial waste (Table 22.6) from nearby settlements, towns, cities, and industries, which often release wastewater that is either untreated or only partially treated directly into the rivers. This wastewater carries organic matter (such as human and food waste), pathogens (including bacteria and viruses), toxic substances (like heavy metals, synthetic chemicals, and dyes), along with suspended solids and thermal pollutants. The Kopai river system consequently faces several problems, such as a reduction in dissolved oxygen (DO), which is essential for aquatic life, an increase in biochemical oxygen demand (BOD), and a deterioration in water quality brought on by high pollutant levels. The basin area is at serious risk for waterborne illnesses like cholera, dysentery, and hepatitis as a result of the tainted water, which is unfit for

Table 22.6 Mechanism and consequences of pollution and deterioration of river health

Mechanism	*Consequences*
Discharge of untreated sewage and industrial waste from nearby settlements	Deterioration of water quality, making it unsafe for drinking, irrigation, or aquatic life
Using chemical fertilizers and pesticides in agriculture leads to runoff into the river	Algal blooms and eutrophication reduce oxygen levels and harm biodiversity

drinking, bathing, or irrigation. Furthermore, the release of industrial waste and untreated sewage disturbs the ecosystem, reduces biodiversity, and threatens regional aquatic food webs.

The Kopai River's health is seriously threatened by using chemical pesticides and fertilizers in agriculture. This practice accelerates algae growth and causes eutrophication due to nutrient overload, especially nitrates and phosphates. This accelerates algae growth and results in algal blooms. When these algae die, their decomposition by bacteria consumes substantial amounts of dissolved oxygen in the water, leading to oxygen depletion. As a result, aquatic life in the Kopai River system increasingly suffers from low oxygen levels (hypoxia), hindering the survival of fish and other species, which ultimately contributes to fish deaths. The biological diversity of the Kopai River ecosystem is deteriorating daily as native aquatic species decline, disrupting the balance of the ecosystem. Furthermore, algal blooms may release toxins (such as those from cyanobacteria), making the water unsuitable for human use and diminishing the overall quality of the river. Nearby agricultural fields use synthetic fertilizers high in nitrogen, phosphorus, and pesticides. During rainfall or irrigation, these chemicals enter waterways through surface runoff or seep into groundwater that feeds into the river.

22.7 Major Findings

22.7.1 Alteration of Channel Morphology

Geospatial observations from 2000 to 2025 reveal significant changes in the main channel of the Kopai River (Figs. 22.7, 22.29, 22.30, and 22.31), including channel narrowing, shifting, and reduced meander intensity. Human activities such as embankments, excessive extraction of natural resources from the surrounding river basin, channelization, and construction projects like brick kilns, bridges, barrages, and sand mining have artificially altered the river's natural path and decreased its fluvial dynamism.

22.7.2 *Impact of Sand Mining*

High-resolution imagery shows sand extraction pits, vehicle access routes, and eroded riverbeds. Field surveys confirmed uncontrolled sand mining along the river, especially close to Kankalitala and Ahmedpur. This practice has altered sediment transport and deposition patterns, causing bank erosion, habitat disruption, and riverbed degradation.

22.7.3 *Decline of Riparian Vegetation*

Satellite images and field surveys reveal a significant decrease in natural vegetation and a decline in plant cover along riverbanks. Urban development, agricultural encroachment, and sand mining have destroyed ecological buffer zones, increasing the river's susceptibility to siltation and flash floods.

22.7.4 *Deterioration of River Health*

Degradation of the water quality in the Kopai River is primarily caused by pollution. Untreated domestic wastewater discharge and solid waste dumping were discovered at multiple locations during ground-truthing. Particularly during the lean season, these activities have lowered water quality, caused eutrophication, and reduced aquatic life.

22.7.5 *Human Encroachment and Urbanization*

The floodplain's built-up areas have significantly increased, especially in the vicinity of Bolpur and Illambazar. The natural storage and infiltration capacities of the fluvial system have been reduced due to the encroachment of illegal construction and urban sprawl on floodplains. Urbanization and sand mining are two important anthropogenic drivers that have been identified.

22.7.6 *Vanishing Traditional Water Bodies*

Numerous historic ponds and canals connected to the river have dried up or filled in. Due to this loss, the main river is under more stress during the monsoon and dry seasons, interfering with seasonal recharge cycles and hydrological connectivity.

22.7.7 *Reduction of Baseflow of the Kopai River and Its Dry Season Fragmentation*

Satellite imagery captured during dry months displays fragmented stretches of water and disconnected pools. Factors such as groundwater depletion, upstream check dams, and climate variability have diminished baseflow, impacting ecological and human water requirements.

22.7.8 *Cultural and Livelihood Impacts*

Local interviews revealed decreased fishing activities and alterations to river-centric festivals, such as the Poush Mela River rituals. The river system's degradation has affected socio-cultural practices and traditional livelihoods dependent on the river.

22.7.9 *Agricultural Pressure*

Expanding paddy fields and vegetable farming along the riverbanks rely heavily on groundwater irrigation. Excessive water extraction and fertilizer runoff have led to chemical pollution and decreased soil fertility in nearby areas. Intensive agricultural practices that depend on the Kopai River have caused significant soil erosion and sediment accumulation in its narrow course. Additionally, runoff from fertilizers and pesticides has harmed the fluvial ecosystem and riparian vegetation.

22.7.10 *Gaps in Governance and Policy Implementation*

There are conspicuously few monitoring stations or checkpoints along the river's most vulnerable areas. This suggests that environmental regulations and river management policies are not sufficiently regulated and enforced, allowing unsustainable practices to continue unchecked.

22.7.11 Vulnerability of the Kopai River

The Kopai River is susceptible to erosion and seasonal floods because it is a rain-fed river with permeable soil. These vulnerabilities have been exacerbated by human activities such as the building of dams, concrete bridges, and seasonal flood control measures, which have disturbed the Kopai River's typically narrow channel's natural meandering and flow.

22.8 Significance of the Study

This study is essential for a comprehensive and scientific understanding of how human activities affect river systems. Encroachment, land-use changes, unregulated sand mining, wastewater discharge, pollution, and agricultural growth along its banks are just a few of the human-induced stresses that the Kopai River, a minor tributary of the Mayurakshi in the Ganga River system, is subject to. The study looks at how erosion, sedimentation, channel changes, and ecological degradation are caused by human activity interfering with natural fluvial processes. High-resolution free satellite data have been used to track changes in the river's morphology, hydrodynamics, morphodynamics, and the state of the floodplain over the past few decades to analyse these effects. Time series analysis using satellites is economical and practical for mapping, recording, and quantifying spatial patterns of human interference.

This study seeks to establish a comprehensive support system for evidence-based environmental assessment and decision-making, utilizing scientific maps and reports. All research findings have been confirmed through on-site verification. The river planform is analysed using remote sensing data, complemented by ground-truthing via GPS, photography, and resident interviews. Therefore, this study sheds light on changes at the micro level, like illicit activity or small land-use conversions that might not be visible in satellite imagery. Furthermore, by helping local administrators identify crucial areas of river degradation, the findings improve the regional environmental management system. NGOs and civil society can use the research's thorough understanding of sand mining hotspots to advocate for conservation and restoration initiatives.

This study is highly relevant to Climate Change and River Health. It holds academic and methodological significance, among other factors.

The study integrates technology and field research within fluvial geomorphology and environmental science. It offers a multidimensional examination of how human activities alter natural river systems and their landscapes. Connecting technology (like remote sensing), fieldwork (including ground-truthing), and applied geography emphasizes environmental issues and provides actionable insights for sustainable river basin management. The present study can provide a replicable model for assessing small river systems under stress. It can be considered a significant contribution to river restoration efforts under administrative water policies and different-scale watershed management. The study can enhance the understanding of human–environment interactions in semi-arid river basins.

22.9 Conclusions and Recommendations

The catchment of the Kopai River is located in the highly erosional plateau fringe region. The Kopai basin is also considered a lateritic riverine landscape. A remarkable transformation in the fluvial system of the Kopai River has occurred due to human interventions. The present study, grounded in integrating high-resolution free satellite imagery and meticulous ground-truthing, reveals that anthropogenic pressures such as indiscriminate sand mining, land-use changes, embankment construction, and urban expansion exist. The morphology, hydrology, catchment process, hydrodynamics, and morphodynamics of the Kopai River have been significantly altered due to human intervention in the fluvial system over the past 50 years. The signature of anthropogenic activity on the ecological integrity of the Kopai basin has been noted. Spatio-temporal analysis of high-resolution free satellite data helps in the detailed visualization and interpretation of channel shifts, sedimentation patterns, and Bankline erosion. Post-field ground-truthing was conducted during field surveys, which validated and enriched these observations in the socio-environmental context. According to the core research findings of the present study, implementing a balanced approach to river basin management harmonizes human developmental needs with the ecological sustainability of the Kopai fluvial system. This study's methodology adaptation aims to show how useful freely available geospatial technologies are for tracking the health of rivers and directing policy interventions at the local level. Fostering community involvement and implementing informed conservation strategies are crucial for ensuring the resilience and restoration of this ecologically and culturally significant river system, as the Kopai River continues to face increasing environmental stress.

The main goal is to restore the Kopai River's natural flow while reducing human interference with its ecosystem. Policymakers and local officials must prioritize habitat conservation, improve water management, enforce pollution controls, encourage sustainable farming practices, and responsibly remove bed materials from the river. Implementing scientifically regulated sand mining in the Kopai River system will help reduce excessive sediment buildup and preserve the river's natural depth and cross-sectional area. Recommendations for maintaining a Cultural-Ecological Balance in the Kopai River are also feasible. Implementing Eco-friendly Idol Guidelines will involve using natural dyes, clay, and biodegradable decorations for idol preparation. Furthermore, specific Immersion Zones along the Kopai River should be designated for safe idol immersion, and volunteer-led cleanup initiatives should be encouraged by involving NSS/NCC students, local youth, and environmental NGOs post-rituals.

Campaigns should be sensitive to these aspects, considering the local culture and incorporating their language into Baul song media. During festival times, it is crucial to establish temporary toilets, waste bins, and water filters. Sustainable management practices regarding the Kopai River basin area are strongly recommended, including preserving vegetation in Riparian Buffer Zones to mitigate runoff and erosion. Moreover, promoting sustainable agriculture is equally important. It can be achieved by endorsing organic farming, integrated pest management (IPM), and water-saving irrigation methods like drip or sprinkler systems. Minimizing mono-cropping ventures, utilizing trees for soil erosion control, and changing employment opportunities are also important. Active encouragement of watershed-based approaches and participatory irrigation management needs to take place. Concerning local panchayats and district governance, enforcing control over unauthorized constructions will make land use planning and pollution management more effective in safeguarding the riverbanks. For community policies, investment in modern waste collection and treatment facilities is crucial to managing pollution in rivers and water bodies.

References

Begum, N., & Mandal, S. (2015). Survey of plants in the rural belt of Birbhum district, West Bengal concerning pollination calendar. *International Journal of current microbiology and applied sciences, 4*, 1118–1131. https://www.researchgate.net/publication/321061178_Survey_of_Plants_in_the_rural_belt_of_Birbhum_District_West_Bengal_with_reference_to_Pollination_Calendar

Bishop, M. P. (2013). Remote sensing and geomorphometry for studying relief production in tectonically active mountain belts. *Geomorphology, 190*, 3–13. https://doi.org/10.1016/j.geomorph.2013.01.020

Biswas, R., Pal, S., Bhattacharyya, S., & Mandal, K. (2020). OLS regression-based causal modelling for flood management: A study on river Mayurakshi at Tilpara barrage section, Suri, Birbhum. *Journal of the Indian Chemical Society*. https://indianchemicalsociety.com/portal/uploads/journal/OctoberB14.pdf

Das, N. (2014). Seasonal hydrological characteristics of Langolhata Wetland of Birbhum district and its impact on agriculture. *Indian Journal of Geography and Environment, 13*, 143–150.

Das, P. (2016). Morphometric diversity on the Kuya River Basin. *International Research Journal of Earth Sciences, 4*(4), 17–28. International Science Community Association. https://www.isca.me/EARTH_SCI/Archive/v4/i4/3.ISCA-IRJES-2016-010.pdf

Das, M., & Mukhopadhyay. S. (2015a). Number, size, distribution, and status of ponds in Kopai River Basin, Eastern India. *The International Journal of Geomatics and Geosciences, 5*.

Das, N., Mukhopadhyay, S. (2015b). Status of Ground Water Hydrology of Labpur block, Birbhum District. *International Journal of Geology, Earth & Environmental Sciences, 5*(3), 19–32. https://www.cibtech.org/J-GEOLOGY-EARTH-ENVIRONMENT/PUBLICATIONS/2015/VOL_5_No_3/03-JGEE-003-NILADRI-STATUS.pdf

Das, T., & Mukhopadhyay, M. (2022). Effect of Brick Field on Riverine Morpho-hydrological Environment: Study in Kopai River of South Bengal Plateau Fringe. *Research Review: International Journal of Multidisciplinary*. https://doi.org/10.31305/rrijm.2022.v07.i08.011

Ghosal, J. (2018). Settlement archaeology of Kopai River, Birbhum District, West Bengal: A preliminary study. *Heritage: Journal of Multidisciplinary Studies in Archaeology, 6*, 575–593. https://www.heritageuniversityofkerala.com/JournalPDF/Volume6/24.pdf

Ghosh, K. G., & Pal, S. (2015). *Impact of dam and barrage on flood trend of the lower catchment of the Mayurakshi River basin, eastern India*. European Water, 50, 3–23. © 2015 E.W. Publications.

Ghosh, K. & Pal, S. (2016). *India Impact of dam and barrage on flood trend of the lower catchment of the Mayurakshi River Basin, Eastern India*, 50, 3–23. https://www.researchgate.net/publication/340446806_India_Impact_of_dam_and_barrage_on_flood_trend_of_lower_catchment_of_Mayurakshi_river_basin_eastern_India

Ghosh, S., Das, R., & Roy, N. (2018). River response to human interventions: Case studies from West Bengal. In S. Ghosh (Ed.), *Anthropogeomorphology of Bhagirathi-Hooghly River System in India* (pp. 1–20). CRC Press. https://doi.org/10.1201/9781003032373

Gregory, K. J. (2006). The human role in changing river channels. *Geomorphology, 79*(3–4), 172–191. https://doi.org/10.1016/j.geomorph.2006.06.018

Islam, A., Barman, S. D., Islam, M., & Ghosh, S. (2020). Role of human interventions in the evolution of forms and processes in the Mayurakshi River Basin. In S. Ghosh (Ed.), *Anthropogeomorphology of Bhagirathi-Hooghly River System in India* (1st ed., 53 pp.). CRC Press. https://doi.org/10.1201/9781003032373

Leopold, L. B., Wolman, M. G., & Miller, J. P. (1964). *Fluvial processes in geomorphology*. W.H. Freeman and Company.

Lillesand, T., Kiefer, R. W., & Chipman, J. (2015). *Remote sensing and image interpretation* (7th ed.). Wiley.

Ministry of Environment, Forest and Climate Change. (2019, October 26). District Survey Report (DSR) of [relevant district if known]. Environment Clearance Portal, Government of India. https://environmentclearance.nic.in/writereaddata/Online/EDS/0_0_26_Oct_2019_1901184331Annexure-DSR.pdf

Mondal, B. K., Mukherjee, I., & Saha, S. (2022). Analyzing morphometric attributes of Kopai River Basin of West Bengal, India, Using Geospatial Technology. In P. K. Shit, B. Bera, A. Islam, S. Ghosh, & G. S. Bhunia (Eds.), *Drainage basin dynamics. Geography of the physical environment*. Springer. https://doi.org/10.1007/978-3-030-79634-1_11

Nag, S., Roy, M. B., Bhattacharya, S., Mondal, S., & Roy, P.K. (2021). Assessment of topographic complexity zone of a drainage basin using geographic information system, In P. K. Roy et al.(Eds.), Advances in water resources management for sustainable use (Chapter 32, Lecture Notes in Civil Engineering 131, pp. 441–456). https://doi.org/10.1007/978-981-33-6412-7_32

Roy, S., & Datta, D. K. (2015). Human impacts on the fluvial system of Rarh Bengal: A geomorphological appraisal. *Transactions, Institute of Indian Geographers, 37*(1), 113–126.

Rudra, K. (2010). The encroaching Ganga and social conflicts: The case of West Bengal, India. In M. A. Stewart & D. K. Sinha (Eds.), *Climate change and water resources* (pp. 217–230). CRC Press.

Schumm, S. A. (1977). *The fluvial system*. John Wiley & Sons.

Santos, J. C. N. D., Palácio, H. A. D. Q., Andrade, E. M. D., Meireles, A. C. M., & Neto, J. O. D. N. (2011). Runoff, soil, and nutrient losses in semiarid uncultivated fields. *Revista Ciencia Agronomica*. https://doi.org/10.1590/s1806-66902011000300030

Walia, H., Rana, A., & Kansal, V. (2018). *A supervised approach to Gurmukhi word sense disambiguation using K-NN method*. https://doi.org/10.1109/confluence.2018.8442616

Wang, Y., Xia, J., Cai, W., Liu, Z., Li, J., Yin, J., Zu, J., & Dou, C. (2023). Response of Fish Habitat Quality to Weir Distribution Change in Mountainous River Based on the Two-Dimensional Habitat Suitability Model. *Sustainability, 15*(11), 8698.

CHAPTER 23

The Interplay of Natural Denudation and Anthropogenic Activities in the Garhbeta Badlands (Silabati River, West Bengal): A Comprehensive Analysis of Geomorphic Changes over the Last Decade

Lopamudra Ganguly, Hasiba Nasrin, Subhadip Mondal, Saptadipa Chakraborty, and Ranjita Mondal

23.1 Introduction

Geomorphology, as the science of landforms and their shaping processes, plays a critical role in understanding and countering environmental degradation, especially in sensitive landscapes such as badlands. These areas, characterized by strong erosion of loose sedimentary deposits, steep slopes, and sparse cover, are extremely vulnerable to natural as well as human-induced disturbances. Garhbeta Badland in West Medinipur,

L. Ganguly (✉) · H. Nasrin · S. Mondal · S. Chakraborty · R. Mondal
Department of Geography, Vedanta College, Kolkata, India
e-mail: ganguly.geography@gmail.com

S. Ghosh (ed.), *Reimagining Indian Rivers for Sustainability*, Palgrave Studies in Environmental Sustainability,
https://doi.org/10.1007/978-3-032-04754-0_23

West Bengal, is such a fragile ecosystem, located within the Rarh Plain, which is an ecotone region between the Chhotanagpur Plateau and the Bengal Basin. The term Rarh (Sanskrit: Roordha, meaning "uneven") is indicative of the terrain's ruggedness, molded by Pleistocene-age lateritic formations and the dense web of rills and gullies provided by west-flowing parallel drainages, such as the Shilabati River (Bandyopadhyay & Das, 1987, 1988; Chakrabarti & Mandal, 2019).

While the dynamic geomorphology of the Shilabati basin has been subjected to natural forces since time immemorial, the denudation rates have been subjected to enormous acceleration by anthropogenic activities like laterite mining, deforestation, unsustainable agriculture, and uncontrolled water diversion. Overgrazing and deforestation for fuelwood have denuded protective vegetation cover and lichen crusts (Gallart, 2013), while mining and inappropriate flood-control works have intensified gully erosion and sediment outflow (Mascherini et al., 2011). Countrywide, India loses about 175 million hectares to erosion, with riverine erosion alone contributing 3.97 million hectares (Tripathi & Singh, 1990); the badlands of the Shilabati basin alone contribute massively to this continuing degradation.

The goal of this research is to quantify anthropogenic disturbances on the Shilabati River's geomorphic dynamics by combining remote sensing (RS), GIS-based terrain analysis, and field measurements for: (1) quantification of spatiotemporal erosion pattern variations (e.g., rill-gully expansion, sediment yield), (2) determination of thresholds of human-altered geomorphic resilience, and (3) development of targeted mitigation solutions for sustainable land management. By linking geomorphic process theory with practical environmental management, this study seeks to inform policy interventions that can stem further degradation in the Garhbeta Badland and similar fluvial systems.

23.2 Significance of Studying Human Impacts on Landscape Evolution

The modern civilization has become the main agent of Earth's surface change, with human activities now dominating over natural geological processes in landscape reshaping (Poesen, 2018a, 2018b). This fact renders the research on anthropogenic geomorphic effects not only academically relevant but also critical to planetary sustainability. West Bengal's Garhbeta badlands provide a grim case study wherein fragile natural systems intersect with human survival requirements, triggering an environmental crisis with wide-ranging implications (Bhattacharya, 2023a, 2023b; Chakrabarti & Mandal, 2019).

23.2.1 The Accelerating Crisis of Landscape Degradation

Where nature operates through slow processes over millennia, human activities squeeze geological time scales into three decades. West Bengal's lateritic badlands, naturally susceptible by virtue of porous soils and weak vegetative cover, are threatened with extinction these days by pervasive mining, felling of forests, and ecologically destructive cultivation (Ghosh et al., 2022a, 2022b). A century's work of soil construction is washed away by each monsoon in weeks, while mines irreversibly change drainage directions that took centuries to create (Bridges & Harding, 2016a, 2016b). These interventions do not just alter the landscape—they literally rewrite the geomorphic evolution rules.

23.2.2 Cascading Ecological and Social Impacts

The disturbance expresses itself through interrelated crises: rivers choked with silt that used to support agriculture now flood unpredictably; centuries-old farming terraces collapse under increased erosion; and microclimates change as vegetation loss modifies evaporation regimes (Nadal-Romero et al., 2018a, 2018b). These environmental shifts for local populations mean lost livelihoods, as farmers and foresters see declining incomes from degraded lands. The loss of biodiversity reaches beyond the easily seen species to small microscopic soil organisms that sustain the very underpinning of terrestrial ecosystems (Gallart, 2013).

Analysis of the impacts of humans on landscape development has become increasingly pertinent within the Anthropocene, given that anthropogenic powers now rival or surpass natural processes in remaking the surface of the Earth (Poesen, 2018a, 2018b). The Garhbeta badlands located on the Shilabati River of West Bengal offer a valuable case where vulnerable lateritic ground meets strong human pressures, forming an excellent natural laboratory for investigating the interactions of humans with geomorphology. Where millennia would be the typical time required for natural erosion processes, human interventions such as mining, deforestation, and extractive agriculture have significantly hastened landscape transformation—monsoon rains now remove centuries of soil build-up in a single growing season, whereas mining activity irreversibly changes drainage patterns that took millions of years to develop (Mascherini et al., 2011). These disturbances cause cascading impacts: elevated sediment loads clog river channels and increase flood hazards, degraded soils lower agricultural yields, and changed microclimates destabilize local ecosystems.

Our study uses an integrated approach that integrates remote sensing, field surveys, and local knowledge to quantify these changes and formulate sustainable solutions. Results show that focused interventions such as native species revegetation, modified agricultural practices, and geomorphology-based zoning can stabilize these vulnerable landscapes. The Shilabati Basin's lessons extend globally, offering insights for managing human–landscape interactions in vulnerable environments worldwide while maintaining the delicate balance between resource needs and environmental preservation. This work ultimately highlights how understanding anthropogenic geomorphic processes is not merely academic, but essential for developing strategies that allow human societies to thrive without compromising the physical systems that sustain them.

23.3 Problems to Be Addressed

Earth incorporates land as its fundamental natural resource since it supports both the establishment of terrestrial ecosystems and human settlements. Soil acts as Earth's upper crust and it enables food production through 90% of the worldwide total. Human-made activities have caused erosion rates to surge 15 to 50 times worldwide which harms the quality of soil and destabilizes ecosystem environments.

Pear-shaped escarpments named Badlands experience extensive erosion through formations such as rills and gullies and develop unique geological land features which renders them highly prone to human-caused environmental damage. The natural destruction of these landscapes occurs through interactions of vegetation absence together with geological structure and climatic elements. Human activities which use improper agricultural practices and engage in deforestation and operate without regulation serve to speed up the decline of these sites.

The Shilabati River in West Bengal hosts the Gangani Badland which currently faces severe environmental crisis. Visitors experience the spectacular lateritic geological wonders of this Pleistocene-era formation which includes escarpments as well as caves and pinnacles and earth pillars. The scientific and educational importance along with geotourist value of Gangani requires attention because various threats are intensifying.

1. The combination of tourist footprints and construction projects leads to quick soil loss of weak geological structures.
2. The economic activities which remove vegetative cover create soil degradation by generating more surface runoff.
3. The landforms that visitors do not regulate get destroyed.

This research demonstrates the essential requirement for development of harmonized conservation plans which:

- Implement sustainable geotourism practices
- Establish erosion control measures
- Promote community-based land management
- Develop policies protecting geoheritage sites

Analyzing Gangani's one-of-a-kind badland system provides essential information about managing geological preservation combined with human participation. These findings show how protected natural landscapes can be safeguarded effectively by committed management which creates sustainable geotourism opportunities.

23.4 Literature Review

Scientific study of badland geomorphology has seen tremendous paradigm changes since the path-breaking studies of Bryan (1922) and Schumm (1956), who initially proposed badlands as natural laboratories for the investigation of erosional processes. Modern studies depict these dynamic landscapes as intricate systems where natural geomorphic processes meet intensifying anthropogenic pressures (Poesen, 2018a, 2018b). Internationally, research proves that human activity has altered badland evolution processes at various spatiotemporal scales worldwide, with the most extreme effects in developing areas where economic necessities tend to trump environmental considerations (Nadal-Romero et al., 2018a, 2018b).

The Mediterranean provides some of the most well-documented instances of anthropogenic badland alteration. Studies in the Spanish Pyrenees (García-Ruiz et al., 2013) and Italian Appennines (Torri et al., 2013) show how intensification of agriculture and deforestation over centuries have increased erosion rates by 200–500% over pre-anthropogenic levels. Likewise, in the North American region, the Badlands National Park is a remarkable example where past overgrazing and mining operations have irrevocably changed drainage patterns and sediment transport regimes (Bridges & Harding, 2016a, 2016b). These international examples define clear patterns of anthropogenic geomorphic change that are essential context for interpreting analogous processes in the Indian subcontinent.

The Garhbeta badlands of West Medinipur district are a key case study in interpreting the complex interaction between anthropogenic processes and geomorphic processes in tropical lateritic landscapes. Located in the more extensive Chotanagpur Plateau periphery, the area has been subject to rapid landscape change that typifies the worldwide trend toward human-driven badland extension. Global research in Mediterranean (García-Ruiz et al., 2013), North American (Bridges & Harding, 2016a, 2016b), and Asian (Joshi & Nagare, 2019) badland environments has shown that human processes generally increase natural rates of erosion 15–50 fold, with especially costly consequences in developing countries where population pressures overlie susceptible ecosystems (Nadal-Romero et al., 2018a, 2018b).

The Pleistocene-aged lateritic badlands of Garhbeta (Das et al., 2020) exhibit distinctive geomorphic responsiveness to human disturbances because of three factors: (1) the existence of a weak duricrust layer that acts as a vital protective cover, (2) high monsoon precipitation regimes that enhance erosional processes, and (3) the socioeconomic environment of the region that leads to intensive land use. Studies published recently have estimated the extent to which certain human activities are remolding the geomorphology of Garhbeta: mining activities in laterite have accelerated gully headcut retreat by 0.8 m/yr to 1.6 m/yr (Bhattacharya, 2023a, 2023b); rainfed agriculture on marginal slopes has increased rill network density by 35–55% since the early 2000s; and fuelwood collection through deforestation has depleted vegetative cover by about 60%, largely reducing the infiltration capacity of the landscape (Ghosh et al., 2022a, 2022b).

These human-induced pressures have significantly changed Garhbeta's denudation system, producing quantifiable changes in sediment dynamics.

Sediment yield monitoring shows a sharp rise from 2.3 t/ha/yr during the 1990s to 8.7 t/ha/yr in recent times (Mandal, 2023a, 2023b), with downstream implications for river systems and agricultural productivity. The altered hydrological regime now has reduced concentration times and increased peak discharges during monsoon events, intensifying gully formation and slope instability. Most worrying are the self-perpetuating degradation cycles, which result from an initial erosion brought about by deforestation or mining to create routes of access to promote further land use change and encroachment.

The Garhbeta example poses special management problems that make it distinct from more intensively studied badland systems. Unlike the semi-arid badlands of Rajasthan or Gujarat, the tropical monsoon climate of Garhbeta induces specific seasonal regimes of erosion and recovery. Also, the lateritic substrates of the area react differently to erosional power than do the alluvial badlands of the Chambal Valley. These considerations, together with West Bengal's extremely high population density and reliance on subsistence agriculture, make up a complex socio-geomorphic system that necessitates context-sensitive solutions.

This research fills some key research gaps in anthropogenic geomorphology: (1) the accurate quantification of tipping points in lateritic badland systems, (2) interaction effects among many concurrent stressors (mining, agriculture, deforestation), and (3) the effectiveness of traditional versus engineered stabilization practices in tropical lateritic environments. Methodological advancements such as high-temporal-resolution UAV surveillance (0.05 m accuracy), innovative sediment fingerprinting methods through the use of rare earth element tracers, and participatory GIS mapping with local stakeholders enable unparalleled recording of human-landscape interactions.

The study adds to theoretical and practical aspects of geomorphology. Theoretically, it provides insight into threshold dynamics in human-altered lateritic systems and the notion of anthropogenic signatures in landscape evolution. In practice, it formulates evidence-based management models that reconcile ecological protection with socioeconomic interests, providing solutions that can be scaled up for similar lateritic badlands elsewhere in Southeast Asia and the global tropics. By combining physical process research with social science methods, the research offers an integrated model for tackling one of the most urgent problems in Anthropocene geomorphology: managing fast landscape change in fragile ecosystems under rising human pressure.

Geomorphology and landscape management applied. The interdisciplinary nature of the study offers a model for explaining human–environment relations in vulnerable landscapes globally, especially in the developing world where economic stress and environmental protection tend to come in opposition. In the end, the research contributes to both scholarly understanding and applied solutions for one of the most urgent challenges facing modern geomorphology: controlling accelerated landscape transformation in the Anthropocene era.

23.5 Study Area

The location map (Fig. 23.1) depicts Garhbeta CD Block-I's geographical position using successive levels of geographical scope which start from national to local administrative divisions. The location map starts with India's national presentation that identifies West Bengal through green illumination to show the positioning of the study area in eastern India. On the top-right section of the display stands a more precise geographic representation of West Bengal specifically showing the West Medinipur district situated within the state limits.

The map of West Medinipur district takes center stage at the bottom-right corner where Garhbeta CD Block-I obtains specific highlighting to depict the most confined study area scope. A purple zoomed-in map of Garhbeta CD Block-I appears in the bottom-left corner to define precisely the researched area where geomorphic and anthropogenic investigations occur. The progressive scale of maps in the figure establishes spatial understanding by showing how Garhbeta CD Block-I fits into national, state, and district territories. The comprehensive cartographic display helps identify the research area from geographical and policy-oriented administrative perspectives.

23.6 Methodology

This study draws upon a multi-perspective methodological framework containing both remote sensing and field-based investigations and GIS-based spatial analyses. The study will focus on understanding how human intervention has influenced the hearing and denudational processes and characteristics of the geomorphology in the Garhbeta Badlands. The remote sensing analysis process utilized various satellite images on the landscape, beginning with Landsat OLI (2022) and progressing to IRS

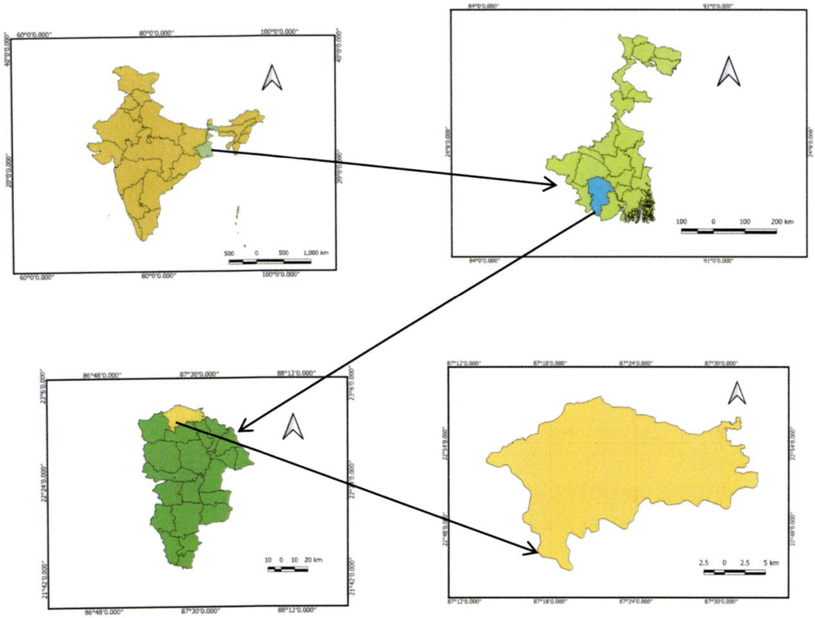

Fig. 23.1 Study area map of Garhbeta-I C.D Block

LISS-III (1995, 2000) and Google Earth to identify multi-temporal land-uses and vegetation cover change. The NDVI values were used to assess vegetation degradation. False Color Composite (FCC) images were utilized to visually identify the changes in the landscape and surface pattern characteristics. Digital elevation models (DEM) using SRTM (2014) and TM DEM datasets were extracted to build slope, aspect, and elevation maps. Terrain metrics were derived to provide TWI and SPI metrics for hydrologic analysis of behavior and erosive potential. The RUSLE (Revised Universal Soil Loss Equation) digital model for estimating soil erosion was obtained by using values of rainfall erosivity (R), soil erodibility (K), slope length and steepness (LS), cover management (C) and conservational practice (P) in GIS analysis. The soil maps were provided by the NBSS & LUP, while climate data and land use data were used as input, to further assist in the RUSLE model construction.

The Revised Universal Soil Loss Equation model (RUSLE) was used in GIS to determine soil erosion rates. Since soil erosion is not an isolated phenomenon each piece of research and physical site search illustrated the biophysical variables involved in calculating annual average soil loss. The research incorporated the biophysical references: (i) the R-factor (rainfall erosivity, the R-factor was derived from Indian Meteorological Department rainfall data from 2020 to 2024; (ii) the K-factor (soil erodibility, K-factor soil erodibility was derived from the NBSS & LUP, soil texture datasets; (iii) LS-factors, based on slope length and slope steepness and taken from 30m SRTM DEM (2014); (iv) C-factor (Cover management) was estimated using classified NDVI taken from Landsat OLI; and (v) P-factor (conservation practices) based on field (on the ground) observations and current land use type (Cartographic). The RUSLE-derived erosion risk map at the gully is validated through sediment yield measurements taken in the field along with visual gully assessments. Gully erosion and morphometry were examined using time series analysis of satellite data and verified with GPS-enabled field surveys and photography over various time codes. Headcut retreat/advancement, gully length, gully depth, gully lateral extent was geo-referenced and recorded from 2010 to 2023. UAV-derived high-resolution topographic data supplemented the visual understanding of slope retreat and scarp degradation following extreme weather stimulation and the visual groundwater incidence. In addition, Digital Elevation Models (DEM) was used to derive slope, aspect, Topographic Wetness Index (TWI) and Stream Power Index (SPI) and represent the assessments of hydrological behaviors and characteristics that suggested erosion susceptibility.

Gully and rill morphometry were studied through multi-date satellite images with field GPS surveys to monitor gully headcut retreat, depth and width changes, and expansion of the overall gully network. Where possible photogrammetric and UAV-derived data was also used to characterize surface morphologies at various stages including scarp retreat and development of rills. To validate the remote sensing analyses, there were extensive field surveys involving slope gradients, sediment yields, soil moisture, and checks for geomorphic features. Semi-structured interviews with local residents provided qualitative detail to illuminate local-scale human land use and understand residents' perspectives on environmental change. Climate data from 2020 to 2024 was obtained from the India Meteorological Department (IMD) and supplemented with field-based meteorological measurements. These data on temperature, humidity and rainfall patterns were analyzed in relation to land degradation metrics to investigate potential microclimatic changes corresponding to human-induced changes in landform geomorphology. In summary, the comprehensive methodology allowed for a detailed understanding of how natural systems of geomorphology are being reshaped under human pressure, while providing scientific evidence that is fundamental to sustainable land management practices in the Shilabati River Basin.

The analysis (Fig. 23.2) method for human-caused activity effects on geomorphic features and erosion processes at the Garhbeta Badland area involved a step-by-step multi-dimensioned procedure. The data collection phase starts with both authentic secondary data combined with remote sensing data. The numerous datasets merge together to produce thematic maps which address geomorphic and environmental parameters including morphology, slope, gully density, soil composition, land use–land cover, and hydro-geomorphology.

The research team subsequently performs quantitative examinations of land degradation characteristics such as scarp retreat and analysis of badlands size and measurements of gully headwalls, rill channel fans and sidewalls alongside gully and rill profiling and soil property and erosion rate evaluation. The chosen metrics deliver detailed awareness about the active procedures shaping the badlands terrain. The examination at level two considers three core elements which include evaluation of both natural and built environment on badland development along with village area influence from gully erosion and human-based causes for heightened soil degradation. The comprehensive evaluation displays both the basic weaknesses present at the site and the additional stress which human activities create to worsen land degradation.

The research uses its collected data to present recommendations about how to control soil erosion while recovering badlands areas. The efforts use environmentally-friendly methods of bioengineering to achieve sustainable environmental management. The new model incorporates ecological factors together with geomorphological elements to develop sustainable resilience for vulnerable badland ecosystems.

23.7 Details of Data

Secondary data (Table 23.1) served as the primary data source for this study which was obtained through remote sensing analysis of various reliable institutional sources. High-resolution Google Earth images and Google Earth View serve as foundational visual references. The land cover

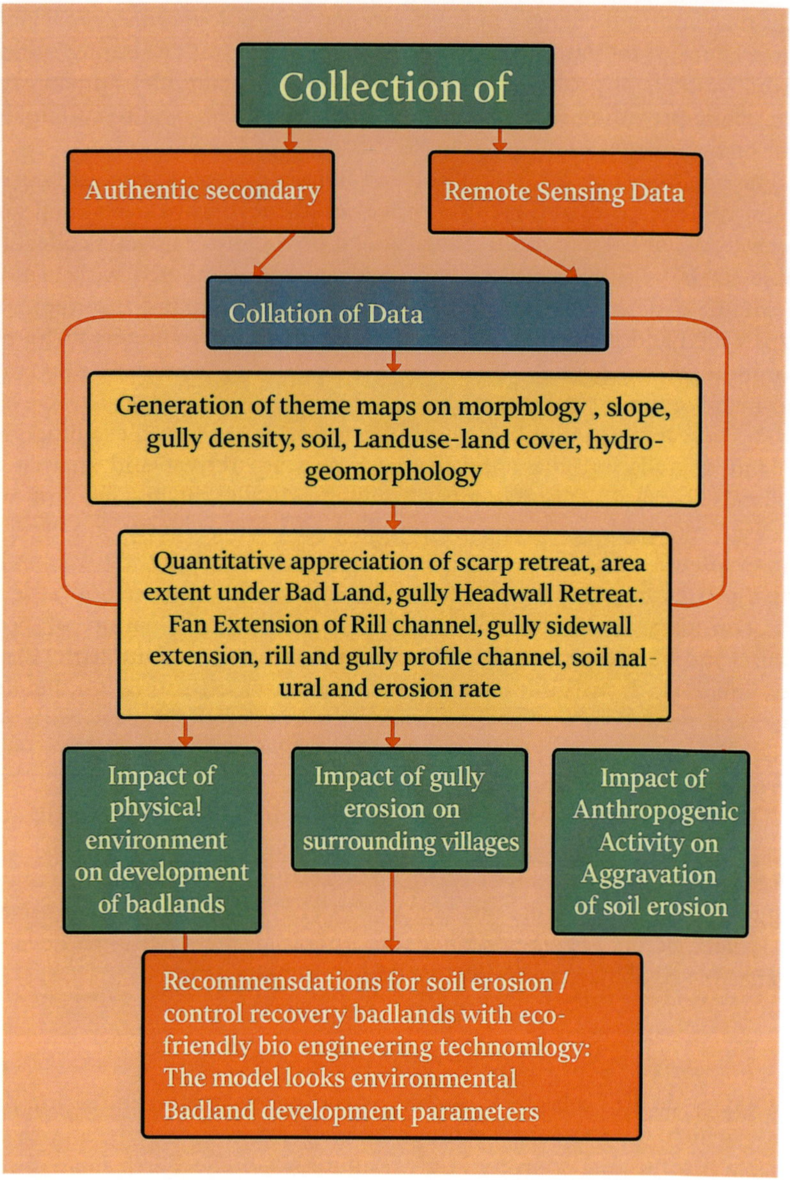

Fig. 23.2 Methodology for the study

analysis benefits from NRSA Hyderabad-provided IRS IC LISS-III geo-coded false color composites (FCC) which show the region in 1995 and 2000. Data acquired from USGS includes topographical and hydrological assessment materials through SRTM data (2014) and TM DEM datasets. The analysis required different types of mapping solutions. The current land use classification process utilized land use maps from Landsat OLI (2022) as well as the Digital Elevation Model (DEM) mapping service provided by Google Earth (2022) for terrain analysis. Environmental data came from the India Meteorological Department (IMD) for analyzing climatic factors. Data about soil quality and erosion risks were obtained through the Bhumi Geoportal from NBSS & LUP resulting in receipt of soil maps and soil loss estimation maps.

Table 23.1 Database for the study

Details of data			
Data type	*Methods*	*Data*	*Source*
Secondary data	Remote sensing	Google image google earth view	Google earth
		1. IRS IC LISS-III Geo coded fcc, 1995	NRSA Hyderabad
		2. IRS IC LISS-III Geo coded fcc, 2000	
		3. SRTM DATA, 2014	USGS SRTM, 2014
		4. TM DEM	USGS SRTM, 2014
	Map information	Elevation DEM Map	Google earth, 2022
		Land use map and information	Landsat OLI, 2022
		Climatological Data	IMD
		Soil map	NBSS & LUP, Bhumi geoportal
		Geology map	Geological survey of India, Bhukosh
		Geomorphology map	Geological survey of India, Bhukosh
		Topographical wet index map	USGS SRTM DEM, 2014
		Stream power index map	USGS SRTM DEM, 2014
		Soil loss estimation map	NBSS & LUP, Bhumi geoportal
		FCC satellite image	Landsat OLI, 2022

The geological and geomorphological features of this region were examined through maps from both the Geological Survey of India (GSI) and Bhukosh so scientists could better understand terrain development patterns. The USGS SRTM DEM (2014) served as a source to generate Topographical Wetness Index and Stream Power Index maps for understanding moisture patterns and erosion susceptibility. Satellite imagery acquired from Landsat OLI during 2022 enabled land cover classification alongside analysis of the area. The collection of various datasets covered a wide range of information to support complete spatial understanding and environmental evaluation of the study region.

23.8 Results

Garhbeta CD Block-I situated in West Medinipur, West Bengal received its slope mapping (Fig. 23.3) from USGS SRTM data collection (2014). The slope map splits the terrain into five categories according to slope incline from 3.63° to 42.3° that shows color transition from light yellow to dark red. Most of the study region features slopes between 3.63° and 13.3° which indicates that it possesses mostly flat areas alongside mildly rolling surfaces. Steeper slopes extending from 23° to 42.3° form scattered regions throughout the block especially in its central and southern areas where they manifest as orange and red-colored areas. The steep slope sections indicate rough terrain containing small hills together with escarpments. The geographical area stretches between the northward orientation at longitude 87°18′0″E to 87°36′0″E and latitude 22°46′0″N to 23°0′0″N. The slope analysis provides vital insights for area development and protection as well as planning for land use and soil erosion prevention within Garhbeta CD Block-I region.

The aspect map (Fig. 23.4) of Garhbeta CD Block-I West Medinipur derives its data from USGS SRTM data (2014) to display slope direction across the geographic area. Slope direction which is measured in standard units affects regional environmental patterns as well as vegetation distribution and site usability potential. The mapping system divides the terrain into nine classes represented by different colors where north, northeast, east, southeast, south, southwest, west, northwest and finally northward-facing slopes receive distinct color shading. The area presents a range of topographic features mainly distributed toward northeast and southeast and southwest directions which indicates the terrain's uneven characteristics. These land areas feature north slopes (with both low and

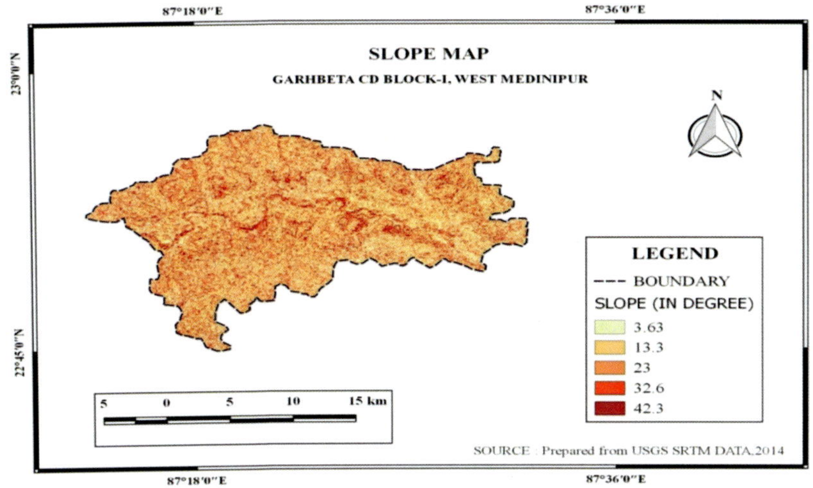

Fig. 23.3 Slope map of Garhbeta C.D Block-I, West Medinipur, West Bengal

high elevation angles) together with west slopes located between 219° and 260.8°. The numerous slope directions across Garhbeta CD Block-I serve as critical factors for studying local climate shifts while they help design agricultural systems and improve soil water retention.

The Digital Elevation Model (DEM) of Garhbeta CD Block-I (Fig. 23.5) in West Medinipur district region of West Bengal was created from the USGS SRTM data (2014) to show the spatial patterns of terrain elevation across the region. The elevation range varies from 27 to 74 meters above mean sea level which is each level appearing in different shades ranging from yellow to red. The northeastern and eastern portions of the block exhibit the lowest elevation range of 27 meters is colored in yellow as these regions have even lowlands which are prone to floods during monsoon seasons as water collects in this region.

The topographic zone with deep red in the map represents areas which slope up to 74 meters and these are found mainly in the southern part with southwestern parts and some parts of the central zones. Apart from being drained, these elevated areas are at risk of soil erosion unless there is sufficient cover of vegetation over the landscape. Such elevation variation is important for land use planning, where elevated land areas are often preferable for settlements and specific agricultural uses and lower

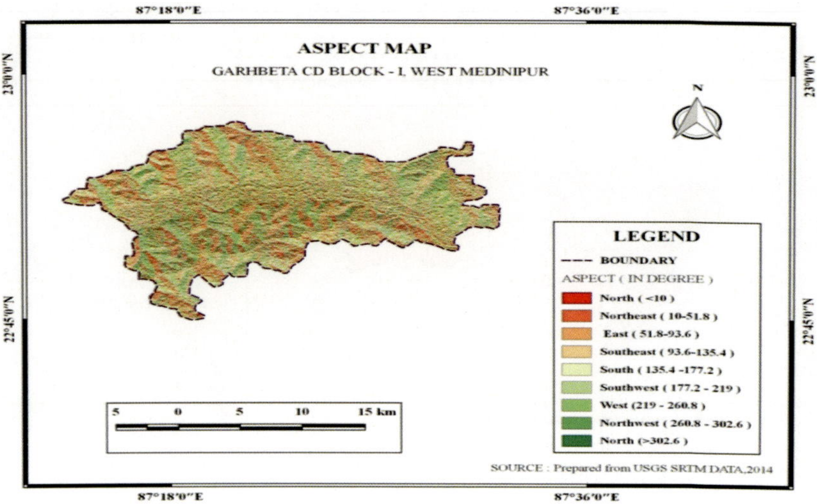

Fig. 23.4 Aspect map of Garhbeta C.D Block, West Medinipur, West Bengal

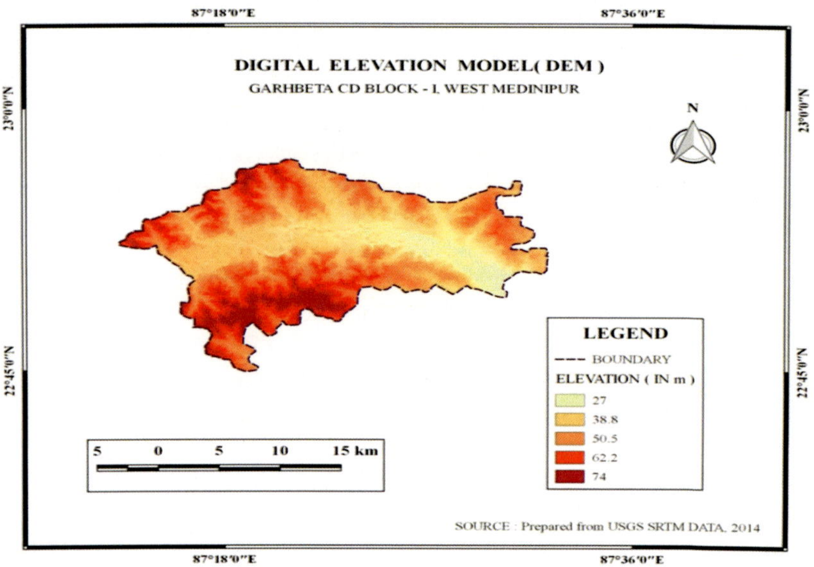

Fig. 23.5 Digital elevation model (DEM) of Garhbeta C.D Block, West Medinipur, West Bengal

areas are more suitable for paddy cultivation, wetlands, or conservation of natural vegetation. Additionally, elevation information is important in environmental management, infrastructure planning (e.g., road and dam building), groundwater prospect evaluation, and disaster risk reduction initiatives, especially in the determination of flood- or erosion-prone areas.

A detailed overview of different land cover patterns (Fig. 23.6) in Garhbeta CD Block-I, West Bengal emerges from the USGS Landsat Legacy 8 OLI data (2022) basis land use and land cover map. The geographical area shows multiple land category divisions which include water bodies, forests, vegetation, flooded vegetation, crops, scrublands, built-up areas and rocky lands.

Cropland stands out as the leading land cover in the block since yellow color dominates the landscape showing agriculture drives both economy and sustenance for residents. Natural and semi-natural vegetation exists predominantly in northwestern and southern areas while built-up areas

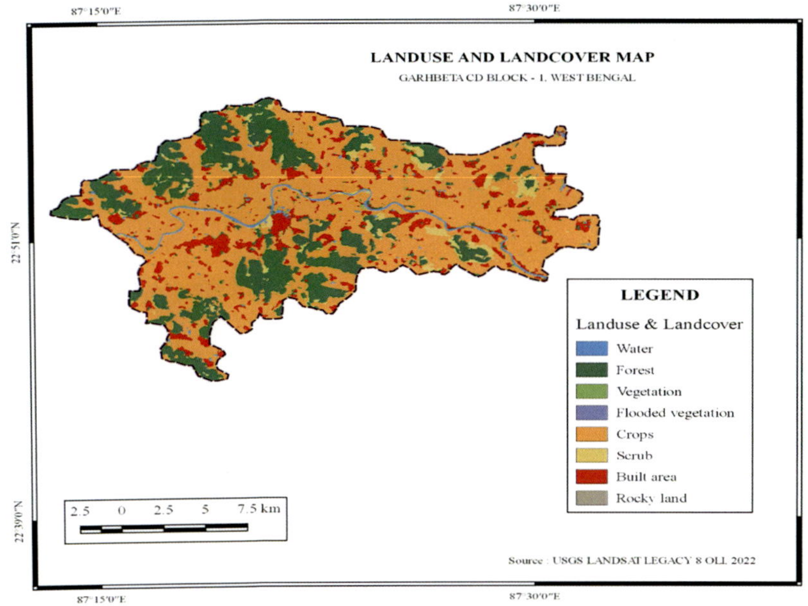

Fig. 23.6 Land use land cover (LULC) map of Garhbeta C.D Block, West Medinipur, West Bengal

stretch mainly in the central and southeastern parts of the study area. Human activities affecting Colorado's land are rapidly growing out of control while urban planning needs proper management for sustainability. Some regions with scrublands (orange color) consistently cross the landscape between different land types and croplands so they represent either lands in recovery or sites that risk damage if humans do not exercise proper management. The presence of flooded vegetation patches primarily occurs next to watercourses while indicating areas that act as wetlands and serve ecological purposes during the monsoon month. The south portion of this territory exhibits minimal gray-colored patches that indicate exposed rocky terrain which cannot be used for farming yet can serve quarrying purposes and various other agendas.

The lithostratigraphic sequence (Table 23.2) covered here provides an overall picture of the geological past of the region, ranging from the Pre-Cambrian through to the Upper Holocene. The base for this sequence is the Archaean Basement Complex, a complex of rocks generated during the Pre-Cambrian period, that forms the earliest rocks in the region. Relying atop this basement complex is the Pre-Tertiary to Early Tertiary Phulkusuma Formation, a major milestone in the geology of the region.

Moving up the stratigraphic column, we find ourselves at the Mio-Pliocene Upper Bhairob Banki Formation. The formation consists of gritty sandstone, clay-stone impressions of leaves, and local conglomerates. Bedded white shale and brownish-white sandstone in this formation indicate a variety of depositional environments, potentially including shallow marine and river systems, indicative of the active geologic history. The formation has signs of climatic and environmental alterations, which are evident from the diverse lithological compositions. This formation is divided from the overlying ones by an unconformity, showing an erosion or non-deposition period. The Lower Pleistocene Lower Lalgarh Formation comprises laterized conglomerates and boulders with petrified wood, implying an erosional setting with major physical weathering, and ancient vegetation. There is another unconformity between this and the Upper to Middle Pleistocene Upper Lalgarh Formation, which is made up of secondary laterite and pebbles and vertebrate fossils and shows a large change in the environment at this time. Fossils of this unit depict changes in the fauna and climate.

The stratigraphy of the Holocene in it comprises the Sijua Formation of Lower Holocene consisting of sandy loam with burrows, showing evidence of depositional environment which might be attributed to deltaic

Table 23.2 Stratigraphic sequence of the Gangani badlands, Garhbeta

Age	*Lithostratigraphy*
Upper Holocene	Kasai formation Bamandiha formation Daintikri formation
..................................*Unconformity*....................................	
Lower Holocene	Sijua formation Sandy loam with burrows
..................................*Unconformity*....................................	
Upper to Middle Pleistocene	Upper Lalgarh Secondary laterite with Pebbles and Vertebrate fossils formation:
Lower Pleistocene	Lower Lalgarh Laterised boulder formation: Conglomerate with petrified wood
..................................*Unconformity*....................................	
Mio Pliocene	Upper Bhairob Gritty Sandstone, clay-stone Banki formation with leaf impressions, grit, local conglomerate. Bedded white shale and Brownish white sandstone
..................................*Unconformity*....................................	
Pre tertiary to early tertiary	Phulkusuma formation
..................................*Unconformity*....................................	
Pre-Cambrian	Archaean Basement complex

or estuarine environments. The Kasai, Bamandiha, and Daintikri Formations mark the Upper Holocene and are representative of newer geological activities and ongoing changes in the environment. The fact that unconformities exist between these units highlights the episodic nature of erosion and deposition. This sequence offers an account of large-scale geological events, climatic changes, and environmental changes over millions of years. For further intensive examination of the stratigraphy of this area, Ghosh and Majumdar (1981) give a detailed explanation of the lithostratigraphic units and their implications regarding the interpretation of the local geologic history ("Lithostratigraphy of Eastern India," Ghosh, R. & Majumdar, 1981).

The map (Fig. 23.7) demonstrates the soil patterns throughout the Garbeta–I Community Development (CD) Block situated in Paschim

Medinipur district within West Bengal state. The layout includes a precise variability key along with an orientation device and measurement indicator to help users navigate areas of space in an effective manner. The map divides the soil types into four main classes where each category receives a specific color to enhance understanding. The main soil category of this region consists of Fine loamy (Typic ustifluvents) which appears as a light brown color in the map. The central belt of the area has this soil type with a balanced combination of sand particles and clay elements. The fertile characteristic of Typic ustifluvents qualifies these alluvial soils to support diverse agricultural production patterns including paddy cultivation as well as vegetable and pulse cultivation.

During the southern central region the coarse loamy soils exist mainly in green areas. The distribution of sand particles in these soils provides good drainage while creating insufficient water retention. The water requirements of drought-resistant crops fit well into coarse loamy soil conditions when combined with appropriate water management strategies and organic matter addition for fertility enhancement. The southern parts

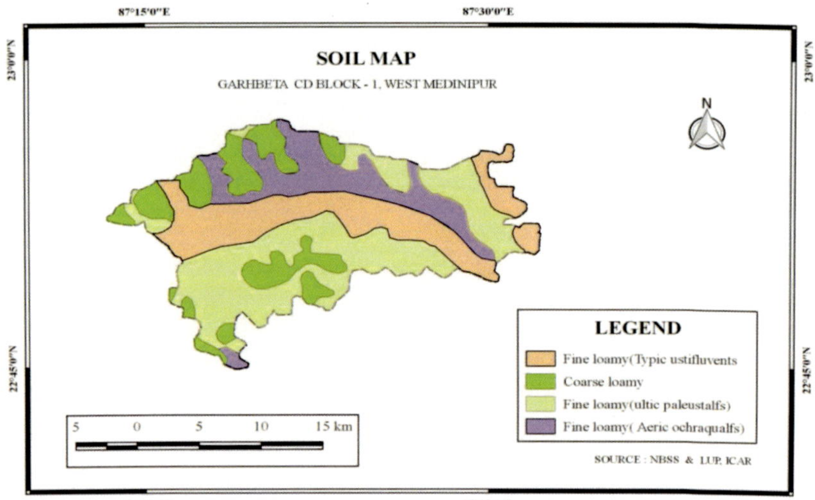

Fig. 23.7 Soil texture map of Garhbeta C.D Block, West Medinipur, West Bengal

of the area show a distribution pattern of Fine loamy soils (slightly palentrophic) which are colored yellow. The fine-textured soils demonstrate initial symptoms of palentrophism despite their textured nature. These soils have an average fertility level but need occasional fertilizer additions for maintaining agricultural yield productivity. Had inadequate soil management policies been applied during a period of prolonged cultivation and leaching these soils advanced toward a serious nutrient deficiency condition.

23.8.1 Microclimatic Shifts Due to Anthropogenic Disturbance

23.8.2 Surface Preparation During Winter Condition (October–February)

The surface humidity decreases immediately following monsoon rains as the atmospheric humidity levels also lower together (Table 23.3). Changes in air moisture adversely affect the local environmental conditions of this area. The region is characterized by wide daily temperature variations which extend from 8 °C to 12 °C. The fast temperature shifts produce detrimental consequences on seasonal plant life mainly in grasslands and badlands because these regions contain delicate vegetation that is strongly affected by climate. Garhbeta experiences cold winter temperatures which drop below 8 °C at this time and the atmospheric humidity reaches levels below 65% in December and January. The surface of the land becomes extremely arid due to this climatic environment. Surface soils become disintegrated because they become loose and powdery in the highlands of the latitudinal region where moisture loss through contraction leads to soil breakage. Environmental stress causes severe land degradation and hastens badlands formation which hinders the recovery of vegetation after the seasonal changes.

23.8.3 Surface Preparation During Summer Condition (March–May)

The summer season brings high daily temperatures reaching 42 °C but temperatures at night reach only 22 °C (Table 23.4). The rapid extremity of environmental temperature changes functions as an important factor which speeds up the process of mechanical weathering on surface materials which remain exposed to the elements. Daytime heat expands the

Table 23.3 Temperature and Humidity condition (2021–2024) during winter season

Year	*2021*			*2022*			*2023*			*2024*		
Month	*Temp* °C		*Hum**	*Temp* °C		*Hum**	*Temp* °C		*Hum**	*Temp* °C		*Hum**
	Max	*Min*	%	*Max*	*Min*	%	*Max*	*Min*	%	*Max*	*Min*	%
December	26.15	5.19	68.5	26.64	4.65	79.75	26.55	6.31	72	25.88	5.3	82.69
January	25.31	5.15	64.56	28.3	6.21	59.62	25.89	5.44	74	28.93	6.42	67.31
February	31.78	10.37	62.25	33.79	8.61	58.25	29.24	7.31	66.12	35.3	6.94	52.88
Average	27.747	6.903333	65.10333	29.577	6.49	65.87333	27.22667	6.3533333	70.7067	30.03667	6.22	67.62667

surface materials but night-time contraction creates repeated stresses which ultimately breaks down the particles on the top layer. The upper surface of the Garhbeta region shows deep-colored granular texture in hollow areas as a sign of severe weathering. Such weathering conditions allow additional physical disintegration to occur. Surface material looseness in this area increases due to irregular "Kalbaishakhi" thunderstorms that occur frequently in this precipitation zone. The material weakness from previous processes makes it easily erodible through flows of surface water or sheet wash that quickly move the dislodged materials. Surface degradation combined with landform changes throughout the area receives extensive support from severe climatic conditions and extreme weather events.

A line chart (Fig. 23.8) represents minimum and maximum temperature measurements collected annually between 2020 and 2024. Two temperature trends are presented in the graph through maximum temperature measurements shown by red squares and minimum temperature measurements represented by blue diamonds. The five-year period shows only slight variations in both maximum and minimum temperature measurements. The recorded maximum temperatures stayed higher than minimum temperatures through each year within a range of 46 °C to 49 °C. The recorded minimum temperatures displayed equivalent movement over this period by varying between 41 °C and 45 °C. The year 2022 became the optimum baseline for both maximum and minimum temperature readings that showed a potential heat maximum peak that year. Both temperature measurements reached their annual minimums in 2023 and 2024 based on the recorded data. After the year 2022 the temperature data appears to remain stable while showing minor decreasing patterns. The data suggests a slight temperature reduction or stabilizing period following this period of thermal peak.

According to the visual plot of monthly rainfall (mm) for the period 2020–2024 (Fig. 23.9), the data reflects both intra-annual and inter-annual variability, indicating seasonal concentration of rainfall and long-term change. The distribution of rainfall for all years always reflects maxima for the monsoon months June to September, and the dry months January to April and November to December are characterized by negligible rain, usually close to or less than 5 mm. This strongly suggests a monsoonal rainfall regime characteristic of the area.

In the years 2020 and 2021, the rains look quite low, with the moderate highs in July and August, but the overall levels of precipitation

Table 23.4 Temperature and humidity condition (2021–2024) during summer season

Year	*2021*			*2022*			*2023*			*2024*		
Month	*Temp* °C		*Hum**	*Temp* °C		*Hum**	*Temp* °C		*Hum**	*TEMP* °C		*Hum**
	Max	*Min*	%	*Max*	*Min*	%	*Max*	*Min*	%	*Max*	*Min*	%
March	37.23	13.91	56.69	38.09	12.18	60.44	36.77	14.74	63.12	42.82	15.59	43.38
April	39.57	19.36	67.81	43.99	20.97	57	40.9	18.25	58.38	41.65	20.28	48.94
May	41.76	23.73	64.75	43.44	21.71	59.81	40.56	23.25	67.62	42.39	23.69	66.06
Average	39.52	19	63.08333	41.84	18.2867	59.08333	39.41	18.746667	63.04	42.28667	19.8533	52.79333

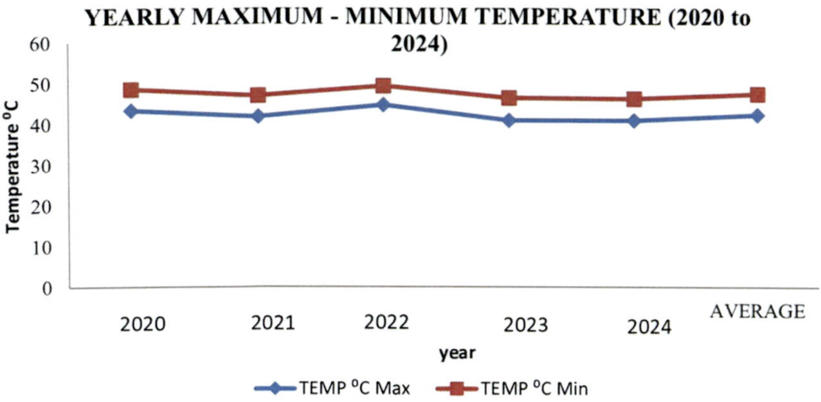

Fig. 23.8 Maximum and minimum temperature of Garhbeta (2020–2024)

are kept low, which indicates either weakened or delayed monsoons of those years. In 2022 and 2023, a low but definite increase is noted, where July and August rainfall goes well beyond 15 mm in a few places, which shows a possible shift toward more stronger monsoonal conditions. Interestingly, 2024 exhibits a clear rise both in intensity and extent, not only peaking well above 15 mm in July but also being more evenly spread from May to September. This may indicate either enhanced monsoon strength or climatic irregularities like premature onset or longer duration of rain (Table 23.5).

The overall pattern also indicates sparse temporal concentration of precipitation, being mostly confined within a 3–4-month period. High-concentration precipitation, particularly if combined with disturbed vegetation cover and enhanced surface exposure (as noticed in the Garhbeta Badlands), potentially leads to higher surface runoff, flash flooding, and enhanced soil erosion. Precipitation-less months during the rest of the year (particularly the winter season) further aggravates drought-like regimes, impacting plant regeneration and preservation of soil moisture. In summary, the rainfall charts as a group indicate that while the monsoon is still the main source of rain, its intensity and reliability change from year to year. The rise in rainfall seen in 2024 will be agriculturally beneficial in the short run but also carry risks of erosion and flooding in a geomorphologically vulnerable area such as Garhbeta. This trend emphasizes the imperative necessity of adaptive land and water management such

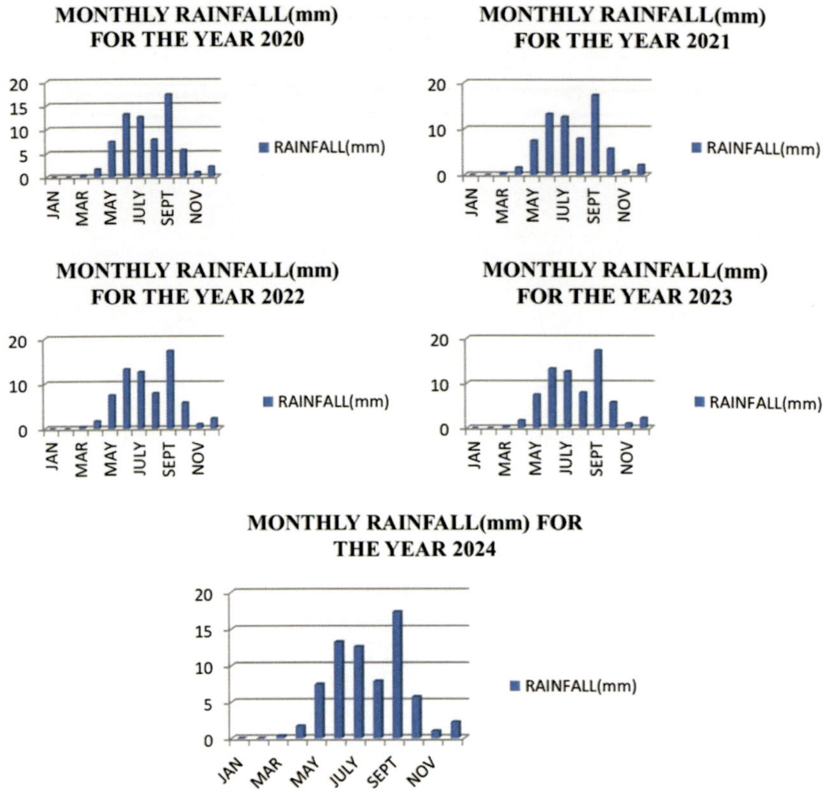

Fig. 23.9 Monthly rainfall of Garhbeta-I CD Block from 2020 to 2024 (mm)

as measures of erosion control, wet-season watering, and measures of soil conservation to mitigate the effects of spotty rainfall.

23.8.4 *Emerging Geomorphic and Climatic Trends in the Shilabati River Basin due to Anthropogenic Pressure*

While the fundamental findings of this study already highlight core trends—e.g., amplified soil erosion within mined areas, gully network expansion, great loss of vegetation, and growing surface runoff—there is also an emerging, and no less important, trend that warrants focused attention: microclimatic change in the Garhbeta Badland environment.

Table 23.5 Monthly rainfall (mm) for the last 5 years (2020–2024) at Garhbeta badlands

Year	*2020*											
Month	*Jan*	*Feb*	*Mar*	*Apr*	*May*	*June*	*July*	*Aug*	*Sept*	*Oct*	*Nov*	*Dec*
RAINFALL (mm)	0.12	0	1.19	1.39	3.72	8.35	22.4	11.71	6.78	8.63	0.69	0.3
Year	*2021*											
Month	*Jan*	*Feb*	*Mar*	*Apr*	*May*	*June*	*July*	*Aug*	*Sept*	*Oct*	*Nov*	*Dec*
RAINFALL (mm)	0	0.08	0.26	4.42	2.1	7.37	10.56	11.17	6.04	1.34	0.12	1.1
Year	*2022*											
Month	*Jan*	*Feb*	*Mar*	*Apr*	*May*	*June*	*July*	*Aug*	*Sept*	*Oct*	*Nov*	*Dec*
RAINFALL (mm)	0	2.5	1.19	1.53	4.06	3.16	6.19	9.92	9.3	5.19	1.07	0.41
Year	*2023*											
Month	*Jan*	*Feb*	*Mar*	*Apr*	*May*	*June*	*July*	*Aug*	*Sept*	*Oct*	*Nov*	*Dec*
RAINFALL (mm)	1.16	0.23	1.77	3.24	6.55	8.09	8.79	10.92	4.63	2.36	0.19	0.01
Year	*2024*											
Month	*Jan*	*Feb*	*Mar*	*Apr*	*May*	*June*	*July*	*Aug*	*Sept*	*Oct*	*Nov*	*Dec*
RAINFALL (mm)	0	0	0.34	1.68	7.46	13.21	12.58	7.89	17.31	5.72	0.98	2.25

This transformation, subtle in everyday observation, is highly important when understood against the background of land surface transformations and seasonal meteorological data between the years 2021 and 2024. From the data, there is a definite trend—a human-induced geomorphic disturbance not only reshaping the physical landscape topography but also modifying the thermal and atmospheric equilibrium, thereby creating new levels of exposure to the ecosystem and human settlements as well.

The initial signal of this microclimatic instability can be traced to the progressive increase in surface temperature maxima, especially in summer months. As reported, the summer average temperature in 2024 reached about 42.28 °C from 39.52 °C in 2021. In particular months such as April and May, the trend is even more dramatic, with recorded maxima exceeding 43 °C in 2022 and remaining similarly high for following years. At the same time, minimum night temperatures have not consistently increased, resulting in an expanding diurnal temperature range (DTR). This inconsistency can be viewed as a disruption of the region's natural microclimatic regulation, previously stabilized by closed vegetation and undisturbed soil moisture regimes. This expansion in temperature range is typical of desertification and aridification mechanisms, usually fueled by deforestation, mining activities, and the exposure of soils' surfaces to air—human-influenced drivers that strongly exist in Garhbeta terrain.

Humidity values during winters and summers spanning from 2021 to 2024 lend more credence to this presumption. The mean relative humidity in summer has fallen significantly, from around 63% in 2021 to 52.79% in 2024. Especially in March and April, the beginning of summer, the trend of decreasing humidity becomes apparent as a function of reduced evapotranspiration rates associated with vegetation loss and soil compaction. Humidity values less than 60% during these pivotal months indicate greater atmospheric dryness, adding to the drying of surface soils and speeding up wind and water erosion. During the winter season, the same trends are evident, with relative humidity varying widely and falling below 65% in several cases. Such climatic changes not only enhance geomorphic instability but also weaken the agro-ecological resilience of the region, endangering rainfed agriculture and natural forest regeneration.

To better comprehend this event calls for an investigation of the spatial correlation between land cover change and thermal flux. As remote sensing data confirms, over 40% of vegetative cover has been lost across the Garhbeta badlands in less than a decade, largely due to laterite mining, overgrazing, and land clearance for agriculture and tourism infrastructure. The loss of tree canopy and the depletion of shrub and grass layers have profound implications for land–atmosphere energy exchanges. Vegetation is a natural temperature buffer that moderates ground heat flux via mechanisms such as shading, transpiration, and root water uptake. Vegetation removal results in enhanced surface albedo and net radiation that warm the land surface more quickly during the day. This serves to increase land surface temperatures, increasing convective processes and, quite possibly, also changing regional rain patterns—despite failing short of full-long-term precipitation model prediction here, the preliminary hints at unusual trends in recent-five-year rainfall foreshadow possibilities like this. Secondly, intense diurnal contrasts also act as a contributing force for exacerbating mechanical weathering of naked surface exposures. During summer, with diurnal temperatures rising above 42 °C and dipping to 22–24 °C, the thermal contraction and expansion of the regolith and rock material intensify. These micro-fractures within the regolith and lateritic crust—augmented by trampling by people, excavation, and traffic—are responsible for increasing the weathering of surface material, particularly along slope margins and escarpment edges. This thermally-activated geomorphic action becomes the prelude for subsequent more accelerated erosion when the monsoon sets in, with rainwater readily mobilizing the loosened sediment to produce flash runoff, sheet wash, and gully incision. Supplementing this rich chain of interaction is the contribution of modified hydrological behavior. The research site has recorded a massive rise in runoff on the surface as a result of both compaction of topsoil from human activities and interference with the natural drainage streams by mining quarries and ad hoc constructions. The runoff goes beyond causing erosion in gully form to an accumulation of ephemeral drainage lines, which subsequently incise within the badland landscape to generate isolated plateaus and fingerlike rill ridges. These are observable in DEM and slope analysis outputs, where there is the appearance of new slope instabilities within previously geomorphically stable areas. With slope instability and rill growth worsening, the feedback loop increases—the more degraded the land, the more likely to degrade further.

What enhances the urgency of this newly emerging microclimatic problem is its impact on society. Local residents, who subsist on traditional farming and seasonal employment, are now subject to greater uncertainty from unpredictable microclimates and diminishing arable land. Soil previously fertile enough to produce millet, pulses, and vegetables is now heat-stressed, eroded, and dried out, making it barren. This has direct implications for food security, water resources, and livelihoods, compelling communities to migrate or exploit further already-degraded land. The experience of "climatic hardship migration"—albeit not central to this paper—is traceable to this accumulating pressure in locations such as Garhbeta, wherein climate and geomorphology are no longer shaped by solely natural forces but by progressively influential human activity.

Significantly, this microclimatic change remains largely unassimilated into the majority of policy-level evaluations and local land management initiatives, which merely address superficial erosion patterns while neglecting atmospheric signals. This represents a significant lacuna in both research literature and practice-oriented environmental governance. Coupling meteorological information with land degradation indicators can provide a multi-scalar, interdisciplinary insight into badland processes, and underpin early-warning for erosion and agricultural failure. There is an urgent need to observe temperature and humidity trends over a longer time period and correlate them with vegetation regeneration, soil preservation, and water use practices. Such an endeavor would not only fill the gap in data but also enable grassroots stakeholders to implement preventive measures prior to irreversible thresholds of desertification.

In conclusion, the temperature, humidity, and fluctuation pattern changes within the last four years of the Shilabati River Basin are not random but highly interrelated with the anthropogenic geomorphic changes of the area. These reflect a transition from a biologically-regulated to a mechanically-dominated landscape system wherein the land is more disturbed by humans than its ability to self-recover. Therefore, any long-term conservation or rehabilitation effort will need to include microclimatic monitoring and restoration techniques—such as reforestation with indigenous species, shielding moisture-retaining vegetation, and zoned land-use regulation. Failing to take this important dimension into account will mean that attempts to curb erosion or restore geomorphic equilibrium may provide only short-term mitigation, leaving the underlying causes of degradation unaddressed.

While the major findings of this study already underscore the main patterns—such as soil erosion within mined areas on the rise, enlargement of gully networks, severe loss of vegetation, and augmentation of surface runoff—there is an arising and no less vital trend which is worthy of elaborate focus: microclimatic change of the Garhbeta Badland landscape. This transition, subtle at the day-to-day level of observation, has far-reaching significance when understood along with land surface alterations and annual meteorological parameters between the periods 2021 and 2024. Evident is a clear pattern from the readings—human intervention geomorphic alteration is not merely reshaping the physical landscape terrain but also manipulating the thermal as well as the atmospheric equilibrium and thereby adding novel layers of sensitivity to the system and human habitat as well.

The first sign of such microclimatic instability resides in the continuously increasing maxima of surface temperature, especially summer ones. As reported, summer average temperature reached about 42.28 °C in 2024, versus 39.52 °C for 2021. For specific months such as April and May, the situation is even more evident, as recorded maxima have exceeded the 43 °C mark in 2022 and have stayed quite high also for the following years. At the same time, minimum night temperatures have not exhibited a uniform rise, resulting in an expanding diurnal temperature range (DTR). This lack of consistency can be viewed as a disruption of the natural microclimatic regulation of the region, previously stabilized by dense cover and undisturbed soil moisture regimes. Such increasing temperature differences mark aridification and desertification processes that typically are promoted by deforestation, mining, and exposure of soils to the surface—anthropogenic conditions most strongly ubiquitous to the Garhbeta region. Cumulative data on humidity conditions during winter to summer periods within 2021 to 2024 substantiates this proposition as well. The mean relative humidity in summer has fallen significantly, from around 63% in 2021 to 52.79% in 2024. Especially in March and April, the beginning of summer, the trend of drying is apparent, indicating reduced evapotranspiration rates due to vegetation loss and soil compaction. Humidity of less than 60% for these key months indicates higher atmospheric dryness, which adds to the drying out of surface soils and hastens wind and water erosion. The same trends can be observed during the winter season, with relative humidity varying sharply and falling below 65% in several instances. These climatic changes not just enhance geomorphic instability but also weaken the agro-ecological resilience of the region, compromising both rainfed agriculture and natural forest regeneration.

A more thorough comprehension of this phenomenon necessitates an exploration of the spatial interplay between land cover change and thermal flux. As remote sensing evidence attests, more than 40% of vegetative cover has been depleted in the Garhbeta badlands in less than a decade due mainly to laterite mining, grazing, and land clearance for tourism infrastructure and agriculture. Tree canopy loss and shrub and grass layer depletion have significant ramifications for land–atmosphere energy exchanges. Vegetation also provides a natural buffering effect for temperature by controlling ground heat flux by processes such as shading, transpiration, and root water uptake. Removal causes increased surface albedo and net radiation, warming the land surface faster during the daytime. This adds to increased land surface temperatures, which in turn increase convective activity and even change local rainfall patterns—while this study does not go so far as long-term precipitation modeling, the early indications of erratic rainfall over the past five years suggest such possibilities. Also, the mechanical weathering of exposed surfaces is worsened by extreme diurnal fluctuations. During summer, as the day temperature hovers above 42 °C and dips to 22–24 °C at night, thermal expansion and contraction of soil and rock material intensify. These micro-fractures within the regolith and lateritic crust, amplified by trampling by humans, excavation, and vehicular traffic, hasten the disintegration of surface material, particularly on slope edges and escarpment margins. This heat-generated geomorphic action becomes the stage-setting for quicker erosion when the monsoon rains set in because rainwater can readily mobilize the loosened sediment, which produces flash runoff, sheet wash, and gully incision. Contributing to this complex web of interaction is the impact of changed hydrological behavior. The study area has produced an extensive surface runoff as a result of human activity-induced compaction of topsoil, as well as disturbance of the natural drainage lines by mining ponds and illegal buildings. The runoff leads not just to gully erosion but to concentration of temporary drainage channels that incise deeper into the badland landscape to produce isolated plateaus and finger-shaped rill features. These characteristics are seen in the DEM and the outputs of slope analysis, revealing new slope instabilities occurring in what were previously regarded as geomorphically stable locations. While slope instability and rill formation become more aggressive, the feedback mechanism becomes stronger—the more degraded the land gets, the more prone it is to degradation.

What compounds the immediacy of this new microclimatic problem is its social significance. Locals, who depend on conventional farming and seasonal employment, now experience heightened uncertainty from erratic microclimates and reducing arable land. Soil that was formerly fertile enough to cultivate millet, pulses, and vegetables is now made barren by heat stress, erosion, and loss of moisture. This has a direct impact on food security, water supply, and livelihoods, driving communities toward migration or further exploitation of already-degraded lands. The process of "climatic hardship migration"—can be traced back to this increasing stress in regions such as Garhbeta, where climate and geomorphology are no longer controlled by natural forces but by more and more prevalent human actions.

Notably, this microclimatic change has not yet been comprehensively incorporated into the majority of policy-level evaluations or local land management plans, which tend to consider only apparent erosion patterns and disregard atmospheric signals. This represents a significant knowledge gap in both scholarly literature and practical environmental governance. The incorporation of meteorological data with land degradation indicators can provide a multi-scalar, interdisciplinary perspective on badland processes, and inform early-warning systems for erosion and crop loss. There is an urgent need to track temperature and humidity patterns over longer time scales and correlate them directly with vegetation rehabilitation, soil conservation, and water management practices. This would not only fill the data gap but also enable local stakeholders to act preventively before irreversible desertification thresholds are reached.

In short, the changes observed in temperature, humidity, and thermal fluctuation regimes over the past four years within the Shilabati River Basin are not coincidental but are strongly integrated with the region's anthropogenic geomorphic processes. They indicate the transition of a biologically-regulated to a mechanically-dominated landscape system under rapid human disturbance compared to the natural recovering capacity of the land. Consequently, any subsequent conservation or rehabilitation activity should involve microclimatic monitoring and restoration planning—such as reforestation with indigenous cover, maintenance of moisture-retentive vegetation, and regulated land-use zoning. Unless this key dimension is tackled, attempts at controlling erosion or geomorphic balancing may only produce temporary relief at the expense of deeper drivers of degradation remaining unsolved.

23.9 Discussion

The Garhbeta badlands of the Shilabati River Basin offer a special and critical geomorphic environment in which the interaction between natural and anthropogenic processes is generating far-reaching changes in landscape form as well as function. The findings of this research unequivocally establish that human-induced pressures not only amplified denudational processes such as soil erosion, gully development, and slope instability, but also initiated more intricate environmental feedbacks involving microclimatic changes and hydrological changes. The discoveries unveil a stratified environmental crisis extending beyond perceptible land degradation and into the sphere of systemic ecological imbalance with far-reaching consequences for the sustainability of the region.

One of the principal findings of the research is the measurement of rates of soil erosion, which present a disturbing contrast between disturbed and undisturbed regions. Natural erosion in vegetation-covered, comparatively stable terrain stays at 8.3 t/ha/year, whereas it increases to about three times as much, that is, 24.6 t/ha/year, in regions influenced by mining and forest clearance. This supports the long-held hypothesis that anthropogenic disturbance, particularly surface mining and agricultural encroachment, strips protective soil covers and leaves unconsolidated sediments to be subjected to accelerated weathering and runoff. The scenario is further intensified during the monsoon season, where intense rainfall produces instantaneous overland flow because of decreased infiltration, causing flash erosion events. The RUSLE model used here confirms this spatial imbalance in erosion risk and offers a useful model for planning future soil conservation. The research also records a 15% increase in gully networks in less than a decade, with average headcut retreat rates of 1.8 meters per annum. These findings highlight a clear pattern: human actions, especially uncontrolled land use at slope margins, are serving as drivers for geomorphic thresholds to be exceeded. Where natural gully systems may consolidate over time, ongoing anthropogenic stress—e.g., trampling, vehicular movement, and road construction—prevents sedimentation and stimulates lateral and vertical gully expansion. These processes transform once-moderate slopes into deeply incised and fragmented terrains, which are not only difficult to restore but also pose hazards to nearby settlements and farmland through increased flood and landslide risk.

The consequences of deforestation are also clearly apparent in both the NDVI analysis and field observations. Over 40% of the original vegetation cover in the study area has been lost in the last 10 years. Such degradation results largely from clearing forests for fuelwood, expansion of rainfed agriculture to marginal lands, and augmented pressure through uncontrolled tourism, especially on geomorphologically fragile areas like the Gangani badlands. Loss of vegetation directly weakens soil surface stability by eroding root networks which anchor sediments, lessening surface irregularities which decelerate runoff, and eliminating cooling by evapotranspiration-driven evaporation, which contributes to regulation of the local temperature and humidity. The cascading impacts of this process not only involve heightened erosion but also soil organic matter loss, reduced fertility, and reduced agricultural potential for local populations.

But one of the most important and yet less investigated outcomes pointed out in this research is microclimatic instability. Climatic data analysis from the years 2021 to 2024 indicates a pattern of increased maximum temperature and reduced relative humidity, particularly in summer." Average peak temperatures rose by almost 3 °C during this brief period, while relative humidity in April and May dipped below 55%, a major departure from the past. These changes can be most directly linked to anthropogenic land-cover change, specifically deforestation and soil exposure. The ground surface takes up additional solar radiation when there is no vegetation, heating the ground during the daytime. During the night, the lack of insulation afforded by canopy cover results in immediate heat loss, and thus greater diurnal temperature ranges. These thermal fluctuations increase physical weathering of the rock and soil surfaces, reducing structural cohesion and facilitating additional geomorphic breakdown.

This discovery highlights an often-neglected aspect of land degradation: the feedback between atmospheric conditions and geomorphic change. As surface temperatures and moisture availability decrease, the region starts to assume characteristics similar to semi-arid conditions, though its general climatic classification as tropical monsoon persists. The change not only endangers local biodiversity and customary cropping systems but also diminishes the ecosystem's capacity to bounce back after disruption. It results in the formation of a self-reinforcing degradation cycle—where land degradation creates microclimatic stress, which further increases degradation. This is extensively reported elsewhere in other badland systems around the world, such as the Mediterranean and the American Southwest, but its reporting in the case of tropical lateritic badlands such as Garhbeta contributes new important findings to the discipline of anthropogenic geomorphology.

The modified hydrological regime also corroborates the same. As infiltration capacity declines as a result of compaction and loss of vegetation, surface runoff becomes higher, causing an increase in the number and intensity of sheet wash and rill erosion events. The lowered water-holding capacity of the degraded soils also implies that rain is less efficient at sustaining vegetation regrowth, thereby perpetuating the cycle of degradation. The LULC map of the study easily identifies a decrease in dense vegetation cover and an expansion of scrubland and degraded cropland—transitional habitats that will either recover if the right conditions are provided or further degrade if present pressures continue.

Socioeconomic impacts of these changes are already being realized on the ground. Farmers cite reduced yields, increased crop failures, and greater difficulty in holding soil moisture in the dry season. The increasing spread of degraded lands has taken cultivation into progressively more marginal areas, frequently with insufficient soil conservation practices. The loss of traditional land-use knowledge is also taking place, as younger generations move toward urban areas because of decreasing land productivity. This rural out-migration results in the abandonment of sparsely populated, resource-poor communities with minimal capacity to adopt sustainable land use, further exacerbating conservation challenges.

The information and evidence introduced above suggest a compelling case for multi-scalar action. At the local level, programs should involve short-term afforestation and revegetation utilizing indigenous plants adapted to the lateritic terrain. Check dams, vegetative strips, and terracing-based erosion control at the community level should be encouraged where gully activity is intense. Regionally, planning involves the need to zone land use to prohibit mining and tourism expansion in geomorphically vulnerable areas. Public awareness campaigns and stakeholder participation maps can assist communities in recognizing and acting upon environmental hazards before irreversible damage is incurred.

Furthermore, future research will need to pursue an interdisciplinary solution, blending geomorphological modeling with socio-environmental studies. Monitoring systems that link remote sensing, ground climate stations, and local knowledge can provide immediate feedback on intervention effectiveness. The policy-level interventions need to be in sync also with national initiatives like India's National Mission on Sustaining Himalayan Ecosystems (NMSHE) and State Action Plan on Climate Change (SAPCC), with finance and technical assistance also customized for non-Himalayan yet highly sensitive areas like the Shilabati Basin.

23.10 Conclusion

The results of this research highlight the significant and complex effect of anthropogenic activities on the geomorphic processes and denudation processes of the Shilabati River Basin, especially in the sensitive lateritic badlands of Garhbeta, West Medinipur. By means of remote sensing analysis, GIS-based modeling, and field observations combined, it has been evidently shown that unchecked human interventions particularly deforestation, mining, unsustainable agriculture, and hasty infrastructural development have been major drivers of natural erosive processes, overhauling not just the physical structure of the landscape but its ecological and climatic harmony as well. The erosion rates of soil in human-disturbed sites, particularly up to 24.6 t/ha/year as against 8.3 t/ha/year in undisturbed sites, indicate the extent of degradation and vulnerability (Bhattacharya, 2023a, 2023b; Mandal, 2023a, 2023b).In addition to physical changes, the research points to an important but little-recognized aspect of degradation: the initiation of microclimatic change due to anthropogenic geomorphic disruption. The rise in temperature extremes, low humidity levels, and increased diurnal temperature ranges that occurred from 2021 to 2024 are directly associated with vegetation degradation and surface exposure resulting from land misuse. These patterns imply that human influences are not just redefining the landscape but also changing the local climatic conditions, resulting in a feedback cycle of accelerated erosion, discouraging plant regeneration, and creating additional land degradation (Gallart, 2013; Ghosh et al., 2022a, 2022b). This trend is replicated in Mediterranean badlands (Nadal-Romero et al., 2018a, 2018b) and North American landscapes (Bridges & Harding, 2016a, 2016b), as it confirms the international applicability of this case study.

In addition, their socio-economic consequences are both short-term and long-term. As agricultural productivity decreases with diminished soil fertility and moisture content, local populations experience growing livelihood insecurity. The degradation of previously fertile land, the extension of gully systems, and changed drainage patterns increase the danger of displacement, food insecurity, and rural outmigration. Here, the Garhbeta badlands are a microcosm of the larger issues confronting numerous tropical and subtropical regions experiencing rapid, uncontrolled land-use change during the Anthropocene (Chakrabarti & Mandal, 2019; Poesen, 2018a, 2018b).This study promotes an integrated, ecosystem-based land management approach, integrating geomorphological knowledge with sustainable development strategies. Short-term measures like native species reforestation, land-use zoning enforcement, controlled mining, erosion control measures, and sustainable geotourism promotion are needed to prevent further degradation. In addition, the integration of community engagement, traditional knowledge, and adaptive policy mechanisms is paramount to maintaining long-term resilience and sustainable landscape use.

References

Bandyopadhyay, S., & Das, J. K. (1987). Lateritic badlands of West Bengal: A geomorphological appraisal. *Indian Journal of Landscape Systems and Ecological Studies, 10*(1), 25–34.

Bandyopadhyay, S., & Das, J. K. (1988). Evolution of badland topography in the lateritic terrain of West Bengal. *Geographical Review of India, 50*(2), 41–50.

Bhattacharya, R. (2023a). Impact of laterite mining on gully headcut dynamics in tropical badlands. *Journal of Geomorphological Studies, 28*(2), 112–125.

Bhattacharya, R. (2023b). Laterite mining and erosion dynamics in Garhbeta badlands. *Journal of Indian Geomorphology, 12*(1), 77–89.

Bridges, A., & Harding, L. (2016a). Geomorphic transformation of the Badlands National Park: A century of anthropogenic influence. *North American Geoscience Review, 14*(1), 55–72.

Bridges, M., & Harding, A. (2016b). Anthropogenic impacts on drainage systems in Badlands National Park. *Geographical Review, 106*(2), 145–161.

Bryan, K. (1922). Erosion and sedimentation in badlands. *Geographical Review, 12*(3), 451–475.

Chakrabarti, S., & Mandal, A. (2019). Lateritic terrain evolution in the Rarh region of West Bengal. *Indian Journal of Regional Geography, 6*(3), 77–91.

Das, S., Ghosh, P., & Sinha, R. (2020). Pleistocene evolution of lateritic terrains in West Bengal: Implications for badland formation. *Earth Surface Processes and Landforms, 45*(5), 939–955.

Gallart, F. (2013). Vegetation crust loss and hydrological response in Mediterranean badlands. *Earth Surface Processes and Landforms, 38*(1), 68–77. https://doi.org/10.1002/esp.3239

García-Ruiz, J. M., Beguería, S., Nadal-Romero, E., González-Hidalgo, J. C., Lana-Renault, N., & Sanjuán, Y. (2013). A meta-analysis of soil erosion rates across the Mediterranean region. *Geomorphology, 196*, 1–11.

Ghosh, D., Saha, R., & Roy, A. (2022a). Vegetation loss and infiltration capacity decline in Garhbeta badlands: A GIS-based analysis. *Environmental Monitoring in Eastern India, 11*(4), 213–226.

Ghosh, S., Bera, K., & Roy, T. (2022b). Vegetation cover decline and hydrological changes in the Garhbeta badlands. *Environmental Monitoring and Assessment, 194*(9), 639.

Joshi, R., & Nagare, S. (2019). Anthropogenic transformation of lateritic badlands in western India: A spatiotemporal assessment. *CATENA, 175*, 512–525.

Mandal, B. (2023a). Trends in sediment yield in Garhbeta's lateritic badlands: A temporal analysis. *Journal of Soil and Water Conservation, 18*(2), 112–126.

Mandal, K. (2023b). Sediment yield dynamics in lateritic badlands of West Bengal: A temporal analysis. *Journal of Soil and Water Conservation India, 9*(1), 98–110.

Mascherini, M., Brandani, F., & Catoni, R. (2011). Anthropogenic influences on gully erosion in Mediterranean badlands: A case study from central Italy. *CATENA, 84*(1–2), 123–131. https://doi.org/10.1016/j.catena.2010.10.001

Nadal-Romero, E., González-Hidalgo, J. C., & Lana-Renault, N. (2018a). Human impacts on badland erosion and geomorphology: A global review. *Earth-Science Reviews, 185*, 302–319. https://doi.org/10.1016/j.earscirev.2018.06.008

Nadal-Romero, E., Poesen, J., Verachtert, E., & Martín-Duque, J. F. (2018b). Human-influenced badlands: An overview of their distribution, processes, and conservation challenges. *Earth-Science Reviews, 185*, 123–138.

Poesen, J. (2018a). Badlands and human-induced geomorphic change: From natural processes to anthropocene signatures. *Geomorphology, 309*, 1–7.

Poesen, J. (2018b). Soil erosion in the Anthropocene: A review of human-induced geomorphic processes. *Earth's Future, 6*(5), 636–652. https://doi.org/10.1002/2017EF000766

Schumm, S. A. (1956). Evolution of drainage systems and slopes in badlands at Perth Amboy, New Jersey. *Geological Society of America Bulletin, 67*(5), 597–646.

Torri, D., Poesen, J., & Borselli, L. (2013). Badland morphology and erosion processes in the Italian Apennines: A long-term analysis. *Geomorphology, 204*, 101–112.

Tripathi, R. P., & Singh, R. P. (1990). Soil erosion and conservation in India. *Soil Conservation Society of India, Bulletin, 8*(1), 12–19.

CHAPTER 24

Integrating Remote Sensing and GIS for Monitoring of Riverine Geohazards: An Overview of Concepts and Applications

Manisha Giri and Anup Das

24.1 Understanding Riverine Geohazards

"Geohazard" this word comes from two (2) words: (1) Geo- from the word "geological" meaning from the Earth, (2) a situation that poses a risk to property, life, also the environment. So, Geohazards, also known as Geological hazards, are the consequences of active natural geological processes that pose risks to human populations and infrastructure. These events can also occur naturally in remote areas without causing harm (see "Geohazards—Alaska Nature and Science"). Geohazards encompass a wide range of phenomena, including volcanic eruptions, earthquakes, landslides, avalanches, glacial surges, outburst floods, tsunamis, and ground collapses due to thawing permafrost. These hazards can appear

M. Giri (✉) · A. Das
North Eastern Space Applications Centre, Umiam, Meghalaya, India
e-mail: manishagiri946@gmail.com

S. Ghosh (ed.), *Reimagining Indian Rivers for Sustainability*, Palgrave Studies in Environmental Sustainability,
https://doi.org/10.1007/978-3-032-04754-0_24

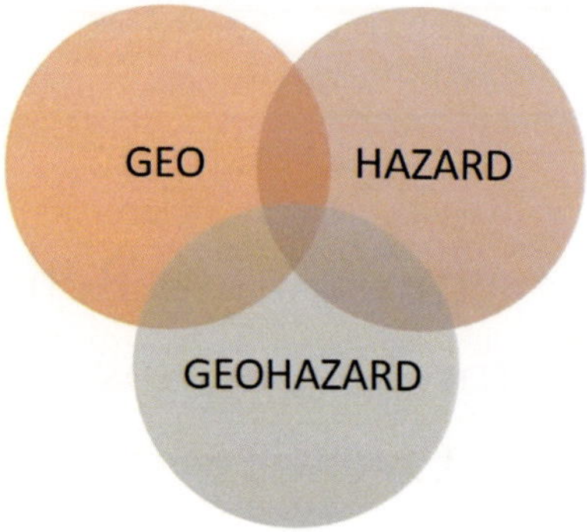

Fig. 24.1 Concept of geohazard

suddenly, emphasizing the need to monitor high-risk areas, such as landslides, debris flows, or permafrost thaw. Preparing for these geohazards is crucial for mitigating their impact (Fig. 24.1).

A geologic hazard is any geological condition that can cause damage or loss of property or life (see " Geologic Hazards"). This includes flooding, erosion, ground failure, earthquakes, tsunamis, and volcanic activity (see " Geologic Hazards"). Cryospheric hazards related to the frozen parts of the Earth, such as sea ice, river ice, glaciers, and permafrost, are particularly significant in northern and alpine areas. Geomorphological hazards, defined as events or processes that can harm humans or their activities, underscore the importance of understanding human and natural interactions in geohazards. Given the vital role of river systems as life sources for numerous communities globally, applying advanced tools becomes essential to address problems related to river hazards, aiming to mitigate the impact of floods while enhancing overall resilience (see "Geohazards—Stantec").

Riverine environments are susceptible to a specific set of geohazards that are intrinsically linked to the dynamic nature of flowing water and its interaction with the surrounding landscape. These hazards can be broadly categorized to provide a structured understanding. Overland flooding, also known as clearwater floods, occurs when a river exceeds its banks and inundates the adjacent floodplain. River migration, or avulsion, is another significant hazard where a river establishes a new primary channel, leading to reduced or ceased flow in its original course. This process can be rapid or unfold over decades. Erosion, the wearing away and removal of rocks and soil by water, poses a continuous threat to riverbanks and adjacent land (Boundary Integrated Watershed Service, 2022). Sedimentation, the deposition of transported material, can lead to land gain in some areas but also contribute to navigational challenges and altered river morphology (Boundary Integrated Watershed Service, 2022). Debris floods and debris flows, characterized by a moving mass of earth, rock, and water, often coincide with heavy rainfall and flash flooding (see "Geohazards—Alaska Nature and Science"). Other notable riverine geohazards include ice jams, which can cause localized flooding (see "Guide to Geologic Hazards") and outburst floods resulting from the sudden breach of natural dams like glacial ice (see "Geohazards—Alaska Nature and Science"). Water-triggered geohazards extend beyond these, encompassing landslides due to saturated soil, changes in groundwater pressure that can potentially lead to slope failures or sinkholes, soil liquefaction in seismically active riverine areas, river scouring that can undermine infrastructure, and karst formation resulting from water dissolving soluble rocks (Kraft, 2024). The presence of river crossings also introduces vulnerabilities, as floods and heavy rains can cause scour, exposing buried pipelines to the erosive forces of the water flow (Five geological hazards to handle with integrity management, n.d.) (Fig. 24.2).

Geospatial technology has become an essential set of tools for effectively evaluating and managing various river-related geohazards. Central to geospatial technology is remote sensing, which is actively utilized in hazard evaluation, including the identification and analysis of floods and landslides. As many geohazards can occur swiftly and unexpectedly, the long-term monitoring capabilities provided by geospatial methods are extremely valuable for monitoring high-risk regions. Additionally, the information gathered and assessed through geospatial technologies is crucial for developing informed strategies to mitigate hazard risk.

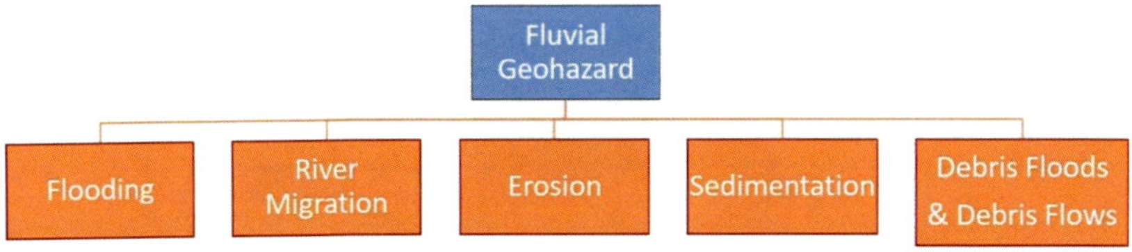

Fig. 24.2 Different types of fluvial geohazard

24.2 Fundamentals of Geospatial Technology for River Geohazards

The effective application of geospatial technology in understanding and addressing riverine geohazards depends upon a comprehensive understanding of its fundamental principles and components, which primarily encompass Remote Sensing (RS), Geographic Information Systems (GIS), and Global Navigation Satellite Systems (GNSS) (Fig. 24.3).

Remote sensing combines science and art to gather information about the Earth's surface without direct contact. This process involves sensing and capturing the energy reflected or emitted from the Earth's surface, then analyzing and applying this data. The fundamental principle of remote sensing is that various objects and surfaces on Earth can be identified and differentiated by the distinct variations in radiant energy they emit or reflect. For example, water, soil, and vegetation each emit energy in different quantities across various electromagnetic bands (Sergieieva, 2025). Several factors influence this interaction, including the object's chemical and physical composition, the type of material, its surface roughness, and the radiant intensity, incidence angle, and wavelength of the

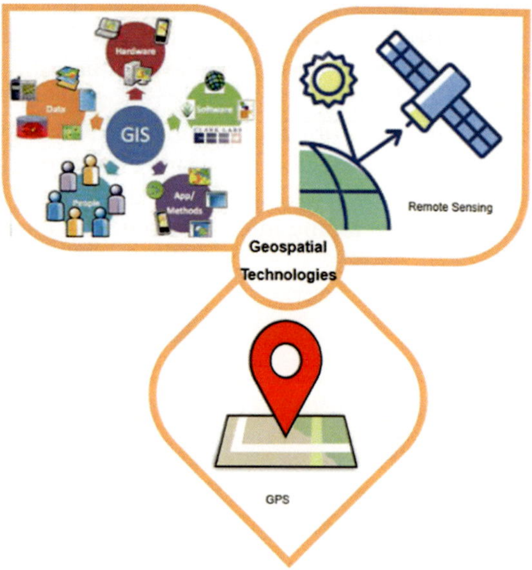

Fig. 24.3 Major three components of geospatial technologies

electromagnetic radiation (Sergieieva, 2025). The remote sensing process typically involves several key stages: the presence of an energy source (often the sun), the interaction of this radiation with the atmosphere, its subsequent interaction with the object of study, the recording of the energy by a sensor, the transmission of data to a processing facility, the analysis and interpretation of the processed data, and finally, its practical application (Sergieieva, 2025).

Electromagnetic radiation (EMR) underpins remote sensing, propagating as waves defined by wavelength, frequency, and velocity (Aggarwal et al., n.d.). The velocity of light (c) equals the product of frequency (ν) and wavelength (λ) ($c = \nu\lambda$) (Aggarwal et al., n.d.). EMR's different wavelengths interact with the atmosphere through scattering (Rayleigh, Mie, non-selective) and absorption (see "Fundamentals of Remote Sensing"), affecting energy reaching the Earth's surface and returning to the sensor. When EMR reaches the surface, it can be transmitted, reflected (specular or diffuse), scattered, or absorbed, quantified by radiometric properties like absorptance, transmittance, and reflectance (Mapping Around, 2023). Remote sensing systems can be broadly classified as passive or active. Passive systems rely on natural energy sources, primarily the sun, and measure the energy reflected or emitted by the target (Mapping Around, 2023). Examples include optical and thermal sensors. Active remote sensing systems, on the other hand, generate their own energy and measure the energy reflected back from the target (Mapping Around, 2023). Radar and LiDAR are examples of active sensors. Sensors used in remote sensing are often mounted on satellites, operating in various regions of the electromagnetic spectrum, including visible, infrared, and microwave (Siddiqui et al., 2021). Satellite missions are characterized by their orbit (e.g., Low Earth Orbit (LEO), Sun-Synchronous Orbit (SSO)), which influences their coverage and revisit time, as well as the spatial resolution (size of the area represented by one pixel), spectral resolution (number and width of spectral bands), and radiometric resolution (sensitivity of the sensor to differences in energy) of the data they collect (see "Fundamentals of Remote Sensing"). Understanding these principles is crucial for selecting and interpreting remote sensing data to analyze riverine geohazards (Table 24.1).

Geographic Information Systems (GIS) provide a framework for managing, analyzing, and visualizing spatial data, playing a pivotal role in the study and management of riverine geohazards (Waldo Tobler, n.d.). A GIS is a system designed to capture, store, manipulate, analyze, manage,

Table 24.1 Details of digital elevation model (DEM)

DEM	*Horizontal resolution*	*Absolute resolution*	*Relative resolution*
SRTM	30 m	~10–16 m	~6–8 m
ASTER DEM	30 m	~17 m	~10–12 m
ALOS PALSAR DEM	12.5 m	~5 m	~3–4 m
CartoDEM v3R1	10 m	~3–5 m	~2–3 m
Copernicus DEM	30 m and 10 m	~1–2 m	<1 m

and present data that are spatial or geographic in nature (Waldo Tobler, n.d.). It integrates location data, which specifies where things are, with descriptive information, which details what those things are like (What is GIS? Geographic Information System Mapping Technology, n.d.). The fundamental concept of a GIS is its ability to relate seemingly unrelated information by using location as a common index variable (Waldo Tobler, n.d.). GIS spatial data can be represented using two primary models: raster and vector (Wikipedia Contributors, 2025). The raster model represents geographic features as a grid of cells, where each cell contains a value representing a specific attribute. The vector model, conversely, uses geometric primitives such as points, lines, and polygons to represent discrete objects (Wikipedia contributors., 2025a). A functional GIS comprises five key components: hardware (the computer system), software (the GIS applications and tools), data (the geographic and attribute information), people (the users and analysts), and methods (the procedures and techniques used) (see "What is GIS? Components of a GIS—City of High Point") (Table 24.2).

The core tasks a GIS performs include data input, manipulation, management, query, and analysis (see "What is GIS? Components of a

Table 24.2 Some GIS *Software* for geohazards applications

List of software for geohazards	
• ArcGIS (Arc Hydro Tool) • QGIS (open source) • ArcGIS Pro • GEE (Cloud Computing Platform) • MODFLOW	• Global Mapper • HEC-RAS • HEC-HMS • HEC-GeoRAS (ArcGIS Extension) • Erdas Imagine etc

GIS—City of High Point"). Spatial analysis techniques are particularly relevant for studying riverine geohazards and include overlay analysis, which integrates different data layers to identify spatial relationships; network analysis, helpful in analyzing river networks and flow paths; and neighborhood analysis, which examines the characteristics of areas surrounding specific locations (see "What is GIS? Components of a GIS—City of High Point"). For spatial data to be effectively used in a GIS, it needs to be accurately georeferenced, meaning its location on the Earth's surface is precisely defined using a coordinate system (Waldo Tobler, n.d.). GIS databases are designed to store both geometric information (location and shape) and geographic phenomena' attributes (properties) (Wikipedia Contributors, 2025). Often, these databases utilize a relational database management system (DBMS) to organize and manage the data efficiently (see "What is GIS? Components of a GIS—City of High Point"). GIS provides the essential framework for integrating diverse geospatial data related to riverine environments, enabling comprehensive spatial analysis to understand geohazard processes and their impacts on vulnerable areas and infrastructure.

Global Navigation Satellite Systems (GNSS) are satellite constellations that provide location and time, enabling ground-based receivers to determine their exact position through trilateration (Patrick. Gindler, n.d.). Current operational GNSS include the U.S. Global Positioning System (GPS), Russia's GLONASS, Europe's Galileo, and China's BeiDou (Patrick. Gindler, n.d.). These systems consist of three segments: the space segment (the satellites), the control segment (ground stations), and the user segment (the receivers) (see https://www.meted.ucar.edu/sign_in.php?go_back_to=/GIS/GNSS_positioning/print.php). GNSS positioning measures distances to at least four satellites by calculating radio signal travel time from satellites to receiver (MOBATIME Swiss Time Systems, 2024). Precise time synchronization between satellites and the receiver is crucial for accurate positioning (MOBATIME Swiss Time Systems, 2024).

Global Navigation Satellite System (GNSS) technology offers essential positioning capabilities across various applications, such as surveying, mapping, and navigation (Patrick. Gindler, n.d.). Within riverine environments, GNSS is indispensable for accurately identifying data acquisition platforms, including aircraft and drones, and for establishing control points necessary for georeferencing remote sensing imagery (Mendez,

2024b). Multiple factors, including atmospheric delays, signal interference, and physical obstructions (Mendez, 2024b) may influence the precision of GNSS positioning. To improve accuracy, methodologies such as Differential GNSS (DGPS) and Real-Time Kinematics (RTK) are utilized, employing reference stations to furnish corrections to the measurements obtained by the receiver (Mendez, 2024a). In addition to positioning, the precise timing capabilities of GNSS are invaluable for synchronizing sensors and monitoring time-dependent geohazard processes within river systems. GNSS acts as a foundational technology, upon which numerous other geospatial data acquisition and analysis techniques pertinent to riverine geohazards are constructed.

24.3 Impact of Natural and Human Interference on the River System

GeoHazards in river systems arise from floods, erosion, landslides, and sedimentation, which result from both natural processes and human activities, posing serious risks to both our lives and livelihoods. When a settlement is situated in an area prone to collapses, the initial edge of a landslide and the spots where debris collects can, unfortunately, lead to injuries and property damage. Roads and bridges are especially at risk, facing threats from collapses, landslides, and debris flows, which can block travel routes and harm infrastructure. Our precious arable land also suffers as a result (Ding et al., 2019). These hazards threaten human settlements and the environment, varying in occurrence and severity based on geography and climate. River-related geohazards stem from hydrological and geological processes. Floods can result from intense rainfall, particularly in tropical, monsoonal, and cold mountainous regions where snowmelt swells rivers beyond their capacity. Coastal areas may experience severe flooding due to storm surges and tsunamis pushing seawater into river channels. Erosion occurs as water gradually wears down riverbanks and bottomlands, especially in high-altitude areas with steep gradients. Landslides, prevalent in mountainous regions, can happen after heavy rainfall destabilizes the soil. Earthquakes may also destabilize river valleys, leading to landslides. Excessive sedimentation complicates river dynamics as rivers carry and deposit sediment, affected by changes in flow patterns due to tectonic uplift (Roy & Vidyasagar, 2025).

Human activities have exacerbated these hazards; urbanization has disrupted river systems by replacing permeable soils with impermeable surfaces, increasing runoff and flooding risk. Poor drainage infrastructure heightens flood risk, while deforestation promotes erosion and landslides. Agriculture, particularly over-cultivation on slopes, causes sediment transport from fields, obstructing river channels. Dam construction alters natural river flow and sediment transport, while trapped sediment erodes riverbeds and disrupts ecosystems, risking severe flooding when maintenance is lacking (He et al., 2022). Mining and quarrying activities near rivers can destabilize the land and contribute to erosion and sedimentation by introducing loose material and undermining slope integrity, thereby heightening landslide risks. Climate change, driven by human activities, influences landslides through changes in rainfall patterns, intense storms, and rapid glacier melt, resulting in severe flooding and erosion. While natural processes cause landslides, their frequency and intensity increasingly relate to human activities. Urban development, deforestation, agriculture, dam construction, and mining all amplify risks in distinct geographical and climatic contexts, necessitating effective hazard management (Ding et al., 2019; Roy & Vidyasagar, 2025).

24.4 Environmental Crisis: Amplifying River Geohazards

The ongoing environmental crisis, primarily driven by climate change, significantly amplifies geohazards in river basins by altering natural processes. More intense and frequent rainfall events, a direct consequence of a warming climate, lead to increased and severe flooding, rapid riverbank erosion due to higher flow velocities and saturated soils, and a greater occurrence of landslides and debris flows as slopes become unstable. Additionally, accelerated glacier melt in mountainous regions can initially cause increased river discharge and dangerous glacial lake outburst floods, while rising sea levels contribute to inundation and altered sediment dynamics in lower river reaches (Chameides, 2017). These climate-induced changes are often exacerbated by human activities such as deforestation and unsustainable river management, creating a feedback loop where environmental degradation and geological hazards intensify one another, posing growing threats to communities and infrastructure within river basins (Gdrc, n.d.).

24.5 Geospatial Data Acquisition Techniques for Riverine Geohazard Analysis

We use several geospatial data acquisition techniques to collect valuable information for analyzing geohazards in river environments. Each technique offers unique data types with their own spatial, temporal, and spectral features, making them perfect for specific applications.

Satellite imagery provides a comprehensive and synoptic view of riverine landscapes, thereby facilitating the mapping and monitoring of extensive areas over time (Byrraju et al., 2020). Both optical and radar imagery hold significant value in studying riverine geohazards. Optical satellite imagery captures reflected sunlight across various spectral bands, such as visible, near-infrared, and shortwave infrared (see "Fundamentals of Remote Sensing"). These bands are particularly effective in differentiating between water, vegetation, and other land cover types (Sergieieva, 2025). For instance, water exhibits strong absorption in the infrared spectrum, making it easily identifiable in optical imagery (Nagel et al., 2023). The spectral signatures associated with vegetation can also yield insights regarding its health and stress levels, which may indicate erosion or other geomorphic processes (see "Fundamentals of Remote Sensing"). Multi-spectral data obtained from missions such as Landsat and Sentinel are extensively utilized for mapping river extents, monitoring changes in land cover within floodplains, and estimating river widths, all of which are crucial for flood hazard analyses (NASA Earth Observatory, n.d.). Moreover, spectrally based techniques have been developed to estimate river bathymetry (water depth) from optical satellite imagery (Legleiter & Niroumand-Jadidi, 2024).

Synthetic Aperture Radar (SAR) imagery is an active remote sensing method that emits microwave radiation to measure backscattered signals. One of its key benefits is the ability to penetrate clouds and function day and night, making it ideal for monitoring fast-changing river conditions like flooding, often in poor weather. Interferometric SAR (InSAR), along with its sophisticated variants such as Differential InSAR (DInSAR) and Persistent Scatterer InSAR (PS-InSAR), serves as robust methodologies for precisely detecting and tracking ground deformation. These techniques have proven effective in monitoring railway deformations, assessing changes in ground moisture near rivers, and identifying landslides triggered by river erosion or other factors. Several satellite missions like ERS-1/2, ENVISAT, RADARSAT-1/2, Sentinel-1, TerraSAR-X,

COSMO-SkyMed, and ALOS supply significant radar data essential for studying riverine geohazards (Cigna, 2018). The temporal resolution of satellite imagery, which varies depending on the satellite and its orbit, is critical for monitoring dynamic river processes. Missions like Sentinel-1 and Sentinel-2 offer high revisit frequencies, enabling near real-time monitoring of events like floods (Byrraju et al., 2020).

Aerial photography is a valuable tool for studying river environments, providing high-resolution imagery to map river morphology and bank changes. Historical collections offer a unique record for understanding long-term river system changes, including channel migration and floodplain evolution. Modern aerial photography and advances in photogrammetry and GIS software enable detailed mapping beyond typical satellite imagery. Softcopy mapping involves scanning high-resolution aerial photographs and processing them with digital elevation models (DEMs) to create three-dimensional views, revealing subtle geomorphic features like small landslides. Unmanned Aerial Vehicle (UAV) methods use orthophotography and high-precision DEMs to identify and monitor geohazards in reservoir bank slopes and riverine environments. These methods provide very high spatial resolution data for localized analysis.

LiDAR (Light Detection and Ranging) is an active remote sensing technology that uses pulsed lasers to measure distances to the Earth's surface, producing accurate three-dimensional data about its shape and features. In riverine studies, two key types of LiDAR are significant: topographic LiDAR, which utilizes a near-infrared laser to map land surfaces, and bathymetric LiDAR, which uses a water-penetrating green laser to measure the riverbed and seafloor elevations. USGS scientists employ lightweight, portable LiDAR systems on boats to collect detailed topographic information along rivers; when combined with bathymetric data, this creates comprehensive digital models of entire river channels. UAV-mounted LiDAR systems excel in densely vegetated alpine gorges by penetrating the canopy to gather terrain and geomorphological information on geohazards. LiDAR-derived Digital Elevation Models (DEMs) are precise and can generate various visual products, including hillshade maps and Red Relief Image Maps (RRIMs), which improve understanding of geomorphic features associated with riverine geohazards such as landslides, rockfalls, and debris flows. Additionally, LiDAR assesses sediment retention from erosion control barriers and quantifies riverbank erosion by comparing datasets over time. LiDAR's ability to provide high-resolution topographic data in vegetated landscapes makes it an essential tool for identifying subtle geomorphic features and understanding riverine geohazard processes.

Drone surveys using UAVs are a flexible, cost-effective method for acquiring high-resolution imagery and video of river environments. Drones capture data on short-lived events like debris flows and enable repeated surveys for monitoring stream erosion. Structure from Motion (SfM) techniques create three-dimensional models from the overlapping images, including high-resolution orthomosaics and Digital Elevation Models (DEMs). These products are vital for analyzing river morphology, bank stability, and geohazards' impacts. For instance, drone surveys examine how hurricanes reshape river channels via landslides and floods. USGS uses drones to map coastal environments and assess changes over time. Drones equipped with multispectral, near-infrared (NIR), and LiDAR sensors provide detailed land and water quality data, enabling precise assessments in remote areas. Some drones even feature radar systems to measure water depth, channel bathymetry, velocity, and river discharge. Their high mobility and rapid deployability make drones ideal for post-disaster riverine geohazard assessments (Fig. 24.4 and Table 24.3).

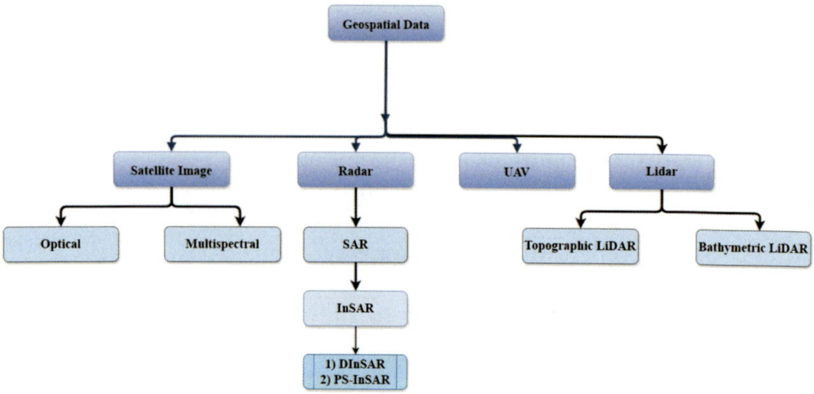

Fig. 24.4 List of data that can be used for geohazard application

Table 24.3 Types and triggers of riverine geohazards

Geohazard type	*Description*	*Common Triggers*
Overland flooding	The river overflows its banks and spills into the floodplain	Heavy rainfall, snowmelt, storm surges
River migration (avulsion)	The river creates a new main channel, reducing flow in the original channel	Changes in sediment load, slope, or flow regime; log jams; human interventions
Erosion (land loss)	Wearing away and removal of rocks and soil by water	Water flow exceeding bank resistance, vegetation removal, changes in flow velocity
Sedimentation (land gain)	Deposition of transported material	Changes in flow velocity, sediment supply, natural barriers, and human interventions
Debris floods/debris flows	Moving mass of earth, rock, and water traveling down a slope	Heavy rainfall, flash flooding, landslides, volcanic eruptions
Bank erosion	Wearing away of riverbanks	High flow, extra silt charge, river channel shift, changes in vegetation
Channel migration	Lateral movement of a river channel over time	Flow discharge, water surface slope, water velocity, sediment transport, bank erodibility
Sedimentation issues	Problems caused by excessive sediment deposition (e.g., navigation issues)	Increased erosion in the watershed, changes in river flow patterns, and human alterations to the river channel
Ice jams	Accumulation of ice that obstructs river flow, causing localized flooding	Rapid temperature changes, breakup of ice cover
Outburst floods	Sudden and massive flood resulting from the breach of a natural dam	Failure of glacial dams, landslide dams, or other natural obstructions
Landslides (river-erosion triggered)	Mass movement of earth or rock down a slope, initiated by river erosion	Undercutting of riverbanks, saturation of slopes by fluvial water, and alterations in slope stability resulting from fluvial processes

24.6 Analyzing River Morphology or Morphometric with Geospatial Technology

Geospatial technology provides a suite of powerful tools for analyzing the morphology of rivers, including their channel migration, bank erosion, and sedimentation patterns. Channel migration, a fundamental process in fluvial systems, can be effectively analyzed using multi-temporal remote sensing data and GIS techniques (Nagel et al., 2023). By comparing river channel boundaries extracted from satellite imagery acquired at different time intervals, it is possible to quantify the rates and patterns of channel movement (Nagel et al., 2023). Various satellite missions, such as Landsat and Sentinel, provide long-term archives of imagery that are invaluable for reconstructing historical channel migration patterns (Nagel et al., 2023). These analyses can reveal trends in channel shifting, identify areas prone to avulsion (sudden channel abandonment), and provide insights into the factors driving these changes, such as discharge and sediment load variations (Manjusree et al., 2002). GIS techniques, including the Digital Shoreline Analysis System (DSAS), allow for the statistical analysis of these changes, quantifying rates of erosion and accretion along riverbanks and tracking the net movement of the channel over time (Nath & Ghosh, 2024). Furthermore, specialized GIS tools and plugins, such as RiverMetrics for QGIS, automate the extraction of key morphological parameters like sinuosity (the degree to which a river meanders) and braiding (the presence of multiple interconnected channels), which are essential indicators of river channel stability and evolution (Rosa et al., 2024). Web-based applications are also being developed to leverage the vast amounts of freely available satellite imagery to continuously monitor river migration risks, particularly near critical infrastructure like bridges (Boothroyd et al., 2021). These advancements in geospatial technology enable a more comprehensive and efficient understanding of river channel dynamics than traditional field-based methods.

Riverbank erosion, a significant geohazard in many river systems, can be effectively assessed using remote sensing and GIS to map and measure erosion and accretion zones. Multi-temporal satellite imagery, particularly from Landsat, provides a historical record that identifies areas where riverbanks have retreated or advanced over time (Jaafar & Daliman, 2022). GIS techniques are used to delineate riverbank lines from imagery acquired at different dates and to calculate the area of land lost or gained due to erosion and accretion (Jaafar & Daliman, 2022). Tools like

DSAS facilitate the quantitative analysis of riverbank changes by calculating erosion and accretion rates at various points along the river (Hasan et al., 2025). Integrating GIS with other analytical techniques, such as machine learning, can enhance the accuracy and predictive capabilities of riverbank erosion assessments by incorporating various influencing factors, including hydraulic parameters, soil texture, and vegetation cover. GIS-based approaches can also be used to model the probability of riverbank erosion by considering factors such as the distance of floodplain areas from the active channel and flood magnitudes, as well as incorporating geomorphic variables like bank morphology and sediment type. The increasing availability of free and open-source geospatial tools and data makes these assessments more accessible, enabling better identification of vulnerable areas and the development of effective mitigation strategies.

Sedimentation patterns in river channels and floodplains can be effectively analyzed using LiDAR and other remote sensing techniques (King, 2017). LiDAR's ability to capture highly accurate three-dimensional topographic data makes it particularly well-suited for measuring and monitoring sediment deposition and erosion (Keane, n.d.). By comparing LiDAR datasets acquired at different time points, it is possible to quantify the volume of sediment that has accumulated or been eroded, providing valuable insights into sediment transport dynamics within river systems (Gupta et al., 2011). Both airborne and terrestrial LiDAR systems are used for these studies, with terrestrial LiDAR offering very high precision for localized measurements of sediment yield deposition (Alexiou et al., 2024). Bathymetric LiDAR can even map sediment deposits below the water surface, providing a more complete picture of sediment distribution in river channels (Keane, n.d.). Beyond LiDAR, analyzing changes in the spectral reflectance of water and land surfaces in satellite imagery can also provide qualitative and semi-quantitative information about sediment concentration and distribution (see "Fundamentals of Remote Sensing"). For instance, high turbidity levels, indicative of high sediment load, can often be detected in optical imagery. The ability to monitor sedimentation patterns is crucial for understanding river channel stability, assessing the impact of erosion control measures, and managing geohazards like flooding, which can be exacerbated by sediment accumulation (Table 24.4).

Table 24.4 Some satellite missions for river monitoring

Satellite mission	*Sensor type*	*Key spectral bands*	*Spatial resolution*	*Temporal resolution*
Landsat	Optical	Visible (Blue, Green, Red), Near-Infrared (NIR), SWIR	30 meters	16 days
Sentinel-1	Radar	C-band SAR	10 meters	6–12 days
Sentinel-2	Optical	Visible, NIR, SWIR	10–20 meters	5 days
SPOT	Optical	Visible, NIR, SWIR	2.5–20 meters	1–26 days
WorldView	Optical	Visible, NIR, SWIR	0.3–1.8 meters	Variable
TerraSAR-X	Radar	X-band SAR	1–16 meters	11 days
COSMO-SkyMed	Radar	X-band SAR	1–100 meters	Variable
ALOS	Radar	L-band SAR	10–70 meters	Variable

24.7 Mapping and Monitoring Specific Riverine Geohazards

Geospatial technology is critical in mapping and monitoring specific geohazards in riverine environments, including floods, landslides triggered by river erosion, and debris flows.

Flood mapping and monitoring have been significantly enhanced by integrating satellite imagery, Digital Elevation Models (DEMs), and GIS (Keane, n.d.). Satellite imagery, both optical and radar, delineates the spatial extent of flooding (Rosa et al., 2024). Optical imagery can effectively identify inundated areas based on the distinct spectral signature of water, particularly in the infrared spectrum (Nagel et al., 2023). Radar imagery, with its ability to penetrate cloud cover, is crucial for mapping floods during adverse weather conditions (Ray, n.d.). Accurate river widths derived from satellite imagery are also valuable for improving the accuracy of flood hazard analyses (NASA Earth Observatory, n.d.). High-resolution topographic information from digital elevation models (DEMs), often generated from light detection and ranging (LiDAR) data, is essential for creating detailed flood inundation models within a geographic information system (GIS) environment (Keane, n.d.). GIS platforms integrate these datasets, allowing for the visualization of flood extents, the analysis of flood frequency, and the assessment of potential impacts on infrastructure and populations (Rosa et al., 2024). Furthermore, the increasing availability of near real-time satellite data and automated GIS processing techniques enable timely flood monitoring, which is critical for issuing early warnings and facilitating effective disaster response (Team et al., 2025).

Landslide mapping and monitoring in riverine environments benefit greatly from applying various remote sensing techniques, including optical imagery, radar, and LiDAR, integrated within a GIS framework (Byrraju et al., 2020). Optical imagery can be used to identify landslide scars, changes in vegetation cover indicative of instability, and the deposition of landslide debris in river channels (Yang et al., 2024). Satellite radar imagery, particularly InSAR techniques, effectively detects subtle ground deformations that may precede or follow landslide events (Byrraju et al., 2020). LiDAR technology, with its ability to penetrate vegetation and provide high-resolution topographic data, is invaluable for mapping landslide morphology, identifying pre-failure conditions, and even detecting concealed landslides under forest cover (Yang et al., 2024). UAV photographic methods are also being used for early identification of potential reservoir geohazards, including landslides on steep riverbanks (Dai et al., 2024). GIS is crucial in integrating these diverse remote sensing datasets with other relevant information, such as geological maps, hydrological data, and historical landslide inventories, to create landslide susceptibility maps and assess the risk to river systems and infrastructure (Team et al., 2025). Monitoring the impact of landslides on rivers, such as changes in sediment load, channel blockage, or the potential for triggering secondary hazards like debris flows, is also facilitated by geospatial technology through post-event assessment using high-resolution imagery and topographic data.

Debris flow identification and tracking in riverine environments rely on applying remote sensing and GIS to map source areas, model runout paths, and monitor events (Boundary Integrated Watershed Service, 2022). High-resolution imagery from drones and satellites is particularly useful for mapping the steep slopes and channels in river catchments where debris flows typically initiate (Drones for the Geosciences: A new eye in the sky, 2018). DEMs derived from LiDAR or Structure from Motion (SfM) techniques applied to drone imagery can be used to model the potential runout paths of debris flows and identify downstream areas at risk (Yang et al., 2024). GIS platforms are essential for integrating topographic data, hydrological information, and land cover data to develop models that simulate the initiation and flow of debris flows (What is GIS? Geographic Information System Mapping Technology, n.d.). Rapid deployment of remote sensing platforms like drones after a debris flow event allows for quick assessment of the damage extent, identification of potential secondary hazards such as river blockage, and monitoring post-event recovery processes (Drones for the Geosciences: A new eye in the sky, 2018). LiDAR-derived DEMs can also aid in identifying the characteristic features of debris flows, such as accumulation fans, which help map their extent and understand their dynamics (Yang et al., 2024) (Table 24.5).

24.8 Geospatial Technology for Risk and Vulnerability Assessment of Riverine Geohazards

Geospatial technology, particularly GIS, provides a robust platform for assessing the risk and vulnerability of riverine geohazards for communities and infrastructure.

Identifying vulnerable communities and infrastructure is a crucial step in risk assessment, and GIS excels at integrating various spatial datasets to achieve this (Team et al., 2025). By overlaying maps of geohazard zones, such as flood inundation areas or landslide susceptibility zones, with spatial data on population density, infrastructure networks (roads, railways, pipelines, power lines), and land use, it becomes possible to pinpoint the specific elements at risk (Team et al., 2025). Overlaying flood maps with residential areas identifies individuals at risk of flooding. Mapping critical infrastructure like hospitals or schools in high-hazard zones allows for targeted preparedness (What is GIS? Geographic Information System Mapping Technology, n.d.). GIS integrates diverse information, enhancing understanding of spatial relationships and supporting informed decision-making in land-use planning and emergency management (What is GIS? Geographic Information System Mapping Technology, n.d.). Examining geohazards alongside socio-economic data in GIS helps identify vulnerable communities, factoring in population density, demographics (such as elderly or low-income groups with limited mobility), and at-risk economic activities (Team et al., 2025) (Fig. 24.5).

Table 24.5 Comparison of geospatial data acquisition techniques for riverine geohazards

Technique	*Spatial resolution*	*Temporal resolution*	*Advantages*	*Limitations*
Satellite imagery	0.3–100 meters	Daily to Monthly	Broad spatial coverage, long-term archives available, relatively cost-effective for large areas, can provide spectral information	Optical imagery affected by cloud cover has lower spatial resolution compared to aerial or drone imagery
Aerial photography	0.05–1 meter	Event-based	High spatial resolution and valuable historical data available can be used for detailed topographic mapping through photogrammetry	It can be expensive to acquire, has limited temporal resolution (usually flown for specific projects or after events), and requires orthorectification and DEM generation processing

(continued)

Table 24.5 (continued)

Technique	*Spatial resolution*	*Temporal resolution*	*Advantages*	*Limitations*
LiDAR	0.1–1 meter	Event-based to Periodic	Very high accuracy and detail in topographic data can penetrate vegetation (especially airborne), and bathymetric LiDAR can map riverbeds	Acquiring and processing can be expensive and require specialized equipment and expertise
Drone surveys (UAVs)	0.01–0.1 meter	On-demand to Periodic	High spatial resolution, highly flexible and rapidly deployable, cost-effective for localized areas, can capture video and oblique imagery	There is a limited coverage area per flight, battery life limitations, regulations that may restrict operations, and processing required for SfM to generate 3D data

Assessing risk levels entails the identification of vulnerabilities as well as the evaluation of the likelihood and intensity of hazards (see "Geohazards—Stantec"). Through the application of spatial analysis techniques, Geographic Information Systems (GIS) integrate factors pertaining to hazard probability, intensity, and vulnerabilities. Hazard probability may be estimated utilizing historical data, statistical analysis, or modeling of physical processes. Furthermore, hazard intensity, which includes metrics such as flood depth or landslide volume, can also be modeled using GIS in conjunction with topographic, hydrological, or geotechnical data. By amalgamating these hazard characteristics with data concerning vulnerable elements, GIS provides estimations of overall risk, frequently expressed in terms of potential damages or losses. For example, GIS has the capability to model flood scenarios (e.g., 100-year flood, 500-year flood) and to overlay inundation maps with property values in order to estimate economic losses. In addition, GIS can model the trajectories of

Fig. 24.5 Flood Tripura on 2024 (*Source* https://morungexpress.com/uploads/2024/08/89226619_1724416982_202408233210677.jpg and https://cf-img-a-in.tosshub.com/lingo/itne/images/story/202408/66cf3aac500b5-tripura-flood-285643192-16x9.jpg%3Fsize=948:533)

landslides and debris flows, thereby assessing which infrastructure and communities are likely to be impacted by these high-intensity hazards (see "Geohazards—Stantec").

Developing risk mitigation strategies is crucial for geospatial technology in riverine geohazards. GIS aids spatial planning by identifying development sites that minimize exposure to high-risk areas. It also helps design protective structures like flood levees by analyzing topographic data and simulating impacts. For disaster preparedness, GIS is vital for planning evacuation routes, locating emergency shelters, and pre-positioning resources. Furthermore, geospatial technology creates early warning systems for riverine hazards such as floods and landslides by integrating real-time data from remote sensing and ground sensors into a GIS. This setup monitors conditions leading to hazards, triggering alerts when thresholds are exceeded to warn communities for timely action.

24.9 Case Studies: Successful Applications of Geospatial Technology in Riverine Geohazards

- The use of satellite imagery and GIS to monitor riverbank erosion and its impact on communities along the Brahmaputra River in India (Figs. 24.6, 24.7 and 24.8).
- The application of LiDAR technology to map flood hazard assessment, improving flood forecasting and emergency response (Fig. 24.9).

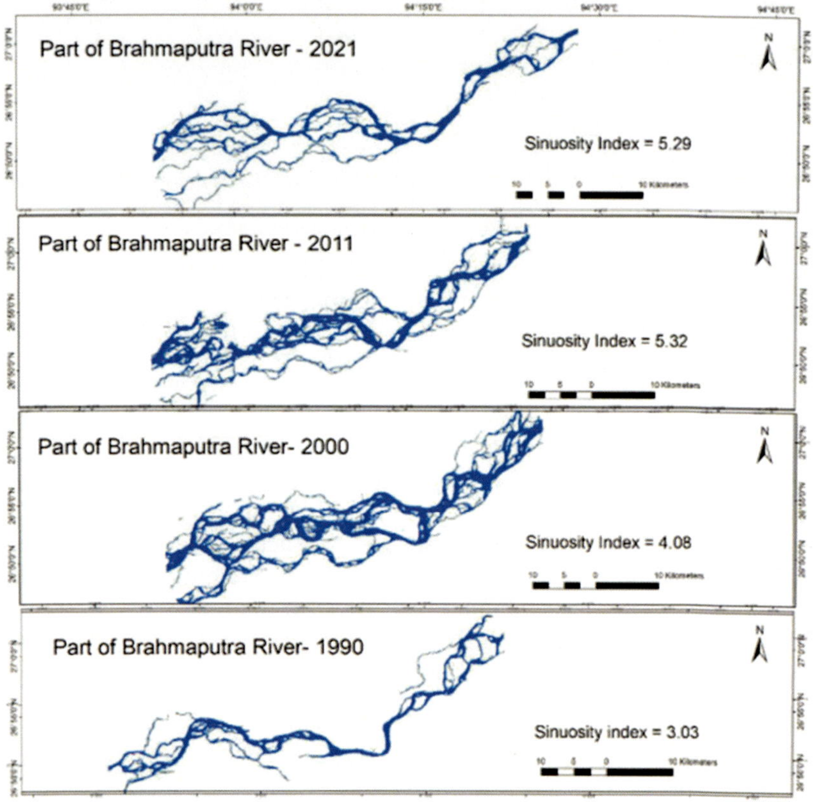

Fig. 24.6 Analysis on changes of River Morphology river bank (*Source* https://ars.els-cdn.com/content/image/1-s2.0-S2665972722000289-gr3.jpg)

Fig. 24.7 Brahmaputra River bank erosion (*Source* https://i0.wp.com/nenow.in/wp-content/uploads/2018/05/river-erosion.jpg?w=870&ssl=1)

24.10 Future Trends and Challenges in Geospatial Technology for Riverine Geohazards

Geospatial technology is increasingly used to understand and address geohazards in rivers, with emerging trends and challenges shaping its future. Remote sensing technologies like hyperspectral imaging provide detailed information about water quality parameters, vegetation stress, and land cover changes. Cloud computing platforms like Google Earth Engine and AWS are transforming how geospatial data is processed and analyzed, enabling large-scale studies and the development of sophisticated algorithms for real-time monitoring and modelling of riverine geohazards.

Geospatial analysis and modelling also improve understanding and management of these hazards. Machine learning techniques are being integrated with GIS for tasks such as automated feature extraction, predictive modelling of geohazard susceptibility, and enhancing risk assessments. The growing volume of geospatial data from various sources necessitates applying big data analytics techniques to extract meaningful insights for understanding complex riverine processes and associated hazards.

Fig. 24.8 River bank erosion in lower gangetic region (Malda and Farakka Region) (*Source* Discharge-driven rapid bank-erosion and its impact on sediment budgeting in the lower Gangetic plains—Scientific Figure on ResearchGate. Available from: https://www.researchgate.net/figure/Field-photographs-a-human-establishments-at-the-eroding-bank-of-the-Ganges-near_fig2_364347484 [accessed 11 Jun 2025])

Integrating GNSS with other geospatial technologies leads to enhanced monitoring capabilities. Real-time kinematic (RTK) GNSS provides centimeter-level accuracy in positioning, making it valuable for high-precision surveying of riverbank movement and ground deformation associated with landslides. Combining GNSS with drone technology allows for accurate tracking of UAV flight paths and precise georeferencing of imagery and topographic data collected, enabling high-resolution monitoring of changes in riverine environments over time.

However, challenges remain in the effective application of geospatial technology for riverine geohazard management, such as data accessibility limitations and technical challenges in integrating data from diverse

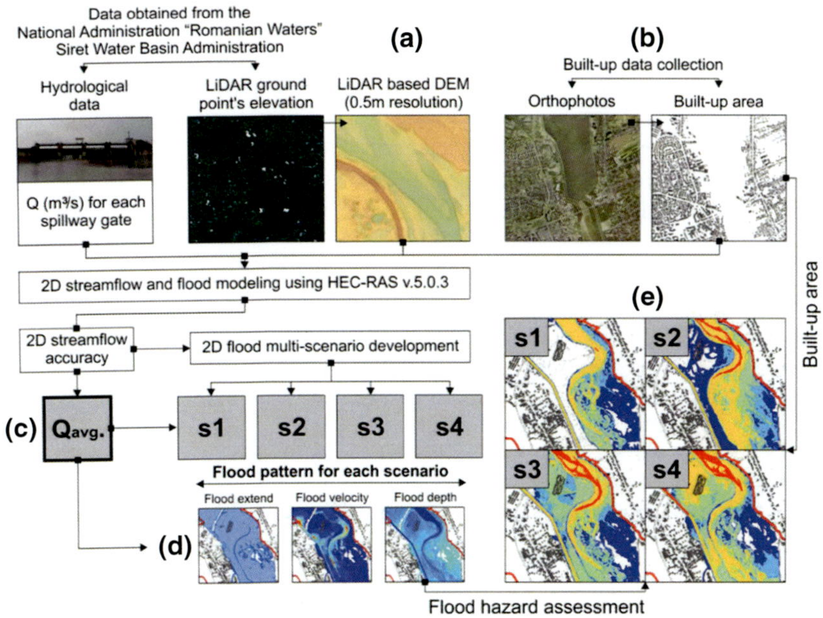

Fig. 24.9 LiDAR-based application for Flood Hazard assessment (*Source* Cîmpianu et al., 2019)

sources. Overcoming these challenges through improved data sharing protocols, standardization of data formats, and robust data integration techniques will be essential for realizing the full potential of geospatial technology in mitigating risks posed by geohazards in rivers.

24.11 Conclusions

Geospatial technology is crucial for mapping and managing geohazards in riverine environments, including floods, landslides, and debris flows. It enhances risk assessment and vulnerability identification, facilitating effective mitigation strategies. Advancements in hyperspectral imaging, satellite constellations, and cloud computing provide deeper insights into riverine geohazards. However, data accessibility, accuracy, and integration must be addressed. Ongoing research and collaborative data-sharing initiatives are vital for fostering resilient communities against rising risks.

Geospatial technology promotes safety and sustainability in riverine environments globally. Geo-hazards in river basins, such as floods, landslides, and erosion, have a significant impact on both the environment and human societies. They cause damage to infrastructure, disrupt ecosystems, displace communities, and lead to economic losses. Factors such as heavy rainfall, tectonic activity, and human interventions, including dam construction and deforestation, often exacerbate these hazards.

References

Admin, O. (2023, December 14). *Understanding the advancements in geospatial technology*. agaiti.com. https://www.agaiti.com/2023/12/understanding-advancements-in.html. Accessed November 14, 2023.

Aggarwal, S., & Photogrammetry and Remote Sensing Division, Indian Institute of Remote Sensing, Dehradun. (n.d.). Principles of remote sensing. In *Satellite remote sensing and GIS applications in agricultural meteorology* (pp. 23–38).

Alexiou, S., Papanikolaou, I., Schneiderwind, S., Kehrle, V., & Reicherter, K. (2024). Monitoring and quantifying soil erosion and sedimentation rates in centimeter accuracy using UAV-photogrammetry, GNSS, and t-LiDAR in a post-fire setting. *Remote Sensing, 16*(5), 802. https://doi.org/10.3390/rs16050802

A review of GNSS constellations · VectorNav. (n.d.). https://www.vectornav.com/resources/inertial-navigation-primer/theory-of-operation/theory-gnss

Australia, G. (2024, August 5). *Historical aerial photography*. Geoscience Australia. https://www.ga.gov.au/scientific-topics/national-location-information/historical-aerial-photography

Backman, L. (n.d.). *ESRI—About GIS*. https://maps.unomaha.community/Workshops/Career/ESRI/comp_gis.html

Boothroyd, R. J., Williams, R. D., Hoey, T. B., University of Glasgow, University of Glasgow, & Brunel University London. (2021). *"InfraRivChange": A Web-Based application to monitor river migration at sites of critical bridge infrastructure in the Philippines* (pp. 213–215). https://www.cdri.world/upload/fellowship/1781441302871997_202311020846infrarivchange%20a%20web%20based%20application%20to%20monitor.pdf

Boundary Integrated Watershed Service. (2022, July 15). *Boundary floods and hazards*. ArcGIS StoryMaps. https://storymaps.arcgis.com/stories/a281cb8f6d2e4d45928a3e1602e9deff

Brosnahan, S., & Sherwood, C. R. (2018). *Unmanned aerial systems capabilities of the U.S. geological survey woods hole coastal and marine science center*. Fact Sheet. https://doi.org/10.3133/fs20183061

Byrraju, S. V., Rizos, D. C., & Qian, Y. (2020). Satellite radar imagery for detection and monitoring of geohazards. *Transportation Research Record Journal of the Transportation Research Board, 2674*(3), 283–292. https://doi.org/10.1177/0361198120910746

Canada Centre for Remote Sensing. (n.d.). Fundamentals of remote sensing. In *Canada centre for remote sensing* (p. 7). https://natural-resources.canada.ca/sites/www.nrcan.gc.ca/files/earthsciences/pdf/resource/tutor/fundam/pdf/fundamentals_e.pdf

Chameides, B. (2017, August 3). *Getting down to the basics of sea level rise—Mississippi River Delta*. Restore the Mississippi River Delta. https://mississippiriverdelta.org/getting-down-to-the-basics-of-sea-level-rise/

Cigna, F. (2018). Observing geohazards from space. *Geosciences, 8*(2), 59. https://doi.org/10.3390/geosciences8020059

Cîmpianu, C. I., Stoleriu, C. C., Pérez, M. N., & Paveluc, L. E. (2019). Using high-density LiDAR data and 2D streamflow hydraulic modeling to improve urban flood hazard maps: A HEC-RAS multi-scenario approach. *Water, 11*(9), 1832. https://doi.org/10.3390/w11091832

Conaway, J., Eggleston, J. R., Legleiter, C. J., Jones, J. W., Kinzel, P. J., & Fulton, J. W. (2019). Remote sensing of river flow in Alaska—New technology to improve safety and expand coverage of USGS streamgaging. *Fact Sheet*. https://doi.org/10.3133/fs20193024

Dai, Z., Zhang, Y., Zhang, C., Fu, X., Zhang, P., & Ye, R. (2024). A photographic method to identify reservoir geohazards induced by rock mass deterioration of hydro-fluctuation belt. *Frontiers in Earth Science*, 12. https://doi.org/10.3389/feart.2024.1365272

Das, T., Sil, B. S., & Ashwini, K. (2024). Spatial modeling and mapping of riverbank erosion through the integration of machine learning application and in-situ data. *Canadian Journal of Civil Engineering*. https://doi.org/10.1139/cjce-2024-0402

De Rosa, P., Fredduzzi, A., & Cencetti, C. (2024). Geographic information system tools for river evolution analysis. *Water, 16*(17), 2512. https://doi.org/10.3390/w16172512

Dekaraja, D., & Mahanta, R. (2021). Riverbank erosion and migration inter-linkage: with special focus on Assam, India. *Environmental Systems Research, 10*(1). https://doi.org/10.1186/s40068-020-00214-0

Dey, S., Basu, A., Banerjee, S. N., & Jain, V. (2022). Discharge-driven rapid bank-erosion and its impact on sediment budgeting in the lower Gangetic plains. *Episodes, 46*(2), 259–267. https://doi.org/10.18814/epiiugs/2022/022027

Ding, M., Tang, C., & Miao, C. (2019). Response analysis of valley settlements to the evolution of debris flow fans under different topographic conditions: A case study of the upper reaches of Min River, China. *Bulletin of Engineering*

Geology and the Environment, 79(3), 1639–1650. https://doi.org/10.1007/s10064-019-01641-9

Drone based River Mapping & Surveying. IG Drones. (n.d.). *IG Drones.* https://igdrones.com/services/river-mapping-services

Drones for the Geosciences: A new eye in the sky. (2018). *New Mexico Earth Matters.* https://geoinfo.nmt.edu/publications/periodicals/earthmatters/18/n1/em_v18_n1.pdf

Five geological hazards to handle with integrity management. (n.d.). https://www.rosen-group.com/en/expertise/experience-center/blog/five-geological-hazards-you-can-handle-with-an-integrity-management-approach

Fundamentals of Remote Sensing. NASA Applied Sciences. https://appliedsciences.nasa.gov/sites/default/files/2022-11/Fundamentals_of_RS_Edited_SC.pdf. Accessed March 27, 2025.

Flood situation slightly improves in Tripura, CM takes aerial survey. https://www.morungexpress.com/flood-situation-slightly-improves-in-tripura-cm-takes-aerial-survey

Fulton, J. W., Anderson, I. E., Chiu, C., Sommer, W., Adams, J. D., Moramarco, T., Bjerklie, D. M., Fulford, J. M., Sloan, J. L., Best, H. R., Conaway, J. S., Kang, M. J., Kohn, M. S., Nicotra, M. J., & Pulli, J. J. (2020). QCAM: SUAS-based doppler radar for measuring river discharge. *Remote Sensing, 12*(20), 3317. https://doi.org/10.3390/rs12203317

Gdrc, H. S. (n.d.). *Cyclical Interlinkages of disasters and the Environment.* https://www.gdrc.org/uem/disasters/disenvi/cyclical.html

Geologic Hazards. Alaska Division of Geological & Geophysical Surveys. https://dggs.alaska.gov/hazards/index.html. Accessed March 27, 2025.

Geohazards—Alaska Nature and Science (U.S. National Park Service). https://www.nps.gov/subjects/aknatureandscience/geohazards.htm. Accessed March 27, 2025.

Geohazards—Stantec. https://www.stantec.com/en/services/geohazards. Accessed March 27, 2025.

Giri, M., & Nongkynrih, J. M. (2024). Comparison utilization of UAV images & satellite images for geospatial analysis. *International Journal of Innovative Science and Research Technology (IJISRT)*, 637–645. https://doi.org/10.38124/ijisrt/ijisrt24may083

Giri, S. (2020). Water quality prospective in twenty first century: Status of water quality in major river basins, contemporary strategies and impediments: A review. *Environmental Pollution, 271*, Article 116332. https://doi.org/10.1016/j.envpol.2020.116332

Global Navigation Satellite System (GNSS). (2005). *In Global Navigation Satellite System (GNSS).* https://www.princeton.edu/~alaink/Orf467F07/GNSS.pdf

Global Navigation Satellite System [Explained]. (2025, March 13). *Advanced navigation.* https://www.advancednavigation.com/tech-articles/global-navigation-satellite-system-gnss-and-satellite-navigation-explained/

Global Navigation Satellite System & Components of GNSS. (n.d.). https://www.onestopgis.com/Remote-Sensing/4-Ground-Truth-Data-and-GPS/3-Global-Navigation-Satellite-Systems/2-Global-Navigation-Satellite-System-Components-of-GNSS.html

Guide to Geologic Hazards. Alaska Division of Geological & Geophysical Surveys. https://dggs.alaska.gov/pubs/hazards-glossary

Guo, C., Xu, Q., Xiujun, D., Li, W.-l, Zhao, K., Lu, H., & Ju, Y. (2021). Geohazard recognition and inventory mapping using airborne LiDAR data in complex mountainous areas. *Journal of Earth Science., 32*, 1079–1091. https://doi.org/10.1007/s12583-021-1467-2

Gupta, S., Kessler, A., Dolliver, H., Univ. of Minnesota, & Univercity of Wisconsin. (2011). *Natural vs. anthropogenic factors affecting sediment production and transport from the Minnesota river Basin to lake Pepin.* https://wrl.mnpals.net/_flysystem/fedora/2023-09/natural-vs.-anthropogenic-factors-affecting-sediment-production-and-transport-from-the-minnesota-river-basin-to-lake-pepin.pdf

Harvard University Graduate School of Design. (2016, July 13). *Fundamentals of geographic information systems: Theory and applications—Harvard Graduate School of Design.* Harvard Graduate School of Design. https://www.gsd.harvard.edu/course/fundamentals-of-geographic-information-systems-theory-and-applications-spring-2003/

Hasan, J., Zaman, M. A. U., & Faridatul, M. I. (2025). Assessing spatiotemporal trends of riverbank erosion using a GIS-DSAS-based integrated approach. *Geology Ecology and Landscapes*, 1–19. https://doi.org/10.1080/24749508.2025.2451450

He, Y., Ding, M., Liu, K., & Lei, M. (2022). The impact of geohazards on sustainable development of rural mountain areas in the upper reaches of the Min River. *Frontiers in Earth Science*, 10. https://doi.org/10.3389/feart.2022.862544

https://www.meted.ucar.edu/sign_in.php?go_back_to=/GIS/GNSS_positioning/print.php

Impact. Collaborative Center for Landslide Geohazards. (n.d.). https://landslides-geohazards.gatech.edu/impact

Islam, Md. T. & Department of Earth Science, University of Gothenburg, Gothenburg, Sweden. (2010). River channel migration: A remote sensing and GIS analysis. In *Proceeding of 'ESA Living Planet Symposium*. Bergen, Norway. https://gup-server.ub.gu.se/v1/asset_data/128129.pdf

Jaafar, E., & Daliman, S. (2022). Analysis of river channel change using remote sensing images at Kelantan river basin. *IOP Conference Series: Earth and Environmental Science., 1102*, Article 012082. https://doi.org/10.1088/1755-1315/1102/1/012082

Kaku, K. (2018). Satellite remote sensing for disaster management support: A holistic and staged approach based on case studies in Sentinel Asia. *International Journal of Disaster Risk Reduction, 33*, 417–432. https://doi.org/10.1016/j.ijdrr.2018.09.015

Kalita, K. K., Kalita, K. K., & Kalita, K. K. (2021, November 11). *River erosion: An enduring problem*. NorthEast Now. https://nenow.in/opinion/river-erosion-enduring-problem.html

Kanickaraj, J. 2. 2. B. (2018, January 2). *Geographic Information Systems (GIS) defined*. ARC Advisory Group. https://www.arcweb.com/blog/geographic-information-systems-gis-defined

Keane, C. (n.d.). What is Lidar and what is it used for? American Geosciences Institute. https://profession.americangeosciences.org/society/intersections/faq/what-lidar-and-what-it-used/

Khatun, M., Rahaman, S. M., Garai, S., Das, P., & Tiwari, S. (2021). Assessing river bank erosion in the Ganges using remote sensing and GIS. In *Advances in Geographic Information Science* (pp. 499–512). https://doi.org/10.1007/978-3-030-75197-5_22

King, J. (2017, December 4). *Power of LIDAR*. Hidden Hydrology. https://www.hiddenhydrology.org/power-of-lidar/

Kraft, K. (2024, May 31). *7 geologic hazards and their water-related triggers*. Teren. https://www.teren4d.com/7-geohazards-water-related-triggers/

Legleiter, C. J., & Niroumand-Jadidi, M. (2024). Maps of water depth derived from satellite images of selected reaches of the American, Colorado, and Potomac Rivers acquired in 2020 and 2021 (ver. 2.0, September 2024): U.S. Geological Survey data release. https://doi.org/10.5066/P1APEJEP

Limited, I. (2019, October 22). *Essential components of GIS*. InfoSys BPM. https://www.infosysbpm.com/blogs/geospatial-data-services/gis-five-essential-components.html

Liu, M., Shao, Y., Li, R., Wang, Y., Sun, X., Wang, J., & You, Y. (2020). Method for extraction of airborne LiDAR point cloud buildings based on segmentation. *PLoS ONE, 15*(5), Article e0232778. https://doi.org/10.1371/journal.pone.0232778

Logan, J. B., & East, A. E. (2024). *Aerial imagery and structure-from-motion derived data products from a UAS survey of the debris flow at South Fork Campground*. U.S. Geological Survey data release. https://doi.org/10.5066/P144KDGN

Manjusree, P., Satyanarayana, P., Bhatt, C., Sharma, S., Srinivasa Rao, G., & National Remote Sensing Centre. (2002). Remote sensing and GIS for river morphology studies. In *Remote sensing applications area*. ISRO.

Mapping Around. (2023, November 30). *Fundamental principals of remote sensing—Mapping around*. https://mappingaround.in/fundamental-principals-of-remote-sensing-geospatial-analysis/

Mapping the river with lasers. (n.d.). *USGS*. https://www.usgs.gov/centers/columbia-environmental-research-center/mapping-river-lasers

Mendez, M. (2024, November 8). *Global Navigation Satellite System (GNSS) and Satellite Navigation explained*. Inertial Labs. https://inertiallabs.com/gnss-and-satellite-navigation-explained/

Mendez, M. (2024, August 29). *An in-depth look at the principles of GNSS*. Inertial Labs. https://inertiallabs.com/an-in-depth-look-at-the-principles-of-gnss/

MOBATIME Swiss Time Systems. (2024, June 20). *GNSS: Global navigation satellite systems—MOBATIME*. MOBATIME Swiss Time Systems. https://www.mobatime.com/article/gnss-global-navigation-satellite-system/

Nagel, G. W., Darby, S. E., & Leyland, J. (2023). The use of satellite remote sensing for exploring river meander migration. *Earth-Science Reviews, 247*, Article 104607. https://doi.org/10.1016/j.earscirev.2023.104607

NASA Earth Observatory. (n.d.). *A satellite view of river Width*. https://earthobservatory.nasa.gov/images/86172/a-satellite-view-of-river-width

Nath, A., & Ghosh, S. (2024). Geo-spatial analyses of meandering rivers, assessing past and future impacts on bank landforms and LULC changes. *Water Policy, 26*(12), 1234–1260. https://doi.org/10.2166/wp.2024.062

O'Leary, D., Harris, R., Piraux, O., Sommerville, A., & Stantec Consulting Ltd. (n.d.). *Geohazard mapping and evaluation using softcopy and LiDAR technologies*. Stantec Consulting Ltd.

Patrick. Gindler. (n.d.). *GNSS*. https://www.unoosa.org/oosa/de/ourwork/psa/gnss/gnss.html

Pelletier, R. E., Griffin, R. H., National Aeronautics and Space Administration, National Space Technology Laboratories, Earth Resources Laboratory, United States Department of Agriculture, Soil Conservation Service, & Southern National Technical Center. (1986). *Utility of remotely sensed data for identification of soil conservation practices*. In National Aeronautics and Space Administration (Technical Report DC-Y5–00400). https://core.ac.uk/download/pdf/42821277.pdf

Piovan, S. E., Hodgson, M. E., Mozzi, P., Porter, D. E., & Hall, B. (2023). LiDAR-change-based mapping of sediment movement from an extreme rainfall event. *GIScience & Remote Sensing, 60*(1). https://doi.org/10.1080/15481603.2023.2227394

Quantifying River Bank Erosion and Accretion Patterns along the Gorai River in Kushtia, Bangladesh: A geospatial analysis utilizing GIS and remote sensing techniques. Scientific Research Publishing. https://www.scirp.org/journal/paperinformation?paperid=131340. Accessed March 27, 2025.

Ray, S. S., Mahalanobis National Crop Forecast Centre, & Department of Agriculture & Cooperation. (n.d.). Basics of remote sensing. In *Mahalanobis National Crop Forecast Centre*. https://www.ncfc.gov.in/publications/p1.pdf

Roy, J., Vidyasagar Teachers' Training College, United Nations Environment Programme, Wohl, Wohl, Wohl, National Transportation Library, & Wikipedia. (2025). Geohazards in river systems: Causes, impacts & mitigation strategies. In *Journal of Emerging Technologies and Innovative Research* (Vol. 12, Issue 4). https://www.jetir.org

Satpalda. (2025, February 19). *Components of remote sensing*. SATPALDA. https://satpalda.co/components-of-remote-sensing/

Sergieieva, K. (2025, March 13). *Remote sensing: Painting a picture of Earth from above*. EOS Data Analytics. https://eos.com/blog/remote-sensing/

Siddiqui, M. A., Jamia Millia Islamia, Said, S., Punia, M. P., BITS, Jamia Millia Islamia, & AMU. (2021). *Basic Principles of Remote Sensing* [Book]. In M. P. Punia & BITS, Remote Sensing, GIS and GPS. https://www.shivajicollege.ac.in/sPanel/uploads/econtent/a04bd35bcdf8b338ba0cc3622571158c.pdf

Team, S., Team, S., & Team, S. (2025, March 26). *Explaining the Six Components of a Remote Sensing System—SkyWatch*. SkyWatch. https://skywatch.com/explaining-the-six-components-of-a-remote-sensing-system/

Toal, A. (2024, April 8). *Soil testing in agriculture: Enhancing sustainable farming*. RhizeBio—Improving Soil Performance. https://rhizebio.com/2023/07/27/the-importance-of-soil-testing-in-agriculture/

Tripura flood causes Rs 306 crore damage to Urban Local Bodies, relief operations underway. https://cf-img-a-in.tosshub.com/lingo/itne/images/story/202408/66cf3aac500b5-tripura-flood-285643192%E216x9.jpg%3Fsize=948:533

Tualcom. (2025, April 29). *Understanding GNSS receivers—TUALCOM*. TUALCOM. https://www.tualcom.com/understanding-gnss-receivers/

Using LIDAR to measure sediment retention by sediment barriers: College of Architecture, Design and Construction. (2021, October 15). *College of architecture*. Design and Construction. https://cadc.auburn.edu/using-lidar-to-measure-sediment-retention-by-sediment-barriers/

Waldo Tobler. (n.d.). *Fundamentals of GIS* (Unit–1). https://www.pmctech.org/images/cs/pdf/course/15.pdf

Wang, C., Zhang, W., Ji, Y., Marino, A., Li, C., Wang, L., Zhao, H., & Wang, M. (2024). Estimation of aboveground biomass for different forest types using data from Sentinel-1, Sentinel-2, ALOS PALSAR-2, and GEDI. *Forests, 15*(1), 215. https://doi.org/10.3390/f15010215

What is GIS? Components of a GIS—City of High Point. https://www.highpointnc.gov/998/What-is-GIS-PDF. Accessed March 27, 2025.

What are the five components of GIS? (2025, January 19). *Kent State Online*. https://onlinedegrees.kent.edu/blog/what-are-the-five-components-of-gis

What are glacial lakes? National Snow and Ice Data Center. (n.d.). https://nsidc.org/learn/ask-scientist/what-are-glacial-lakes

What is GIS? Geographic Information System Mapping Technology. (n.d.). https://www.esri.com/en-us/what-is-gis/overview

Wikipedia Contributors. (2025, April 1). *Geological structure measurement by LiDAR*. Wikipedia. https://en.wikipedia.org/wiki/Geological_structure_measurement_by_LiDAR

Wikipedia contributors. (2025, April 1). *Geographic information system*. Wikipedia. https://en.wikipedia.org/wiki/Geographic_information_system

Winterbottom, S. J., & Gilvear, D. J. (2000). A GIS-based approach to mapping probabilities of river bank erosion: Regulated River Tummel. *Scotland. Regulated Rivers Research & Management, 16*(2), 127–140. https://doi.org/10.1002/(sici)1099-1646(200003/04)16:2

Wu, T., Xu, Z., Chen, R., Wang, S., & Li, T. (2023). Channel activity remote sensing retrieval model: A case study of the lower yellow river. *Remote Sensing, 15*(14), 3636. https://doi.org/10.3390/rs15143636

https://www.highpointnc.gov/998/What-is-GIS-PDF#:~:text=A%20working%20GIS%20integrates%20five,data%2C%20people%2C%20and%20methods.&text=Hardware%20is%20the%20computer%20on,stand%2Dalone%20or%20networked%20configurations

Yang, Y., Zhao, Z., Zhou, D., Lai, Z., Chang, K., Fu, T., & Niu, L. (2024, June 21). Identification and analysis of the geohazards located in an alpine valley based on multi-source remote sensing data. *Sensors (Basel), 24*(13): 4057. PMID: 39000836; PMCID: PMC11244093. https://doi.org/10.3390/s24134057

CHAPTER 25

From Flowing to Fragmented River: A Geospatial Study of Channel Decay and Anthropogenic Influence on the River Churni of West Bengal, India

Debika Ghosh and Shreyasee Das Mistry

25.1 Introduction

Water resource is the key element of our life. From the ancient past, human civilizations have been developed along the banks of river like the Harappan civilization was evolved along the bank of Indus, Mesopotamia civilization was evolved along the banks of river Tigis-Euphates. There are so many examples of development of human civilizations along the river banks to take advantages of river as a resource. World's first civilizations were developed along River Indus, Nile, Huang He, Tigris and Euphrates

D. Ghosh (✉) · S. Das Mistry
Post Graduate Department of Geography, Krishnagar Government College, Nadia, West Bengal, India
e-mail: idebikaghosh@gmail.com

S. Ghosh (ed.), *Reimagining Indian Rivers for Sustainability*, Palgrave Studies in Environmental Sustainability,
https://doi.org/10.1007/978-3-032-04754-0_25

(Mackline & Lewin, 2015; Wang & He, 2022). There are so many advantages to live along the banks of a river. Rivers are not just body of water but it also acts as a medium of transportation, source of water to the agricultural field, domestic uses, habitat for many species. Therefore, it can be said that river is a lifeline to the organic community for their survival. Humans depend on rivers for various essential needs like drinking water, irrigation, and industrial use, crop irrigation, transportation, food sources, hydroelectric power generation, recreational activities, etc. But unfortunately, various human activities are continuously affecting very badly the entire river system. Various human activities like heavy construction along the river banks, unplanned settlement growth, construction of dams, deforestation, disposal of waste into the river water etc. lead to various problems in river system. Anthropogenic activities play crucial role in destruction of channel, pollution of river water, flow destruction or simply it alters the entire geomorphic processes associated with river. Decay of river is a burning problem in today's world. Development of human civilizations along the river banks and different anthropogenic activities continuously act as an influential factors for decaying river ecosystem (Ekka et al., 2019; Li et al., 2022). Rivers are being badly affected by pollutant water in industrial region (Crini et al., 2022; Sahu, 2023; Sekharan et al., 2022). Near about 99% of river in the world is affected by various human activities like intensive cultivation, expansion of urban area, use of fertilizer and pesticides and other contaminants (Li et al., 2022; Meter et al., 2016). Different anthropogenic activities like urbanization, population growth, future development of hydro power and dam have very much bad effects on rivers and it leads to destroy the water quality (Sidabutar et al., 2017; Winton et al., 2021). Rivers have got affected badly by different national water ways (Roy Choudhury, 2020).Decay of river denotes the shrinking of river channels mainly caused by various anthropogenic activities. The prime objective of this research is to study the problem of river decay and tries to find out human impacts on the river decay using RS and GIS technique. After disintegration, this research focuses

- To study the spatio-temporal change of decaying river over long period of time
- To find out the change of land use land cover within 1 Kilometer (km) buffer zone of both sides of the decaying river over long period of time.

- This study tries to identify connection between decay of channel and anthropogenic activities.
- To focus on the management of decaying rivers.

To achieve the objectives of the study, river Churni of Nadia district, West Bengal, India, have been taken into consideration. The present study has its importance at local to international levels. This research shows scenario of conversion of a living river into a decaying river. This is an attempt to show how various human activities gradually destroyed a living river. This study established direct connection between various human activities and a living river scientifically. And, the study also focuses on the management of decaying rivers which will be definitely fruitful for the management of river decay and this will be beneficial for the society. The preconceived idea has been taken in this research that **"Decay of riveris highly caused by various human activities".**

25.2 Materials and Methods

25.2.1 Study Area

River Churni is one of the most important streams in Nadia district of West Bengal. River Churni is the created distributary of river Mathabhanga (District Census Hand Book Nadia, 2011a, b). The river Mathabhanga divides into two rivers Ichhamati and Churni near Majhdia at Krishnaganj Community Development (C.D.) block of Nadia district (District census Handbook, Nadia, 2011a, b; District Survey Report of Nadia District, Government of West Bengal, 2022). It is flowing from the source of river Mathabhanga near Majdia of Krishnaganj Community Development (C.D.) block to its mouth at river Bhagirathi near Shibpur at Ranaghat-II C.D. block (District Census Hand Book Nadia, 2011a, b). The river channel is situated from 23.404449 N, 88.710259 E at its source at Majhdia of Krishnaganj C.D. block to 23.132949 N, 88.502289 E at its confluence near Shibpur of Ranaghat-II C.D. block. The river Churni is flowing at the eastern to middle part of Nadia district. This is the Moribund part of Gangetic delta (District Census Hand Book Nadia, 2011a, b). River Churni passes through krishnaganj, Hashkhali, Ranaghat-I and Ranaghat-II C.D. blocks (Fig. 25.1). In micro level administrative division passes through Araghata, Chakda,

Hijuli 1 and 2 Gram Panchayat, Shivnibus Gram Panchayat, krishnaganj, Ramnagar I, Kalinarayanpur and Anulia which entirely fall under Nadia district administrative area. River Churni is an important distributary of river Mathabhanga and river Mathabhanga is originated from the riverbank of Padma at Munshigunaj near Kustia district in Bangladesh. Therefore, it can be said that river Churni is an artificial canal (Thakur, 2022). River Churni is seriously affected by various human activities. And, the river Churni is decaying day by day. The river Churni provides huge support to the community residing along the banks of river Churni as a resource (Thakur, 2022). The total number of households in the four community development (C.D.) blocks are 289,308 where the total number of households are 78,432, 56,507, 85,687, and 68,682 in Kaliganj Ranaghat-I, Ranaghat-II, Hanskhali community development (C.D.) blocks respectively (District Census Hand Book Nadia, 2011a, b).

Remote sensing and GIS based methodology has been accomplished in this study (Fig. 25.2).

25.2.2 Data Collection

The present work has been done based on satellite images which have been collected from reliable sources (Table 25.1). Google Earth images of 2010 and 2024 have also been consulted to do the present work.

25.2.3 Channel Width (W), Length, and Sinuosity Index of River Churni from 2010 to 2024

To calculate the width of the river Churni in 2010 and 2024, total sixteen (16) cross sections have been drawn across the river Churni. (Fig. 25.3) and here, total stretch of river Churni has been considered (fragmented part of river also included). The name of the cross sections has been given like A, B, C, D, E, F, G, H, I, J, K, L, M, N, O, P, in both years. Total four (04) cross sections in every community development block; have been drawn to measure the width of river Churni in 2010 and 2024 (Table 25.2) by using GIS technique. The length of the river has also been measured using GIS technique in 2010 and 2024 (Fig. 25.4).

Sinuosity index (Eq. 25.1) of river Churni for the year 2010 and 2024 have been measured based on the Sinuosity index proposed by Schumm (1963). Sinuosity is the deviation of channel that is the observed path from its expected path (Charlton, 2008). The formulae for the calculation

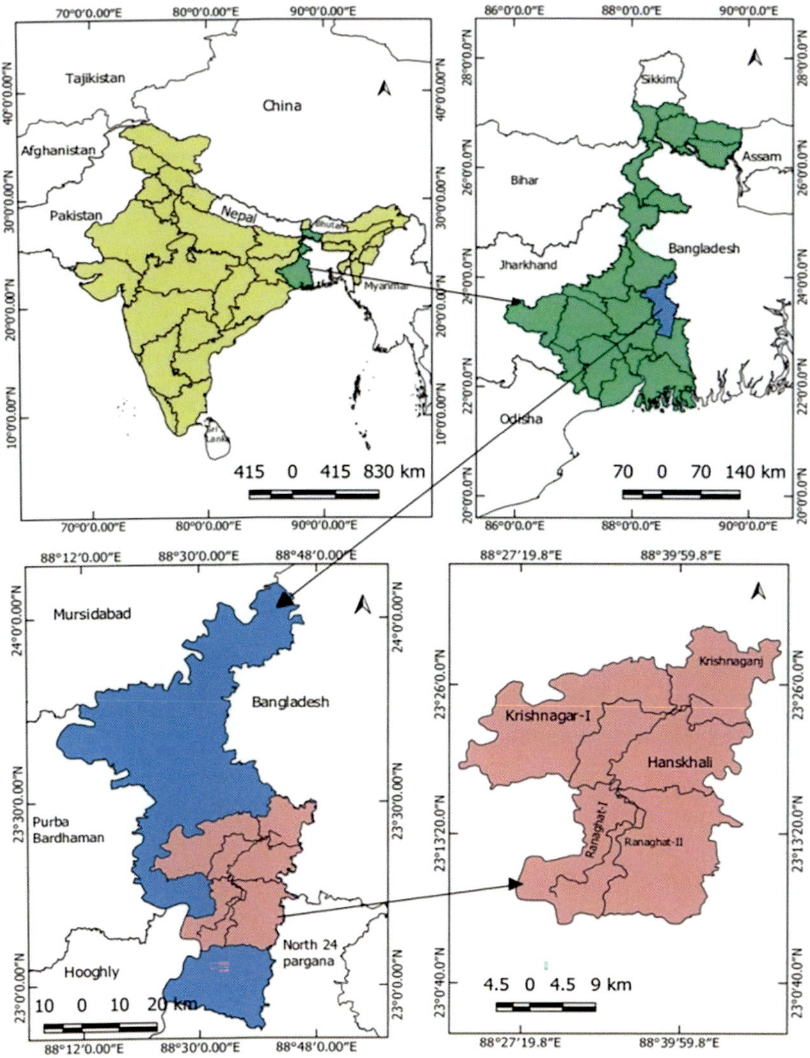

Fig. 25.1 Location map of the study area

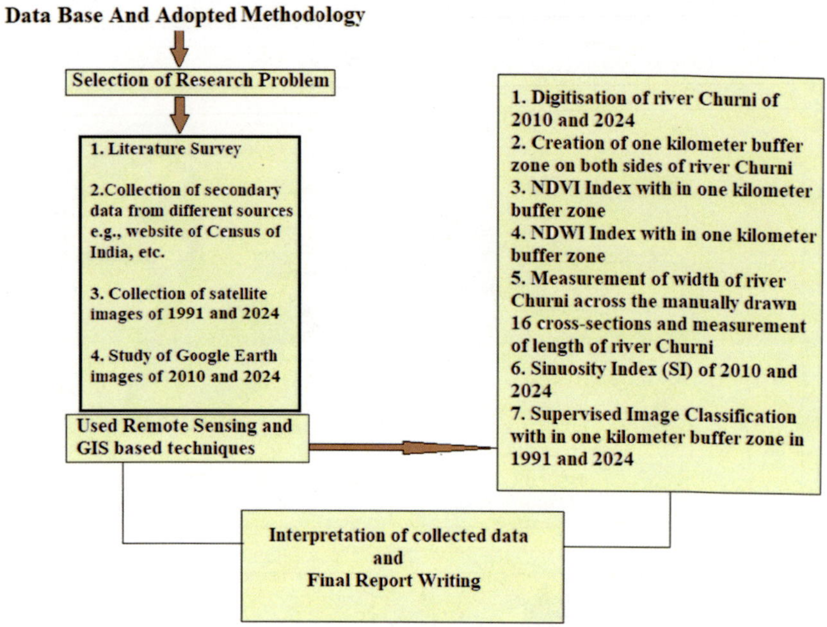

Fig. 25.2 Data base and adopted methodology used in this research

Table 25.1 Details of used satellite images

Date of acquisition imageries	*Path/row*	*Data set*	*Producer*
09.05.1991	138/44	MSS (Landsat-5)	USGS
29.02.2024	138/44	OLI_TIRS (Landsat-8)	USGS

of Sinuosity index (SI) is as follow:

$$\mathbf{SI} = \mathbf{OL}/\mathbf{EL} \tag{25.1}$$

(Where, SI = Sinuosity index, OL = Observed path/Actual path of the river, EL = expected path of the river). The SI value 1.0 indicates straight channel, SI value 1.0–1.5 indicates sinuous channel and SI value greater than 1.5 indicates meandering channel.

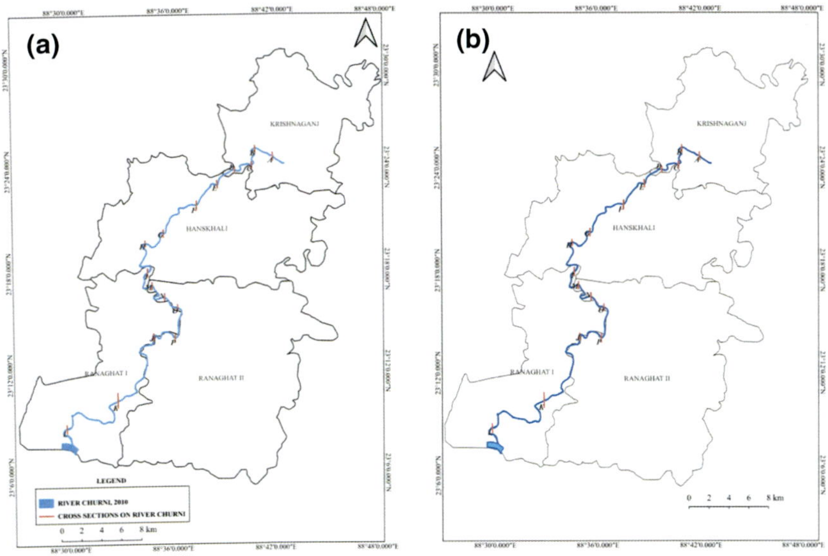

Fig. 25.3 Sixteen (16) cross sections across the river Churni in 2010 (**a**) and 2024 (**b**)

In this study sinuosity index has been used to detect the deviation of river Churni from its expected path. And, here, only the water containing parts of river Churni have been considered to compute SI. The change in SI indicates changes the in channel planform where human activities can not be ignored.

To gather knowledge about the area of the study, district census handbook, Nadia, 2011, and district survey report of Nadia district, November 2022, have been consulted.

25.2.4 Buffer Zone Creation

One kilometer buffer zone on the both sides of river Churni has been created using GIS technique. NDVI, NDWI, and Supervised Image Classification have been done within the buffer zone which has been created on the both sides of river Churni for the identification of changes of different features like vegetation cover, water bodies, settlement areas etc.

Table 25.2 Measuring the width (in meters) of river Churni in 2010 and 2024

C.D. blocks	*Name of the cross sections*	*Width of river Chiurni (2010)*	*Width of river Chiurni (2024)*	*Average width of river Chiurni (2010)*	*Average width of river Chiurni (2024)*
Krishnaganj	A	52.092	52.113	42.334	38.5125
	B	37.682	33.676		
	C	34.759	34.873		
	D	44.803	33.388		
Hanskhali	E	41.591	20.176	51.29225	32.5805
	F	58.485	37.331		
	G	55.586	35.553		
	H	49.507	37.262		
Ranaghat I	I	61.638	61.028	75.2425	64.665
	J	105.415	95.729		
	K	57.913	50.662		
	L	76.004	51.241		
RANAGHAT II	M	43.924	36.714	52.29275	37.39775
	N	59.594	51.307		
	O	55.592	36.054		
	P	50.061	25.516		

25.2.5 Normalized Difference Vegetation Index (NDVI)

The Normalized Difference Vegetation Index (NDVI) (Eq. 25.2) is one of the important indexes used in remote sensing and GIS. It is used to clarify the complicated multi-spectral imagery and it is the most popular index for vegetation assessment. NDVI is basically calculated with any multispectral sensor with a visible and a near-IR band (Huang et al., 2021). In this present study, NDVI has been used to know the change in vegetation cover within one kilometer buffer zone in the year 1991 and 2024. The formulae of NDVI (Bhatta, 2021) is as follow:

$$\mathbf{NDVI} = (\mathbf{NIR} - \mathbf{RED})/(\mathbf{NIR} + \mathbf{RED}) \tag{25.2}$$

where: NIR is the reflectance in the near-infrared band (700–1100 nanometers), RED is the reflectance in the red band (600–700 nanometers).

The values of NDVI range from 0.3 to 0.8 indicate a dense and healthy vegetation canopy. Negative values indicate clouds, water, and snow fields.

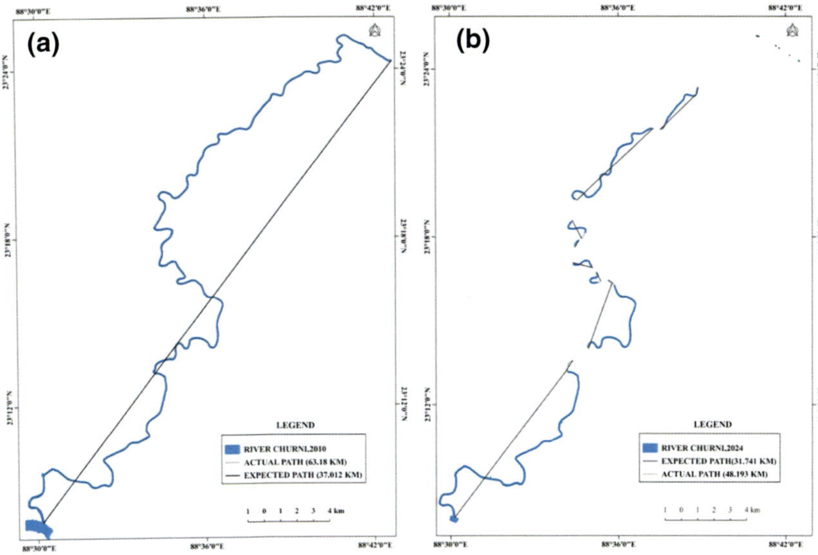

Fig. 25.4 Calculated Sinuosity Index of river Churni from 2010 (**a**) to 2024 (**b**)

The values of NDVI range from 0.1 to 0.2 indicate soil. The value of NDVI ranges from 0.3 to 0.6 indicates stressed vegetation (Bhatta, 2021). In this present research, NDVI index has been used to detect the condition of vegetation within one kilometer buffer zone from both side of river Churni because in maintenance of stream corridor healthy vegetation play very important role.

25.2.6 Normalized Difference Water Index (NDWI)

Normalized different water index (NDWI) (Eq. 25.3) is calculated from satellite images and this index is commonly and successfully used to identify surface water body and to detect the quality of water body (Ozelkan, 2019). In this present study, NDWI has been used to know the change in surface water body within one kilometer buffer zone in the year 1991 and 2024. The formulae of NDWI (Bhatta, 2021) is as follow:

$$\mathbf{NDWI = (NIR - SWIR) / (NIR + SWIR)} \tag{25.3}$$

where: NIR stands for the near-infrared band (700–1100 nanometers), SWRI stands for Short-Wave Infrared (900 to 1700 nanometers).

NDWI values ranges from −1 to +1.

Generally, the values of NDWI range from −1.0 to +1.0. The values of NDWI from 0.5 and more than 0.5 indicate water bodies. Negative value (−0.2) of NDWI indicates built up areas. Values of NDWI toward −0.1 indicates water less dry areas. Vegetation has very smaller values in NDWI which distinguishing vegetation easily from water bodies. In this present research, NDWI index has been used to detect the condition of water body within one kilometer buffer zone from the both sides of river Churni which is very necessary to know.

25.2.7 Land Use/Land Cover Change

Supervised image classification has been done using maximum likelihood method using QGIS software for the year 1991 and 2024 within the one kilometer buffer zone on the both sides of river Churni to identify the changes in land use and land cover. In the years 1991 and 2024, the features like river, agricultural use in the channel, agricultural land, vegetation or forest area have been identified within one kilometer buffer area to find out changes in land use and land cover. The percentage of area under the above selected parameters for land use and land cover change has been calculated using QGIS software. The land use and land cover change within one kilometer buffer area has been done to identify the positive or negative changes in selected parameters for land use and land cover in this study. The changes may help to draw a connection between increased human activities and loss of natural coverage.

25.3 Results and Discussion

25.3.1 Spatio-Temporal Changes in River Churni

25.3.1.1 Change in the Width and Length of River Churni from the Year 2010 to 2024

The width of river Churni has been changed drastically across the drawn cross sections. It has been observed from Figs. 25.5 and 25.6 that the widths of river Churni have been decreased across 16 cross sections. The average width (in meter) of river Churni were 42.33, 51.29, 75.24, and

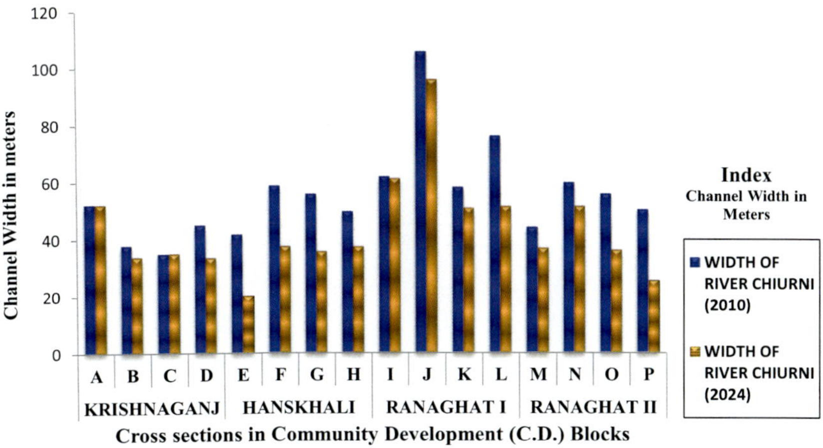

Fig. 25.5 Changing width of river Churni in C.D. Blocks across the sixteen cross sections of from the year 2010 to 2024

52.29 at Krishnaganj, Hanskhali, Ranaghat-I, and Ranaghat-II community development blocks respectively in the year 2010 (Fig. 25.6). The average width (in meter) of river Churni are 38.51, 32.58, 64.66, and 37.39 at Krishnaganj, Hanskhali, Ranaghat-I, and Ranaghat-II community development blocks respectively in the year 2024 (Fig. 25.6). Thus, it can be said that there are massive change in the average width of river Churni from the year 2010 to 2024. The reduction in width of river Churni indicates gradual decay of river Churni across the cross sections.

It has observed that the length of river Churni has been decreased in the year 2024 from the year 2010 (Fig. 25.4). The measured length of river Churni was 63.18 kilometers in the year 2010 and surprisingly, it is 48.19 kilometers in 2024. Therefore, river Churni has lost 14.99 kilometers length from the year 2010 to 2024.

25.3.2 *Change in the Sinuosity of River Churni from the Year 2010 to 2024*

The calculated value of sinuosity index shows that there is a change in the sinuosity index of river Churni from the year 2010 to 2024. In the year 2010, the value of sinuosity index was 1.84 and in the year 2024,

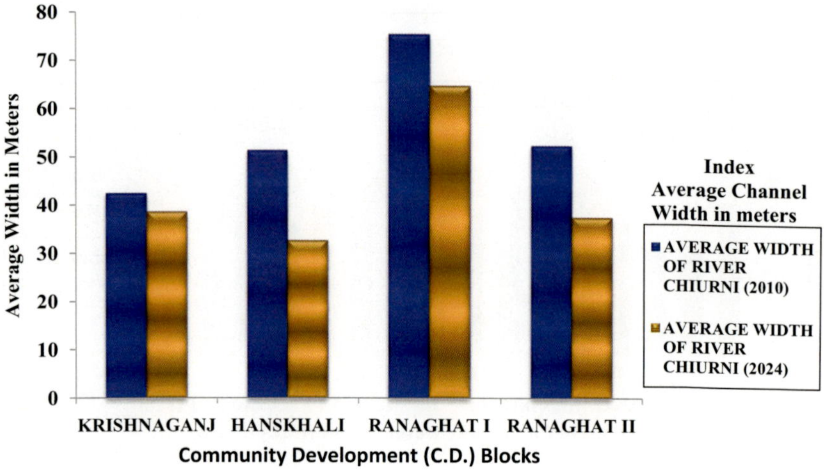

Fig. 25.6 Changing of average width of river Churni in C.D. Blocks across the sixteen cross sections of from the year 2010 to 2024

the value of sinuosity index is 1.33 (Fig. 25.7). According to sinuosity value of SI proposed by Schumm (1963), it can be said that path of river Churni in the year 2010 was meandering and the path of river Churni is sinuous in the year 2024. So, there is a change in channel planform. The river Churni turned into sinuous from meandering path in the year 2024 and the anthropogenic activities cannot be ignored. Specially, uses of channel bed for different purposes.

25.3.3 Changes in the Vegetation Cover from the Year 1991 to 2024

The calculated values of NDVI clearly show that in the year 1991, the highest value of NDVI was 0.0263 and the lowest value of NDVI was −0.221(Fig. 25.8a). The highest value of NDVI is 0.0463 and the lowest value of NDVI was −0.056 (Fig. 25.8b) in the year 2024. It can be easily said from the values of NDVI in both years that there are no healthy and dense vegetation cover within one kilometer buffer zone on the both sides of river Churni.

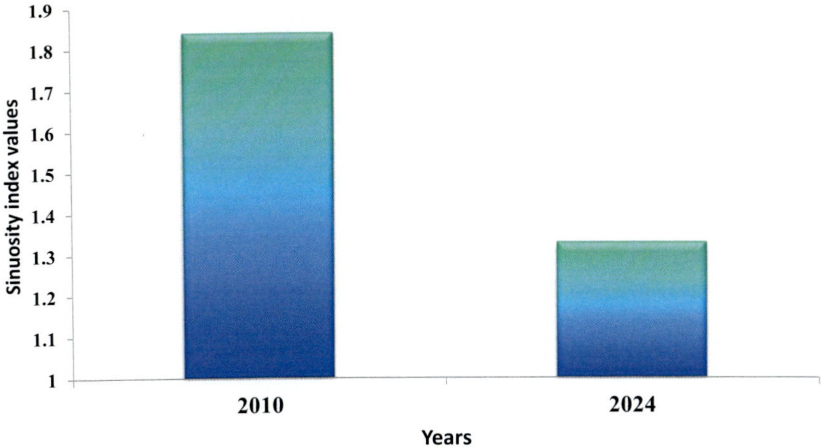

Fig. 25.7 The sinuosity index of river Churni from the year 2010 to 2024

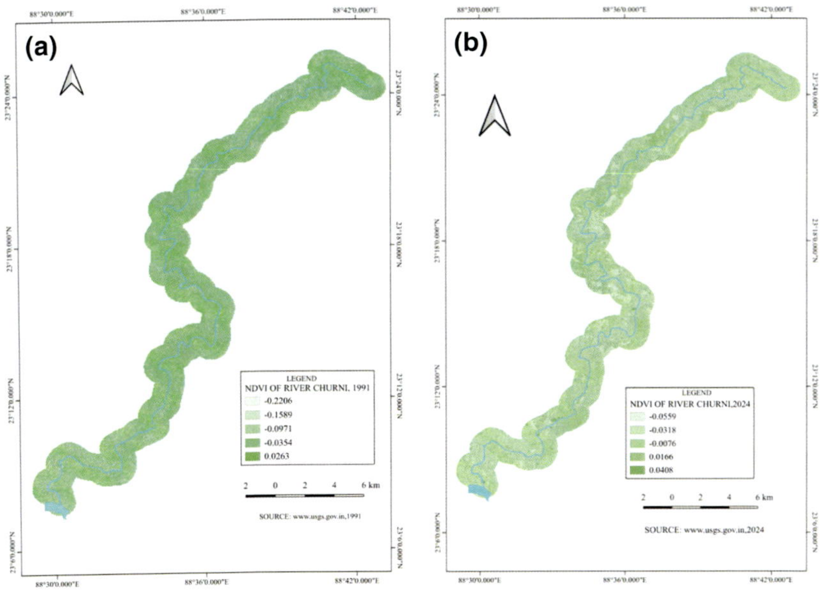

Fig. 25.8 Normalized Difference Vegetation Index within one kilometer buffer zone of river Churni from the year 1991 (**a**) to 2024 (**b**)

25.3.4 Changes in the Water Body from in the Year 1991 to 2024

The calculated values of NDWI clearly show that in the year 1991, the highest value of NDWI was 0.305 and the lowest value of NDWI was – 0.257 (Fig. 25.9a). The highest value of NDWI is 0.040 and the lowest value of NDWI was −0.150 (Fig. 25.9b) in the year 2024. It can be easily said from the values of NDWI in both years that water bodies are degraded or no fresh water coverage within one kilometer buffer zone on the both sides of river Churni. The values of NDWI clearly show that the water quality of the river Churni is not healthy. Water pollution has significant impact on river decay and it destroys the entire environment and channel network of the particular area where it flows. Pollution destroys water quality, deteriorating stream ecosystems, and can destroy biodiversity and it brings water related diseases. Chatterjee (2013) stated about the deteriorating water quality of river Churni and gave emphasis on various unethical anthropogenic activities on decay and deteriorating condition of river Churni.

25.3.5 Changes in the Land Use/Land Cover from the Year 1991 to 2024

There are drastic changes in land use/land cover within one kilometer buffer area on all sides of river Churni. It can be said that the five selected parameters (viz. water body/river, agricultural use on the river channel, settlement cover, agricultural land, and vegetation cover) are drastically changed from the year 1991 to 2024 (Fig. 25.10a–b). The percentage of area under water body/river in the year 1991 was 13.62 and the area under water body/river in the year 2024 is 5.26% (Fig. 25.11). Thus, 8.36% area of water body/river had been decreased from the year 1991 to 2024. The percentage of area under agricultural use on the river channel was 11.36 in the year 1991 and the area under agricultural use on the river channel is 16.80% in the year 2024 (Fig. 25.11). Thus, 5.44% area under agricultural use on the river channel had been increased from the year 1991 to 2024. The scenario clearly gives indication of channel decay of river Churni. The percentage of area under settlement in the year 1991 was 21.15 and the area under settlement in the year 2024 is 32.11% (Fig. 25.11). Thus, 10.96% area under settlement had been increased from the year 1991 to 2024.

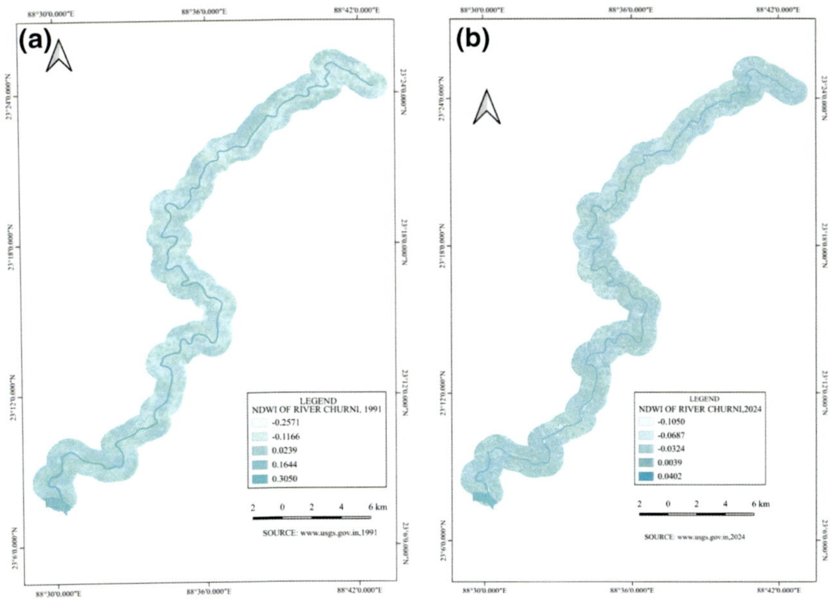

Fig. 25.9 Normalized Difference Water Index within one kilometer buffer zone of river Churni from the year 1991 (**a**) to 2024 (**b**)

The percentage of area under agricultural land in the year 1991 was 25.45 and the area under agricultural land in the year 2024 is 37.32% (Fig. 25.11). Thus, 11.87% area under agricultural land had been increased from the year 1991 to 2024. The percentage of area under vegetation cover in the year 1991 was 28.43 and the area under vegetation cover in the year 2024 is 8.56% (Fig. 25.11). Thus, 19.87% area under vegetation cover had been decreased from the year 1991 to 2024. Thus, it has been clearly proved that drastic changes in land use/land cover have been observed within one kilometer buffer area from all sides of river Churni. Anthropogenic activities cannot be ignored in these changes of land use/land cover. Conversion of river bed into agricultural land, growing settlement areas, reduction in vegetation cover, reduction in water coverage, all these parameters give clear indication of decay of river Churni and unhealthy condition of river Churni.

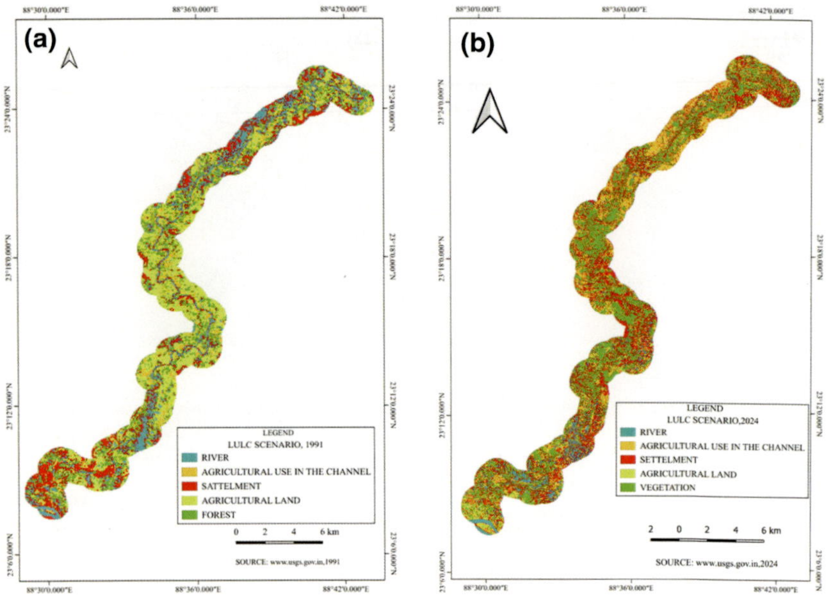

Fig. 25.10 Land use/land cover map within one kilometer buffer zone of river Churni from the year 1991 (**a**) to 2024 (**b**)

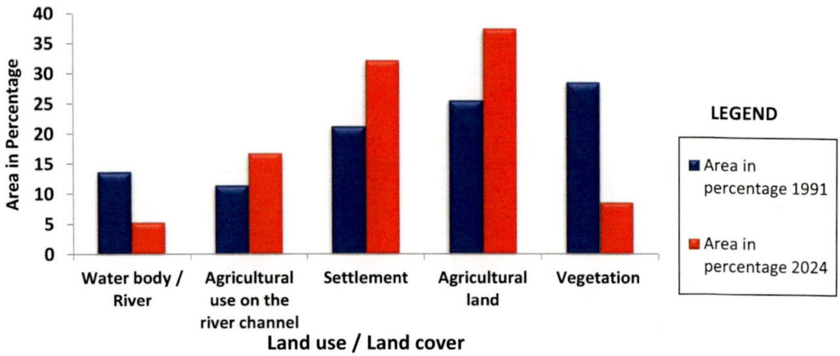

Fig. 25.11 Land use/land cover change within one kilometer buffer zone of river Churni from the year 1991 to 2024

25.3.6 Management Strategies

It is very needful to take prompt management strategies for the survival of decaying river because the existence of all biotic communities very much depends on water body. If proper management strategies can be implemented then the problem of river decay can be handled well (Fig. 25.12) and it will be very much essential and beneficial for the human society. If proper management strategies can be implemented then it again will become a healthy river just like before. Again, it will be a good source of resource. Based on this river, very good fishing ground can be developed which will again support many people economically.

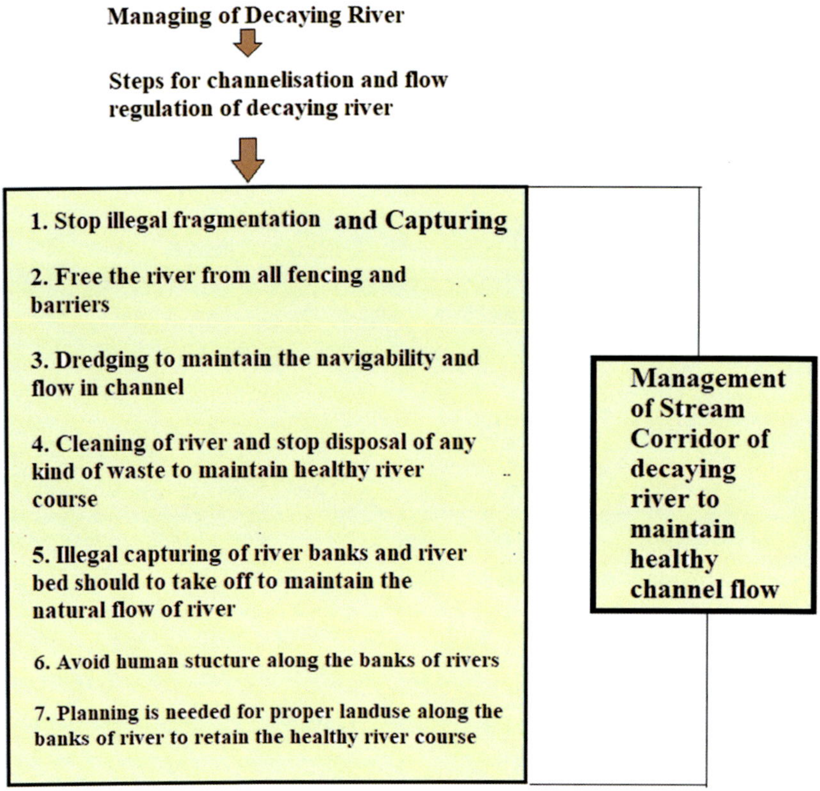

Fig. 25.12 Management strategies to prevent channel decay

For management of decaying river, first of all, illegal capturing (Fig. 25.13) and fragmentation of river should to stop. It is very necessary to free the river from all its barriers. Dredging is needed to maintain the course of river in the affected areas. River encroaching has introduced siltation on river bed. Illegal capturing of river banks and river bed should to take off to maintain the natural flow of decaying river. To maintain the healthy river flow, management of stream corridor or watershed management is very much necessary (Fig. 25.12).

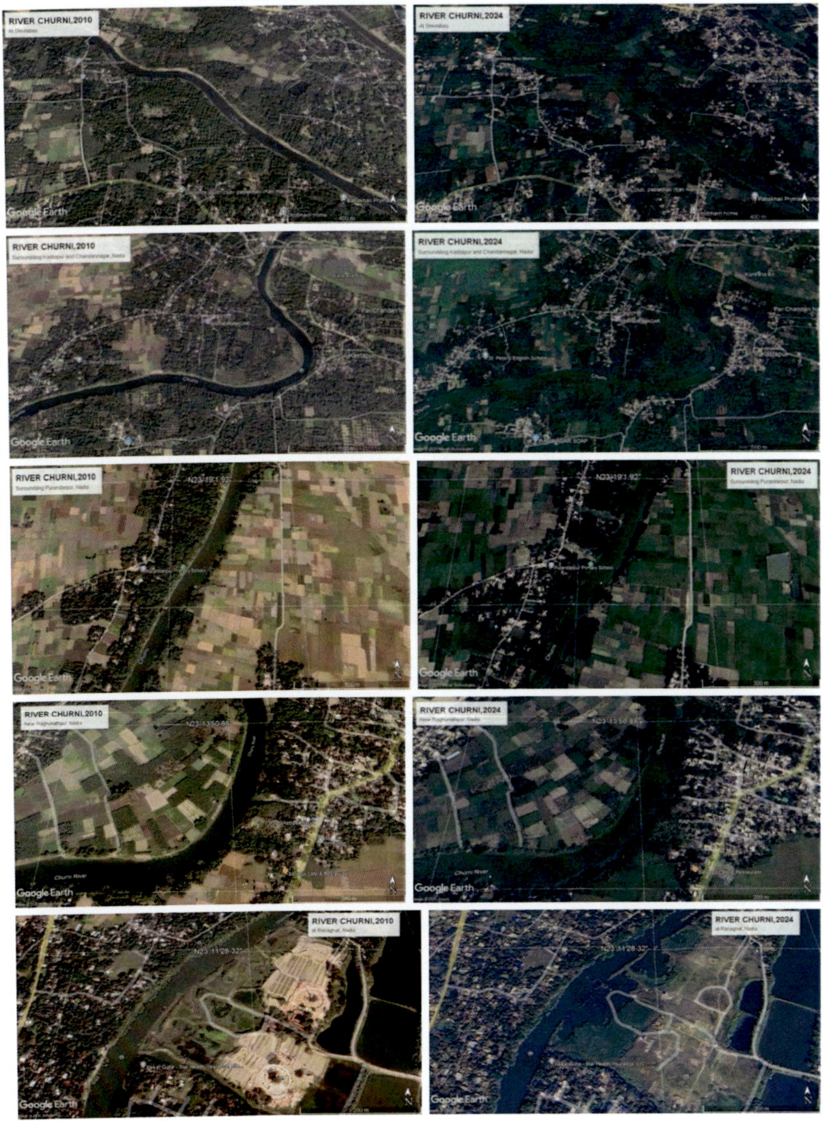

Fig. 25.13 Capturing and fragmentation of river Churni in some selected stretches in 2024 based on comparative views of satellite images in 2010 and 2024

25.4 Conclusion

It can be concluded that river decay is a burning problem for the human society. It is continuously bringing various problems in human livelihood and in natural environment. The preconceived idea in this research is **"Decay of river is highly caused by various anthropogenic activities"** and the application of different methods like width measurement to find out the shrinking width of river, supervised image classification to identify drastic changes in land use and land cover specially decreasing area of water bodies, increasing areas of settlements, increasing use of area on the river channel, etc. NDWI and NDVI both have in most of the cases negative values which indicates not good condition in respect of water bodies and vegetation cover respectively. All the used Indies have scientifically supported the taken hypothesis. The main reason of decay of river is various anthropogenic activities which is clear from the foregoing discussion. The case study of river Churni scientifically proved that positive change in the area under settlement from the year 1991 to 2024 has significant impact on encroaching of river width of Churni from 1991 to 2024. The problem of river decay or encroaching river can be stopped by the concerned authority. Use of land along the river or the management of watershed needs to be maintained properly. Construction of houses, other concrete structures need to avoid along the river banks. Decaying river again can be reformed in its previous living condition if suggested strategies can be implemented by concern authority.

Acknowledgements The authors are thankful to USGS and Google Earth Pro for providing satellite images and the authority related to making QGIS software. We want to convey our deep gratitude to **Mr. Samiron Das** for his endless support until the completion of the study.

References

Bhatta, B. (2021). *Remote sensing and GIS India*. Oxford University Press.

Charlton, R. (2008). *Fundamentals of fluvial geomorphology*. London.

Chatterjee, M. (2020). An enquiry into the evolution and impact of human interference on the Churni river of Nadia district, West Bengal. *International Journal of Current Research, 5*(5), 1088–1092.

Crini, M. N., Lichtfouse, E., Liu, G., Balaram, V., & LadoRibeiro, A. R. L. (2022). Worldwide cases of water pollution by emerging contaminants: A review. *Environmental Chemistry Letters, 20*, 2311–2338. https://doi.org/10.1007/s10311-022-01447-4ff

Census of India. (2011a). *District census hand book: Village and town dictionary part-XII-A (20) Nadia West Bengal*. Retrieved January 2018, from https://censusindia.gov.in/census.website/1

Census of India. (2011b). *District census hand book: Village and town dictionary part-XII-A (20) Nadia West Bengal*. Retrieved January 2018, from https://censusindia.gov.in/census.website/1

District Survey Report of Nadia District. (2022). Department of Industry, Commerce and Enterprises Government of West Bengal. http://www.dmm.gov.in/pdfs/DSR/DSR_Nadia.pdf

Ekka, A., Pande, S., Jiang, Y., & Zaag, P. V. D. (2019). Understanding the impact of anthropogenic activities on river ecosystem services in India: A hydro-economic approach. *Geophysical Research Abstracts, 21*, 2019–8753.

Huang, S., Tang, L., Hupy, J. P., et al. (2021). A commentary review on the use of normalized difference vegetation index (NDVI) in the era of popular remote sensing. *Journal of Forest Research, 32*, 1–6. https://doi.org/10.1007/s11676-020-01155-1

Li, P., Fang, L., Yuanzhu, W., Mi, W., Ji, L., Uixiang, Z. G., Yang, P., Chen, Z., & Bi, Y. (2022). Anthropogenic activities accelerated the evolution of river trophic status. *Ecological Indicators*. https://doi.org/10.1016/j.ecolind.2022.108584

Mackline, M. G., & Lewin, J. (2015). The rivers of civilization. *Quaternary Science Reviews, 114*, 228–244. https://doi.org/10.1016/j.quascirev.2015.02.004

Meter, K.V., Thompson, S., & Basu, N. B. (2016). Human impacts on stream hydrology and water quality. In B. Jeremy & E. H. S. Jones (Ed.), *Stream ecosystems in a changing environment* (pp. 441–490). Academic Press. https://doi.org/10.1016/B978-0-12-405890-3.00011-7

Özelkan E (2020) Water body detection analysis using NDWI Indices derived from landsat-8 OLI. *Polish Journal of Environmental Studies, 29*(2), 1759–1769. https://doi.org/10.15244/pjoes/110447

Roy Choudhury, S. (2020). *Human activities put dolphins at survival risk*. Environmental sciences Ecology conservation Biology and Research.

Sidabutar, N. V., Namara, I., Hartono, D. M., & Soesilo, T. E. B. (2017). The effect of anthropogenic activities to the decrease of water quality. *IOP Conference Series Earth and Environmental Science*, *67*(1), 12–34. https://doi.org/10.1088/1755-1315/67/1/012034

Sahu, S. (2023). *River water pollution, sources, and human health effect in India*. Sustainability Agri Food and Environmental Research. 12:2024 Special issue: Climate change.

Thakur, P. (2022). River Churni: assessing trans boundary river pollution and options for livelihood protection. *Electronic Journal of Social and Strategic Studies, 3*(1), 21–40. https://doi.org/10.47362/EJSSS.2022.3102

Winton, R. S., Teodoru, C. R., Calamita, E., Kleinschroth, F., Banda, K., Nyambe, I., & Wehrli, B. (2021). Anthropogenic influences on Zambian water quality: Hydropower and land-use change. *Environmental Science: Processes & Impacts, 23*, 981–994.

Wang, H., & He, G. (2022). Rivers: Linking nature, life, and civilization. *River, 1*(1), 25–36. https://doi.org/10.1002/rvr2.7

CHAPTER 26

Geographical Dilemma of River Bank Erosion: Empirical Observations in the Dwarakeswar River Basin (West Bengal, India)

Arka Prava Ray

26.1 Introduction

River bank erosion is a natural process that occurs when the force of water, wind, and some other exogenic and endogenic process eroded the banks of river (Islam & Guchhait, 2017). In India this phenomenon is particularly dominated in the eastern and northern states where rivers like Ganges and Bramhaputra flows. This type of bank erosion can lead to loss of agricultural land, homestead, property, and displacement of community etc. West Bengal, a state crisscrossed by rivers like the Ganges, Hooghly, Jalangi, and Ichamati, faces recurrent issues of river bank erosion (Das & Samanta, 2023; Thakur et al., 2012). These rivers,

A. P. Ray (✉)
Mirjapur High School, Bankura, West Bengal, India
e-mail: ararayl@gmail.com

S. Ghosh (ed.), *Reimagining Indian Rivers for Sustainability*, Palgrave Studies in Environmental Sustainability,
https://doi.org/10.1007/978-3-032-04754-0_26

particularly in their lower reaches, carry massive sediment loads and meander significantly, leading to frequent bankline shifts. The problem is most severe in the districts of Malda, Murshidabad, Nadia, North 24 Parganas, and Dakshin Dinajpur. Erosion not only causes land loss but also severely affects agriculture, infrastructure, and local livelihoods (Islam & Guchhait, 2017).

Although river bank erosion is a natural phenomenon, it can cause serious problems and difficulties when it intensifies or is hastened by human activity or environmental changes. A summary of the main problems and difficulties related to river bank erosion is provided below:

- Environmental Impact—Erosion can disrupt the ecological balance by destroying fish, mammal, and plant habitats. Water turbidity is increased by sediment from erosion, which hinders light penetration and has an impact on aquatic life. Species that depend on vital wetland regions near riverbanks may be impacted if they disappear.
- Agricultural Loss—Reduced agricultural production results from farmers losing arable land beside waterways. Deposition of eroded dirt in other areas might impact drainage and soil quality.
- Infrastructure and Property Damage—Houses next to deteriorating riverbanks may collapse or become dangerous. Roads and bridges are damaged. Progressive erosion may necessitate the evacuation of entire villages.
- Economic Consequences—Communities that depend on fishing or agriculture can have financial difficulties. It may be expensive to build revetments, embankments, and other defences. Properties in places that are prone to erosion lose value and are difficult to insure.
- Social and Political Challenges—Land ownership conflicts may occur when riverbanks change. It can be politically delicate to decide who is responsible and what should be done for impacted populations. There are emotional, sociological, and practical difficulties with relocating communities.
- Management Challenges—Planning is more challenging since river flow patterns are impacted by climate unpredictability. an increase in the number of severe events. Numerous erosion control techniques, such concrete embankments, might have negative downstream effects. Frequently, infrastructure needs ongoing maintenance and observation. Extensive data and modelling on the dynamics of erosion are lacking in many areas.

River bank erosion in West Bengal is a multifaceted issue, driven by both natural and human factors. While structural interventions provide temporary relief, long-term sustainability lies in integrated planning, public participation, and ecological approaches. Without timely and effective actions, erosion will continue to displace communities, degrade environments, and hinder economic development in the region. In this chapter the issue is more related to the Dwarakeswar which is a major river in the western part of West Bengal. The Dwarakeswar runs for about 202.2 km (length of main stream) (Sinha, 2016). Dwarakeswar River (Dhalkisor) originates from Tilboni hill in Purulia District and enters Bankura District near Chhatna P.S. It flows through the district Bankura, part of Hooghly and Midnapur districts. The river Dwarekeshwar meets the river Silai near Ghatal (Midnapore West, West Bengal, India) and the confluence is known as Rupnarayan River. This river originates from area where drought is a common factor and flows through the area where flood has major historical background. Here, the main objective of this work is to demonstrate the loss of agricultural and homestead land due to riverbank erosion along the Dwarakeshwar River in the small mouza of Madhuban, located in the drought-prone Bankura District of West Bengal.

26.2 Study Area

The study area (Fig. 26.1) lies in the southern eastern portion of the district of Bankura in West Bengal of India (Figs. 26.1, 26.2). The District Bankura consists of 22 C.D. Blocks. The Block Bishnupur is one of the prosperous blocks of Bankura. The Madhuban mouza (J.L.No. 78) lies within it is severely affected by erosion. The study area covers a geometrical extension from 23° N to 23° 5′ 24″ N latitude and from 87° 20′ E to 87° 20′30‴ E longitude. It is relatively a small area which has investigated intensive field-study, a detailed analysis and discussion of micro-level data (village wise). The area is covered mainly by four geological units belonging to four geological eras beginning from the Archaean to Holocene. It is noted here that the study area is unique from its adjoining eastern and southern portions. The eastern and southern portions have monotony in geological formation which is dominated by alluvial tracts of the Holocene period.

In Purulia district, the Dwarkeswar River rises from the Panjonia peak (322 m) close to the Tilaboni hill of the extended Chhotanagpur Plateau

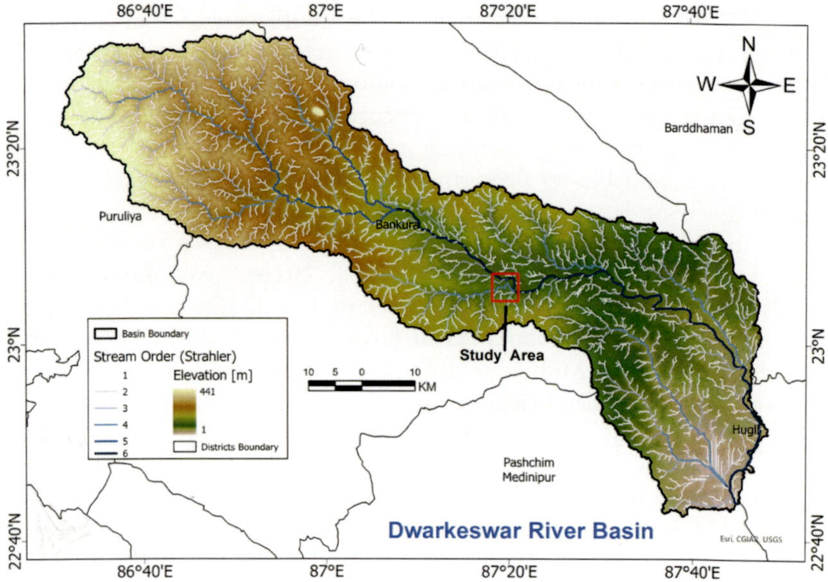

Fig. 26.1 Elevation map of Dwarkeswar river basin

(Roy, 2019). It then flows into the Bankura district, which is close to Chhatna. In summer, the average monthly temperature is 32.55 °C, while in winter, it is 18.98 °C. Rainfall is 117.76 mm annually on average. The temperature ranges from 4 to 9 °C during the day. There is a severe shortage of drinking water in the higher catchments from April to May. The dug wells (8–11 wells per 200 m linear distance) overcrowd the upper portion, while the deep tube well irrigation (down to 50 m) overcrowds the middle and bottom portions. The most striking feature of the Dwarekeswar River (200.5 km long) is that it concurrently exhibits the tidal impact at its lower confluence, the monsoonal flood in the middle, and flash floods in its higher catchments (Malik et al., 2020; Nag & Lahiri, 2011).

26.2.1 Geographic Dilemma

A difficult or complicated circumstance brought on by geographical elements or places is referred to as a geographic issue or dilemma. These

Fig. 26.2 Field photographs shwoing acutal condition of bank failure in the river valley

conundrums can have many facets, including anything from the physical landscape to human growth and repercussions. The Global South is disproportionately affected by the climate catastrophe, having been least responsible for its emergence yet most affected by its effects. For geographers and policymakers, this presents moral and practical conundrums. Complex conundrums arise when the demands of territorial growth and sustainability are balanced with the necessity of local self-determination. In hazard research, the term "geographic dilemma" describes the intricate relationship between human activity and natural dangers that results in varying effects in various geographic regions. Analysing the physical processes of risks as well as the socioeconomic vulnerabilities that increase a person's susceptibility to damage is necessary to comprehend this conundrum. Hazards can be made worse by human activities including urbanization, deforestation, and irresponsible land use. Building in landslide-prone places or on floodplains, for example, makes a person more vulnerable. The effects of hazards are not consistent. Topography, soil composition, and climate are examples of local factors that might affect the severity and distribution of risks. Additionally, the damage might be increased when several dangers are present (for example, a flood after bank erosion). Human activity, climate change, and hazards are all related. For instance, the frequency and severity of extreme weather occurrences may change due to climate change. A geographical viewpoint on hazards promotes a comprehensive method of risk assessment and mitigation. This entails fostering sustainable land management techniques, creating efficient early warning systems, and taking into account the interaction of human and environmental variables.

The term "bank erosion" describes the geomorphic processes that cause river banks to erode and retreat, affecting river ecosystems, human activities associated with rivers, and the loss of land and habitats. A natural process known as "bank erosion" happens when water flows through a river or stream, wearing away the banks and frequently carrying sand and debris with it. Even though this process is natural, human activity has a big impact on how quickly and how much it happens. Deforestation and dam construction are two examples of human actions that might worsen the natural process of bank erosion. Effective control of bank erosion requires an understanding of its geomorphic environment. Bank erosion can result in infrastructure destruction, population displacement, and agricultural land loss, which can fuel social discontent and economic hardship. Communities that have been displaced may experience health hazards, social isolation, and economic instability. Conventional bank protection techniques do not always work or last, and they might have unforeseen effects on the ecosystem of the river and its environs. It is imperative that management strategies be comprehensive, coordinated, and take into account the larger watershed environment. River bank erosion is a complex geographical dilemma in India, requiring a multifaceted approach to mitigate its impacts. By understanding the causes and consequences of erosion, implementing sustainable land use practices, and investing in disaster preparedness, India can reduce the vulnerability of its riverine communities and protect its rich natural resources.

26.3 Bank Erosion in Madhuban Mouza

Tropic of Cancer passes through the bank of Dwarakeswar. More over Vindya—Pleistocene rest of Chhotonagpur Plateau is connected with the birth place of the river Dwarakeswar. For this reason, the Dwarakeswar is carrying stone and sand in a huge amount. So, the sediments mainly sand are stored on the wombs of Dwarakeswar and impede in the main current of the river. So, the flow had and has to change its course in other direction being obstructed by the heap of sand. So, the river has to make its way in other direction. That causes bank erosion. Few cultivated lands of several mouzas have been severely eroded and the Madhuban mouza of Bishnupur Block is one of them. Damdama and ainchbari and their surroundings are severely affected by this erosion. From the maps as well as from field observation it is found that Maduban mouza is mostly affected by bank erosion. From rough estimation, about 400 acres of cultivated lands has been devoured by the river Dwarakeswar in last hundreds of years. This river bank erosion has created an endemic and recurrent natural hazard. It is not the phenomenon of present but the riverbank erosion is exaggerated by the anthropogenic activity.

26.3.1 Causes of Erosion

- The Madhuban mouza lies in Tropical monsoonal climatic zone. Basin area is lateritic with bare surfaces. Early monsoonal rainfall responsible for rain drop erosion of soil. This type of top soil with bare surface is very much vulnerable to bank erosion.
- When Dwarakeshwar enters into the stage of maturity, it becomes sluggish and meanders or braid. This change in velocity causes massive riverbank erosion.
- Erosion is caused due to the extensive over bank spills.
- Gradual migration of Dwarakeshwar River towards the north is responsible for river bank erosion
- Madhuban is situated in the southern bank of river and just of its opposite bank (Shantipur) has been embanked so the velocity is being chased to the bank of Madhuban where the erosion is being done gradually. When the huge water run in the rainy season. Then whole water is being chased in the embankment and forcibly returned to the opposite bank where erosion takes place.

- Loosely packed recent deposited materials consisting of silt and sands are highly susceptible to erosion. Here the flash flood accelerates the rate of bank erosion.
- Changes in riparian vegetation pattern and sand mining escalate poor drainage during monsoon and promote bank failures.

Field observations and hydro-geomorphic study (Fig. 26.2 and 26.3) reveal the processes of bank failure during monsoon flow or flood. By increasing the bank's overall instability and causing seepage and fluvial erosion, increased water flow can cause bank failure directly. The velocity at which water flows have a direct impact on the removal of soil particles through fluvial erosion. Through increased pore water pressure, decreased erosion resistance, and the possibility of mass failure when the bank's weight surpasses its strength, increased flow further destabilizes the bank. Particularly near the toe (the bottom of the bank), the power of the stream immediately erodes the bank (Das et al., 2021). The bank may become more prone to collapse as a result of overhanging blocks or a steep bank angle. Seepage erosion is another effect of increased water flow, in which groundwater runs out of the bank with sufficient force to erode the bank's substance. If concentrated, this might be referred to as piping, in which a soil "pipe" is eroded. In addition, high water levels and flow rates can make the bank more unstable overall by (a) decreasing resistance to erosion, (b) raising pore water pressure, and (c) causing mass failure. An important issue is that sparse vegetation, or the lack of it, can significantly contribute to bank failure by reducing the overall stability of riverbanks. Vegetation, especially with deep roots, helps to bind soil particles together, increasing the soil's strength and resistance to erosion. Sparse or absent vegetation weakens the bank's ability to withstand the forces of water flow. If vegetation is replaced with shallower-rooted plants like grass, it can actually weaken the soil compared to the deep roots of mature vegetation. Sparse vegetation can lead to increased water flow and velocity in the channel, further exacerbating erosion.

26.4 Socio Economic Impacts

Riverbank erosion has disastrous socio-economic effect of this small mouza, which totally depends on agriculture. The majority of the affected people perceive riverbank erosion as a natural phenomenon but in many

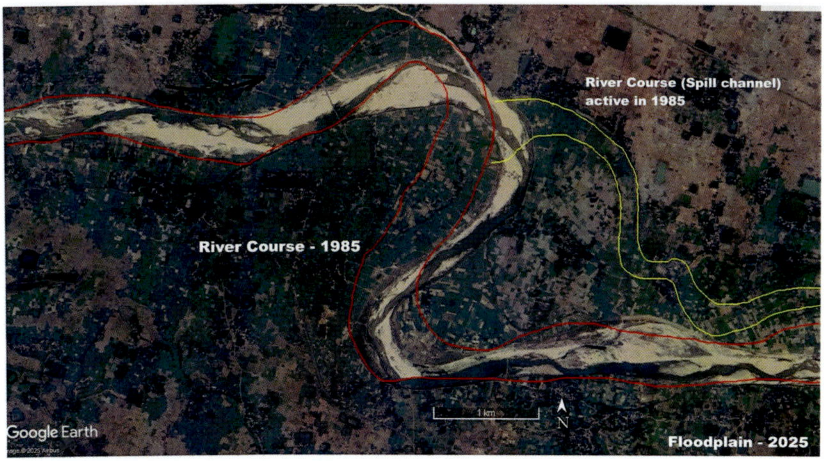

Fig. 26.3 Channel shifting between 1985 and 2025 instigate bank erosion near Madhuban mouza

cases the people believe erosion to be the "will of God". However, these days, riverbank erosion is seen as one of the major causes for national poverty. The degree of economic loss and vulnerability of population due to bank erosion has dramatically increased in recent years. The impact of land loss involves primarily the loss of homestead land, housing structures, crops, cattle, trees. Loss of homesteads forces people to move to new places without any option and puts them in disastrous situations. About one million people are directly affected each year by bank erosion in the country. Bank erosion affects people, irrespective of farm sizes. Riverbank erosion causes setback for village agriculture. Along with homestead settlements, it erodes farmland, infrastructure, and the communication system. It affects the crop income of vulnerable groups. The big farmers are the worst affected, followed by medium farmers, and marginal groups. The affected people lose their assets and are forced to draw on savings and often fall into further debt. Researchers found that the land lost is much more than the land that rises out of riverbed through accretion. Displacement is the immediate impact of riverbank erosion. The displaced usually move to nearby areas but migration to distant places is not uncommon. In erosion-prone areas, most families have witnessed a displacement in their lifetime. This involuntary movement can go up to 10 times or even more. A survey conducted in Madhuban mouza has revealed that villages lost 400 acres of land in last 100 years (Table 26.1, Fig. 26.4).

Table 26.1 Records of the eroded land of Madhuban Mouza that permanently lost

Name of the owner	*Dag number*	*Area of land (in acre)*	*Nature of land*
			Agricultural or homestead
Jabbal Majhi,Sakti Majhi, & others	22	1	Agricultural
Aditya Sit	115	0.25	Agricultural
Sitanath Majhi	1112, 135, 136, 236, 239, 240, 244, 169	1.12	Agricultural
Kamal Baul	145	1.25	Agricultural
Bibhuti Bauri	446	0.33	Agricultural
Banku dhak	420	0.33	Agricultural
Madev Dhak	113	0.41	Agricultural
Sankar Basaki	129	0.5	Agricultural
Gopal Baul	112	0.33	Homestead
Achinta Baul,Pares Baul, & others	150–158	2.33	Homestead
Ashini Majhi	445	0.33	Agricultural
Bagala dhak	446	0.5	Agricultural
Sanatan Pal	431, 432	0.41	Agricultural
Chandi Majhi	468	0.25	Agricultural
Madhusudan Dey	445	0.16	Agricultural
Bhakta Dey,Nabagopal Dey Kalipada Dey & others	557, 558, 560, 561	2	Homestead
Bhabotosh Dey,Madhusudan Dey	562	1	Homestead
Khandu Lohar	714, 715, 716, 717	0.75	Agricultural
Kalipada Choudhury	784	1.33	Agricultural
Rabindra Faujdar	800	1	Agricultural

Note 1 acre = 4046.86 m^2

Fig. 26.4 The photographs shwoing destruction of home during high water flow level, people particpation to recover the eroded bank and intalation of sand bags as protection strategies

26.5 Management Programme

26.5.1 Low-cost Treatment of River Bank

The erosion causes a lot of damage at Madhuban Mouza. Table 26.1 shows that in last 100 years it swallowed a good tract of cultivated land. Now the management question is very important. There are various steps towards the management of riverbank erosion but here researcher concluded the plan which are most effective in this study area.

In study area most of the people are cultivators and if the river bank erosion management programme will very costly, then the Government and as well as poor cultivators are not able to adopt these sophisticated technologies and for that they feel inferior and neglect conventional agronomic practices. But the main objective of civilization is the full-fledged development of society. From this point of view, the farmers of rural Bankura will be able to afford this technological management, if input costs are reduced. The only way to reduce the cost of management is the use of the tools and components which are derived from conventional practices, which are regionally suitable.

Soft engineering or sometimes-called bio engineering is a management technology that controls riverbank erosion which is more nature and compatible with environment and also low cost.

Construction of little civil structures have created pockets of slow-moving water along the banks, which not only halts the erosive action but also throws sediment in the area forming toes needed to support the yielding banks. The newly formed toes gradually develop into berms, which are stabilized latter and will provide support the river bank. In urban areas this structure is made from masonry, stone, iron, etc. Those will bad influence to the vegetation growth and habitat for species that are living around water. Because those are costly, this kind of erosion control is only conducted to protect important place or populous and costly urban area, which is enough costly. While, in rural area where the value of land or property is low, cannot be protected by above structure, because that is not feasible if compare the cost of structure.

World Bank and Directorate Water Resources Management have developed bio engineering since year 1996. Bio engineering is low-cost riverbank protection, easy to construct, using natural and local materials, so that will harmonies with environment. Bio Engineering, it is combination bamboo, and local environment friendly product, the construction was design to protect riverbank from current attack.

Bank protection measures, which are matching with the spirit and soul of eco-hydraulic, must be:

– Using natural materials, which can be found around, that will be low cost.
– Labour intensive, lessening unemployment so that as according to condition of our country.
– Can provide habitat for water animal and crop.

Few eco-friendly methods which are low cost are discussed below:

- **Vertical Timber Posts**
  Wooden posts or pile driven vertically into the stream bed can be used to protect they lower part of a bank. That is recommended at least half of the post should be driven into the ground.
- **Woody Debris**
  In areas where trees are removed or the bare river bank, wood debris like brush helps to protect bank erosion. Debris can also be used to slow the water motion, trap silt and sediment and permit the growth of river-edge plants. Cut logs and branches can be secured against a bank protection or for filling scour holes.
- **Wicker Spiling and Hurdles**
  Spiling can be used to protect the base of steep banks and to create dikes on smaller stream. Stakes are driven into the ground and

Table 26.2 Remedial method to prevent bank erosion of Madhuban Mouza

Dag. no	*Method of prevent erosion*
11, 12, 135, 136, 236, 239, 240, 244, 169	Plantation of bamboo groves, and plantation of long root Grass
145, 446, 420, 130, 129, 152, 153, 154, 155, 156	Knitting by bamboo sticks
157, 158, 445, 446, 434, 432	
468, 150, 151	
558, 560, 561, 562, 714, 715, 717, 784, 800, 834	Knitting by bamboo sticks and alternatively post can be staggered in two rows and brush piled between them

branches woven between them. Alternatively, post can be staggered in two rows and brush piled between them. The distance between the posts should be consistent with the length and thickness of the weaving materials.

Here in Madhuban mouza is eroded extensively by the river. Here few essential remedies include as Table 26.2.

26.6 Conclusion

Bank erosion is a complex issue with both natural and human-induced drivers. By understanding the human role in exacerbating and mitigating erosion, we can develop effective strategies to protect riverbanks and the communities and ecosystems that depend on them. Human activity has a significant impact on whether bank erosion failure is exacerbated or lessened. Activities such as deforestation, building close to riverbanks, and inappropriate land management techniques are examples of how humans may exacerbate the issue. They may, however, also put solutions into action, such as replanting plants, erecting barriers, and implementing sensible land use plans to stop erosion. Following observations highlight the positive direction of river bank management:

- Bank protection that commonly use is made from masonry or concrete; this type is expensive and not natural or not environmentally friendly.

- Bio engineering is a bank erosion control structure which more nature and compatible with environment and also low cost.
- Directorate of Water resources Development have developed bio engineering since year 1996. This is very much effective in this area.
- Meeting, and seminar for dissemination of bio engineering must be conducted. The activities can change the thinking of river engineer that erosion control have not to use massif and expensive structure.

References

Das, R., & Samanta, G. (2023). Impact of floods and river-bank erosion on the riverine people in Manikchak Block of Malda District, West Bengal. *Environment, Development and Sustainability, 25*, 13595–13617.

Das, V. K., Debnath, K., Roy, S., et al. (2021). Effect of turbulent structures on the riverbank erosion due to tidal influence: A case study from the Rupnarayan River, eastern India. *Journal of Earth System Science, 130*, 62. https://doi.org/10.1007/s12040-021-01554-w

Islam, A., & Guchhait, S. K. (2017). Search for social justice for the victims of erosion hazard along the banks of river Bhagirathi by hydraulic control: A case study of West Bengal, India. *Environment, Development and Sustainability, 19*, 433–459.

Malik, S., Pal, S. C., Chowduri, I., et al. (2020). Prediction of highly flood prone areas by GIS based heuristic and statistical model in a monsoon dominated region of Bengal Basin. *Remote Sensing Applications: Society and Environment, 19*, 100343. https://doi.org/10.1016/j.rsase.2020.100343

Nag, S. K., & Lahiri, A. (2011). Morphometric analysis of Dwarakeswar watershed, Bankura district, West Bengal, India, using spatial information technology. *International Journal of Water Resources and Environmental Engineering, 3*(10), 212–219.

Roy, S. K. (2019). Flood as a catastrophic hazards on Dwarakeswar river basin. *International Journal of Scientific Development and Research, 4*(8), 6–13.

Sinha, M. (2016). Dwarekeshwar River Basin and Anthropogenic Intervention as Sand Mines. *International Journal of Research in Geography, 2*(2), 48–56.

Thakur, P. K., Laha, C., & Aggarwal, S. P. (2012). River bank erosion hazard study of river Ganga, upstream of Farakka barrage using remote sensing and GIS. *Natural Hazards, 61*, 967–987.

Index

S. Ghosh (ed.), *Reimagining Indian Rivers for Sustainability*, Palgrave Studies in Environmental Sustainability,
https://doi.org/10.1007/978-3-032-04754-0

FSC
www.fsc.org
MIX
Papier | Fördert
gute Waldnutzung
FSC® C083411

Kolibri 360 GmbH
Ferdinand-Jühlke-Straße 7
99095 Erfurt, Deutschland
produktsicherheit@kolibri360.de